ANNALS OF
THE NEW YORK ACADEMY
OF SCIENCES

Volume 956

EDITORIAL STAFF

Executive Editor
BARBARA M. GOLDMAN

Managing Editor
JUSTINE CULLINAN

Associate Editors
MARION L. GARRY
ANGELA FINK
LINDA HOTCHKISS MEHTA

The New York Academy of Sciences
2 East 63rd Street
New York, New York 10021

THE NEW YORK ACADEMY OF SCIENCES
(Founded in 1817)

BOARD OF GOVERNORS, September 2001 – September 2002

TORSTEN N. WIESEL, *Chairman of the Board*
JOHN F. NIBLACK, *Vice Chairman of the Board*
BILL GREEN, *Past Chairman*

Honorary Life Governors
WILLIAM T. GOLDEN   JOSHUA LEDERBERG
JOHN T. MORGAN, *Treasurer*

Governors

ELEANOR BAUM	D. ALLAN BROMLEY	KAREN E. BURKE
LAWRENCE B. BUTTENWIESER	PRAVEEN CHAUDHARI	
JOHN H. GIBBONS	MICHAEL GOLDEN	RONALD L. GRAHAM
JACQUELINE LEO	WILLIAM J. McDONOUGH	SANDRA PANEM
RICHARD RAVITCH	RICHARD A. RIFKIND	JOHN J. ROCHE
SARA LEE SCHUPF	JAMES H. SIMONS	LEE VANCE

HELENE L. KAPLAN, *Counsel* [ex officio]   NANCY B. EISENBERG, *Interim Secretary* [ex officio]

NEUROBIOLOGY OF EYE MOVEMENTS
FROM MOLECULES TO BEHAVIOR

ANNALS OF THE NEW YORK ACADEMY OF SCIENCES
Volume 956

NEUROBIOLOGY OF EYE MOVEMENTS FROM MOLECULES TO BEHAVIOR

Edited by Henry J. Kaminski and R. John Leigh

The New York Academy of Sciences
New York, New York
2002

Copyright © 2002 by the New York Academy of Sciences. All rights reserved. Under the provisions of the United States Copyright Act of 1976, individual readers of the Annals are permitted to make fair use of the material in them for teaching or research. Permission is granted to quote from the Annals provided that the customary acknowledgment is made of the source. Material in the Annals may be republished only by permission of the Academy. Address inquiries to the Permissions Department (permissions@nyas.org) at the New York Academy of Sciences.

Copying fees: For each copy of an article made beyond the free copying permitted under Section 107 or 108 of the 1976 Copyright Act, a fee should be paid through the Copyright Clearance Center, Inc., 222 Rosewood Drive, Danvers, MA 01923 (www.copyright.com).

♾ The paper used in this publication meets the minimum requirements of the American National Standard for Information Sciences—Permanence of Paper for Printed Library Materials, ANSI Z39.48-1984.

Library of Congress Cataloging-in-Publication Data

Neurobiology of eye movements : from molecules to behavior / edited by Henry J. Kaminski and R. John Leigh.
    p. cm. — (Annals of the New York Academy of Sciences ; v. 956)
Includes bibliographical references and index.
ISBN 1-57331-394-7 (cloth: alk. paper) — ISBN 1-57331-395-5 (paper: alk. paper)
    1. Eye—Movements—Regulation—Congresses. 2. Neurobiology—Congresses. I. Kaminski, Henry J. II. Leigh, R. John. III. Series.

Q11.N5 vol. 956
[QP477.5]
500 s—dc21
[612.8]
                                          2002002393

GYAT/B-M
Printed in the United States of America
ISBN 1-57331-394-7 (cloth)
ISBN 1-57331-395-5 (paper)
ISSN 0077-8923

ANNALS OF THE NEW YORK ACADEMY OF SCIENCES
Volume 956
April 2002

NEUROBIOLOGY OF EYE MOVEMENTS
FROM MOLECULES TO BEHAVIOR

Editors and Conference Organizers
HENRY J. KAMINSKI AND R. JOHN LEIGH

[This volume is the result of a conference entitled **Neurobiology of Eye Movements: From Molecules to Behavior**, co-sponsored by the New York Academy of Sciences, Case Western Reserve University Schoool of Medicine, and University Hospitals of Cleveland, held on October 4–6, 2001, in Cleveland, Ohio.]

CONTENTS

Foreword. *By* HENRY J. KAMINSKI AND R. JOHN LEIGH xii

A Personal Introduction to Eye Movements. *By* ROBERT B. DAROFF 1

Part I. The Extraocular Muscle and Orbital Tissues

Extraocular Muscle: Cellular Adaptations for a Diverse Functional Repertoire. *By* JOHN D. PORTER 7

The Orbital Pulley System: A Revolution in Concepts of Orbital Anatomy. *By* JOSEPH L. DEMER 17

Mechanics of Eye Movements: Implications of the "Orbital Revolution". *By* THOMAS HASLWANTER 33

Differential Susceptibility of the Ocular Motor System to Disease. *By* HENRY J. KAMINSKI, CHELLIAH R. RICHMONDS, LINDA L. KUSNER, AND HIROSHI MITSUMOTO 42

Applications of Molecular Genetics to the Understanding of Congenital Ocular Motility Disorders. *By* ELIZABETH C. ENGLE 55

Calcium Channelopathy Mutants and Their Role in Ocular Motor Research. *By* JOHN S. STAHL 64

Part II. Brainstem Mechanisms for Controlling Gaze

Modern Concepts of Brainstem Anatomy: From Extraocular Motoneurons to Proprioceptive Pathways. *By* J.A. BÜTTNER-ENNEVER, A.K.E. HORN, W. GRAF, AND G. UGOLINI 75

Studies of the Role of the Paramedian Pontine Reticular Formation in the Control of Head-Restrained and Head-Unrestrained Gaze Shifts. *By* DAVID L. SPARKS, ELLEN J. BARTON, NEERAJ J. GANDHI, AND JON NELSON .. 85

The Contribution of Midbrain Circuits in the Control of Gaze. *By* ULRICH BÜTTNER, JEAN A. BÜTTNER-ENNEVER, HOLGER RAMBOLD, AND CHRISTOPH HELMCHEN 99

Contribution of the Superior Colliculus and the Mesencephalic Reticular Formation to Gaze Control. *By* DAVID M. WAITZMAN, JAY PATHMANATHAN, RACHEL PRESNELL, AMANDA AYERS, AND STACY DEPALMA 111

Neural Discharge in the Superior Colliculus during Target Search Paradigms. *By* EDWARD L. KELLER AND ROBERT M. MCPEEK 130

Midbrain Disorders of Vertical Gaze: A Quantitative Re-evaluation. *By* JAMES A. SHARPE AND JI SOO KIM 143

Part III. Cerebellar Mechanisms of Controlling Gaze

Cerebellar Influences on Saccade Plasticity. *By* F.R. ROBINSON, A.F. FUCHS, AND C.T. NOTO 155

Distributed Model of Collicular and Cerebellar Function during Saccades. *By* LANCE M. OPTICAN AND CHRISTIAN QUAIA 164

The Cerebellar Contribution to Eye Movements Based upon Lesions: Binocular Three-Axis Control and the Translational Vestibulo-Ocular Reflex. *By* DAVID S. ZEE, MARK F. WALKER, AND STEFANO RAMAT 178

Spatial Orientation of Caloric Nystagmus. *By* YASUKO ARAI, SERGEI B. YAKUSHIN, MINGJIA DAI, MIKHAIL KUNIN, THEODORE RAPHAN, JUN-ICHI SUZUKI, AND BERNARD COHEN 190

Part IV. Cerebral and Basal Ganglionic Influences on Gaze Control

The Role of the Lateral Intraparietal Area of the Monkey in the Generation of Saccades and Visuospatial Attention. *By* MICHAEL E. GOLDBERG, JAMES BISLEY, KEITH D. POWELL, JACQUELINE GOTTLIEB, AND MAKOTO KUSUNOKI 205

Effects of Cortical Lesions on Saccadic Eye Movements in Humans. *By* CH. PIERROT-DESEILLIGNY, C.J. PLONER, R.M. MÜRI, B. GAYMARD, AND S. RIVAUD-PÉCHOUX 216

Visual-Vestibular and Visuovisual Cortical Interaction: New Insights from fMRI and PET. *By* THOMAS BRANDT, STEFAN GLASAUER, THOMAS STEPHAN, SANDRA BENSE, TAREK A. YOUSRY, ANGELA DEUTSCHLÄNDER, AND MARIANNE DIETERICH 230

Scanpaths: The Path to Understanding Abnormal Cognitive Processing in
Neurological Disease. *By* CHRISTOPHER KENNARD 242

Antisaccades and Task Switching: Studies of Control Processes in Saccadic
Function in Normal Subjects and Schizophrenic Patients. *By* JASON J. S.
BARTON, MARIYA V. CHERKASOVA, KRISTEN LINDGREN, DONALD C.
GOFF, JAMES M. INTRILIGATOR, AND DARA S. MANOACH 250

Part V. Binocular Aspects of Gaze Control

Neural Mechanisms for the Control of Vergence Eye Movements. *By*
PAUL D.R. GAMLIN 264

Neural Basis of Disjunctive Eye Movements. *By* W.M. KING AND WU ZHOU . 273

Population Coding in Cortical Area MST. *By* A. TAKEMURA, K. KAWANO,
C. QUAIA, AND F.A. MILES 284

Adaptive Control of Vergence in Humans. *By* CLIFTON M. SCHOR, JAMES S.
MAXWELL, JEFREY MCCANDLESS, AND ERICH GRAF 297

Part VI. Vertigo, Nystagmus, and Other Disorders that Disrupt Clear Vision

Inferior Vestibular Neuritis. *By* G.M. HALMAGYI, S.T. AW, M. KARLBERG,
I.S. CURTHOYS, AND M.J. TODD 306

Otolith Function: Basis for Modern Testing. *By* GARY D. PAIGE 314

Vergence-Mediated Modulation of the Human Horizontal Angular VOR
Provides Evidence of Pathway-Specific Changes in VOR Dynamics. *By*
DAVID M. LASKER, STEFANO RAMAT, JOHN P. CAREY, AND
LLOYD B. MINOR 324

Genetics of Familial Episodic Vertigo and Ataxia. *By* ROBERT W. BALOH AND
JOANNA C. JEN 338

Animal Models for Visual Deprivation-Induced Strabismus and Nystagmus.
By RONALD J. TUSA, MICHAEL J. MUSTARI, VALLABH E. DAS, AND
RONALD G. BOOTHE 346

Development of New Treatments for Congenital Nystagmus. *By* LOUIS F.
DELL'OSSO 361

A Neurobiological Approach to Acquired Nystagmus. *By* R. JOHN LEIGH,
VALLABH E. DAS, AND SCOTT H. SEIDMAN 380

Poster Papers

Extraocular Muscle Gene Expression and Function after Dark Rearing. *By*
FRANCISCO H. ANDRADE, ANITA P. MERRIAM, AND JOHN D. PORTER .. 391

Conservation of Synapse-Signaling Pathways at the Extraocular Muscle
Neuromuscular Junction. *By* SANGEETA KHANNA AND JOHN D. PORTER 394

Extraocular Muscle Fatigue. *By* HENRY J. KAMINSKI AND CHELLIAH R.
RICHMONDS 397

Nitric Oxide and cGMP Modulation of Extraocular Muscle Contraction. *By*
HENRY J. KAMINSKI AND CHELLIAH R. RICHMONDS 399

Ocular Motor Myotonic Phenomenon in Myotonic Dystrophy. *By* MAURIZIO VERSINO, SILVIA COLNAGHI, GIORGIO SANDRINI, AND VITTORIO COSI . . 401

Midbrain Reticular Formation Circuitry Subserving Gaze in the Cat. *By* PAUL J. MAY, SUSAN WARREN, BINGZHONG CHEN, FRANCES J.R. RICHMOND, AND ETIENNE OLIVIER 405

Activity in the Primate Rostral Superior Colliculus during the "Gap Effect" for Pursuit and Saccades. *By* RICHARD J. KRAUZLIS, NATALIE DILL, AND KRISTA KORNYLO 409

Time-Optimality and the Spectral Overlap of Saccadic Eye Movements. *By* M.R. HARWOOD AND C.M. HARRIS 414

Knowledge of Future Target Position Influences Saccade-Associated Head Movements. *By* JOHN S. STAHL 418

Normal and Abnormal Slowing of Saccades: Are They One and the Same Phenomenon? *By* R.C. MCGIVERN, J.M. GIBSON, D.W. JENNINGS, K. LAVERY, AND C. MONTGOMERY 421

Smooth Pursuit Tracking: Saccade Amplitude Modulation during Exposure to Microgravity. *By* J.T. SOMERS, M.F. RESCHKE, A. BERTHOZ, AND L.C. TAYLOR 426

Saccadic Palsy after Cardiac Surgery: Visual Disability and Rehabilitation. *By* ROBERT L. TOMSAK, BRUCE T. VOLPE, JOHN S. STAHL, AND R. JOHN LEIGH 430

Small Vertical Saccades Have Normal Speeds in Progressive Supranuclear Palsy (PSP). *By* L. AVERBUCH-HELLER, C. GORDON, A. ZIVOTOFSKY, C. HELMCHEN, H. RAMBOLD, U. BÜTTNER, J. BÜTTNER-ENNEVER, AND R. J. LEIGH 434

Saccadic and Vestibular Abnormalities in Multiple Sclerosis: Sensitive Clinical Signs of Brainstem and Cerebellar Involvement. *By* DEBORAH L. DOWNEY, JOHN S. STAHL, ROONGROJ BHIDAYASIRI, JOY DERWENSKUS, NANCY L. ADAMS, ROBERT L. RUFF, AND R. JOHN LEIGH 438

A Form of Inherited Cerebellar Ataxia with Saccadic Intrusions, Increased Saccadic Speed, Sensory Neuropathy, and Myoclonus. *By* BARBARA E. SWARTZ, MARGIT BURMEISTER, JEFFREY T. SOMERS, KLAUS G. ROTTACH, IRINA N. BESPALOVA, AND R. JOHN LEIGH 441

Anticompensatory Eye Position ("Contraversion") in Optokinetic Nystagmus. *By* S. GARBUTT, M.R. HARWOOD, AND C.M. HARRIS 445

Adaptive Control of Saccades in Children with Dancing Eye Syndrome. *By* LAURA E. MEZEY AND CHRISTOPHER M. HARRIS 449

Integration of Motion Signals for Smooth Pursuit Eye Movements. *By* RICHARD T. BORN AND CHRISTOPHER C. PACK 453

Visually Driven Eye Movements Elicited at Ultra-short Latency Are Severely Impaired by MST Lesions. *By* AYA TAKEMURA, YUKA INOUE, AND KENJI KAWANO 456

Neuronal Activity in Monkey Cortical Area 6 during the Initial Phase of Smooth Pursuit against a Stationary Background. *By* YASUSHI KODAKA, SOHEI CHIMOTO, AND KENJI KAWANO 460

Perisaccadic Occipital EEG Changes Quantified with Wavelet Analysis. *By* I. BODIS-WOLLNER, H. VON GIZYCKI, M. AVITABLE, Z. HUSSAIN, A. JAVEID, A. HABIB, A. RAZA, AND M. SABET 464

Human Gaze Shifts to Acoustic and Visual Targets. *By* L.C. POPULIN, D.J. TOLLIN, AND J.M. WEINSTEIN 468

A Bayesian Approach to Change Blindness. *By* MATTHIAS NIEMEIER, J. DOUGLAS CRAWFORD, AND DOUGLAS B. TWEED 474

Sensory Processing Delays Measured with the Eye-Movement Correlogram. *By* J.B. MULLIGAN 476

Inhibitory Control of Saccade Generation Mediated by Various Oculomotor Regions in Response to Reading Difficulty. *By* SHUN-NAN YANG AND GEORGE W. MCCONKIE 479

Botulinum Toxin Therapy for Apraxia of Lid Opening. *By* DAN BOGHEN, VIORIKA TOZLOVANU, ANDREEA IANCU, AND ROBERT FORGET 482

Distinguishing Progressive Supranuclear Palsy from Other Forms of Parkinson's Disease: Evaluation of New Signs. *By* SANDRA KUNIYOSHI, DAVID E. RILEY, DAVID S. ZEE, STEPHEN G. REICH, CHRISTINA WHITNEY, AND R. JOHN LEIGH 484

Binocular Eye Movement Responses to Dichoptically Presented Horizontal and/or Vertical Stimulus Steps. *By* JOHANNES VAN DER STEEN AND RYOTA KANAI 487

Visual Processing in Disparity Vergence Control. *By* SCOTT B. STEVENSON .. 492

Anticipatory Saccadic–Vergence Responses in Humans. *By* ARUN N. KUMAR, YANNING HAN, STEFANO RAMAT, AND R. JOHN LEIGH 495

Vergence Eye Movements in Strabismus Patients. *By* HORACE R. LEE AND MOSHE EIZENMAN 499

Vergence Disorders in Progressive Supranuclear Palsy. *By* KITTISAK KITTHAWEESIN, DAVID E. RILEY, AND R. JOHN LEIGH 504

A New Way to Find the Primary Position in Three-Dimensional Measurements. *By* J. YGGE, M. BENASSI, AND R. BOLZANI 508

Neural Mechanisms of Three-Dimensional Eye and Head Movements. *By* ELIANA M. KLIER, HONGYING WANG, AND J. DOUGLAS CRAWFORD ... 512

The Visuomotor Transformation for Arm Movement Accounts for 3-D Eye Orientation and Retinal Geometry. *By* D.Y.P. HENRIQUES, J.D. CRAWFORD, AND T. VILIS 515

Implementation of Listing's Law in Patients with Unilateral Sixth Nerve Palsy. *By* AGNES M.F. WONG, DOUGLAS TWEED, AND JAMES A. SHARPE .. 520

Hyperdeviation and Static Ocular Counterroll in Unilateral Abducens Nerve Palsy. *By* AGNES M.F. WONG, DOUGLAS TWEED, AND JAMES A. SHARPE 523

Vestibulo-Ocular Responses during Mirror-Viewing. *By* Y. HAN, A. KUMAR, J. SOMERS, J.I. KIM, AND R.J. LEIGH 527

The Effects of Horizontal Head Position (Yaw axis) and Step Velocity on the Vestibulo-Ocular Reflex. *By* PHILLIP KRAMER, ELLIOT M. FROHMAN, DANIELE NUTI, AND DAVID S. ZEE 530

The Vertical Vestibulo-Ocular Reflex, and Visual-Vestibular Interaction during Active Head Motion. *By* JI SOO KIM AND JAMES A. SHARPE 533

A Three-Channel Model for Generating the Vestibulo-Ocular Reflex in Each Eye. *By* LAURENCE R. HARRIS, KARL A. BEYKIRCH, AND MICHAEL FETTER 537

Rectified Cross-Axis Adaptation of the Vestibulo-Ocular Reflex in Rhesus Monkey. *By* M.F. WALKER AND D.S. ZEE 543

Three-Dimensional Eye-Movement Responses to Surface Galvanic Vestibular Stimulation in Normal Subjects and in Patients: A Comparison. *By* H.G. MACDOUGALL, A.E. BRIZUELA, I.S. CURTHOYS, AND G.M. HALMAGYI 546

Translational VOR Responses to Abrupt Interaural Accelerations in Normal Humans. *By* S. RAMAT AND D.S. ZEE 551

Rapid Adaptation of Translational Vestibulo-Ocular Reflex: Time Course, Consolidation, and Specificity. *By* W. ZHOU, P. WELDON, B. TANG, AND W.M. KING 555

Rapid Adaptation of Translational Vestibulo-Ocular Reflex: Independence of Retinal Slip. *By* W. ZHOU, P. WELDON, B. TANG, AND W.M. KING 558

Effects of a Large-Field Visual Scene on the Vergence Response to Naso-Occipital Linear Motion in Monkeys. *By* YOSHIRO WADA, YASUSHI KODAKA, AND KENJI KAWANO 561

Dynamic Changes of Eye Cyclo Position during Head Tilt. *By* TONY PANSELL, JAN YGGE, AND HERMANN D. SCHWORM 564

The Myth of Static Ocular Counter-Rolling: The Response of the Eyes to Head Tilt. *By* ROBERT S. JAMPEL 568

Otolith Effect on Torsional Quick Phases of Vestibular Nystagmus in Humans. *By* A.A. KORI, A. SCHMID-PRISCOVEANU, AND D. STRAUMANN 572

Impaired Linear Vestibulo-Ocular Reflex Initiation and Vestibular Catch-Up Saccades in Older Persons. *By* JUN-RU TIAN, BENJAMIN T. CRANE, GERALD WIEST, AND JOSEPH L. DEMER 574

A New Differential Diagnosis for Spontaneous Nystagmus: Lateral Canal Cupulolithiasis. *By* ALEXANDRE R. BISDORFF AND DAMIEN DEBATISSE 579

Changes in the Angular Vestibulo-Ocular Reflex after a Single Dose of Intratympanic Gentamicin for Ménière's Disease. *By* J.P. CAREY, T. HIRVONEN, G.C.Y. PENG, C.C. DELLA SANTINA, P.D. CREMER, T. HASLWANTER, AND L.B. MINOR. 581

Unilateral Rebound Nystagmus: One Manifestation of Two Different Pathologic Processes. *By* M.L. ROSENBERG AND D.S. ZEE 585

Delayed-Onset Seesaw Nystagmus Posttraumatic Brain Injury with Bitemporal Hemianopia. *By* ERIC R. EGGENBERGER 588

Main Sequence of Convergence Retraction Nystagmus Indicates a Disorder of Vergence. *By* HOLGER RAMBOLD, DETLEF KÖMPF, AND CHRISTOPH HELMCHEN 592

Effects of Intravenous Opioids on Eye Movements in Humans: Possible Mechanisms. *By* K.G. ROTTACH, A.E. DZAJA, W.A. WOHLGEMUTH, T. EGGERT, AND A. STRAUBE 595

Evaluation of Current Optical Methods for Treating the Visual Consequences of Nystagmus. *By* STACY S. YANIGLOS, JOHN S. STAHL, AND R. JOHN LEIGH 598

Latency of Dynamic and Gaze-Dependent Optotype Recognition in Patients with Infantile Nystagmus Syndrome versus Control Subjects. *By* RICHARD W. HERTLE, MITRA MAYBODI, GEORGE F. REED, AMIR H. GUERAMI, DONGSHENG YANG, AND EDMOND J. FITZGIBBON 601

A Robust, Normal Ocular Motor System Model with Latent/Manifest Latent Nystagmus and Dual-Mode Fast Phases. *By* JONATHAN B. JACOBS AND LOUIS F. DELL'OSSO 604

A Hypothetical Fixation System Capable of Extending Foveation in Congenital Nystagmus. *By* JONATHAN B. JACOBS AND LOUIS F. DELL'OSSO ... 608

Congenital Periodic Alternating Nystagmus: Response to Baclofen. *By* DAVID SOLOMON, NEIL SHEPARD, AND ANUPAM MISHRA 611

Index of Contributors 617

Financial assistance was received from:

Co-sponsors
- CASE WESTERN RESERVE UNIVERSITY SCHOOL OF MEDICINE
- UNIVERSITY HOSPITALS OF CLEVELAND

Major Funder
- EVENOR ARMINGTON FUND

Supporters
- CLEVELAND FOUNDATION
- NATIONAL INSTITUTE OF NEUROLOGICAL DISORDERS AND STROKE, NIH

Contributors
- ALLERGAN CORPORATION
- CASE WESTERN RESERVE UNIVERSITY AND UNIVERSITY HOSPITALS OF CLEVELAND VISUAL SCIENCES RESEARCH CENTER, VISUAL SCIENCES TRAINING PROGRAM, AND DEPARTMENT OF OPHTHALMOLOGY
- MERCK & CO. INC.

The New York Academy of Sciences believes it has a responsibility to provide an open forum for discussion of scientific questions. The positions taken by the participants in the reported conferences are their own and not necessarily those of the Academy. The Academy has no intent to influence legislation by providing such forums.

Contributors to the conference **Neurobiology of Eye Movements: From Molecules to Behavior**, held in Cleveland, Ohio, October 4–6, 2001, to honor Robert B. Daroff, M.D., pictured on the steps of Severance Hall, home of the Cleveland Orchestra

Foreword

This volume of the *Annals* is the fruit of a conference, *Neurobiology of Eye Movements: From Molecules to Behavior*, held at Case Western Reserve University, Cleveland, Ohio, USA on October 4-6, 2001. More than 50 basic and clinical scientists contributed to this conference, held to honor Dr. Robert B. Daroff, who has been influential in the fields of eye movements and neurology over the past 30 years. The presentations highlighted the range of current knowledge about the control of eye movements, but also featured discoveries—some revolutionary—that have opened up new areas of research.

We are grateful to many individuals for making this project a success. First, we thank the conference contributors, who traveled from around the world, at a difficult time, to present their work. The smooth running of the meeting and prompt publication of this volume were made possible by the staff of the New York Academy of Sciences, especially Sherryl Usmani, Dr. Rashid Shaikh, Marion Garry, and Sheila Kane. The staff of the Cleveland Museum of Art made it possible to hold the poster session in an ambience of high art, and also to reproduce the face of the jester Calabaza from the portrait by Diego Velázquez on the cover of this volume. Thanks go out to Henry L. Meyer III and the Key Corporation for sponsorship of the reception at the Museum of Art. The staff of the Cleveland Orchestra made possible an outstanding evening of music for conference participants. Elizabeth Ciemins, Patricia Anderson, and Dr. Joy Derwenskus made special contributions, and we thank them. We appreciate the help of Dr. Nathan Berger, Dean of the Case Western Reserve University School of Medicine, and Farah Walters, President and Chief Executive Officer of University Hospitals of Cleveland, in obtaining grant support.

We are pleased to acknowledge major support by the Evenor Armington Fund, and sponsorship by Case Western Reserve University School of Medicine, University Hospitals of Cleveland, Dorothy R. Blair, the Cleveland Visual Sciences Research Center, Visual Sciences Training Program and the Department of Ophthalmology of Case Western Reserve University, the Cleveland Foundation, the National Institute of Neurological Disorders and Stroke, the Allergan Corporation, and Merck & Company, Inc.

— HENRY J. KAMINSKI
— R. JOHN LEIGH

A Personal Introduction to Eye Movements

ROBERT B. DAROFF

University Hospitals of Cleveland, 11100 Euclid Avenue, Cleveland, Ohio 44106-5015, USA

ABSTRACT: The presentations at this meeting demonstrate that the control of eye movements is probably the best understood motor system from the level of genes and molecules to the neural orchestration of complex behaviors. Here I review my early involvement in this field, with personal impressions of how our current concepts developed and the individuals who shaped them.

KEYWORDS: eye movements; ocular motor; oculomotor; control systems

I am pleased to write an opening paper for a conference that aims, for the first time, to deal with the ocular motor system from the level of molecules to that of complex behavior. The ability to address such a range of issues is an exciting commentary on how much progress has been made in our field. My main focus in this paper concerns some of the individuals behind these advances over the past 30 years, especially those who influenced me, and others whom I have been privileged to help. As a starting point, I turned to some reflections I made in 1996 both on my personal development and advances in ocular motor research.[1]

HOW I CAME TO STUDY EYE MOVEMENTS

Why study eye movements? This question can be restated and expanded several ways. Why did eye movements evolve and why are they necessary? Why are eye movement abnormalities so common in neurological patients and a knowledge of them important in neurological diagnoses? Why is eye movement research a growth industry? And how did I become more interested in eye movements than other areas of clinical neurology? I will answer these questions but not in the order presented.

First, let's take Daroff from boyhood to a neurologist with a primary interest in eye movements. I decided to become a physician at age 13, in the eighth grade, in the New York City public school system, largely because of the influence of a science teacher, Herman Horn. As a freshman at the University of Chicago (1952–1953), while rooming with Carl Sagan, I became engrossed in Freud and was determined to be come a psychiatrist-psychoanalyst. My experience as a third-year medical student at the University of Pennsylvania made me reluctant to give up organic medicine for psychoanalysis, and the exposure to excellent clinical neurological educators—Gabriel Schwarz, James Toole, Abraham Ornsteen, Wilmer Anderson, and

Address for correspondence: Robert B. Daroff, M.D., University Hospitals of Cleveland, 11100 Euclid Avenue, Cleveland, OH 44106-5015. Voice: 216-844-3809; fax: 216-844-7611.
rbd2@ po.cwru edu

John Bevilacqua—steered me to neurology. During my internship, I saw a copy of David Cogan's *Neurology of the Ocular Muscles,* second edition,[2] in an ophthalmologist's living room; I was told it was a good book, so I bought it but didn't read it until the next summer (1962). I was a beginning neurology resident at Yale, on a three-month rotation at the West Haven Veterans Administration Hospital, when we admitted a patient with a supranuclear ophthalmoplegia. The attending neurologist, Lewis Levy, suggested that one of the house officers become authoritative about eye movements and give a talk to the group. There were four of us on service and three were rotators (two from psychiatry and one from medicine). As the only neurologist, I had to "volunteer" and then spent the next two months reading about eye movements, almost to the exclusion of other areas of neurology. My primary source was Cogan,[2] supplemented by Walsh's *Clinical Neuro-Ophthalmology,* second edition,[3] Duke-Elder's *Textbook of Ophthalmology,*[4] and reprints from Goodwin Breinin on ocular electromyography lent to me by a fellow resident (James Pritchard) who had learned of my assignment. At the end of my rotation at the VA in September 1962, I presented a report to the ward team based upon a triple-spaced, 26-page manuscript entitled, "Conjugate Eye Movements," that remained in my file cabinet until 1996, when I filed it with the National Auxiliary Publication Service.[5]

Later in 1962, my chief, Gilbert H. Glaser, recommended Kestenbaum's *Clinical Methods of Neuro-Ophthalmalogic Examination,* second edition,[6] but I found it dry and didn't read it until the spring of 1964 while doing a neuro-ophthalmology elective at the University of Miami under J. Lawton Smith. During a week that Smith was out of town, I went to the library and started reading Walsh. Fortunately, Edward Norton (neuro-ophthalmologist–turned–retinal surgeon) came by and asked what I was reading in Walsh; when I replied that my intent was to read it completely, Norton said something like "Walsh isn't a good idea for cover-to-cover reading; Kestenbaum is better." I followed this advice and never regretted it. Kestenbaum's observations and intuition were incredibly perceptive. Those thinking they uncovered a "new" ocular sign should first check Kestenbaum,[6] as we experienced with periodic alternating nystagmus and the shifting null.[7] Later in 1964, Bender's *The Oculomotor System*[8] was published, and I was introduced to contemporary ocular motor physiology. "Contemporary" is important because my major previous source, Cogan's second edition,[2] was published in 1956, and most of the references were from the early part of the century.

Throughout my residency and neuro-ophthalmology fellowship (1967–1968) with William Hoyt, I envisioned myself as an eye movement clinical phenomenologist. Although I collaborated with Donald Higgins in a control system analysis of ocular dysmetria,[9] the methodology and conceptualization were all Higgins'. I didn't have the background to consider sophisticated eye movement research, and when Arnold Starr published his seminal paper on the eye movement abnormalities in Huntington's chorea in 1967,[10] I was distinctly ambivalent. Although excited about this first application of quantitative oculography to a clinical eye movement problem discussed in modern terms, I regretted being unable to do such a study myself.

I should express my indebtedness to William (Bill) Fletcher Hoyt, my neuro-ophthalmology mentor at the University of California, San Francisco. When I returned to the states in 1966 after a year in Vietnam,[11] I imagined that life would henceforth be stress-free. Then came my fellowship with Hoyt. Although intense,

Hoyt's example led me to fixate not only with the fovea, but compulsively on a discipline—which for Hoyt was the totality of neuro-ophthalmology and which for me was eye movements.

In early 1968, I was negotiating by mail with Peritz Scheinberg to join his faculty at the University of Miami. He asked what research project I planned, and I replied that I didn't do research but was quite productive as a descriptive phenomenologist. Prior to joining the faculty in Miami, I had written 22 journal articles with only the previously mentioned paper with Higgins[9] constituting real "research." Scheinberg wrote back, "Everyone on my faculty does systematic research. Submit a project." Thus, forced, I wrote a proposal relating phenytoin blood levels to nystagmus (recorded by oculography), to supplement Kutt's clinical observations.[12] This required eye movement recording instrumentation and, fortunately, the Miami VA Hospital and my friend and colleague, Noble David, had the resources to purchase DC oculographic instrumentation (I had learned from an electrical engineer that eye position cannot be recorded accurately with AC coupled instrumentation.)

We started recording eye movements in 1969 at the Miami VA in the expansively titled Ocular Motor Neurophysiology Laboratory. (I explained elsewhere[13] my preference for "ocular motor," rather than "oculomotor." It simply prevents confusion between the entire system and one of its components—the third cranial [oculomotor] nerve.) Our recorder had eight channels and, wanting to use them all, we recorded from each eye simultaneously. To our surprise, horizontal eye movements in normal subjects were not always conjugate; one or the other eye would frequently overshoot or undershoot. Unwittingly, we were to be the first to record each eye separately with DC electro-oculography. Previous researchers used bi-temporal electrodes (averaging the movements of the two eyes) or recorded only from one eye, assuming that the other eye was doing the same thing. Heavily engaged in clinical activities, I didn't have the time to pursue research aggressively, but our chief resident in neurology, Ronald Weber, wanted to spend a year as a Fellow in my laboratory. I tried to dissuade him but he was insistent, and we studied the metrics of normal horizontal eye movements.[14,15] There were two additional lucky breaks. First, Louis Dell'Osso, an electrical engineer, who did his Ph.D. thesis on the analysis of his own congenital nystagmus, was recruited to Miami by the Department of Biomedical Engineering. Because my laboratory was equipped and operational, Dell'Osso became my partner and added sophisticated engineering/control system expertise. Second, Todd Troost, having finished his neuro-ophthalmology fellowship with Hoyt, wanted to study eye movements, and Hoyt told him that we had the best clinical eye movement recording laboratory in North America. That started a very productive Dell'Osso-Troost-Daroff collaboration. Not unexpectedly, we never did perform the study of phenytoin and nystagmus that I proposed to Scheinberg during my recruitment to Miami.

We also had a spectacular group of Fellows over the years, including Larry Abel, currently in Australia, and Jim Sharpe, a participant in this conference. Our success in Miami was enhanced by the Bascom Palmer Eye Institute, with the neuro-ophthalmology duo of Lawton Smith and Joel Glaser feeding us patients and providing us the luxury of concentrating on the efferent system, much as Robert Tomsak currently does for us in Cleveland. And then there was Thomas Brandt, a bright young neurologist trained by Richard Jung in Freiburg, whose visit to our laboratory led to a fruitful collaborative relationship.

I arrived in Cleveland in 1980 with Dell'Osso, Troost, and Abel, and had the good fortune of recruiting John Leigh from Hopkins and, later, John Porter from Kentucky. Two of our residents, Henry Kaminski and John Stahl, stayed on to bolster our ocular motor team.

EYE MOVEMENT BASICS

Eye movements subserve vision. The primate ocular motor system consists of several subsystems that generate different types of eye movements to different stimuli, but *all* in the service of vision. Walls wrote the classic paper on the phylogenetic evolution of eye movements.[16] Afoveate animals, such as the rabbit, needed eye movements primarily to stabilize the eyes in space during head movement; their dominant eye movement subsystems were vestibular and optokinetic. With the development of the fovea—an area of the retina with the best visual discriminatory function—eye movement subsystems evolved both to foveate eccentric targets and maintain foveation when the target moved. Saccades are fast eye movements that refixate the fovea. The great American psychologist William James captured this function in the late nineteenth century when he said, "The peripheral retina is like a sentinel and when an object of regard falls upon it, it shouts 'hark, who goes there' and calls the fovea to the spot." Pursuit (tracking) eye movements maintain fixation when objects move. Vergence movements both capture and maintain fixation in the Z axis. In addition, there is a fixation subsystem that holds the eyes on intended targets. Leigh and Zee[17] summarized all these eye movement types, their anatomical substrates, physiological properties, and their dysfunction secondary to neurological disease.

The ease of recording is a major reason for the glut of eye movement studies in humans. Quantitative analysis of limb movements is difficult because of the multiple joints, vectors, afferent inputs, and variable mechanical loads. With eye movements, however, the inputs can be isolated by stabilizing the head (eliminating vestibular influences) and controlling the other major input, the visual targets and tasks. Moreover, eye movements can be measured in all vectors (horizontal, vertical, diagonal, and torsional) with precision down to minutes of arc.[18]

I previously reviewed[19] the prevalent notion in the 1960s that the ocular motor system was organized from cortex to eye muscle with parallel fast (saccadic) and slow (pursuit) pathways and how this concept evolved, based upon the study of motoneuronal firing patterns in primates, into our current belief that the same basic anatomical substrate (from cortex to muscle) subserves all eye movements. Over the years, as a contributor to and reviewer for neurological journals edited by Fred Plum (*Archives of Neurology* and *Annals of Neurology*), I was repeatedly asked by Plum what the study of eye movements tells us about movements in general. It was a difficult question to answer other than, "I don't know, but I can measure eye movements precisely and I cannot do the same with limb movements. Therefore, I study eye movements."

In 1986, David Robinson addressed this issue in a paper, "Is the Oculomotor System a Cartoon of Motor Control?"[20] The eye has only one "joint," whereas the limbs have multiple joints. Eye movement forces are linear whereas limb movements are nonlinear. Reciprocal innervation is always present with eye movements but not in the limbs. Limb muscles have stretch reflexes; the eye muscles do not. The eyes are not encumbered by external forces that add to the complexity of studying limbs. Different types of eye movements (saccades, pursuit, vestibular, vergence) can be studied separately. Most importantly, the eye movement motoneurons are precisely identified, allowing a "bottom up" study. Eye movement physiologists can begin at the brainstem motoneurons and then go progressively higher to the cerebral cortex, moving from the simplest to the more complex premotor neuronal pools. With limb movements, a "top down" approach is required, beginning with premotor structures. All this explains why eye movements are much easier to study, and understand, than limb movements.

Robinson answered Fred Plum's question by stating, "The simplicity of the oculomotor system may allow us to penetrate deeper and more quickly into the problems of central planning and execution of movements in general."[20] Personally, I am less optimistic and have always studied eye movements as a separate and distinct neural system without expecting any insights into limb-movement control. In the same paper, Robinson[20] listed the five major factors responsible for the technological advances in the understanding of eye movements over the past three decades:

(1) The ability to record from single neurons in alert behaving animal
(2) The development of tracers to determine afferent connections of nuclear structures
(3) Computers that permit rapid data analysis
(4) Precise recording techniques to perform noninvasive measurements in all planes, down to minutes of arc, in humans and experimental animals
(5) The systems approach where models provide the necessary hypotheses to focus basic and clinical research.

The proceedings of this current conference extend this list with several new techniques including the ability to study the effects of genetic deletions on eye movements in mice, identification of neurotransmitters in the ocular motor system by measuring the effects of pharmacological inactivation, and mapping of cerebral activity during specific behaviors with functional imaging.

WHERE EYE MOVEMENTS HAVE LED ME

I was fortunate to have had supportive mentors, excellent research environments, and superb collaborators. Moreover, the David Robinson–types began unraveling the ocular motor system as I was beginning my research. At the time when I was deciding upon a fellowship in neuro-ophthalmology, with an emphasis on studying eye movements and nystagmus, Frank Elliott of the University of Pennsylvania wrote to me, "A lot of intensely intelligent and highly dedicated workers have given their lives to this subject of nystagmus and very little has come out of it." I was reminded that Wilbrand, a famous German neuro-ophthalmologist, told the young Robert Wartenberg in 1921, "Never write about nystagmus, it will lead you nowhere."[21] As the reader will realize from the chapters that follow, the study of eye movements has led me into the company and friendship of the many brilliant and scholarly contributors to this conference, and the excitement generated by their biological curiosity and perseverance.

ACKNOWLEDGMENTS

I am grateful to Georg Thieme Verlag for allowing me to reproduce, almost in its entirety, a previously published chapter.[1]

REFERENCES

1. DAROFF, R.B. 1996. Eye movements in health and disease. *In* Challenges in Medicine. W. Siegenthaler & R. Haas, Eds.: 59–63. George Thieme Verlag. Stuttgart.
2. COGAN, D.G. 1956. Neurology of the Ocular Muscles, 2nd edit. C.C. Thomas. Springfield, IL.
3. WALSH, F.B. 1957. Clinical Neuro-Ophthalmology, 2nd edit. Williams & Wilkins. Baltimore, MD.
4. DUKE ELDER, S. 1932. Textbook of Ophthalmology, Vol. 1., C.V. Mosby. St. Louis, MO.
5. NATIONAL AUXILLARY PUBLICATION SERVICE. Document no. 05275. Microfiche Publications, P.O. Box 3513, Grand Central Station, New York, NY 10163–3513.
6. KESTENBAUM, A. 1961. Clinical Methods of Neuro-Ophthalmologic Examination, 2nd edit. Grune & Stratton. New York.
7. DAROFF, R.B. & L.F. DELL'OSSO. 1974. Periodic alternating nystagmus and the shifting null. Can. J. Otol. **3:** 367–371.
8. BENDER, M.B. 1964. The Oculomotor System. Harper and Row, New York.
9. HIGGINS, D. & R.B. DAROFF. 1966. Overshoot and oscillation in ocular dysmetria. Arch. Ophthalmol. **75:** 742–745.
10. STARR, A. 1967. A disorder of rapid eye movements in Huntington's chorea. Brain **90:** 545–564.
11. DAROFF, R.B. 1999. Neurology in a combat zone. J. Neurol. Sci. **170:** 131–137.
12. KUTT, H., H. WINTERS, R. KOKENGE & F. MCDOWELL. 1964. Diphenylhydantoin metabolism, blood levels, and toxicity. Arch. Neurol. **11:** 642–648.
13. DAROFF, R.B. 2000. Random comments: neurologists and neuro-ophthalmology; the "ocular motor" system; update on ophthalmoplegic migraine. Semin. Neurol. **20:** 145–149.
14. WEBER, R.B. & R.B. DAROFF. 1971. The metrics of horizontal saccadic eye movements in normal humans. Vision Res. **11:** 921–928.
15. WEBER, R.B. & R.B. DAROFF. 1972. Corrective movements following refixation saccades: type and control system analysis. Vision Res. **12:** 467–475.
16. WALLS, G.L. 1962. The evolutionary history of eye movements. Vision Res. **2:** 69–80.
17. LEIGH, R.J. & D.S. ZEE. 1999. The Neurology of Eye Movements, 3rd edit. Oxford University Press. New York.
18. DELL'OSSO, L.F. & R.B. DAROFF. 1999. Eye movement characteristics and recording techniques. *In* Neuro-Ophthalmology, 3rd edit. J.S. Glaser, Ed.: 327–343. Lippincott Williams & Wilkins. Philadelphia.
19. DAROFF, R.B. & A. NEETENS. 1989. Neurological Organization of Ocular Movement [Forward]. Bull. Soc. Belge Ophthalmol. **237:** iii–ix.
20. ROBINSON, D.A. 1986. Is the oculomotor system a cartoon of motor control? Prog. Brain Res. **64:** 411–417.
21. WARTENBERG, R. 1953. Diagnostic Tests in Neurology. Year Book. Chicago. p.42.

Extraocular Muscle: Cellular Adaptations for a Diverse Functional Repertoire

JOHN D. PORTER

Departments of Ophthalmology, Neurology, and Neurosciences, Case Western Reserve University and The Research Institute of University Hospitals of Cleveland, Cleveland, Ohio 44106-5068, USA

ABSTRACT: Oculomotor control systems are considerably more complex and diverse than are spinal skeletomotor systems. Moreover, individual skeletal muscles are frequently functional role-specific, while all extraocular muscles operate across a very wide dynamic range. We contend that the novel phenotype of the extraocular muscles is a direct consequence of the functional demands imposed upon this muscle group by the central eye movement controllers. This review highlights five basic themes of extraocular muscle biology that set them apart from more typical skeletal muscles, specifically, the (a) novel innervation pattern, (b) heterogeneity in contractile proteins, (c) structural and functional compartmentalization of the rectus and oblique muscles, (d) diversity of extraocular muscle fiber types, and (e) relationship between the novel muscle phenotype and the differential response of these muscles in neuromuscular and endocrine disease. Finally, new data from broad genome-wide profiling studies are reviewed, with global gene expression patterns lending substantial support to the notion that the extraocular muscles are fundamentally different from traditional skeletal muscle. This novel eye muscle phenotype represents an adaptation that exploits the full range of variability in skeletal muscle to meet the needs of visuomotor systems.

KEYWORDS: extraocular muscle; muscle phenotype; muscle development; neuromuscular; gene profiling; microarray; SAGE; review

INTRODUCTION

For over 125 years, color has been a durable concept in skeletal muscle biology. The notion of red and white muscles continues to be invaluable as a unifying concept in muscle development, structure, function, and dysfunction. Muscle role specificity then is achieved simply on the basis of the proportional content of a few, three to four, highly conserved myofiber types that represent gradations in color from red to white. For example, in a red muscle (e.g., soleus) there is a predominance of red (type I) and intermediate (type IIA) fiber types that mediate its slow-twitch, but fatigue-resistant role in maintenance of upright posture. By contrast, an immediate neighbor of soleus, the gastrocnemius, acts phasically upon movements of the ankle

---

Address for correspondence: John D. Porter, Ph.D., Department of Ophthalmology, University Hospitals of Cleveland and Case Western Reserve University, 11100 Euclid Avenue, Cleveland, OH 44106-5068. Voice: 216-844-7053; fax: 216-844-4792.

jdp7@po.cwru.edu

through the same tendon as the soleus, achieving fast-twitch contractions but with an inability to long-sustain force, due to its predominately white (type IIB) muscle fiber content.

Some skeletal muscles diverge from the muscle fiber prototype, instead exhibiting fundamental differences in fiber types, such that the traditional colors of muscle biology cannot be applied to them. The clearest example is the muscles that move the eyes, the extraocular muscles (EOMs). The EOMs represent the final common pathway for a very complex and diverse set of movement control systems. This review organizes the evidence that EOM must utilize the full range of options in muscle biology to serve the functional roles assigned to them. In particular, contemporary gene-profiling techniques firmly establish that EOM has diverged from other skeletal muscles, employing a wide variety of traits used in both skeletal and cardiac musculature, as well as breaking some of the established "rules" for skeletal muscle biology.

THE WORKING ENVIRONMENT FOR EXTRAOCULAR MUSCLE

The functional goals of the EOMs are maintenance of stable gaze position and generation of target-acquiring eye movements. In this task, EOM serves as the effector of the final common pathway for five distinct neural control systems that mediate the well-identified eye movement classes discussed elsewhere in this volume. Because the area of high-acuity vision, the fovea, subtends a very small angle of visual space, the task of gaze control must be accomplished with high precision through the coordinated activity of the six EOMs. Because it is the pattern of usage that is a key determinant of skeletal muscle properties, the complex and diverse neural output patterns of the oculomotor system can shape the eye muscle phenotype.

Taken together, this working environment for EOM is much more sophisticated than that of any other skeletal muscle. The use of traditional tools in muscle biology, from electron microscopy to gene expression analysis, has identified several specializations that have been used to generate a comprehensive EOM fiber type classification scheme. Known adaptations are reviewed, emphasizing several major themes that have emerged in EOM biology. This discussion is followed by consideration of data from our most recent efforts to using genome-wide profiling techniques for development of a comprehensive understanding of EOM biology.

THEME 1. THE NOVEL INNERVATION PATTERN OF EXTRAOCULAR MUSCLE

It is well established that skeletal muscle characteristics are influenced by the patterned activity of the motoneurons that innervate them. Thus, understanding of motoneuron discharge patterns produced by oculomotor control systems is germane to knowledge of the diversity of EOM biology. Recordings from oculomotor motoneurons in chronic alert monkeys making eye movements to visual targets identified a highly stereotyped discharge pattern.[1] Studies report several important findings, including (a) tight linkage between sustained motoneuron activity and eye position, including smooth changes in this activity during slow eye movements and a step

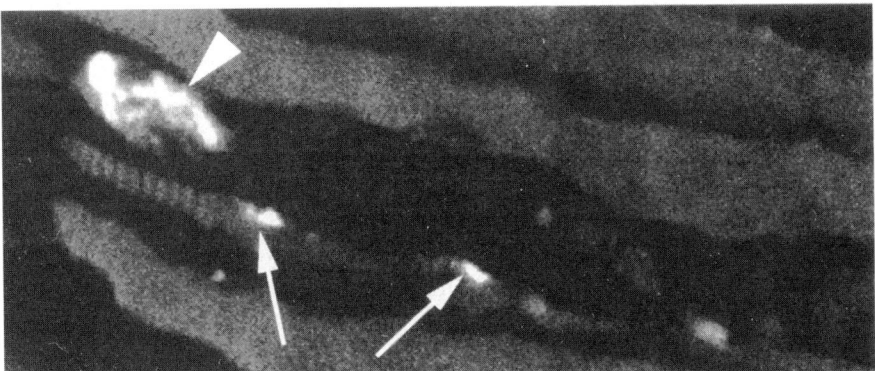

FIGURE 1. Confocal photomicrograph of a longitudinal section of rodent EOM stained with Texas red-α-bungarotoxin to label acetylcholine receptor aggregations at neuromuscular junctions. Distinctive SIF and MIF junctional morphologies are identified. *Arrowhead* indicates en plaque junction on an EOM SIF. *Arrows* indicate two en grappe junctions on a single EOM MIF. (Courtesy of S. Khanna.)

change in activity during rapid eye movements, (b) rapid and large pulses of motoneuron discharge associated with saccadic eye movements, and (c) an overall high activity of oculomotor motoneurons, often exceeding that of spinal motoneurons by an order of magnitude. At the level of oculomotor motoneuron activity patterns, EOM fibers must be responsive over an unprecedented dynamic range that requires adaptations for contraction speed and fatigue resistance well beyond that experienced by the more typical skeletal muscles.

Adult mammalian skeletal muscle exhibits a stereotypical pattern of a motor axon contacting each muscle fiber roughly in the center of the fiber length. Acetylcholine released from nerve terminals interacts with muscle surface receptors, leading to an opening of sodium channels, generation of a propagated action potential along the muscle fiber length, and an all-or-none twitch contraction. While EOM has a high percentage (80–85%) of such singly innervated muscle fiber (SIF) types, these muscles are one of a few in mammals that also exhibit multiply innervated muscle fibers (MIF) that do not propagate action potentials (FIG. 1).[2] Instead of a rapid twitch, these fibers undergo slow, graded contractions that increase muscle force only slowly and to a small degree. Moreover, motoneurons that innervate MIFs are small and spatially segregated from SIF motoneurons in the oculomotor, trochlear, and abducens nucleus, suggesting that they may receive premotor signals that are, at least in part, different from those of SIF motoneurons.[3] The function(s) subserved by a phylogenetically primitive MIF system (nearly identical to that in amphibians) in eye movements are, as yet, speculative. There are suggestions that the EOM MIFs either may act to finely increment muscle force around primary position[4] to maintain fixation or may simply be part of a sophisticated proprioceptive apparatus adapted for oculomotor control systems.[5]

THEME 2. HETEROGENEITY OF MYOSIN EXPRESSION IN EXTRAOCULAR MUSCLE

The contractile function of striated muscle is mediated by the sliding of actin and myosin filaments relative to one another, such that overall muscle shortening is achieved via the additive effect of incremental length changes in sarcomere units arranged end-to-end along the length of individual muscle fibers. While skeletal muscle actin is conserved across all skeletal muscle fiber types, myosin heavy chains are represented by a multi-gene family and are differentially expressed in the various skeletal muscle fiber types. Myosin isoform content is functionally tuned to the specific muscle fiber types because the distinct myosin heavy chain isoforms exhibit diversity in contraction speed and energetics, and skeletal muscle fiber types typically express only one heavy chain gene suited to the contractile properties of the particular fiber type. During development, skeletal muscles transiently express embryonic and neonatal myosin heavy chain isoforms, while adult muscles express only the adult myosin heavy chain types I, IIA, IIB, and IIX/D.

By contrast, EOM exhibits considerable heterogeneity in myosin heavy chain expression.[6–13] First, expression of more than one myosin isoform in single muscle fibers is frequently observed in EOM. Second, embryonic and neonatal myosin isoforms are only incompletely downregulated in EOM fiber types, because some fibers retain developmental isoforms while also expressing at least one adult isoform.[6, 9,12] This finding suggests that functional properties of the developmental myosins are, at least in part, better suited than adult isoforms to the demands placed upon some of the EOM fiber types. Third, one of the two EOM MIF types expresses α-cardiac myosin, an isoform otherwise seen only in the heart and muscles of mastication, and the typical slow-twitch (type I) myosin.[10] Finally, the EOMs express a novel EOM-specific myosin heavy chain isoform, seen only in EOM and select laryngeal muscles.[6,9,14,15] Thus, EOM expresses nearly all known striated muscle myosin heavy chain isoforms (the only exception is a superfast myosin found only in carnivore jaw muscles; but because carnivore EOM has not been evaluated for this isoform, the possibility that it is expressed in EOM in a species-specific fashion cannot be excluded). In sensory deprivation studies, we have established that the EOM-specific myosin is developmentally regulated by maturing visuomotor and vestibular systems,[16,17] consistent with the hypothesis that some EOM properties are emergent traits of the oculomotor system in general. Finally, the pattern of differential expression of contractile protein isoforms also extends to a contractile regulatory protein, troponin, for which isoforms that are normally rare in skeletal muscle predominate in EOM.[18]

THEME 3. EXTRAOCULAR MUSCLE COMPARTMENTALIZATION

Skeletal muscles frequently exhibit heterogeneous distribution of muscle fiber types across the muscle cross-sectional area, often having specific layering or compartmentalization in fiber-type distribution. The functional consequences of such compartmental organization have been addressed previously in several muscle groups, including the muscles of mastication, dorsal neck, and leg musculature.[19,20]

It has been long recognized that the rectus and oblique EOMs of all mammalian species exhibit compartmentalization into two distinctive layers. A c-shaped orbital layer is adjacent to the bony walls of the orbit, while the global layer is immediately adjacent to the optic nerve and the globe itself. This feature is highly conserved in phylogeny as many sub-mammalian species also exhibit the compartmentalized organization. The two layers contain different muscle fiber types (see below) with distinct activation patterns during eye movements. The two layers also significantly differ in myosin isoform composition (see above); the functional properties of the predominant orbital layer isoforms (EOM-specific and embryonic) are incompletely understood, so functional inferences for the orbital layer are not easily based upon myosin composition. It has been recognized that the orbital layer terminates early, prior to the formation of a distinct muscle tendon, while the global layer clearly gives rise to the tendon of insertion into the globe. Recently, accumulated morphologic and high-resolution imaging evidence has established that the EOM layers do indeed have functionally discrete insertion points,[21,22] with the global layer inserting upon the eye proper while the orbital layer inserts into expansions of orbital connective tissues known as EOM pulleys. A review of the participation of EOM in an orbital functional model, the active pulley hypothesis, is discussed by Demer in this volume.

THEME 4. EXTRAOCULAR MUSCLE-SPECIFIC FIBER TYPES

Patterned differences in skeletal muscle fiber types define three to four fiber types that are conserved in nearly all muscle groups. While the concept of distinct muscle fiber types has been invaluable, caution must be exercised in declaring any new skeletal muscle fiber type without robust supporting data across a wide range of muscle structural/functional traits. Beginning with the basic differences in innervation pattern and myosin gene expression patterns discussed above, the notion that EOM fiber types do not respect the traditional fiber-type classification schemes has evolved. It is now accepted that the EOMs contain six distinct fiber types with a broad spectrum of differences from other skeletal muscle fiber types; the EOM fiber types are highly conserved across mammalian species. The most accepted scheme for EOM fiber

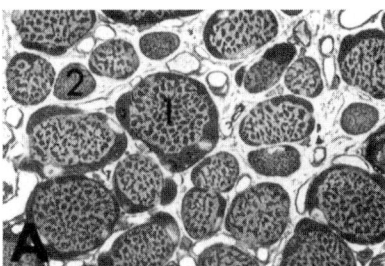 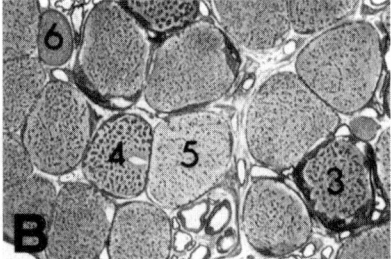

FIGURE 2. Phase-contrast light photomicrographs of EOM fiber types in monkey orbital (**A**) and global (**B**) layers. The six EOM fiber types are (1) orbital SIF, (2) orbital MIF, (3) global red SIF, (4) global intermediate SIF, (5) global white SIF, and (6) global MIF.

types includes six distinct types with terminology based upon location (orbital or global), innervation pattern (SIF or MIF), and fatigue properties, or color (red, intermediate, or white).[2,5,23] Light photomicrographs illustrating EOM fiber types for rhesus monkey are shown in FIGURE 2. From single-unit physiology studies of oculomotor motoneurons[1] and intraoperative EMG data,[24] it is thought that all EOM fiber types participate in all types of eye movement. However, although considerable information is available from motor unit studies in acute physiologic studies,[25] a key outstanding question in EOM biology is the potential and likely relationship of EOM fiber types to oculomotor motoneurons with different on-positions and firing rate/eye position plot slopes. These data will be critical to establishing firm relationships between the diversity of EOM fiber types and the functional demands of eye movement control systems.

THEME 5. DIFFERENTIAL RESPONSE OF EXTRAOCULAR MUSCLE IN DISEASE

The conservation of three to four structural and functional fiber types found in most skeletal muscles correlates with the generalized response of skeletal muscles in a variety of neuromuscular and autoimmune disorders. Some diseases may, for example, preferentially affect the slow (type I) or the fast (type II) fibers across many skeletal muscles. Duchenne muscular dystrophy is a progressive myopathic disorder that ultimately results in cyclic degeneration and regeneration of nearly all skeletal muscles, but with type II fiber preference. By contrast, there are isolated disorders, such as the distal myopathies, that affect some muscle groups, but not others, with no clear scientific understanding of the mechanisms behind the heterogeneous response.

In contrast, the selective sparing or targeting of the EOMs in various diseases is a frequent occurrence (see Kaminski, this volume).[23] The novel phenotype of EOM makes this muscle group either more or less susceptible to disease than other skeletal muscles. Prominent examples in which linkages between EOM phenotype and disease responses are being studied include muscular dystrophy[26–31] and myasthenia gravis (see Kaminski, this volume). Establishment of relationships between the EOM phenotype and preferential sparing or targeting of muscle groups in disease may not only advance knowledge of EOM biology, but might provide insight into new treatment regimens.

THE PENULTIMATE THEME: GENE PROFILING AND THE EXTRAOCULAR MUSCLE PHENOTYPE

Taken together, evidence to date establishes that the biology of EOM is clearly different from that of other skeletal muscle, but the notion of EOM-specific traits has been given little attention by general muscle biologists, perhaps because of the highly specialized nature of this muscle group. Given the limited resources of a relatively small number of investigators in this field, it is difficult to meet the goal of a comprehensive understanding of EOM adaptations for the complex oculomotor repertoire using a "one-gene-at-a-time" approach. A contemporary solution to this problem is application of genome-wide expression profiling techniques that allow simultaneous determination of expression levels for thousands of genes—DNA microarray and serial analysis of gene expression (SAGE). Microarray technology relies upon hybridization of gene products of a given tissue to gene-specific probes that are arrayed on a "chip." Microarray is rapid, acutely sensitive, and can detect small quantities of expressed genes, but is limited to a current maximum of the 12,000 or so genes with probes on each array. By contrast, SAGE is both time- and labor-intensive and has difficulty detecting genes expressed at low levels, but is not constrained to a specific set of gene probes, and data are more portable for use in comparisons with findings from other laboratories. Our lab is currently using both techniques to profile EOM gene expression in health and disease.

By DNA microarray, we have compared three presumptive classes of skeletal muscle—EOM, masticatory, and hindlimb. Pairwise comparisons using conservative criteria have identified 287 genes that are differentially expressed in EOM in comparison to the other skeletal muscle groups (FIG. 3).[32] Among the expression differences, EOM is low in enzymes and regulators related to glycogen metabolism; this finding is consistent with low glycogen content of EOM and suggests that there are important differences in energy metabolism in EOM. By contrast, we observed no differences in gene expression between masticatory and hindlimb muscles. These EOM data contrast with the rather limited number of genes (30–40) that were differ-

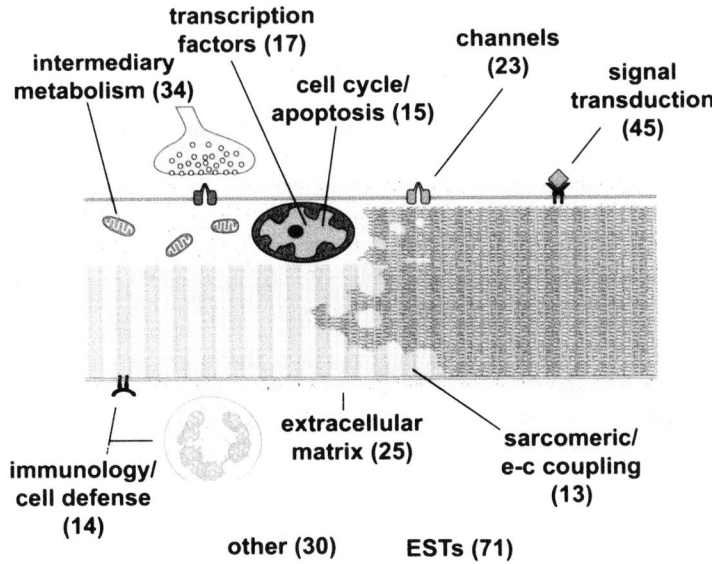

FIGURE 3. Muscle fiber schematic illustrating the breadth of differentially expressed genes in EOM versus other skeletal muscles, as identified using Affymetrix DNA microarrays. In all, 287 genes exhibited EOM-specific expression patterns; these are distributed across major function categories important to muscle biology (number of genes per category is shown in parentheses). EST denotes expressed sequence tags or putative genes. The full list of differentially regulated genes is available at http://www.pnas.org/content/vol98/issue21/index.shtml.

entially expressed in comparisons of traditional red and white skeletal muscles.[33] Thus, it is clear that EOM is more different from any other skeletal muscle than the extremes of typical skeletal muscle, red and white, are from each other. Our SAGE data[34] reinforce the distinctive gene expression pattern of EOM, including EOM-specific patterns in contractile, energy metabolism, ribosomal, and cytoskeletal protein gene expression. Collectively, these patterned differences in gene expression define EOM as a distinct muscle class and ultimately may explain the unique response of these muscles in neuromuscular diseases.

CONCLUSIONS

The historical argument that EOM represents a fundamentally different class of skeletal muscle is strong. Recent gene expression data confirms and greatly extends knowledge of the depth and breadth of gene expression differences between EOM and the more traditional skeletal muscles. EOM utilizes the full latitude of striated muscle structural and functional traits, and then some. Moreover, there is compelling evidence that the EOM phenotype is determined by, and thereby exquisitely adapted for, the demands of a diverse and complex eye movement control system. In efforts to understand fully the biology of eye movements, a comprehensive understanding of oculomotor system components from cerebral cortex to EOM is necessary at all levels from the molecular to the behavioral.

ACKNOWLEDGMENTS

Studies were accomplished through invaluable efforts of past and present members of the Porter laboratory and the Cleveland EOM group (F.H. Andrade, H.J. Kaminski, J.S. Stahl, and J.D. Porter). Relocation of the Porter laboratory to Cleveland has opened doors to previously unimagined opportunities in EOM biology; Henry Kaminski, John Leigh, Jon Lass, and Robert Daroff made this move possible. The original work behind this review was supported by grants from The Research Institute of University Hospitals of Cleveland, Research to Prevent Blindness (Departmental and Senior Scientific Investigator Awards); the Muscular Dystrophy Association USA; The Evenor Armington Fund; NIH grants EY09834, EY12779, and P30 EY11373; and the Carl F. Asseff, M.D. Professorship in Ophthalmology.

For his nearly 25 years of mentorship and friendship, this review is dedicated to the memory of Robert F. Spencer, Ph.D. (deceased March 24, 2001).

REFERENCES

1. ROBINSON, D.A. 1970. Oculomotor unit behavior in the monkey. J. Neurophysiol. **33:** 393–403.
2. SPENCER, R.F. & J.D. PORTER. 1988. Structural organization of the extraocular muscles. Rev. Oculomotor Res. **2:** 33–79.
3. BÜTTNER-ENNEVER, J.A., A.K. HORN, H. SCHERBERGER & P. D'ASCANIO. 2001. Motoneurons of twitch and nontwitch extraocular muscle fibers in the abducens, trochlear, and oculomotor nuclei of monkeys. J. Comp. Neurol. **438:** 318–335.

4. ROBINSON, D.A. 1978. The functional behavior of the peripheral ocular motor apparatus: a review. *In* Disorders of Ocular Motility. G. Kommerell, Ed.: 43–61. Bergman. Munich.
5. PORTER, J.D., R.S. BAKER, R.J. RAGUSA & J.K. BRUECKNER. 1995. Extraocular muscles: Basic and clinical aspects of structure and function. Surv. Ophthalmol. **39:** 451–484.
6. BRUECKNER, J.K., O. ITKIS & J.D. PORTER. 1996. Spatial and temporal patterns of myosin heavy chain expression in developing rat extraocular muscle. J. Muscle Res. Cell Motil. **17:** 297–312.
7. WASICKY, R., F. ZIYA-GHAZVINI, R. BLUMER, *et al.* 2000. Muscle fiber types of human extraocular muscles: a histochemical and immunohistochemical study. Invest. Ophthalmol. Visual Sci. **41:** 980–990.
8. WINTERS, L.M., M.M. BRIGGS & F. SCHACHAT. 1998. The human extraocular muscle myosin heavy chain gene (MYH13) maps to the cluster of fast and developmental myosin genes on chromosome 17. Genomics **54:** 188–189.
9. WIECZOREK, D.F., M. PERIASAMY, G.S. BUTLER-BROWNE, *et al.* 1985. Co-expression of multiple myosin heavy chain genes, in addition to a tissue-specific one, in extraocular musculature. J. Cell Biol. **101:** 618–629.
10. RUSHBROOK, J.I., C. WEISS, K. KO, *et al.* 1994. Identification of alpha-cardiac myosin heavy chain mRNA and protein in extraocular muscle of the adult rabbit. J. Muscle Res. Cell Motil. **15:** 505–515.
11. ASMUSSEN, G., I. TRAUB & D. PETTE. 1993. Electrophoretic analysis of myosin heavy chain isoform patterns in extraocular muscles of the rat. FEBS Lett. **335:** 243–245.
12. JACOBY, J., K. KO, C. WEISS & J.I. RUSHBROOK. 1990. Systematic variation in myosin expression along extraocular muscle fibres of the adult rat. J. Muscle Res. Cell Motil. **11:** 25–40.
13. RUBINSTEIN, N.A. & J.F. HOH. 2000. The distribution of myosin heavy chain isoforms among rat extraocular muscle fiber types. Invest. Ophthalmol. Visual Sci. **41:** 3391–3398.
14. BRIGGS, M.M. & F. SCHACHAT. 2000. Early specialization of the superfast myosin in extraocular and laryngeal muscles. J. Exp. Biol. **203:** 2485–2494.
15. LUCAS, C.A., A. RUGHANI & J.F. HOH. 1995. Expression of extraocular myosin heavy chain in rabbit laryngeal muscle. J. Muscle Res. Cell Motil. **16:** 368–378.
16. BRUECKNER, J.K., L.P. ASHBY, J.R. PRICHARD & J.D. PORTER. 1999. Vestibulo-ocular pathways modulate extraocular muscle myosin expression patterns. Cell Tissue Res. **295:** 477–484.
17. BRUECKNER, J.K. & J.D. PORTER. 1998. Visual system maldevelopment disrupts extraocular muscle-specific myosin expression. J. Appl. Physiol. **85:** 584–592.
18. BRIGGS, M.M., J. JACOBY, J. DAVIDOWITZ & F.H. SCHACHAT. 1988. Expression of a novel combination of fast and slow troponin T isoforms in rabbit extraocular muscles. J. Muscle Res. Cell Motil. **9:** 241–247.
19. EASON, J.M., G. SCHWARTZ, K.A. SHIRLEY & A.W. ENGLISH. 2000. Investigation of sexual dimorphism in the rabbit masseter muscle showing different effects of androgen deprivation in adult and young adult animals. Arch. Oral Biol. **45:** 683–690.
20. ENGLISH, A.W. & W.D. LETBETTER. 1982. A histochemical analysis of identified compartments of cat lateral gastrocnemius muscle. Anat. Rec. **204:** 123–130.
21. DEMER, J.L., S.Y. OH & V. POUKENS. 2000. Evidence for active control of rectus extraocular muscle pulleys. Invest. Ophthalmol. Visual Sci. **41:** 1280–1290.
22. KHANNA, S. & J.D. PORTER. 2001. Evidence for rectus extraocular muscle pulleys in rodents. Invest. Ophthalmol. Visual Sci. **42:** 1986–1992.
23. PORTER, J.D. & R.S. BAKER. 1996. Muscles of a different 'color': the unusual properties of the extraocular muscles may predispose or protect them in neurogenic and myogenic disease. Neurology **46:** 30–37.
24. COLLINS, C.C. 1971. Orbital mechanics. *In* The Control of Eye Movements. P. Bach-y-Rita & C.C. Collins, Eds.: 283–325. Academic Press. New York.
25. GOLDBERG, S.J. & M.S. SHALL. 1999. Motor units of extraocular muscles: recent findings. Prog. Brain Res. **123:** 221–232.
26. KARPATI, G. & S. CARPENTER. 1986. Small-caliber skeletal muscle fibers do not suffer deleterious consequences of dystrophic gene expression. Am. J. Med. Genet. **25:** 653–658.

27. KHURANA, T.S., R.A. PRENDERGAST, H.S. ALAMEDDINE, et al. 1995. Absence of extraocular muscle pathology in Duchenne's muscular dystrophy: role for calcium homeostasis in extraocular muscle sparing. J. Exp. Med. **182:** 467–475.
28. RAGUSA, R.J., C.K. CHOW, D.K. ST. CLAIR & J.D. PORTER. 1996. Extraocular, limb and diaphragm muscle group-specific antioxidant enzyme activity patterns in control and mdx mice. J. Neurol. Sci. **139:** 180–186.
29. PORTER, J.D. & P. KARATHANASIS. 1998. Extraocular muscle in merosin-deficient muscular dystrophy: cation homeostasis is maintained but is not mechanistic in muscle sparing. Cell Tissue Res. **292:** 495–501.
30. PORTER, J.D., J.A. RAFAEL, R.J. RAGUSA, et al. 1998. The sparing of extraocular muscle in dystrophinopathy is lost in mice lacking utrophin and dystrophin. J. Cell Sci. **111:** 1801–1811.
31. PORTER, J.D., A.P. MERRIAM, A.A. HACK, et al. 2001. Extraocular muscle is spared despite the absence of an intact sarcoglycan complex in gamma- or delta-sarcoglycan-deficient mice. Neuromuscul. Disord. **11:** 197–207.
32. PORTER, J.D., S. KHANNA, H.J. KAMINSKI, et al. 2001. Extraocular muscle is defined by a fundamentally distinct gene expression profile. Proc. Natl. Acad. Sci. USA **98:** 12062–12067.
33. CAMPBELL, W.G., S.E. GORDON, C.J. CARLSON, et al. 2001. Differential global gene expression in red and white skeletal muscle. Am. J. Physiol. Cell Physiol. **280:** C763–C768.
34. CHENG, G. & J.D. PORTER. 2002. Transcriptional profile of rat extraocular muscle by serial analysis of gene expression. Invest. Ophthalmol. Visual Sci. **43:** In press.

The Orbital Pulley System: A Revolution in Concepts of Orbital Anatomy

JOSEPH L. DEMER

Departments of Ophthalmology and Neurology, and the Jules Stein Eye Institute, University of California, Los Angeles, USA

ABSTRACT: Magnetic resonance imaging (MRI) now enables precise visualization of the mechanical state of the living human orbit. Resulting insights have motivated histological re-examination of human and simian orbits, providing abundant consistent evidence for the *active pulley hypothesis*, a re-formulation of ocular motor physiology. Each extraocular muscle (EOM) consists of a global layer (GL) contiguous with the tendon and inserting on the eyeball, and a similar-sized orbital layer (OL) inserting on a connective tissue ring forming the EOM *pulley*. The pulley controls the EOM path and serves as the EOM's functional origin. Activity of the OL positions the pulley along each rectus EOM to assure that its pulling direction shifts by half the change in ocular orientation, the half-angle behavior characteristic of a linear ocular motor plant. Half-angle behavior is equivalent to Listing's law of ocular torsion, and makes 3-D ocular rotations effectively commutative. Pulleys are configured to maintain oblique EOM paths orthogonal to half-angle behavior, and violate Listing's law during the vestibulo-ocular reflex. Rectus pulley positions shift during convergence, facilitating stereopsis.

Innervations, fiber types, and metabolism of the OL and GL differ, consistent with the elastic loading of the former, and viscous loading of the latter. Disorders of the location and stability of rectus pulleys are associated with predictable patterns of incomitant strabismus that may mimic cranial nerve palsies. Surgical interventions improve defective pulley function. Understanding of ocular motor control requires characterization of the behavior of the EOM pulleys as well as knowledge of angular eye orientation.

KEYWORDS: active pulley hypothesis; extraocular muscles; magnetic resonance imaging; pulleys

INTRODUCTION

Implicit in efforts to study central ocular motor control is a concept of EOM mechanics. Until recently, a qualitative and intuitive notion of EOM mechanics prevailed, broadly underlying much traditional thinking. This notion characterized EOM actions by the simple geometric relationships between their origins in the annulus of Zinn and their scleral insertions. Even the earliest quantitative modeling efforts uncovered consequent anomalies related to the absence of specific constraints

Address for correspondence: Joseph L. Demer, M.D., Ph.D., Jules Stein Eye Institute, 100 Stein Plaza, UCLA, Los Angeles, CA 90095-7002, USA. Voice: 310-825-5931; fax: 310-206-7826.

jld@ucla.edu

on EOM paths,[1,2] leading to the supposition that fundamental orbital anatomy might have been incompletely understood.[3,4] The resulting re-examination of EOM anatomy and physiology has been so revealing as to motivate a fundamental paradigm shift[5] that has broad basic and clinical implications. The new paradigm will be termed the *active pulley hypothesis*.

OCULAR KINEMATICS AS A UNIFYING CONCEPT

An initial study of the ocular motor system examined the simple case of movements in a single degree of freedom, usually horizontal rotation. This initial simplification made many ocular motor phenomena tractable for study, but it allowed investigators to ignore the serious kinematic problem of commutativity. Rotations of a three-dimensional (3-D) object such as the eyeball are not mathematically commutative; that is, final eye orientation depends on the order of rotations.[6] Angular velocity of the eye is not equal to the rate of change of its orientation, but rather is related both to the time derivative and to instantaneous eye orientation.[7,8] Each combination of horizontal and vertical positions, for an arbitrary sphere, could be associated with infinitely many torsional positions.[9] A noncommutative peripheral ocular motor apparatus would require control by a central nervous system that kept track of the sequence of previous eye movements, a problem of daunting complexity!

While the ocular motor system cannot ever be entirely commutative, a fortunate arrangement of EOMs and orbital connective tissues makes it appear, for all practical purposes, commutative to the brain.[10] This occurs if the velocity axis of any ocular rotation shifts by half of the shift in ocular orientation,[10] which is one mathematical formulation of Listing's law,[8] the "half-angle rule." A more familiar formulation of Listing's law states that, with the head upright and stationary, any eye orientation can be reached from any other by rotation about a single axis, and that all such possible axes lie in a single plane, Listing's plane. When Listing's law is observed, 3-D eye position is closely approximated by the time integral of 3-D eye velocity, simplifying the problem for neural integrators.[10]

ORBITAL MECHANICS AND COMMUTATIVITY

Previously, Listing's law was believed to be implemented entirely by complex neural commands. However, no neural substrate for Listing's law has been identified. In the superior colliculus, saccades are encoded as the 2-D rate of change of eye orientation, implying that any computation of the third dimension, torsion, is accomplished downstream.[10,11] Even in the oculomotor nucleus and rostral interstitial nucleus of the medial longitudinal fasiculus, saccadic burst commands are better correlated with rate of change of 3-D eye position than with angular eye velocity.[11,12]

A central tenet of the active pulley hypothesis is the existence of connective tissue structures in the orbit that constrain EOM paths to conformity with Listing's half-angle rule.[10,13,14] FIGURE 1 (top) is a side view of a diagrammatic globe showing a horizontal rectus EOM. The EOM's rotational axis is perpendicular to the line con-

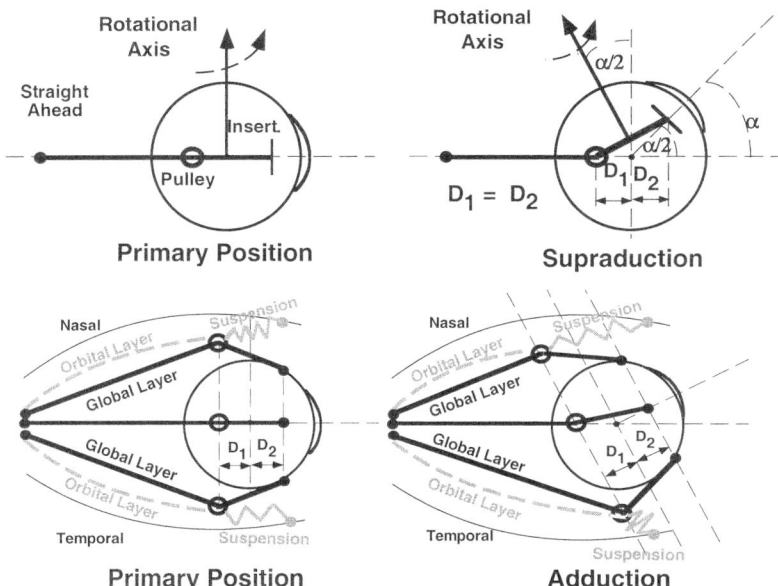

FIGURE 1. Top: Lateral view of orbit. Rotational axis for rectus EOM is perpendicular to the line connecting its pulley to the scleral insertion. With the eye in supraduction of (small) angle α, Listing's law is satisfied if distance D_1 from pulley (dark ring) to globe center is equal to distance D_2 from insertion to globe center. **Bottom:** Superior view of orbit showing shifts in horizontal rectus pulley position required to maintain Listing's half-angle relationship in tertiary positions of adducted elevation and adducted depression. Pulleys are depicted as dark rings.

necting its pulley with its scleral insertion, thus vertical in primary gaze. Now consider viewing of a horizontally centered target at supraduction angle α. If the distance from the pulley to globe center D_1 is equal to the distance from the insertion to globe center D_2, then (for small angles typical of the oculomotor range) the rotational axis will tilt posteriorly by α/2, as required by Listing's law.

In tertiary (oblique) gaze positions, pulley behavior can still explain Listing's law (FIG. 1, bottom).[14] Beginning in adduction, the tertiary positions of adducted supraduction and adducted infraduction can be attained in conformity to Listing's law if the distance from the pulley to globe center D_1 remains equal to the distance from the insertion to globe center D_2. Because these distances must remain constant relative to the globe, Listing's law requires that pulleys move in the orbit to follow the movements of the scleral EOM insertions.

Since all rectus EOMs have pulleys implementing Listing's half-angle behavior, the oblique EOMs would appear to be unnecessary when the head is upright and stationary. However, when the head is rotated, Listing's law is violated by the vestibuloocular reflex (VOR), for which the velocity axis rotates by about 25%[15] to as little as 0%[16,17] of the ocular angle. A kinematically ideal VOR would have a rotation axis coincident with the head and independent of eye position in order to optimally stabilize retinal images. The active pulley hypothesis proposes that this violation of

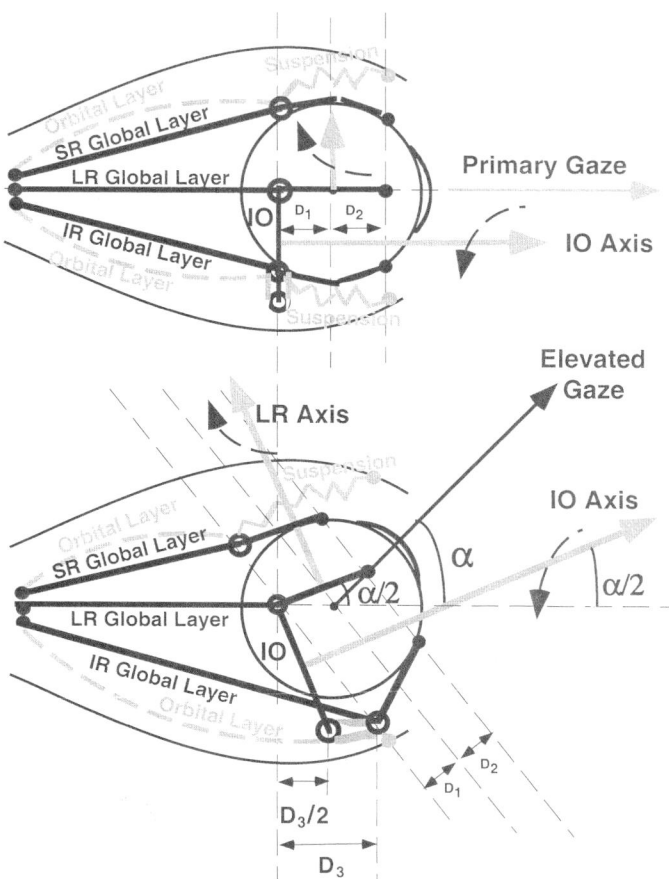

FIGURE 2. Lateral diagrammatic view of inferior oblique (IO) and rectus pulleys. **Top:** Rotational axis of lateral rectus (LR, *vertical gray arrow*) is perpendicular to segment from its pulley to its insertion. Since the IO pulley is coupled to the IR and LR pulleys, the IO rotational axis (*horizontal gray arrow*) parallels primary gaze. **Bottom:** Since distance D_1 from the LR pulley to globe center is equal to distance D_2 from globe center to the LR insertion, supraduction to angle α causes LR rotational axis (*short gray arrow*) to tilt by angle $\alpha/2$, implementing Listing's law. Relaxation of the IR orbital layer allows the IR pulley to move anteriorly by D_3, and partial coupling of the IO pulley to the IR pulley moves the IO pulley anteriorly by $D_3/2$. This maintains the terminal IO path perpendicular to the LR path anterior to its pulley, so that the IO rotational axis (*long gray arrow*) remains perpendicular to the LR rotational axis.

Listing's law is mediated by the oblique EOMs, whose velocity axes are actively maintained orthogonal to the Listing's velocity axis of the rectus EOMs (FIG. 2). The situation appears clearest for the inferior oblique (IO) muscle, whose pulley is half coupled to that of the inferior rectus (IO), and whose OL inserts, among other sites, on the LR pulley.[18] In primary gaze (FIG. 2, top), this arrangement forces the terminal segment of the IO to be perpendicular to the gaze direction so that the IO produces a purely torsional action around the gaze line. In supraduction (FIG. 2, bottom), the IO pulley moves anteriorly by half the distance that the IR pulley moves; the IO rotational axis is then no longer perpendicular to the gaze direction, but it remains perpendicular to the Listing's velocity axis of the rectus EOMs. The IO's action then always violates Listing's law by producing a torsional velocity axis orthogonal to it. The amount of ocular axis shift during the VOR should, as observed,[16] then be inversely related to torsional VOR gain, and ideally zero when torsional VOR gain is unity.

The superior oblique (SO) muscle, which has a rigid pulley in the superonasal orbit and a very broad scleral insertion, is thus a special case. It is likely that the breadth of the anatomical SO insertion allows its functional insertion at the point of tangency to shift with gaze to maintain a rotational axis parallel to that of the IO. The oblique EOMs then act in concert to violate Listing's law and to confer the noncommutativity observed for the VOR.[19] Because, however, the kinematic requirement for the VOR is to rotate the eye exactly opposite the head, this instance of noncommutativity does not cause problems for the brain as would noncommutativity in visually guided eye movements. Another instance of the VOR is static ocular counterrolling in response to maintained head tilt relative to gravity. Recordings from burst neurons in monkeys are compatible with torsional shift of rectus pulleys in the direction of ocular counterroll induced by static head tilt.[20] This is an example of the importance of understanding pulley actions in interpreting neuronal behavior in pre-motor structures.

HISTOLOGICAL EVIDENCE FOR ACTIVE PULLEY HYPOTHESIS

FIGURE 3 summarizes human and monkey histological data obtained by histochemical and immunohistochemical staining of serial sections.[14,21,22] Rectus and IO pulleys consist of encircling collagen rings, stiffened by elastin and smooth muscle (SM). Anterior to these rings, rectus EOMs are invested by collagen slings convex toward the orbital walls. Posterior to the pulley rings, pulley slings are convex toward the globe. Pulley rings are suspended anteriorly from the orbital rim by dense collagenous bands, stiffened by elastin. Elastin has the property of reversible extensibility, and thus confers a spring-like property. Pulley suspensions also contain SM, particularly in a band running from the medial rectus (MR) pulley to the nasal aspect of the IR pulley (Muller's peribulbar muscle). This SM receives a rich autonomic innervation: sympathetic projections employing norepinephrine from the superior cervical ganglion, cholinergic parasympathetic projections probably from the ciliary ganglion, and nitroxidergic innervation from the pterygopalatine ganglion.[22]

Each EOM consists of two layers, a global layer (GL) contiguous with the scleral insertion, and an orbital layer (OL) that inserts on pulley tissues. Quantitative analysis in humans and monkeys indicates that the OLs of the four rectus EOMs contains roughly similar numbers of fibers, averaging about 55% of total EOM fibers.[23] The remaining fibers comprise the GL, varying numerically in proportion to abundance of connective tissue in each pulley. For the rectus and IO EOMs, the OL is C-shaped and lies on the EOM's orbital surface. The OL fibers are not contiguous with the scleral insertions, but instead insert on the pulleys so that OL contraction stretches the anteriorly anchored pulley suspensions and translates the pulleys posteriorly.

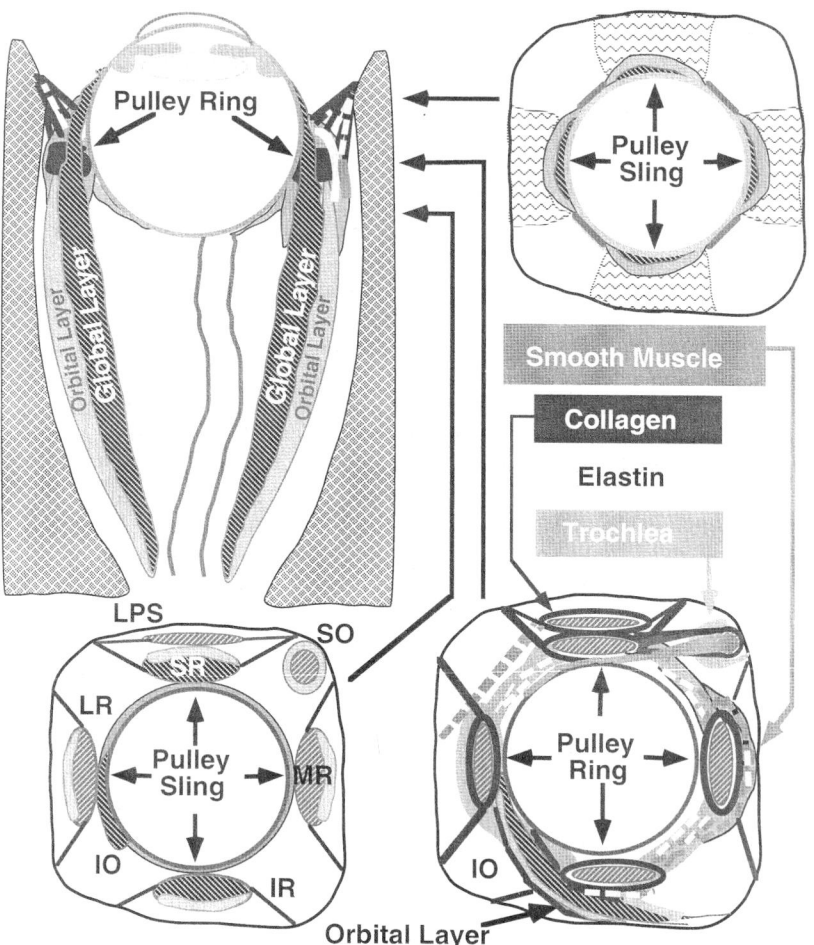

FIGURE 3. Diagrammatic representation of structure of orbital connective tissues and their relationship to the fiber layers of the rectus EOMs. Coronal views represented at levels indicated by arrows in horizontal section. IR, inferior rectus muscle; LLA, lateral levator aponeurosis; LPS, levator palpebrae superioris muscle; LR, lateral rectus muscle; MR, medial rectus muscle; SO, superior oblique muscle; SR, superior rectus muscle.

The IR and IO pulleys are mechanically coupled via Lockwood's ligament. The OL of the IR inserts on the IO pulley, the collagenous sleeve of the IO temporally, and also on the LR pulley.[18] Contraction of the IO OL thus would translate the IR pulley nasally and the LR pulley inferiorly. In lower animals such as rabbit, the SO contains EOM fibers that directly insert on the sclera. In rabbit, both the OL and GL pass through the cartilaginous trochlea; the OL inserts on the pulley apparatus around the SR, whereas the GL inserts on the sclera. In humans and monkeys, the SO OL inserts coaxially posterior to the trochlea on a dense collagenous sheath enveloping its ex-terior, whereas slightly more anteriorly the GL becomes contiguous with the SO tendon. Both the SO sheath and tendon pass through the trochlea. The SO sheath inserts on the nasal border of the SR pulley, while the tendon thins and passes inferior to the SR pulley to insert broadly on the posterior lateral sclera. Contraction of the SO OL thus would translate the SR pulley nasally, while contraction of the SO GL rotates the globe.

FUNCTIONAL IMAGING EVIDENCE FOR ACTIVE PULLEY HYPOTHESIS

Magnetic resonance imaging (MRI) using surface coils has been valuable in testing the predictions of the active pulley hypothesis.[24] While MRI can occasionally demonstrate pulley structures directly, it has been more useful to characterize the discrete inflections in EOM paths produced by pulleys in secondary[25,26] and tertiary[27] gaze positions. Such data, obtained from 2-mm-thick coronal image planes, have strongly supported the active pulley hypothesis (FIG. 4). For example, the MR and LR pulleys both move posteriorly during EOM contraction, by amounts corresponding to movements of their scleral insertions.[27] When the anteroposterior positions of the MR and LR pulleys are plotted as functions of horizontal gaze angle, the fits should intersect at Listing's primary position (FIG. 4). As predicted by the active pulley hypothesis, this intersection occurs 8 mm posterior to the globe center, roughly the distance D_1 equal to distance D_2 from the globe center to the MR and LR scleral insertions. This intersection occurred under these conditions in slight adduction, suggesting an apparent nasal tilt of Listing's plane of 5 deg. Because the MR inserts 1.3 mm anterior to the LR, D_1 must be larger for the LR than MR so that Listing's primary position is almost exactly straight ahead. The SR and IR pulleys behave similarly, with position plots also intersecting as predicted, but with a small superior tilt.[27]

Imaging shows a straight IO path from its origin on the inferonasal orbit to its pulley near the IR crossing. The IO pulley moves anteriorly with supraduction, and posteriorly with infraduction (FIG. 5), by 53% of the IR insertion's travel.[28] This is consistent with the active pulley hypothesis. The separation between the IR GL and OL can be directly visualized (FIG. 5), making it obvious that while the GL inserts on the eyeball, most of the contraction in sustained infraduction occurs in the OL as it displaces both the IR and IO pulleys.

Coronal plane MRI demonstrates highly stereotypic locations of normal rectus pulleys in central gaze[29] as seen in FIGURE 6. The normal MR pulley is stable during all eye movements, conjugate and vergence. The other rectus pulleys move somewhat in the coronal plane. The LR and MR pulleys, which receive insertions from the IO OL, move 2–3 mm inferiorly and nasally, respectively, during IO contraction in supraduction. The IR pulley's nasal shift may be assisted by Muller's peribulbar muscle, a 1×5 mm SM band extending from the IR to the MR pulleys.[22]

Listing's planes for the two eyes are known to rotate temporally during convergence by about twice the vergence angle,[30] corresponding to excyclotorsion in depression and incyclotorsion in elevation.[31,32] This is necessary to maintain alignment of corresponding retinal meridia during near viewing,[33] and is greater during stereopsis than with disparity alone. Thus, during binocular viewing of near

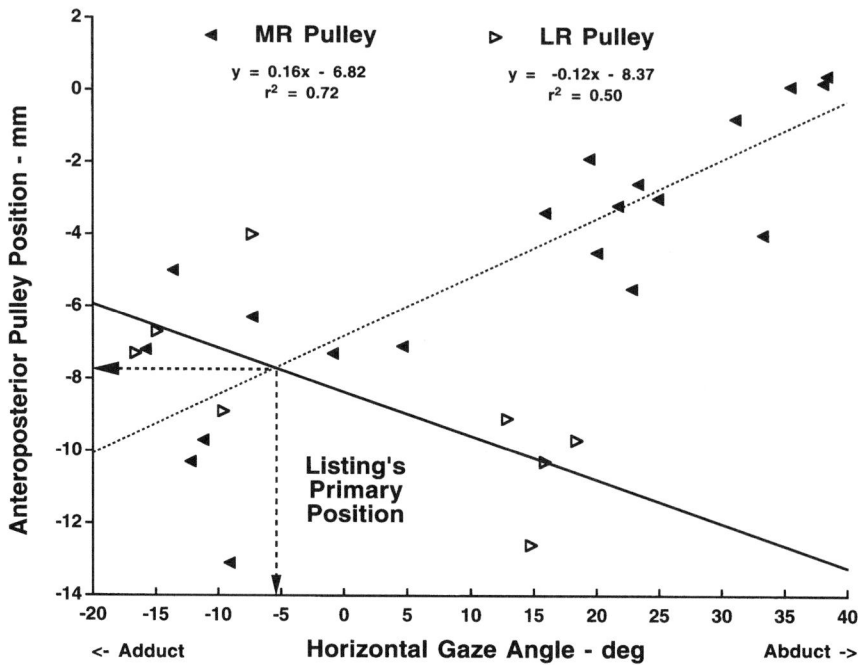

FIGURE 4. Anteroposterior positions (referenced to globe center) of the medial (MR) and lateral rectus (LR) pulleys as a function of craniotopic horizontal gaze angle in eight normal subjects, based on path inflections produced by superimposed supraduction and infraduction. Note anterior travel of MR pulley in abduction, and posterior travel in adduction, with mirror image behavior by the LR. Linear fits to the pulley positions for the two EOMs intersect ~8 mm posterior to globe center, the same as the distance from globe center to the EOM insertions. This corresponds to Listing's primary position 6 deg temporal to the craniotopic central direction.

and far targets aligned on one eye, the Listing plane for that unmoving eye is nevertheless tilted in association with the vergence movement of the other eye. Orbital MRI in this situation demonstrates that pulleys correspondingly move in the aligned eye.[34] In craniotopic central gaze, this corresponds to a lateral shift of the SR, inferior shift of the LR, and nasal shift of the IR, altogether amounting to an excyclorotation of the rectus pulley array in the orbit. This is appropriate to accomplish the extorsion of the globe observed in convergence when gaze is below primary position, which MRI evidence suggests to be 13 deg above craniotopic central gaze for supine subjects.[27] It has been proposed by van Rijn and van den Berg that a form of Hering's law of equal innervation exists for vergence, such that both eyes receive symmetric version commands for remote targets, and mirror symmetric vergence commands for near targets.[31] We have extended this suggestion to propose that symmetric control may be applied to the pulleys via the oblique OLs and peribulbar SM, so that pulleys in both orbits are configured with mirror symmetry during convergence regardless of superimposed conjugate gaze.[14] Quantitative analysis of MRI images suggests

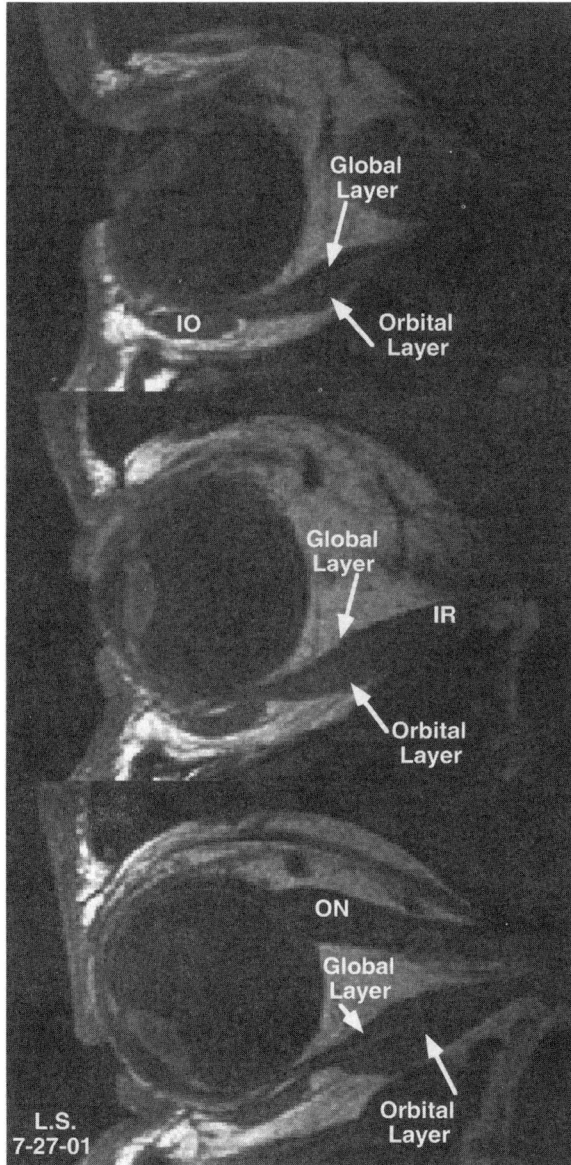

FIGURE 5. Sagittal MRI of human orbit showing that the inferior oblique (IO) moves anteriorly in supraduction and posteriorly in infraduction. The global layer (GL) of the inferior rectus (IR) contracts modestly in infraduction and inserts on the globe, while the orbital layer contracts more vigorously and inserts on the IO/IR pulley. ON, optic nerve.

that the SO does not contribute to this reconfiguration for convergence because its cross-sectional area is unchanged in the aligned eye, which exhibits an increase in IO cross section suggestive of IO contraction (Kono, Wright, and Demer, unpublished data). The IO also moves posteriorly in convergence. Consistent with recordings of horizontal rectus EOM tensions during the same experiment in monkey,[35] analysis of human EOM cross sections did not indicate co-contraction in convergence.

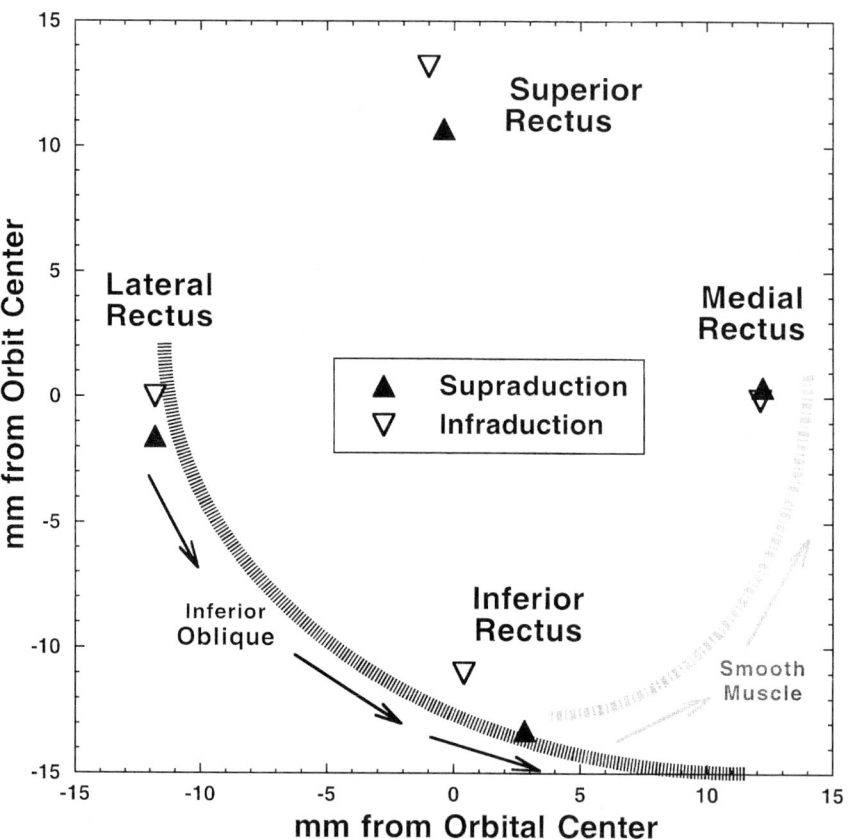

FIGURE 6. Effect of vertical gaze position on mean locations of rectus pulleys in 11 normal adults, redrawn from Clark, Miller & Demer.[29] Note the shift of the LR and IR pulleys along the path of the IO muscle in supraduction, where the IO contracts. Nasal shift of the IR could also result from smooth muscle contraction. The MR pulley is stable over the entire range of gaze.

IMPLICATIONS FOR MUSCLE BIOLOGY

The active pulley hypothesis may partially account for the complex variety of EOM fiber types, starting from the premise that fiber characteristics are adapted to physiologic demands.[36] About 80% of OL fibers of each EOM are fast, twitch-generating, singly innervated fibers (SIFs) resembling mammalian skeletal muscle fibers, while 20% are multiply innervated fibers (MIFs) that either do not conduct action potentials or do so only in their central portions.[37] Orbital SIFs are specialized for oxidative metabolism and fatigue resistance.[37] The vascular supply in the OL is high,[38] about 50% greater in humans than the GL.[39] The high metabolism, fatigue resistance, and luxurious blood supply of the numerous OL SIFs are tailored to their continuous elastic loading by pulley suspensions. Expression of unique myosin isoforms in OL SIFs may also be related to the requirement for fast twitch against continuous loading because alterations in EOM activity patterns can change EOM-specific myosin gene expression.[40] The function of the sparse and primitive OL MIFs remains unclear.

About 90% of GL fibers are fast, twitch-generating SIFs, while 10% are slow, non-twitch MIFs resembling those of amphibians.[37] The SIFs are often divided into three types: red, intermediate, and white, distinguished by their density of mitochondria and fatigue resistance.[37] The largest and most granular red SIFs, constituting about 33% of GL fibers, are very similar to OL SIFs and are highly fatigue resistant, while intermediate and white SIFs are progressively less resistant.[37] The predominant static GL loading by the moderate contractile force of antagonist EOM accounts for the GL's higher overall recruitment threshold than that of the OL and the lesser oxidative, vascular, and fatigue-resistant features of orbital SIFs. However, during saccades the high viscous GL loading by the relaxing antagonist EOM requires the high transient force of the intermediate and white SIFs. The function of the rare GL MIFs remains obscure.

IMPLICATIONS FOR ELECTROPHYSIOLOGY

Electromyographic (EMG) recordings in the human LR GL demonstrated both a pulse and step of activity during saccades, the former required to drive the formidable viscous load imposed by the relaxing antagonist EOM, and the latter to oppose the lesser elastic load as position is maintained.[41] Recordings of insertional tension of simian horizontal rectus EOMs have confirmed both saccadic pulses and steps.[42] In the human GL, EMG showed both pulses and steps during saccades, while in the OL there was only a step.[41] Fibers in the OL have lower recruitment thresholds than GL fibers,[41,43,44] prompting Collins to propose that the OL might have a special role in fixation.[41] The insight that the OL inserts on the pulley permits an alternative interpretation. The mechanical load on the OL is dominated by elasticity of the attached pulley suspension. Collins has pointed out that the main load on an EOM attached to the globe is viscosity due to the relaxing antagonist EOM.[41] A force pulse in the OL is unnecessary to achieve brisk pulley motion against an elastic load. However, elastic loading by connective tissue requires that the OL sustain active tension throughout the oculomotor range. In contrast, the GL remains under tension even when relaxed due to stretching by the antagonist EOM. The active pulley hypothesis predicts that motor neurons preferentially innervating fibers in the OL should exhibit step but not pulse changes in activity during saccades. Many such "tonic" motor neurons have been found in the abducens and oculomotor nuclei.[45]

PULLEY DISEASE AS A CAUSE OF STRABISMUS

The orbital SIF is the last EOM fiber type to mature, occurring post-natally during establishment of binocular alignment.[46,47] Nemestrina monkeys, a macaque species with a high prevalence of naturally occurring strabismus, suggest evidence for a role for an OL abnormality in the pathogenesis of strabismus. These monkeys transiently exhibit tubular aggregates only in orbital SIFs during the first six months of

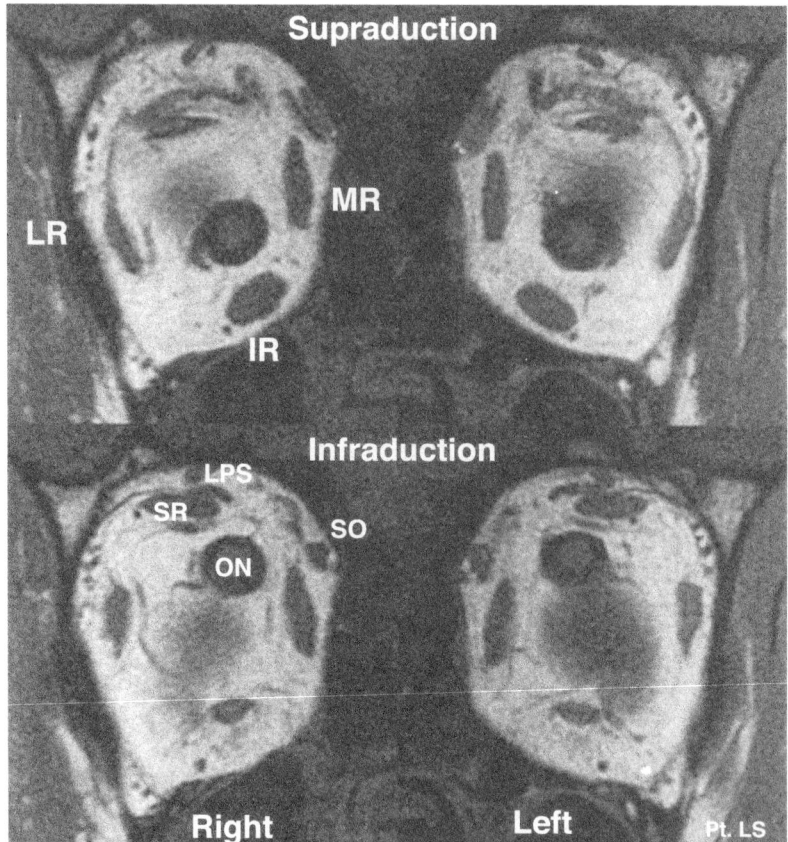

FIGURE 7. Coronal MRI of patient with Y-pattern exotropia in elevation due to extorsion of the SR, LR, and IR pulleys in this gaze position. ON, optic nerve.

life, while fasicularis monkeys exhibit neither OL tubular aggregates nor naturally occurring strabismus.[46]

MRI has demonstrated the coronal plane locations of rectus EOM pulleys to be stereotypic in normal [29,48] and most strabismic subjects.[29] The 95% confidence intervals of coronal plane pulley coordinates are $< \pm 1$ mm.[24] A computational model of binocular alignment based on the static force balance approach of Robinson[1,2] incorporates fixed pulleys,[49] and is now available as the Macintosh application *Orbit*. The expected effect of coronal plane heterotopy (malpositioning) of pulleys is readily computed using *Orbit*.[50] Many cases of incomitant cyclovertical strabismus are associated with heterotopy of one or more rectus EOM pulleys >2 SDs from normal. Patterns of incomitance in individual patients consistently match those predicted by *Orbit* simulation based on the measured pulley locations, suggesting that pulley heterotopy *caused* the strabismus.[24,51,52] In most of these cases the strabismus had an A- or V-pattern. These clinical findings mimic all features of what has been heretofore regarded as "oblique" EOM dysfunction[51] and suggest that clinical nosology be significantly revised.[53] Several lines of evidence suggest that ocular torsion due to strabismus did not cause the pulley heterotopy: (1) typically only one or two pulleys were heterotopic; (2) the amount of ocular torsion was insufficient to account for the amount of pulley heterotopy; and (3) patients with similar ocular torsion due to SO palsy lack this sort of pulley heterotopy.[54] Extreme pulley heterotopy is associated with esotropia and hypotropia in axial high myopia.[24,55]

Although normal pulleys shift little with gaze changes (FIG. 6), one or several pulleys may become unstable and shift with gaze to markedly change EOM action. Inferior shift of the LR pulley mimics the restrictive hypotropia in adduction traditionally attributed to SO tendon sheath pathology (Brown syndrome), or X-pattern exotropia.[56] An exaggeration of the physiologic excyclorotation of the rectus pulley array in convergence may produce a marked Y-pattern exotropia, one present only in elevated gaze (FIG. 7).

IMPLICATIONS FOR STRABISMUS—TREATMENT

Strabismus treatment probably affects the action of the OLs on their pulleys. Botulinum toxin treatment of strabismus produces its most lasting effects on orbital SIFs[38] and might therefore alter active pulley behavior. Strabismus surgery that alters the relationships between EOM insertions and pulleys probably produces unintended effects that should be considered in surgical planning. In particular, any pulley manipulation compromising the orderly relationship between eye orientation and EOM rotational axes would compromise neural control of eye movement,[10] and would be expected to produce at least dynamic binocular misalignments in tertiary gaze positions. Noncommutativity of ocular rotations occurs in some strabismic patients with pulley abnormalities, complicating clinical assessment of ocular versions.

Specific treatment of pulley disorders requires knowledge of specific abnormalities. Orbital imaging is clinically helpful in many cases of incomitant strabismus.[24] For example, pulleys may be stabilized by posterior fixation for treatment of restrictive hypotropia in adduction (Brown syndrome)[57] or esotropia,[58] anchored to the sclera to restrict ocular rotation in the field of EOM action,[59] or shifted to correct heterotopy or enhance the action of transposed EOM insertions in paralytic strabismus.[60]

IMPLICATIONS FOR OCULAR MOTOR RESEARCH

The active pulley hypothesis represents a paradigm shift for understanding of the ocular motor system. It is now clear that the ocular motor system is overdetermined in the mathematical sense: there are far too many internal variables than can be ascertained by measurement of the three angular degrees of freedom of each eye. As demonstrated for convergence, the same ocular orientation can be attained by differing EOM configurations, presumably representing different neural command sets. It therefore follows that specific clinical diagnoses of the mechanisms of strabismus will generally require more information than can be obtained even in theory from external examination or recording of eye movements. It is highly likely that separate neural command schemes prevail for the OLs that rotate the eye, and the GLs that position the pulleys and thus control the directions of EOM actions. In future ocular motor research, it will no longer be sufficient to correlate angular eye position with neuronal discharges. It will be necessary to study EOM pulley behavior as well as eye orientation; MRI is one way to do this in humans and possibly monkeys. Many classical ocular motor experiments deserve reconsideration in light of the active pulley hypothesis. Such reconsideration is likely to clarify our basic and clinical understanding of the neurobiology of eye movements and to demand correlative studies of the biomechanics and developmental biology of the orbital connective tissues.

ACKNOWLEDGMENTS

This work was supported by National Institutes of Health grant EY8313. J.L.D. was recipient of an unrestricted grant from Research to Prevent Blindness and is Larraine and David Gerber Professor of Ophthalmology.

REFERENCES

1. ROBINSON, D.A. 1975. A quantitative analysis of extraocular muscle cooperation and squint. Invest. Ophthalmol. **14:** 801–825.
2. MILLER, J.M. & D.A. ROBINSON. 1984. A model of the mechanics of binocular alignment. Comput. Biomed. Res. **17:** 436–470.
3. MILLER, J.M. & D. ROBINS. 1987. Extraocular muscle sideslip and orbital geometry in monkeys. Vision Res. **27:** 381–392.
4. MILLER, J.M. 1989. Functional anatomy of normal human rectus muscles. Vision Res. **29:** 223–240.
5. KUHN, T.S. 1996. The Structure of Scientific Revolutions. University of Chicago Press. Chicago, IL.
6. HASLWANTER, T. 1995. Mathematics of three-dimensional eye rotations. Vision Res. **35:** 1727–1739.
7. TWEED, D. & T. VILIS. 1987. Implications of rotational kinematics for the oculomotor system in three dimensions. J. Neurophysiol. **58:** 832–849.
8. TWEED, D. & T. VILIS. 1990. Geometric relations of eye position and velocity vectors during saccades. Vision Res. **30:** 111–127.
9. VAN DEN BERG, A.V. 1995. Kinematics of eye movement control. Proc. R. Soc. Lond. **260:** 191–197.
10. QUAIA, C. & L.M. OPTICAN. 1998. Commutative saccadic generator is sufficient to control a 3-D ocular plant with pulleys. J. Neurophysiol. **79:** 3197–3215.
11. HEPP, K. 1994. Oculomotor control: Listing's law and all that. Curr. Opinion Neurobiol. **4:** 862–868.
12. VAN OPSTAL, J., K. HEPP, Y. SUZUKI & V. HENN. 1996. Role of the monkey nucleus reticularis tegmenti pontis in the stabilization of Listing's plane. J. Neurosci. **16:** 7284–7296.
13. RAPHAN, T. 1998. Modeling control of eye orientation in three dimensions. I. Role of muscle pulleys in determining saccadic trajectory. J. Neurophysiol. **79:** 2653–2667.
14. DEMER, J.L., S.Y. OH & V. POUKENS. 2000. Evidence for active control of rectus extraocular muscle pulleys. Invest. Ophthalmol. Visual Sci. **41:** 1280–1290.
15. MISSLISCH, H., D. TWEED, M. FETTER, *et al.* 1994. Rotational kinematics of the human vestibuloocular reflex. III. Listing's law. J. Neurophysiol. **72:** 2490–2502.

16. MISSLISCH, H. & B.J. HESS. 2000. Three-dimensional vestibuloocular reflex of the monkey: optimal retinal image stabilization versus Listing's law. J. Neurophysiol. **83:** 3264–3276.
17. PALLA, A., D. STRAUMANN & H. OBZINA. 1999. Eye-position dependence of three-dimensional ocular rotation axis orientation during head impulses in humans. Exp. Brain Res. **129:** 127–133.
18. DEMER, J.L., S.Y. OH & V. POUKENS. 2001. Orbital layers of the oblique extraocular muscles (EOMs) insert on the orbital connective tissue system. Invest. Ophthalmol. Visual Sci. (ARVO Abstr.) **42:** S517.
19. TWEED, D.B., T.P. HASLWANTER, V. HAPPE & M. FETTER. 1999. Non-commutativity in the brain. Nature **399:** 261–263.
20. SCHERBERGER, H., J.-H. CABUNGCAL, K. HEPP, et al. 2001. Ocular counterroll modulates the preferred direction of saccade-related pontine burst neurons in the monkey. J. Neurophysiol. **86:** 935–949.
21. DEMER, J.L., J.M. MILLER, V. POUKENS, et al. 1995. Evidence for fibromuscular pulleys of the recti extraocular muscles. Invest. Ophthalmol. Visual Sci. **36:** 1125–1136.
22. DEMER, J.L., V. POUKENS, J.M. MILLER & P. MICEVYCH. 1997. Innervation of extraocular pulley smooth muscle in monkeys and humans. Invest. Ophthalmol. Visual Sci. **38:** 1774–1785.
23. OH, S.Y., V. POUKENS & J.L. DEMER. 2001. Quantitative analysis of extraocular muscle global and orbital layers in monkey and human. Invest. Ophthalmol. Visual Sci. **42:** 10–16.
24. DEMER, J.L. & J.M. MILLER. 1999. Orbital imaging in strabismus surgery. *In* Clinical Strabismus Management: Principles and Techniques. A.L. Rosenbaum & P. Santiago, Eds.: 84–98. Mosby. New York.
25. CLARK, R.A., A.L. ROSENBAUM & J.L. DEMER. 1999. Magnetic resonance imaging after surgical transposition defines the anteroposterior location of the rectus muscle pulleys. J. Am. Assoc. Pediatr. Ophthalmol. Strabismus **3:** 9–14.
26. CLARK, R.A., J.M. MILLER & J.L. DEMER. 2000. Three-dimensional location of human rectus pulleys by path inflections in secondary gaze positions. Invest. Ophthalmol. Visual Sci. **41:** 3787–3797.
27. KONO, R., R.A. CLARK & K.L. DEMER. 2002. Active pulleys: magnetic resonance imaging of rectus muscle paths in tertiary gazes. Invest. Ophthalmol. Visual Sci. Submitted.
28. DEMER, J.L., R.A. CLARK & J.L. MILLER. 1999. Magnetic resonance imaging (MRI) of the functional anatomy of the inferior oblique (IO) muscle. Invest. Ophthalmol. Visual Sci. (ARVO Abstr.) **40:** S772.
29. CLARK, R.A., J.M. MILLER & J.L. DEMER. 1997. Location and stability of rectus muscle pulleys inferred from muscle paths. Invest. Ophthalmol. Visual Sci. **38:** 227–240.
30. KAPOULA, Z., M. BERNOTAS & T. HASLWANTER. 1999. Listing's plane rotation with convergence: role of disparity, accommodation, and depth perception. Exp. Brain Res. **126:** 175–186.
31. VAN RIJN, L.J. & A.V. VAN DEN BERG. 1993. Binocular eye orientation during fixations: Listing's law extended to include eye vergence. Vision Res. **33:** 691–708.
32. SOMANI, R.A.B., J.F.X. DESOUZE, D. TWEED & T. VILIS. 1998. Visual test of Listing's law during vergence. Vision Res. **38:** 911–923.
33. TWEED, D. 1997. Visual-motor optimization in binocular control. Vision Res. **37:** 1939–1951.
34. DEMER, J.L., R. KONO & W. WRIGHT. 2002. Reconfiguration of human rectus extraocular muscle pulleys during binocular convergence demonstrated by MRI. Invest. Ophthalmol. Visual Sci. (ARVO Abstr.) **43:** In press.
35. MILLER, J.M., C.J. BOCKISH & D.S. PAVLOVSKI. 1999. No co-contraction during convergence. Invest. Ophthalmol. Visual Sci. (ARVO Abstr.). **40:** S772.
36. PORTER, J.D. & K.F. HAUSER. 1993. Diversity and developmental regulation of extraocular muscle: progress and prospects. Acta Anat. **147:** 197–206.
37. PORTER, J.D., R.S. BAKER, R.J. RAGUSA & J.K. BRUECKNER. 1995. Extraocular muscles: basic and clinical aspects of structure and function. Surv. Ophthalmol. **39:** 451–484.

38. SPENCER, R.F. & J.D. PORTER. 1988. Structural organization of the extraocular muscles. *In* Neuroanatomy of the Oculomotor System. J. Buttner-Ennever, Ed.: 33–79. Elsevier. Amsterdam.
39. OH, S.Y., V. POUKENS, M.S. COHEN & J.L. DEMER. 2001. Structure-function correlation of laminar vascularity in human rectus extraocular muscle. Invest. Ophthalmol. Visual Sci. **42:** 17–22.
40. BRUECKNER, J.K., L.P. ASHBY, J.R. PRICHARD & J.D. PORTER. 1999. Vestibulo-ocular pathways modulate extraocular muscle myosin expression patterns. Cell Tissue Res. **295:** 477–484.
41. COLLINS, C.C. 1975. The human oculomotor control system. *In* Basic Mechanisms of Ocular Motility and Their Clinical Implications. G. Lennerstrand & P. Bach-y-Rita, Eds.: 145–180. Pergamon. New York.
42. MILLER, J.M. & D. ROBINS. 1992. Extraocular muscle forces in alert monkey. Vision Res. **32:** 1099–1113.
43. BARMACK, N.H. 1978. Laminar organization of the extraocular muscles of the rabbit. Exp. Neurobiol. **59:** 304–321.
44. SCOTT, A.B. & C.C. COLLINS. 1973. Division of labor in human extraocular muscle. Arch. Ophthalmol. **90:** 319–422.
45. HENN, V. & B. COHEN. 1972. Eye muscle motor neurons with different functional characteristics. Exp. Brain Res. **45:** 561–568.
46. PORTER, J.D. & R.S. BAKER. 1993. Developmental adaptations in the extraocular muscles of Macaca nemestrina may reflect a predisposition to strabismus. Strabismus **1:** 173–180.
47. SPENCER, R.F. & K.W. MCNEER. 1988. Morphology of the extraocular muscles in relation to the clinical manifestation of strabismus. *In* G. Lennerstrand, G.K. von Noorden & E.C. Campos, Eds.: 37–46. Plenum Press. New York.
48. KRZIZOK, T.H. & B.U. SCHROEDER. 1999. Measurement of recti eye muscle paths by magnetic resonance imaging in highly myopic and normal subjects. Invest. Ophthalmol. Visual Sci. **40:** 2554–2560.
49. MILLER, J.M., D.S. PAVLOVSKI & I. SHAEMEVA. 1999. Orbit 1.8 Gaze Mechanics Simulation. Eidactics. San Francisco, CA.
50. MILLER, J.M. & J.L. DEMER. 1999. Biomechanical modeling in strabismus surgery. *In* Clinical Strabismus Management: Principles and Techniques. A.L. Rosenbaum & P. Santiago, Eds.: 99–113. Mosby. St. Louis, MO.
51. CLARK, R.A., J.M. MILLER, A.L. ROSENBAUM & J.L. DEMER. 1998. Heterotopic rectus muscle pulleys or oblique muscle dysfunction? J. Am. Assoc. Pediatr. Ophthalmol. Strabismus **2:** 17–25.
52. DEMER, J.L. 1999. Heterotopy of extraocular muscle pulleys causes incomitant strabismus. *In* Advances in Strabismology. G. Lennerstrand, Ed.: 91–94. Aeolus Press. Buren, the Netherlands.
53. DEMER, J.L. 2001. Clarity of words and thoughts about strabismus. Am. J. Ophthalmol. **132:** 757–759.
54. CLARK, R.A., J.M. MILLER & J.L. DEMER. 1998. Displacement of the medial rectus pulley in superior oblique palsy. Invest. Ophthalmol.Visual Sci. **39:** 207–212.
55. KRZIZOK, T., D. WAGNER & H. KAUFMANN. 1996. Elucidation of restrictive motility in high myopia by magnetic resonance imaging. Arch. Ophthalmol. **115:** 1019–1027.
56. OH, S.Y., R.A. CLARK, F. VELEZ, *et al.* 2002. Incomitant strabismus caused by instability of rectus pulleys. Invest. Ophthalmol. Visual Sci. Submitted.
57. ROSENBAUM, A.L., S.Y. OH, R.A. CLARK & J.L. DEMER. 2001. A new mechanism for Brown syndrome. Abstracts of 27th Annual Meeting of the American Association for Pediatric Ophthalmology and Strabismus. Orlando, FL. 1p.
58. KRZIZOK, T.H., H. KAUFMANN & H. TRAUPE. 1997. New approach in strabismus surgery in high myopia. Br. J. Ophthalmol. **81:** 625–630.
59. CLARK, R.A., S.J. ISENBERG, S.J. ROSENBAUM & J.L. DEMER. 1999. Posterior fixation sutures: a revised mechanical explanation for the fadenoperation based on rectus extraocular muscle pulleys. Am. J. Ophthalmol. **128:** 702–714.
60. CLARK, R.A. & J.L. DEMER. 2002. Surgical transposition of rectus extraocular muscle pulleys in treatment of paralytic strabismus. Am. J. Ophthalmol. **133:** 119–128.

Mechanics of Eye Movements: Implications of the "Orbital Revolution"

THOMAS HASLWANTER

Department of Neurology, University Hospital Zurich and Institute of Theoretical Physics, ETH Zurich, Switzerland

ABSTRACT: Our understanding of the functional structure of extraocular muscles has undergone a profound change: while these muscles used to be represented by strings running straight from their origin in the posterior orbita to their insertion on the globe, we now know that their paths and pulling directions are dominated by fibromuscular pulley structures, keeping them close to the orbital wall for most of their path. An overview is presented of recent models that have been developed to understand the implications of muscle pulleys for the neural control of eye movements and the applications of such models to the interpretation of experimental data.

KEYWORDS: muscle pulleys; modeling; extraocular muscles; orbital plant; Listing's plane; Listing's law; orbit

THE "ORBITAL REVOLUTION"

The first "golden age" of oculomotor research, around 1860, profoundly changed our understanding of the eye, its movements and its mechanics. The experiments by Listing, Donders, and Helmholtz shaped the subsequent understanding of the kinematics of eye movements. At the same time, Volkmann's anatomical measurements of the ventral origin of the extraocular muscles (EOMs), and of their insertion points on the globe, provided the basics of the understanding of the mechanics of the oculomotor plant.[1] This picture of EOMs as strings running straight from their origin to the insertion on the globe dominated the view of the orbital mechanics for a long time. About 100 years later, a combination of modeling[2,3] and imaging work[4–6] sparked off the "orbital revolution": researchers realized that the EOMs do not run freely in the orbit, but for most of their path stay close to the orbital wall. Initially attributed to a combination of retrobulbar pressure on the fat tissues behind the orbit[7] and attachments of the EOMs to the orbital wall by connective tissue,[8] the EOM paths are now thought to be dominated by pulley-like sleeves, which consist of connective tissue as well as smooth muscles.[9] Ample evidence now also exists that these muscle pulleys are actively innervated[10] (Demer, this volume). In this paper I will refer to these fibromuscular structures simply as "pulleys." Imaging work has shown that irregularities in these pulleys can lead to incomitant strabismus[11] and that their

---

Address for correspondence: Dr. Thomas Haslwanter, Department of Neurology, Frauenklinikstr. 26, CH- 8091 Zürich, Switzerland. Voice: +41-1-255 3996; fax: +41-1-255 4507.
haslwant@neurol.unizh.ch

locations in turn are affected by muscle palsies.[12] Nevertheless, the detailed implications of these pulleys for the control of eye movements are still controversial.

IMPLICATIONS OF THE "ORBITAL REVOLUTION"

The experimental data that have revolutionized our understanding of the anatomy and physiology of the EOMs have been provided mainly by the work of Demer, Miller, and co-workers (Demer, this volume, and references therein). To understand the implications of these findings for the control signals required to move and hold the eye in eccentric positions, a number of models have been developed. Two types of models can be distinguished:

(1) Models with a simplified plant that focus on the information processing and control aspects under idealized conditions. By simplifying the mechanics of the oculomotor plant, these models can investigate static as well as dynamic situations.
(2) Biomechanically correct models that incorporate all known anatomical and physiological details of the oculomotor plant. Because such models include many measured parameters, they are usually quite complex. This has restricted them so far to the simulations of fixations only.

I will present an overview of the most important models and their applications to experimental data.

IDEALIZED-PLANT MODELS AND THEIR IMPLICATIONS

"Idealized-plant models" dramatically simplify the mechanical and physiological properties of the oculomotor plant, in order to better simulate its effects on the control of eye movements. These models commonly assume three idealized muscle pairs (*horizontal*, *vertical*, and *torsional*), which are perfectly orthogonal to each other. The forces generated by these muscle pairs are strictly linear to their innervation. In these models, the location of the pulleys is usually described by a free, adjustable parameter.

The first such model that used the mechanical properties of the oculomotor plant to explain experimental findings was by Schnabolk and Raphan.[13] Using an oculomotor plant without pulleys but with idealized EOMs, they showed that a two-dimensional horizontal-vertical torque command to the oculomotor plant would nicely explain Listing's law. (Note that we talk about "torque" here and not about "force," because a torque is necessary to cause a rotation. In all idealized-plant models, only rotational movements of the eye are considered.) Listing's law states that for static subjects all rotation vectors describing the eye rotation from the primary position to any eye position align along a plane. Using this plane to determine the coordinate system in which data are described, this can be rephrased as "all eye positions have zero torsion."[14] In an idealized plant the elastic-restoring torque T, exerted by the elasticity of the EOMs and the orbital tissue onto the eye, is related to the rotation vector $\vec{rot}$ describing the current eye position by

$$\vec{T} = -a * \vec{rot}. \qquad (1)$$

Therefore, a zero torsional component in $\vec{rot}$ automatically implies a zero torsional component in the corresponding torque vector $\vec{T}$. In other words, in this model a two-dimensional, horizontal–vertical torque command is sufficient to generate all eye positions in Listing's plane.

Tweed et al. quickly pointed out that such a model predicts "blips" in the torsional eye position during eccentric saccadic eye movements, which are not found experimentally.[15] (A later study by Straumann et al. did find small but consistent torsional position blips for eccentric horizontal eye movements.[16] But they were much smaller than the ones predicted by Schnabolk and Raphan, and had idiosyncratic directions for vertical eye movements.) The reason for these torsional blips lies in the noncommutativity of rotations.

This noncommutativity entails a paradox: eye *positions* in Listing's plane have zero torsion, but eye *movements* require a nonzero torsional velocity to stay in Listing's plane.[14] FIGURE 1 shows this effect graphically: the circles at the center of each surface mark a starting position, and the surrounding surfaces show the amount of torsion that is necessary to move to an adjacent position in Listing's plane. For example, starting from the primary position (0/0), any surrounding position in Listing's plane can be reached with no torsion (FIG. 1, flat surface lower left). Also, a purely vertical movement from a vertical starting position requires no torsion (upper left surface, moving "in the groove"). However, a horizontal eye movement from the same starting position needs a torsional component proportional to the size of the movement in order to stay in Listing's plane (upper left surface, moving perpendicular to the groove).

The change of the surface shape with eye position shows that an eye position–dependent torsional eye-velocity component is necessary to keep the eye in Listing's plane. This torsional *velocity* component was absent from the Schnabolk/Raphan model, which caused the torsional *position* blips during saccades. Raphan corrected for this by suggesting that pulleys on the EOMs might implement the necessary eye position–dependent torsion during eye movements. FIGURE 2 illustrates the idea: for a horizontal rectus muscle (lateral or medial rectus), a pulley located exactly on the horizontal rotation axis would lead to an axis of eye velocity that tilts with the eye (FIG. 2, top and left). In contrast, if there were no pulleys, the axis of eye velocity would always be (almost) earth-vertical, independent of the eye position (FIG. 2, bottom and left).

Parametrizing these extremes with $k = 1$ (for eye-fixed rotation axes, FIG. 2 top) and $k = 0$ (for head-fixed rotation axis, FIG. 2 bottom), Raphan showed that a value of $k = 0.5$ would correctly implement Listing's law for fixations as well as for saccades.[17] This is not too surprising because it had been pointed out previously that Listing's law entails the "half-angle rule" for eye velocity: the axis of eye velocity has to lie halfway between a head-fixed and an eye-fixed axis in order to follow Listing's law.[18] Quaia and Optican generalized this result for all eye movements, and showed mathematically that EOM pulleys have a different effect on the pulse and on the step components of saccadic innervation.[19] The innervation step generates the

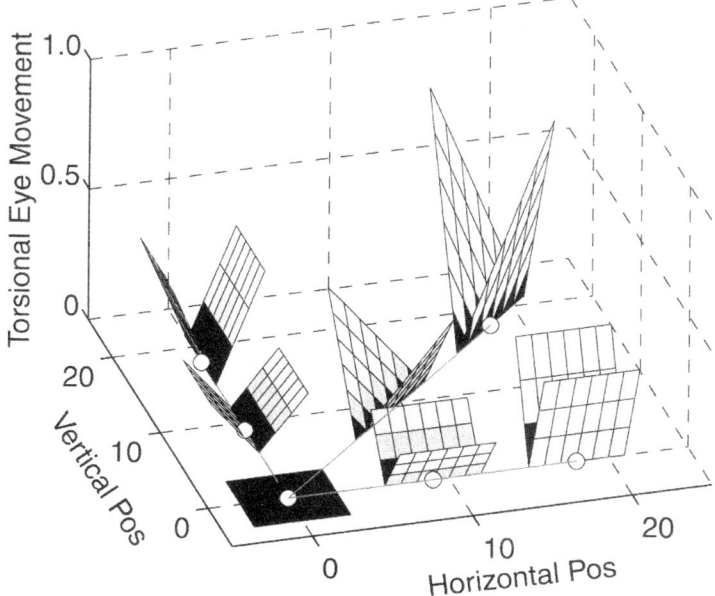

FIGURE 1. The paradox of Listing's plane and torsional eye movements: for different starting points in Listing's plane (*circles at the center of each surface*), a position-dependent, nonzero amount of torsional eye movement is required to move the eye to adjacent horizontal/vertical eye positions in Listing's plane (i.e., with zero torsion!). All eye positions are in Listing's plane, and the angles are in degrees.

torque that keeps the eye in an eccentric eye position by compensating for the elastic restoring forces of the oculomotor plant that pull the eye back toward the straight-ahead position. This torque, which corresponds to the torque $\vec{T}$ in Eq. 1, is independent of muscle pulleys. In contrast, the saccadic pulse is necessary to move the eye and must compensate for the torque generated by the viscous force induced by the eye movement. This torque is proportional to the angular eye velocity, which—as illustrated in FIGURE 2—depends on the pulley locations. Quaia and Optican also showed that with "correctly" located pulleys ($k = 0.5$) the saccadic step is almost exactly the same as the mathematical integral of the saccadic pulse. This allows a simple online computation of the saccadic step from the pulse signal, and solves the problem of the 3-dimensional (3-D) velocity–position transformation, without encountering the noncommutativity of 3-D rotations. Such an arrangement would explain, for example, the lack of torsional eye movement signals in the superior colliculus[20,21] and shows the importance of ocular plant mechanics for the understanding of the neural control of eye movements. However, as Quaia pointed out, more experiments are needed to test the predictions of these models.[19]

The discussion about the implications of EOM pulleys sometimes gets mixed up with the controversy about commutativity or noncommutativity in the control of eye

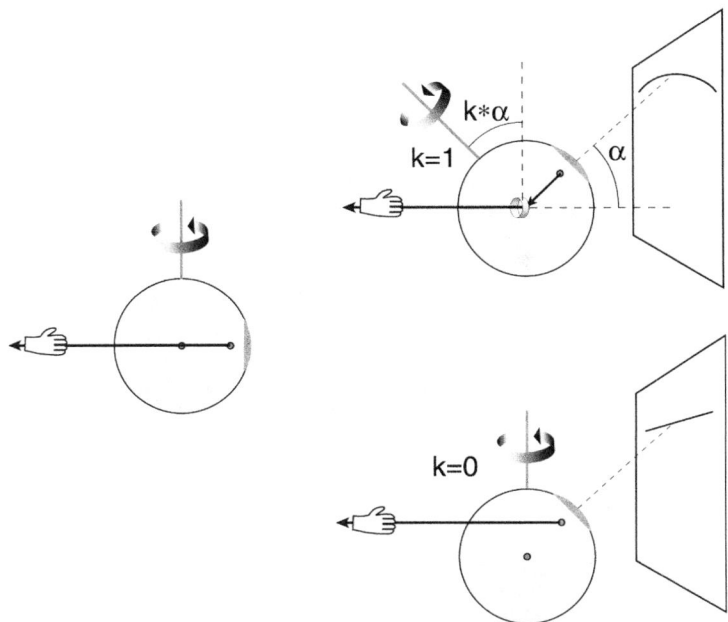

FIGURE 2. Sketch of the effect of a muscle pulley. (*Left*) Pulling on a horizontal rectus muscle exerts a torque about an earth-vertical axis. (*Top right*) If the pulley for the horizontal muscle were located at the center of the globe, the torque that rotates the eye: would always be perpendicular to the line of sight: an eye elevation of α would lead to a tilt of the eye velocity axis of $k*\alpha$, with $k = 1$. (*Bottom right*) If there were no pulley, the torque would always be about an earth-vertical axis, independent of the vertical eye position ($k = 0$, assuming a very long muscle).

movements. A "commutative controller" means that the neural command for a horizontal eye movement is independent of the vertical eye position, and vice versa. The models described above have shown that such commutative controllers, which only use a horizontal-vertical but no torsional eye position signal, can implement Listing's law, if the EOM pulleys are properly located.

Contrary to a widely held opinion, many researchers who believe in a *non-commutative*, 3-D controller for eye movements do not deny the implications of pulleys, and sometimes also incorporate muscle pulleys in their models.[22] However, they have shown that under dynamic conditions (VOR or combined eye-head movements) the torsional component of eye position is actively controlled, something that would not be possible with a horizontal-vertical control system.[22,23] The proposed noncommutative control systems that can explain these observations are not necessarily complex and can be implemented in a simple, plausible way.[22] Furthermore, the visuomotor transformation from retinal information into kinematically correct eye movements requires accurate control over horizontal, vertical, *and* torsional eye position.[24] Therefore, the often heated argument about commutativity or noncommutativity in the control of eye movements is not about pulleys, but about whether the central nervous system uses a 2-D (horizontal/vertical) or a 3-D (horizontal/vertical/torsional) controller for eye movements.

PULLEY MODELS AND EXPERIMENTAL DATA

Models that simplify the mechanics of the orbital plant have also been used to interpret experimental data. For example, Thurtell *et al.* found intriguing differences between active and passive rapid horizontal head movements.[25] To investigate whether EOM pulleys could be responsible for these differences, they applied the model by Raphan[17] to the vestibulo-ocular reflex.[26] They found that to reproduce the eye movements during the first 50 ms of active head impulses, their model required a pulley coefficient k of 0.5. In contrast, fitting eye movements during passive head impulses led to a pulley coefficient closer to zero (0.2) However, it is arguable how much this reveals about pulleys: it simply shows that during active movements the eye stays in Listing's plane ($k = 0.5$), whereas for passive movements—where the central nervous system does not know about the upcoming head rotation—the central nervous system makes a compromise between Listing's plane ($k = 0.5$) and optimal retinal image stabilization ($k = 0$). No pulleys would be necessary to implement such a control system.

EOM pulleys play a role in all experiments in which the pulling direction of the EOMs is critical for the interpretation of the experimental data. For example, Rambold *et al.* wanted to determine which muscle groups are involved in the generation of nystagmus after lesions in the interstitial nucleus of Cajal. Using a geometric muscle model that accounts for the change of the EOM pulling directions by pulleys, their data suggested a co-activation of eye muscles similar to the effects of electrical stimulation of the anterior canal nerve.[27]

The recent finding that the locations of EOM pulleys are not constant but can be actively changed by innervation of the global layers of the EOMs has been used to explain changes in the orientation of Listing's plane, such as the lateral tilt of Listing's plane during convergence (Demer, this volume). This mechanism has also been invoked to motivate the lateral tilt of Listing's plane during light sleep.[28]

Single-cell recordings in the paramedian pontine reticular formation (PPRF) have suggested that during torsional movements of the eye, pulleys may also change their location in the coronal plane. Scherberger *et al.* determined the preferred on-direction of neurons in the PPRF while orienting the monkey in different roll positions, and showed that these on-directions are consistent with pulleys moving in the coronal plane by half the amount of the eye movement.[29]

BIOMECHANICAL MODELS

All models mentioned so far dramatically idealize the orbital plant. They do not take into consideration the functional differences between the individual EOMs. For example, although the functional origin of the rectus muscles is determined by EOM pulleys that can change their location,[10] the functional origin of the superior oblique is fixed by the trochlea. In addition, idealized-plant models use two gross simplifications. For one thing, they assume that the tension in the muscles is linearly proportional to the innervation. In contrast, the force exerted by real EOMs has a complicated, nonlinear length-tension relationship,[30] which is ignored by these models. The second simplification is the assumption of three perfectly aligned push-pull muscle pairs. This assumption avoids the complication of an overdetermined system, which is present in the real oculomotor plant: for each eye we have six EOMs, but only three rotational degrees of freedom. To investigate the actual effects of the individual components of the oculomotor plant, *biomechanical models* incorporating our current knowledge about the anatomical and physiological properties of the orbital plant are indispensable.

The first biomechanical model by Robinson—which also inspired all subsequent modeling efforts—already indicated that the EOMs could not run in a straight line from their origin to their insertion.[2] This would cause too much "bridle" effect (horizontal rectus muscles becoming elevators in up-gaze, for instance), and too much twist of the muscle at the insertion. Imaging work then indicated that the EOMs in fact stay close to the orbital wall for most of their path.[4–6] This finding was included in two independent models, both based on the original Robinson model. Simonsz developed a program that was optimized for a handheld computer, aimed at assisting in complicated strabismus surgeries.[7] In parallel, Miller refined his SQUINT model of binocular alignment[3] and made it commercially available as *Orbit*. While Simonsz tried to make his model as straightforward as possible, Miller's model was characterized by the anatomical accuracy, a user-friendly interface, and superb graphics. *Orbit* now also includes passive pulleys and allows simulation of a number of ophthalmologic pathologies (e.g., muscle palsies, Duane's retraction syndrome) through modification of parameters characterizing the anatomic and physiologic properties of the orbital plant.

Most recent developments of biomechanical models are related to *Orbit*. Since the source code of *Orbit* is proprietary, Porrill *et al.* implemented many *Orbit* ideas in the MATLAB model *EyeLab*.[31] Parameters were thereby chosen such that the behavior approximated that of *Orbit*. While the idealized-plant models mentioned above only indicated that properly placed pulleys would *facilitate* the implementation of a commutative controller for eye movements, *EyeLab* demonstrated that the actual structure of the orbital plant is in fact such that commutative horizontal-vertical control signals would implement Listing's law.[31]

OUTLOOK

Because of the computational complications arising from an overdetermined system (six EOMs, but only three degrees of freedom), current biomechanical models are restricted to the investigation of fixations only. They also cannot easily incorporate recent anatomical findings like the active-pulley hypothesis with its layered view of the EOMs, or additional orbital structures like Lockwood's ligament (coupling between the inferior oblique and the inferior rectus pulley). To incorporate these elements, the development of a new, strongly object-oriented model of the oculomotor plant is necessary and is currently under development by Miller and coworkers. Also, an extension of biomechanical models to include eye movement dynamics is desirable. This requires a better understanding of the force development in

EOMs during actual eye movements[32] and inclusion of the complex properties of muscle force generation.[33,34]

REFERENCES

1. VOLKMANN, A.W. 1869. Zur Mechanik der Augenmuskeln. Ber. Sächs Ges. Wiss. Math. Naturwiss. Klasse **21:** 28–69.
2. ROBINSON, D.A. 1975. A quantitative analysis of extraocular muscle cooperation and squint. Invest. Ophthalmol. **14:** 801–825.
3. MILLER, J.M. & D.A. ROBINSON. 1984. A model of the mechanics of binocular alignment. Comput. Biomed. Res. **17:** 436–470.
4. SIMONSZ, H.J., F. HARTING, B.J. DE WAAL & B.W. VERBEETEN. 1985. Sideways displacement and curved path of recti eye muscles. Arch. Ophthalmol. **103:** 124–128.
5. MILLER, J.M. & D. ROBINS. 1987. Extraocular muscle sideslip and orbital geometry in monkeys. Vision Res. **27:** 381–392.
6. MILLER, J.M. 1989. Functional anatomy of normal human rectus muscles. Vision Res. **29:** 223–240.
7. SIMONSZ, H.J. 1996. Robinson's computerized strabismus model comes of age. Strabismus **4:** 25–40.
8. DEMER, J.L., J.M. MILLER, V. POUKENS, et al. 1995. Evidence for fibromuscular pulleys of the recti extraocular muscles. Invest. Ophthalmol. Visual Sci. **36:** 1125–1136.
9. MILLER, J.M., J.L. DEMER & A.L. ROSENBAUM. 1993. Effect of transposition surgery on rectus muscle paths by magnetic resonance imaging. Ophthalmology **100:** 475–487.
10. DEMER, J.L., S.Y. OH & V. POUKENS. 2000. Evidence for active control of rectus extraocular muscle pulleys. Invest. Ophthalmol. Visual Sci. **41:** 1280–1290.
11. DEMER, J.L., R.A. CLARK & J.M. MILLER. 1999. Heterotopy of extraocular muscle pulleys causes incomitant strabismus. *In* Proceedings of the 8th Congress of the International Strabismological Association. G. Lennerstrand, Ed.: 91–94. Aeolus Press. Buren, the Netherlands.
12. CLARK, R.A., J.M. MILLER & J.L. DEMER. 1998. Displacement of the medial rectus pulley in superior oblique palsy. Invest. Ophthalmol. Visual Sci. **39:** 207–212.
13. SCHNABOLK, C. & T. RAPHAN. 1994. Modeling three-dimensional velocity-to-position transformation in oculomotor control. J. Neurophysiol. **71:** 623–638.
14. HASLWANTER, T. 1995. Mathematics of three-dimensional eye rotations. Vision Res. **35:** 1727–1739.
15. TWEED, D., H. MISSLISCH & M. FETTER. 1994. Testing Models of the Oculomotor Velocity-to-Position Transformation. J. Neurophysiol. **72:** 1425–1429.
16. STRAUMANN, D., D.S. ZEE, D. SOLOMON, et al. 1995. Transient torsion during and after saccades. Vision Res. **35:** 3321–3334.
17. RAPHAN, T. 1998. Modeling control of eye orientation in three dimensions. I. Role of muscle pulleys in determining saccadic trajectory. J. Neurophysiol. **79:** 2653–2667.
18. TWEED, D., W. CADERA & T. VILIS. 1990. Computing three-dimensional eye position quaternions and eye velocity from search coil signals. Vision Res. **30:** 97–110.
19. QUAIA, C. & L.M. OPTICAN. 1998. Commutative saccadic generator is sufficient to control a 3-D ocular plant with pulleys J. Neurophysiol. **79:** 3197–3215.
20. HEPP, K., A.J. VAN OPSTAL, D. STRAUMANN, et al. 1993. Monkey superior colliculus represents rapid eye movements in a two-dimensional motor map. J. Neurophysiol. **69:** 965–979.
21. VAN OPSTAL, A.J., K. HEPP, Y. SUZUKI & V. HENN. 1995. Influence of eye position on activity in monkey superior colliculus. J. Neurophysiol. **74:** 1593–1610.
22. TWEED, D.B., T.P. HASLWANTER, V. HAPPE & M. FETTER. 1999. Non-commutativity in the brain. Nature **399:** 261–263.
23. TWEED, D., T. HASLWANTER & M. FETTER. 1998. Optimizing gaze control in three dimensions. Science **281:** 1363–1366.
24. KLIER, E.M. & D.J. CRAWFORD. 1998. Human oculomotor system accounts for 3-D eye orientation in the visual-motor transformation for saccades. J. Neurophysiol. **80:** 2274–2294.

25. THURTELL, M.J., R.A. BLACK, G.M. HALMAGYI, et al. 1999. Vertical eye position-dependence of the human vestibuloocular reflex during passive and active yaw head rotations. J. Neurophysiol. **81:** 2415–2428.
26. THURTELL, M.J., M. KUNIN & T. RAPHAN. 2000. Role of muscle pulleys in producing eye position-dependence in the angular vestibuloocular reflex: a model-based study. J. Neurophysiol. **84:** 639–650.
27. RAMBOLD, H., C. HELMCHEN & U. BUTTNER. 2000. Vestibular influence on the binocular control of vertical-torsional nystagmus after lesions in the interstitial nucleus of Cajal. Neuroreport **11:** 779–784.
28. CABUNGCAL, J.H., H. MISSLISCH, H. SCHERBERGER, et al. 2001. Effect of light sleep on three-dimensional eye position in static roll and pitch. Vision Res. **41:** 495–505.
29. SCHERBERGER, H., J.H. CABUNGCAL, K. HEPP, et al. 2001. Ocular counterroll modulates the preferred direction of saccade-related pontine burst neurons in the monkey. J. Neurophysiol. **86:** 935–949.
30. SIMONSZ, H.J. 1994. Force-length recording of eye muscles during local-anesthesia surgery in 32 strabismus patients. Strabismus **4:** 197–218.
31. PORRILL, J., P.A. WARREN & P. DEAN. 2000. A simple control law generates Listing's positions in a detailed model of the extraocular muscle system. Vision Res. **40:** 3743–3758.
32. MILLER, J.M. & D. ROBINS. 1992. Extraocular muscle forces in alert monkey Vision Res. **32:** 1099–1113.
33. GOLDBERG, S.J., M.A. MEREDITH & M.S. SHALL. 1998. Extraocular motor unit and whole-muscle responses in the lateral rectus muscle of the squirrel monkey. J. Neurosci. **18:** 10629–10639.
34. GOLDBERG, S.J. & M.S. SHALL. 1999. Motor units of extraocular muscles: recent findings. Prog. Brain Res. **123:** 221–232.

Differential Susceptibility of the Ocular Motor System to Disease

HENRY J. KAMINSKI,[a,b] CHELLIAH R. RICHMONDS,[a] LINDA L. KUSNER,[c] AND HIROSHI MITSUMOTO[d]

Departments of Neurology,[a] Neurosciences,[b] and Biophysics and Physiology,[c] Case Western Reserve University, University Hospitals of Cleveland, Cleveland, Ohio, USA

Louis Stokes Veterans Affairs Medical Center, Cleveland, Ohio, USA

Department of Neurology,[d] Columbia-Presbyterian Medical Center, New York, New York, USA

ABSTRACT: This review summarizes an alternative approach to the understanding of neuromuscular disease. By contrasting disease susceptibility of extraocular muscle and ocular motor neurons, it is hoped that unique insights into disease mechanisms may be identified. Disorder of eye movements leads to dramatic symptoms for patients and the ocular motor system is relatively limited in its ability to compensate rapidly for such disruptions. However, more profound reasons exist as to why myasthenia gravis compromises neuromuscular transmission at ocular muscle synapses as well as why Graves' ophthalmopathy exists. In contrast, muscular dystrophies spare the eye muscles while devastating all other skeletal muscles; the same is true for motor neuron diseases. It is hoped that this review will encourage others to view the world of neuromuscular diseases as delineated into those that spare the ocular motor system and those that do not.

KEYWORDS: extraocular muscle; skeletal muscle; ocular motor neuron; neuromuscular disease

INTRODUCTION

Among the most striking properties of the peripheral ocular motor system is its varied involvement by disorders of nerve and muscle. The degree of this differential susceptibility occurs across a continuum from near-complete sparing of ocular motor nerve (OMN) and extraocular muscles (EOM) by devastating disorders, such as Duchenne muscular dystrophy and motor neuron diseases, to isolated damage in the case of Graves' ophthalmopathy. There are no disorders in which the pathophysiology of this differential targeting is understood well. This review provides a broad overview of several disorders where insight exists into ocular motor system disease. John Porter provides in this volume the background of EOM biology to appreciate discussions of pathology. The goal of this presentation, in keeping with the rest of

Address for correspondence: Henry J. Kaminski, M.D., Department of Neurology, University Hospitals of Cleveland, 11100 Euclid Avenue, Cleveland, Ohio 44106. Voice: 216-368-0250; fax: 216-368-0249.

hjk3@po.cwru.edu

this conference, is to foster creative interactions among various disciplines in order to better understand the ocular motor system and its involvement by disease.

GENERAL REASONS FOR SUSCEPTIBILITY OF THE OCULAR MOTOR SYSTEM TO DISEASE

Some straightforward reasons exist for the apparent differential susceptibility for disorders that compromise ocular motor nerve and muscle function. The most simple, and least scientifically exciting, is that dysfunction leads to patient complaints of double or blurred vision.[1] Such problems are usually not ignored by individuals, and rather prompt medical attention is sought. Because of the precision required in maintaining alignment of the visual axes, even a minor reduction of force generation produces dramatic symptoms. This is in marked contrast to a small reduction in force generation of a limb muscle, which may not be readily appreciated, and is more easily ignored by a patient.

Although the importance of proprioception in ocular motor control is increasingly becoming appreciated (see Dell'Osso, this volume), its role in maintaining ocular alignment by dynamic control of muscle force generation is unlikely to be as important as in other skeletal muscles. The lack of tendon organs in EOM supports a less important role for proprioception. The importance of this is that in a deltoid muscle reduced force generation could be compensated by feedback mechanisms, which recruit additional motor units. Such a mechanism appears not to be in place for the ocular motor system.

DISEASES THAT COMPROMISE THE OCULAR MOTOR SYSTEM

Although the reasons provided above are attractive in their simplicity, there appear to be deeper bases for EOM and OMN to be specifically targeted by certain neuromuscular disorders. The following sections detail diseases where some insights are available.

Myasthenia Gravis

Myasthenia gravis (MG) is an autoimmune disorder that causes a reduction of skeletal muscle acetylcholine receptors, which produces neuromuscular transmission failure.[2] Weakness of the EOM and the levator palpebrae are the initial signs of MG in the majority of patients and ultimately occurs in nearly all; it may remain restricted to these muscles in 10–15% of patients.[3] One of the authors (H.J.K.) has recently provided a detailed review of the preferential involvement of the EOM neuromuscular junctions by myasthenia gravis,[4] and the arguments are only briefly recapitulated here.

In addition to the general reasons for ocular motor susceptibility to disease, several factors specific to EOM place them at risk for myasthenia gravis involvement. The observation that genetic defects of neuromuscular transmission lead to ocular misalignment in many patients suggests that the unique physiological properties of the ocular motor system may be of paramount importance in targeting of EOM by neuromuscular transmission abnormalities. The high firing frequencies of OMN may contribute to neuromuscular transmission fatigue. All disorders of neuromuscular transmission share in common with myasthenia gravis a compromise in the safety factor. Repetitive stimulation leads to a relative reduction of acetylcholine release from the nerve terminal. At a normal junction, this is inconsequential because of a high safety factor, but at a myasthenic junction the safety factor is reduced. Thus, the EOM junctions subject to high rates of stimulation would be expected to be particularly susceptible to a reduced safety factor. One question to be answered is how normal EOM junctions are adapted to endure such high rates of neuronal stimulation.

EOM singly innervated fibers possess anatomical characteristics that may make them more susceptible to neuromuscular transmission block. These fibers have less prominent synaptic folds, and therefore one would predict fewer AChRs and sodium channels on the post-synaptic membrane.[5] A reduction in AChR, sodium channels, and quantal content would reduce the safety factor predisposing EOM to neuromuscular transmission failure. However, the EOM miniature end-plate potential amplitudes are similar to those of leg muscle junctions, indicating that AChR density is similar at these synapses.[6] An estimation of the safety factor at EOM neuromuscular junctions has only been published in abstract form.

Certain multi-innervated fibers of EOM are similar to reptile tonic and intermediate fibers that have tonic contractile characteristics, with the functional consequence that force generation is directly proportional to the membrane depolarization caused by the endplate potential. Therefore, a safety factor does not exist for tonic fibers, and any reduction of end-plate potential induced by a loss of acetylcholine receptors would decrease contractile force of these fibers.[7,8] Also, the neuromuscular junctions have a scarcity to complete absence of junctional folds. These structures in singly innervated fibers serve to increase the concentration of acetylcholine receptors and to concentrate current on to the depths of the folds where sodium channels are highly concentrated.

The autoimmune process of patients with ocular myasthenia does differ from patients with generalized disease. Ocular myasthenia patients have lower levels of acetylcholine receptor autoantibodies.[9,10] The intensity of T cell responses to AChR epitopes of patients with ocular myasthenia are lower than that of generalized patients, and they fluctuate over time.[11] The combination of mild autoimmune disease and physiologic susceptibility may be the explanation for the existence of purely ocular myasthenia. Patients with purely ocular myasthenia may not develop general disease because these patients may have residual regulatory mechanisms that suppress the autoimmune reaction. T cells (CD4+) may not be sufficiently activated to drive a pathogenic antibody response.[11] Although EOM is unique in its expression of the fetal acetylcholine receptor at their neuromuscular junctions,[12,13] no evidence exists that there is specific immunological targeting of the receptor by the autoimmune disease.[11,14,15]

A new insight has been made since the publication of the review by Ubogu and Kaminski[4]—the identification of the differential expression of complement regulatory proteins in EOM (TABLE 1).[16] Destruction of the neuromuscular junction by myasthenia gravis is mediated by complement. If EOM were to express low levels of decay accelerating factor (daf), which functions to inhibit complement deposition, in the vicinity of the neuromuscular junction, then this could be an additional factor contributing to the development of ocular manifestations with a low level of disease intensity.

TABLE 1. Differences in gene expression of complement regulatory protein genes between extraocular muscle and jaw or leg muscle

Gene	EOM vs. Leg muscle	EOM vs. Masseter
Decay accelerating factor 1 (Daf1)	−5	−7
Decay accelerating factor 2 (Daf2)	ns	−10.7
CD59 antigen (protectin)	ns	2
Complement factor H-related protein[a]	12.3	5.8
Complement factor H-related protein[a]	ns	5.4
Complement factor H-related protein h (Cfh)	ns	4.3

[a]Denotes results from two different probes in the Affymetrix microarray.
ns, not significant. Data from Porter et al.[16]

Lambert-Eaton Myasthenic Syndrome

Lambert-Eaton myasthenic syndrome is a presynaptic disorder, which is produced by antibodies directed at calcium channels at the nerve terminal leading to a compromise of synaptic vesicle release. In contrast to myasthenia gravis, only mild to moderate ptosis is observed among these patients, and eye movement disturbances are not as prominent.[17] Given the pathogenesis of the disorder, one may expect EOM function to be severely compromised since ocular motor neurons, because of their high firing frequencies, would not have a significant reserve of readily releasable synaptic vesicles. Miniature end-plate potential frequency is greater at EOM junctions, indicating greater calcium influx to the nerve terminal;[6] and this property may be relatively protective from the defect produced by Lambert-Eaton myasthenic syndrome.

Botulism

Double vision is an early and nearly universal manifestation of botulism.[18] Botulinum toxin injections to the face for treatment of craniofacial dystonias may produce misalignment of the eyes despite preservation of the normal peak velocities during saccades.[19] In their clinical study, Stahl and colleagues proposed that the effects of botulinum toxin are twofold—an initial effect related to blockade of the neuromuscular junction, which is partially relieved by edrophonium, and a delayed effect due to fiber atrophy. Unlike the short-term and reversible effect that botulinum toxin has on most muscle fibers, the orbital singly innervated fibers of EOM undergo atrophy.[20,21] Once again, the reasons for their degeneration is a mystery. The specific degeneration of these EOM fibers is all the more surprising because EOM do not show significant morphological alteration with axotomy.[22,23]

Graves' Ophthalmopathy

Mild ocular disease of Graves' is produced by β-adrenergic hypersensitivity produced by excessive amounts of thyroid hormone, while more severe involvement results from inflammatory edema of the orbital contents. The spectrum of orbital inflammation ranges from mild to severe leading to exophthalmos, which may lead to compression of the optic nerve. As observed with myasthenia gravis, muscle involvement may be restricted to single muscles or one orbit. The EOM, lacrimal gland, and orbital connective tissue enlarge to several times their normal size; this arises from enlargement of the extracellular space by increased amounts of connective tissue and hydrophilic extracellular-matrix compounds.[24,25] Acidic glycoproteins accumulate in the ground substance, bind water, and produce swelling of the orbital contents.[26] The extent of inflammation is variable. Plasma cells, mast cells, and macrophages infiltrate retrobulbar fat and the extraocular muscles. Adipocytes may be atrophied or slightly enlarged with thyrotoxicosis.[27] Some EOM biopsies reveal prominent T-cell infiltrates with degenerating muscle fibers,[28] while others have found little inflammation and no muscle fiber damage.[29] Ultrastructural studies indicate that the primary change is an interstitial inflammatory edema.[27]

Graves' ophthalmopathy is now considered a separate autoimmune disorder from immune thyrotoxicosis.[30] In Graves' disease thyroid-stimulating antibodies are thought to initiate immune-mediated destruction of the thyroid and thyrotoxicosis.[31] Some of these antibodies were thought to cross-react with orbital membranes, suggesting a direct link between the thyroid disease and the ophthalmopathy. The pathogenesis of ophthalmopathy may be related to an immune-mediated process involving a shared antigen between thyroid and eye muscle protein.[32,33] Antibodies in patients with ophthalmopathy react with EOM.[34,35] Retroorbital connective tissue also binds immunoglobulins from patients with ophthalmopathy.[36] T-lymphocytes of Graves' patients are cytotoxic toward EOM-derived myotubes in cell culture.[37] The pathogenesis and propagation of thyroid-associated ophthalmopathy appears to be primarily T-cell mediated.[38] It is not known whether a specific antigen to orbital tissue initiates Graves' ophthalmopathy, although thyrotropin receptor or a structurally similar protein is a candidate.[39] Another factor that may direct the immune response to EOM is the interconnections between the lymphatic drainage of the thyroid and the orbit. Consequently, orbital tissue may be exposed to high concentrations of thyroglobulin and its anti-thyroid autoantibodies.[40] Variations in lymphatic anatomy could then account for the variable involvement of specific muscles or more predominant involvement of a single orbit. Complement regulatory proteins may modulate the severity of autoimmune thyroid disease.[41] Whether the differential expression of these proteins by EOM (TABLE 1) may influence development of Graves' ophthalmopathy is unknown.

Acute Demyelinating Neuropathy

Peripheral neuropathies generally do not involve the OMN with the dramatic exception of the acute demyelinating neuropathy, Guillain-Barré syndrome. Patients who develop ocular motor paresis as part of wide spread autoimmune demyelination of peripheral nerves possess antibodies against the GQ1b ganglioside.[42,43] Such antibodies are also identified in patients with Miller Fisher syndrome, which manifests with ophthalmoparesis and ataxia with loss of deep tendon reflexes, and may have coincident central nervous system pathology. Human OMN contains a large amount of GQ1b ganglioside, significantly more than the ventral and dorsal roots of the spinal cord. The GQ1b ganglioside is specifically localized to the paranodal regions of the extramedullary portion of oculomotor, trochlear, and abducens nerves. These data, taken together, indicate that specific immunologic targeting explains the development of eye movement abnormalities in certain cases of demyelinating neuropathy.

DISEASES THAT SPARE THE OCULAR MOTOR SYSTEM

More perplexing problems arise from the lack of involvement of the ocular motor system by disease. These lie on the muscle and the nerve side. As in the case of myasthenia gravis, much effort has been expended to understand differential involvement of EOM by muscular dystrophies, but an answer has not been found.[44] For motor neuron diseases tantalizing observations are available, but as yet little focused research time has been expended.

Muscular Dystrophies

Although the genetic defects of the transmembrane dystrophin-glycoprotein complex (DGC) have been identified to underlie the majority of muscular dystrophies, the link between the gene defect and muscle fiber degeneration remains unknown with a variety of hypotheses under active investigation. A novel approach to understanding muscular dystrophy pathogenesis has been the investigation of the sparing of EOM from degeneration in dystrophin-deficient muscular dystrophy (Duchenne/Becker).

Patients with dystrophin-deficiency, who are severely compromised, have essentially normal eye movements as measured by infrared oculography.[45] A single patient with Becker muscular dysrophy has been reported who had slowed saccadic velocities, and this exception suggests that other modifying influences lead to EOM compromise.[46] Pathological studies confirm the sparing of EOM. The structure of EOM is spared in animal models of dystrophin deficiency, including the *mdx* mouse and *cmdx* dog.[47–50] There are no signs of myofiber degeneration, connective tissue accumulation, or central nuclei characteristic of regenerated muscles in EOM in any of the animal models. Even among aged mice, e no alterations in EOM are found.[50] Other forms of muscular dystrophy caused by mutations of laminin-2 (congenital muscular dystrophy) and sarcoglycan (limb-girdle type dystrophy) also have no pathological abnormalities of EOM.[51,52]

The sparing of EOM may be explained by the expression of alternative proteins that could compensate for lost proteins, or by the functional or metabolic characteristics of these muscles, which may protect them from the consequences of the primary defects. Andrade *et al.* have recently reviewed this subject in detail and only major points are recapitulated here.[44]

The DGC could differ in normal EOM, but thus far no variations in its structure from other skeletal muscle have been identified. EOM appears to have the capacity to increase utrophin expression in response to dystrophin deficiency, and this ability may contribute to sparing in Duchenne muscular dystrophy. Mice deficient in both dystrophin and utrophin exhibit substantial EOM pathology, thereby supporting a putative sparing role by utrophin.[50] However, the orbital layer fibers that show the greatest upregulation of utrophin in mdx mice are spared in the dystrophin/utrophin double knockout mouse, suggesting additional mechanisms are at work.

Disruption of the DGC is thought to lead to elevations of intracellular calcium by destabilization of the sarcolemma. The EOM could be spared in these disorders by their apparent constitutively high calcium sequestration capacity. The EOM are fast contracting and constantly active muscles, also capable of sustained tetanus. The small myofibers, unique troponin T isoform composition, abundant mitochondria and sarcoplasmic reticulum (SR), and high SR calcium and calcium-ATPase content of extraocular muscles are consistent with these capabilities. Relations of force generation to calcium concentration suggest EOM increment force in a more graded manner and over a wider range of cytosolic calcium concentrations.[5,53] The ability to modulate free cytosolic calcium precisely and above the levels typically seen in skeletal muscles is a normal function for EOM. Therefore, enhanced calcium homeostasis may be protective from calcium overload caused by muscular dystrophy. The calcium sequestration capacity of EOM appears to protect the EOM from toxic agents that act via increases in intracellular calcium concentration. The aminoacyl local anesthetics increase intracellular calcium via frank sarcolemmal breaks and are myotoxic to the extent that they have been used to induce necrosis prior to the study of muscle regeneration. By contrast, the local anesthetics elicit only mild and limited damage when applied to EOM.[54] Also, calcium ionophore A-23187 does not kill EOM fibers at concentrations that are lethal to other skeletal muscle.[48]

The EOM have enhanced antioxidant capacity, including high levels of nitric oxide synthase activity, which may serve an antioxidant function.[55,56] There is evidence of increase oxidative stress in patients and animal models of muscular dystrophy; however, free radical scavengers do not effectively treat the disease. EOM of mdx mice are similar to other skeletal muscle in their reduction of nitric oxide synthase.[57,58] However, controversy remains on the role of nitric oxide and other free radicals in the pathogenesis of muscular dystrophies and EOM's oxidative capacity still prove to play a role in its sparing by these disorders.[44,59-61]

Motor Neuron Disease

OMNs are generally spared in human motor neuron diseases, including infantile and juvenile spinal muscular atrophy, Kennedy's disease, amyotrophic lateral sclerosis (ALS), and poliomyelitis.[62] The OMNs must differ from spinal and bulbar motor neurons, which are selectively affected in these diseases. The ocular motoneurons may have specific resistance to degeneration, or spinal and bulbar motor neurons may be selectively vulnerable.

Although the routine clinical examination fails to identify eye movement disturbance, when ocular motility is investigated formally, the velocity of saccadic or smooth pursuit movement is found to be decreased in about 50% of patients with typical amyotrophic lateral sclerosis;[63-65] however, these may be related to extrapyramidal or supranuclear dysfunction.[66,67] OMN dysfunction may contribute to these abnormalities. In histopathological studies, OMN degeneration was found in only 4 of 53 amyotrophic lateral sclerosis patients, but essentially all cases had degeneration of neurons in the hypoglossal nuclei.[68] OMN of amyotrophic lateral sclerosis patients rarely show inclusion bodies that are characteristic of ALS, indicating that the OMN are not completely spared.[69]

Rarely, otherwise typical motor neuron disease is associated with external ophthalmoplegia, neuronal loss, and gliosis in the ocular motor nuclei.[66,70] Amyotrophic lateral sclerosis patients who survive for long periods because of artificial ventilation develop a predominantly supranuclear ophthalmoplegia.[71] Pathological evaluation of such patients demonstrates lesions not only in the ocular nuclei but also extensive neuron degeneration outside the motor system.[72] Therefore, histopathological studies in such patients indicate that OMNs are not completely spared. This appears to occur in patients with variants of typical motor neuron disease or when patients survive for extreme periods.

Differences in OMNs from spinal and bulbar motor neurons exist that could influence their being spared by motor neuron diseases. These include expression of calcium-binding proteins and androgen receptor, excitatory neurotransmission, and trophic factors; brief discussions of each follow.

The ratio of excitatory to inhibitory synapses on OMNs is approximately 1.0.[73], higher than found in spinal motoneurons, and this suggests a paucity of inhibitory input on OMN somata, probably a lack of Renshaw cell and spindle afferent inhibition, which permits relatively undamped firing rate of OMN. Glutamate is the predominant excitatory neurotransmitter in the oculomotor system.[74-76] Because OMNs have more glutamatergic excitatory afferent and much higher neuronal firing rates than spinal motor neurons, they could be experience greater excitotoxic insult than spinal motoneurons.[77] Perhaps, the OMNs have mechanisms that protect them against excitototoxic insults, which are critical for their resistance to involvement by motor neuron diseases.

OMNs and other motor neurons differ in the expression of calcium-binding proteins, which may influence their susceptibility to calcium-mediated mechanisms of degeneration. Calbindin D-28k and parvalbumin are not found in spinal motoneurons, but are present in the OMN, preganglionic sympathetic neurons of the anteromediolateral spinal nuclei, and the Onuf nucleus, all spared in ALS.[78-80] Intraneuronal calcium binding proteins may provide selective resistance to the ALS disease process (see also discussion of Lambert-Eaton myasthenic syndrome). Again, these observations suggest a calcium-mediated mechanism for OMN sparing, but evidence for this is lacking.

EOM differs in trophic factors for survival and are likely to depend on different growth factors and transcriptional regulation (see Engel, this volume). Spinal motor neurons can support only the early stages of EOM myocyte growth in co-cultures. Neonatal EOM explants developed myotubes that are immunoreactive for myosin, became innervated, and matured in parallel with hindlimb muscle explants. However, in long-term cultures (after the third week *in vitro*), EOM that is innervated by spinal motor neurons degenerate.[81] When EOM primordia are innervated by midbrain explants, many of which contain OMNs, the cultures thrive for more than 60 days. This experiment suggests that there may be cell-to-cell specificity and trophic interactions between a specific type of motor neuron and its target muscle fibers. Such differences in trophic influences offer another mechanism for sparing of OMN by motor neuron diseases.

Kennedy's disease, which selectively affects bulbar and spinal motoneurons, is caused by an abnormal trinucleotide repeat in the gene encoded for the androgen receptor protein.[82] Androgen receptors are predominantly identified in spinal and bulbar motoneurons, but were thought to be rarely found in OMNs;[83] however, more recent work found that OMNs have moderate androgen receptor immunoreactivity.[84] The presence or absence of the androgen receptor alone may not be sufficient to ex-

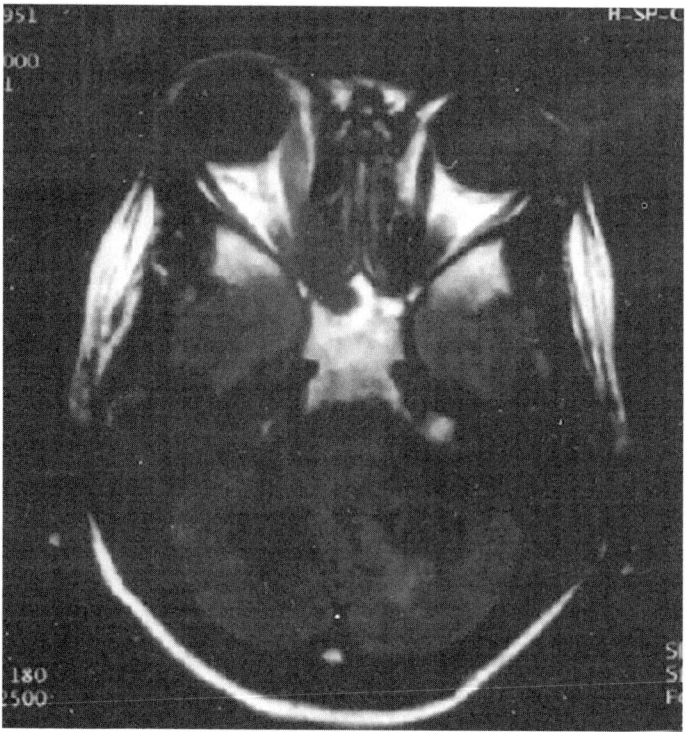

FIGURE 1. Magnetic resonance image of a patient with orbital myositis. Note enlargement of right medical rectus (left side of scan).

plain the selective resistance of OMNs. Androgens appear to act as growth factor for spinal motoneurons, and androgen receptors appear to be regulated by brain-derived neurotrophic factor, which may provide a link to the influence of growth factors on the course of motor neuron diseases.[62]

CONCLUSION

This review has concentrated on those diseases in which insights exist as to the differential susceptibility to EOM and OMN. However, other examples are readily appreciated. The inflammatory myopathies diffusely involve skeletal muscle, and may involve the heart, yet EOM is spared. In contrast, an isolated inflammatory condition of EOM, and likely other orbital tissue, termed orbital myositis is observed without evidence of systemic muscle involvement (FIG. 1). Some types of mitochondrial diseases may involve EOM at the exclusion of other skeletal muscle, while others do not. No meaningful insights into differential susceptibility exist in these conditions. In the last decade, studies of the differential susceptibility of EOM to neuromuscular pathology have provided considerable insight into disease mechanisms, and it is hoped this unique approach will expand to understanding ocular motor neurons and other disorders.

ACKNOWLEDGMENTS

The Office of Research and Development, Medical Research Service of the Department of Veterans Affairs, and NIH grants EY-11998, EY-13238, and P30 EY-11373 supported this work.

REFERENCES

1. LEIGH, R.J. & D.S. ZEE. 1999. The Neurology of Eye Movements. 3 edit. F.A. Davis. Philadelphia, PA.
2. ENGEL, A.G. 1999. The Myasthenic Syndromes. Oxford University Press. New York.
3. GROB, D., E.L. ARSURA, N.G. BRUNNER & T. NAMBA. 1987. The course of myasthenia gravis and therapies affecting outcome. Ann. N.Y. Acad. Sci. **505:** 472–499.
4. UBOGU, E.E. & H.J. KAMINSKI. Preferential involvement of extraocular muscle by myasthenia gravis. Neuroophthalmology. In press.
5. SPENCER, R.F. & J.D. PORTER. 1988. Structural organization of the extraocular muscles. In Neuroanatomy of the oculomotor system. J. Buttner-Ennever, Ed.: 33–79. Elsevier. Amsterdam.
6. MOSIER, D.R., L. SIKLÓ & S. APPEL. 2000. Resistance of extraocular motoneuron terminals to effects amyotrophic lateral sclerosis sera. Neurology **54:** 252–255.
7. RUFF , R.L., H.J. KAMINSKI, E. MAAS & P. SPIEGEL. 1989. Ocular muscles: physiology and structure-function correlations. Bull. Soc. Belge Ophtalmol. **237:** 321–352.
8. JACOBY, J., D.J. CHIARANDINI & E. STEFANI. 1989. Electrical properties and innervation of fibers in the orbital layer of rat extraocular muscles. J. Neurophysiol. **61:** 116–125.
9. ODA, K. & Y. ITO. 1981. Myasthenia gravis: antibodies to acetylcholine receptor in ocular myasthenia gravis. J. Neurol. **225:** 251–258.
10. LIMBURG, P.C., T.C. THE, E. HUMMEL-TEPPEL & H. OOSTERHUIS. 1983. Anti-acetylcholine receptor antibodies in myasthenia gravis. I. Relation to clinical parameters in 250 patients. J. Neurol. Sci. **58:** 357–370.
11. WANG, Z., B. DIETHELM-OKITA, D. OKITA, et al. 2000 T-cell recognition of muscle acetylcholine receptor in ocular myasthenia gravis. J. Neuroimmunol. **108:** 29–39.
12. KAMINSKI, H.J., L.L. KUSNER & C.H. BLOCK. 1996. Expression of acetylcholine receptor isoforms at extraocular muscle endplates. Invest. Ophthalmol. Visual Sci. **37:** 345–351.
13. HORTON, R.M., A.A. MANFREDI & B.M. CONTI-TRONCONI. 1993. The "embryonic" gamma subunit of the nicotinic acetylcholine receptor is expressed in adult extraocular muscle. Neurology **43:** 983–986.
14. MACLENNAN, C., D. BEESON, A.-M. BUIJS, et al. 1997. Acetylcholine receptor expression in human extraocular muscles and their susceptibility to myasthenia gravis. Ann. Neurol. **41:** 423–431.
15. KAMINSKI, H. & R. RUFF. 1997. Ocular muscle involvement by myasthenia gravis. Ann. Neurol. **41:** 419–420.
16. PORTER, J.D., S. KHANNA, H.J. KAMINSKI, et al. 2001. Extraocular muscle is defined by a fundamentally distinct gene expression profile. Proc. Natl. Acad. Sci. USA **98:** 12062–12067.
17. O'NEILL, J.H., N.M.F. MURRAY & J. NEWSOM-DAVIS. 1998. The Lambert-Eaton myasthenic syndrome. A review of 50 cases. Brain **111:** 577–596.
18. BLECK, T. 1995. Clostridium botulinum. In Principles and Practice of Infectious Disease. G. Mendell, J. Bennett & R. Dolin, Eds.: 2178–2182. Churchill Livingstone. New York.

19. STAHL, J., L. AVERBUCH-HELLER, B.F. REMLER & R.J. LEIGH. 1998. Clinical evidence of extraocular muscle fiber-type specificity of botulinum toxin. Neurology **51:**1093–1099.
20. PORTER, J.D., S. STREBECK & N.F. CAPRA. 1991. Botulinum-induced changes seen in monkey eylid muscle. Comparison with changes seen in extraocular muscle. Arch. Ophthalmol. **109:** 396–404.
21. SPENCER, R.F. & K.W. MCNEER. 1987. Botulinum toxin paralysis of adult monkey extraocular muscle: structural alterations in orbital, singly innervated muscle fibers. Arch. Ophthalmol. **105:** 1703–1711.
22. PORTER, J.D., L.A. BURNS & E.J. MCMAHON. 1989. Denervation of primate extraocular muscle: a unique pattern ofstructurral alterations. Invest. Ophthalmol. Visual Sci. **30:** 1894–1908.
23. CHRISTIANSEN, S.P., R.S. BAKER, M. MADHAT & B. TERRELL. 1993. Type specific changes in fiber morphometry following denervation of caninine extraocular muscle. Exp. Mol. Pathol. **56:** 87–95.
24. SMITH, T., R.S. BAHN & C.A. GORMAN. 1989. Connective tissue, glycosaminoglycans, and diseases of the thyroid. Endocr. Rev. **10:** 366–391.
25. HUFNAGEL, T.J., W.F. HICKEY, W.H. COBBS, *et al.* 1984. Immunohistochemical and ultrastructural studies on the exenterated orbital tissues of a patient with Graves' disease. Ophthalmology **91:** 1411–1419.
26. FELLS, P. 1978. Orbital pathology. *In* The Thyroid. 4th edit. S. Werner & S. Ingbar, Eds.: 660. Harper & Row. New York.
27. RILEY, F. 1972. Orbital pathology in Graves' disease. Mayo Clin. Proc. **47:** 975.
28. WEETMAN, A.P., S. COHEN, K.C. GATTER, *et al.* 1989. Immunohistochemical analysis of the retrobulbar tissues in Graves' ophthalmopathy. Clin. Exp. Immunol. **75:** 222–227.
29. TALLSTEDT, L. & R. NORBERG. 1988. Immunohistochemical staining of normal and Graves' extraocular muscle. Invest. Ophthalmol. Visual Sci. **29:**175–184.
30. FELDON, S. 1990. Graves' ophthalmopathy. Is it really thyroid disease? Arch. Intern. Med. **150:** 948–950.
31. WALL, J.M., N. SALVI, A. BERNARD, *et al.* 1991. Thyroid-associated ophthalmopathy—a model for the association of organ-specific autoimmune disorders. Immunol. Today **12:** 150–153.
32. YAMADA, M., A.W. LI & J.R. WALL. 2000. Thyroid-associated ophthalmopathy: clinical features, pathogenesis, and management. Crit. Rev. Clin. Lab. Sci. **37:** 523–549.
33. WALL, J. 1995. Extrathyroidal manifestations of Graves' disease. J. Clin. Endocrinol. Metab. **80:** 3427–3429.
34. AHMAN, A., J. BAKER, A. WEETMAN, *et al.* 1987. Antibodies to porcine eye muscle in patients with Graves' ophthalmopathy: identification of serum immunoglobins directed against unique determinants by immunoblotting and enzyme-linked immunosorbent assay. J. Clin. Endocrinol. Metab. **64:** 454–460.
35. HIROMATSU, Y., H. FUKAZAWA, F. GUINARD, *et al.* 1988. A thyroid cytotoxic antibody that cross-reacts with an eye muscle cell surface antigen may be the cause of thyroid-associated ophthalmopathy. J. Clin. Endocrinol. Metab. **67:** 565–570.
36. SCHIFFERDECKER, E., U. KETZLER-SASSE, O. BOEHM, *et al.* 1989. Re-evaluation of eye muscle autoantibody determination in Graves' ophthalmopathy: failure to detect a specific antigen by use of enzyme-linked immunosorbent assay, indirect immunofluorescence, and immunoblotting techniques. Acta Endocrinol. (Copenh.) **121:** 643–650.
37. BLAU, H., I. KAPLAN, T. TAO & J. KRISS. 1983. Thyroglobulin-independent, cell-mediated cytotoxicity of human eye muscle cells in tissue culture by lymphocytes of a patient with Graves' ophthalmopathy. Life Sci. **32:** 45–53.
38. PAPPA, A., V.CALDER, R. AJJAN, *et al.* 1997. Analysis of extraocular muscle-infiltrating T cells in thyroid-associated ophthalmopathy (TAO). Clin. Exp. Immunol. **109:** 362–369.
39. PASCHKE, R., G. VASSART & M. LUDGATE. 1995. Current evidence for and against the TSH receptor being the common antigen in Graves' disease and thyroid-associated ophthalmopathy. Clin. Endocrinol. (Oxf.) **42:** 565–569.
40. KRISS, J. 1975. Studies on the pathogenesis of Graves' ophthalmopathy (with some related observations regarding therapy). Recent Prog. Horm. Res. **31:** 3533.

41. TANDON, N., S.L. YAN, B.P. MORGAN & A.P. WEETMAN. 1994. Expression and function of multiple regulators of complement activation in autoimmune thyroid disease. Immunology **81:** 643–647.
42. CARPO, M., R. PEDOTTI, F. LOLLI, et al. 1998. Clinical correlate and fine specificity of anti-Q1b antibodies in peripheral neuropathy. J. Neurol. Sci. **155:** 186–191.
43. CHIBA, A., S. KUSUNOKI, H. OBATA, et al. 1993. Serum anti-GQ1b IgG antibody is associated with ophthalmoplegia in Miller Fisher syndrome and Guillain-Barre syndrome: clinical and immunohistochemical studies. Neurology **43:** 1911–1917.
44. ANDRADE, F.H., J.D. PORTER & H.J. KAMINSKI. 2000. Eye muscle sparing by the muscular dystrophies: lessons to be learned? Microsc. Res. Tech. **48:** 192–203.
45. KAMINSKI, H.J., M. AL-HAKIM, R.J. LEIGH, et al. 1992. Extraocular muscle are spared in advanced Duchenne dystrophy. Ann. Neurol. **32:** 586–588.
46. SCELSA, S., D. SIMPSON, B. REICHLER & M. DAI. 1996. Extraocular muscle involvement in Becker muscular dystrophy. Neurology **46:** 564–566.
47. KARPATI, G. & S. CARPENTER. 1986. Small-caliber skeletal muscle fibers do not suffer deleterious consequences of dystrophic gene expression. Am. J. Med. Genet. **25:** 653–658.
48. KHURANA, T.S., R.A. PRENDERGAST, H.S. ALAMEDDINE, et al. 1995. Absence of extraocular muscle pathology in Duchenne's muscular dystrophy: role for calcium homeostasis in extraocular muscle sparing. J. Exp. Med. **182:** 467–475.
49. RAGUSA, R. & J. PORTER. 1994. Extraocular muscle is highly protected from dystrophic changes in the mdx mouse. Invest. Ophthalmol. Visual Sci. **35:** 2198.
50. PORTER, J.D., J.A. RAFAEL, R.J. RAGUSA, et al. 1998. The sparing of extraocular muscle in dystrophinopathy is lost in mice lacking utrophin and dystrophin. J. Cell. Sci. **111:** 1801–1811.
51. PORTER, J.D. & P. KARATHANASIS. 1998. Extraocular muscle in merosin-deficient muscular dystrophy: cation homeostasis is maintained but is not mechanistic in muscle sparing. Cell Tissue Res. **292:** 495–501.
52. PORTER, J.D., A.P. MERRIAM, A.A. HACK, et al. 2001. Extraocular muscle is spared despite the absence of an intact sarcoglycan complex in gamma- or delta-sarcoglycan–deficient mice. Neuromuscul. Disord. **11:** 197–207.
53. SCHACHAT, F.H., M.S. DIAMOND & P.W. BRANDT. 1987. Effect of different troponin T-tropomyosin combinations on thin filament activation. J. Mol. Biol. **198:** 551–554.
54. PORTER, J.D., D.P. EDNEY, E.J. MCMAHON & L.A. BURNS. 1988. Extraocular myotoxicity of the retrobulbar anesthetic bupivacaine hydrochloride. Invest. Ophthalmol. Visual Sci. **29:** 163–174.
55. RAGUSA, R.J., C.K. CHOW, D.K. ST. CLAIR & J.D. PORTER. 1996. Extraocular, limb and diaphragm muscle group-specific antioxidant enzyme activity patterns in control and mdx mice. J. Neurol. Sci. **139:** 180–186.
56. RICHMONDS, C.R. & H.J. KAMINSKI. 2001. Nitric oxide synthase expression and nitric oxide effects on contractility in rat lateral rectus. FASEB J. **15:** 1764–1770.
57. KAMINSKI, H.J. & F.H. ANDRADE. 2001. Nitric oxide: biologic effects on muscle and role in muscle diseases. Neuromuscul. Disord. **11:** 517–524.
58. WEHLING, M., J.T. STULL, T.J. MCCABE & J.G. TIDBALL. 1998. Sparing of mdx extraocular muscles from dystrophic pathology is not attributable to normalized concentration or distribution of neuronal nitric oxide synthase. Neuromuscul. Disord. **8:** 22–29.
59. CROSBIE, R.H. 2001. NO vascular control in Duchenne muscular dystrophy. Nat. Med. **7:** 27–29.
60. WEHLING, M., M.J. SPENCER & J.G. TIDBALL. 2001. A nitric oxide synthase transgene ameliorates muscular dystrophy in mdx mice. J. Cell. Biol. **155:** 123–131.
61. CROSBIE, R.H., V. STRAUB, H.Y. YUN, et al. 1998. mdx muscle pathology is independent of NOS perturbation. Hum. Mol. Genet. **7:** 823–829.
62. MITSUMOTO, H., D.A. CHAD & E.P. PIORO. 1997. Amyotrophic lateral sclerosis. F.A. Davis/Oxford Press. New York.
63. JACOBS, L., D. BOZIAN, R.R. HEFFNER JR. & S.A BARRON. 1981. An eye movement disorder in amyotrophic lateral sclerosis. Neurology **31:** 1282–1287.
64. LEVEILLE, A., J. KIERNAN, J.A. GOODWIN & J. ANTEL. 1982. Eye movements in amyotrophic lateral sclerosis. Arch. Neurol. **39:** 684–686.

65. MARTI-FABREGAS, J. & C. ROIG. 1993. Oculomotor abnormalities in motor neuron disease. J. Neurol. **240:** 475–478.
66. AVERBUCH-HELLER, L., C. HELMCHEN, A.K.E. HORN, et al. 1998. Slow vertical saccades in motor neuron disease: correlation of structure and function. Ann. Neurol. **44:** 641–648.
67. GIZZI, M., A. DIROCCO, M. SIVAK & B. COHEN. 1992. Ocular motor function in motor neuron disease. Neurology **42:** 1037–1046.
68. LAWYER, J.R. & M.G. NETSKY. 1953. Amyotrophic lateral sclerosis. Arch. Neurol. Psychiatry **69:** 171–192.
69. OKAMOTO, K., S. HIRAI, M. AMARI, et al. 1993. Oculomotor nuclear pathology in amyotrophic lateral sclerosis. Acta Neuropathol. **85:** 458–462.
70. HARVEY, D.G., R.M. TORACK & H.E. ROSENBAUM. 1979. Amyotrophic lateral sclerosis with ophthalmoplegia. Arch. Neurol. **36:** 615–617.
71. HAYASHI, H., S. KATO, T. KAWADA & T. TSUBAKI. 1987. Amyotrophic lateral sclerosis: oculomotor function in patients in respirators. Neurology **37:** 1431–1432.
72. MIZUTANI, T., M. AKI, R. SHIOZAWA, et al. 1990. Development of ophthalmoplegia in amyotrophic lateral sclerosis during long-term use of respirators. J. Neurol. Sci. **99:** 311–319.
73. SPENCER, R. & P. STERLING. 1977. An electron microscopic study of motoneurones and interneurones in the cat abducens nucleus identified by retrograde transport of horseradish peroxidase. J. Comp. Neurol. **176:** 65–86.
74. DURAND, J. 1993. Synaptic excitation triggers oscillations during NMDA receptor activation in rat abducens motoneurons. Eur. J. Neurosci. **5:** 1389–1397.
75. DOI, K., K. TSUMOTO & T. MATSUNAGA. 1990. Actions of excitatory amino acid antagonists on synaptic inputs to the rat medial vestibular nucleus: an electrophysiological study *in vitro*. Exp. Brain Res. **82:** 254–262.
76. KOPYSOVA, I., S. KOROGOD, J. DURAND & S. TYC-DUMONT. 1996. Local mechanisms of phase-dependent postsynaptic modifications of NMDA-induced oscillations in the abducens motoneurons: a stimulation study. J. Neurophysiol. **76:** 1015–1024.
77. BROWN, R.H. 1995. Amyotrophic lateral sclerosis: recent insights from genetics and transgenic mice. Cell **80:** 687–692.
78. INCE, P., N. STOUT, P. SHAW, et al. 1993. Parvalbumin and calbindin D-28k in the human motor system and in motor neuron disease. Neuropathol. Appl. Neurobiol. **19:** 291–299.
79. ALEXIANU, M.E., B.-K. HO, H. MOHAMED, et al. 1994. The role of calcium-binding proteins in the selective motorneuron vulnerability in amyotrophic lateral sclerosis. Ann. Neurol. **36:** 846–858.
80. REINER, A., G. MEDINA, G. FIGUEREDO-CARDENAS & S. ANFINSON. 1995. Brainstem motoneuron pools that are selectively resistant in amyotrophic lateral sclerosis are preferentially enriched in parvalbumin: evidence from monkey brainstem for a calcium-mediated mechanism in sporadic ALS. Exp. Neurol. **131:** 239–250.
81. PORTER, J.D. & K.F. HAUSER. 1993. Survival of extraocular muscle in long-term organotypic culture: differential influence of appropriate and inappropriate motoneurons. Dev. Biol. **160:** 39–50.
82. LA SPADA, A.R., E.M. WILSON, D.B. LUBAHN, et al. 1991. Androgen receptor gene mutations in X-linked spinal and bulbar muscular atrophy. Nature **352:** 77–79.
83. SAR, M. & W.E. STUMPF. 1977. Androgen concentration in motor neurons of cranial nerves and spinal cord. Science **197:** 77–79.
84. MATSUURA, T. 1996. [Study of androgen receptor expression and neuronal vulnerability in X-linked spinal and bulbar muscular atrophy]. Hokkaido Igaku Zasshi **71:** 785–799.

Applications of Molecular Genetics to the Understanding of Congenital Ocular Motility Disorders

ELIZABETH C. ENGLE

Neurology and Pediatrics (Genetics) Children's Hospital, Boston, Massachusetts, USA and Department of Neurology, Harvard Medical School, Boston, Massachusetts, USA

ABSTRACT: The congenital fibrosis syndromes (CFS), including congenital fibrosis of the extraocular muscles (CFEOM) and Duane syndrome (DS), are rare congenital strabismus syndromes that present with nonprogressive restrictive ophthalmoplegia with or without ptosis. Although historically believed to result from primary extraocular muscle (EOM) fibrosis, our laboratory's work is based on the hypothesis that these disorders result from distinct, but analogous, developmental defects of the oculomotor (nIII), trochlear (nIV), and abducens (nVI) nuclei. We have defined three inherited CFEOM phenotypes (CFEOM1–3) and have mapped each phenotype to a distinct genetic locus (*FEOM1–3*). Individuals with CFEOM1 are born with bilateral ptosis and both eyes fixed in a downward position with absent upgaze and aberrant horizontal gaze. This disorder maps to the *FEOM1* locus on chromosome 12cen.[1,2] Neuropathology studies of CFEOM1 reveal the absence of the superior division of oculomotor nerve and its corresponding alpha motor neurons in the midbrain, with abnormalities of target EOMs.[3] These neuropathology findings parallel those previously identified in Duane syndrome, in which there is an absence of nVI and the abducens nerve.[4,5] Individuals with CFEOM2 are born with bilateral ptosis and exotropia. This atypical form of CFEOM maps to the *FEOM2* locus on chromosome 11q13 and results from mutations in ARIX (PHOX2A).[6,7] ARIX encodes a homeodomain transcription factor protein previously shown to be required for nIII/nIV development in mouse and zebrafish.[8,9] Together, these findings support the hypothesis that the congenital fibrosis syndromes result from parallel defects in nIII, nIV, and nVI nuclear development. Functional studies of the CFEOM genes should provide additional insight into the unique features of the extraocular lower motor neuron axis in health and disease. (For full refs. 1–9, see reference list of the main paper.)

KEYWORDS: strabismus; ophthalmoplegia; genetics; congenital fibrosis; extraocular muscle; oculomotor nuclei

Isolated strabismus affects 1–5% of the general population.[10] Most forms of strabismus are multifactorial in origin and, while there is probably an inherited component, the genetics of these disorders remain unclear. A group of isolated strabismic disor-

Address for correspondence: Elizabeth C. Engle, M.D., Enders 551, The Children's Hospital, 300 Longwood Avenue, Boston, MA 02115. Voice: 617-355-8371; fax: 617-277-0496.
engle@enders.tch.harvard.edu

Ann. N.Y. Acad. Sci. 956: 55–63 (2002). © 2002 New York Academy of Sciences.

ders referred to as the "congenital fibrosis syndromes," however, can be inherited as autosomal dominant or recessive traits. These disorders provide an opportunity to employ genetic techniques to investigate the molecular basis of strabismus and the unique aspects of the EOM lower motor neuron unit.

The congenital fibrosis syndromes are relatively rare forms of strabismus that were categorized together in 1950 based on common clinical features of congenital, nonprogressive restrictive ophthalmoplegias with active limitation and passive restriction of globe movement.[11] These disorders included Duane syndrome, CFEOM, strabismus fixus, vertical retraction syndrome, and Brown syndrome. The restrictive ophthalmoplegia, "tight" feel of EOMs at surgery, and finding of connective tissue on EOM biopsies led investigators to propose that these disorders resulted from primary EOM fibrosis. Our laboratory's work is based on the hypothesis that at least two of these disorders, CFEOM and Duane syndrome, result from distinct, but analogous, developmental defects of the nIII, nIV and/or nVI.

The specific aim of our laboratory's work has been to define the clinical phenotypes of the inherited forms of the congenital fibrosis syndromes and to determine the neuropathology and gene mutation underlying each phenotype. Once identified, the function of each gene in normal and abnormal neurodevelopment can be investigated. Thus far, we have defined three genetically distinct forms of CFEOM (*FEOM1–3*) and identified the *FEOM2* gene as reviewed below.

CFEOM SYNDROMES

CLASSIC CFEOM (CFEOM1)

CFEOM is frequently inherited, and many large autosomal dominant and a few recessive families have been reported in the literature (reviewed in Ref. 10). Most reports of families and sporadic individuals with CFEOM describe a stereotypical phenotype we refer to as "classic CFEOM" or CFEOM1.[3] An individual with classic CFEOM demonstrates (1) congenital nonprogressive bilateral external ophthalmoplegia and ptosis; (2) an infraducted primary position of both eyes with the inability to raise either eye above the horizontal midline; and (3) forced duction testing positive for restriction, if testing is performed. Of note, the horizontal position of each eye may be midline, eso- or exotropic; horizontal movements can be full to none; and aberrant movements may be present, but are not required. These individuals typically lack binocular vision, have significant bilateral refractive errors with high astigmatism, and frequently have amblyopia in the non-fixing eye. The pupillary reaction and remaining neurological exam are normal in most cases.

FEOM1 (MIM #135700)

The first family enrolled in our study was a large CFEOM family in which all affected members met classic CFEOM criteria. We now refer to such families as CFEOM1 pedigrees. We performed a genome-wide linkage screen and mapped the family's phenotype to the centromeric region of chromosome 12,[2] now referred to as the FEOM1 locus. Subsequently, we ascertained many additional CFEOM1 pedigrees and confirmed linkage to the FEOM1 locus, defined by the flanking recombinant markers D12S1584 and D12S1668.[1,2] In addition, D'Esposito *et al.*[12] reported a classic Italian family whose phenotype mapped to the FEOM1 locus, and Black et al.[13] reported three classic families, two of whose phenotypes were consistent with linkage to FEOM1. Thus far, virtually all CFEOM1 families demonstrate autosomal dominant inheritance with full penetrance. We are currently analyzing candidate genes in the FEOM1 region to identify the CFEOM1 gene.

There has been an ongoing debate in the literature as to whether CFEOM1 is a primary myopathic or neurogenic disorder. On the one hand, the restrictive nature of the ophthalmoplegia and the frequent finding of fibrotic tissue on EOM biopsy led to the belief that CFEOM was primarily due to a myopathic process. On the other hand, clinical observations such as synergistic convergence and divergence, globe retraction, and the association of CFEOM with Marcus-Gunn jaw winking,[14–20] as well as EMG studies,[21,22] suggest aberrant innervation. We reported the first postmortem examination of an affected member of a CFEOM1 pedigree whose disease gene maps to the *FEOM1* locus (FIG. 1b).[3] The autopsy study revealed the absence of the superior division of the oculomotor nerve, which normally innervates the superior rectus (SR) and levator palpabrae superioris (LPS) muscles. There was an absence of the corresponding alpha motor neurons of the oculomotor nucleus and loss of large motor axons in the proximal portion of the oculomotor nerve. The SR and LPS muscles were a diminutive structure containing fat, connective tissue, and scant myofibers. The remaining EOMs were not fibrotic. These abnormalities of the LPS and SR would account for the bilateral ptosis and infraducted globes found in CFEOM1 individuals.

Of note, these postmortem findings are not consistent with some reports of CFEOM IR surgical biopsies that have been read as pathologic fibrous tissue[17,18,20,23] or islands of normal myofibers within areas of fibrosis.[3,18,22,24] However, because EOMs have very long tendonous insertions of fibrous connective tissue,[25] the tendon or transitional region from tendon to muscle, rather than the muscle belly, may be erroneously sampled. This sampling error has been shown to occur in a large series of non-CFEOM biopsies,[26] as well as in biopsies from CFEOM patients.[24] We have performed serial sections from IR tendonous insertion from a CFEOM1 patient biopsy and age-matched control; the tendon and transition zones were of equivalent length and both were indistinguishable from the published "islands of normal muscle" and "pathologic fibrosis" reported in CFEOM.[3] Nonetheless, we did find extensive fibrous tissue in SR biopsies. We hypothesize that this fibrosis occurs secondary to lack of innervation, similar to the patchy fibrosis reported as a secondary process in Duane retraction syndrome where abnormalities of innervation have also been documented.[4,5]

ATYPICAL CFEOM (CFEOM2, CFEOM3)

A small number of reports describe CFEOM phenotypes that do not meet classic criteria. Both atypical families and atypical sporadic individuals have been reported under many names (reviewed in Ref. 10). By our classification, individuals with atypical CFEOM have restrictive ophthalmoplegia in the nIII and/or nVI distribution but, unlike those with classic CFEOM, are able to raise one or both eyes above the horizontal midline or have unilateral rather than bilateral ptosis or ophthalmoplegia. Unlike with classic CFEOM, there is marked phenotypic variability between these atypical cases.

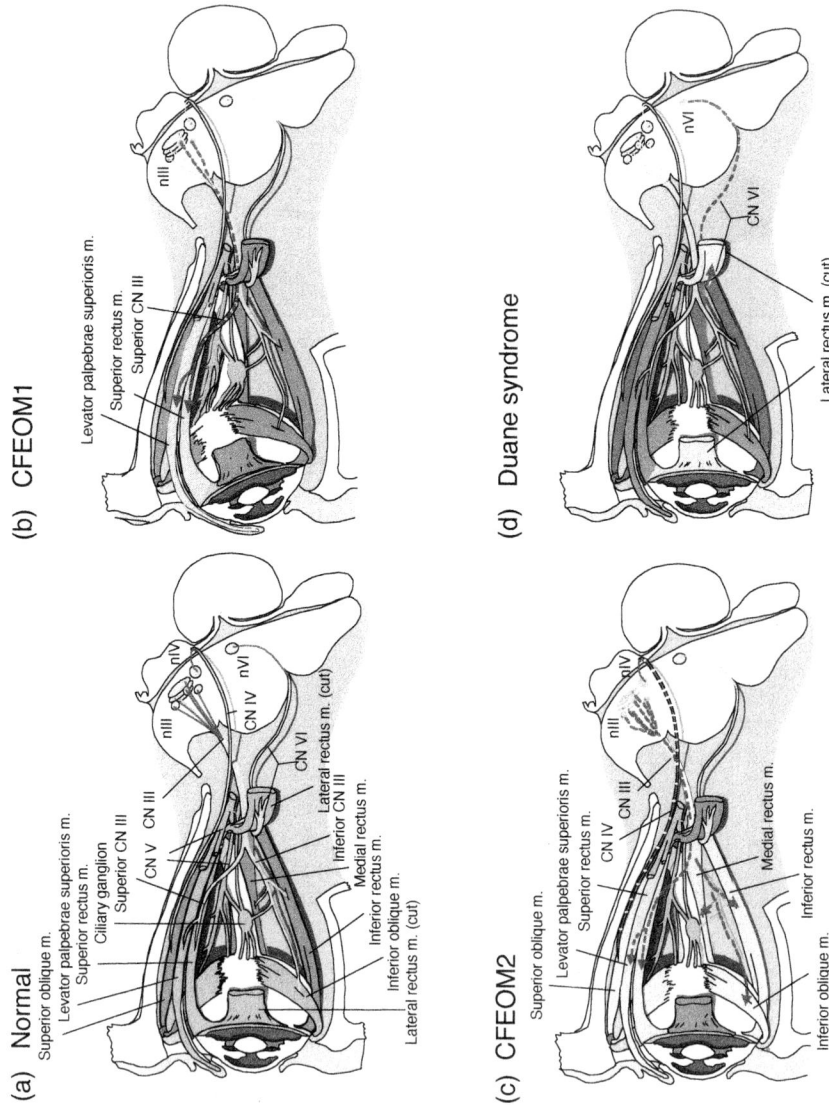

FIGURE 1. *See following page for caption.*

CFEOM2 (MIM #602078), FEOM2 Locus

We have identified three Saudi Arabian and one Turkish pedigrees with an atypical CFEOM phenotype, which we refer to as CFEOM2. Affected individuals are born with bilateral ptosis with their eyes fixed primarily in an exotropic position, with or without secondary hypertropia or hypotropia,[6,27,28] leading to our hypothesis that CFEOM2 results from maldevelopment of both nIII and nIV (FIG. 1c).[6] CFEOM2 can also be considered an inherited form of Brown's "strabismus fixus,"[11] which had been described previously only as a sporadic disorder.[17,24] Unlike CFEOM1, the families with CFEOM2 are consanguineous and demonstrate autosomal recessive inheritance. We conducted a genome-wide linkage screen and mapped the gene for CFEOM2 to the *FEOM2* locus on chromosome 11q13.[6] Through linkage analysis and homozygosity mapping, we reduced the *FEOM2* critical region to less than a 1-cm region delimited by *D11S4162* and *D11S1369*, and constructed a bacterial artificial chromosome physical map that allowed us to identify candidate genes in the *FEOM2* region. Sequence analysis of one of these genes, *ARIX* (*PHOX2A*), revealed three distinct mutations in the CFEOM2 pedigrees.[7] Two mutations disrupt splicing of *ARIX*, and the third mutation results in a conserved amino acid substitution. *ARIX* encodes a homeodomain transcription factor that plays a primary role in the development of nIII and nVI alpha motor neurons in mice and zebrafish.[8,9] This finding corresponds well to the predicted neuropathology of CFEOM2,[6] and provides the first genetic proof that CFEOM2 results from aberrant development of these motor nuclei (FIG. 1c).

In mice and zebrafish, *ARIX* is also essential to the generation and survival of adrenergic neurons and a larger population of brainstem motor neurons, as well as in the determination of the noradrenergic neurotransmitter phenotype.[29–35] Mice and zebrafish with *ARIX* mutations die at birth and have much more extensive pathology than what we anticipate in CFEOM2 individuals. In addition to the loss of nIII and nIV, these animal models also lack the locus coeruleus and the pterygopalatine (CN VII) and otic (CN IX) parasympathetic ganglia of the head, and have atrophic geniculate (CN VII), petrosal (CN IX), and nodose (CN X) cranial sensory ganglia.[8,9,30,33] In contrast, patients with CFEOM2 have normal life expectancies and do not exhibit obvious signs or symptoms of these abnormalities. We hope that future studies of *ARIX* function will determine why, in humans, its function in nIII and nIV appears to be dissociated from its function in the locus coeruleus and the autonomic nervous system.

FIGURE 1. Schematic drawings of the lateral orbit and brainstem in (**a**) normal, (**b**) CFEOM1, (**c**) CFEOM2, and (**d**) Duane syndrome subjects demonstrating the established or proposed neuropathologies. The lateral rectus has been cut in order to expose the medial aspects of the orbit. Diseased muscles are pale, absent nerves are depicted as *dashed lines*, and absent brainstem nuclei are *dotted* with absence of a black border. In addition, the abnormal structures in **b–d** are labeled. The schematic neuropathology findings of CFEOM1 and Duane syndrome are based on autopsy findings[5,34] and of CFEOM2 is based on the function of Arix in mice and zebrafish.[8,9] (FIG. 1a,b,d from Engle.[10] Reprinted with permission of Oxford University Press. FIG. 1c from Nakano *et al.*[7] Reprinted with permission of *Nature Genetics*.)

CFEOM3 (MIM #604361), FEOM3 and FEOM1 Loci

We identified a second atypical CFEOM phenotype in a large Canadian family. The affected members of this family demonstrate marked variability of clinical presentation. The most severely affected individuals meet classic CFEOM criteria, but, in all cases, their eyes are fixed in both a downward (hypotropic) and outward (exotropic) position. The phenotype of mildly affected family members is not classic because these individuals have normally positioned eyes with only a slight limitation of vertical gaze. We conducted a genome-wide screen and mapped this atypical dominant CFEOM phenotype with reduced penetrance to a third *FEOM* locus, referred to as *FEOM3*, located within an ~ 5-cm region at the telomere of chromosome 16q.[36] We propose that this form of CFEOM results from a variable error in the development of nIII. Subsequently, we have mapped a family with the CFEOM3 phenotype to the FEOM1 locus.[36a]

DUANE SYNDROME

Duane syndrome (DS) accounts for 1–5% of strabismus cases[37,38] and is the most common form of the congenital fibrosis syndromes. Individuals with DS typically have limited abduction, variably limited adduction, and globe retraction on attempted adduction. DS can occur in isolation or in association with other congenital anomalies, particularly of the skeleton, ear, eye, and kidney.[10,39] Early studies of EOM in DS[40–42] reported fibrosis and abnormal insertions of the LR and/or MR muscles, suggesting a primary myopathic etiology. Subsequently, EMG studies revealed that the simultaneous co-contraction of electrically active MR and LR muscles correlated with globe retraction.[43,44] Postmortem examinations of isolated patients with DS revealed the absence of the nVI and the abducens nerve (CNVI) on the affected side(s), and partial innervation of the LR muscle(s) by branches of CNIII.[4,5] These studies support the hypothesis that DS is caused by a primary neurodevelopmental anomaly of nVI (FIG. 1d).

Duane syndrome is typically sporadic and therefore is not as amenable to molecular genetic studies as the more frequently inherited CFEOM. Cytogenetic analyses of several sporadic individuals with DS, however, have revealed deletions of chromosome 8q13 (*DURS1* locus).[45–48] Linkage analysis of rare families with autosomal dominant DS mapped the *DURS2* locus to chromosome 2q31.[49–51] Identification of the *DURS* genes should elucidate the molecular basis of DS and may provide insight into the development of the abducens nucleus and/or the lateral rectus muscle.

CONCLUSIONS

Identification of *ARIX* as the CFEOM2 disease gene combined with the neuropathology findings in CFEOM1 and Duane syndrome provide mounting evidence that at least some forms of the congenital fibrosis syndromes result from errors in nIII, nIV, and nVI development (summarized in TABLE 1). We anticipate that the various forms of CFEOM will result from mutations in genes essential to nIII and/or nIV development and that the more restricted forms of CFEOM, such as CFEOM1 and CFEOM3, may result from mutations in ARIX-regulated genes. In contrast, we an-

TABLE 1. Fibrosis syndrome genetics

Disorder	Phenotype	Neuropathology	Locus	Gene
CFEOM1	Classic	nIII superior division	FEOM1 12cen	?
CFEOM2	Atypical	nIII + nIV	FEOM2 11q13	ARIX
CFEOM3	Atypical	Variable nIII?	FEOM3/FEOM1 16qter/12cen	?
Duane syndrome 1		nVI	DURS1 8q13	?
Duane syndrome 2		nVI	DURS2 2q31	?

ticipate that the various genes responsible for DS will result from mutations in genes essential to nVI development. Together, the identification and functional studies of the congenital fibrosis genes should elucidate the targeting of EOM in these disorders as well as contribute to our understanding of the signaling cascade required for normal cranial motor nuclear development.

REFERENCES

1. ENGLE, E.C. et al. 1995. Congenital fibrosis of the extraocular muscles (autosomal dominant congenital external ophthalmoplegia): genetic homogeneity, linkage refinement, and physical mapping on chromosome 12. Am. J. Hum. Genet. **57:** 1086–1094.
2. ENGLE, E.C. et al. 1994. Mapping a gene for congenital fibrosis of the extraocular muscles to the centromeric region of chromosome 12. Nat. Genet. **7:** 69–73.
3. ENGLE, E.C. et al. 1997. Oculomotor nerve and muscle abnormalities in congenital fibrosis of the extraocular muscles. Ann. Neurol. **41:** 314–325.
4. HOTCHKISS, M.G. et al. 1980. Bilateral Duane's retraction syndrome: a clinical-pathological case report. Arch. Ophthalmol. **98:** 870–874.
5. MILLER, N.R. et al. 1982. Unilateral Duane's retraction syndrome (type 1). Arch. Ophthalmol. **100:** 1468–1472.
6. WANG, S. et al. 1998. Congenital fibrosis of the extraocular muscles type 2 (CFEOM2), an inherited exotropic strabismus fixus, maps to distal 11q13. Am. J. Hum. Genet. **63:** 517–525.
7. NAKANO, M. et al. 2001. Homozygous mutations in *ARIX (PHOX2A)* result in congenital fibrosis of the extraocular muscles type 2. Nat. Genet.. **29:** 315–320.
8. PATTYN, A. et al. 1997. Expression and interactions of the two closely related homeobox genes Phox2a and Phox2b during neurogenesis. Development **124:** 4065–4075.
9. GUO, S. et al. 1999. Development of noradrenergic neurons in the zebrafish hindbrain requires BMP, FGF8, and the homeodomain protein soulless/Phox2a. Neuron **24:** 555–566.
10. ENGLE, E. 1998. The genetics of strabismus: Duane, Moebius, and fibrosis syndromes. *In* Genetic diseases of the eye: a textbook and atlas. E. Traboulsi, Ed.: 477–512. Oxford University Press. New York.
11. BROWN, H.W. 1950. Congenital structural muscle anomalies. *In* Strabismus Ophthalmic Symposium. J.H. Allen, Ed.: 205–236. C.V. Mosby Co. St. Louis, MO.
12. D'ESPOSITO, D. et al. 1995. Oftalmoplegia congenita esterna: conferma della presenza di un locus genetico sul chomosoma 12. *In* Atti 75 Congresso S.O.I.: 273–274. Rome.

13. BLACK, G. *et al.* 1998. Locus heterogeneity in autosomal dominant congenital external ophthalmoplegia. J. Med. Genet. **35:** 985–988.
14. HEUCK, G. 1879. Ueber angeborenen vererbten Beweglichkeits—Defect der Augen. Klin. Monatsbl. Augenheilkd. **17:** 253–278.
15. VOSSIUS, A. 1982. Zwei Falle von angeborener fast vollstandiger Unbeweglichkeit beider Augen und der oberen Augenlider. Beitr. Augenheilkd. **5:** 1–10.
16. RUMPH, M. 1974. Fibrose du muscle droit inferieur, anomalies d'insertions et aplasies musculaires, une cause rare de troubles hereditaires non progressifs et congenitaux de la motilite oculaire. Ann. Oculistique **207:** 831–829.
17. HARLEY, R.D. *et al.* 1978. Congenital fibrosis of the extraocular muscles. J. Pediatr. Ophthalmol. Strabismus **15:** 346–358.
18. BRODSKY, M.C. *et al.* 1989. Neural misdirection in congenital ocular fibrosis syndrome: implications and pathogenesis [see comments]. J. Pediatr. Ophthalmol. Strabismus **26:** 159–161.
19. POLLOCK, S.C. *et al.* 1990. Congenital fibrosis syndrome [response to letter; comment]. J. Pediatr. Ophthalmol. Strabismus **27:** 329.
20. BRODSKY, M. 1998. Hereditary external ophthalmoplegia synergistic divergence, jaw winking, and oculocutaneous hypopigmentation: a congenital fibrosis syndrome caused by deficient innervation to extraocular muscles. Ophthalmology **105:** 717–725.
21. CIBIS, G. *et al.* 1984. Electromyography in congenital familial ophthalmoplegia. *In* Strabismus II. R.D. Reinecke, Ed.: 379–390. Grune & Stratton. New York.
22. HOUTMAN, W.A. *et al.* 1986. Hereditary congenital external ophthalmoplegia. Ophthalmologica **193:** 207–218.
23. LAUGHLIN, R.C. 1956. Congenital fibrosis of the extraocular muscles; a report of six cases. Am. J. Ophthalmol. **41:** 432–438.
24. APT, L. & R.N. AXELROD. 1978. Generalized fibrosis of the extraocular muscles. Am. J. Ophthalmol. **85:** 822–829.
25. DRACHMAN, D.A. *et al.* 1969. Experimental denervation of ocular muscles. A critique of the concept of "ocular myopathy." Arch. Neurol. **21:** 170–183.
26. MARTINEZ, A.J. *et al.* 1980. Structural features of extraocular muscles of children with strabismus. Arch. Ophthalmol. **98:** 533–539.
27. TRABOULSI, E. *et al.* 1993. Congenital fibrosis of the extraocular muscles: report of 24 cases illustrating the clinical spectrum and surgical management. Am. Orthoptic J. **43:** 45–53.
28. ASSAF, A. 1997. Bilateral congenital vertical gaze disorders: congenital muscle fibrosis or congenital central nervous abnormalities. Neuroophthalmology **17:** 23–30.
29. ZELLMER, E. *et al.* 1995. A homeodomain protein selectively expressed in noradrenergic tissue regulates transcription of neurotransmitter biosynthetic genes. J. Neurosci. **15:** 8109–8120.
30. MORIN, X. *et al.* 1997. Defects in sensory and autonomic ganglia and absence of locus coeruleus in mice deficient for the homeobox gene Phox2a. Neuron **18:** 411–423.
31. YANG, C. *et al.* 1998. Paired-like homeodomain proteins, Phox2a and Phox2b, are responsible for noradrenergic cell-specific transcription of the dopamine beta-hydroxylase gene. J. Neurochem. **71:** 1813–1826.
32. PATTYN, A. *et al.* 1999. The homeobox gene Phox2b is essential for the development of autonomic neural crest derivatives. Nature **399:** 366–370.
33. PATTYN, A. *et al.* 2000. Specification of the central noradrenergic phenotype by the homeobox gene Phox2b. Mol. Cell. Neurosci. **15:** 235–243.
34. SWANSON, D.J. *et al.* 2000. The homeodomain protein Arix promotes protein kinase A-dependent activation of the dopamine beta-hydroxylase promoter through multiple elements and interaction with the coactivator cAMP-response element-binding protein. J. Biol. Chem. **275:** 2911–2923.
35. ADACHI, M. *et al.* 2000. Paired-like homeodomain proteins Phox2a/Arix and Phox2b/NBPhox have similar genetic organization and independently regulate dopamine beta-hydroxylase gene transcription. DNA Cell Biol. **19:** 539–554.
36. DOHERTY, E. *et al.* 1999. CFEOM3: a new extraocular congenital fibrosis syndrome that maps to 16q24.2-q24.3. Invest. Ophthalmol. Visual Sci. **40:** 1687–1694.

36a. SENER, E.C. et al. 2000. New clinical variation of a fibrosis syndrome in a Turkish family maps to the CFEOM1 locus on chromosome 12. Arch. Ophthalmol. **118:** 1090–1097.
37. DANIS, P. 1948. Sur les anomalies congenitale de la motilite oculaire d'origine musculaire et en particular sur le sundrome de Stilling-Turk-Duane. Ann. Ocul. **1811:** 148.
38. KIRKHAM, T. 1970. Inheritance of Duane's syndrome. Br. J. Ophthalmol. **54:** 323–329.
39. PFAFFENBACH, D. et al. 1972. Congenital anomalies in Duane's retraction syndrome. Arch. Ophthalmol. **88:** 635.
40. APPLE, C. 1939. Congenital abducens paralysis. Am. J. Ophthalmol. **22:** 169–173.
41. DUANE, A. 1905. Congenital deficiency of abduction, associated with impairment of adduction, retraction movements, contraction of the palpebral fissure and oblique movements of the eye. Arch. Ophthalmol. **34:** 133–159.
42. GIFFORD, H. 1926. Congenital defects of abduction and other ocular movements and their relation to birth injuries. Am. J. Ophthalmol. **9:** 3.
43. GUNDERSON, T. & B. ZEAVIN. 1956. Observations on the retraction syndrome of Duane. Arch. Ophthalmol. **55:** 576.
44. HUBER, A. 1984. Duane's retraction syndrome; consideration on pathophysiology and etiology. In Strabismus II. R. Reinecke, Ed.: 345–361. Grune & Stratton. Orlando, FL.
45. VINCENT, C. et al. 1994. A proposed new contiguous gene syndrome on 8q consists of branchio-oto-renal (BOR) syndrome, Duane syndrome, a dominant form of hydrocephalus and trapeze aplasia; implications for the mapping of the BOR gene. Hum. Mol. Genet. **3:** 1859–1866.
46. CALABRESE, G. et al. 1998. Detection of an insertion deletion of region 8q13-q21.2 in a patient with Duane syndrome: implications for mapping and cloning a Duane gene. Eur. J. Hum. Genet. **6:** 187–193.
47. CALABRESE, G. et al. 1998. Narrowing the Duane syndrome critical region at chromosome 8q13. Am. J. Hum. Genet. **63(Suppl.):** A1417.
48. CALABRESE, G. et al. 2000. Narrowing the Duane syndrome critical region at chromosome 8q13 down to 40 kb. Eur. J. Hum. Genet. **8:** 319–324.
49. APPUKUTTAN, B. et al. 1999. Localization of a gene for Duane retraction syndrome to chromosome 2q31. Am. J. Hum. Genet. **65:** 1639–1646.
50. CHUNG, M. et al. 2000. Clinical diversity of hereditary Duane's retraction syndrome. Ophthalmology **107:** 500–503.
51. EVANS, J.C. et al. 2000. Confirmation of linkage of Duane's syndrome and refinement of the disease locus to an 8.8-cm interval on chromosome 2q31. Hum. Genet. **106:** 636–638.

Calcium Channelopathy Mutants and Their Role in Ocular Motor Research

JOHN S. STAHL

Departments of Neurology, Case Western Reserve University and Cleveland Veterans Affairs Medical Center, Cleveland, Ohio 44106, USA

ABSTRACT: Thanks to technical advances in eye movement recording, the mouse is destined to become increasingly important in ocular motor research. An advantage of this species is the wide range of existing mutant strains and techniques to generate new mutations affecting specific cell types. Mutations of ion channels may be used to modulate the intrinsic properties of neurons, and this approach may generate insight into the degree to which neuronal computations depend upon those intrinsic properties as opposed to the properties of circuits of neurons. Dendritic calcium currents carried by P-type voltage-activated calcium channels have been widely postulated to perform important computational functions in cerebellar Purkinje cells. Mutations of this channel lead to human diseases, and several ataxic strains of mice are now known to harbor mutations of this calcium channel. Murine P-channel mutants such as *rocker* are ataxic, but have normal or near-normal numbers of cerebellar Purkinje cells and thus offer the opportunity to study the effects of biophysical perturbations as opposed to outright cell destruction or inactivation. Initial studies of *rocker* mice reveal an array of ocular motor abnormalities, including static hyperdeviation of the eyes and an attenuation of vestibulo-ocular reflex gains at high stimulus frequencies. The pattern of gain and phase abnormalities is entirely different in *lurcher*, an ataxic mutant in which Purkinje cells degenerate. The ocular motor abnormalities of *rocker* progress with animal age, underscoring the importance of careful attention to animal age when performing ocular motor studies in this short-lived species.

KEYWORDS: video oculography; vestibulo-ocular reflex; nystagmus; migraine; cerebellum

INTRODUCTION

The cerebellar Purkinje cell is remarkable for its tremendous dendritic arbor and its unique active conductances, which enable the most distal parallel fibers to influence the Purkinje cell's axonal firing rate while also allowing input effects to be tightly restricted within the dendrite.[1–4] A consensus has emerged that active dendritic currents are critical to the computations being carried out within Purkinje cells, but the exact nature of those computations remains elusive.[5] Recent advances in the biology of calcium channels offer a new approach to the problem. The genes encod-

Address for correspondence: Dr. John Stahl, Department of Neurology, University Hospitals of Cleveland, 11100 Euclid Avenue, Cleveland, OH 44106-5040. Voice: 216-791-3800 ext. 5235; fax: 216-421-3040.

jss6@po.cwru.edu

ing the protein subunits of the neuronal calcium channel have been cloned, and mouse strains carrying mutations of those genes have been identified.[6] Studying the vestibular and ocular motor function in such mutants, by determining the effects of perturbing the normal calcium currents, can elucidate the role of the currents.

The main route of calcium influx in Purkinje cell dendrites is the P-type calcium channel, one of a family of high-voltage–activated calcium channels that now includes P, Q, N, R, and L subtypes.[1,6,7] Each of the channels consists of an α_1 subunit, which contains the ion pore, and accessory proteins β, γ, and $\alpha_2\delta$, which exert modulatory effects.[8] Multiple genes exist for each of the subunits, and the channel subtype is determined by the gene from which the channel's α_1 subunit is derived. The pore unit of P- and Q-type channels is derived from the same gene by alternative mRNA splicing,[9] and anatomical studies aimed at localizing the channels do not necessarily distinguish between the two types. P/Q channels are expressed in multiple brain regions, but within the regions related to vestibular and ocular motor control they are concentrated on a relatively limited number of cell types—Purkinje cells, cerebellar granule cells, and inferior olive neurons.[10–14] In most of these cell types, the P/Q channel function is thought to relate to synaptic transmission.[15] An exception is the Purkinje cell, where the channel is situated to play roles both in dendritic calcium spiking and synaptic transmission.[10,13,14]

Mouse geneticists have long recognized a set of mutant mice characterized by cerebellar ataxia with and without seizures. Many of these are now known to harbor mutations of the P/Q-type calcium channel. These mutants are generating great interest, in part because they have been proposed as animal models for human P/Q calcium channelopathies such as familial hemiplegic migraine, episodic ataxia type 2, and spinocerebellar ataxia type 6.[16,17] Of the mouse mutants, *rocker*, *tottering*, *rolling-Nagoya*, *tottering^{4J}*, *tottering^{5J}*, *tottering3^J*, and *leaner* (listed in rough order of the severity of their ataxia) have mutations of the gene encoding the $\alpha1$ subunit.[18–20] Four of the mutants—*rocker*, *tottering*, *tottering^{4J}*, and *tottering^{5J}*—are reported to have histologically normal or near-normal cerebella,[19–23] which indicates that the ataxia of these mutants is potentially attributable to abnormal signal processing within the cerebellar circuitry, rather than degeneration and loss of the neuronal elements. In contrast, the *leaner* and *rolling* mutations are associated with cerebellar degeneration, particularly of the anterior folia.[24,25]

The electrophysiologic properties of the mutant channels have been investigated in various central nervous system *in vitro* preparations. Calcium entry through P-channels is reduced in the Purkinje cells of *leaner*,[26,27] *tottering*,[28] and *rolling*,[29] as well as in the hippocampal Schaffer collaterals of *tottering*.[30] Studies using the HEK293 expression system have shown that mutations in the human α_{1A} gene can result in P-channel currents that are either abnormally high or low.[31,32] Alterations in calcium currents have been speculated to lead to abnormal modulation of cellular machinery and negative trophic effects,[27,28] but there has been little discussion of the extent to which the abnormal calcium currents might immediately affect neuronal signaling.

Although the first study of mouse eye movement recordings dates to the 1970s, relatively little work has been done in this species because of the technical difficulties imposed by the animal's small size. Early studies using video and EOG techniques[33,34] were limited by a calibration technique that was based upon the unrealistic assumption that the gain of the optokinetic reflex is unity for low-speed stimuli. In more recent video studies,[35,36] the conversion of linear pupil position to eye angle was accomplished using a trigonometric transformation that incorrectly assumed that the pupil's radius of rotation equals the radius of the eyeball. Magnetic search-coil recordings do not suffer from the calibration problems of video and EOG techniques, but early search-coil studies[37,38] reported low vestibulo-ocular reflex gains and/or bizarre phase relationships, raising the question of whether the mouse was a suitable model for studying eye movements. However, a recent study that combined video and improved search-coil techniques introduced a more realistic method of calibrating video recordings; it demonstrated that compensatory eye movements in the mouse were typical of other afoveate mammals, and indicated that some of the atypical features reported in earlier studies were attributable to the implanted search coils interfering with free rotation of the eyes.[39]

We now report initial results of a study of the eye movements of the P/Q calcium channel mutant, *rocker*. *Rocker* is an autosomal recessive mutation that is maintained on a genetic background closely related to the C57BL/6 inbred mouse strain.[20] The mutation results in a single amino acid substitution in the α_1 subunit near the mouth of the ion pore. For purposes of comparison, we also present limited data from *lurcher*, a strain whose glutamate channel mutation leads to postnatal degeneration of all cerebellar Purkinje cells.[40]

METHODS

Rocker and *lurcher* mutants of either sex were obtained from the breeding colonies of Dr. Karl Herrup at Case Western Reserve University. Three types of animals served as controls for the *rockers*—C57BL/6J animals purchased from Jackson Laboratories, animals drawn from the C57BL/6J-derived colony in which *rocker* was developed, and animals heterozygous for the *rocker* mutation. Animals were prepared for chronic video recording by surgical implantation of a head fixation pedestal as previously described.[39] Prior to each recording session, animals were treated topically with 0.5% physostigmine salicylate to prevent excessive dilation of the pupil in darkness. Without such treatment the pupil becomes so large as to render reliable tracking of its center impossible.

Even with pharmacologic treatment, pupil size still varies as a function of illumination. This variation must be accounted for in the conversion of pupil position to eye angle; the iris rides over the strongly curved surface of the lens, and thus dilation of the pupil is accompanied by a reduction of the distance from the plane of the pupil to the apex of corneal curvature. Calibration of the video recordings was accomplished using the previously reported method, with a refinement designed to compensate for the effect of changes of pupil size.[39] We recorded pupil diameter as well as the horizontal positions of the pupil and a reflection of a reference illuminator mounted immediately adjacent to the camera lens. As in the previous method, we rotated the video camera ±10° about the stationary animal and determined Δ_1 and Δ_2, the linear horizontal distances between the pupil and reference reflection, taken at the two extremes of camera position. We calculated Rp, the distance from the plane of the pupil to the origin of corneal curvature as $(\Delta_1 - \Delta_2)/(20\pi/180)$ and regressed this value against the pupil diameter. The regression coefficients give Rp as a function of pupil diameter. FIGURE 1 shows an example of the relationship. Note that Rp

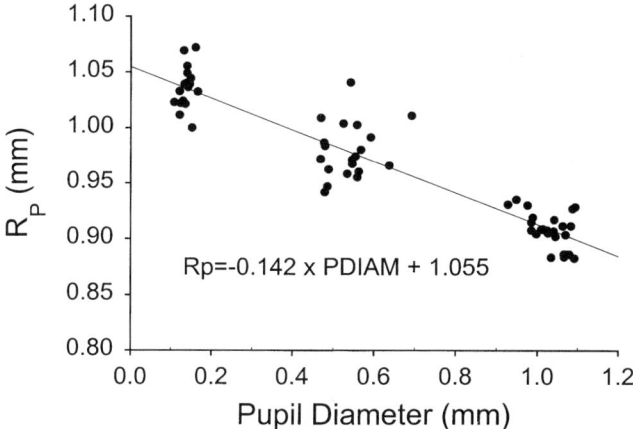

FIGURE 1. Effect of pupil diameter on Rp, the distance from the plane of the pupil to the origin of corneal curvature. Rp decreases as the pupil enlarges. The linear regression curve and equation are superimposed.

varied by approximately 15% of its mean value as pupil diameter varied over a 1.0 mm range. Horizontal eye angle is obtained from the video signal by the equation $E = \arcsin(\Delta_X/Rp)$, where Δ_X is the horizontal distance between the pupil and corneal reflection. This calibration technique assumes that the surface of the cornea is spherical and that pupil vertical position remains within a few degrees of the position it held during calibration. For recordings of static vertical elevation, the reference illuminator was placed on the horizontal plane of the center of the camera lens. In this position the corneal reflection appears at the equator of the eye. The static elevation was then determined as $\arctan(\Delta_Y/Rp)$, where Δ_Y is the difference in the vertical positions of the pupil and corneal reflection.

Static elevations were determined with the animal stationary in the light. Compensatory eye movement dynamics were determined as the animal was oscillated sinusoidally at 0.1 Hz 10° (0–peak) amplitude; 0.2 Hz 10°; 0.4 Hz ~10°; 0.8 Hz ~9.2°; 0.8 Hz ~4.6°; and 1.6 Hz ~3.9° in darkness (VOR) or light (VVOR). Eye movement gains and phases were obtained by Fourier analysis.[39] Average gain and phase curves for each animal type were created by averaging all data within a session to obtain each animal's session curves, then averaging 2–3 session curves to obtain all-session gain and phase curves for each animal, and then averaging individual animal curves to obtain curves for each animal type.

RESULTS

Static eye elevation was determined in 19 control mice, 12 *rocker* homozygotes, 3 *rocker* heterozygotes, and 2 *lurcher* mutants. FIGURE 2 plots elevation as a function of animal age. Some animals were recorded multiple times and their data points are joined by lines. In the youngest age group (<200 days), static elevation was greater

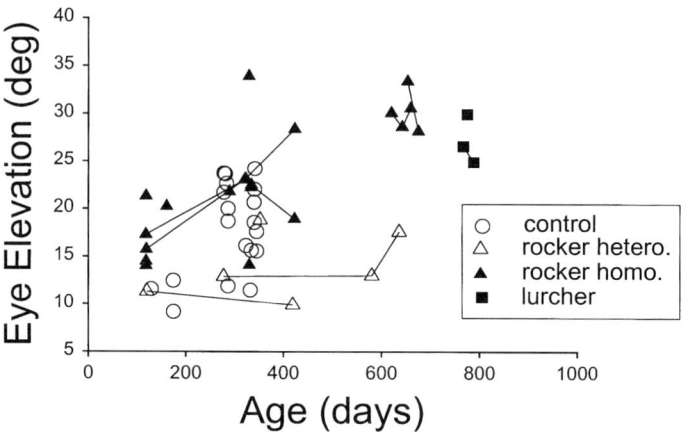

FIGURE 2. Absolute elevation of the pupil recorded with the animal stationary in the light, plotted as a function of animal age in days. Where individual animals were recorded multiple times in their lives, the measurements are joined by lines.

for *rocker* homozygotes than for any of the three C57BL/6 controls, or the single *rocker* heterozygote. Static elevation averaged 17.2 ± 3.0° for the six *rockers* and 11.1 ± 1.4° for the pool of three control animals and one heterozygote. The difference in elevations was statistically significant ($p = 0.006$, two-tailed t test). More data was obtained at the 200–450-day age range, and there the distributions of elevation overlapped. Nevertheless, the average elevation for the pooled control and heterozygous animals differed significantly from the average for the *rocker* homozygotes (respectively, 18.2 ± 4.4° versus 23.0 ± 5.8°, $p = 0.035$). Inspection of the data in FIGURE 2 suggests a tendency for eye elevation to increase with animal age in both the *rocker* and control animals, with average elevation reaching 30.3° in the two homozygous mutants recorded beyond 600 days of age. Static elevation was also recorded in two elderly *lurcher* mutants. Both of these exhibited striking elevation, closely approximating that of the elderly *rocker* mutants. Humans with cerebellar degeneration often exhibit downbeat nystagmus, i.e., an upward bias of slow-phase velocity. FIGURE 3 shows a 40-s recording of horizontal and vertical positions of the eye in one homozygous *rocker* mutant. While there are slow vertical drifts in both light and darkness, there is no upward directional preponderance. The possibility of a connection between the static elevation in mutant mice and human downbeat nystagmus is discussed further below.

VOR and VVOR dynamics were assessed in two 4-month-old *rockers*, two 19-month-old *rockers*, and three control animals (aged 2, 2, and 8 months). The gain and phase plots for the young and old *rockers* appear in FIGURE 4. The control animal data is repeated in the two sets of plots for ease of reference. Normal mice exhibited excellent VVOR, with gains approaching unity and phase leads near 0 degrees across the tested frequency range. VOR gain declined and phase advanced as stimulus frequency decreased, as expected, based upon the biophysics of the vestibular end-organ.[41] Average VOR gain peaked at 0.90 at 0.8 Hz. This value exceeds the search-

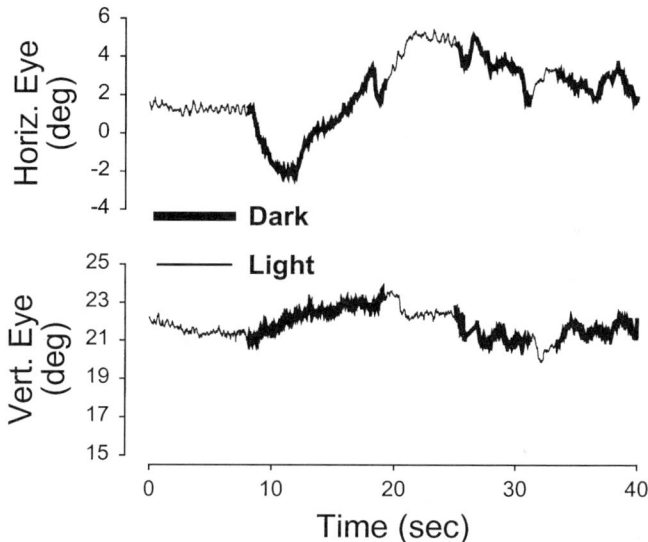

FIGURE 3. Example of horizontal and vertical positions of the eye of a *rocker* recorded with the animal stationary. Periods of darkness are indicated by heavy lines. Note absence of downbeat nystagmus.

coil–derived values reported for a well-studied afoveate mammal, the rabbit.[42] It also exceeds the best values that have been obtained for the mouse using search-coil recordings,[43] as well as those reported in a recent video study in which calibration was based upon the radius of the eyeball.[44] Of note, the VVOR values at low-stimulus frequencies also exceed those reported in our own previous video study.[39] In the current experiments we enclosed the animals in a stationary drum painted with a highly contrasted pattern, and presumably this strong visual input more effectively countered the visual effects of apparatus moving with the animal (the illuminators, camera, and restraint structure) that tend to drive down the VVOR gain.

Young *rocker* animals exhibited a decline in VVOR gain as stimulus frequency increased. This decline likely reflects the flattening of the underlying VOR gain curve; *rocker* VOR gain failed to increase with increasing stimulus frequency as sharply as it did in control animals. The high-frequency attenuation of the VVOR and plateauing of VOR gain were still more pronounced in the older animals. Phase lead of normal and mutant animals was similar at high frequencies. At lower frequencies the phase lead of the VOR increased more rapidly in the mutants. The greater phase lead may reflect dysfunction of the velocity-to-position neural integrator, which is deficient following flocculectomy and in human patients with cerebellar degeneration.[45,46]

FIGURE 5 shows the gain and phase plots for two *lurcher* animals, recorded at 690–695 days of age. Again, the average curves for the three control animals are included for reference purposes. The *lurcher* exhibits gain abnormalities differing

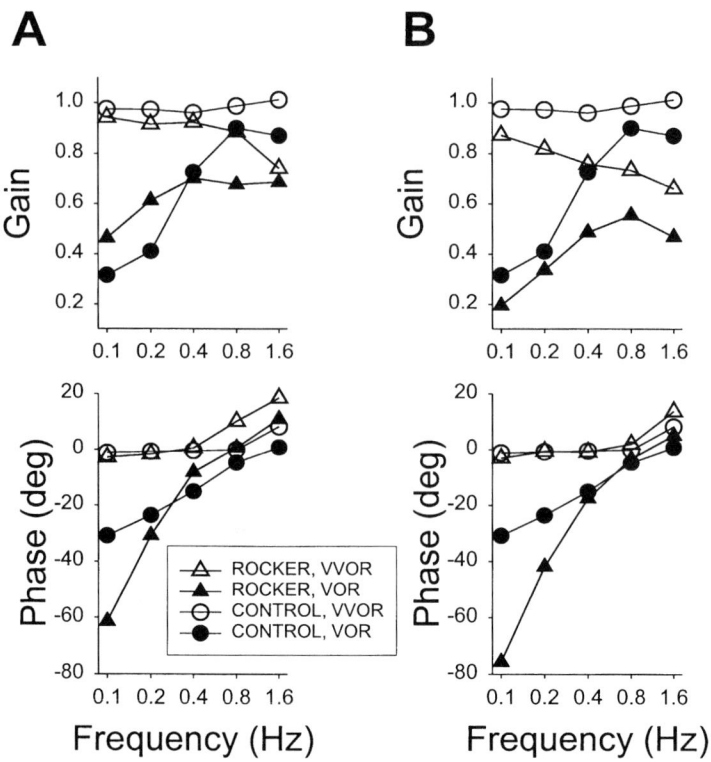

FIGURE 4. Average gain and phase plots for *rocker* mutants and normal controls during VOR and VVOR. (**A**) 4-month-old *rockers;* (**B**) 19-month-old *rockers*.

markedly from those of *rocker*. VVOR gains were subnormal at low frequencies and became supernormal at 0.8 Hz. VOR gains were similar to those of the control animals at low frequencies, but exceeded the normal animals as frequency increased. The VVOR gain pattern can be explained by a poor optokinetic response at intermediate frequencies, resulting in the inability of the animals to use vision to increase gain at 0.1–0.4 Hz, and a similar inability to use vision to suppress an excessive VOR gain at 0.8 Hz. As in *rocker*, VOR phase lead is abnormally large, consistent with deficiency of the neural integrator. The greater severity of this deficiency is not surprising in that Purkinje cells undergo complete degeneration in *lurcher*, whereas in rocker they are physically present but altered in their electrophysiological properties. The pronounced phase lead of the VVOR at 0.2–1.6 Hz is consistent with the deficient optokinetic high-frequency response inferred from the gain data; visual inputs available during VVOR are unable to compensate for the phase lead of the underlying VOR. Increased VOR gains and decreased optokinetic responses have also been demonstrated in *lurcher* using search-coil recordings.[47]

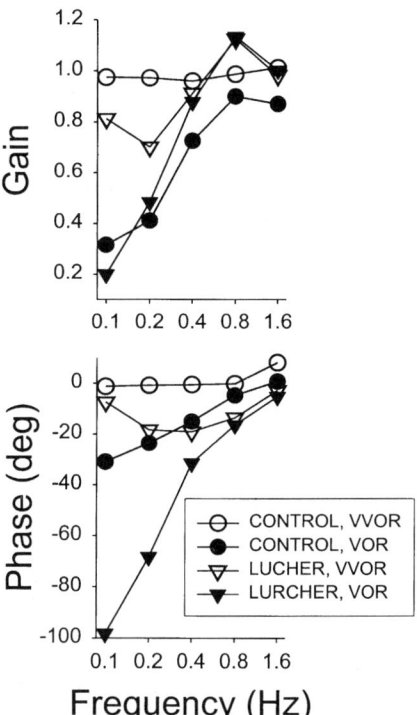

FIGURE 5. Average gain and phase plots for VOR and VVOR in *lurcher* mutants compared to same control curves shown in FIGURE 4.

DISCUSSION

These initial results support the feasibility and potential of using murine calcium channelopathy mutants to study the contribution of normal calcium biophysics to ocular motility. The *rocker* mutant exhibits a range of behavioral defects, with both similarities to and differences from the deficiencies observed in a mutant (*lurcher*) that undergoes Purkinje cell degeneration. The similarities (e.g., abnormal ocular elevation and deficient velocity-to-position integration) may indicate that some of the behavioral abnormalities observed following cerebellar lesions or degeneration can be produced by merely altering Purkinje cell signals.

Our description of abnormal eye elevation in ataxic mice is, to our knowledge, the first report of its kind. Because of the technical difficulties of determining absolute eye position in untrained animals, most studies report eye position relative to an arbitrary zero point, and thus little is known about the control of absolute eye position in afoveate mammalian species. Fortunately, simply placing the reference illuminator in the horizontal plane of the video camera's optical axis allows us to obtain the absolute elevation of the pupil above the horizon. The upward deviation of the eyes is reminiscent of the upward drifts seen in patients with cerebellar degeneration, and is attributed to disinhibition of anterior semicircular canal pathways.[45,48] The lack of an upward drift in the dark may indicate that the upward deviation in the mouse stems from another cause, such as aberrant signaling within otolithic circuits. However, this conclusion is not inescapable. Lacking a fovea or even a visual streak,[49] the mouse may lack the need to maintain a particular orientation of its eyes with respect to the environment. Saccadic movements intended to correct a misorientation would be absent, and thus the eye would slowly drift upward until the upward velocity bias comes into equilibrium with elastic forces tending to recenter the eye. The tendency for the mouse eye to recenter is strong, owing to the short time constant of its velocity-to-position integrator.[43] This tendency would be further enhanced in the ataxic mutants, which have, based upon the VOR phase data described in RESULTS, even shorter integrator time constants, at least in the horizontal plane. Thus, the static upward deviation could reflect the balance of an upward canal signal and a tendency of orbital elasticity to recenter the eye, together with the animal failing to use the saccadic system to correct the vertical position offset.

Both the static elevation and the alterations of VOR/VVOR dynamics in *rocker* appeared to progress with age. The abnormalities of the P-type calcium currents are presumably present and fixed from birth, barring age-related changes in the expression or post-translational modifications of the channel's components. In that case, the progressive alterations in behavior may relate to the progressive alterations of Purkinje cell dendritic architecture described in this mutant.[20] Because of the progressive deficits, it is important that age be carefully controlled when studying these mutants. As demonstrated in FIGURE 2, the passage of only a few months was associated with increases in the average eye elevation. It should be noted that, given a life span of only two years, a few months represents a sizeable percentage of a mouse's life. In humans, a variety of indices of ocular motor performance are known to change with age.[50–53] Ocular motor senescence is similarly probable in the normal mouse and, given its future importance in vestibular and ocular motor research, it will be important to perform longitudinal studies in this species.

ACKNOWLEDGMENTS

We gratefully acknowledge Dr. Karl Herrup and Dr. Theresa Zwingman for providing the mutants used in this study. This work was partially supported by core center grant EY11373.

REFERENCES

1. LLINAS, R. & M. SUGIMORI. 1980. Electrophysiological properties of *in vitro* Purkinje cell dendrites in mammalian cerebellar slices. J. Physiol. **305:** 197–213.
2. JAEGER, D., E. DE SCHUTTER & J.M. BOWER. 1997. The role of synaptic and voltage-gated currents in the control of Purkinje cell spiking: a modeling study. J. Neurosci. **17:** 91–106.
3. DE SCHUTTER, E. & J.M. BOWER. 1994. An active membrane model of the cerebellar Purkinje cell. II. Simulation of synaptic responses. J. Neurophysiol. **71:** 401–419.
4. LLINAS, R. & H. MORENO. 1998. Local Ca^{2+} signaling in neurons. Cell Calcium **24:** 359–366.
5. YUSTE, R. & D.W. TANK. 1996. Dendritic integration in mammalian neurons, a century after Cajal. Neuron **16:** 701–716.

6. ASHCROFT, F. 2000. Ion Channels and Disease. Academic Press. San Diego, CA.
7. LLINAS, R., M. SUGIMORI, J. LIN, et al. 1989. Blocking and isolation of a calcium channel from neurons in mammals and cephalopods utilizing a toxin fraction (FTX) from funnel-web spider poison. Proc. Natl. Acad. Sci. USA **86:** 1689–1693.
8. WALKER, D. & M. DE WAARD. 1998. Subunit interaction sites in voltage-dependent Ca^{2+} channels: role in channel function. TINS **21:** 148–154.
9. BOURINET, E., T. SOONG, K. SUTTON, et al. 1999. Splicing of alpha 1A subunit gene generates phenotypic variants of P- and Q-type calcium channels. Nat. Neurosci. **2:** 407–415.
10. HILLMAN, D., S. CHEN, T. AUNG, et al. 1991. Localization of P-type calcium channels in the central nervous system. Proc. Natl. Acad. Sci. USA **88:** 7076–7080.
11. TANAKA, O., H. SAKAGAMI & H. KONDO. 1995. Localization of mRNAs of voltage-dependent Ca(2+)-channels: four subtypes of alpha 1– and beta-subunits in developing and mature rat brain. Brain Res. Mol. Brain Res. **30:** 1–16.
12. STEA, A., W. TOMLINSON, T. SOONG, et al. 1994. Localization and functional properties of a rat brain alpha 1A calcium channel reflect similarities to neuronal Q- and P-type channels. Proc. Natl. Acad. Sci. USA **91:** 10576–10580.
13. WESTENBROEK, R., T. SAKURAI, E. ELLIOTT, et al. 1995. Immunochemical identification and subcellular distribution of the alpha1A subunits of brain calcium channels. J. Neurosci. **15:** 6403–6418.
14. CRAIG, P., A. MCAINSH, A. MCCORMACK, et al. 1998. Distribution of the voltage-dependent calcium channel alpha1A subunit throughout the mature rat brain and its relationship to neurotransmitter pathways. J. Comp. Neurol. **397:** 251–267.
15. CATTERALL, W. 1998. Structure and function of neuronal Ca^{2+} channels and their role in neurotransmitter release. Cell Calcium **24:** 307–323.
16. OPHOFF, R., G. TERWINDT, M. VERGOUWE, et al. 1996. Familial hemiplegic migraine and episodic ataxia type-2 are caused by mutations in the Ca(2+) channel gene CACNL1A4. Cell **87:** 543–552.
17. ZHUCHENKO, O., J. BAILEY, P. BONNEN, et al. 1997. Autosomal dominant cerebellar ataxia (SCA6) associated with small polyglutamine expansions in the alpha1A-voltage-dependent calcium channel. Nat. Genet. **15:** 62–69.
18. DOYLE, J., X. REN, G. LENNON, et al. 1997. Mutations in the Cacnl1a4 calcium channel gene are associated with seizures, cerebellar degeneration, ataxia in tottering and leaner mutant mice. Mamm. Genome **8:** 113–120.
19. ZWINGMAN, T., W. FRANKEL & K. HERRUP. 2000. Two new mouse variants of the voltage-dependent calcium channel subunit gene alpha1A. Soc. Neurosci. Abstr. **24:** 316.14.
20. ZWINGMAN, T.A., P.E. NEUMANN, J.L. NOEBELS, et al. 2001. Rocker is a new variant of the voltage-dependent calcium channel gene Cacna1a. J. Neurosci. **21:** 1169–1178.
21. ISAACS, K.R. & L.C. ABBOTT. 1992. Development of the paramedian lobule of the cerebellum in wild-type and tottering mice. Dev. Neurosci. **14:** 386–393.
22. ISAACS, K. & L. ABBOTT. 1995. Cerebellar volume decreases in the tottering mouse are specific to the molecular layer. Brain Res. Bull. **36:** 309–314.
23. NOEBELS, J.L. & R.L. SIDMAN. 1979. Inherited epilepsy: spike-wave and focal motor seizures in the mutant mouse tottering. Science **204:** 1334–1336.
24. MURAMOTO, O., I. KANAZAWA & K. ANDO. 1981. Neurotransmitter abnormality in Rolling mouse Nagoya, an ataxic mutant mouse. Brain Res. **215:** 295–304.
25. HERRUP, K. & S. WILCZYNSKI. 1982. Cerebellar cell degeneration in the leaner mutant mouse. Neuroscience **7:** 2185–2196.
26. DOVE, L.S., L.C. ABBOTT & W.H. GRIFFITH. 1998. Whole-cell and single-channel analysis of P-type calcium currents in cerebellar Purkinje cells of leaner mutant mice. J. Neurosci. **18:** 7687–7699.
27. LORENZON, N.M., C.M. LUTZ, W.N. FRANKEL, et al. 1998. Altered calcium channel currents in Purkinje cells of the neurological mutant mouse *leaner*. J. Neurosci. **18:** 4482–4489.
28. WAKAMORI, M., K. YAMAZAKI, H. MATSUNODAIRA, et al. 1998. Single tottering mutations responsible for the neuropathic phenotype of the P-type calcium channel. J. Biol. Chem. **25:** 34857–34867.

29. MORI, Y., M. WAKAMORI, S. ODA, et al. 2000. Reduced voltage sensitivity of activation of P/Q-Type Ca^{2+} channels is associated with the ataxic mouse mutation rolling Nagoya (tg(rol)). J. Neurosci. **20:** 5654–5662.
30. QIAN, J.& J. NOEBELS. 2000. Presynaptic Ca^{2+} influx at a mouse central synapse with Ca^{2+} channel subunit mutations. J. Neurosci. **20:** 163–170.
31. TORU, S., T. MURAKOSHI, K. ISHIKAWA, et al. 2000. Spinocerebellar ataxia type 6 mutation alters P-type calcium channel function. J. Biol. Chem. **275:** 10893–10898.
32. HANS, M., S. LUVISETTO, M.E. WILLIAMS, et al. 1999. Functional consequences of mutations in human alpha1A calcium channel subunit linked to familial hemiplegic migraine. J. Neurosci. **19:** 1610–1619.
33. MITCHINER, J., L. PINTO & J.J. VANABLE. 1976. Visually evoked eye movements in the mouse (Mus musculus). Vision Res. **16:** 1169–1171.
34. GRUSSER-CORNEHLS, U. & P. BOHM. 1988. Horizontal optokinetic ocular nystagmus in wildtype (B6CBA+/+) and weaver mutant mice. Exp. Brain Res. **72:** 29–36.
35. MANGINI, N., J.J. VANABLE, M. WILLIAMS, et al. 1985. The optokinetic nystagmus and ocular pigmentation of hypopigmented mouse mutants. J. Comp. Neurol. **241:** 191–209.
36. KATOH, A., H. KITAZAWA, S. ITOHARA, et al. 1998. Dynamic characteristics and adaptability of mouse vestibulo-ocular and optokinetic response eye movements and the role of the flocculo-olivary system revealed by chemical lesions. Proc. Natl. Acad. Sci. USA **95:** 7705–7710.
37. DE ZEEUW, C.I., C. HANSEL, F. BIAN, et al. 1998. Expression of a protein kinase C inhibitor in Purkinje cells blocks cerebellar LTD and adaptation of the vestibulo-ocular reflex. Neuron **20:** 495–508.
38. KOEKKOEK, S.K.E., A.M. ALPHEN, J. VAN DER BURG, et al. 1997. Gain adaptation and phase dynamics of compensatory eye movements in mice. Genes & Function **1:** 175–190.
39. STAHL, J., A. VAN ALPHEN & C. DE ZEEUW. 2000. A comparison of video and magnetic search coil recordings of mouse eye movements. J. Neurosci. Methods **99:** 101–110.
40. HEINTZ, N. & P. DE JAGER. 1999. GluR delta 2 and the development and death of cerebellar Purkinje neurons in lurcher mice. Ann. N.Y. Acad. Sci. **868:** 502–514.
41. WILSON, V.J. & G. MELVILL JONES. 1979. Mammalian Vestibular Physiology. Plenum. New York.
42. COLLEWIJN, H. 1981. The Oculomotor System of the Rabbit and Its Plasticity. Springer-Verlag. New York.
43. VAN ALPHEN, A., J. STAHL, S. KOEKKOEK, et al. 2001. The dynamic characteristics of the mouse vestibulo-ocular and optokinetic response. Brain Res. **890:** 296–305.
44. IWASHITA, M., R. KANAI, K. FUNABIKI, et al. 2001. Dynamic properties, interactions and adaptive modifications of vestibulo-ocular reflex and optokinetic response in mice. Neurosci. Res. **39:** 299–311.
45. STRAUMANN, D., D. ZEE & D. SOLOMON. 2000. Three-dimensional kinematics of ocular drift in humans with cerebellar atrophy. J. Neurophysiol. **83:** 1125–1140.
46. ZEE, D.S., A. YAMAZAKI, P.H. BUTLER, et al. 1981. Effects of ablation of flocculus and paraflocculus on eye movements in primate. J. Neurophysiol. **46:** 878–899.
47. VAN ALPHEN, A., C. DE ZEEUW & J. STAHL. 1999. Eye movement recordings in Lurcher mice. Soc. Neurosci. Abstr. **25:** 661.13.
48. WALKER, M.F. & D.S. ZEE. 1999. Directional abnormalities of vestibular and optokinetic responses in cerebellar disease. Ann. N.Y. Acad. Sci. **871:** 205–220.
49. JEON, C.-J., E. STRETTOI & R.H. MASLAND. 1998. The major cell populations of the mouse retina. J. Neurosci. **18:** 8936–8946.
50. MUNOZ, D., J. GROUGHTON, J. GOLDRING, et al. 1998. Age-related performance of human subjects on saccadic eye movement tasks. Exp. Brain Res. **121:** 391–400.
51. SHARPE, J. & D. ZACKON. 1987. Senescent saccades. Effects of aging on their accuracy, latency, and velocity. Acta Otolaryngol. **104:** 422–428.
52. TIAN, J., I. SHUBAYEV, R. BALOH, et al. 2001. Impairments in the initial horizontal vestibulo-ocular reflex in older humans. Exp. Brain Res. **137:** 309–322.
53. PAIGE, G. 1994. Senescence of human visual–vestibular interactions: smooth pursuit, optokinetic, and vestibular control of eye movements with aging. Exp. Brain Res. **98:** 355–372.

Modern Concepts of Brainstem Anatomy

From Extraocular Motoneurons to Proprioceptive Pathways

J.A. BÜTTNER-ENNEVER,[a] A.K.E. HORN,[a] W. GRAF,[b] AND G. UGOLINI[c]

[a]*Institute of Anatomy, Ludwig-Maximilian University of Munich, 80336 Munich, Germany*

[b]*CNRS-Collège de France, Paris, France*

[c]*CNRS, Gif-sur-Yvette, France*

ABSTRACT: The extraocular muscles, unlike the skeletal muscles, contain non-twitch muscle fibers. Recent experiments have located the non-twitch motoneurons. They lie around the periphery of the oculomotor, trochlear and abducens nuclei, separate from the more usual twitch motoneurons that cluster within the boundaries of the classical motor nuclei. The premotor inputs to non-twitch neurons were traced by the injection of rabies virus into the distal tip of the lateral rectus muscle. Retrogradely labeled cells were found in areas associated with the neural integrator, vergence and smooth pursuit premotor areas, but *not* the saccadic premotor burst neurons or the direct vestibulo-ocular pathways. The rabies tracing emphasizes for the first time that the central mesencephalic reticular formation (cMRF) and the supraoculomotor area exert direct premotor control over the non-twitch motoneurons. Because the two sets of motoneurons do not receive the same afferents, they must have different functions; these are not yet clarified. These results are not compatible with the concept of a single final common pathway from motoneurons to eye muscles.

Putative sensory receptors, palisade endings, are located at the tips of non-twitch muscle fibers reminiscent of an inverted muscle spindle, which would make the non-twitch motoneurons, γ-motoneurons. We propose that twitch motoneurons are the major source of tension used for eye movements, whereas non-twitch motoneurons are more important for fine alignment of the eyes. Furthermore, the non-twitch motoneurons could be controlled through sensory feedback networks (including perhaps proprioceptive signals from the palisade endings) that are relayed through the superior colliculus and via cMRF to the non-twitch motoneurons. The clinical repercussions of these hypotheses are discussed.

KEYWORDS: non-twitch motoneurons; palisade endings; superior colliculus; oculomotor nuclei; central mesencephalic reticular formation; premotor neurons; supraoculomotor area; near-response area; neural integrator; marginal zone; medial vestibular nucleus; eye movements; medial rectus C-group; final common pathway

---

Address for correspondence: Prof. J.A. Büttner-Ennever, Institute of Anatomy, Ludwig-Maximilian University, Pettenkoferstr. 11, D-80336 Munich, Germany. Voice: (*)49 89 5160 4851/4876; fax: (*)49 89 5160 4857.

buettner@anat.med.uni-muenchen.de

INTRODUCTION

There are no fast stretch-reflexes from eye muscles when they are pulled,[1] and so it is often assumed that there is no proprioceptive input from the extrinsic eye muscles to the brain. The puzzling lack of muscle spindles from the eye muscles of monkey and cat, but the abundance of them in man and sheep, has compounded the problem,[2] and consequently the topic of eye muscle proprioception has been to a large extent ignored. However, recent reviews have repeatedly emphasized that much evidence exists for the use of sensory information from eye muscles by the brain.[2–5] At least six different types of muscle fibers are found in the eye muscles of vertebrates;[6] some are fast-twitch, others are intermediate- or slow-twitch, which are also found in skeletal muscles. In the extraocular muscles there are also non-twitch muscle fibers (Felderstruktur), which are highly unusual and not a component of skeletal muscles in mammals. Taking as a starting point the recent discovery in the monkey of a new separate group of extraocular motoneurons that innervate the non-twitch extraocular muscle fibers,[7] we will then go on to describe the premotor inputs to the non-twitch motoneurons, which differ from those of the classical twitch motoneurons.[8] Finally, we will consider the function of these "non-twitch brainstem pathways," and propose that theoretically they could play a role in the fine sensory control of eye alignment.

METHODS

Experiments were carried out on macaque monkeys (*M. mulatta* or *fascicularis*). Animal care and experimental procedures conformed with European Union standards. Under general anesthesia, an eye muscle was exposed and injected with a retrograde tract tracer [cholera toxin subunit B (CT, 1–10 µl, 1%; from List Campbell, CA), or wheat-germ agglutinin conjugated to horseradish peroxidase (WGA.HRP, 5–30 µl, 2.5% Sigma, St. Louis, MO), or iodinated wheat-germ agglutinin ($[^{125}I]$ WGA, 4–50 µl, Büttner-Ennever and Akert 1981), or rabies virus (CVS strain). The propagation of the tracers took 2–3 days, after which the animals were killed by an overdose of anesthesia and perfused with phosphate-buffered saline, followed by 4% paraformaldehyde and 10% sucrose at pH 7.4. Brains were dissected out and cut stereotaxically in two blocks, gelatin embedded, and cut in frozen sections (50 µm). The WGA.HRP, CT and $[^{125}I]$WGA were visualized using the standard methods.[7] Rabies virus was visualized immunohistochemically using monoclonal antibodies (anti-rabies P protein) and the peroxidase anti-peroxidase method.[9,10]

RESULTS

The location of motoneurons supplying the slow, non-twitch and the fast-twitch muscle fibers of the six extraocular muscles of monkeys was achieved by injections of simple retrograde tracers (CT, WGA.HRP, $[^{125}I]$ WGA) into eye muscles that were placed either (1) centrally, within the central end-plate zone of the twitch fibers. In these injections the tracer would theoretically be taken up by both the large end-plates of the twitch motoneurons and the end-plates of the multiply innervated

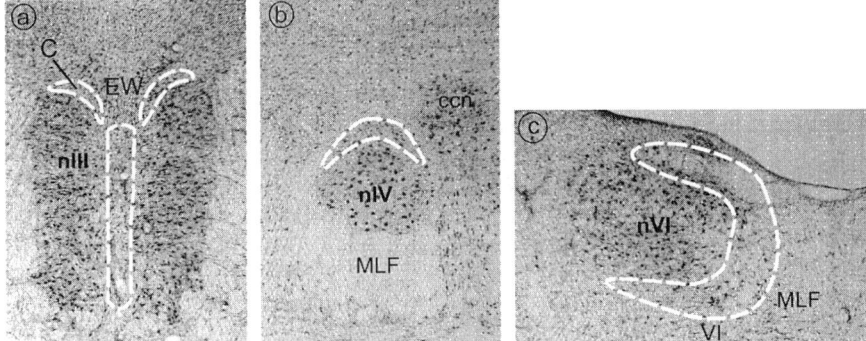

FIGURE 1. Transverse sections stained with cresyl violet show the oculomotor (nIII), trochlear (nIV), and abducens (nVI) nuclei of the monkey. The location of the non-twitch motoneurons, retrogradely filled with retrograde tracer injected into the distal tip of the extraocular eye muscles, is indicated by the white dashed lines: they lie around the periphery of the nuclei, separate from the twitch motoneurons in the classical subgroups. Technically it is very difficult to fill all the regions of the eye muscles with tract tracers, e.g., proximal end of muscles. Therefore, the plots of the location of the slow non-twitch motoneurons are probably still incomplete. C, C-group; ccn, central caudal nucleus; EW, Edinger-Westphal nucleus; MFL, medial longitudinal fasciculus; VI, abducens nerve.

non-twitch motoneurons that cover the length of the fiber. (2) Alternatively, the tracer injection was placed distally, near the myotendinous junction where only motor end-plates of the multiply innervated non-twitch muscle fibers are found. This injection would theoretically fill only non-twitch motoneurons. The myotendinous region also contains the putative sensory receptors called palisade endings.

Injections into the central end-plate zone of the muscles labeled large motoneurons within the abducens, trochlear or oculomotor nucleus, and smaller motoneurons lying mainly around the periphery of the motor nuclei. Injections into the distal tip of the muscle labeled small and medium–large-sized peripheral neurons strongly, and almost exclusively. The peripheral neurons labeled from the lateral rectus muscle surround the medial half of the abducens nucleus: from the superior oblique they form a cap over the dorsal trochlear nucleus; from the inferior oblique and superior rectus they are scattered bilaterally around the midline, between the oculomotor nucleus; from both medial and inferior rectus they lie mainly (but not exclusively) in the C-group, on the dorsomedial border of the oculomotor nucleus (FIG. 1).[7] The neurons of the C-group lie in close association with the Edinger-Westphal nucleus (EW),[11] and their labeled dendrites extended out into the supraoculomotor area (SOA).

The injection of a *transsynaptic* retrograde tracer, rabies virus, into the distal tip of the lateral rectus muscle has been reported previously.[8] The virus labeled motoneurons predominantly in the periphery of the abducens nucleus, in the same way as in the simple tracer experiments; and injections into the central end-plate zone of the muscle labeled motoneurons throughout the abducens nucleus. Uptake by sensory endings in the lateral rectus muscle does not occur with rabies virus: no labeling was found in Gasser's ganglion in either experiment, in keeping with our previous findings in another model.[10] After a survival time of 2.5 days, without any clinical symptoms in the monkey, the virus had replicated itself and passed retrogradely over the synapses on the motoneurons into the immediate premotor neurons. Following a rabies injection into the distal tendon portion of the lateral rectus muscle, transneuronally labeled premotor neurons, with a monosynaptic input to the non-twitch abducens motoneurons, were abundant bilaterally in the medial vestibular nuclei (predominantly in the parvocellular division), in the nucleus prepositus hypoglossi and its marginal zone, in the caudal SOA and the adjacent central mesencephalic reticular formation (cMRF). In addition, a crucial observation was that there was *no significant labeling* of horizontal saccade generator networks in the paramedian pontine reticular formation (PPRF). After distal injections no tracer was found in the excitatory burst neurons (EBNs) in nucleus reticularis pontis caudalis or in the dorsal paragigantocellular nucleus containing the inhibitory burst neurons (IBNs). Neither were the direct vestibulo-oculomotor pathways from the magnocellular medial vestibular nucleus (MVN_{mag}) labeled.[12] In contrast, rabies injections into the central end-plate zone of the lateral rectus muscle strongly labeled the EBNs in PPRF, the IBN and MVN_{mag}, and the oculomotor internuclear neurons.[13] The labeling also included all regions that were traced from the distal injections, as is to be expected from the methodology of this experiment. Additional regions labeled, but not considered further here, are the superior vestibular nuclei, the y-group, the olivary pretectal nucleus, an area lateral to the rostral tip of the trigeminal motor nucleus bilaterally, and a few neurons in the ipsilateral Scarpa's ganglion.

DISCUSSION

Motoneurons

Plotting the location of extraocular motoneurons with simple retrograde tracers in the eye muscles shows that the fast-twitch and slower non-twitch motoneurons lie separately: the twitch motoneurons within the boundaries of the classical motor nuclei, and the non-twitch motoneurons around the periphery. The organization of the subgroups for individual eye muscles in the oculomotor nucleus is completely different for the two motoneuron types.[7] The twitch motoneurons of inferior rectus, medial rectus, superior rectus, and inferior oblique have relatively separate subgroups within the classical oculomotor nucleus. In contrast, the non-twitch motoneurons of medial and inferior rectus lie together in the dorsomedial C-group, and those of inferior oblique and superior rectus are mixed along the midline. Excitatory inputs to the C-group would cause the eyes to converge and move down, and those to the midline cluster would move the eyes upwards.

No recordings have been made from identified extraocular non-twitch motoneurons, apart from one study in the frog[14] where the neuronal activity was tonic and correlated with intended eye position. It is not clear how much the non-twitch units contribute to the development of extraocular eye muscle tension. However, many vertebrates have non-twitch muscle fibers in their extraocular eye muscles. The fibers do not have propagated action potentials, produce weak tetanic tensions, and are extremely fatigue resistant.[15] Although at present their function is unknown, they certainly serve a basic purpose in the control of eye movements. We will propose below that they could participate in the control of eye alignment associated with sensory feedback pathways.

Premotor Neurons

The results using the transsynaptic tracer, rabies virus, confirm previous reports[7] that there are independent premotor circuits supplying "fast, twitch" and "slow, non-twitch" motoneurons. They show for the first time that the premotor input from the saccade generator (EBNs and IBNs in PPRF), and direct vestibulo-ocular reflex pathways from MVN_{mag}, project directly to the twitch, but *not* the non-twitch motoneurons.[8] This is in direct conflict with the concept of a single, final common pathway from motoneurons to eye muscles; there appears to be a dual motor control of extraocular eye muscles.

The premotor regions targeting the non-twitch motoneurons directly are, with few exceptions, bilateral. The inputs come from the nucleus prepositus hypoglossi, a part of the neural integrator which serves to maintain gaze;[16] the marginal zone which plays a role in visual control of gaze through smooth pursuit eye movements;[12,17] and the SOA, a clearly defined region dorsal to the oculomotor nucleus, which is associated with the premotor control of vergence[11,18] and the near-response.[19,20] Interestingly, all these regions subserve eye movement types (vergence, smooth pursuit, gaze holding) that use visual feedback to control the eye position. In view of these synaptic connections, it seems that the non-twitch motoneurons may be more suited to the fine motor control of eye alignment, possibly controlled by sensory feedback pathways.

The Central Mesencephalic Reticular Formation (cMRF)

The rabies tracing results also emphasize the role of the cMRF and SOA as premotor regions to extraocular motoneurons, which has up until now been largely overlooked. The reason for this is not entirely clear, but probably reflects the extraordinary power of the "rabies virus technique" compared to all other tract tracing methods previously used. The cMRF and SOA input to non-twitch motoneurons is firmly established by the rabies results. But because they are also labeled by central injections into eye muscles, it is not clear from these results whether the cMRF and SOA also project to twitch motoneurons. Other experiments indicate that their input is predominantly to the non-twitch motoneurons.[21,22]

The cMRF lies lateral to the oculomotor nucleus and projects monosynaptically onto non-twitch motoneurons. It is a specific part of MRF which was first defined by Cohen and colleagues as an area from which horizontal saccades can be evoked by electrical stimulation.[23] Since then the region has been carefully studied both anatomically and physiologically.[24] The cMRF contains neurons that encode various aspects of saccades. Reversible chemical lesions lead to hypermetric saccades, destabilization of gaze fixation, and shifting of the position of primary gaze.[25,26] Neuroanatomically the most prominent relationship of cMRF is its intense reciprocal interconnections with the superior colliculus (SC).[27]

Hypothetical Sensory Feedback Pathways

How is it possible to integrate the function of a set of slow non-twitch motoneurons with a unique set of premotor inputs into the oculomotor neural networks that have been studied and modeled so thoroughly over the last 25 years? One particular aspect of eye movements has been severely neglected—that is, the sensory control

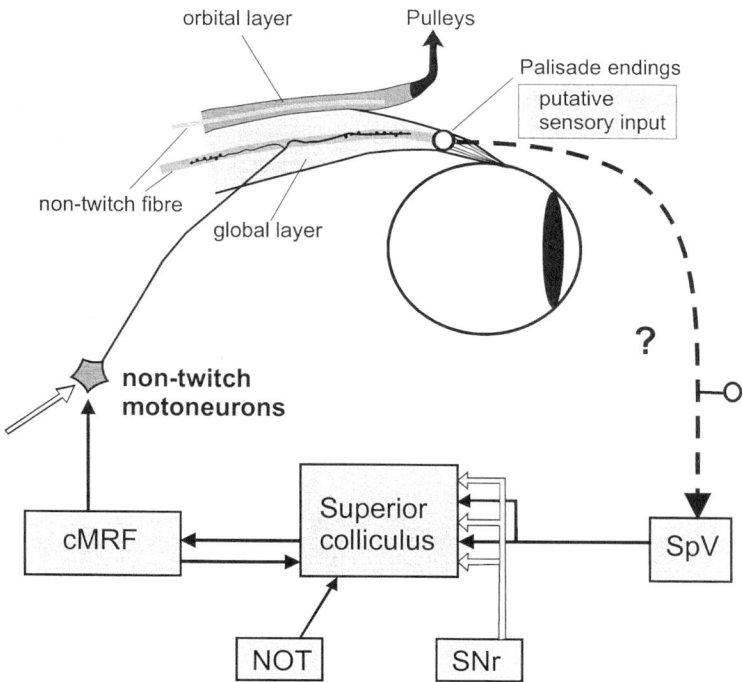

FIGURE 2. A sensory feedback hypothesis: the non-twitch muscle fibers of the global layer have motoneurons in the oculomotor nucleus, and palisade endings at the myotendinous junctions. The orbital muscle layer branches off and inserts on the pulleys, which leaves the palisade endings in an ideal position to monitor globe movements, and could provide a putative proprioceptive signal to the brain, perhaps to the spinal trigeminal nucleus (SpV). Premotor inputs to the non-twitch motoneurons come from several areas concerned with visual control of eye movements (*open arrow*): neural integrator, the marginal zone, and near response regions, also from the central mesencephalic reticular formation (cMRF), which is tightly connected to the superior colliculus (SC), reciprocally. If the SC acts as an interface between sensory and motor signals, it could relay spinal trigeminal signals concerning extraocular muscle tension back to the non-twitch motoneurons via the SC and cMRF. The hypothesis is not proven, but may stimulate new ideas on the sensory control of eye movements. NOT, nucleus of the optic tract; SNr, substantia nigra pars reticulata.

of eye movements, including eye muscle proprioception. Is it possible that the non-twitch muscle fibers are a type of giant fusiform fiber, with sensory receptors at the ends—an enormous "inverted muscle spindle," as Robinson suggested.[5] We will argue through this possibility, even in the knowledge that there is no proof for it. However, on the other hand, numerous facts fit well with the existence of sensory feedback pathways through SC and cMRF controlling the activity of non-twitch motoneurons, and surprising little evidence contradicts the hypothesis (FIG.2).

If the contraction of the non-twitch muscle fibers is related to the stabilization of the eye position with respect to sensory feedback signals, such as vision, fused images, or eye and neck proprioception, then the question arises as to which neural pathways are involved. The non-twitch motoneurons receive inputs from the neural integrator, premotor vergence areas, and regions related to smooth pursuit eye movements (see FIG. 2, open arrow), as well as the cMRF. The cMRF has already been suggested as a structure carrying sensory information to the oculomotor system.[28,29] The results of mechanical lesions on optokinetic nystagmus[28] and chemical lesions causing defects in fixation and saccade size[25,26] could be related to disrupting a sensory feedback pathway to the eye muscles.

The cMRF is powerfully interconnected with the SC,[27,30] and furthermore the fibers giving rise to this projection are branches of the tectoreticular neurons.[30,31] The exact layers of the SC giving rise to the tectoreticular pathways are targeted by two interdigitating groups of terminals[32]: one from the substantia nigra (pars reticulata) carrying a motor command (SNr, FIG. 2),[33] and the other from the spinal trigeminal nucleus (SpV, FIG. 2). If the non-twitch motoneuron premotor pathways structures interact in the SC, this interface could be influenced by descending afferents to the SC from higher centers and could modulate the alignment (positional error) signals at this level. For example, a structure such as nucleus of the optic tract (NOT) in the pretectum has been shown to influence the latent nystagmus in monkeys reared with visual deprivation[34,35]: in accordance with the proprioceptive hypothesis NOT, a recipient of visual and optokinetic inputs[36] projects massively to the rostral half of SC[37] where the trigeminal input is concentrated (FIG. 2).

The proposal that SC, at least the rostral part, carries a positional error network, providing a position signal rather than a motor signal, has been recently made by Krauzlis and colleagues.[38,39] It lends support to the proprioceptive hypotheses set out here, as do the conclusions of Grantyn et al.,[40] who studied the slow post-saccadic eye movements that can result from the excitation of tectoreticular neurons, and proposed that a component of the drifts motor signal circumvents the saccadic generator. The circumventing signal could be the pathway from SC through the cMRF and onto the non-twitch motoneurons.

What information does the SpV send to the SC? It is certainly of a sensory nature and may come from eye proprioceptors.[41,42] However, the pathway from specific functional sensory receptors of the eye muscle to the SpV has not been satisfactorily demonstrated, partly on account of the enormous variation between species.[2,3]

There seems to be no simple correlation between the possession of muscle spindles in the eye muscles and the oculomotor repertoire of any species. Golgi tendon organs are very rare, too; palisade endings, however, are abundant in man, monkey, and cat. So it seems at present that palisade endings are the most likely candidates for providing a sensory signal to the brainstem. The palisade endings lie only at the myotendinous junctions of the slow non-twitch fibers of the *global layer* (FIG.2), and the non-twitch muscle fibers of the orbital layer do not possess the palisade endings.[6,43] Demer and colleagues[44] have shown that the *orbital layers* of the eye muscles terminate on the pulleys or fibrous connective tissue of the orbit, which provide sleeves to direct the pulling direction of the eye muscles. This leaves the global layer to insert on the sclera and direct the position of the eye itself. The palisade endings lie in an opportune position to monitor this activity (FIG. 2). However, it is totally unclear whether the palisade endings project to the SpV.[43]

Clinical Implications

If the palisade endings at the myotendinous junction do provide a sensory signal used for fine alignment of the eye, then damage to the tip of the eye muscles, or an operation to reinsert the muscle on the sclera, could cause fatal damage to this system.[5] The result could be a residual strabismus or nystagmus. This hypothesis is supported by experiments on dogs with congenital nystagmus: the resection of all the eye muscles and subsequent reinsertion at exactly the original position led to a significant improvement of their nystagmus.[45] This rather surprising result could be explained on the basis of damaging a defective sensory signal from palisade endings in the dogs' extrinsic eye muscles.

In conclusion, we have considered the afferent and efferent pathways of the slow non-twitch motoneurons, which innervate the unique non-twitch muscle fibers of the eye attached to palisade endings. Because it is possible that the non-twitch muscle fibers are fusiform fibers and could function as a sensory receptor, we have attempted to integrate this proprioceptive hypothesis into the well-studied circuitry of the oculomotor system. The proposals, which suggest an integration of sensory feedback signals passing through the SC and via cMRF onto the non-twitch motoneurons, are new and still entirely hypothetical; it is hoped they will lead to discussion of new approaches in the analysis of experimental and clinical data.

ACKNOWLEDGMENT

This research was supported by the European Union (Grant BIO4-CT98-0546).

REFERENCES

1. KELLER, E.L. & D.A. ROBINSON. 1971. Absence of a stretch reflex in extraocular muscles of the monkey. J. Neurophysiol. **34:** 908–919.
2. RUSKELL, G.L. 1999. Extraocular muscle proprioceptors and proprioception. Prog. Retin. Eye Res. **18:** 269–291.
3. DONALDSON, I.M.L. 2000. The functions of the proprioceptors of the eye muscles. Philos. Trans. R. Soc. Lond. [Biol.] **355:** 1685–1754.
4. WEIR, C.R., P.C. KNOX & G.N. DUTTON. 2000. Does extraocular muscle proprioception influence oculomotor control? Br. J. Ophthalmol. **84:** 1071–1074.
5. STEINBACH, M.J. 2000. The palisade ending: an afferent source for eye position information in humans. *In* Advances in Strabismus Research: Basic and Clinical Aspects. G. Lennerstrand, J. Ygge & T. Laurent, Eds.: 33–42. Portland Press. London.
6. SPENCER, R.F. & J.D. PORTER. 1988. Structural organization of the extraocular muscles. Rev. Oculomotor Res. **2:** 33–79.
7. BÜTTNER-ENNEVER, J.A., A.K.E. HORN, H. SCHERBERGER & P. D'ASCANIO. 2001. Motoneurons of twitch and nontwitch extraocular muscle fibers in the abducens, trochlear, and oculomotor nuclei of monkeys. J. Comp. Neurol. **438:** 318–335.
8. UGOLINI, G., J.A. BÜTTNER-ENNEVER, M. DOLDAN, *et al.* 2001. Horizontal eye movement networks in primates: differences in monosynaptic input to slow and fast abducens motoneurons. Soc. Neurosci. Abstr. **27:** 403.13.
9. UGOLINI, G. 1995. Specificity of rabies virus as a transneuronal tracer of motor networks: transfer from hypoglossal motoneurons to connected second-order and higher-order central nervous system cell groups. J. Comp. Neurol. **356:** 457–480.

10. TANG, Y., O. RAMPIN, F. GIULIANO & G. UGOLINI. 1999. Spinal and brain circuits to motoneurons of the bulbospongiosus muscle: retrograde transneuronal tracing with rabies virus. J. Comp. Neurol. **414**: 167–192.
11. MAY, P.J., N.F. WRIGHT, R.C.S. LIN & J.T. ERICHSEN. 2000. Light and electron microscopic features of medial rectus C-subgroup motoneurons in macaques suggest near triad specializations (Abstr.). Invest. Ophthalmol. Visual Sci. **41:** 4353.
12. BÜTTNER-ENNEVER, J.A. 1992. Patterns of connectivity in the vestibular nuclei. Ann. N.Y. Acad. Sci. **656:** 363–378.
13. CLENDANIEL, R.A. & L.E. MAYS. 1994. Characteristics of antidromically identified oculomotor internuclear neurons during vergence and versional eye movements. J. Neurophysiol. **71:** 1111–1127.
14. DIERINGER, N. & W. PRECHT. 1986. Functional organization of eye velocity and eye position signals in abducens motoneurons of the frog. J. Comp. Physiol. **158:** 179–194.
15. NELSON, J.S., S.J. GOLDBERG & J.R. MCCLUNG. 1986. Motoneuron electrophysiological and muscle contractile properties of superior oblique motor units in cat. J. Neurophysiol. **55:** 715–726.
16. FUKUSHIMA, K., C.R. KANEKO & A.F. FUCHS. 1992. The neuronal substrate of integration in the oculomotor system. Prog. Neurobiol. **39:** 609–639.
17. MCFARLAND, J.L. & A.F. FUCHS. 1992. Discharge patterns in nucleus-prepositus-hypoglossi and adjacent medial vestibular nucleus during horizontal eye movement in behaving macaques. J. Neurophysiol. **68:** 319–332.
18. MAY, P.J., J.D. PORTER & P.D.R. GAMLIN. 1992. Interconnections between the primate cerebellum and midbrain near-response regions. J. Comp. Neurol. **315:** 98–116.
19. GAMLIN, P.D.R., Y.H. ZHANG, R.A. CLENDANIEL & L.E. MAYS. 1994. Behavior of identified Edinger-Westphal neurons during ocular accommodation. J. Neurophysiol. **72:** 2368–2382.
20. MAYS, L.E. & P.D.R. GAMLIN. 1995. Neuronal circuitry controlling the near response. Curr. Opin. Neurobiol. **5:** 763–768.
21. BÜTTNER-ENNEVER, J.A., P. GROB, K. AKERT & B. BIZZINI. 1981. A transsynaptic autoradiographic study of the pathways controlling the extraocular eye muscles, using B-IIb tetanus toxin fragment. Ann. N.Y. Acad. Sci. **374:** 157–170.
22. BÜTTNER-ENNEVER, J.A., P. GROB, K. AKERT & B. BIZZINI. 1981. Transsynaptic retrograde labeling in the oculomotor system of the monkey with tetanus toxin BIIb fragment. Neurosci. Lett. **26:** 233–238.
23. COHEN, B., D.M. WAITZMAN, J.A. BÜTTNER-ENNEVER & V. MATSUO. 1986. Horizontal saccades and the central mesencephalic reticular formation. Brain Res. **64:** 243–255.
24. WAITZMAN, D.M., V.L. SILAKOV & B. COHEN. 1996. Central mesencephalic reticular formation (cMRF) neurons discharging before and during eye movements. J. Neurophysiol. **75:** 1546–1572.
25. WAITZMAN, D.M., V.L. SILAKOV, S. DEPALMA-BOWLES & A.S. AYERS. 2000. Effects of reversible inactivation of the primate mesencephalic reticular formation. I. Hypermetric goal-directed saccades. J. Neurophysiol. **83:** 2260–2284.
26. WAITZMAN, D.M., V.L. SILAKOV, S. DEPALMA-BOWLES & A.S. AYERS. 2000. Effects of reversible inactivation of the primate mesencephalic reticular formation. II. Hypometric vertical saccades. J. Neurophysiol. **83:** 2285–2299.
27. CHEN, B. & P.J. MAY. 2000. The feedback circuit connecting the superior colliculus and central mesencephalic formation: a direct morphological demonstration. Exp. Brain Res. **131:** 10–21.
28. KOMATSUZAKI, A., J. ALPERT, H.E. HARRIS & B. COHEN. 1972. Effects of mesencephalic reticular formation lesions on optokinetic nystagmus. Exp. Neurol. **34:** 522–534.
29. MOSCHOVAKIS, A.K., C.A. SCUDDER & S.M. HIGHSTEIN. 1996. The microscopic anatomy and physiology of the mammalian saccadic system. Prog. Neurobiol. **50:** 133–133.
30. MOSCHOVAKIS, A.K., A.B. KARABELAS & S.M. HIGHSTEIN. 1988. Structure-function relationships in the primate superior colliculus. II. Morphological identity of presaccadic neurons. J. Neurophysiol. **60:** 263–301.
31. MOSCHOVAKIS, A.K., A.B. KARABELAS & S.M. HIGHSTEIN. 1988. Structure-function relationships in the primate superior colliculus. I. Morphological classification of efferent neurons. J. Neurophysiol. **60:** 232–262.

32. HARTING, J.K. & D.P. VAN LIESHOUT. 1991. Spatial relationships of axons arising from the substantia nigra, spinal trigeminal nucleus, and pedunculopontine tegmental nucleus within the intermediate gray of the cat superior colliculus. J. Comp. Neurol. **305:** 543–558.
33. HIKOSAKA, O., Y. TAKIKAWA & R. KAWAGOE. 2000. Role of the basal ganglia in the control of purposive saccadic eye movements. Physiol. Rev. **80:** 953–978.
34. MUSTARI, M.J., R.J. TUSA, A.F. BURROWS, et al. 2001. Gaze-stabilizing deficits and latent nystagmus in monkeys with early-onset visual deprivation: role of the pretectal NOT. J. Neurophysiol. **86:** 662–675.
35. TUSA, R.J., M.J. MUSTARI, A.F. BURROWS & A.F. FUCHS. 2001. Gaze-stabilizing deficits and latent nystagmus in monkeys with brief, early-onset visual deprivation: eye movement recordings. J. Neurophysiol. **86:** 651–661.
36. HOFFMANN, K.P., C. DISTLER & U. ILG. 1992. Callosal and superior temporal sulcus contributions to receptive field properties in the macaque monkey's nucleus of the optic tract and dorsal terminal nucleus of the accessory optic tract. J. Comp. Neurol. **321:** 150–162.
37. BÜTTNER-ENNEVER, J.A., B. COHEN, A.K.E. HORN & H. REISINE. 1996. Efferent pathways of the nucleus of the optic tract in monkey and their role in eye movements. J. Comp. Neurol. **373:** 90–107.
38. KRAUZLIS, R.J., M.A. BASSO & R.H. WURTZ. 2000. Discharge properties of neurons in the rostral superior colliculus of the monkey during smooth-pursuit eye movements. J. Neurophysiol. **84:** 876–891.
39. BASSO, M.A., R.J. KRAUZLIS & R.H. WURTZ. 2000. Activation and inactivation of rostral superior colliculus neurons during smooth-pursuit eye movements in monkeys. J. Neurophysiol. **84:** 892–908.
40. GRANTYN, A.A., Y. DALEZIOS, T. KITAMA & A.K. MOSCHOVAKIS. 1996. Neuronal mechanisms of two-dimensional orienting movements in the cat. 1. A quantitative study of saccades and slow drifts produced in response to the electrical stimulation of the superior colliculus. Brain Res. Bull. **41:** 65–82.
41. PORTER, J.D. 1986. Brainstem terminations of extraocular muscle primary afferent neurons in the monkey. J. Comp. Neurol. **247:** 133–143.
42. NELSON, J.S., M.A. MEREDITH & B.E. STEIN. 1989. Does an extraocular proprioceptive signal reach the superior colliculus? J. Neurophysiol. **62:** 1360–1374.
43. LUKAS, J.R., R. BLUMER, M. DENK, et al. 2000. Innervated myotendinous cylinders in human extraocular muscles. Invest. Ophthalmol.Visual Sci. **41:** 980–990.
44. DEMER, J.L., S. Y. OH & V. POUKENS. 2000. Evidence for active control of rectus extraocular muscle pulleys. Invest. Ophthalmol. Visual Sci. **41:** 1280–1290.
45. DELL'OSSO, L.F., R.W. HERTLE, R.W. WILLIAMS & J.B. JACOBS. 1999. A new surgery for congenital nystagmus: effects of tenotomy on an achiasmatic canine and the role of extraocular proprioception. J. Am. Acad. Pediatr. Ophthalmol. Strabismus **3:** 166–182.

Studies of the Role of the Paramedian Pontine Reticular Formation in the Control of Head-Restrained and Head-Unrestrained Gaze Shifts

DAVID L. SPARKS,[a] ELLEN J. BARTON,[a] NEERAJ J. GANDHI,[a] AND JON NELSON[b]

[a]*Division of Neuroscience, Baylor College of Medicine, Houston, Texas 77030, USA*

[b]*Department of Physical Therapy, The College of St. Scholastica, Duluth, Minnesota 55811, USA*

ABSTRACT: Results of three experiments related to the role of the paramedian pontine reticular formation (PPRF) in the control of gaze are described. (1) Chronic unit recording methods, used to study the on-directions of short-lead burst neurons in head-restrained monkeys, and (2) reversible inactivation techniques confirmed the traditional view of the importance of PPRF in the control of horizontal eye movements. Reversible inactivation of neurons in the vicinity of identified short-lead burst neurons produced dramatic reductions in the speed of saccades to horizontal target displacements. The reductions in velocity were largely compensated for by an increase in saccade duration. Only minor, if any, effects were observed upon the velocity, duration, and amplitude of saccades to upward target displacements. (3) Microstimulation was applied to omnipause neurons to gate activity of excitatory burst neurons that discharge during coordinated eye-head movements. The microstimulation failed to noticeably slow (prevent) head movements when stimulation was applied during (prior to onset of) gaze shifts, suggesting that signals relayed to motoneurons innervating the neck muscles are not inhibited by the omnipause neurons. In other words, the desired gaze signal is parsed into eye and head pathways upstream of the excitatory burst neurons.

KEYWORDS: saccades; PPRF; reversible inactivation; gaze; OPNs; microstimulation

INTRODUCTION

Most models of the neural control of saccadic eye movements assume that neurons in the paramedian zone of the pontine reticular formation (PPRF) generate motor command signals that effect changes in the horizontal positions of the eyes during saccades and the quick phases of nystagmus. This assumption is based upon accumulated evidence obtained from microstimulation, lesions, clinical, anatomical, and chronic single-unit recording data (see Refs. 1–4 for reviews).

Address for correspondence: David L. Sparks, Ph.D., Division of Neuroscience, Baylor College of Medicine, One Baylor Plaza, Houston, TX 77030.

sparks@cns.bcm.tmc.edu

Recently, the traditional view of PPRF function has been challenged in two ways. First, one model[5] assumes that saccade commands are not decomposed into horizontal and vertical commands until the level of the motoneurons. The concept was motivated by data indicating that individual neurons in PPRF discharge before vertical saccades[6] and by findings suggesting that the "on-direction" of PPRF neurons may be distributed continuously, rather than clustered around horizontal (or vertical) directions.[7] Second, recent papers have suggested that neurons in the PPRF send both eye-velocity and head-velocity signals to extraocular muscle motoneurons. This suggestion emerged from studies of the activity of inhibitory burst neurons (IBNs) during eye, head, and gaze movements generated by monkeys without head restraint. Cullen and Guitton[8] found that for most of the IBNs studied the number of spikes in a burst was better correlated with gaze amplitude than with the amplitude of either the eye or head components of the gaze shift. Below, we report data relevant to the issues raised by these challenges to the traditional view of PPRF function.

EXTRACTION OF HORIZONTAL AND VERTICAL COMMAND SIGNALS

Best Direction of Short-Lead Burst Neurons in PPRF

It is surprising that issues about the on-direction of cells in PPRF are still unresolved 30 years after chronic unit recording methods allowed the first descriptions of pontine cells with saccade-related and eye-position–related activity.[9–12] However, untrained subjects were used in early experiments, and the range of saccade directions and amplitude generated while the activity of a cell was being studied was not under experimental control. Also, the stability, accuracy, and precision of the methods used to record eye position in these early studies were less than ideal.

Descriptions of the on-direction of burst neurons in PPRF found in more recent papers are usually secondary to the major purpose of the paper. For example, the evidence that the on-direction short-lead burst neurons (SLBNs) can have a large vertical component was obtained, primarily, from intra-axonal recordings performed in experiments designed to trace anatomical pathways following the intracellular injection of tracer substances.[7,13] The number of saccade directions and amplitudes was, of necessity, small. Moreover, these experiments were also performed in untrained animals, and the distribution of saccade directions and amplitudes used for estimates of on-direction was limited. Below, we present data obtained from 18 SLBNs under experimental conditions that allowed greater control over saccade direction and amplitude.

The activity of SLBNs in the PPRF was recorded while rhesus monkeys, with their heads restrained, generated saccades to small (0.1°) visual targets for a liquid reward. The targets consisted of an array of LEDs located on a tangent screen. The array spanned 48° horizontally and 40° vertically, with a 2° or 4° spacing. Eye positions were recorded using the search-coil method.[14] Horizontal and vertical eye positions were sampled every 2 ms, and the digitized interspike intervals were recorded with 100 μs resolution. Quantitative analysis was done off-line. Eye velocity was computed from the horizontal and vertical eye position data using a 5-point central difference differential algorithm. Saccade onset and offset were defined by a velocity criterion, usually 30°/s. Burst onset and offset, defined as the first spike and the last

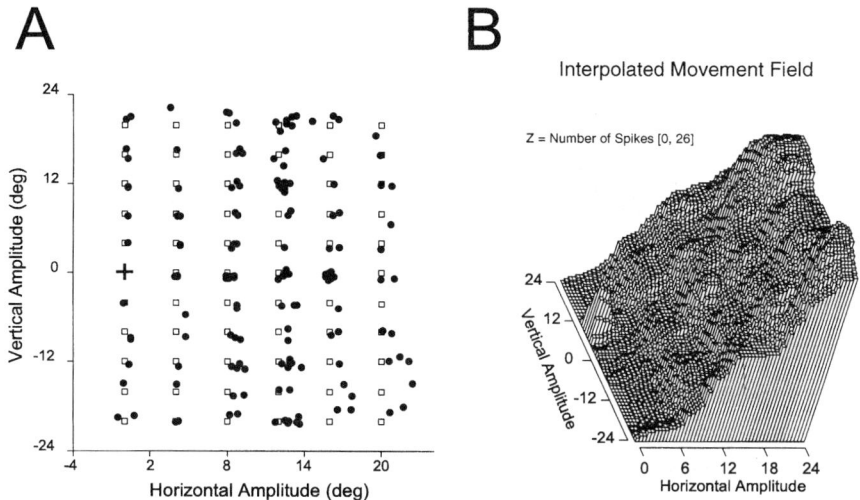

FIGURE 1. Direction and amplitude of saccades used to construct a three-dimensional movement field of an SLBN (# spikes as a function of the horizontal and vertical amplitude). (**A**) Locations of visual targets (*open squares*) with respect to the initial fixation target (+) and the end points (*filled circles*) of 160 primary saccades made to acquire these targets. The saccade-related discharge was recorded concurrently with the behavioral data shown. (**B**) The number of spikes (z axis, ranges from 0 to 26) is plotted as a function of horizontal and vertical amplitude for the same trials shown in **A**. The plot was obtained by interpolating from the 160 discrete values.

spike, respectively, in a saccade-related burst, were determined by the investigator. Linear and curvilinear regression analyses were used to examine the relationship between SLBN burst and saccade metrics. The best direction of a given SLBN was determined using the method described by Henn and Cohen.[15] For each saccade in a data set, a component amplitude (DPOS) along a specified projection plane was calculated as

DPOS = saccade vector amplitude*cos (saccade direction angle−projection plane)

for projection planes ranging from 0° to 360° in 1° increments. The plane showing the highest correlation between number of spikes and component amplitude (DPOS) along the given projection plane was defined as the best direction.

FIGURE 1A illustrates the range of saccade amplitudes and directions generated during behavioral trials while we were recording the activity of one SLBN. Panel B is an interpolated movement field plot of the number of spikes in the saccade-related burst of activity as a function of the horizontal and vertical amplitude of the saccade. Visual inspection of the movement field plot indicates that much of the variance in the number of spikes in the saccade-related burst can be attributed to the amplitude of the horizontal component of the saccade. The Henn and Cohen analysis indicated that DPOS was 0 for this neuron.

Results of the Henn and Cohen analysis for a different SLBN are illustrated in FIGURE 2. Panel A plots the number of spikes in the saccade-related burst as a func-

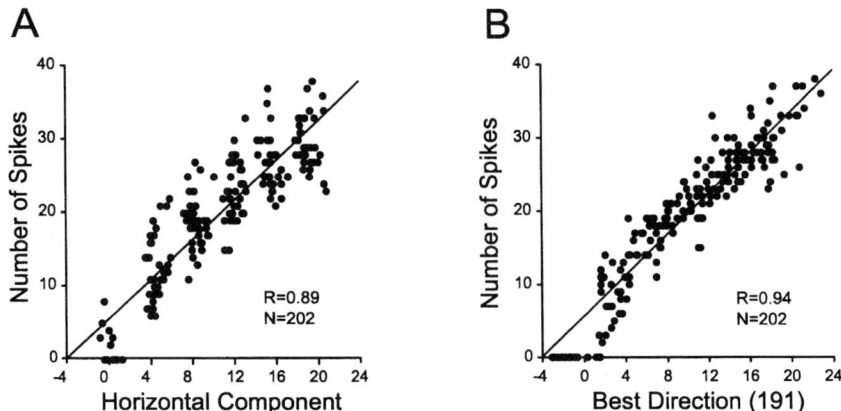

FIGURE 2. The number of spikes in the discharge of a SLBN as a function of the horizontal component versus the best direction component. (**A**) The number of spikes plotted as a function of the amplitude of the horizontal component (correlation coefficient = 0.89). (**B**) The number of spikes as a function of the amplitude of the best direction component (correlation coefficient = 0.94). The best direction = 191° as determined by the technique of Henn and Cohen (see METHODS). The trials in **A** and **B** are identical ($n = 202$).

TABLE 1. "Best direction" analysis for SLBNs in PPRF

Best Direction	R – Best Direction	R – Horizontal or Vertical	Direction	N
0	.91	.91	Right	120
0	.94	.94	Right	244
1	.97	.97	Right	147
2	.76	.75	Right	205
3	.96	.95	Right	293
5	.97	.96	Right	241
355	.88	.86	Right	128
353	.80	.77	Right	242
8	.97	.94	Right	128
11	.96	.90	Right	609
50	.76	.55	Right	304
181	.89	.88	Left	158
185	.96	.94	Left	174
191	.92	.85	Left	187
193	.85	.73	Left	228
276	.93	.89	Down	113
83	.95	.91	Up	325
101	.88	.83	Up	296

By column: (**1**) Best direction determined using the method of Henn and Cohen. (**2**) Correlation coefficient for the relationship between number of spikes and component amplitude along a projection plane of the best direction of the cell. (**3**) Correlation coefficient for the relationship between number of spikes and the amplitude of the movement along cardinal (horizontal or vertical) directions. (**4** and **5**) Optimal saccade direction (up, down, left or right) and the number of trials used in computing the correlation coefficients.

tion of the horizontal component of the movement. Panel B plots the same data as a function of the component amplitude along a projection plane of 191°, the best direction for this cell. The scatter of points about the line of best fit is reduced, and the correlation coefficient increases from 0.89 to 0.94. TABLE 1 presents results of the "best direction" analysis for 18 PPRF SLBNs, including three that discharged most vigorously in association with vertical saccades. The best direction of 9 of the 18 cells was within 5° of either the horizontal or vertical axes; the best direction of 13 of the 18 cells was within 10°; and the best direction of 17 of the 18 cells was within 20° of horizontal or vertical axes. The best direction of one cell was 50°.

These results are in good agreement with two descriptions of the best direction of IBNs. Scudder and colleagues[16] noted that although the best directions of IBNs were seldom purely horizontal, most clustered near the horizontal axes. The extent of the spread, measured as root-mean-square deviation from horizontal, was only 14.2° for SLBNs and 9.9° for long-lead burst neurons. Cullen and Guitton[17] reported that the preferred directions in their sample of IBNs varied from 29° upward to 30° downward. The mean preferred direction was −1°, closely aligned with the horizontal plane. Based upon these results and the data presented in this paper, models in which information about horizontal components of saccades is not extracted until the level of extraocular motoneurons is not necessary. Moreover, the assumptions of the Quaia and Optican model[5] that the on-directions are evenly distributed and equally represented are not supported by available data. It is true, however, that the on-directions of the pontine neurons having monosynaptic connections with motoneurons are unknown.

Reversible Inactivation of Neurons in Rostral PPRF

Most models of the saccadic system assume that neurons in PPRF extract a signal of horizontal eye displacement and that a similar signal specifying the vertical component of saccades resides in the rostral intrastitial nucleus of the medial longitudinal fasciculus (rIMLF) and other nuclei in rostral midbrain. These models predict that reversible inactivation of cells in PPRF should affect primarily horizontal movements and that alterations in the accuracy or speed of saccades to vertical targets would be small.

Reversible lesions were produced using the recording/injection probe developed by Malpeli and Schiller[18] and was accomplished by pressure injection of 50–200 nl of 2% lidocaine. In these initial experiments, we used lidocaine because, based upon results of our experiments in the SC, the size of the affected area is smaller with lidocaine than with muscimol injections. Also, lidocaine injections produced behavioral effects that lasted for 20–40 min, whereas the effects of muscimol could persist for several hours. Thus, with lidocaine injections it was possible to make multiple injections during an experimental session. Also, the extent to which performance deficits could be attributed to behavioral strategies rather than disruption of neural function could be assessed by observing the pattern and time course of behavioral recovery.

FIGURE 3 illustrates the effect of consecutive inactivations on saccades to targets presented either 20° to the left or 20° above the fixation target, with target location randomly interleaved. The lidocaine was injected twice into the rostral PPRF at the site of SLBN activity. Separate panels plot the peak velocity, duration, and amplitude

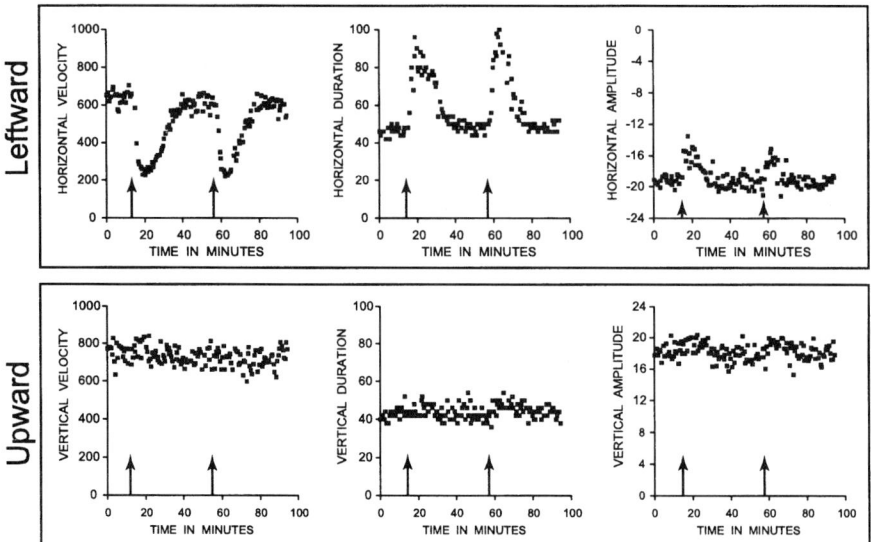

FIGURE 3. Upward vs. leftward movements: the effect of consecutive injections of lidocaine on saccades to horizontal and vertical target displacements. **Top panel:** peak velocity, duration, and amplitude of the horizontal component of saccades (target = −20, 0) as a function of time. **Bottom panel:** peak velocity, duration, and amplitude of the vertical component of saccades (target = 0,20) as a function of time. Each point represents a measurement from a single trial and the arrows indicate time of injections. The measurements shown in the two panels are synchronous because the targets were randomly interleaved. Amount of lidocaine: injection #1, 150 nl; #2, 200 nl.

of movements to horizontal (top) and vertical (bottom) target displacements measured during a 90-min period beginning about 15 min before the first of two injections (upward arrows). Before the injections, peak velocity of leftward movements was about 600°/s, but dropped to about 250°/s after the injection and gradually recovered over the next 20–30 min. With a similar time course, the duration of the saccades increased from control values of about 45 ms to as much as 95 ms, and then returned to control. The lengthened duration counterbalanced the reduced velocity, but not entirely, as saccades remained hypometric (see figure legend for more details). Referring to the bottom panels, note that the inactivation of neural activity in rostral PPRF had only minor, if any, effects upon the velocity, duration, and amplitude of saccades to upward target displacements.

In summary, all eight inactivations caused an increase in the duration and a decrease in the peak velocity of horizontal saccades in the ipsilesional direction, while the results for upward saccades were inconsistent and relatively small. For saccades of both directions, the amount of lidocaine and the location of the injection within the PPRF influenced the magnitude of the effect. Across the eight injections, the average maximal reduction in horizontal peak velocity was 389.4°0/s for saccades to targets that were displaced 20° horizontally. For the same saccades, the average maximal increase in duration was 32.6 ms. The prolonged duration compensated for the diminished velocity, but usually not completely, with the saccades still hypometric for six of eight injections. For saccades to targets that were displaced 20° vertically, no significant influences occurred on the vertical peak velocity of six of eight injections. For the remaining two injections, the significant effects were minor and in opposite directions (maximal change: 32.9°/s and −54.8°/s). Six of the eight injections produced a relatively small but significant increase in vertical duration (average maximal change 5.1 ms), while the other two caused no effect and a decrease in duration (3.4 ms). Because the duration of the vertical component increased slightly without an associated decrease in peak velocity for five of eight injections, the amplitude of the vertical component increased for these five injections (average maximal change 1.7°), while the other three injections induced no effect ($n = 2$) and a decrease (−1.7°) in amplitude.

It is not possible to attribute the behavioral consequences of the lidocaine injections to disruption of the activity of a particular functional class of PPRF neurons. Lidocaine injections into the rostral PPRF are presumed to affect the activity of all neural elements in the area of the injection. This includes (a) the SLBNs described in the previous section—cells generating a burst of activity primarily associated with the horizontal component of saccades and the quick phases of nystagmus; (b) long-lead burst neurons that may generate less discrete saccade-related bursts and have peak activity associated with movements that are not purely horizontal; (c) a smaller number of cells discharging maximally before vertical movements; and (d) fibers passing through the region of the injection. Despite the many types of neurons affected by the injections, the reversible lesions in rostral PPRF produce large changes in the velocity and duration of saccades to horizontal targets without having major effects upon the velocity, duration, or amplitude of saccades to randomly interleaved vertical targets. The deficits we observed were similar to those described by other researchers following permanent and more extensive lesions in the same regions of the brainstem.[4,19,20] Although not studied in detail, we did note that injections at more caudal sites disrupted both horizontal and vertical rapid eye movements. Caudal injections involving the vestibular nuclei and nucleus prepositus hypoglossi produced gaze-holding deficits and gaze-evoked nystagmus similar to what has been previously reported.[21-23]

Collectively, the recording and reversible inactivation data presented in this paper add to the large body of evidence indicating that cells in the PPRF specialize in the generation of commands for horizontal eye movements. Because we did not measure torsional rotations of the eye, our data do not address the suggestion[24] that the PPRF contains a three-dimensional arrangement of signals similar to what has been observed in the rostral midbrain.

STIMULATION OF PONTINE OMNIPAUSE NEURONS DURING AND BEFORE COORDINATED MOVEMENTS OF THE EYES AND HEAD

The PPRF contains neurons involved in the control of movements of the eyes and the head. Anatomical studies have shown that projections to both neck and eye motoneuron pools arise from both the oral and caudal divisions of the PPRF.[25] Signals carried by reticulospinal neurons are related to both neck muscle activity and eye position. Single pontine neurons send axons to the spinal cord and to the motor nuclei of extraocular muscles, allowing the same signals to be sent to neck and eye muscles (see Grantyn and Berthoz[26]). Microstimulation at the site of putative excitatory burst neurons (EBNs), which are a subset of the SLBNs embedded within the complicated pontine network of signals for gaze control, produce slow, constant velocity changes in gaze accomplished by coordinated movements of the eyes and head. The ratio of eye/head contribution to the slow change in gaze direction depends upon the initial positions of the eyes in the orbits.[27]

According to traditional views of PPRF function, the EBNs issue a saccadic eye movement command that is delivered, monosynaptically, to the extraocular motoneurons (eye pathway, FIGURE 4A). More specifically, the number of spikes in the saccade-related burst generated by the EBNs is tightly correlated with saccade amplitude. During coordinated eye-head movements, however, the spike count of EBNs is better correlated with gaze amplitude than with the eye component.[28] This observation also holds for the putative inhibitory burst neurons,[8,29] which are thought to reflect the activity of EBNs.

Omnipause neurons (OPNs) located within *nucleus raphe interpositus,* a discrete collection of cells found along the midline of the PPRF, discharge tonically during fixation and are silent during saccades. They monosynaptically inhibit the EBNs[30] and, thus, gate the activity of cells generating a command for an eye movement. Selectively activating OPNs using microstimulation interrupt ongoing saccades or delays their onset.[12,31,32] Paré and Guitton[33] stimulated the OPN region in cats during ongoing eye-head gaze shifts. They state that stimulation interrupted transiently both eye and head movements and that head deceleration was seen less frequently than gaze deceleration. It is difficult to assess the validity of these statements because data from only six trials are presented.

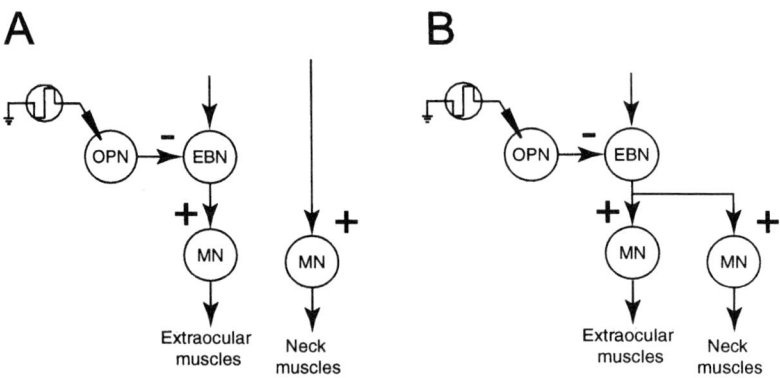

FIGURE 4. Rationale of the OPN stimulation experiment. Based on head-restrained experiments, the OPNs gate activity of EBNs that generate a command for an eye movement (pathway leading to extraocular muscles). For coordinated eye-head movements, if the command conveyed to neck motoneurons arises from structures above the level of EBNs (**A**), stimulation of the OPNs during a gaze shift will perturb only the eye component; the head movement should remain unattenuated. If the command relayed to the neck motoneurons is parsed downstream of the EBNs (**B**), stimulation of the OPNs is predicted to attenuate both eye and head trajectories.

To distinguish between the traditional and recently suggested roles of the EBNs in control of gaze, we exploited the antagonistic relationship of EBN and OPNs. If EBNs encode eye movements only, then the command signals to motoneurons innervating neck muscles must arise from structures upstream of the EBNs (FIG. 4A). Hence, stimulation of the OPNs during a gaze shift will perturb only the eye component; the head movement should remain unattenuated. If, on the other hand, the command relayed to the neck motoneurons is parsed downstream of the EBNs (FIG. 4B), stimulation of the OPNs is predicted to attenuate both eye and head movements.

FIGURE 5A illustrates the typical effects of OPN stimulation triggered on gaze onset. After a short latency, OPN microstimulation halted the change in gaze, typically until the end of stimulation, but had little, if any, effect on the head movement. Thus, the effect was mediated via a perturbation in the eye trajectory: the ocular component stopped in mid-flight and the eyes immediately began to counterrotate to stabilize gaze (gain of vestibulo-ocular reflex was 1). Stimulation prior to the onset of the gaze shift (FIG. 5B) delayed the onset of gaze, but the head movement was typically initiated during the microstimulation and, therefore, led gaze onset. The direction of gaze was stable during the stimulation because the eyes counterrotated in the orbits, thereby compensating for the ongoing head movement. At the end of the microstimulation train, the gaze shift was initiated by a saccadic eye movement in the same direction as the ongoing head movement.

To identify potential perturbations in head movements, head velocity traces were examined after aligning them on onset of head movement. FIGURE 6 superimposes head velocities of control (thin lines) and stimulation (thick lines) trials; both individual trials (left panels), and their mean ± 2 standard errors (right panels) are plotted. For trials when stimulation was triggered after gaze onset (top panels), the traces from stimulation trials are plotted until the onset of the resumed movement. For trials when stimulation was applied prior to gaze onset (bottom panels), the traces from stimulation trials are plotted until the approximate onset of the gaze shift. If EBNs transmit movement commands to the neck motoneurons as well as the extraocular motoneurons, the head velocity traces should deviate from the control traces and should reach zero velocity if the head movement were completely arrested. When stimulation is applied prior to gaze onset, the head velocity should remain at zero if the EBNs are the sole source of neck muscle motoneuron activation associated with the command for a gaze shift. The head velocity traces did not differ significantly during control and stimulation conditions regardless of when the OPNs were stimulated. These results provide strong support for the hypothesis that the EBNs participate in the eye pathway and that commands to motoneurons innervating the neck muscles originate from structures upstream of the EBNs.

Why then does the relationship between measures of the neural discharge of EBNs and saccade parameters deteriorate during combined eye-head movements? One possibility that has been suggested[34] is illustrated in FIGURE 7. Assume the EBNs generate commands for eye movements and that commands for moving the head originate in an independent, parallel pathway, as illustrated in Panel A. Movements of the head can influence the innervation signals reaching extraocular muscles via the vestibulo-ocular reflex (VOR) which is assumed to have a gain of 1 during the hypothetical eye and head movements illustrated. Panels B and C plot eye position and velocity, respectively, as a function of time for four movements. A signal to

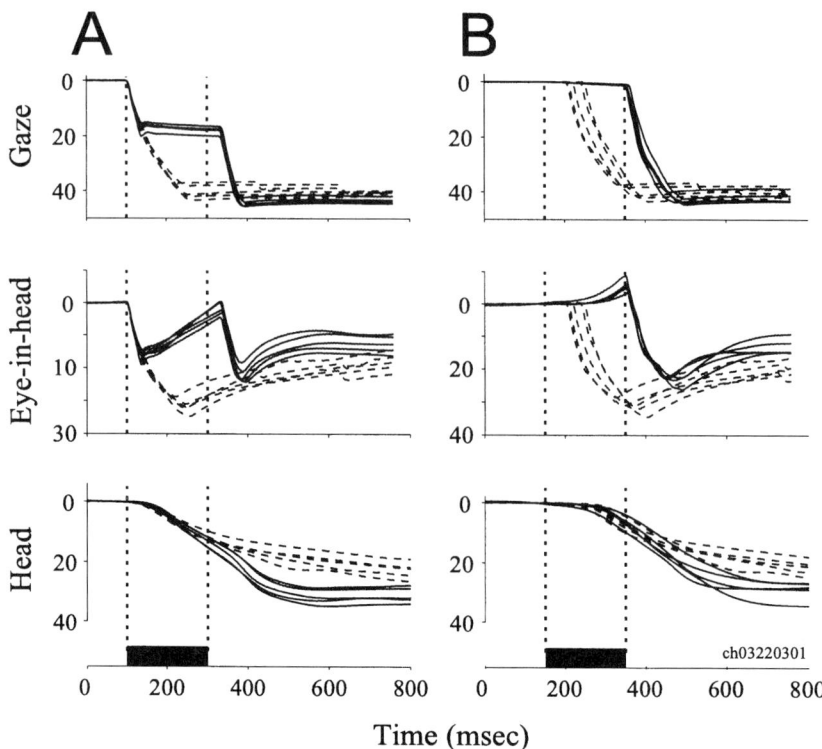

FIGURE 5. Stimulation of an OPN region during head-unrestrained gaze shifts attenuates only the eye component. Plots of control (*dashed lines*) and stimulation (*solid lines*) traces of horizontal gaze (*top*), eye-in-head (*middle*), and head (*bottom*) movements to a briefly flashed target. Stimulation was triggered either during (**A**) or before (**B**) gaze onset. Data was aligned on gaze onset in **A** and on cue to initiate movement in **B**. Stimulation onset always occurred at 150 ms for the trials displayed in **B**. Stimulation duration: 200 ms.

move the head did not occur during two of the movements (solid lines; head-restrained condition) when commands to move the eyes 20° and 10° to the right were issued. Commands to move the head 30° to the right and the eyes 20° to the right occurred on the other two trials (dashed lines; head-unrestrained condition).

In the absence of a head movement, measures of executed eye movements can serve as *estimates of the saccadic commands* transmitted to the extraocular motoneurons from the EBNs. This is possible because the motoneurons received no additional inputs from other oculomotor subsystems (e.g., vergence, pursuit, vestibular) during the time when the saccade command is being implemented. Thus, the amplitude (peak velocity) of the movement is tightly correlated with the number of spikes (peak discharge rate) in the saccade-related burst, allowing us to examine how information about the parameters of saccadic movements may be represented in the activity of EBNs.

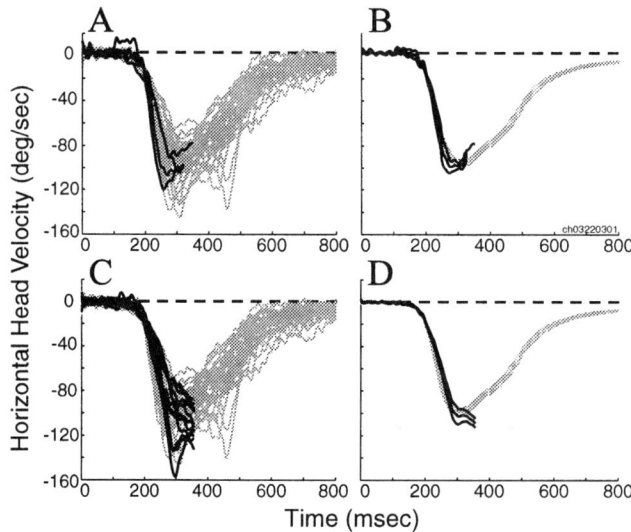

FIGURE 6. Lack of effects of stimulation of an OPN region on head movements. Plots display horizontal head velocity as a function of time for control (*thin, gray lines*) and stimulation (*thick, black lines*) trials. All traces, aligned on head onset, are from same stimulation site as that shown in FIGURE 4. Individual control and stimulation trials (**A,C**) as well as their means and ± 2 standard errors from the mean (**B,D**) are illustrated. (**A,B**) Stimulation was triggered on gaze onset. Stimulation trials are shown until the onset of the resumed movement. (**C,D**) Stimulation was triggered prior to gaze onset. One hundred milliseconds before head onset and 280 ms of the head movement are plotted for the stimulation trials; this corresponds roughly to the time period prior to the onset of the gaze shift. Note that the control traces show head movements when a gaze shift was executed.

When the motoneurons are receiving inputs from the other oculomotor subsystems while a saccade is being generated, measurements of the amplitude or velocity of the executed saccade will be *unreliable and misleading estimates* of the command issued by EBNs. Correlations of EBN activity with the amplitude, velocity, and duration of the executed eye movement cannot be used to make inferences about the motor signals being generated by EBNs because of the dissociation between the command that is issued and the movement that is executed. This dissociation is illustrated by the two simulated movements shown with dashed lines (FIG. 7, B and C). The EBNs generated identical commands for a 20° movement on both trials. Similarly, the command for a 30° head movement was the same except that it was issued 60 ms earlier for one of the movements. Because the head was moving and the VOR was active with a gain of 1, the executed eye movements had amplitudes of 18° and 15° and peak velocities of 350°/s and 275°/s instead of the 20° amplitude and 375°/s peak velocity that would have occurred if the head had not been moving. Because of the dissociation between the issued command and the executed movement, relationships between movement parameters and the activity of any neuron issuing an oculomotor command above the level of the motoneurons are not informative in the absence of accurate information about other influences on

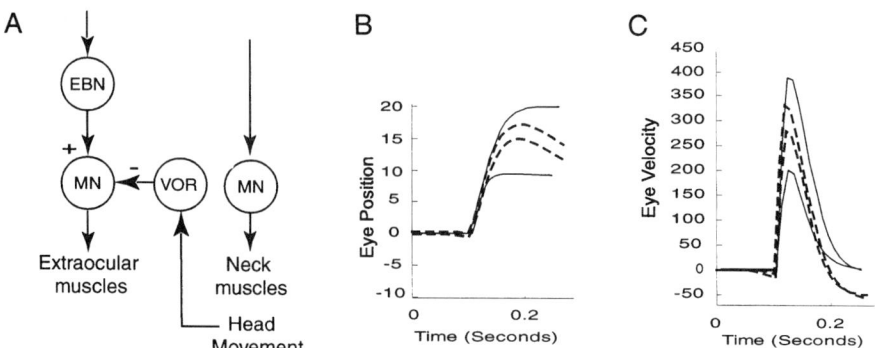

FIGURE 7. (**A**) Excitatory burst neurons (EBNs) generate commands for eye movements that are transmitted monosynaptically to extraocular muscle motoneurons (MN). Signals to neck muscle motoneurons are assumed to originate in an independent parallel pathway. Movements of the head can influence the innervation of extraocular muscles via the vestibulo-ocular reflex (VOR) which is assumed to have gain of 1. (**B**) Plots of eye position as a function of time for four hypothetical eye movements. *Solid lines*: movements occurring when commands to move the eye 20° to the right or 10° to the right were generated by EBNs and when a command to move the head was not issued. *Dashed lines*: movements occurring when a command to move the eye 20° to the right was issued in conjunction with a command to move the head 30° to the right. The difference in the amplitude of the two eye movements occurs because the command to move the head occurred 60 ms earlier in one case. (**C**) Plots of eye velocity as a function of time for the same four movements.

motoneuron activity. Thus, the data obtained by correlating the activity of neurons above the level of motoneurons with various movement parameters has limited usefulness.

ACKNOWLEDGMENT

This project was supported by NIH grants EY-01189 (D.L.S.), EY-02520 and EY-07009 (N.J.G.).

REFERENCES

1. RAPHAN, T. & B. COHEN. 1978. Brainstem mechanisms for rapid and slow eye movements. Annu. Rev. Physiol. **40:** 527–552.
2. FUCHS, A.F., C.R.S. KANEKO & C.A. SCUDDER. 1985. Brainstem control of saccadic eye movements. Annu. Rev. Neurosci. **8:** 307–337.
3. MOSCHOVAKIS, A.K. & S.M. HIGHSTEIN. 1994. The anatomy and physiology of primate neurons that control rapid eye movements. Annu. Rev. Neurosci. **17:** 465–488.
4. MOSCHOVAKIS, A.K., C.A. SCUDDER & S.M. HIGHSTEIN. 1996. The microscopic anatomy and physiology of the mammalian saccadic system. Prog. Neurobiol. **50:** 133–254.
5. QUAIA, C. & L.M. OPTICAN. 1997. Model with distributed vectorial premotor bursters accounts for the component stretching of oblique saccades. J. Neurophysiol. **78:** 1120–1134.

6. VAN GISBERGEN, J.A.M., D.A. ROBINSON & S. GIELEN. 1981. A quantitative analysis of generation of saccadic eye movements by burst neurons. J. Neurophysiol. **45:** 417–442.
7. STRASSMAN, A., S.M. HIGHSTEIN & R.A. MCCREA. 1986. Anatomy and physiology of saccadic burst neurons in the alert squirrel monkey. II. Inhibitory burst neurons. J. Comp. Neurol. **249:** 358–380.
8. CULLEN, K.E. & D. GUITTON. 1997. Analysis of primate IBN spike trains using system identification techniques. II. Gaze versus eye movement based models during combined eye-head gaze shifts. J. Neurophysiol. **78:** 3283–3306.
9. SPARKS, D.L. & R.P. TRAVIS. 1971. Firing patterns of reticular neurons during horizontal eye movements. Brain. Res. **33:** 477–481.
10. COHEN, B . & V. HENN. 1972. Unit activity in the pontine reticular formation associated with eye movements. Brain Res. **46:** 403–410.
11. LUSCHEI, E.S. & A.F. FUCHS. 1972. Activity of brain stem neurons during eye movements of alert monkeys. J. Neurophysiol. **35:** 445–461.
12. KELLER, E.L. 1974. Participation of medial pontine reticular formation in eye movement generation in monkey. J. Neurophysiol. **37:** 316–332.
13. STRASSMAN, A., S.M. HIGHSTEIN & R.A. MCCREA. 1986. Anatomy and physiology of saccadic burst neurons in the alert squirrel monkey. I. Excitatory burst neurons. J. Comp. Neurol. **249:** 337–357.
14. ROBINSON, D.A. 1963. A method of measuring eye movement using a scleral search coil in a magnetic field. IEEE Trans. Biomed. Eng. **10:** 137–145.
15. HENN, V. & B. COHEN. 1976. Coding of information about rapid eye movements in the pontine reticular formation of alert monkeys. Brain Res. **108:** 307–325.
16. SCUDDER, C.A., A.F. FUCHS & T.P. LANGER. 1988. Characteristics and functional identification of saccadic inhibitory burst neurons in the alert monkey. J. Neurophysiol. **59:** 1430–1454.
17. CULLEN, K.E. & D. GUITTON. 1997. Analysis of primate IBN spike trains using system identification techniques. I. Relationship to eye movement dynamics during head-fixed saccades. J. Neurophysiol. **78:** 3259–3282.
18. MALPELI, J. & P. SCHILLER. 1979. A method of reversible inactivation of small regions of brain tissue. J. Neurosci. Methods **1:** 143–151.
19. HENN, V., W. LANG, K. HEPP & H. REISINE. 1984. Experimental gaze palsies in monkeys and their relation to human pathology. Brain **107:** 619–636.
20. HEPP, K., V. HENN, T. VILIS & B. COHEN. 1989. Brainstem regions related to saccade generation. *In* The Neurobiology of Oculomotor Research. R.H. Wurtz & M.E. Goldberg, Eds.: 105–212. Elsevier. Amsterdam.
21. UEMURA, T. & B. COHEN. 1973. Effects of vestibular nuclei lesions on vestibulo-ocular reflexes and posture in monkeys. Acta Otolaryngol. Suppl. **315:** 1–71.
22. CANNON, S.C. & D.A. ROBINSON. 1987. Loss of the neural integrator of the oculomotor system from brain stem lesions in monkey. J. Neurophysiol. **57:** 1383–1409.
23. CHERON, G., E. GODAUX, J.M. LAUNE & B. VANDERKELEN. 1986. Lesions in the cat prepositus complex: effects on the vestibulo-ocular reflex and saccades. J. Physiol. **372:** 75–94.
24. CRAWFORD, J.D. & T. VILIS. 1992. Symmetry of oculomotor burst neuron coordinates about Listing's plane. J. Neurophysiol. **68:** 432–448.
25. ROBINSON, F.R., J.O. PHILLIPS & A.F. FUCHS. 1994. Coordination of gaze shifts in primates: brainstem inputs to neck and extraocular motoneuron pools. J. Comp. Neurol. **346:** 43–62.
26. GRANTYN, A. & A. BERTHOZ. 1985. Burst activity of identified tecto-reticulo-spinal neurons in the alert cat. Exp. Brain Res. **57:** 417–421.
27. GANDHI, N.J. & D.L. SPARKS. 2000. Microstimulation of the pontine reticular formation in monkey: effects on coordinated eye-head movements. Soc. Neurosci. Abstr. **26.**
28. LING, L., A.F. FUCHS, J.O. PHILLIPS & E.G. FREEDMAN. 1999. Apparent dissociation between saccadic eye movements and the firing patterns of premotor neurons and motoneurons. J. Neurophysiol. **82:** 2808–2811.
29. CULLEN, K.E., D. GUITTON, C.G. REY & W. JIANG. 1993. Gaze-related activity of putative inhibitory burst neurons in the head-free cat. J. Neurophysiol. **70:** 2678–2683.

30. CURTHOYS, I.S., C.H. MARKHAM & N. FURUYA. 1984. Direct projection of pause neurons to nystagmus-related excitatory burst neurons in the cat pontine reticular formation. Exp. Neurol. **83:** 414–422.
31. KELLER, E.L. 1977. Control of saccadic eye movements by midline brain stem neurons. *In* Control of Gaze by Brain Stem Neurons. R. Baker & A. Berthoz, Eds.: 327–336. Elsevier/North-Holland. Amsterdam.
32. KING, W.M. & A.F. FUCHS. 1977. Neuronal activity in the mesencephalon related to vertical eye movements. *In* Control of Gaze by Brain Stem Neurons. R. Baker & A. Berthoz, Eds.: 319–326. Elsevier/North-Holland. Amsterdam.
33. PARÉ, M. & D. GUITTON. 1998. Brain stem omnipause neurons and the control of combined eye-head gaze saccades in the alert cat. J. Neurophysiol. **79:** 3060–3076.
34. SPARKS, D.L. 1999. Conceptual issues related to the role of the superior colliculus in the control of gaze. Curr. Opin. Neurobiol. **9:** 698–707.

The Contribution of Midbrain Circuits in the Control of Gaze

ULRICH BÜTTNER,[a] JEAN A. BÜTTNER-ENNEVER,[b] HOLGER RAMBOLD,[c] AND CHRISTOPH HELMCHEN[c]

[a]*Department of Neurology, Ludwig Maximilians University, D-81377 Munich, Germany*

[b]*Department of Neurology and Institute of Anatomy, Ludwig Maximilians University, D-81377 Munich, Germany*

[c]*Department of Neurology, Medical School of Lübeck, D-23538 Lübeck, Germany*

ABSTRACT: The midbrain contains several structures important for the generation of torsional and vertical eye movements including the rostral interstitial nucleus of the MLF (riMLF) and the interstitial nucleus of Cajal (iC). While the riMLF is the immediate premotor structure for the generation of torsional and vertical saccades, the iC is considered a major part of the neural integrator for torsional and vertical eye movements. Experiments in monkeys show that a unilateral inactivation of the riMLF with muscimol leads to spontaneous contralesional torsional nystagmus, whereas an iC inactivation causes ipsilesional torsional nystagmus. In addition, inactivation of either structure leads to a tonic ocular torsion to the contralesional side. While the deficits after a riMLF lesion are thought to result from an imbalance of the saccade generator, a vestibular imbalance probably causes the deficits after an iC lesion. Contralesional and ipsilesional torsional nystagmus is also found in patients with unilateral mesencephalic lesions. A detailed analysis of the lesions from MRI scans shows a preferential involvement of the riMLF for patients with contralesional torsional nystagmus, and a major involvement of iC in cases with ipsilesional torsional nystagmus. Thus, the direction of torsional nystagmus appears to be a valuable topodiagnostic sign for patients with midbrain lesions.

KEYWORDS: rostral interstitial nucleus of the MLF; interstitial nucleus of Cajal; midbrain; torsional eye movements; vertical eye movements

INTRODUCTION

The midbrain contains a number of structures, which are involved in the control of gaze.[1,2] Whereas pontine structures mainly control horizontal aspects of eye movements, the midbrain is preferentially involved in the control of vertical and torsional movements. In the midbrain a number of specific structures have been outlined including the rostral interstitial nucleus of the medial longitudinal fasciculus (riMLF), the interstitial nucleus of Cajal (iC), the nuclei of the posterior commissure (nPC), and the posterior commissure (PC). Other structures are the superior colliculus (sc) and the mesencephalic reticular formation (MRF), which lies adjacent to the

Address for correspondence: Prof. Dr. U. Büttner, Neurologische Klinik, Marchioninistr. 15, D-81377 Munich, Germany. Voice: 49 089-7095-2560; fax: 49 089-7095-5761.
ubuettner@brain.nefo.med.uni-muenchen.de

third nucleus and iC. These structures will be treated in a separate chapter (see Waitzman, this volume) and will not be considered here.

The riMLF is considered as the immediate premotor structure for the generation of vertical and torsional saccades. It is a winglike structure that lies dorsomedial to the anterior pole of the red nucleus (rn). A useful landmark is the posterior thalamosubthalamic artery, which runs dorsomedially to riMLF.[3] The thalamosubthalamic artery arises between the bifurcation of the basilar artery and the origin of the posterior communicating artery. Sometimes it is a single vessel supplying both riMLF.[4] Caudally riMLF is separated from iC by the tractus retroflexus. In humans, riMLF lies 1–2 mm from the midline and has a mediolateral diameter of 6 mm and a ventrodorsal diameter of 2–3 mm. riMLF receives major afferents from the paramedian pontine reticular formation (PPRF), particularly from the omnipause neurons, sc, nPC, fastigial nucleus (medial deep cerebellar nucleus), and the contralateral riMLF.[3]

riMLF projects directly onto motoneurons in the third and fourth nucleus. One excitatory upward burst neuron sends axon collaterals to both, superior rectus and inferior oblique subnuclei, that is, to motoneurons with the same pulling direction.[5] Correspondingly, one downward excitatory burst neuron innervates inferior rectus and superior oblique motoneurons on the same side.[6] riMLF also projects to iC on both sides, the vestibular nuclei and the nuclei prepositus hypoglossi (ppH). riMLF fibers cross to the contralateral riMLF and iC below the aqueduct and not through the PC (FIG. 1).[5,6]

Single-unit recordings in riMLF of the alert-behaving monkey reveal that saccade-related burst neurons have either an upward or downward on-direction.[7] With vestibular stimulation in the roll plane that leads to torsional saccades, it can be demonstrated that neurons always encode ipsitorsional saccades. Thus, during recordings in the right riMLF, neurons are active during torsional saccades with extorsion of the right eye and intorsion of the left eye.[8]

Consequently, a unilateral lesion that inactivates the burst neurons in the riMLF leads to a loss of all ipsitorsional saccades. In addition, there is also a tonic torsional deviation of both eyes to the contralateral side combined with torsional nystagmus, the fast phase beating also to the contralateral side.[9] This finding has been interpreted as a result of an imbalance of the saccade generators on both sides of the brainstem. A unilateral PPRF lesion leads to a similar deficit in the horizontal plane—a loss of ipsilateral saccades and a tonic deviation to the contralateral side combined with a horizontal nystagmus beating contralesionally.[10]

Experimentally, a unilateral riMLF lesion only leads to slowing of downward saccades, whereas saccade velocity in both vertical directions is slowed after bilateral lesions. The vertical and torsional VOR are normal after a unilateral lesion. In addition to the tonic contralesional eye deviation, a contralesional head tilt can be observed after a unilateral lesion.[9]

The iC is considered as the neural integrator for vertical and torsional eye movements.[11,12] It is a compact nucleus and lies dorsolateral to the third nucleus embedded in the fascicles of the MLF just outside the central gray.[3] In humans it has a diameter of 4 mm and is located 1–3 mm lateral to the midline. It receives afferents from the vestibular nuclei, the y-group, the nPC, riMLF, and the contralateral iC. It projects to the third and fourth nuclei, the vestibular nuclei, and the nucleus ppH. In contrast to

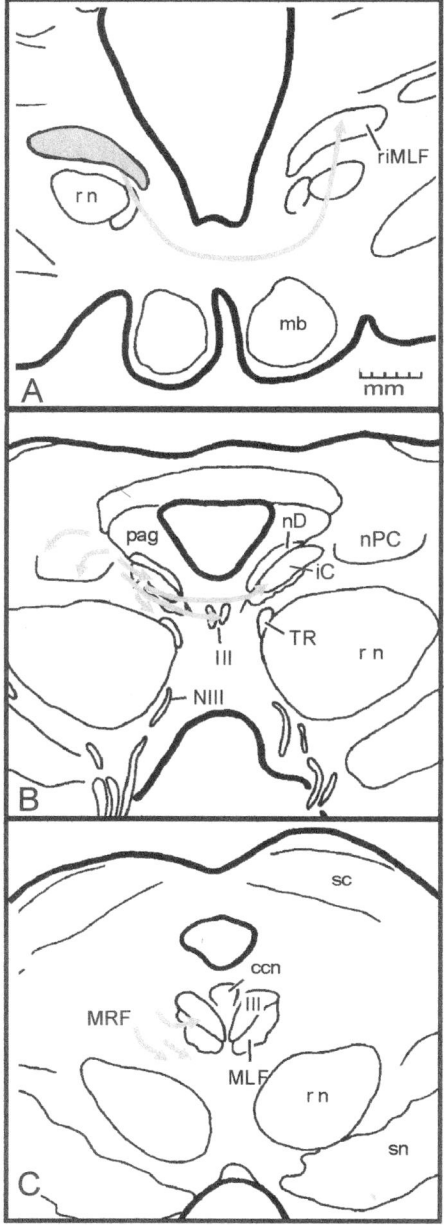

FIGURE 1. Transverse sections of the human mesencephalon from rostral to caudal showing schematically structures related to torsional and vertical eye movements and efferents (*arrows*) of the rostral interstitial nucleus of the medial longitudinal fasciculus (riMLF) (**A**). The interstitial nucleus of Cajal (iC), the posterior commissure (PC), and the nuclei of the posterior commissure (nPC) are shown in **B**; and the mesencephalic reticular formation (MRF), the medial longitudinal fasciculus (MLF), and the third nucleus (III) in **C**. riMLF projects to contralateral riMLF and iC below the aqueduct. It also sends axons to III and IV (not shown), the ipsilateral iC, nPC, and MRF. *Abbreviations*: ccn, central caudal nucleus; mb, mammillary bodies; nd, nucleus of Darkschewitsch; pag, periaqueductal gray; rn, red nucleus; sc, superior colliculus; sn, substantia nigra.

the riMLF, it contains different cell groups including burst neurons similar to those found in the riMLF.[13] In addition, burst-tonic neurons with vertical and torsional on-directions, tonic neurons, and vestibular neurons have been encountered.[14,15]

In comparison to riMLF and iC, little is known about the specific function of the nPC. Axons of nPC are distributed in the posterior commissure PC. NPC projects to the contralateral npc, iC, and riMLF. The oculomotor-related fibers cross mainly in the ventral part of the PC.[3] There are no systematic single-unit studies of nPC neurons in the alert-behaving monkey. A lesion of PC leads to an upgaze-paralysis and to convergent-retraction nystagmus.

The experimental studies over the last years increased our knowledge enormously about the neuronal mechanisms underlying vertical and torsional gaze. This applies especially for torsional eye movements, the study of which has been made possible mainly by recording three-dimensional eye movements as a routine, generally using the precise magnetic search-coil technique, as well as advanced vestibular stimulation paradigms, particularly in the roll plane.

It is now also possible in many places to record patients' three-dimensional eye movements with the search-coil technique and relate their deficits to lesions shown in high-resolution MRI. Experimental results from the monkey and patient data will be presented below. The results will be discussed particularly in relation to the topo-diagnostic value of these oculomotor deficits.

METHODS

Experimental Studies in the Monkey

Monkeys (*Macaca mulatta*) were chronically prepared for single-unit recordings (for details see Helmchen *et al.*[12]). During the experiments, the monkeys sat with the head fixed and erect in a primate chair. For three-dimensional eye position recording, two search coils were attached to the eye ball, approximately orthogonal to each other (see Bartl *et al.*[16] for further details). Eye movements were recorded using a two-field search-coil system (Skalar 3020 eye-posistion meter). Signals were digitized with a 12-bit AD converter (DAP 1200), sampled at 200–1000 Hz and low-pass filtered with cutoff frequencies of 100–200 Hz. For eye movement calibration an *in vivo* procedure was used as described previously,[16] which considered magnetic-field inhomogeneities, offset voltages, and real and apparent cross talk. Monkeys were also trained to fixate defined vertical and horizontal target positions. Prior to each injection experiment, the primary position and the width of Listing's plane were determined. These data were compared with postinjection values. Microinjection sites in iC were determined by characteristic single-unit recordings.[13] A muscimol–NaCl solution (0.3–0.5 µl; 1 µg muscimol in 1 µl 0.9% NaCl) was delivered slowly over a period of 45 s. Spontaneous eye movements in light and darkness were recorded prior and at least 60 min after the injection. Analysis was performed offline using the following parameters: torsional, vertical, and horizontal slow phases of nystagmus, range of stable eye positions, saccade amplitude, and peak velocity.

Patient Investigations

All patients had MRI-identified mesencephalic lesions and clinical evidence of torsional and vertical nystagmus. For eye movement recordings three-dimensional scleral search coils (Skalar, Delft, NL) were placed under topical anesthesia in one eye. The patient sat with the head erect and fixed in a cubic magnetic frame with a side length of 1.8 m. The search-coil system had three orthogonal magnetic fields and was frequency modulated (Remmel, Ashland, MA, USA). Calibration was performed as described previously.[16] Search coils were corrected for nonlinearities and offsets. To calibrate the torsional coil, a precision gimbal system was rotated in small steps through a range of torsional, vertical, and horizontal positions. Eye movement channels were sampled with 500–1000 Hz. Patients were asked to perform spontaneous eye movements and visually guided saccades (5–20 deg horizontally and 5–10 deg vertically).

For eye movement analysis, torsional, vertical, and horizontal eye-position signals were differentiated to obtain eye velocity using a low-pass gaussian filter. Saccade velocity was compared with age-matched control subjects. To exclude effects of vigilance, vertical saccade velocity was also compared intraindividually. Because many patients had a limited oculomotor range and generated only small saccades in the vertical plane, only saccades of 4–10 deg were analyzed and used for comparison. Time constants for the exponential decay of nystagmus slow phases were determined as described previously.[12]

To determine tonic ocular torsion, fundus photographs (laser scanning ophthalmoscope) were taken from both eyes with the head erect, and the deviation of the maculopapillary meridian from the horizontal plane was measured.[17]

Definitions

Direction of eye rotation was defined according to the right-hand rule with positive values for leftward and downward movements. Positive torsion refers to extorsion of the right and intorsion of the left eye.

Magnetic Resonance Imaging

For midbrain images a T1-weighted spin-echo sequence and a T2-weighted turbo-spin-echo sequence were performed in sagittal, axial, and coronal slices on a 1.5 Tesla MRI unit (Siemens Magnetom Vision and Symphony). High-resolution images with 3-mm contiguous slices and an in-plane resolution of less than 0.5 mm were performed. For tilt reconstruction the anterior-posterior commissure (AC–PC) was used on axial slices as a line of reference. riMLF and iC cannot yet be identified in MRI. Therefore, neighboring structures were used for anatomical identification.[18,19] They included the mamillary tract, mamillary bodies, fornix, tractus retroflexus, superior colliculus, posterior commissure, oculomotor and trochlear nucleus, red nucleus, substantia nigra, and crus cerebri.

RESULTS

Experimental Studies in the Monkey

Muscimol injections in iC lead with a latency of <5 min to a vertical-torsional spontaneous nystagmus during gaze straight ahead (FIG. 2). Direct inspection and video recordings demonstrate that the direction of the nystagmus is conjugate on both eyes, particularly in the vertical direction. Thus, there is no seesaw nystagmus. However, torsional and vertical amplitudes are different on both eyes with the torsional component larger on the contralesional eye and the vertical component larger

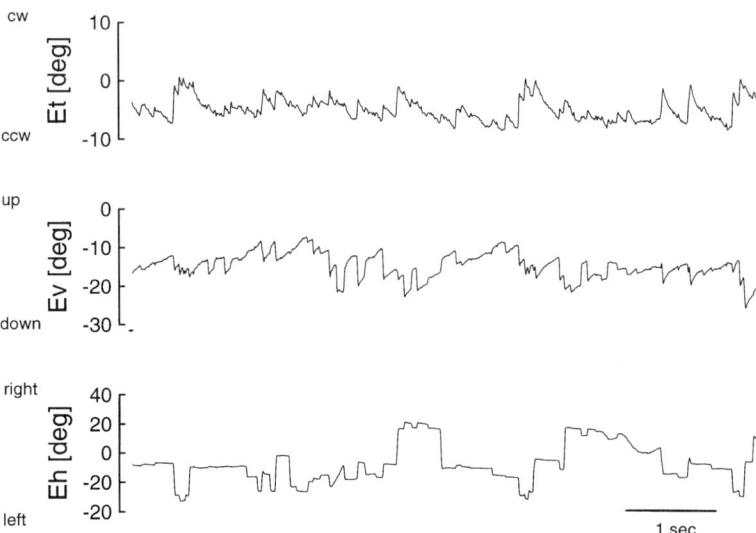

FIGURE 2. Ipsilesional (clockwise, cw) torsional-vertical spontaneous nystagmus in the monkey after a unilateral muscimol injection into the right interstitial nucleus of Cajal (iC). Traces from above: torsional (Et), vertical (Ev), and horizontal (Eh) eye position. In addition, there is a vertical downbeating component. Monocular three-dimensional search-coil registration.

on the ipsilesional eye. These aspects will not be considered further.[20,21] The torsional nystagmus component is always ipsilesional. The slow-phase velocity increases on down gaze but not on lateral gaze. Slow phases always show an exponential decay, with time constants ranging from 200 to 500 ms.

The vertical downbeat component increases on down and lateral gaze. Upbeat gaze evoked-nystagmus is generally not seen after unilateral injections. Only occasionally is it present more than 20 min after injection on extreme upgaze. In these instances the downbeat nystagmus is still present during gaze straight ahead. Thus, stable vertical eye positions were shifted upward. Time constants of the vertical slow phases also range from 200–500 ms in the light.

In addition to the torsional-vertical nystagmus, there is also a tonic ocular torsion of both eyes to the contralesional side which can reach values of up to 20 deg and is on average 13 deg. In the monkey this can be determined precisely by measuring the shift of Listing's plane after the injection. The slow phase of the torsional nystagmus is always in the same direction as the tonic torsional shift with the fast phase beating in the opposite direction.

After unilateral iC injections and in the presence of torsional-vertical nystagmus, hardly any torsional saccades are found in the contralesional direction. The velocity of ipsilesional torsional saccades is not altered compared to preinjection values obtained during torsional VOR. Large vertical saccades are missing after the injection, and the average amplitude is reduced from 8–9 to 4–5 deg. Up- and downward sac-

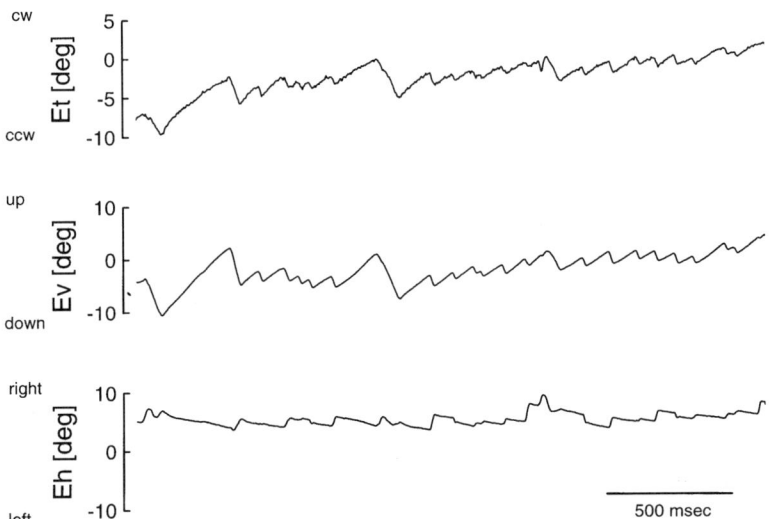

FIGURE 3. Contralesional torsional spontaneous nystagmus recorded in patient 1, who had a unilateral mesencephalic lesion on the right side affecting mostly the rMLF and possibly also the rostral pole of iC. Traces as in FIGURE 2. There is also a vertical downbeating component.

cades are affected equally. Within this limited amplitude range the amplitude velocity relationship remains normal.

Patients

Case histories, oculomotor and MRI findings of patients are presented below. Patient 1 had contralesional torsional nystagmus and patient 2 had ipsilesional torsional nystagmus. Both had unilateral midbrain lesion.

Patient 1 (Contralesional nystagmus)

A 45-year-old woman complained of a sudden onset of vertigo, oscillopsia, vertical diplopia, and a tendency to fall to the left. On neurological examination, a mild left-sided hemiparesis was found. Apart from a head tilt to the left (contralesional), neuroophthalmological examination revealed a conjugate torsional nystagmus (TN) with fast phases beating in a negative torsional direction. It did not change under Frenzel's goggles or in darkness. In addition, a conjugate vertical downbeating component was present (FIG. 3). On gaze straight ahead, there was a skew deviation with a left hypotropia (about 10 deg, on right eye fixation) or right hypertropia (about 10 deg, on left eye fixation). TN was more pronounced on downward and leftward gaze, but did not change its direction. Vision was normal. There was a normal pupil reaction to light and accomodation, but vergence response was diminished. There was no convergence-retraction nystagmus (CRN) and lid movements appeared normal. Vertical saccades were slowed, downward more than upward, and vertical oculomotor range was limited (upward and downward about 15 deg), but vertical vestibulo-ocular reflex (VOR) elicited a full range of vertical eye movements. Fast phases during torsional VOR were slowed to the right side, i.e., in a positive torsional direction. Vertical pursuit was moderately saccadic, more for downward than upward pursuit.

Fundus photography revealed 13.5 deg of right-eye incyclorotation and 18 deg of left-eye excyclorotation, that is, a conjugate negative ocular torsion. The subjective visual vertical (SVV) was tilted to the left by 14 deg for both eyes. Caloric irrigation showed normal responses.

Clinical follow-up examinations of this patient were performed for more than one year. Oscillopsia ceased after about six months. The TN on gaze straight ahead was always present and did not change its direction. Vertical saccade slowing improved slightly but was still present. MRI scan showed a unilateral midbrain lesion on axial T2-weighted images extending from the right paramedian thalamus to the rostral median midbrain. The lesion clearly involved the right riMLF and the nPC, possibly also the rostral pole of the iC. It spared the posterior commissure and the oculomotor nucleus. Because of its sudden onset, the lesion was suspected to be an ischemic unilateral infarction of the paramedian thalamopeduncular deep penetrating midbrain vessels despite the fact that vascular diagnostic workup (Doppler and duplex, echocardiography) was normal. The patient was treated with aspirin.

Patient 2 (ipsilesional nystagmus)

A 71-year-old woman presented with a history of an acute onset of vertical diplopia, vertigo, and an unsteady gait early in the morning of the day of admission. Clinical examination revealed vertical gaze palsy with slowing of downward saccades more than upward saccades and a reduced vertical oculomotor range, but full range of vertical slow eye movements during vertical VOR, suggesting a supranuclear lesion. In addition, there was a vertical-torsional nystagmus on gaze straight ahead with different amplitudes for both eyes: the left eye showed a larger vertical amplitude than the right eye, with conjugate torsional components beating positive (ipsilesional). It did not change with Frenzel's goggles or in darkness. Vertical gaze evoked nystagmus (GEN) was found on upward gaze with a small but conjugate torsional amplitude beating in the same direction as in gaze straight ahead. Downward gaze did not elicit GEN. Lid-saccade velocity seemed to be normal; there was neither CRN nor lid retraction. Thus, lid saccades seemed to be faster than eye saccades. A left and right hypertropia of about 5 deg was inconsistently found. Ocular counterroll could be elicited only poorly without a clear asymmetry. Likewise, fast phases of TN could not be consistently elicited. There was a slight head tilt to the left of 2–3 deg that, once again, was inconsistent. In addition, a falling tendency toward the left side was noted. Horizontal eye movements, including vergence and pupillary reactions to light and accomodation, were normal as was visual acuity. Otherwise there were no abnormal neurological signs. A general medical examination was normal.

Fundoscopy revealed excyclorotation (positive torsion) of the right eye of 7 deg and incyclorotation (positive torsion) of the left eye of 1 deg, indicating positive (ipsilesional), but inconsistent and asymmetric ocular torsion. SVV was tilted to the right by 2.6 deg (right eye) and 2.9 deg (left eye).

Follow-up examinations were performed twice within the first 10 days and after three months. Three days after admission, spontaneous nystagmus during gaze straight ahead decreased in amplitude, and vertical saccades seemed to be faster.

There was, as a new sign, CRN during slightly upward gaze. After six days vertical saccades had a full oculomotor range and were normal during upward but still slowed during downward gaze. There was no longer skew deviation. After three months there was neither nystagmus nor slowing of vertical saccades nor any other oculomotor abnormality.

High-resolution axial T2-weighted MR images (3-mm contiguous slices, in-plane resolution of <0.5 mm, one week after admission) revealed a small unilateral signal-intense lesion in the right midbrain extending from the ventral to ventrolateral border of the periaqueductal gray to the dorsomedial border of the red nucleus, most likely involving the iC. Its rostral border almost reached the floor of the third ventricle. It possibly involved the most rostral part of the third nucleus, but clearly spared the majority of the posterior commissure (PC) and its nucleus (NPC) as well as the riMLF. Apart from the midbrain, no other CNS lesions were found.

Because of the sudden onset of clinical symptoms, the signal intensity was suspected to be an infarction. The patient was treated with ticlopidine after tests to determine the underlying cause were negative.

COMMENTS

Experimental studies in the monkey and patient investigations show clearly that a unilateral midbrain lesion can lead to contralesional or ipsilesional torsional nystagmus (FIG. 4). In experimental studies with defined lesions, contralesional nystagmus is due to an riMLF lesion and ipsilesional nystagmus due to an iC lesion. The patient data

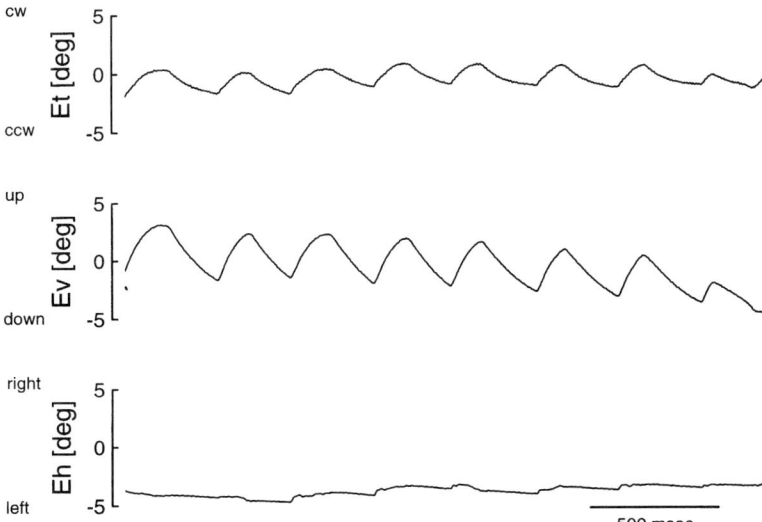

FIGURE 4. Ipsilesional torsional spontaneous nystagmus recorded from a patient with a small unilateral mesencephalic lesion on the right side, which included the iC and extended only partially into the riMLF rostrally. The vertical component is upbeating.

presented demonstrate that even small lesions usually affect both structures, the riMLF and iC. Although it is still not possible to directly outline riMLF and iC on present MRIs, the detailed analysis shows that riMLF was mostly affected in patient 1 compared to patient 2, in which the lesion almost exclusively involved iC. This anatomical evaluation is in agreement with the oculomotor findings of contralesional nystagmus in patient 1 and an ipsilesional nystagmus in patient 2. Thus, based on the knowledge of animal experiments, it can be concluded that also in humans a riMLF lesion leads to contralesional nystagmus, and an iC lesion to ipsilesional nystagmus.

In comparing experimental animal studies and patient investigations, one must consider that the more recent experimental studies generally used microinjections with neuropharmacological agents like muscimol, which inactivate cells but not fibers of passage. FIGURE 1 underlines the intimate vicinity of premotor structures and fibers connecting the premotor structures and the oculomotor nuclei. Experimentally little is known about the specific effect of small destructive lesions compared to microinjections.

In an earlier study,[22] it was suggested that an iC lesion leads to seesaw nystagmus with the vertical fast phases going in opposite directions. Neither in the experimental studies nor in the patient investigations did we find any evidence for this. The vertical fast phases were always conjugate in the same direction. There is, however, a difference in amplitude, both for the vertical and torsional component.[20,21] The reasons for the difference of seesaw and conjugate nystagmus are not yet clear. The lesions in the study by Halmagyi *et al.* were generally large and caused by hemorrhages in contrast to infarctions. In relation to hemorrhages, excitatory mechanisms rather than lesion effects have also been discussed.[23] In this case excitation of riMLF structures would cause the ipsilesional nystagmus rather than an iC lesion.

All experimental and clinical studies show that unilateral midbrain lesions also lead to a tonic contralesional torsion.[9,24] As a distinguishing oculomotor sign, however, it is not useful because it is seen after both riMLF and iC lesions.[2] The cause of this deficit is quite different. In the case of iC, it is most likely due to a vestibular imbalance, whereas with riMLF it is related to an imbalance of the torsional saccade generators.[10] Large lesions including both riMLF and iC should actually lead to less nystagmus—assuming that both nystagmus directions cancel each other out—and more pronounced tonic contralesional torsion. There is actually some experimental evidence for this.[25] This question has not thus far been studied in patients.

ACKNOWLEDGMENTS

This work was supported by the Deutsche Forschungsgemeinschaft (DFG) and the German-Israel Foundation (GIF). The authors wish to thank S. Langer for technical assistance, I. Wendl for preparing the manuscript, and C. Rapaport for editing the English text.

REFERENCES

1. BHIDAYASIRI, R., G.T. PLANT & R.J. LEIGH. 2000. A hypothetical scheme for the brainstem control of vertical gaze. Neurology **54:** 1985–1993.

2. BÜTTNER, U. & C. HELMCHEN. 2000. Eye movement deficits after unilateral mesencephalic lesions. Neuro-ophthalmology **24:** 469–484.
3. BÜTTNER-ENNEVER, J.A. & U. BÜTTNER. 1988. Neuroanatomy of the oculomotor system. The reticular formation. Rev. Oculomotor Res. **2:** 119–176.
4. PERCHERON, G. 1973. The anatomy of the arterial supply of the human thalamus and its use for the interpretation of the thalamic vascular pathology. J. Neurol. **20:** 1–13.
5. MOSCHOVAKIS, A.K., C.A. SCUDDER & S.M. HIGHSTEIN. 1991. The structure of the primate oculomotor burst generator. I. Medium-lead burst neurons with upward on-directions. J. Neurophysiol. **65:** 203–217.
6. MOSCHOVAKIS, A.K., C.A. SCUDDER, S.M. HIGHSTEIN & J.D.WARREN. 1991. Sculture of the primate oculomotor burst generator, II. Medium-lead burst neurons with downward on directions. J. Neurophysiol. **65:** 218–229.
7. BÜTTNER, U., J.A. BÜTTNER-ENNEVER & V. HENN. 1977. Vertical eye movement related activity in the rostral mesencephalic reticular formation of the alert monkey. Brain Res. **130:** 239–252.
8. VILIS, T., K. HEPP, U. SCHWARZ & V. HENN. 1989. On the generation of vertical and torsional rapid eye movements in the monkey. Exp. Brain Res. **77:** 1–11.
9. SUZUKI, Y., J.A. BÜTTNER-ENNEVER, et al. 1995. Deficits in torsional and vertical rapid eye movements and shift of Listing's plane after uni- and bilateral lesions of the rostral interstitial nucleus of the medial longitudinal fasciculus. Exp. Brain Res. **106:** 215–232.
10. HENN, V. 1992. Pathophysiology of rapid eye movements in the horizontal, vertical and torsional directions. In Ocular Motor Disorders of the Brain Stem. U. Büttner & T. Brandt, Eds.: 373–391. Bailliere Tindall. London.
11. CRAWFORD, J.D., W. CADERA & T. VILIS. 1991. Generation of torsional and vertical eye position signals by the interstitial nucleus of Cajal. Science **252:** 1551–1553.
12. HELMCHEN, C., H. RAMBOLD, L. FUHRY & U. BÜTTNER. 1998. Deficits in vertical and torsional eye movements after uni- and bilateral muscimol inactivation of the interstitial nucleus of Cajal (IC) of the alert monkey. Exp. Brain Res. **119:** 436–452.
13. HELMCHEN, C., H. RAMBOLD &U. BÜTTNER. 1996. Saccade-related burst neurons with torsional and vertical on-directions in the interstitial nucleus of Cajal of the alert monkey. Exp. Brain Res. **112:** 63–78.
14. KING, W.M., A.F. FUCHS & M. MAGNIN. 1981. Vertical eye movement-related responses of neurons in midbrain near interstitial nucleus of Cajal. J. Neurophysiol. **46:** 549–562.
15. FUKUSHIMA, K., C. HARADA, J. FUKUSHIMA & Y. SUZUKI. 1990. Spatial properties of vertical eye movement-related neurons in the region of the interstitial nucleus of Cajal in awake cats. Exp. Brain Res. **79:** 25–42.
16. BARTL, K., C. SIEBOLD, S. GLASAUER, et al.1996. A simplified calibration method for three-dimensional eye movement recordings using search-coils. Vision Res. **36:** 997–1006.
17. DIETERICH, M. & T. BRANDT. 1992. Cyclorotation of the eyes and the subjective visual vertical. In Ocular Motor Disorders of the Brain Stem. U. Büttner & Th. Brandt, Eds.: 301–315. Bailliere Tindall. London.
18. HELMCHEN, C., S. GLASAUER, K. BARTL & U. BÜTTNER. 1996. Contralesionally beating torsional nystagmus in a unilateral rostral midbrain lesion. Neurology **47:** 482–486.
19. RIORDAN-EVA, P. et al. 1996. Abnormalities of torsional fast phase eye movements in unilateral rostral midbrain disease. Neurology **47:** 201–207.
20. RAMBOLD, H., C. HELMCHEN & U. BÜTTNER. 1999. Unilateral muscimol inactivations of the interstitial nucleus of Cajal in the alert rhesus monkey do not elicit seesaw nystagmus. Neurosci. Lett. **272:** 75–78.
21. RAMBOLD, H., C. HELMCHEN & U. BÜTTNER. 2000. Vestibular influence on the binocular control of vertical-torsional nystagmus after unilateral lesions of the interstitial nucleus of Cajal (iC) in the alert monkey. Neuroreport **11:** 779–784.
22. HALMAGYI, G.M. et al. 1994. Jerk-waveform see-saw nystagmus due to unilateral meso-diencephalic lesion. Brain **117:** 789–803.
23. BENTLEY, C.R. et al. 1998. Fast eye movement initiation of ocular torsion in mesodiencephalic lesions. Ann. Neurol. **43:** 729–737.

24. DIETERICH, M. & T. BRANDT. 1993. Ocular torsion and tilt of subjective visual vertical are sensitive brainstem signs. Ann. Neurol. **33:** 292–299.
25. MABUCHI, M. 1970. Rotatory head response evoked by stimulating and destroying the interstitial nucleus and surrounding region. Exp. Neurol. **27:** 175–193.

Contribution of the Superior Colliculus and the Mesencephalic Reticular Formation to Gaze Control

DAVID M. WAITZMAN, JAY PATHMANATHAN, RACHEL PRESNELL, AMANDA AYERS, AND STACY DePALMA

Department of Neurology, University of Connecticut Health Center, Farmington, Connecticut 06030-3974, USA

ABSTRACT: Converging lines of evidence support a role for the intermediate and deep layers of the superior colliculus (SC) and the mesencephalic reticular formation (MRF) in the control of combined head and eye movements (i.e., gaze). Recent microstimulation, single-cell recording, and lesion experiments are reviewed in which monkeys are free to move their heads. Cells in the SC discharge in advance of combined head and eye movements and most likely provide a gaze error signal to downstream structures. In contrast, the neurons in the MRF are of at least two types. Eye cells have features that are similar to neurons in the rostral portion of the SC, but fire before the onset of horizontal eye movments. A second group of MRF neurons begin to fire after the onset of the gaze shift and are most closely associated with movements of the head. The peak discharge of these late-onset MRF neurons occurs near the peak head velocity. Stimulation in the rostral SC generates eye movements with fixed amplitude and direction. A similar response is noted after stimulation of the more dorsal portion of the caudal MRF. Stimulation in the caudal portion of the SC produces combined head and eye movements of fixed amplitude. Electrical activation of the more ventral portions of the caudal MRF generates goal-directed and centering eye movements. Temporary inactivation of the SC with the GABA agonist muscimol generated hypometria and curved trajectories of contralateral eye movements. Inactivation of the caudal MRF produced contralateral hypermetria and ipsilateral hypometria of saccades. Release of the monkey's head demonstrated a profound contralateral head tilt. Taken together, these data suggest that the gaze signal generated in the SC is filtered by neurons in the MRF to generate a feedback signal of eye motor error. The head signal found in the MRF could cancel a portion of the gaze signal coming from the SC in the form of head velocity feedback.

Keywords: saccades; head movement; oculomotor system; tecto-reticular-spinal; feedback

---

Address for correspondence: David M. Waitzman, MD, PhD, Department of Neurology, University of Connecticut Health Center, 263 Farmington Avenue, Farmington, CT 06030-3974. Voice: 860- 679-4313 (Lab), 860- 679-8011; fax: 860-679-4446.
waitzman@nso2.uchc.edu

INTRODUCTION

For more than a century the superior colliculus has been known to participate in the visual guidance of eye movement. It has also been evident in lower mammals, particularly rodents, that the colliculus participates in the visual guidance of combined head, eye, and body movements.[43] Over the last three decades various investigators have demonstrated that the superior colliculus of the cat and primate provides a combined head and eye or *gaze* signal for lower brainstem structures.[15,17,24,28,31] Because the permanent removal of the superior colliculus produces only minimal changes in the amplitude or direction of eye movements, it is not the sole mediator of gaze. Rather, the colliculus must coordinate with a number of other midbrain structures including the mesencephalic reticular formation (MRF), with which it forms dense reciprocal connections. In contrast to collicular damage, loss of the MRF produces a profound hemi-neglect and an ipsi-lesion gaze preference in both patients and monkeys. In order to understand the differences and similarities in these two midbrain regions in the control of combined head and eye movement, two types of experiments will be reviewed: chronic single-neuron unit recordings and electrical microstimulation. We will use the results of these experiments as a tool to address the question of how the collicular gaze signal is transformed into separate head and eye signals.

METHODS

All experiments were approved by the Animal Care and Use Committee of the University of Connecticut Health Center. In brief, rhesus monkeys (*Macaca mulatta*) were surgically prepared for eye movement recording, chronic single-unit recording, electrical microstimulation, and microinjections under isoflurane anesthesia.[38] An eye coil was implanted under the conjunctiva.[20] The coil wires were led to a dental acrylic cap in which a head fixation device and a recording chamber were stereotactically embedded to allow access to the MRF. In some monkeys, a second recording chamber was positioned over the superior colliculus. Monkeys free to move their head were trained to follow rear-projected, jumping spots of light 1°–2° in diameter. Since eye and head position could be monitored, the monkey was rewarded for orienting gaze within a window around the target. The oculomotor nuclei were localized with the head restrained through the MRF cylinder positioned over the parietal cortex tilted 15° from the sagittal plane. The MRF was then localized 2 mm lateral to the oculomotor nuclei (FIG. 1A and B). Through a second chamber tilted 38° caudally, a microelectrode was passed perpendicular to the surface of the superior colliculus permitting identification of the characteristic visual responses of the superficial collicular layers. The portion of the visual field that caused superficial layer cells to fire was used as a guide for the generation of eye movements of appropriate amplitude and direction to activate the underlying cells of the intermediate and deep collicular layers.

Well-isolated single-neuron discharges were fed to a window discriminator whose acceptance pulses, in addition to head (re: space) and gaze (re: space) signals, were then recorded on a PC running REX software. Off-line analysis permitted calculation of eye position (re: head) by subtracting head position from gaze position.

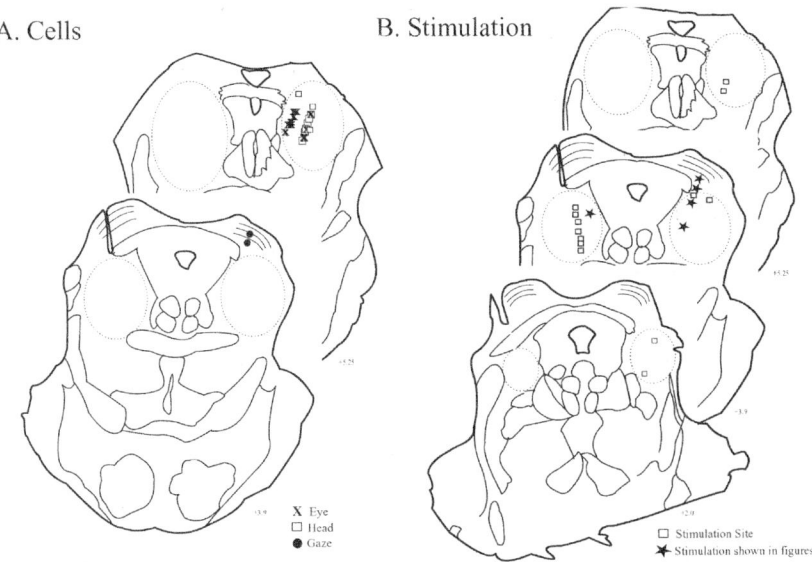
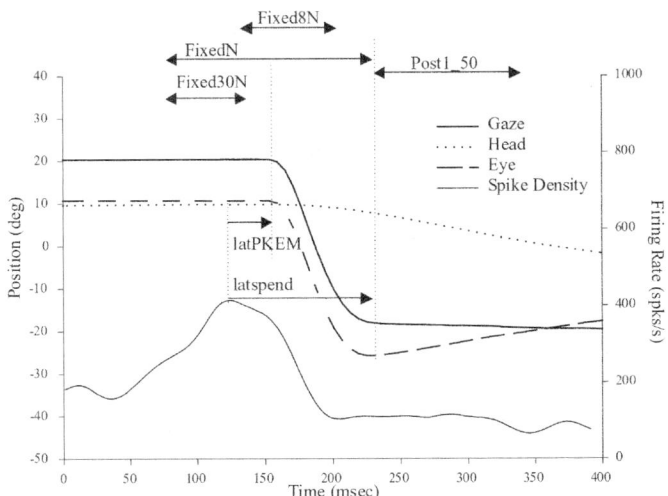

FIGURE 1. (**A**) Location of cells recorded in the MRF and superior colliculus (SC). (**B**) Stimulation sites in the MRF and superior colliculus of two head-unrestrained monkeys. *Stars* indicate stimulation sites displayed in the FIGURES 5 and 6. *Upper star* in the superior colliculus was located at the same depth in a section 0.5 mm more rostral than the one illustrated. *Dotted lines* in **A** and **B** outline the region of the MRF. (**C**) A 40° gaze movement executed across the midline. Traces are from *top to bottom:* gaze, eye, head, and spike density. Spike-counting intervals: FixedN: 30 ms before the start to the end of the gaze movement; Fixed30N: 30 ms before to 8 ms before the onset of the gaze movement; Fixed8N: 8 ms before the onset to 8 ms before the end of the gaze movement; Post 1_50: the first 50 ms after end of gaze movement. Latency measures: latPKEM, time in ms from peak of spike density to the start of the gaze movement; latspend, time in ms from peak of spike density to the end of the gaze movement.

Gaze fields were obtained by having the monkey make repetitive centrifugal movements from primary position to a circle of eight visual targets, positioned at six different amplitudes (5°, 10°, 15°, 20°, 25°, 30°). Large saccades across the midline were used to increase the range of gaze shifts tested. Spike counts over several specific intervals (FIG. 1C) were used to identify the gaze field for each neuron. The group of saccades of preferred amplitude and direction that caused the neuron to generate the largest spike counts were used to perform an analysis of the temporal aspects of the neuronal response with parameters of gaze movement.[6–8,35] We refer to this as "dynamic analysis" in the text below.

At the conclusion of all experiments, injections of WGA-HRP or fluorescent tracers were placed in either the superior colliculus or the MRF. After the appropriate time interval to permit the transport of the respective agents, the monkeys were euthanized via cardiac perfusion under deep barbiturate anesthesia.

RESULTS

Monkeys were trained to perform large gaze movements across the midline as shown in FIGURE 1C. Initially the monkey fixates a target 20° to the right. This 20° rightward gaze position was achieved by use of a 10° rightward head turn combined with a 10° rightward eye displacement (relative to the head). The LED target then jumped 40° to the left. The monkey followed this target shift using the combination of a 35° leftward eye movement and a 5° leftward head movement. Note that after the initial rapid shift of gaze (using primarily the eyes), the vestibular ocular reflex was activated to center the eyes in the head. In order to understand the analysis of MRF neuron discharge, the spike density response of a sample neuron recorded in the MRF is shown below the gaze movement. Four different spike-counting intervals were used: (1) an interval before the saccade (Fixed30N); (2) an interval during the saccade (Fixed8N); (3) an interval that included both of these (FixedN); and (4) the 50-ms interval after the end of the gaze shift (Post1_50).

Neurons in the Mesencephalic Reticular Formation

Twenty mesencephalic reticular formation (MRF) neurons have been recorded in two monkeys free to move their heads (FIGURE 1A). These neurons discharged during gaze movements made in both light and dark. An example of an MRF neuron that discharged primarily during small amplitude, 5°–10° saccades, is shown in FIGURE 2 (Eye Neurons, $n = 11$). The discharge began 30–40 ms before and reached a peak of 500 spikes/s about 10 ms before saccade onset (FIG. 2A and 2B). The greatest number of spikes occurred *before* (Fixed30N) small, horizontal saccades, producing a gaze field that had its peak near 5° to the left (contraversive) and 2° down (FIG. 2C, shaded box). This neuron fired at a much-reduced rate during large amplitude gaze shifts. Therefore, while measuring the total spike number associated with each saccade (FixedN) (FIG. 2D), a large response was found for saccades smaller than 10° and for saccades greater than 40°. Note that the increased response was found only when measuring the spikes *during* (Fixed8N) and not before (Fixed30N) larger saccades. Moreover, while the peak discharge of this neuron led small 5°–10° saccades, it occurred at a consistent time before the end of (that is, during) large saccades (FIG.

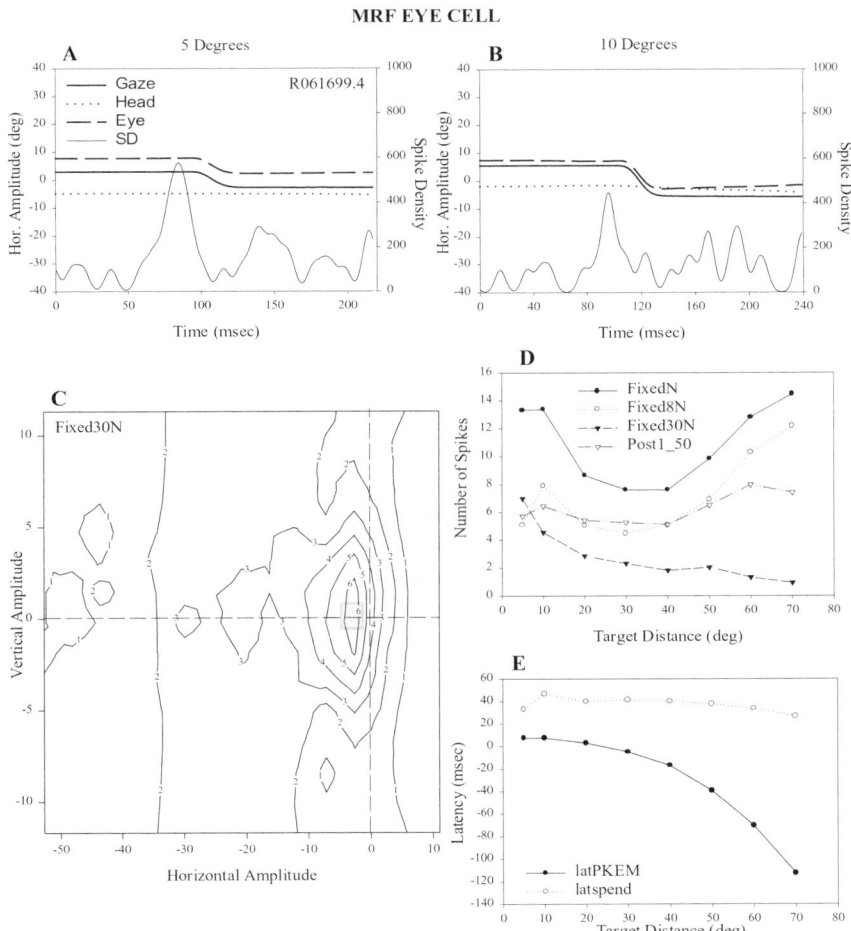

FIGURE 2. An eye movement related neuron recorded in the right MRF. (**A** and **B**) Traces from *top to bottom* are eye, gaze, head, and spike density for an individual trial. This cell fired for gaze movements smaller than 14°. The peak discharge occurred 20 ms before gaze onset. (**C**) Movement field was generated using Fixed30N (presaccadic) spike-counting interval. *Shaded box* indicates the region of peak discharge and the preferred direction and amplitude used in the dynamic analysis. (**D**) Spike counts for various intervals around the gaze shift. The highest spike counts were observed when the entire discharge before and during the saccade was included. The increased discharge for large gaze shifts (FixedN) was the result of a corrective saccade that occurred at the end of the movement. (**E**) Positive numbers indicate that the peak discharge occurred before the eye movement. The peak discharge occurred before small saccades and during the gaze shift for larger movements. The constant latency of peak discharge before saccade end (*open symbols*) suggests that this cell was related to residual eye motor error.

2E, open circles). This suggested that this eye cell was associated with residual motor error, and not saccade amplitude per se.[38] Recent work has emphasized the difficulty of associating particular saccade characteristics with the temporal response of a single neuron, the so-called neural uncertainty problem.[34] To circumvent this problem, the dynamic responses of this MRF neuron was analyzed for just those saccades that brought final gaze position to within ±3° of the saccade amplitude that would elicit the cell's peak response (Spike Count Isobar labeled 6, FIG. 2C). The variance accounted for (VAF) was greatest when an eye velocity model was correlated with the spike discharge of the preferred saccades (VAF: 0.22; latency: 19 ms). Other models that were considered included position and a combination of higher-order terms.

A different response was found for a second group of MRF neurons (Head Neurons, $n = 9$, FIGS. 1A and 3). These cells had a small background discharge associated with contralateral gaze shifts that were less than 40° in amplitude (FIG. 3A). However, these neurons had higher firing rates when the head began to move during contraversive gaze shifts >45° (FIG. 3B). This particular neuron had its highest discharge near the point of maximum head velocity during a 70° gaze shift (arrow). The gaze field of this neuron included all leftward (contraversive) combined movements of the head and eyes greater than 20°. The number of spikes analysis showed that the spikes collected before and during the gaze shift (FixedN) increased with gaze amplitude (FIG. 3D). Only a minimal response (and thus no gaze field) was found when the Fixed30N interval was used, which counted only the presaccadic spikes (FIG. 3D, Fixed30N). Moreover, the peak discharge of these types of neurons was most often fixed with respect to the end of the gaze shift (FIG. 3E). Dynamic analysis of the peak neuronal activity associated with leftward gaze shifts of 65° showed the highest VAF for a model that combined head position and head velocity (VAF: 0.3; latency: 32 ms). A second model that combined gaze position and gaze velocity produced a VAF that did not equal that using the head parameters (VAF: 0.19; latency: 37 ms). The impression that head and not gaze was coded by this neuron is confirmed by inspection of the neural response, which was greatest for the head and not the eye portion of the gaze shift (FIG. 3B). Clearly, additional experiments in which initial eye position was deviated in the ipsilateral direction of movement would be helpful in distinguishing gaze from head responses.

Neurons in the Intermediate and Deep Layers of the Superior Colliculus

Neurons in the intermediate and deep layers of the superior colliculus have been examined in great detail by a number of investigators. We have studied a number of collicular neurons utilizing the same techniques used to study neurons in the MRF. One such neuron is shown in FIGURE 4. This neuron is typical of most build-up neurons located in the more caudal portions of the superior colliculus.[26] An increase in the background discharge began 100 to 130 ms before the onset of the gaze shift. About 30 ms before the beginning of the saccade, the cell produced a burst of discharge associated with an eye movement of 10° (FIG. 4A). This discharge was cut off or clipped with the end of the saccade.[37] Little or no head movement was associated with this 10° gaze shift. With larger gaze shifts of 30° to 60°, the activity of the cell continued throughout the gaze shift (FIG. 4B). This more prolonged discharge returned to the background discharge rate by the time the gaze shift had been

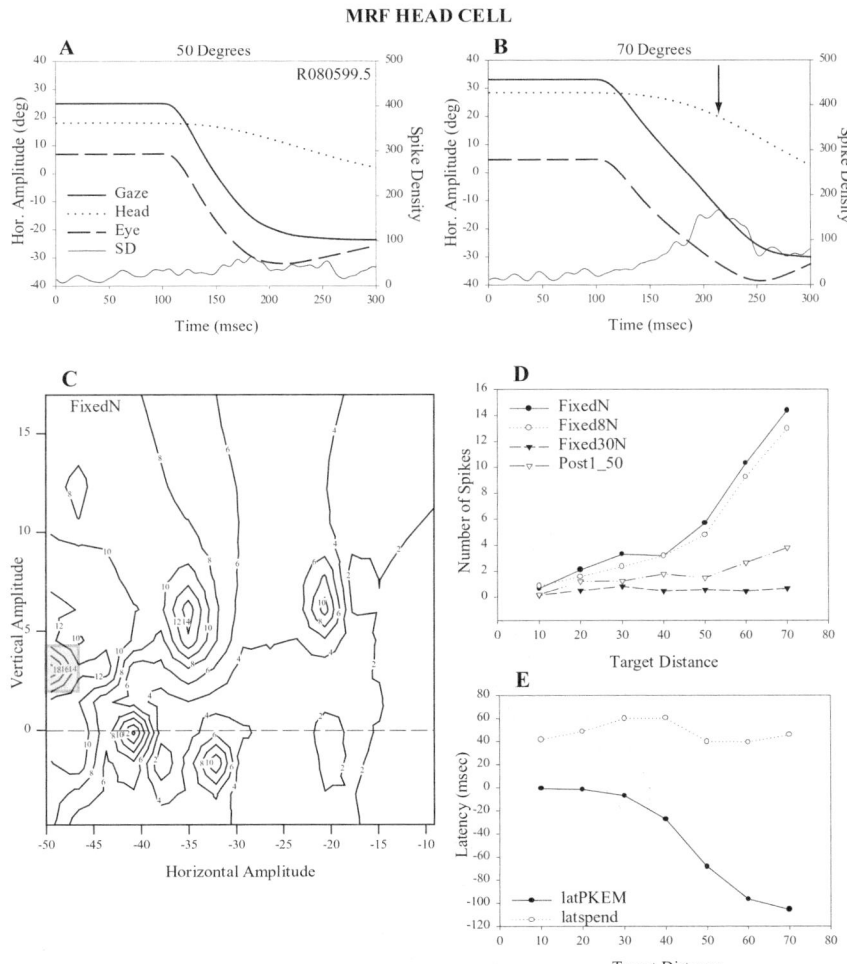

FIGURE 3. A head-associated neuron recorded in the right MRF. Conventions are the same as in FIGURE 2, except for **C**. (**A** and **B**) This neuron had the highest discharge during the execution of a 70° gaze shift. The peak discharge of this neuron coincided with peak head velocity indicated by the *arrow*. (**C**) Movement field generated using the FixedN (total) spike counting interval. *Shaded box* indicates the region of peak discharge for this cell and the preferred direction and amplitude used for the dynamic analysis. (**D**) The highest discharge rate occurred in association with larger gaze shifts (FixedN). (**E**) The constant latency of peak discharge before saccade end (*open symbols*) suggests that this cell was related to residual head motor error or head velocity.

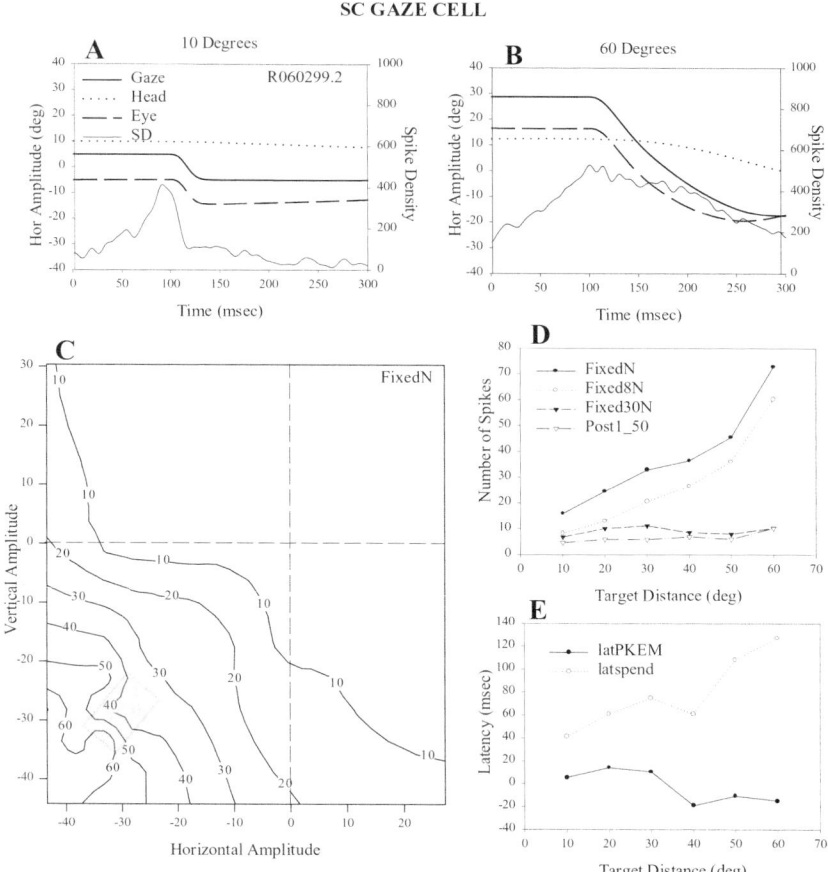

FIGURE 4. A neuron recorded in the right superior colliculus related to gaze displacement. Conventions are the same as in FIGURE 3. (**A** and **B**) Note that the discharge of this neuron increased both in duration and discharge rate for gaze shifts up to at least 60°. The peak discharge occurred 20 ms before gaze saccade onset and the tail of the discharge increased with gaze duration. In **A** the discharge was clipped (cut off) just before saccade end. This neuron would have been previously classified as an eye movement–related cell in a head-restrained monkey. (**C**) Movement field generated using FixedN (total) spike counting interval. *Shaded box* indicates the region of peak discharge for this cell and the preferred direction and amplitude used for dynamic analysis. (**D**) Spike counts for various intervals around the gaze shift. The bulk of the discharge occurred during the gaze shift (FixedN and Fixed8N) and not before (Fixed30N). (**E**) The latency of the peak discharge was fixed with respect to the onset of the gaze saccade and increased with respect to the end of the gaze saccade.

completed (i.e., clipped discharge). The movement field of this neuron showed that it preferred gaze movements of 25° to 35° amplitude (FIG. 4C; shaded box). However, there was a nearly linear increase in the number of spikes with increased gaze amplitude (FIG. 4D, FixedN). In contrast to the MRF neurons, dynamic analysis showed that the collicular neurons were more closely associated with a gaze velocity, not head velocity model. Again, only gaze shifts that landed within ±5° of the amplitude that produced the peak response of this collicular neuron were used in the dynamic analysis (VAF: 0.34; latency: 32 ms). This confirms earlier reports that neurons in the superior colliculus provide a signal of gaze error to downstream oculomotor structures.[12]

Electrical Microstimulation in the MRF

Previous microstimulation experiments in head-restrained monkeys have demonstrated that fixed vector (retinotopic) saccades can be elicited from dorsal portions of the MRF, and variable amplitude saccades—dependent upon initial eye position—could be evoked from the ventral portions of the MRF.[5] We have now electrically microstimulated these same regions in monkeys whose heads were unrestrained. At dorsal sites in the MRF, contraversive fixed amplitude (retinotopic) saccades were elicited at low current (25 µA) and short duration (50 to 100 ms) (FIG. 5A and 5B). The saccades displayed are unselected so that initial eye position (eye position re: head) was allowed to vary. As a result, shifts in initial gaze position represent a mixture of initial eye (re: head) and head (re: space) positions. The variation in initial *horizontal* eye position is represented by the heavy portion of the number line below each column of gaze movements (FIG. 5A). For example, in the first column on the left, initial eye position (re: head) varied from −12.57° to +2.65°. Notice that despite the significant variation in initial eye position the elicited movements were of similar amplitude and direction. Furthermore, when stimulus duration exceeded 50 ms, two saccades of similar amplitude were elicited at short latency (i.e., "staircase saccades"). Four stimulation sites in the dorsal MRF elicited fixed vector saccades.

More ventral locations elicited two types of variable amplitude saccades. The horizontal component of the more common type of variable amplitude saccade elicited with the head restrained was initial position dependent, similar to previously reported results. However, these elicited saccades had a vertical component of movement, which was also dependent upon initial eye position (FIG. 5C). In combination, the dependence of the horizontal and vertical components in the initial position produced "goal-directed saccades" that reversed direction if initial eye position was deviated away from (i.e., contralateral to) the side of stimulation. This particular site was not stimulated with the head unrestrained. At the most caudal portions of the MRF, the reversal of saccade direction was so great that actual "centering saccades" were generated (FIG. 5E and F). We encountered MRF centering locations at about the same frequency ($n = 9$) as the "goal-directed" MRF site ($n = 7$). One point to note is that such centering saccades could have been easily missed in previous stimulation studies if initial eye position had not been varied systematically (away from primary position) at all locations. Similar centering saccades have not been demonstrated during stimulation of any portion of the superior colliculus. Additional experiments are under way to compare the variable amplitude saccades and gaze reversals during head-restrained and unrestrained conditions.

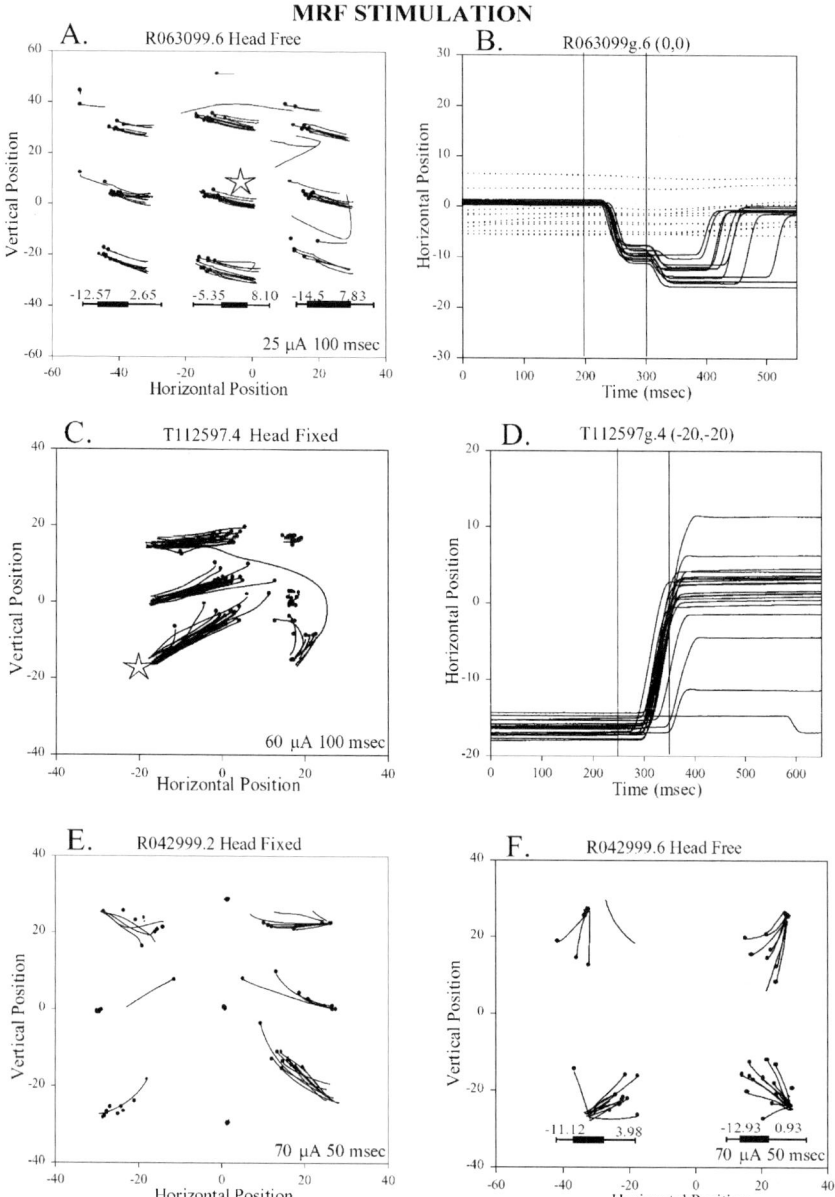

FIGURE 5. Stimulation of the MRF. **(A)** Fixed vector saccades were elicited in the right dorsal MRF of a head-unrestrained monkey. *Star* indicates specific group of trials shown in **B**. Variation in eye (re: head) horizontal initial position is shown by the *heavy bars* on the number lines below each of the initial gaze positions (−30°, 0°, and +30°). Numbers above each bar indicate the range of initial eye position sampled. **(B)** Horizontal gaze (*solid line*) and horizontal head positions (*dotted lines*) for the position labeled by a *star*. **(C)** Goal-

Electrical Microstimulation in the Intermediate and Deep Layers of the Superior Colliculus

To compare the results of microstimulation used in the MRF with those used in the superior colliculus, we repeated the same experiment in the same monkey with the electrode positioned in the intermediate or deep layers of the superior colliculus (FIG. 6). With the head restrained, fixed vector (retinotopic) saccades were elicited from the rostral portions of the superior colliculus (FIG. 6A). The direction and amplitude of these saccades remained the same after the head was released despite an increase in stimulus duration from 50 to 200 ms (FIG. 6B). This was not the case for stimulation sites in the more caudal portions of the superior colliculus where more goal-directed, variable amplitude saccades were elicited (FIG. 6C and D). With the head restrained, these saccades appeared directed toward a region left and up (note the different directions of saccades starting from superior and inferior initial eye positions in FIG. 6C). Even more striking was the lack of a stimulation response at locations ipsilateral to the direction of movement (i.e., variable-amplitude); this was evident at eight caudal superior colliculus sites tested extensively and was different from the MRF stimulation, where ipsiversive initial eye positions often resulted in reversal of saccade direction. When the head was released (FIG. 6D and F), the amplitude of the gaze shifts initiated when gaze was deviated in the direction of movement increased compared to the head fixed condition. As a result, *gaze* shifts had constant amplitudes regardless of their initial position. This finding confirms previous observations of electrical microstimulation in unrestrained monkeys.[10] However, in contrast to previous reports, considerable differences in the direction of saccades elicited with initial positions above, at, or below the horizontal plane persisted (FIG. 6E). For example, saccades elicited from a gaze position 20° above the horizontal were directed down and to the left (200° angle). The direction of gaze shifts elicited from a gaze position 20° below the horizontal was directed up and to the left (160° angle). At the same time, the amplitudes of these two gaze shifts were similar. This shift in direction could not occur from any shift in the position of the electrode in the superior colliculus.

DISCUSSION

Anatomic Connections of the MRF with the Superior Colliculus and Other Brain Stem Structures

Anatomically, the MRF (FIG. 1A; also referred to as nucleus cuneiformis) is closely linked to other oculomotor structures. It is reciprocally interconnected with neurons located in the intermediate and deep layers of the superior colliculus. These projections are topographically organized.[4] Dorsal regions of the MRF associated

---

directed saccades elicited from the left MRF of a headrestrained monkey. The *star* represents the group of trials shown in **D** (Horizontal eye position versus time). (**E** and **F**) Horizontal and vertical eye (**E**) or gaze (**F**) position of centering saccades elicited from the right, caudal MRF of a monkey whose head was restrained (**E**) and unrestrained (**F**). Bar conventions are the same as described in **A**.

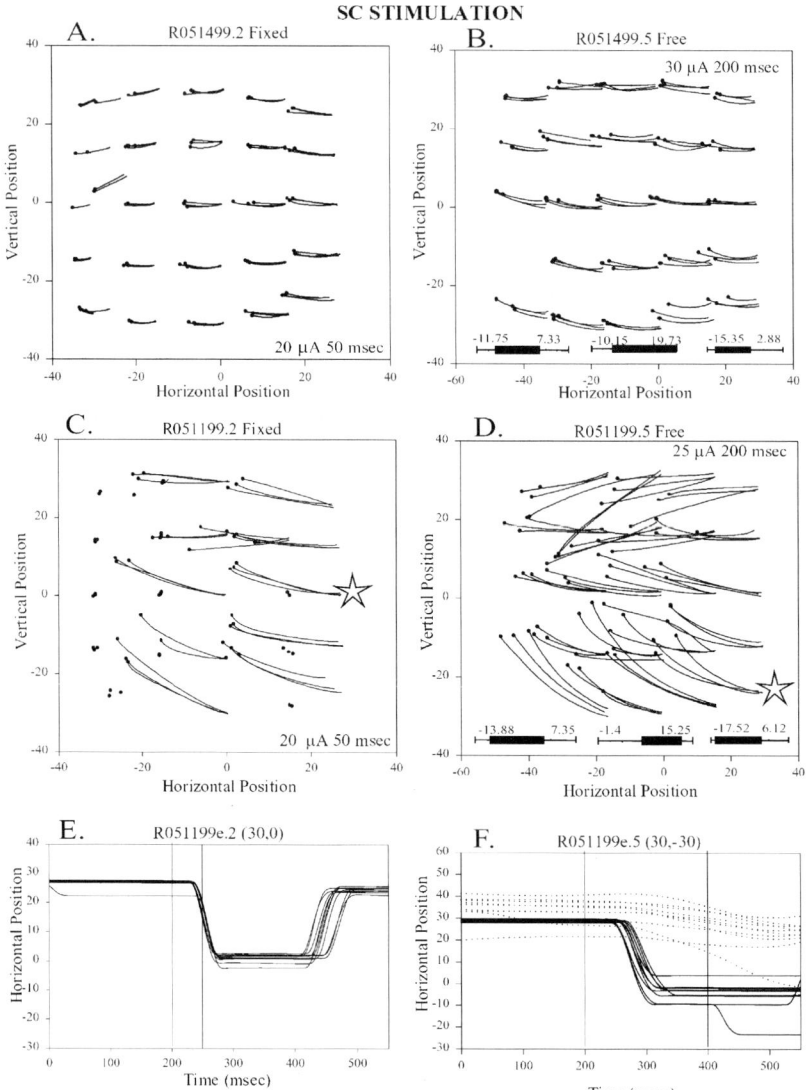

FIGURE 6. Stimulation of the superior colliculus. (**A**) Fixed vector saccades elicited from the right, rostral superior colliculus. (**B**) Similar fixed vector saccades elicited from the same site in the rostral superior colliculus as in **A** when the monkey's head was released. *Bars* below each of the initial gaze positions have same conventions as in FIGURE 5. (**C**) Goal-directed site found in the right caudal superior colliculus. *Star* indicates the specific group of trials shown in **E**. (**D**) The saccades elicited from the same site shown in **C** were less goal-directed when the monkey's head was released. *Star* indicates the specific group of trials shown in **F**. Initial eye (re: head) position bars have the same conventions as in FIGURE 5. (**E**) Horizontal eye position versus time of the saccades shown by the *star* in **C**. (**F**) Horizontal gaze (re: space) (*solid line*) and horizontal head position (*dotted line*) versus time for the trials shown by the *star* in **D**.

with small fixed vector saccades receive projections from the rostral portions of the superior colliculus that are associated with small saccades. On the other hand, the more ventral portion of the MRF (variable saccade region) receives projections from the more caudal portion of the superior colliculus that is associated with large gaze shifts.[3,19] The MRF also has a moderate projection to the excitatory burst region of the paramedian pontine reticular formation (PPRF), but sends a much heavier, bilateral projection to the omnipause region (i.e., raphe interpositus) of the PPRF where the cells pause for each saccade.[9,14,22,32] The MRF also provides contralateral descending projections to the nucleus reticularis tegmenti pontis (one of the major sources of oculomotor input to the cerebellum) and very strong ipsilateral (and weak contralateral) projections to the cervical spinal cord. En route to the cervical spinal cord these axons produce collateral branches that bilaterally innervate the head-associated nuclei of the medulla including nucleus gigantocellularis, nucleus reticularis ventralis, and nucleus lateralis reticularis.[9] The MRF also provides ipsilateral ascending projections to the intramedullary lamina of the thalamus, ventral lateral geniculate nucleus, and hypothalamus. The afferents to the MRF from the cervical spinal cord and the PPRF may provide a conduit for the feedback of head information from the spinal cord and saccade-associated activity of the excitatory burst region of the PPRF destined for the superior colliculus.[2,29]

Comparison of the Activity and Function of Neurons in the MRF and Superior Colliculus

The evidence that one of the major sources of afferents to the MRF arise from the intermediate and deep layers of the superior colliculus would suggest that the discharge of neurons in the MRF would be similar to those found in the superior colliculus. In fact, many of the low-background MRF "eye" neurons showed similar burst and build-up characteristics as those of the neurons in the superior colliculus. Similar to the SC, activity of MRF low-background, build-up neurons began 120 to 150 ms before the onset of contraversive saccades.[18,38] Following this build-up of activity, these neurons would produce a tightly coupled burst of spikes that started before and continued into the saccade. The burst of MRF neurons rostral to the posterior commissure could begin as early as 45 ms before saccade onset,[18] while the peak discharge of more caudal MRF neurons (behind the posterior commissure; FIGS. 3 and 4) would occur on average 6 ms before saccade onset.[38] The activity of both groups of neurons was cut off or clipped with the end of the saccade. Often these cells, like those in the superior colliculus, responded to a visual stimulus, especially when it was used as the object of a saccade. Burst neurons (without a build-up) make up the remainder of the cMRF low-background saccade-related neurons. Again similar to the burst neurons in the superior colliculus, these cells began to fire 30 to 40 ms before the onset of contraversive saccades discharged in association with visual stimuli. In addition, the distal edge of the movement fields of these neurons was closed.

One significant difference between the collicular and reticular saccade-associated neurons was that MRF neurons discharge during spontaneous saccades in the dark.[37,38] Furthermore, the majority of MRF neurons have high-background activity. High-background neurons have a continuous discharge of greater than 10 spikes per second between eye movements. This background activity was often modulated by task parameters and could increase or decrease during periods of active fixation.[38] High-background MRF neurons also produce a saccade-associated burst of spikes that continues during the saccade and can be clipped or partially clipped with saccade end. Thus, the movement fields of these neurons were either closed (clipped cells) or open (partially clipped). An interesting feature of high-background MRF neurons was that their activity was often inhibited before and during ipsilateral saccades. The latency of the peak discharge of low- and high- background MRF neurons located caudal to the posterior commissure was slightly shorter than that measured to a similar group of burst and build-up neurons from the superior colliculus.[38] MRF neurons located rostral to the posterior commissure are somewhat different. They have bursts associated with vertical (rather than horizontal) saccades, and their discharge can actually precede the collicular burst, beginning up to 45 ms before the onset of eye movement.[18] In sum, when examined in head-restrained monkeys, collicular and MRF neurons have many similarities including burst onset, latency, and movement field characteristics. In contrast to neurons in the SC, MRF neurons typically had higher background discharge, rostral MRF neurons had earlier latency, and most MRF neurons fired during spontaneous eye movements in the dark.

When the monkey's head is allowed to move, clear differences emerge between these two structures. A new class of MRF head-associated neurons not previously described in either the reticular formation or superior colliculus was found. These neurons fired for gaze movements larger than 45°. The peak discharge occurred during the head-movement portion of the gaze shift and was most closely associated with peak head velocity. The discharge was greatest for the largest horizontal gaze shift that could be elicited. The head neurons were not activated by oblique gaze movements directed more than 30° from the horizontal plane. Analysis of the discharge of these neurons showed a close relationship to head position and head velocity. The latency of these head cells would be most appropriate to provide a feedback signal to the superior colliculus about the current position and velocity of the head during an ongoing combined head and eye movement. The second difference between the superior colliculus and the MRF was the apparent lack of gaze-associated neurons in the remainder of the MRF neurons recorded to date.

Additional differences between the superior colliculus and the MRF in head-unrestrained animals were demonstrated using electrical microstimulation. Regions of the MRF and superior colliculus where small amplitude, fixed-vector saccades were generated, continued to produce fixed-vector saccades of similar amplitude when the head was unrestrained. Stimulation of more caudal regions of the superior colliculus that produced goal-directed saccades with the head restrained continued to produce convergent saccades with the head unrestrained (FIG. 6). The amplitudes of these gaze-evoked movements were relatively fixed, but their direction was not. This is not characteristic of a region providing a gaze displacement command.[10] An alternative explanation has been suggested by Klier and colleagues that the persistent convergence of gaze following release of the head is most likely secondary to the superior colliculus providing a signal of desired gaze *direction* in retinal coordinates, not gaze displacement.[21] We have not been able to precisely reproduce this experiment in the MRF, but most goal-directed MRF sites were associated with some reversal of eye direction, which was not often noted after stimulation in the superior colliculus. In addition, a number of MRF stimulation sites were associated with centering saccades. Stimulation of centering regions after the monkey's head was released continued to produce centering eye movements. Little if any head movement was noted. These stimulation results suggest that while the output of the superior colliculus is primarily in gaze or retinal coordinates, the output of the caudal MRF reflects a mixture of gaze and spatial coordinates, possibly from simultaneous activation of dissimilar components.

An alternative method for determining the function of the MRF and the superior colliculus in the control of combined head and eye movements is to remove these cells from the circuit. Previous lesion experiments in which either the MRF or superior colliculus was electrolytically or surgically damaged were fraught with multiple drawbacks, including damage to fibers of passage, inadvertent damage to adjacent structures or blood supply, and a severe inability to determine the immediate effect of the lesion since a period of recovery following the lesion was required before the eye movements could be observed. To better understand the function of the MRF and superior colliculus in the control of head and eye movements, recent experiments have used pressure injections of microliter quantities of muscimol, a $GABA_A$ agonist, to temporarily inactivate the neurons and not the axons in passage of each region. Selective injections of muscimol into the superior colliculus produced saccades that were hypometric and whose trajectories were curved.[1] Muscimol injections into the MRF produced two primary effects that depended upon the location of the injection. Injections in the more rostral portion of the MRF adjacent to the interstitial nucleus of Cajal (InC) produced hypometric vertical saccades.[41] On the other hand, injections in the MRF caudal to the posterior commissure adjacent to the oculomotor nucleus, produced hypermetric saccades to the contralateral side.[40] Other observations following caudal injections of muscimol included macrosaccadic square-wave jerks. The direction and amplitude of these spontaneous saccades brought the eyes to a particular position or "goal" in space. When the monkey's head was released following the caudal muscimol injections, there was a contralateral head tilt. Changes in the generation of gaze shifts have not been studied following muscimol injection in either the superior colliculus or MRF of head-unrestrained monkeys.

Taken together, the findings from single-unit, electrical microstimulation, and inactivation experiments have significant implications for the processing of combined head and eye movement information in the midbrain. Despite the close anatomic linkage between these two regions, they clearly do not perform the same function. On the basis of single-unit recording, the superior colliculus provides a signal corresponding to desired gaze displacement to downstream structures for moving the head and eyes. This would correlate well with the results of saccade hypometria produced by inactivating the colliculus with muscimol because it would reduce the size of the outgoing signal. The single-unit recordings in the MRF suggest a multifaceted function for this structure. Low-background MRF burst neurons could participate in the feed-forward function of separating the horizontal component of movement from the oblique output of the superior colliculus. This signal could be directed to the nucleus reticularis pontis (a precerebellar nucleus) or the PPRF. The group of MRF burst neurons associated with vertical saccades and located rostral to the posterior commisure could perform a similar function for the vertical component of gaze.[18] A second purpose for saccade-associated MRF activity is to update neurons in the superior colliculus with a feedback signal corresponding to the current change in eye position and eye velocity.[40] The timing of MRF head neurons suggests that they also could provide collicular feedback, but for current head position and velocity. The high background activity of the majority of cMRF neurons between saccades could be used to suppress the occurrence of eye movement via connections to the omnipause region of the raphe interpositus in the PPRF. Evidence of saccade hypermetria and repetitive square-wave jerks following muscimol injection in the MRF support these ideas.

Separate or Combined Head and Eye Control in the Midbrain

Combined movements of the head and eye pose a formidable control problem for the brain stem where most of the circuitry resides for the generation of such coordinated movements. The eyes have low inertia, twelve muscles, and can move at very high velocities. The head on the other hand has very high inertia, more than 30 muscles, and moves at lower velocity than the eye. How do the superior colliculus and the MRF interact with these disparate structures? At least two different models have been proposed. The *separate head and eye hypothesis* proposes that the gaze signal from a place-coded structure such as the superior colliculus is transformed into separate head and eye signals, which then target the portions of the brain stem controlling the head and eye.[30] In this scenario, the authors divide the gaze system into two pieces: a static portion that provides the command signal, and a second, dynamic portion that actually carries out the execution of the head and eye movements. The input to the model utilizes target position (re: space), which is immediately converted to a static gaze error signal by subtracting two feedback signals: static eye position (re: head) and static head position (re: space). One major drawback of this idea is that while an effect of initial eye position has been demonstrated in the stimulation results of the superior colliculus and MRF (centering and goal-directed movements), only a subset of the MRF sites (i.e., centering) showed a persistence of this effect after releasing the head. At the remaining sites in the MRF and superior colliculus, shifts in gaze were associated with moderate changes in gaze direction, and little change in gaze amplitude, suggesting that these regions are coded in gaze coordinates. As a result, an alternative, *gaze hypothesis,* may better fit the findings.[16] In this model, gaze signals originating from the SC would target both head and eye regions of the brain stem and spinal cord. The inappropriate head or eye portion of the gaze signal would then be subtracted at the particular target region, eye, or head. At the same time, this schema would require a gaze feedback signal directed to the superior colliculus to shutdown the gaze signal output with the end of the gaze movement. This could be accomplished using a single gaze feedback, or separate eye and head plant efference feedback loops that summed as they reached the superior colliculus.[16] If the gaze hypothesis is correct, then cells in the tectorecipient regions such as the cMRF should be associated with gaze if they are in the feedforward path, or either head, eye, or gaze if they function in providing collicular feedback. To date, MRF neurons have been most closely associated with head or eye signals and not gaze. In addition, the temporal pattern of firing for many of the MRF head neurons and some eye neurons would be appropriate for performing a feedback function. Loss of eye feedback signals would produce the hypermetric, goal-directed saccades that have been recently described following temporary inactivation of the MRF with muscimol.[40] In sum, the anatomic and physiologic evidence suggests that the transformation of gaze signals from the superior colliculus into the activity necessary to move the head and eyes requires a number of steps. Most likely, many of the neurons in the closely allied MRF participate in a feedback role to insure the accuracy of the gaze shift.

REFERENCES

1. AIZAWA, H. & R.H. WURTZ. 1998. Reversible inactivation of monkey superior colliculus. I. Curvature of saccadic trajectory. J. Neurophysiol. **79:** 2082–2096.
2. BJORKLAND, M. & J. BOIVIE. 1984. An anatomical study of the projections from the dorsal column nuclei to the midbrain in cat. Anat. Embryol. (Berl.) **170:** 29–43.
3. CHEN, B. & P.J. MAY. 2000. The feedback circuit connecting the superior colliculus and central mesencephalic reticular formation: a direct morphological demonstration. Exp. Brain Res. **131:** 10–21.
4. COHEN, B. & J.A. BUTTNER-ENNEVER. 1984. Projections from the superior colliculus to a region of the central mesencephalic reticular formation (cMRF) associated with horizontal saccadic eye movements. Exp. Brain Res. **57:** 167–176.
5. COHEN, B. et al. 1985. Horizontal saccades induced by stimulation of the central mesencephalic reticular formation. Exp. Brain Res. **57:** 605–616.
6. CULLEN, K.E. & D. GUITTON. 1997. Analysis of primate IBN spike trains using system identification techniques. I. Relationship To eye movement dynamics during head-fixed saccades. J. Neurophysiol. **78:** 3259–3282.
7. CULLEN, K.E. & D. GUITTON. 1997. Analysis of primate IBN spike trains using system identification techniques. II. Relationship to gaze, eye, and head movement dynamics during head-free gaze shifts. J. Neurophysiol. **78:** 3283–3306.
8. CULLEN, K.E. & D. GUITTON. 1997. Analysis of primate IBN spike trains using system identification techniques. III. Relationship to motor error during head-fixed saccades and head-free gaze shifts. J. Neurophysiol. **78:** 3307–3322.
9. EDWARDS, S.B. 1975. Autoradiographic studies of the projections of the midbrain reticular formation: descending projections of nucleus cuneiformis. J. Comp. Neurol. **161:** 341–358.
10. FREEDMAN, E.G., T.R. STANFORD & D.L. SPARKS. 1996. Combined eye-head gaze shifts produced by stimulation of the superior colliculus in rhesus monkeys. J. Neurophysiol. **76:** 927–952.
11. FREEDMAN, E.G. & D.L. SPARKS. 1997. Eye-head coordination during head-unrestrained gaze shifts in rhesus monkeys. J. Neurophysiol. **77:** 2328–2348.
12. FREEDMAN, E.G. & D.L. SPARKS. 1997. Activity of cells in the deeper layers of the superior colliculus of the rhesus monkey: evidence for a gaze displacement command. J. Neurophysiol. **78:** 1669–1690.
13. FREEDMAN, E.G. & D.L. SPARKS. 1999. Coordination of the eyes and head: movement kinematics. Exp. Brain Res. **131:** 22–32.
14. GRANTYN, A. & R. GRANTYN. 1982. Axonal patterns and sites of termination of cat superior colliculus neurons projecting in the tecto-bulbo-spinal tract. Exp. Brain Res. **46:** 243–256.
15. GUITTON, D., M. CROMMELINCK & A. ROUCOUX. 1980. Stimulation of the superior colliculus in the alert cat. I. Eye movements and neck EMG activity evoked when the head is restrained. Exp. Brain Res. **39:** 63–73.
16. GUITTON, D., D.P. MUNOZ & H.L. GALIANA. 1990. Gaze control in the cat: studies and modeling of the coupling between orienting eye and head movements in different behavioral tasks. J. Neurophysiol. **64:** 509–531.
17. GUITTON, D. & D.P. MUNOZ. 1991. Control of orienting gaze shifts by the tectoreticulospinal system in the head-free cat. I. Identification, localization, and effects of behavior on sensory responses. J. Neurophysiol. **66:** 1605–1623.
18. HANDEL, A. & P.W. GLIMCHER. 1997. Response properties of saccade-related burst neurons in the central mesencephalic reticular formation. J. Neurophysiol. **78:** 2164–2175.

19. HUERTA, M.F. & J.K. HARTING. 1984. The mammalian superior colliculus: studies of its morphology and connections. *In* Comparative Neurology of the Optic Tectum. E.H. Vanegas, Ed.: 687–773. Plenum. New York.
20. JUDGE, S.J., B.J. RICHMOND & F.C. CHU. 1980. Implantation of magnetic search coils for measurement of eye position: an improved method. Vision Res. **20**: 535–538.
21. KLIER, E.M., H. WANG & J.D.CRAWFORD. 2001. The superior colliculus encodes gaze commands in retinal coordinates. Nat. Neurosci. **4**: 627–632.
22. MOSCHOVAKIS, A.K. & A.B. KARABELAS. 1985. Observations on the somatodendritic morphology and axonal trajectory of intracellularly HRP-labeled efferent neurons located in the deeper layers of the superior colliculus of the cat. J. Comp. Neurol. **239**: 276–308.
23. MOSCHOVAKIS, A.K., A.B. KARABELAS & S.M. HIGHSTEIN. 1988. Structure-function relationships in the primate superior colliculus. II. Morphological identity of presaccadic neurons. J. Neurophysiol. **60**: 263–302.
24. MUNOZ, D.P, D. GUITTON & D. PELISSON. 1991. Control of orienting gaze shifts by the tectoreticulospinal system in the head-free cat. III. Spatiotemporal characteristics of phasic motor discharges. J. Neurophysiol. **66**: 1642–1666.
25. MUNOZ, D.P. & D. GUITTON. 1991. Control of orienting gaze shifts by the tectoreticulospinal system in the head-free cat. II. Sustained discharges during motor preparation and fixation. J. Neurophysiol. **66**: 1624–1641.
26. MUNOZ, D.P. & R.H. WURTZ. 1995. Saccade-related activity in monkey superior colliculus. I. Characteristics of burst and buildup cells. J. Neurophysiol. **73**: 2313–2333.
27. MUNOZ, D.P. & R.H. WURTZ. 1995. Saccade-related activity in monkey superior colliculus. II. Spread of activity during saccades. J. Neurophysiol. **73**: 2334–2348.
28. PARE, M., M. CROMMELINCK & D. GUITTON. 1994. Gaze shifts evoked by stimulation of the superior colliculus in the head-free cat conform to the motor map but also depend on stimulus strength and fixation activity. Exp. Brain Res. **101**: 123–139.
29. PECHURA, C.M. & R.P. LIU. 1986. Spinal neurons which project to the periaqueductal gray and the medullary reticular formation via axon collaterals: a double-label fluorescence study in the rat. Brain Res. **374**: 357–361.
30. PHILLIPS, J.O. *et al.* 1995. Rapid horizontal gaze movement in the monkey. J. Neurophysiol. **73**: 1632–1652.
31. ROUCOUX, A., D. GUITTON & M. CROMMELINCK. 1980. Stimulation of the superior colliculus in the alert cat. II. Eye and head movements evoked when the head is unrestrained. Exp. Brain Res. **39**: 75–85.
32. SCUDDER, C.A. *et al.* 1996. Anatomy and physiology of saccadic long-lead burst neurons recorded in the alert squirrel monkey. I. Descending projections from the mesencephalon. J. Neurophysiol. **76**: 332–352.
33. SILAKOV, V.L., D.M. WAITZMAN & S.R. DEPALMA. 1999. Combined head and eye movements evoked by microstimulation of the primate mesencephalic reticular formation. Soc. Neurosci. Abstr. **25**: 1651.
34. SPARKS, D.L. 1999. Conceptual issues related to the role of the superior colliculus in the control of gaze. Curr. Opinion Neurobiol. **9**: 698–707.
35. SYLVESTRE, P.A. & K.E. CULLEN. 1999. Quantitative analysis of abducens neuron discharge dynamics during saccadic and slow eye movements. J. Neurophysiol. **82**: 2612–2632.
36. WAITZMAN, D.M., T.P. MA, L.M. OPTICAN & R.H. WURTZ. 1988. Superior colliculus neurons provide the saccadic motor error signal. Exp. Brain Res. **72**: 649–652.
37. WAITZMAN, D.M., T.P. MA, L.M. OPTICAN & R.H. WURTZ. 1991. Superior colliculus neurons mediate the dynamic characteristics of saccades. J. Neurophysiol. **66**: 1716–1737.
38. WAITZMAN, D.M., V.L. SILAKOV & B. COHEN. 1996. Central mesencephalic reticular formation (cMRF) neurons discharging before and during eye movements. J. Neurophysiol. **75**: 1546–1572.
39. WAITZMAN, D.M., V.L. SILOKOV & S.R. DEPALMA. 1998. Goal-directed movements following electrical microstimulation of the mesencephalic reticular formation in primates. Soc. Neurosci. Abstr. **24**: 145.
40. WAITZMAN, D.M., V.L. SILAKOV, S. DEPALMA-BOWLES & A.S. AYERS. 2000. Effects of reversible inactivation of the primate mesencephalic reticular formation. I. Hypermetric goal-directed saccades. J. Neurophysiol. **83**: 2260–2284.

41. WAITZMAN, D.M., V.L. SILAKOV, S. DEPALMA-BOWLES & A.S. AYERS. 2000. Effects of reversible inactivation of the primate mesencephalic reticular formation. II. Hypometric vertical saccades. J. Neurophysiol. **83:** 2285–2299.
42. WAITZMAN, D.M. *et al.* 2001. Signals related to movements of the head and eyes in the mesencephalic reticular formation (MRF) of primates. Soc. Neurosci. Abstr. **27:** 1073.
43. YEOMANS, J.S. & E.J. TEHOVNIK. 1988. Turning responses evoked by stimulation of visuomotor pathways. Brain Res. **472:** 235–259.

Neural Discharge in the Superior Colliculus during Target Search Paradigms

EDWARD L. KELLER AND ROBERT M. McPEEK

The Smith-Kettlewell Eye Research Institute, 2318 Fillmore Street, San Francisco, California 94115, USA

ABSTRACT: Neural studies of oculomotor function in the past have been conducted with the use of very simple visual stimuli. More recently there has been a new emphasis on using more natural stimuli to extend our knowledge of oculomotor organization. Visual search paradigms are an example of the use of these more natural visual surrounds. In search a subject must locate and saccade to a target that appears simultaneously with an array of distractors. When monkeys are used in this paradigm, it is possible to record from neurons located in various central structures in the brain while the initial visual response, subsequent discrimination processes and final saccadic movement unfold. In the present study we used an array of four visual stimuli, and the target was distinguished by its odd color from three distractors of uniform color. Location of the target within the array and its color were randomly selected on each trial. Neurons located in the deeper layers of the superior colliculus (SC) were recorded by standard methods in blocks of search trials.

We found several new features in the discharge of SC neurons using this search paradigm that have not previously been reported in studies using single-target visual displays. (1) In contrast to the "winner take all" behavior previously reported for the SC, we found evidence of concurrent processing of alternative movement vectors. When incorrect movements were made to distractor locations, this concurrent activity was associated with significantly shorter intersaccadic intervals. (2) The discharge profile of the visual response in many units was modified by the appearance of a second prominent peak which followed the initial phasic visual response, but which was clearly differentiated from a third burst in activity associated with a saccade into the cell's response field. In some neurons, the activity in this second peak was discriminatory for the impending saccade vector. That is, it was larger when the target appeared in the response field of the cell than when it contained a distractor. This target selection signal was thus distinct from the burst normally associated with saccades into the movement fields of SC neurons. (3) Some saccades in search had a curved trajectory bowing toward the location of a distractor. These saccades were accompanied by an elevated discharge of neurons coding that distractor.

KEYWORDS: visual search; superior colliculus; saccades; target selection; concurrent processing

Address for correspondence: Edward L. Keller, The Smith-Ketterwell Eye Research Institute, 2318 Fillmore Street, San Francisco, CA 94115. Voice: 415-345-2102; fax: 415-345-8845.
elk@ski.org

Ann. N.Y. Acad. Sci. 956: 130–142 (2002). © 2002 New York Academy of Sciences.

INTRODUCTION

Saccades are rapid movements of the eye that realign the visual axes on new objects of interest in one's visual surround. The neural basis underlying the production of saccades has been studied in the past, for the most part, with simple visual stimuli. Typically, these stimuli consisted of a fixation point followed by the presentation of a single target or sequentially presented single targets. Based on the results of these studies, the superior colliculus (SC) has been identified as a pivotal neural structure involved in the control of saccades.[1] This laminated structure, located on the dorsal surface of the midbrain, receives direct retinal input and projections from a number of cortical areas implicated in saccade production and projects to the brainstem saccadic burst generator, the immediate premotor structure involved in the temporal control of saccades.

Recently there has been a new emphasis on using more natural visual stimuli to extend our knowledge of the organization of the saccadic system.[2] Visual search paradigms are another example of the use of more natural visual surrounds. In search a subject must locate and saccade to a target that appears simultaneously with an array of distractors.[3] Neural recordings have been made in the cortical frontal eye fields of monkeys preforming search tasks.[4] We have recently carried out a series of experiments in monkeys working on a visual search task while recording from neurons in the deeper layers of the SC because this structure has been implicated in several of the various saccade preparatory processes.[5]

By use of a search task, we have found several features in the discharge of SC neurons that have not been previously reported in studies using single-target stimuli. The results allow us to make several new speculations about the organization of the dynamic neural networks for saccade control that reside in the SC.

METHODS

Three adolescent, male rhesus monkeys (*Macaca mulatta*) were used in these studies. A scleral eye-coil and a head-restraint device were implanted on the animals' skulls under isofluorane anesthesia and aseptic surgical conditions. After 2–3 months of training in behavioral tasks, the animals were prepared for chronic single-unit recording in a second aseptic surgery. A stainless steel recording chamber tilted 38 deg posterior from the vertical stereotaxic plane, was positioned on the midline above the SC. Antibiotics (Cefazolin) and analgesics (Buprenex) were administered as needed during the recovery period under the direction of a veterinarian. After the animals had recovered from surgery, recording sessions were made with microelectrodes advanced into the SC with a miniature hydraulic drive system.

Data collection and storage were controlled by a custom real-time program running on a PC. Eye position and velocity were sampled at 1 kHz and digitally stored on disc. A Macintosh computer, which was interfaced with the PC, generated the visual displays using software constructed with the Video Toolbox library.[6] Visual stimuli were presented on a 29-inch color CRT (Viewsonic GA29) in synchronization with the monitor's vertical refresh. The monitor had a spatial resolution of 800 × 600 pixels and a non-interlaced refresh rate of 75 Hz. The monitor was positioned 33 cm in front of the monkey and allowed stimuli to be presented in a field of view of approximately ±32° along the horizontal meridian and ±30° along the vertical meridian.

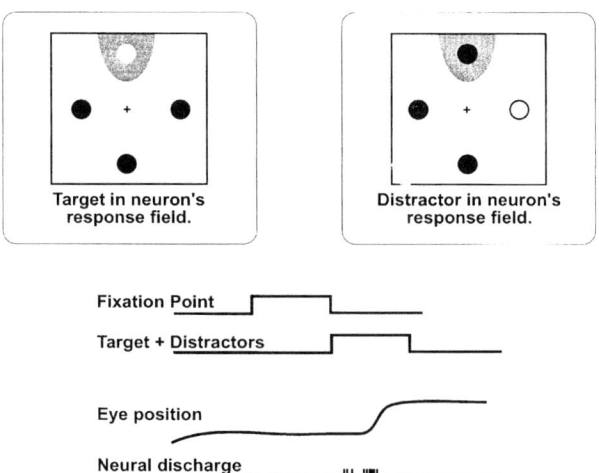

FIGURE 1. Schematic view of the search task. The cartoons at the top represent the spatial arrangement of the visual stimulus array on two example trials. The animal fixates straight ahead (*small cross*) and four stimuli appear simultaneously. One, the target, is of an odd color, while the others, distractors, are of a uniform color. The location and color of the target is randomized on each trial. The array is arranged such that one stimulus on each trial appears in the response field (*region shaded gray*) of the neuron being recorded. The other three stimuli appeared at the same eccentricity, but at 90-deg intervals from the stimulus in the response field. The temporal arrangement of fixation point, stimulus array, eye movements, and neural discharge are shown below.

The monkeys were seated in a primate chair with their heads restrained for the duration of the testing sessions. They executed behavioral tasks for liquid reward, and were allowed to work to satiation. Records of each animal's weight and health status were kept, and supplemental water was given as necessary.

Delayed-Saccade Task

At the beginning of each trial, a white fixation spot appeared in the central position against a homogenous dim background. The monkeys were required to keep their eyes within 1.5–2° of the fixation point during an initial fixation interval of 450–650 ms. At the end of this interval, a single-target stimulus was presented at a peripheral location while the fixation point remained illuminated. Monkeys were required to maintain central fixation until the disappearance of the fixation point 500–700 ms later. Once the fixation point disappeared, they were rewarded for making a saccade to the peripheral stimulus within 70–400 ms. Early or late responses were not rewarded. The target was either a red or green disc.

Search Task

Details for this task are shown in FIGURE 1. Trials began with a 450–650 ms fixation period as for the delayed-saccade task. However, at the end of the fixation period, the fixation point disappeared. Simultaneously, one target and three distractor stimuli were presented at equal eccentricity from the fixation point, but separated by angles of 90° in direction. The stimuli were red or green discs, which were chosen to be approximately equiluminant, with measured luminances of 0.90 and 0.92 cd/m^2, respectively. The distractors were all of the same color and the target differed from them only by virtue of its odd color. In each trial, the colors of the target and distractors were randomly chosen.

The locations of the stimuli were adjusted for each neuron so that on every trial, either the target or a distractor was presented near the center of the neuron's response field, as determined using single stimuli in the delayed-saccade task. The stimuli were M-scaled in order to keep their salience constant across different eccentricities.[7] At an eccentricity of 15°, each stimulus element subtended 2° of visual angle. Monkeys were given a liquid reward for bringing their eyes to the location of the target within 275 ms of the onset of the stimuli. This allowed the possibility of a reward being given if the initial saccade was incorrect, but was followed very rapidly by a correct second saccade. In practice, second saccades usually reached the target too late for the monkey to receive a reward. However, the occasional rewarded two-saccade response encouraged the monkeys not to give up after an initial incorrect response.

Single-Unit Recording

We used standard methods to record single neurons in the deeper layers of the SC of three rhesus monkeys. The microelectrode signal was amplified, bandpass filtered, and displayed on a digital storage oscilloscope. Action potentials were discriminated and converted into TTL pulses using a time-amplitude window discriminator. The computer data acquisition system registered the occurrence of spikes with a resolution of 1 kHz, and the neural data was stored in register with the behavioral measurements.

Data Analysis

Off-line analysis of the eye movement data was performed by algorithms using velocity and acceleration criteria to detect the beginning and end of saccades. The algorithm's identification of saccades was visually inspected for every trial to verify its accuracy. We calculated the curvature of saccade responses using a curvature metric described by Smit and Van Gisbergen.[8] In the summary analyses of the rapid two-saccade responses, we selected error saccades to one of the distractor positions for analysis based on which distractor had the greatest number of error trials. Thus, the responses for each cell all had initial saccades directed to the same distractor, followed by a second saccade to the target positioned in the neuron's response field.

RESULTS

Discrimination of the Target in a Cell's Response Field

We first examined trials in which a correct, single saccade was made to the location of the target in the search array, which might be at a position in or out of the response field of the neuron. The data is aligned on the appearance of the stimulus array (distractors and odd-colored target). Earlier studies using other paradigms had found that the discharge of a type of SC movement-related neuron having tonic ac-

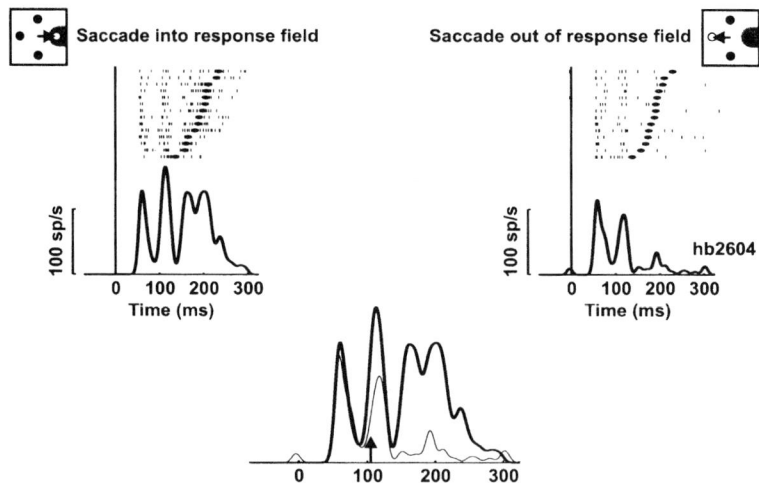

FIGURE 2. Multipeaked response of a sample collicular burst neuron in the search task. Neural activity is represented by raster plots which show individual discharges of the cell on each trial. Mean activity across trials is plotted as average spike density (sigma = 6 ms).[19] Plots are aligned on appearance of the stimulus array (*vertical lines*). Only data from correct single saccades to the target are included. Rasters are arranged on the basis of saccade latency (shortest at the *bottom*). Times of saccade onsets for each trial are indicated on the rasters by the *filled circles*. *Left plate* shows trials in which the target was located in the response field of the cell. *Right plate* shows trials in which the target was located out of the field at the array location in the opposite direction. *Inset* below shows superimposed average spike densities for the two types of trials (*thick curve* for saccades into field and *thin curve* for saccades out of field). The curves are magnified so the time of separation during the second peak of activity can be seen more clearly (*arrow*).

tivity ("build-up" or "prelude" neurons) evolves over time to indicate which stimulus has been selected by the monkey as the saccade goal.[9,10] We found a similar pattern of activity in these cells in our search task. The search task also revealed a more unexpected finding in a different class of SC neuron, the visuomotor (VM) burst neuron, which will be the focus of this section. Previously, it was thought that burst cells do not represent target selection, but rather only serve to trigger saccade initiation. However, our search task uncovered a discrete phase in the discharge of some of these cells which seems to be related to target selection.

Typical results are shown in FIGURE 2 for one such cell. The plate on the left shows a series of trials in which the target appeared in the center of the response field of the neuron and a correct saccade was made to its location. The discharge for individual trials shown on the raster are arranged by saccade latency with the shortest latency at the bottom. The multiphasic nature of the cell's response, when aligned on search array onset, can be seen in the discharge pattern on the individual trials and even more clearly in the averaged spike density trace below the raster. After a short latency from the onset of the search array (~50 ms for this cell), the cell discharges a short burst of several spikes sharply aligned on the onset of the stimulus array. This is the initial visual response of the cell; it is transitory and is followed by a short period of silence or occasional temporally scattered spikes. A second burst of spikes begins at about 100 ms and declines yet again before the final upturn in spike production that constitutes the saccade-related discharge associated with the movement into the response field. The second burst of spikes appears also to be a visual response because it is better aligned with visual stimuli onset than with the saccades to the location of the target.

The plate on the right in FIGURE 2 shows the discharge of the same cell, but now for trials in which the target appeared at a location outside its response field. In this example, only the responses for target appearances in the direction directly opposite the cell's response field are shown; responses, however, to the other two array locations were also collected in the same block of trials, and similar results were obtained for these two locations. The cell again shows a double-peaked visual response in the target aligned data, but lacks the saccade-related response as the movement goes to a location outside the cell's response field.

The double-peaked responses for target locations in and out of the response field are directly compared in the inset below the main figure. In this inset, the averaged spike-density traces from the two cases shown above are magnified and superimposed. It is important to remember that a visual stimulus appeared in the response field of the cell on every trial. Furthermore, because the colliculus is not sensitive to differences in color among visual stimuli, the stimulus in all trials, from the point of view of the SC, is the same.[11] This is reflected in the initial visual responses that are nearly the same in the superimposed traces. However, early on in the development of the second peak of the visual response, the mean response traces separate and become significantly different for responses into and out of the cell's field. For the cell illustrated in this figure, this time of separation was 108 ms. This time of discrimination was typical for most of the cells we analyzed.

Some cells showed a clearer second burst than others. We used a trial-by-trial Poisson spike train analysis to select the VM burst cells having the clearest evidence for two discrete bursts of visual activity in individual trials.[12] We focused our subsequent analyses on the 14 neurons so identified, out of the 29 burst cells in our sample. Thirteen of these 14 cells showed a stronger burst when the monkey chose the stimulus inside the cell's response field versus the opposite stimulus. The mean increase over this population was 63%, and this increase was significant at the $p < .001$ level (t test). Although the second burst was visual, it was modulated by the monkey's choice of saccade goal. Even when short-latency saccades (< 175 ms) were excluded, the second burst was still modulated by the monkey's future execution of saccade vector.

Concurrent Processing of Alternative Movement Vectors

When strongly competing saccade goals are present in tasks such as the search paradigm, very short intersaccadic fixation intervals have been reported in both humans and monkeys.[13–15] We discovered a neural correlate of this foreshortened intersaccadic delay in the discharge of some saccade-related neurons recorded in the deeper layers of the SC during our search task. FIGURE 3 shows the discharge and eye movements for one such neuron and, in contrast to the previous section, the data here are aligned on saccade onset. In FIGURE 3A data are shown for the case where single saccades were made to the center of the cell's response field, which for this

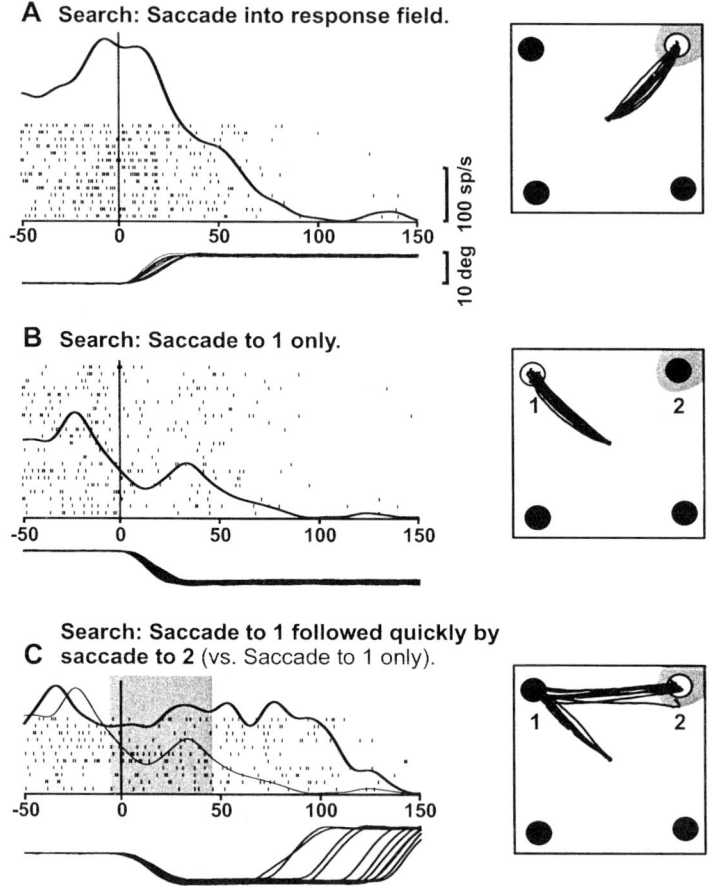

FIGURE 3. Activity of a superior colliculus neuron in the search task. Neural activity is represented by raster plots which show individual discharges of the cell on each trial. Mean activities across trials are plotted as average spike density (sigma = 6 ms). Plots are aligned on saccade onset (*vertical lines*). Eye movements for each raster are shown below. *Insets to the right* show the spatial arrangements of the eye movements superimposed on the stimulus array. The target is always indicated by an *open circle* while the distractors are *filled circles*. The response field of the neuron is the *region shaded gray*. (**A**) Single saccade to the target when it was located in the response field. (**B**) Single saccade to the target when it appeared outside the response field at location 1. (**C**) Initial saccade to a distractor located at 1, followed immediately by a correct saccade to the target at 2. The target at 2 was originally located in the response field. The thick curve superimposed on the raster shows the average spike density for the two-saccade sequences. The spike density for the single-saccade trials from **B** is superimposed (*thinner continuous curve*) on the one for the two-saccade sequence so that the activities can be directly compared. The gray rectangle indicates the perisaccade time interval over which the quantitative analyses were conducted.

neuron was located in the upper right quadrant of the animal's visual field. The cell shows a high frequency burst of discharge that peaks just before saccade onset for movements into the response field of the cell, and then declines during the saccade. FIGURE 3B shows the case when the target is located outside the response field, and single saccades are made to this location. Only data for movements to target location 1 are shown for clarity, but similar results were obtained when the single saccade went directly to the target when it appeared at either of the other two potential locations for search stimuli outside the response field of the cell. The cell's initial presaccadic discharge is high because the cell is a visuomotor neuron and a visual stimulus (a distractor) has appeared in the cell's response field on each trial. However, the cell's discharge declines rapidly before saccade onset and is practically quenched by saccade end. FIGURE 3C compares the mean discharge during a single saccade made to the target at location 1 (from FIG. 3B) to the mean discharge present when an identical saccade is made to a distractor at location 1 followed, after a short intersaccadic fixation interval, by a saccade to the target at location 2. The discharge rates are nearly the same until just before first saccade onset. Then, as the discharge associated with single saccades to location 1 begins to decline (thin curve), the discharge of the cell is maintained during this initial saccade when it will be followed quickly by a second saccade to the target (heavy curve). We hypothesize that it is this maintained activity during the first saccade, which is out of the response field and is not, in the single-saccade case, associated with significant discharge, that represents advance selection or planning for the second saccade (when it will occur at short latency after the first movement). In this exposition we will term this activity concurrent processing.

In order to quantify the difference in maintained discharge, we computed a mean rate by counting spike occurrences during a perisaccadic interval and dividing by the duration of the interval. We defined the "perisaccadic" interval, our epoch of interest, as the 50-ms time period centered on the midpoint in time of the execution of the initial saccade. For the cell shown in FIGURE 3, this perisaccade interval is indicated by the region shaded gray. We obtained sufficient data to quantitatively measure concurrent processing activity during this interval of time in 43 cells. Across our sample population of neurons, there was a significant increase in activity for the rapid two-saccade case over the single-saccade case (Wilcoxon signed-rank test, $p < .001$).

To further support our hypothesis that this activity during the first saccade represents concurrent processing of target selection for the second saccade, we computed an enhancement index defined as $(f_{two} - f_{one})/(f_{best})$, where f_{two} is the mean perisaccadic discharge rate for the initial saccade to a point outside the response field, when it was followed by a second saccade to a target originally in the response field; f_{one} is mean rate for similar single saccades outside the response field, but not followed by a second saccade; and f_{best} is the rate for single saccades when the target was presented inside the cell's response field. If the enhanced discharge seen in the perisaccade interval represents concurrent processing, one would expect this index to be larger when the intersaccadic interval was shorter. In FIGURE 4 we show the results of plotting the enhancement index as a function of the intersaccadic interval between the first and second saccade. This figure shows that enhancement only occurs when the intersaccadic interval is less than about 150 ms, and the most vigorous effect occurs for the very shortest intervals.

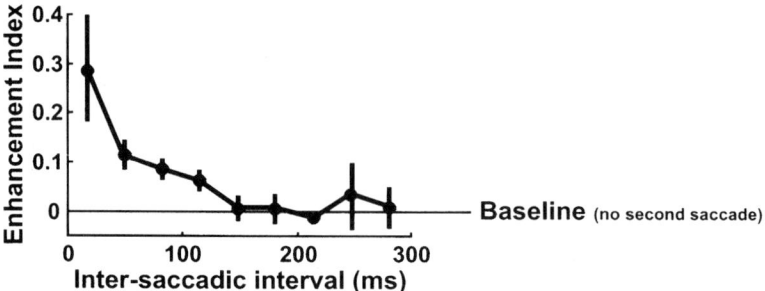

FIGURE 4. The amount of enhancement is greatest for the shortest intersaccadic intervals in two-saccade sequences. The computed enhancement index for all two-saccade movements was divided on the basis of their intersaccadic interval into bins of 30-ms duration. The mean values of enhancement in each bin are shown as *filled circles*. The *vertical bars* represent the SEM. The baseline enhancement when no second saccade occurred (as in FIG. 3B) was zero by definition.

Thus, it appears that this elevated activity represents the advance selection of the second saccade goal, paving the way for the second saccade to be executed with minimal intersaccadic delay.

Discharge in Superior Colliculus Neurons for Curved Saccade Trajectories

We have previously reported that strong competition among alternative targets that is present in the search task results in the more frequent occurrence of curved saccade trajectories that are not often observed when single targets are presented at similar locations.[15] We hypothesized that this trajectory curvature was caused by concurrent activity present on a neural motor map coding the desired location of saccade endpoints. We further showed that curved trajectories occurred most frequently when first saccades were made in search followed by a second saccade to another stimulus location in the search array. In these cases we could predict the direction of curvature (clockwise or counter clockwise) in the first saccade based on the direction of a second subsequent saccade (clockwise or counter clockwise with respect to the endpoint of the first saccade).

In the present series of experiments, we compared the discharge of SC neurons during straight saccades and during curved saccades made to the same target location. We found that we could predict the occurrence and direction of trajectory curvature from the pre-saccadic pattern of activity of neurons located on the SC at sites coding distractors. FIGURE 5 shows an example for one SC cell in which a distractor appeared in the response field of the neuron (upper right quadrant), but the saccade was correctly made to the target located in the lower right quadrant. Four examples of curved trajectories can be seen in FIGURE 5A. In each case the curvature tends to be convex toward the distractor. Other saccades to the same spatial location of the target and in the same block of trials had straight trajectories, and four examples are shown in FIGURE 5A. In FIGURE 5B we compare the spike density trace for the trial with the most curvature in FIGURE 5A to the average spike density obtained for the

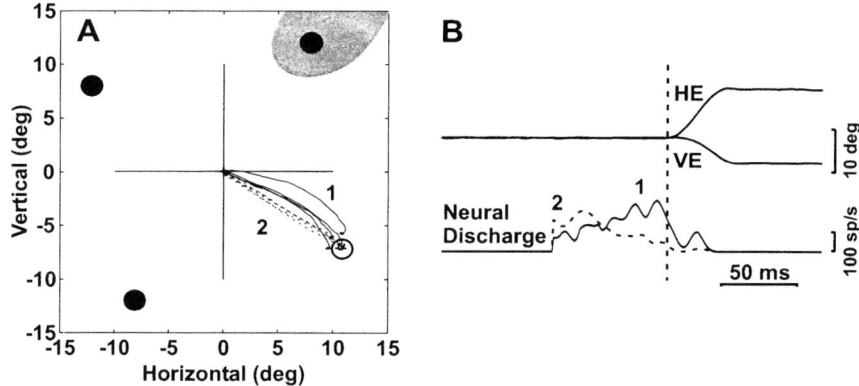

FIGURE 5. Curved saccades are associated with higher presaccadic discharge in collicular neurons located at sites coding distractors. **(A)** Spatial arrangement of the eye movements superimposed on the stimulus array. The target is indicated by an *open circle*, while the distractors are *filled circles*. The response field of the neuron is the *region shaded gray*. Only correct saccades to the target when it was located in lower right quadrant are shown. Some of these saccades are curved (*solid curves*) and others are straight (*dashed curves*). The most curved saccade in the set is indicated by the numeral 1, and the set of straight movements is indicated by a 2. **(B)** Temporal representation of the eye movement (HE is horizontal eye movement and VE vertical eye movement) for the curved saccade 1. Shown below are the associated spike density for saccade 1 (*solid curve*) and the average spike density for the straight movements 2 (*dashed curve*). The spike densities are aligned on the onset of the saccades (*vertical dashed line*). The earlier visual responses of the cell for each trial are removed for clarity.

saccades with straight trajectories shown in FIGURE 5A. The saccade with a curved trajectory was associated with an accelerating discharge that peaked at a relatively high level (~200 spikes/s) just before saccade onset and then was rapidly silenced just after saccade onset. In contrast, for the set of saccades with straight trajectories, the discharge began to decline early—well before saccade onset—and was silenced by the time of this saccade. The other curved saccades included in the example in FIGURE 5A also showed a higher discharge rate in the presaccadic period.

Based on this difference in discharge patterns, we hypothesize that saccade trajectory curvature results when strong competition exists between two populations of SC neurons simultaneously encoding two different saccade goals. Apparently, the site representing the target became dominant in the immediate presaccadic period as the activity at the distractor site declined rapidly. Nevertheless, because ocular muscle activity lags SC drive by perhaps 8 ms or more,[16] this competitive activity at two SC sites leads to a later saccade whose initial direction is pointed in between the visual stimuli coded by the competing sites. The subsequent sharp drop in activity at the distractor site allows the saccade trajectory to curve back toward the actual location of the target. We make this hypothesis more explicit with the schematic shown

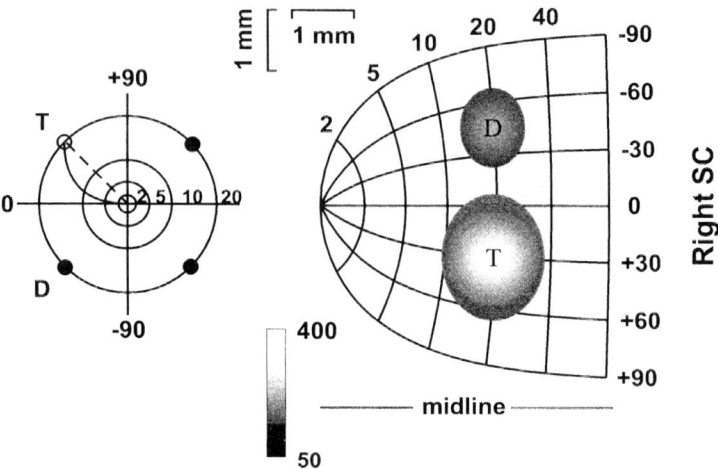

FIGURE 6. Hypothesized population discharge in the colliculus just before the onset of the curved saccade shown in the *inset on the left*. The target (T) appears in the *upper-left quadrant,* and the straight-line path for a movement to the target is shown by the *dashed line in the inset*. Instead, a saccade with an initial direction in between the distractor D and T is made. The straight left direction is indicated by a zero, up by +90 and down by −90. Small numerals on the *circles* indicate eccentricity in the visual field. The schematic on *the right* represents the right superior colliculus, which codes for saccades with a left component. The motor map of the colliculus is indicated by the numerals along its upper edge for eccentricity and elevation along the rightward edge (modified from Ottes *et al.*).[20] Hypothesized extent and intensity of the population discharge in the right colliculus just before the curved saccade onset are shown in *gray scale*. Target-related activity (centered at location marked with a T) is larger, more intense and growing, and that for the distractor (centered at location D) is smaller, less intense and declining rapidly.

in FIGURE 6, which shows the hypothesized distribution of population activity on the SC in the presaccadic period before movement onset. It should be noted in this schematic, and in the actual recordings shown in FIGURE 5, that the final direction of the curved saccades swings well beyond the initial retinotopic direction coded by the target, and hence the final trajectory of the curved saccades is not coded directly by either of the two active populations shown in FIGURE 6.

DISCUSSION

In the first series of experiments, we found a previously unreported second peak in the discharge of many SC visuomotor neurons. This response was visual in nature, but its intensity was modulated by the monkey's choice of saccade goals. A similar early discriminatory visual response in some cortical neurons in the frontal eye fields (FEF) has been reported previously.[4] The timing of the differentiation in discharge for subsequent movements either into or out of the response field of the neuron was similar for our SC cells and the FEF cells, indicating that both structures have neurons whose discharge differentiates whether a distractor or the target is located in their response fields at about the same time. In contrast, the FEF VM cells did not display the two-phase visual response that we observed in many of our SC VM cells, but instead showed a more sustained response that gradually evolved into statistically different distributions for the two search conditions. We found a similar response in the sample of prelude burst cells that we recorded in the SC.

It has been frequently shown that the temporal interval between successive saccades can be considerably shortened in situations where strong competition exists between alternate saccade goals. We and others have hypothesized that the brevity of these intersaccadic intervals are accomplished by the programming of the two movements concurrently.[13-15] In the present communication we report a neurological correlate of this concurrent processing of two saccadic goals. Visuomotor neurons in the deeper layers of the monkey SC show maintained activity during the first saccade when a short-latency second saccade will be made to the location of a target originally in the response field of the cells. This concurrent activity just before and during the first saccade represents target selection processing, because it is greatly diminished or absent by the time of the second movement. After the completion of the first saccade, the saccade vector for the second movement no longer falls in the response field of the neuron. Further experiments will have to be conducted to determine whether cells in the SC at the site coding the saccade vector for the second movement to the target become active just before this movement as did the cells we show here at the site coding the target location before the first movement decline in activity. Alternatively, neural sites below the SC could produce the correct second movement by combining the target selection memory provided by the cells we report here with information about the change in eye position occurring in the first movement.

In addition to the foreshortened intersaccadic intervals that occur frequently in situations in which competition for alternative saccadic goals exist, we have also observed that curved saccade trajectories occur with greater frequency than in trials in which only single-target choices are present. We report here a neural correlate for these curved saccadic movements. We recorded neurons at sites in the SC coding the location of distractors. Lack of significant presaccadic activity at this site following the transitory visual response was associated with a saccade with a straight trajectory to the location of the target. In contrast, when an accelerating and relatively high level of activity was present at a distractor site just before saccade onset to the target, the initial direction of the closely following saccade was pointed toward a location between the target and distractor. We were not able to record simultaneously from neurons in the SC at sites coding the target and a distractor, but we assume that activity at the former site was approximately the same as for saccades to the single target located there, and thus was higher than that recorded at the distractor site. We speculate that this higher activity at the site of the target eventually silences the activity at the distractor site. The final curved trajectory toward the target is not coded by simple retinotopically organized activity at either SC site. We and others have speculated that either a shift in activity on the SC to a third site representing the final direction of approach to the target, or downstream processing to account for eye position changes, could produce the final curved trajectory.[17,18]

ACKNOWLEDGMENTS

This research was supported by National Institutes of Health grants EY08060 and EY06881.

REFERENCES

1. WURTZ, R.H. 1996. Vision for the control of movement. The Friedenwald Lecture. Invest. Ophthalmol. Visual Sci. **37:** 2130–2145.
2. BURMAN, D.D. & M.A. SEGRAVES. 1994. Primate frontal eye field activity during natural scanning eye movements. J. Neurophysiol. **71:** 1266–1271.
3. SCHALL, J.D. 2001. Neural basis of deciding, choosing and acting. Nature Rev. **2:** 33–42.
4. THOMPSON, K.G. *et al.* 1996. Perceptual and motor processing stages identified in the activity of macaque frontal eye field neurons during visual search. J. Neurophysiol. **76:** 4040–4055.
5. SPARKS, D.L. 1999. Conceptual issues related to the role of the superior colliculus in the control of gaze. Curr. Opin. Neurobiol. **9:** 698–707.
6. PELLI, D.G. 1997. The VideoToolbox software for visual psychophysics: transforming numbers into movies. Spat. Vision **10:** 437–442.
7. ROVAMO, J. & V. VIRSU. 1979. An estimation and application of the human cortical magnification factor. Exp. Brain Res. **37:** 495–510.
8. SMIT, A.C. & J.A.M. VAN GISBERGEN. 1990. An analysis of curvature in fast and slow human saccades. Exp. Brain Res. **81:** 335–345.
9. GLIMCHER, P.W. & D.L. SPARKS. 1992. Movement selection in advance of action in the superior colliculus. Nature **355:** 542–545.
10. BASSO, M.A. & R.H. WURTZ. 1998. Modulation of neuronal activity in superior colliculus by changes in target probability. J. Neurosci. **18:** 7519–7534.
11. OTTES, F.P. *et al.* 1987. Collicular involvement in a saccadic colour discrimination task. Exp. Brain Res. **66:** 465–478.
12. HANES, D.P. *et al.* 1995. Relationship of presaccadic activity in frontal eye field and supplementary eye field to saccade initiation in macaque: Poisson spike train analysis. Exp. Brain Res. **103:** 85–96.
13. VIVIANI, P. & R.G. SWENSSON. 1982. Saccadic eye movements to peripherally discriminated visual targets. J. Exp. Psychol. Hum. Percept. Perform. **8:** 113–126.
14. MCPEEK, R.M. *et al.* 2000. Concurrent processing of saccades in visual search. Vision Res. **40:** 2499–2516.
15. MCPEEK, R.M. & E.L. KELLER. 2001. Short-term priming, concurrent processing, and saccade curvature during a target selection task in the monkey. Vision Res. **41:** 785–800.
16. MIYASHITA, N. & O. HIKOSAKA. 1996. Minimal synaptic delay in the saccadic output pathway of the superior colliculus studied in awake monkey. Exp. Brain Res. **112:** 87–196.
17. MCPEEK, R.M. & E.L. KELLER. 2000. Competition between saccadic goals in the superior colliculus or frontal eye fields results in curved saccades. Soc. Neurosci. Abstr. **26:** 291.
18. PORT, N.L. & R.H. WURTZ. 2000. Two electrode recordings in monkey superior colliculus during curved saccades. Soc. Neurosci. Abstr. **26:** 297.
19. RICHMOND, B.J. *et al.* 1987. Temporal encoding of two-dimensional patterns by single units in primate inferior temporal cortex. I. Response characteristics. J. Neurophysiol. **57:** 132–146.
20. OTTES, F.P. *et al.* 1986. Visuomotor fields of the superior colliculus: a quantitative model. Vision Res. **26:** 857–873.

Midbrain Disorders of Vertical Gaze

A Quantitative Re-evaluation

JAMES A. SHARPE AND JI SOO KIM

Division of Neurology, University Health Network, University of Toronto, Toronto, Ontario, Canada

ABSTRACT: The mesodiencephalic junction is the site of the prenuclear control of vertical eye motion. We measured vertical saccades, smooth pursuit (SP), the vertical vestibulo-ocular reflex (VOR), and its interactions with vision during active head motion in 21 patients with midbrain lesions causing palsy of vertical saccades, upward, downward, or in both directions. Most patients with limited slow or slowed saccades in one direction on clinical examination had slowed saccades in the opposed direction. SP gain was decreased in both directions in most patients, and decreased upward or downward in few. VOR gain was subnormal in both directions in many patients, and upward only in one; phase lead of the VOR was recorded in 33% of them. Subnormal SP and VOR gains were often dissociated. Visually enhanced VOR gains were subnormal in both directions in many patients. Cancellation of the VOR was impaired in many patients, both upward and downward in most and upward in few patients. Gaze (eye plus head) tracking gain was subnormal in 29% of patients. Defective SP and defective cancellation of the VOR during head free tracking were often dissociated. We conclude that VOR and SP gains are usually subnormal in patients with paresis of vertical saccades. Impairment of pursuit and the VOR are often dissociated. Phase lead of the VOR implicates damage to velocity-to-position neural integrator for vertical eye motion. These associations and dissociations of impaired vertical eye motion signify discrete structural and functional effects of supranuclear midbrain damage that are undetected by examination of saccades.

KEYWORDS: supranuclear; saccades; smooth pursuit; vertical gaze; eye-head tracking; vestibulo-ocular reflex; vestibulo-ocular reflex cancellation; midbrain; brainstem

Vertical gaze palsy is characteristic of damage to the mesodiencephalic junction.[1,2] This area contains the rostral interstitial nucleus of the medial longitudinal fasciculus (riMLF), the interstitial nucleus of Cajal (INC), the mesencephalic reticular formation (MRF), and the posterior commissure (PC), which are involved in the premotoneuron control of vertical eye movements. Vertical gaze palsies consist of impairment of vertical saccades and variable degradation of vertical smooth eye movements. Neuropathological correlations have demonstrated midline or paramedian

Address for correspondence: James A. Sharpe, M.D., Division of Neurology, University Health Network, 399 Bathurst St., ECW 5-042, Toronto, Ontario M5T 2S8, Canada. Voice: 416-603-5950; fax: 416-603-55596.

sharpej@uhnres.utoronto.ca

damage in the rostral midbrain tegmentum in patients with palsy of vertical saccades.[3-5] Oculography of vertical gaze palsy from focal midbrain damage has been documented in few studies,[3-6] being limited in all but one[5] by the techniques of electro-oculography or infrared reflection. We performed quantitative magnetic search coil oculographic investigations of vertical saccades, smooth pursuit, head-free tracking, and the VOR and its visual enhancement and cancellation during active head motion in patients with palsy of vertical saccades from midbrain damage.

METHODS AND SUBJECTS

Patients

Twenty-one patients with vertical saccadic palsy were selected for study. Their mean age was 48 ± 18 (median 47, range 19 to 85 years). Seventeen patients were young (age < 65, mean 42 ± 12, median 44; 9 men) and four were elderly (age ≥ 65, mean 76 ± 7, median 75; 2 men). Sixteen patients had vascular lesions (eleven infarcts and five hemorrhages). Other lesions included neoplasm in two, demyelination in one, and infection in one. Brain MRI or CT confirmed unilateral lesions in nine and bilateral lesions in four patients. Imaging correlations with deficits will be considered in a future full report. Eight patients had normal MRI or CT; five had infarcts, one had infiltration of the midbrain by acute myeloblastic leukemia, one had herpetic brainstem encephalitis, and one had an idiopathic, longstanding lesion. Control values for vertical saccades were obtained from 39 normal volunteers[7]: 23 young (9 men) and 16 elderly (6 men). Vertical smooth eye movements were recorded in 31 normal subjects: 21 younger and 10 elderly (6 men).[8]

Oculography

Vertical motion of one eye was recorded with a magnetic scleral contact ring and head motion with a magnetic coil secured to the mid-brow, while subjects sat with the head erect (C-N-C Engineering, Seattle, WA). Eye, head, and target positions were digitized at 200 Hz and analyzed as previously reported.[7,8] Statistical analyses with p values of < 0.05 were considered significant. Head and eye movements were recorded under five conditions:

(1) Saccades: Saccades were measured in response to vertical target steps of 5, 10, 20, 30, and $40°$ at intervals from 1 to 5 seconds. Head position was stabilized in the erect position. Target steps elicited centrifugal (toward orbital midposition) and centripetal (from orbital midposition) saccades in each orbital hemirange of vertical motion as well as saccades crossing midposition.

(2) Smooth pursuit: With the head immobile in the erect position, subjects tracked a vertical sine-wave target with peak-to-peak amplitudes of $10°$, $20°$, and $40°$ at 0.125 to 2.0 Hz.

(3) Combined eye-head tracking: Subjects tracked a target moving vertically in sine waves of $20°$ at 0.25, 0.5, 1.0 and 2.0 Hz. Head tracking and eye tracking were verbally encouraged, but subjects were able to select varied combinations of head and eye motion, according to their preference. This head-free pursuit paradigm requires cancellation of the VOR.

(4) VOR: Active pitch head movements in darkness were performed in sinusoidal fashion at an intended peak-to-peak amplitude of $20°$ from the erect position at 0.25, 0.5, 1.0 and 2.0 Hz, paced to the sound of a periodic tone. Subjects attempted to fixate an imaginary stationary target in total darkness, using their memory of the position of the previously illuminated target.

(5) Visual enhancement of the VOR (VVOR): Vertical smooth eye and head movements were recorded while subjects fixated on an earth-fixed stationary laser dot target 1 m from the cornea and actively moved the head, as in condition 4.

RESULTS

Clinical Examinations

Of the 21 patients, five were unable to generate any vertical saccades (four in both directions and one downward) and 15 had limitation of the range of vertical saccades (10 in both directions and 5 selectively upward). One patient had slowing only of upward saccades without limitations of their range. Seventeen patients had limited ranges of smooth pursuit, 14 in both directions and 3 upward. Among the five patients with no voluntary vertical saccades, two could also not generate vertical smooth pursuit. Smooth pursuit excursions were larger than saccade excursions in 11 patients, either upward or downward. In four patients, pursuit was full in range in both directions.

Oculocephalic maneuvers elicited full vertical smooth eye movement excursions in 13 of the 21 patients, and limited excursions in both directions in 8 patients. Compared with saccade excursions, oculocephalic maneuvers during fixation of an earth-fixed target, which elicits the visually enhanced VOR (VVOR), increased vertical excursion in 18 patients, either upward or downward. In 12 patients, ocular excursions with oculocephalic maneuvers were larger than those of pursuit in either direction. Thus the VVOR, as assessed with oculocephalic maneuvers, increased the range of motility in most patients.

Horizontal eye movements were defective in nine patients. Eight patients had saccadic pursuit and six patients had slowed or limited saccades. Two patients had internuclear ophthalmoplegia. Convergence-retraction oscillations (10 patients), light-near dissociation of the pupillary reflex (8 patients), limited convergence (5 patients), skew deviation, or ocular tilt reaction (12 patients) were commonly associated findings. Unilateral fascicular third-nerve palsy occurred in 5 patients. These patients did not have contralateral ptosis or iridoplegia to indicate a vertical opthalmoplegia of oculomotor nucleus origin. Three patients had fourth-nerve palsy (one unilateral and two bilateral).

Saccades

Four patients could not generate any vertical saccades in either direction. In one patient with slowed upward saccades, no downward saccades were elicited. Two patients had slowed upward saccades and normal downward saccades. The other 14 patients had slowed saccades in both directions (FIG. 1). Of the six patients with selective upward saccadic palsy on clinical examination, four patients had slowed

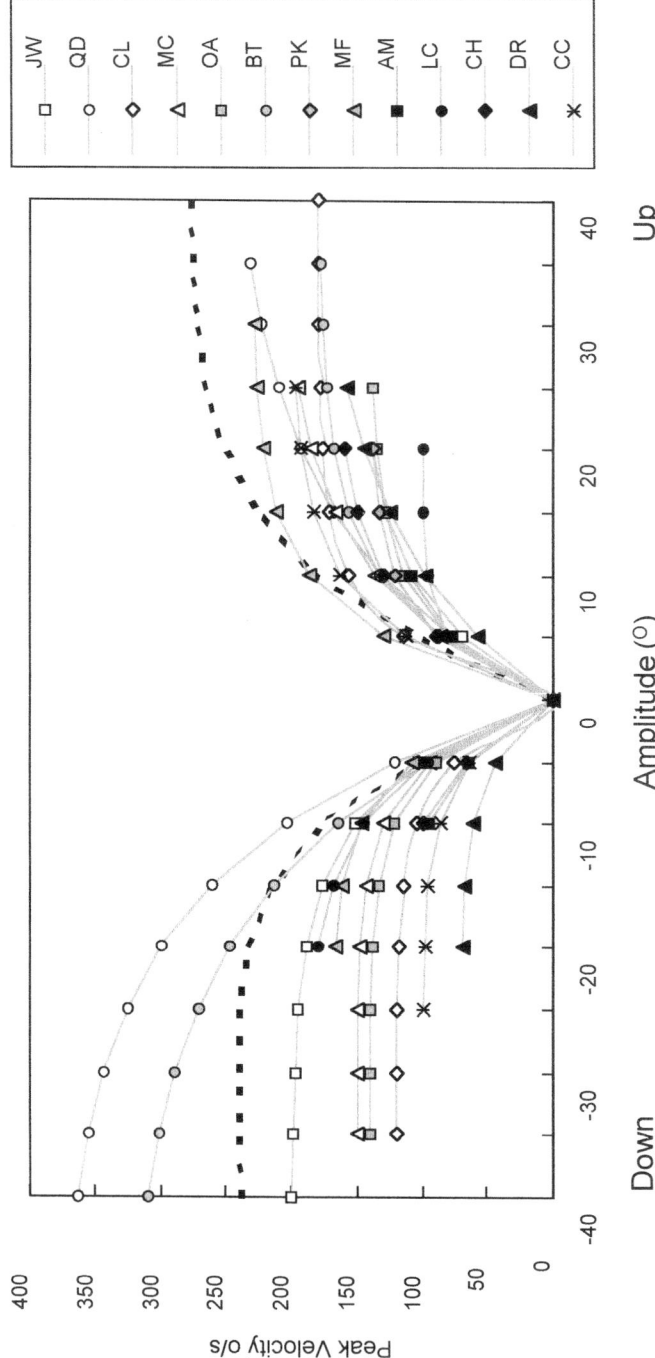

FIGURE 1. Amplitude–peak velocity relationship of all centrifugal, centripetal, and center-crossing vertical saccades in 13 young patients who could make vertical saccades. Each plot is truncated at the maximal amplitude that each patient could generate in corresponding direction during the recording. *Dashed line* indicates the lower limit of normal range (mean − 2 SD).[7] Upward saccades are plotted rightward, and downward saccades are leftward on ordinate.

downward saccades despite full downward excursion. One patient, diagnosed with selective limitation of downward range, also had slowed upward saccades.

Amplitude-velocity relationships between centrifugal and centripetal saccades in each orbital hemirange from midposition did not differ in either upward or downward direction in any patient. Centrifugal upward refixation movements to targets of large amplitude were often composed of multiple-step hypometric saccades, while centripetal upward saccades (toward orbital midposition) were typically accomplished by one- or two-step saccades.

Smooth Pursuit

Pursuit gain was subnormal in both directions in 9 patients, upward in 2 patients, and downward in 3 patients. Pursuit gain was normal in only 7 patients. Eight patients with limited ranges of pursuit (3 upward and 4 in both directions) had normal gains. Three patients with full-pursuit excursions in either direction showed subnormal gain, and excursion was achieved by segments of smooth pursuit and catch-up saccades. All patients with subnormal smooth-pursuit gain, upward or downward, also had slowed or absent saccades in the corresponding direction. Ten patients with slowed or limited saccades in either direction showed normal smooth-pursuit gain, (7 in both directions and 3 patients upward).

Combined Eye-Head Tracking

Head-free pursuit was recorded in 15 patients. Group means of gaze (eye plus head) tracking gains were higher than those of SP at all frequencies in both directions, with significant difference at 0.25 and 0.5 Hz downward, and at 1.0 Hz upward (paired t-test, $p < 0.05$). Gaze gain was subnormal both upward and downward in five patients and upward in one patient. One patient with subnormal gaze gain in both directions had normal upward smooth-pursuit gain. The other five patients with subnormal gaze gain had also decreased smooth-pursuit gain. Two patients with subnormal smooth-pursuit gain in either direction had normal gaze gain. Two patients who moved the head in larger amplitudes than the target amplitudes showed abnormally high gaze gains.

Vestibulo-Ocular Reflex

VOR gains were subnormal in both directions in 14, and subnormal upward in 1 of the 21 patients (FIG. 2). Phase lead of the VOR was recorded in five patients who also had subnormal VOR gain. One elderly patient had phase lag. Only six of the 21-patient group had normal VOR gain and phase. Among 13 patients who had full ranges of eye movements with oculocephalic maneuvers, subnormal VOR gain occurred in eight patients (upward in 1 and in both directions in 7). In contrast, one patient with limited "doll's eye movements" to oculocephalic maneuvers in both directions showed normal VOR gain within the restricted range.

Impairment of smooth pursuit and the VOR was dissociated in 13 patients. Eight patients with normal smooth pursuit in either direction showed subnormal VOR gain in the corresponding direction, and five patients with normal VOR gain had lowered smooth-pursuit gain.

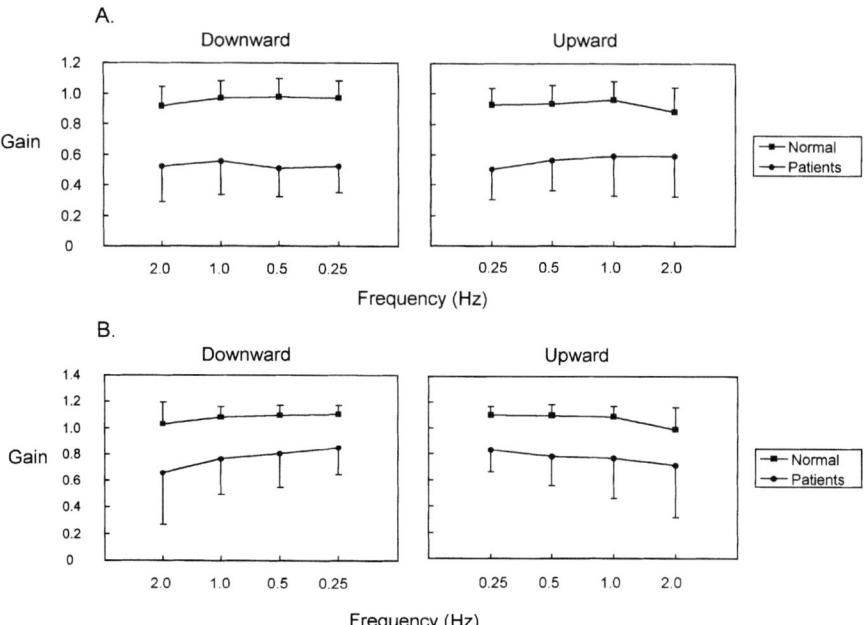

FIGURE 2. Mean group gains (ordinate) of the vertical VOR (**A**) and visually enhanced VOR (VVOR) (**B**) for young control subjects[8] and young patients with subnormal VOR or VVOR gain. Error bars indicate 1 SD.

Visually Enhanced VOR (VVOR)

When patients fixated an earth-fixed target during active head pitch, visually enhanced VOR gains were subnormal in both directions in 14 patients. Six patients with full ocular excursions with oculocephalic maneuvers had subnormal VVOR gains in both directions. Two patients had normal gains of the VVOR despite having subnormal gains of smooth pursuit and the VOR. In two patients, VVOR gains were subnormal while the gains of pursuit and the VOR were within reference ranges. All other patients with subnormal VVOR gains had subnormal pursuit or VOR gains.

Cancellation of the VOR

Cancellation was measured in 14 patients during head-free tracking and was impaired in 9 patients (upward in 2 and both upward and downward in 7). Impairment of smooth pursuit and of VOR cancellation did not correlate well (FIG. 3). Two patients with subnormal upward cancellation quotient (CQ) (impaired VOR cancellation) had normal upward smooth-pursuit gain. Of the seven patients with subnormal CQ in both directions, five patients had subnormal smooth pursuit gain (4 in both directions and 1 upward). The other two patients had normal smooth-pursuit gain in both directions. Three patients with lowered SP gain showed normal CQ, either upward or downward.

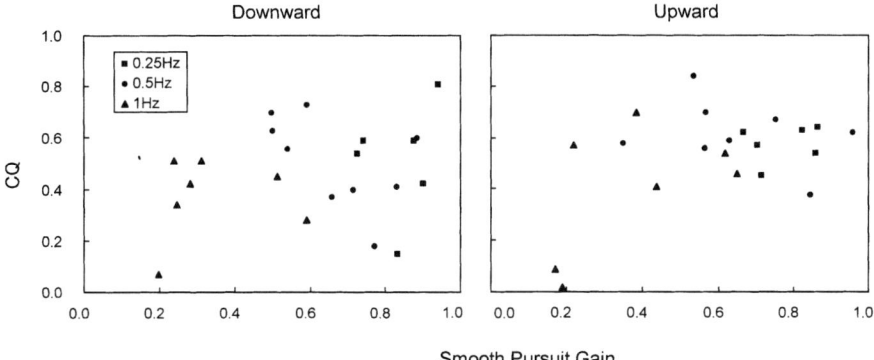

FIGURE 3. Correlation between cancellation quotient (CQ) and smooth-pursuit gain. Impairment of smooth pursuit and of VOR cancellation did not correlate well. CQ = (VOR gain – canceled VOR gain) / VOR gain, where VOR gain is in darkness, and canceled VOR gain is the ratio of vestibular smooth eye movement to head movement velocity during combined eye-head tracking. Each point is plot of an individual patient at a frequency.

DISCUSSION

Lesions in the pretectum often paralyze upward saccades. Quantitative investigation of vertical eye movements in our patients with paresis of vertical saccades after midbrain damage refutes the clinical aphorism that supranuclear palsy of vertical eye motion spares the vertical VOR. Our results indicate that VOR and smooth-pursuit gains are frequently subnormal, although the range of vertical smooth eye movement is typically not limited to the range of vertical saccades. We also found that abnormal pursuit and VOR responses are often dissociated.

Isolated paralysis of downward gaze or combined upward and downward gaze palsy is usually associated with bilateral lesions in the mesencephalic periaqueductal gray matter. Vertical saccade palsy in both directions has been identified after unilateral lesions.[5,9,10] In monkeys, unilateral inactivation of the rostral MRF causes conjugate slowing and hypometria of vertical saccades in both directions.[11] Selective paresis of upward saccades results from unilateral or bilateral lesions of the INC and riMLF or unilateral lesions in or adjacent to the PC.[3,4,12] Our results indicate[5,13] that unilateral lesions can cause selective upward or combined upward and downward vertical gaze palsy and are as effective as bilateral lesions in impairing the VOR, its visual enhancement, or its visual cancellation.

Saccades

Clinical examination showed selective palsy of upward saccades in six patients and selective palsy of downward saccades in one patient, but recordings revealed slowed saccades in the opposite direction in five of them. No patient had isolated paresis of downward saccades confirmed by oculography. This implies that normality of vertical saccades cannot be assured without recording of eye movements. Most of our patients had slowed or limited vertical saccades in both directions while two patients showed selective slowing of upward saccades. Four patients were unable to generate any vertical saccades. When saccades were limited in either direction, saccade speed was typically subnormal. Only one patient with limited saccades in both directions showed normal velocity of upward saccades within the restricted range.

Four patients who were unable to generate any vertical saccades had bilateral thalamomesencephalic infarcts. Complete abolition of vertical saccades can be explained by lesions involving both riMLFs, or their proximate connections, or by unilateral lesions damaging one riMLF and interrupting crossed fibers from the contralateral riMLF. Two of our 21 patients had selective upward saccadic palsy and MRI revealed unilateral lesions in the upper midbrain. Selective upward saccadic palsy after unilateral lesions may be explained by damage to the projection fibers from the riMLF mediating upward saccades. Burst neurons with upward on-direction project bilaterally to oculomotor nucleus neurons, whereas neurons with downward on-directions project ipsilaterally to motoneurons of the oculomotor and trochlear nuclei[14–16]; this makes isolated riMLF lesions more likely to impair downward saccades. Unilateral pretectal lesion near the PC may destroy axons from both riMLF, selectively abolishing upward saccades.

The INC participates in generating vertical saccades. *Unilateral* INC inactivation in monkeys reduces the range of vertical saccades up to 50%, but the saccadic amplitude–peak velocity relationship (main sequence) remains intact within the restricted range, unlike the effect of riMLF lesions.[17] After *bilateral* INC inactivation, the vertical range of saccades decreases when compared to the effect of unilateral lesions, but again the main sequence remains normal.[17] One of our patients had limited upward saccades but normal peak velocities within the restricted range of saccades. Normal saccade speed in the presence of limited saccade range may be explained by selective damage to the INC.

No difference was found in the amplitude–velocity relationship between centripetal and centrifugal saccades in either direction in our patients. We found the asymptotic peak velocity of vertical saccades to be often dissociated from the limitation of their range. Limitation of the oculomotor range could arise from several mechanisms: damage to the burst cells, damage to the neural integrator, or damage to gaze error commands delivered to the burst cells. The dissociation between the degrees of restriction and preservation of peak velocities might arise from involvement of the INC, where damage in monkeys limits the range of vertical saccades whilst preserving their peak velocities.[17] However, integrity of eccentric gaze-holding near the limit of the restricted range, argues against damage to the INC as an explanation in these patients. We attribute the restricted range with relative preservation of speed in such patient to disruption of descending cerebral and collicular gaze commands to the riMLF and INC.

Smooth Pursuit

Gains of smooth pursuit were subnormal in most patients; in five of them, upward and downward impairment was dissociated. The dissociation of impairment in each direction indicates that the pathways mediating upward or downward pursuit take different courses to ocular motor nuclei, just as do riMLF fibers that mediate saccades. The INC contains neurons that modulate their firing during vertical pursuit[18] and project to the contralateral INC, the oculomotor nucleus, and the trochlear nucleus through the PC.[19] INC efferent fibers near or in the PC modulate their discharge in concert with the instantaneous vertical position and the velocity of vertical smooth eye motion,[20] but a role of the INC, if any, may be subordinate to other pursuit pathways that descend or ascend in the tegmentum. Midsagittal splitting of the PC in monkeys reduces the range of upward pursuit (as well as saccades), implying that fibers mediating upward pursuit decussate in this structure.[21] Impairment or abolition of vertical smooth pursuit in the face of normal ranges and gains of the VOR may be explained by damage to descending corticofugal circuits carrying commands for vertical pursuit.

Vestibulo-Ocular Reflex

Subnormal VOR or VVOR gains occurred in 18 of our 21 patients. In 10 of those 18 patients, vertical ocular excursions were full with oculocephalic maneuvers. Since the range of smooth eye motion in response to oculocephalic stimulation is often full in the face of limited vertical saccades, supranuclear palsies of vertical gaze might be interpreted to spare the vertical VOR.[2] As tested in the clinic, oculocephalic maneuvers assess the range of the visually enhanced VOR (VVOR). Our results indicate that normality of the vertical VOR, or its visual enhancement, cannot be assured by a full range of oculocephalic responses.

The INC receives vestibular inputs from the medial, superior, and y cell group of vestibular nuclei.[22] *Unilateral* INC lesions in animals have virtually no effect on vertical VOR gain and phase, whereas bilateral lesions reduce vertical VOR gain by 50% and cause a phase lead of 10–20 degrees.[17,23,24] In monkeys, lesions of the PC prevent maintenance of eccentric vertical eye positions (up or down), reduce the gain of the vertical VOR, and cause phase lead of the eyes before the head, particularly at lower frequencies of head motion.[25] Vertical gaze-evoked nystagmus was not present in our patients. We recorded VOR phase lead in five patients with subnormal VOR gain. Abnormal phase lead had been identified in another patient after a discrete unilateral lesion involving the INC.[5] This phase lead can be explained by relative preservation of vestibular velocity commands that ascend to the midbrain, and impaired transmission of integrated velocity-to-position commands in the indirect VOR pathway. Damage to the INC or its projections, which participate in velocity-to-position integration, can account for the phase lead of the VOR in our patients.

Dissociations between Smooth Pursuit and the VOR

Subnormal gain of the VOR occurred in the presence of normal smooth pursuit in many of our patients. The simian MLF carries neural signals for vertical smooth pursuit and the VOR,[26,27] and bilateral destruction of the MLF in monkeys abolishes vertical VOR and impairs vertical smooth pursuit.[28] Markedly reduced vertical VOR gain, in the presence of only mild impairment of vertical smooth pursuit, in patients with internuclear ophthalmoplegia suggests that pathways extrinsic to the MLF carry substantial vertical smooth pursuit commands in humans.[29] Focal midbrain lesions that disrupt ascending vertical VOR signals in the MLF to the pre-motoneuron nuclei while preserving inputs for vertical smooth pursuit in pathways outside the MLF may result in subnormal gains of the vertical VOR without impairment of vertical smooth pursuit.

Several mechanisms may account for subnormal pursuit gain in the presence of normal VOR gain. Slippage of retinal images can induce plastic changes in VOR gain. Paresis of vertical tracking creates image retinal slip, and when VOR gain is reduced during fixation of earth-fixed targets, the retinal image error may serve as a stimulus to adaptive increase in VOR gain. Either damage to cerebral corticofugal pathways subserving vertical pursuit or selective midbrain involvement of ascending fibers carrying vertical smooth-pursuit signals may impair vertical smooth pursuit while sparing the vertical VOR.

Combined Eye-Head Tracking and Cancellation of the VOR

In our patients, cancellation of the VOR was often impaired during upward and downward head-free tracking, but eye-head tracking gain was normal in most patients. Normal gaze gain in the presence of a poorly canceled VOR is explained by an increased amplitude of head movement to compensate for the opposed vestibular smooth eye movements; increased amplitudes of vertical head motion compensated for impaired cancellation of the VOR and achieved normal gaze gain.

Correlation of impairment of smooth pursuit and VOR cancellation was poor. Five patients with normal smooth pursuit gain showed defective cancellation of the VOR and three patients with normal VOR cancellation had subnormal pursuit gain, either upward or downward. Head-free tracking is thought to be achieved by a central mechanism that cancels the opposed VOR. This dissociation implicates a nonvisual suppression of the VOR during head-free tracking. Rather than a smooth pursuit signal, a head-tracking command may be employed to down-modulate the opposed VOR during combined eye-head tracking.[30]

ACKNOWLEDGMENTS

This work was supported by Canadian Institutes of Health Research (CIHR) Grants ME 5509 and MT 15362 (J.A.S.), and by an Elizabeth Barford Award, University of Toronto (J.S.K.).

REFERENCES

1. CHRISTOFF, N.C. 1974. A clinicopathologic study of vertical eye movements. Arch. Neurol. **31**: 1–8.
2. DAROFF, R.B. & W.F. HOYT. 1971. Supranuclear disorders of ocular control systems in man. *In* The Control of Eye Movements. P.C.C. Bach y Rita & J.E. Hyde, Eds.: 117–235. Academic Press. New York.
3. BÜTTNER-ENNEVER, J.A. *et al.* 1982. Vertical gaze paralysis and the rostral interstitial nucleus of the medial longitudinal fasciculus. Brain **105**: 125–149.
4. PIERROT-DESEILLIGNY, C.H. *et al.* 1982. Parinaud's syndrome: electro-oculographic and anatomical analyses of six vascular cases with deductions about vertical gaze organization in the premotor structure. Brain **105**: 667–696.
5. RANALLI, P.J., J.A. SHARPE & W.A. FLETCHER. 1988. Palsy of upward and downward saccadic, pursuit, and vestibular movements with a unilateral midbrain lesion: pathophysiologic correlations. Neurology **38**: 114–122.
6. BALOH, R.W., J.M. FURMAN & R.D. YEE. 1985. Dorsal midbrain syndrome: clinical and oculographic findings. Neurology **35**: 54–60.

7. HUAMAN, A.G. & J.A. SHARPE. 1993. Vertical saccades in senescence. Invest. Ophthalmol. Visual Sci. **34:** 2588–2595.
8. KIM, J.S. & J.A. SHARPE. 2001. The vertical vestibulo-ocular reflex, and its interaction with vision during active head motion: effects of aging. J. Vestib. Res. **11:** 3–12.
9. RIORDAN-EVA, P. *et al.* 1996. Abnormalities of torsional fast phase eye movements in unilateral rostral midbrain disease. Neurology **47:** 201–207.
10. HELMCHEN, C., S. GLASAUER, K. BARTL & U. BÜTTNER. 1996 Contralesionally beating torsional nystagmus in a unilateral rostral midbrain lesion. Neurology **47:** 482–486.
11. WAITZMAN, D.M. *et al.* 2000. Effects of reversible inactivation of the primate mesencephalic reticular formation. II. Hypometric vertical saccades. J. Neurophysiol. **83:** 2285–2299.
12. AUERBACH, S.H. *et al.* 1982. Sylvian aqueduct syndrome caused by unilateral midbrain lesion. Ann. Neurol. **11:** 91–94.
13. KIM, J.S. & J.A. SHARPE. 2000. Supranuclear palsies of vertical saccades, pursuit and the vestibulo-ocular reflex and its interaction with vision: quantitative associations and dissociations after focal midbrain damage [abstract]. Neurology **54** (Suppl. 3): A170.
14. MOSCHOVAKIS, A.K., C.A. SCUDDER & S.M. HIGHSTEIN. 1991. Structure of the primate oculomotor burst generator. I. Medium-lead burst neurons with upward on-directions. J. Neurophysiol. **65:** 203–217.
15. MOSCHOVAKIS, A.K. *et al.* 1991. Structure of the primate oculomotor burst generator. II. Medium-lead burst neurons with downward on-directions. J. Neurophysiol. **65:** 218–229.
16. BHIDAYASIRI, R., G.T. PLANT & R.J. LEIGH. 2000. Medical hypothesis: a hypothetical scheme for the brainstem control of vertical gaze. Neurology **54:** 1985–1993.
17. HELMCHEN, C., H. RAMBOLD, L. FUHRY & U. BÜTTNER. 1998. Deficits in vertical and torsional eye movements after uni- and bilateral muscimol inactivation of the interstitial nucleus of Cajal of the alert monkey. Exp. Brain Res. **119:** 436–452.
18. KING, W.M., A.F. FUCHS & M. MAGNIN. 1981. Vertical eye movement-related responses of neurons in midbrain near interstitial nucleus of Cajal. J. Neurophysiol. **46:** 549–562.
19. KOKKOROYANNIS, T. *et al.* 1996. Anatomy and physiology of the primate interstitial nucleus of Cajal. I. Efferent projections. J. Neurophysiol. **75:** 725–739.
20. DALEZIOS, Y. *et al.* 1998. Anatomy and physiology of the primate interstitial nucleus of Cajal. II. Discharge pattern of single efferent fibers. J. Neurophysiol. **80:** 3100–3111.
21. PASIK, P., T. PASIK & M.B. BENDER. 1969. The pretectal syndrome in monkeys. I. Disturbances of gaze and body posture. Brain **92:** 521–534.
22. CRAWFORD, J.D. & T. VILIS. 1993. Modularity and parallel processing in the oculomotor integrator. Exp. Brain Res. **96:** 443–456.
23. ANDERSON, J.H., W. PRECHT & C. PAPPAS. 1979. Changes in the vertical vestibulo-ocular reflex due to kainic acid lesions of the interstitial nucleus of Cajal. Neurosci. Lett. **14:** 259–264.
24. FUKUSHIMA, K.T. OHASHI & J. FUKUSHIMA. 1994. Effects of chemical deactivation of the interstitial nucleus of Cajal on the vertical vestibulo-collic reflex induced by pitch rotation in alert cats. Neurosci. Res. **20:** 281–286.
25. PARTSALIS, A., S.M. HIGHSTEIN & A.K. MOSCHOVAKIS. 1994. Lesions of the posterior commissure disable the vertical neural integrator of the primate oculomotor system. J. Neurophysiol. **71:** 2582–2585.
26. KING, W.M., S.G. LISBERGER & A.F. FUCHS. 1976. Responses of fibres in medial longitudinal fasciculus (MLF) of alert monkeys during horizontal and vertical conjugate eye movements evoked by vestibular or visual stimuli. J. Neurophysiol. **39:** 1135–1149.
27. POLA, J. & D.A. ROBINSON. 1978. Oculomotor signals in medial longitudinal fasciculus of the monkey. J. Neurophysiol. **41:** 245–259.
28. EVINGER, L.C., A.F. FUCHS & R. BAKER. 1977. Bilateral lesions of the medial longitudinal fasciculus in monkeys: effects on the horizontal and vertical components of voluntary and vestibular induced eye movements. Exp. Brain Res. **28:** 1–20.

29. RANALLI, P.J. & J.A. SHARPE. 1988. Vertical vestibulo-ocular reflex, smooth pursuit and eye-head tracking dysfunction in internuclear ophthalmoplegia. Brain **111:** 1299–1317.
30. HUEBNER, W.P. *et al.* 1993. An investigation of horizontal combined eye-head tracking in patients with abnormal vestibular and smooth pursuit eye movements. J. Neurol. Sci. **116:** 152–164.

Cerebellar Influences on Saccade Plasticity

F.R. ROBINSON,[a,c] A.F. FUCHS,[b,c] AND C.T. NOTO[a,c]

[a]*Department of Biological Structure,* [b]*Department of Physiology and Biophysics, and* [c]*Regional Primate Research Center, University of Washington, Seattle, Washington 98195, USA*

ABSTRACT: Inaccurate saccades adapt to become more accurate. In this experiment the role of cerebellar output to the oculomotor system in adapting saccade size was investigated. We measured saccade adaptation after temporary inactivation of saccade-related neurons in the caudal part of the fastigial nucleus which projects to the oculomotor brain stem. We located caudal fastigial nucleus neurons with single unit recording and injected 0.1% muscimol among them. Two monkeys received bilateral injections and two monkeys unilateral injections. Unilateral injections made ipsiversive saccades hypermetric (gains >1.5) and contraversive saccades hypometric (gains ~0.6). Bilateral injections made both leftward and rightward saccades hypermetric (gains >1.5). During unilateral inactivation neither ipsiversive nor contraversive saccade size adapted after ~1,000 saccades. During bilateral inactivation, adaptation was either small or very slow. Most intact monkeys completely adapt after ~1,000 saccades to similar dysmetrias produced by intrasaccadic target displacement. After the monkeys receiving bilateral injections made >1,000 saccades in each horizontal direction, we placed them in the dark so that the muscimol dissipated without the monkeys receiving visual feedback about its saccade gain. After the dark period, 20-degree saccades were adapted to be 12% smaller, and 4-degree saccades to be 7% smaller. We expect this difference in adaptation because during caudal fastigial nucleus inactivation, monkeys made many large overshooting saccades and few small overshooting saccades. We conclude from these results that: (1) caudal fastigial nucleus activity is important in adapting dysmetric saccades; and (2) bilateral caudal fastigial nucleus inactivation impairs the relay of adapted signals to the oculomotor system, but it does not stop all adaptation from occurring.

KEYWORDS: monkey; caudal fastigial nucleus; dysmetria; muscimol; inactivation

INTRODUCTION

The motor commands that produce movements must change if our movements are to remain accurate as our bodies change throughout life. Indeed, when a movement is consistently inaccurate, the commands that produce that movement adapt to make the movement more accurate. Several types of evidence indicate that the cerebellum must function normally for motor commands to adapt effectively. For example, people consistently throw balls to one side of a target immediately after they don

---

Address for correspondence: Dr. F.R. Robinson, Department of Biological Structure, University of Washington, Seattle, WA 98195. Voice: 206-685-0614; fax: 206-543-1524.
robinsn@u.washington.edu

goggles that deflect their line of gaze laterally. Whereas normal subjects adapt within ~30 throws so that their throws again send the balls near the target, patients with damage to their cerebellum or its input from the inferior olive do not adapt their throws.[1]

Eye movements also adapt if they are inaccurate. The most thoroughly studied example is the vestibulo-ocular reflex which, when it does not accurately compensate for head rotation, undergoes a change in gain to minimize retinal slip.[2] Although they are less studied, voluntary rapid eye movements also adapt if they are inaccurate. In the laboratory we can make voluntary rapid eye movements, called saccades, seem to be too large if, during each saccade, we move the target in the opposite direction.[3] After a human[3-5] or monkey[6] makes many saccades that seem to be too large, saccades become smaller. Saccade size falls significantly within ~30 saccades in humans and within ~300 saccades in monkeys.

Like vestibulo-ocular reflex adaptation, which depends on regions of the cerebellar cortex called the flocculus and ventral paraflocculus,[7] saccade adaptation also depends on particular parts of the cerebellum. Large lesions including both the medial cerebellar cortex and the underlying fastigial nucleus[8] render monkeys unable to adapt their saccades. Smaller lesions of the cerebellar cortex centered on lobule VII of the vermis also abolish the ability to adapt saccades.[9,10]

Lobule VII projects heavily to the caudal part of the fastigial nucleus (CFN),[11] so that it is reasonable to suppose that the CFN also participates in saccade adaptation. However, no current data indicate what role it might play. Therefore, the goal of the experiments described here was to characterize the role of the CFN in saccade adaptation. We investigated this in two ways. First, we measured adaptation when CFN activity was temporarily compromised by injections of the $GABA_a$ agonist muscimol. Second, we tested whether saccades were adapted after the muscimol dissipated while the monkey was in the dark to prevent any visual feedback about the accuracy of its saccades.

METHODS

The subjects for these experiments were four juvenile male rhesus monkeys (monkeys M, B, R, and H). In sterile operations we implanted each monkey with a three-turn eye coil[12] and used stainless steel screws and dental acrylic to attach three acrylic appliances to the monkey's skull to hold the head steady as we recorded eye movements and single unit activity. In addition, we implanted a stainless steel recording chamber over a circular hole cut through the skull on the midline centered 8 mm posterior to ear bar zero. We attached the recording chamber, oriented directly dorsoventral, to the skull with stainless steel screws and dental acrylic.

After the monkey had recovered at least a week from surgery we trained it to use saccades to track the movements of a small red target spot generated with a laser diode projected onto a tangent screen 57 cm in front of the animal. The image of the spot reflected off two intervening computer-controlled mirror galvanometers which directed the spot to any part of the screen. A monkey received a dollop of applesauce from a feeding tube near its mouth when it kept its eyes within ~1 degree of the target spot for ~1 second. We measured the monkey's eye position with the search coil technique.[13]

After the monkey reliably tracked >2,000 target movements with saccades we used standard extracellular single unit recording techniques to locate the CFN whose neurons discharge a burst of action potentials for nearly every saccade whatever its direction or size.[14–16] Once we located saccade-related CFN neurons we injected muscimol among them with a pipette. The pipette was a 32-gauge hypodermic tube with a pulled-glass tip (ID $\cong$ 25 µm) glued over its end. Polyethylene tubing connected a solenoid valve to the back of the pipette. We filled the pipette and several centimeters of the connected tube with a solution of 1 mg/ml (8.75 mM) muscimol in normal saline solution. We positioned the tip of the pipette in the dorsoventral center of the saccade-related region and injected the muscimol solution using brief air pressure pulses. We left the pipette in place for 5 minutes after each injection and then withdrew it.

Before each injection we recorded saccades of several sizes in each horizontal direction. After an injection we had the monkey track at least 1,000 target steps in each horizontal direction. We recorded voltages proportional to horizontal and vertical eye and target positions on videotape using a pulse code modulation (PCM) adapter.

Monkeys R and H received unilateral injections and monkeys M and B received bilateral injections. Unilateral CFN inactivation makes contraversive saccades hypometric and ipsiversive saccades hypermetric, whereas bilateral inactivation makes saccades in both horizontal directions hypermetric.[17] After the monkey made ~1,000 saccades in each horizontal direction we released its head from restraint and turned out the lights in the booth. We monitored the monkey with an infrared camera and frequently offered it water via the feeding tube. Monkeys were in the dark for at least 10 hours. We then immobilized the monkey's head and recorded saccades of several sizes; there was no evidence of muscimol-induced saccade dysmetria.

We digitized the records of each experiment and analyzed them with custom software that measured many features of each saccade including its gain, that is, saccade size divided by distance from the eye to the target before the saccade. We omitted from analysis any saccade whose direction deviated >10 degrees from horizontal. These were mostly small corrective saccades or saccades away from the target. We graphed saccade gains during CFN inactivation as a function of the number of saccades that the monkey made and fit the plots with an exponential curve. The rate constant and asymptote of the curve indicated the speed and size of adaptation.[6] We also used a t test to compare the mean gain of the first and last 50 saccades made during inactivation and treated any $p < 0.05$ as significant.

All surgical and behavioral training procedures were approved by the Animal Care and Use Committee at the University of Washington. The animals were cared for by the veterinary staff of the Regional Primate Research Center. They were housed under conditions that comply with National Institutes of Health standards as stated in the *Guide for the Care and Use of Laboratory Animals* (DHEW Publication NIH85-23, 1985) and with recommendations from the Institute of Laboratory Animal Resources and the American Association for Accreditation of Laboratory Animal Care.

RESULTS

Before injection each monkey had saccade gains of ~1 in both directions. The injections into each monkey made saccades in both horizontal directions dysmetric and also impaired the adaptation that would normally have corrected a similar apparent dysmetria induced by intrasaccade target movements. Unilateral injection into monkey R gave the first 50 contraversive and ipsiversive saccades mean gains of 0.82 and 2.48, respectively. After 1,830 contraversive saccades the mean gain of the last 50 contraversive saccades was still 0.82 ($p = 0.8$). Following 717 ipsiversive saccades, the mean gain of the last 50 saccades, 3.04, was significantly larger ($p < 0.01$) than that of the first 50, but this gain change was in the wrong direction, away from 1.

Unilateral injection into monkey H gave the first 50 contraversive and ipsiversive saccades mean gains of 0.77 and 2.38, respectively. FIGURE 1A is an example of these dysmetrias as monkey H tracked an ipsiversive and then a contraversive target movement. FIGURE 1B shows the gains of contraversive and ipsiversive saccades after unilateral CFN inactivation. After 2,992 contraversive saccades the mean gain of the last 50 contraversive saccades, 0.73, was smaller than that of the first 50. Following 1,179 ipsiversive saccades the mean gain of the last 50 saccades, 2.19, was smaller than that of the first 50. Neither change was significant ($p = 0.3$ and 0.1, respectively). Thus, after unilateral CFN inactivation, the gain of dysmetric saccades did not change significantly towards 1.0 in either monkey.

Bilateral injections into both monkeys M and B made horizontal saccades hypermetric. In both of these monkeys the gains of saccades ipsiversive to the side that we injected first, left in both monkeys, were smaller and less variable than the gains of saccades in the other direction.

In monkey M the mean gains of the first 50 leftward and rightward saccades after CFN inactivation were 1.69 and 2.79, respectively. This caused alternating hypermetric saccades as monkey M tried to track a single target movement (FIG. 2A). FIGURE 2B shows the gain of leftward and rightward saccades after bilateral CFN inactivation in monkey M. After >1,600 overshooting leftward and rightward saccades the mean gain of the last 50 leftward saccades, 1.25, was significantly smaller ($p < 0.001$) than that of the first 50 saccades. The gain of the last 50 rightward saccades, 2.45, was not significantly different ($p = 0.06$) from that of the first 50.

The significant reduction in the gain of monkey M's leftward saccades exhibited a time constant that was ~10 × larger than normal. The change in leftward saccade gain during inactivation had a rate constant of 5,096 saccades. When its CFNs were normal, monkey M adapted with rate constants of ~500 saccades to apparent saccade overshoots of similar size induced by intrasaccade target movements.

In monkey B the mean gains of the first 50 leftward and rightward saccades after CFN inactivation were 1.29 and 3.60, respectively. After >1,200 overshooting saccades in each direction the mean gain of the last 50 leftward saccades, 1.24, was smaller than that of the first 50, and the mean gain of the last 50 rightward saccades, 2.88, was smaller than that of the first 50. Neither decrease was significant ($p = 0.3$ and 0.4, respectively). Thus, when both CFNs were inactivated in two monkeys, saccade size either did not change significantly within the number of saccades we tested or it changed with an abnormally long rate constant.

After the bilateral CFN inactivations in monkeys M and B, we let the muscimol dissipate while the monkeys were in the dark. Thereafter, saccades *were* adapted, some sizes more than others. In monkey M the gains of saccades tracking 4-, 8-, 12-, and 16-degree leftward target movements were moderately adapted by about the same amount (FIG. 3B). Saccades to 20-degree leftward target movements were most adapted (FIG. 3B). Saccades to rightward target movements of 4, 8, and 12 degrees

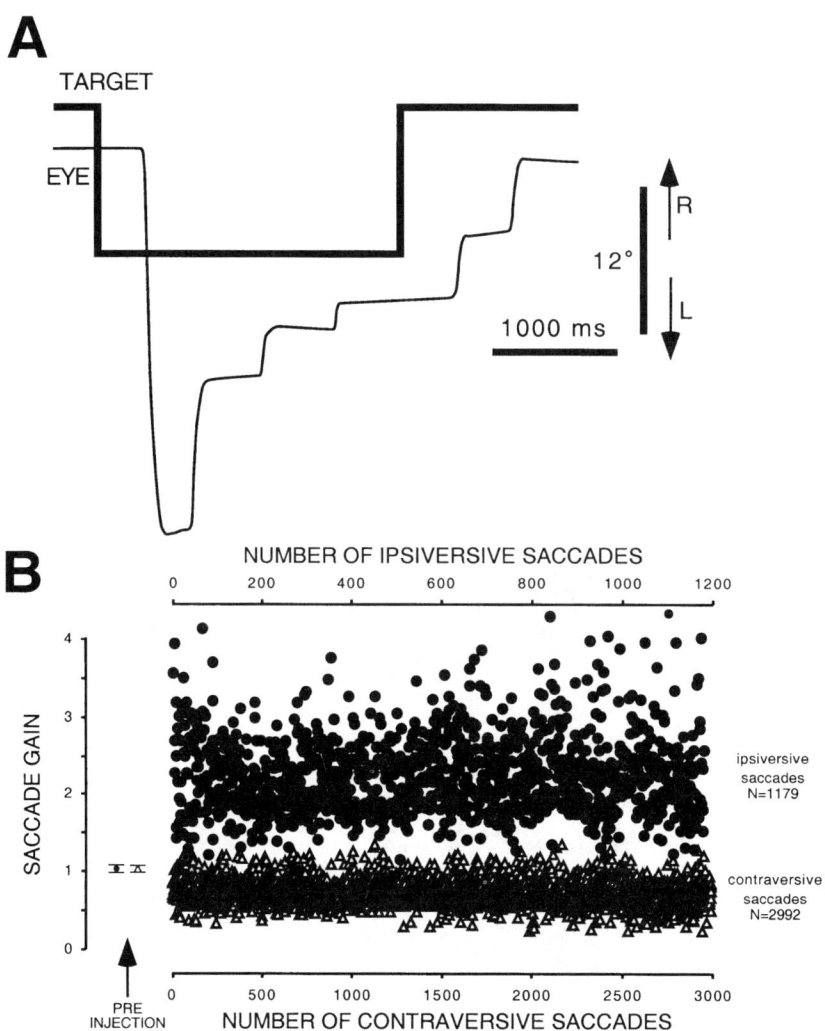

FIGURE 1. Consequences of inactivating monkey H's left caudal fastigial nucleus (CFN). (**A**) Eye and target position as the monkey tracks first a leftward and then a rightward 12-degree target movement with hypermetric leftward (ipsiversive) saccades and hypometric rightward (contraversive) saccades. Note that the monkey makes many more contraversive than ipsiversive saccades. The leftward offset of the eye from the target, visible on the left and right sides of the figure, is a consequence of inactivating the left CFN.[17] (**B**) Saccade gains as a function of the number of saccades. Separate scales for the number of ipsiversive and contraversive saccades appear at the *top* and *bottom* of the the graph, respectively. The mean gains (± SD) before inactivation are indicated with *symbols to the left*.

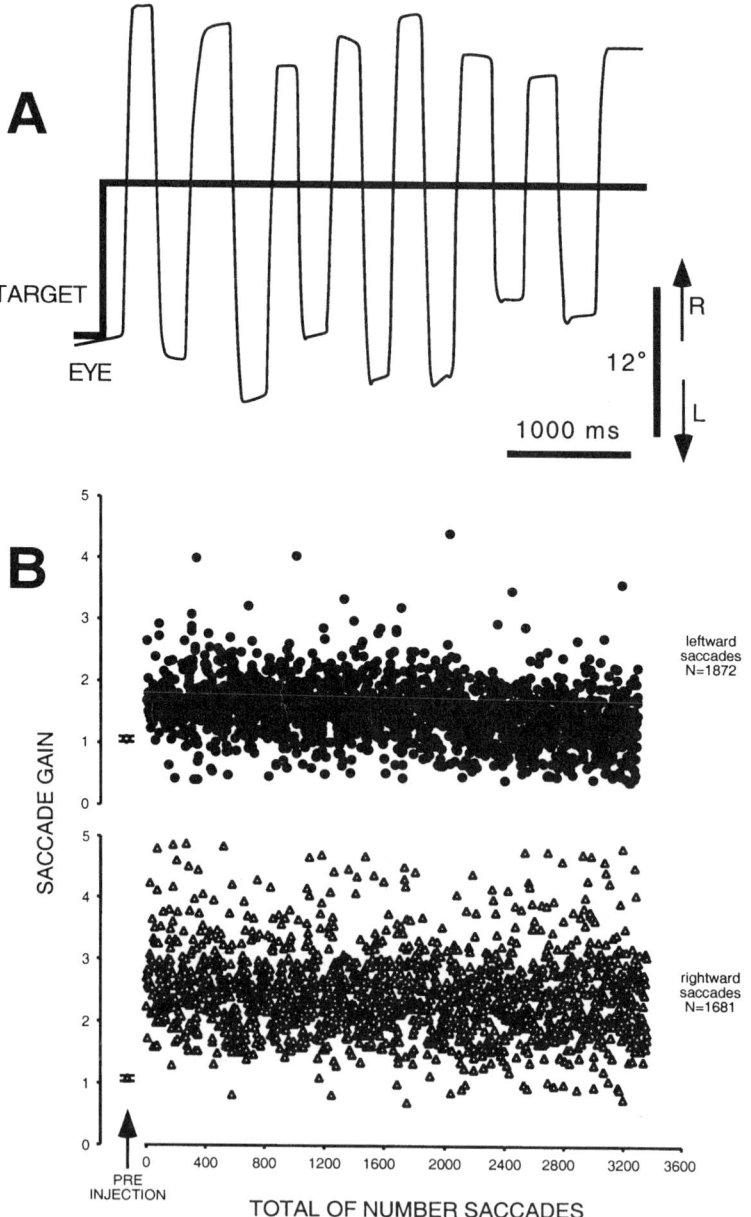

FIGURE 2. Consequences of inactivating both the left and right caudal fastigial nucleus (CFN) in monkey M. (**A**) Eye and target position as the monkey tracks a rightward target movement with alternating hypermetric leftward and rightward saccades. (**B**) Saccade gain as a function of the number of saccades. Saccades to the left are shown on *top*, and those to the right on the *bottom*. The mean gains (± SD) before inactivation are indicated with *symbols to the left*.

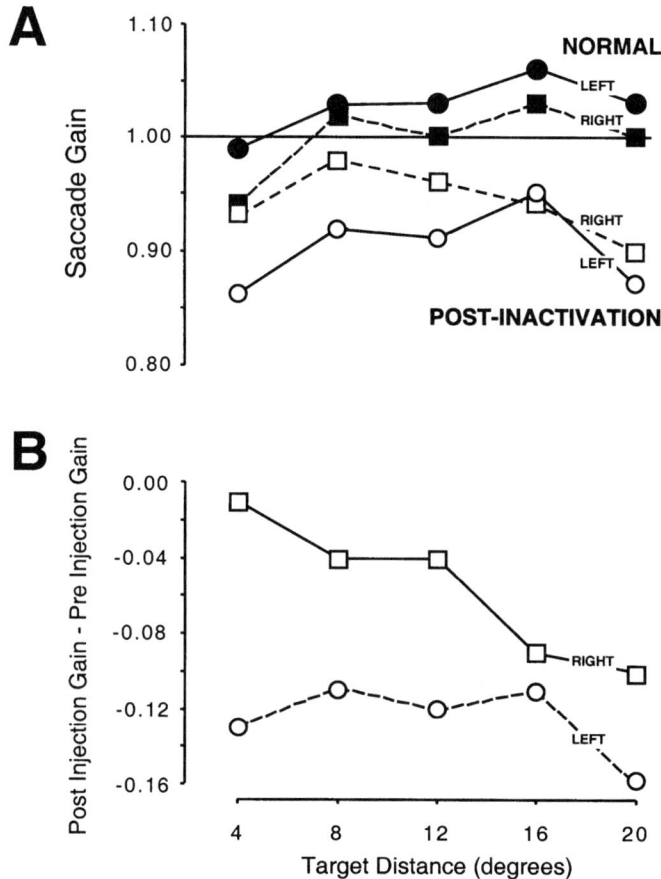

FIGURE 3. Evidence for adaptation during bilateral caudal fastigial nucleus (CFN) inactivation. (**A**) Gains of saccades to leftward (*circles*) and rightward (*squares*) target movements of five different sizes before (*filled symbols*) and after (*open symbols*) bilateral CFN inactivation. (**B**) Differences in gain between saccades recorded before and after bilateral CFN inactivation.

were adapted very little, but those to 16- and 20-degree rightward target movements were more adapted (FIG. 3B).

In monkey B (not shown), saccades tracking leftward target movements of 20 degrees were the most adapted, whereas saccades tracking rightward target movements of 12 degrees were the most adapted.

DISCUSSION

This work produced two major findings. First, CFN inactivation disrupts saccade adaptation. Saccade size did not change significantly in the adaptive direction during unilateral CFN inactivation. Bilateral inactivation also stopped these changes or dramatically increased their rate constant. Second, bilateral CFN inactivation does not prevent all adaptation from occurring. After the muscimol had dissipated following bilateral inactivation, saccades in both horizontal directions are at least partly adapted.

Some features of the pattern of adaptation after bilateral CFN inactivation are consistent with what we know about the specificity of adaptation. Adapting saccades of a particular size alters saccades of that size the most. The effect on saccades that are not adapted directly depends on how similar they are to the adapted size; the more similar, the more adapted they are.[18] During bilateral CFN inactivation, monkeys made many saccades ≥20 degrees in both directions (e.g., FIG. 2A). Therefore we expect that saccades ≥20 degrees will be more adapted than small saccades. This was true of monkey M's saccades in both directions and monkey B's saccades to the left. However, we cannot currently explain why saccades to rightward target movements < 20 degrees in monkey M were so well adapted or why those to 12-degree rightward target movements were the most adapted in monkey B.

Our current data do not allow us to conclude whether the adaptation of saccade gains during bilateral CFN inactivation was as large as it would have been if the CFNs had been functioning normally. To do this we must measure the adaptation elicited in intact monkeys by conditions similar to those faced by an injected monkey. Then we must determine how much of this adaptation persists after 10 hours in the dark and compare this to the adaptation evident after bilateral inactivation followed by 10 hours in the dark. The results of these experiments will demonstrate either that CFN activity makes a significant contribution to adapting saccade size or that adaptation occurs entirely upstream of the CFN.

ACKNOWLEDGMENTS

We are grateful to Bob Cent for writing and maintaining the software for saccade analysis and to Bruce Brown and the veterinary care staff at the Regional Primate Research Center, University of Washington, for their excellent care of the animals. This study was supported by National Institutes of Health Grant RR00166 and National Eye Institute Grants EY10578 and EY00745.

REFERENCES

1. MARTIN, T.A., J.G. KEATING, H.P. GOODKIN, et al. 1996. Throwing while looking through prisms. I. Focal olivocerebellar lesions impair adaptation. Brain **119:** 1183–1198.
2. GONSHOR, A. & G. MELVILL JONES. 1976. Short-term adaptive changes in the human vestibulo-ocular reflex arc. J. Physiol. **256:** 361–379.
3. MCLAUGHLIN, S.G. 1967. Parametric adjustment in saccadic eye movements. Percept. & Psychophys. **2:** 359–362.
4. MILLER, J.M., A. ANISTIS & W.B. TEMPLETON. 1981. Saccadic plasticity: parametric adaptive control by retinal feedback. J. Exp. Psychol. **7:** 356–366.
5. DEUBEL, H., W. WOLF & G. HAUSKE. 1986. Adaptive gain control of saccadic eye movements. Human Neurobiol. **5:** 245–253.
6. STRAUBE, A., A.F. FUCHS, S. USHER & F.R. ROBINSON. 1997. Characteristics of saccadic gain adaptation in rhesus macaques. J. Neurophysiol. **77:** 874–895.

7. Ito, M., P.J. Jastreboff & Y. Miyashita. 1982. Specific effects of unilateral lesions in flocculus upon eye movements in albino rabbits. Exp. Brain Res. **79:** 249–260.
8. Optican, L.M. & D.A. Robinson. 1980. Cerebellar-dependent adaptive control of primate saccadic system. J. Neurophysiol. **44:** 1058–1076.
9. Takagi, M., D.S. Zee & R.J. Tamara. 1998. Effects of lesions of the oculomotor vermis on eye movements in primate: saccades. J. Neurophysiol. **80:** 1911–1931.
10. Barash, S., A. Melikyan, A. Sivakov, et al. 1999. Saccadic dysmetria and adaptation after lesions of the cerebellar cortex. J. Neurosci. **19:** 10931–10939.
11. Yamada, J. & H. Noda. 1987. Afferent and efferent connections of the oculomotor cerebellar vermis in the macaque monkey. J. Comp. Neurol. **265:** 224–241.
12. Fuchs, A.F. & D.A. Robinson. 1966. A method for measuring horizontal and vertical eye movement chronically in the monkey. J. Appl. Physiol. **21:** 1068–1070.
13. Robinson, D.A. 1963. A method of measuring eye movement using a scleral search coil in a magnetic field. IEEE Trans. Biomed. Eng. **10:** 137–145.
14. Ohtsuka, K. & H. Noda. 1990. Direction selective saccadic-burst neurons in the fastigial oculomotor region of the macaque. Exp. Brain Res. **81:** 659–662.
15. Ohtsuka, K. & H. Noda. 1991. Saccadic burst neurons in the oculomotor region of the fastigial nucleus of macaque monkeys. J. Neurophysiol. **65:** 1422–1434.
16. Fuchs, A.F., F.R. Robinson & A. Straube. 1993. Role of the caudal fastigial nucleus in saccade generation. I. Neuronal discharge patterns. J. Neurophysiol. **70:** 1723–1740.
17. Robinson, F.R., A. Straube & A.F. Fuchs. 1993. Role of the caudal fastigial nucleus in saccade generation. II. Effects of muscimol inactivation. J. Neurophysiol. **70:** 1741–1758.
18. Noto, C.T., S. Watanabe & A.F. Fuchs. 1999. Characteristics of simian adaptation fields produced by behavioral changes in saccade size and direction. J. Neurophysiol. **81:** 2798–2813.

Distributed Model of Collicular and Cerebellar Function during Saccades

LANCE M. OPTICAN[a] AND CHRISTIAN QUAIA[b]

[a]*Laboratory of Sensorimotor Research, National Eye Institute, NIH, Bethesda, Maryland 20892, USA*

[b]*DEEI, Università degli Studi di Trieste, Trieste, Italy*

ABSTRACT: How does the brain tell the eye where to go? Classical models of rapid eye movements are lumped control systems that compute analogs of physical signals such as desired eye displacement, instantaneous error, and motor drive. Components of these lumped models do not correspond well with anatomical and physiological data. We have developed a more brain-like, distributed model (called a neuromimetic model), in which the superior colliculus (SC) and cerebellum (CB) play novel roles, using information about the desired target and the movement context to generate saccades. It suggests that the SC is neither sensory nor motor; rather it encodes the *desired sensory consequence* of the saccade in retinotopic coordinates. It also suggests a non-computational scheme for motor control by the cerebellum, based on *context learning* and a novel spatial mechanism, the *pilot map*. The CB learns to use contextual information to initialize the pilot signal that will guide the saccade to its goal. The CB monitors feedback information to steer and stop the saccade, and thus replaces the classical notion of a displacement integrator. One consequence of this model is that no desired eye movement signal is encoded explicitly in the brain; rather it is *distributed* across activity in both the SC and CB. Another is that the transformation from spatially coded sensory information to temporally coded motor information is implicit in the velocity feedback loop around the CB. No explicit spatial-to-temporal transformation with a normalization step is needed.

KEYWORDS: saccades; control system; modeling; movement; brain; superior colliculus; cerebellum; vermis; fastigial nucleus

INTRODUCTION

Historically, models of the neural control of movement have been designed with classical systems or control theory principles in mind.[1–7] All of these models require the computation of a desired eye displacement signal (e.g., target eccentricity on the retina processed by a sensory-motor transformation).[8] Desired displacement is then fed into an inverse kinematic controller that computes the innervation required to make that eye movement.[1,9–12] Models also must include a spatial-to-temporal transformation (STT)[13] because sensory information is present in the brain on spa-

Address for correspondence: Dr. Lance M. Optican, Bldg. 49, Room 2A50, Laboratory of Sensorimotor Research, National Eye Institute, NIH, Bethesda, MD 20892-4435. Voice: 301-496-3549.

LOptican@NIH.GOV

tially distributed maps (e.g., on the retina), whereas movements are represented by the firing rates of motor neurons. In most models the STT is performed by a special block that uses an explicit normalization procedure.[14]

Other approaches to the control of movement have been proposed that do not depend upon all the features of classical, hierarchical models. Houk et al.[15] proposed that movement depended upon motor learning in the cerebellum to modulate reverberatory, positive feedback pathways to control movements. Others proposed distributed models of the superior colliculus that enclose its movement map in a feedback loop.[4,16–19] However, none of these distributed models have discussed how the desired eye movement or the STT might be represented in the brain.

Recently, we proposed a distributed model of the saccadic pulse generator[20,21] that used distributed maps in the frontal eye fields (FEF), lateral intraparietal cortex (LIP), superior colliculus (SC) and the cerebellum (CB) to control movements. Study of this model reveals that although it reproduces saccades and neural activity very well, it does not compute any of the signals present in classical models. Instead, it uses cooperation between areas that learn to recognize specific movement contexts and areas that generate motor drive signals.

This paper will explain how our model achieves saccadic accuracy without computing a desired eye movement, motor error or inverse or forward kinematic signals. It also shows how an essential function of the brain, converting sensory information on maps into firing rates of motor neurons, can be carried out without an STT module that performs an explicit normalization step.

METHODS

Simulations were produced with a program written in Matlab (The MathWorks, Inc., Natick, MA) and the C programming language, modified from a previous simulation of our neuromimetic model.[20,21] The modifications allowed the initial locus of activity on the target map (a 33×33 element matrix) in the SC and on the pilot map (a 33×33 element matrix) in the CB to be set independently. For other details, see Lefèvre et al.[20]

RESULTS

Neuromimetic Model

This section presents an overview of the saccadic system model we recently proposed,[20,21] emphasizing those elements of the model needed to generate the pulse of innervation for a visually guided saccade. A functional diagram of our model of the saccadic system is shown in FIGURE 1. The model consists of cerebral cortical regions (FEF and LIP) that provide information to both the SC and the CB about the retinal location of the target to be foveated. The outputs from the SC and the CB project to the brain stem premotor nuclei that drive the eye.

The SC produces three signals. The first is a veto signal from fixation cells in the rostral SC, which are tonically active between, and silent during, saccades. This signal directly excites the omnipause neurons (OPNs), which strongly inhibit the sac-

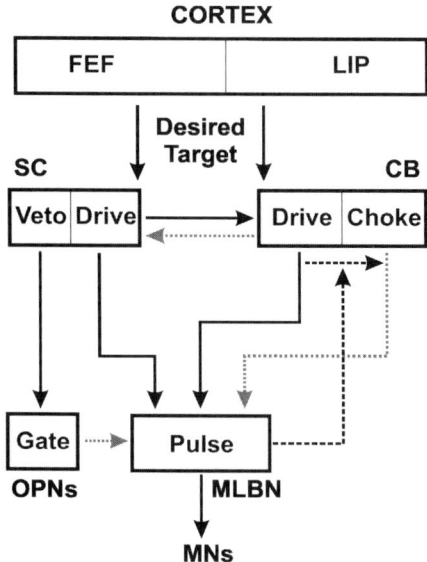

FIGURE 1. Schematic diagram of a neuromimetic model of the saccadic system. Regions of cerebral cortex (e.g., frontal eye fields, FEF; lateral intraparietal sulcus, LIP) provide target location information to both the SC and the CB. The SC and CB have multiple functions. The caudal SC drives the eye toward the target in retinal coordinates (Drive); the rostral SC contains fixation neurons, which prevent saccades from occurring (Veto). The contralateral CB drives the eye toward the target in motor coordinates (Drive); the ipsilateral CB stops the movement by choking off the drive signal (Choke). Omnipause neurons (OPNs) act as a gate to prevent saccades. Medium lead burst neurons (MLBN) provide the pulse of innervation to the motor neurons needed for saccades. (*Dotted gray arrows* indicate inhibitory pathways, *black arrows* show excitatory pathways; *black dashed arrows* show velocity feedback and spread of activity across CB).

cadic premotor, medium-lead, burst neurons (MLBNs) carrying the pulse of innervation required to make a saccade. The SC also provides part of the pulse of innervation to the MLBNs. This signal is probably conveyed through the long-lead burst neurons.[22] As this drive signal is simply a weighted sum of all the activity on the SC map, this pathway has strong convergence. Originally we called this a directional drive, but it could also be called a *retinal drive*, as it simply generates a change in muscle tension determined by the locus of activity on a retinotopic map. This interpretation is consistent with studies showing that saccades evoked by electrical stimulation are better correlated with movements in retinal coordinates.[23,24] Note that retinotopic information alone is, in general, insufficient to specify the eye movement. Knowledge of eye position is also needed, because of the noncommutativity of rotations in three dimensions.[8] The retinal drive's direction is determined by the lateromedial location of the activated site, whereas its intensity is a function of both the rostrocaudal location of the active site and the level of activation.

The retinal drive signal is excitatory, peaks at saccade onset, decays throughout the saccade, but is not over by saccade end. Finally, the SC provides to the CB (through the nucleus reticularis tegmenti pontis, NRTP) information about which target was selected. This information is distributed, with the projection of SC cells diverging widely in the cerebellum. This pathway provides a signal encoding the location of the desired, or selected, target in retinal coordinates (*what* to look at). This role is consistent with recent studies showing that the activity of the SC is related to the likelihood that a stimulus will be selected for a saccade.[25–27]

The role of the CB is more complex, and it varies during the movement. The CB learns to use contextual information to choose the initial locus of activity on a spatial map, which determines the pilot signal that will guide the saccade to its goal. During the first part of the saccade, the contralateral CB drives the excitatory MLBNs ipsilateral to the movement, increasing the initial acceleration of the eyes. This signal can also be considered a directional drive; the main difference between the collicular and the cerebellar drives is that the direction of the former is fixed, whereas the direction of the latter can be modified during a saccade. This allows the CB to compensate for different contexts (e.g., initial eye positions) and movement errors, and guide a saccade to its final position. The output that steers the eye is in motor coordinates, and we call it a *pilot drive*. Toward the end of the saccade, the ipsilateral CB begins activating the inhibitory MLBNs contralateral to the movement, choking off the drive to the motoneurons and stopping the movement. Thus, the overall cerebellar contribution is positive at the beginning of the movement, but negative at the end. The movement is over when this negative contribution offsets the positive drive coming from the SC and the contralateral CB. In this scheme, the CB is the structure that monitors feedback information, steers the movement, and decides when it is time to terminate the movement, determining its exact metrics. Note that this model guarantees accuracy by using local feedback information. However, that information is not fed to a comparator, the cornerstone of classic feedback control systems.[12] Instead, it causes activity to spread through the CB (see below). This is one of the major differences between our model and most lumped models. Thus, the CB replaces the classical notions of a displacement, or resettable integrator[3] and a comparator[1] with the dynamically evolving state of the pilot map.

In our model the speed with which the transition from driving to choking occurs is a function of the intensity of the pulse. As there is a direct proportionality between the pulse and the speed of the movement,[28] the faster the movement, the faster the transition. Accordingly, there is going to be a correlation between the duration of the saccade and the population activity of the CB (which has been confirmed by recent studies of Purkinje cells in the vermis[29]).

Finally, in our model the CB also provides a signal to the rostral SC that encodes, in a very approximate way, the progress of the saccade toward the target. This signal reduces the activity in the caudal SC because of reciprocal inhibition between rostral and caudal SC neurons. One effect of this reduction is that during the movement the SC drive is gradually weighted less than the CB drive, facilitating steering. The other effect is that at the end of the saccade the fixation neurons in the rostral SC can resume firing immediately, exciting the OPNs and preventing ocular oscillations.

Achieving Saccade Accuracy

The goal of the saccadic system is to produce fast and accurate movements. As noted above, in our model the NRTP sends a signal encoding the location of the tar-

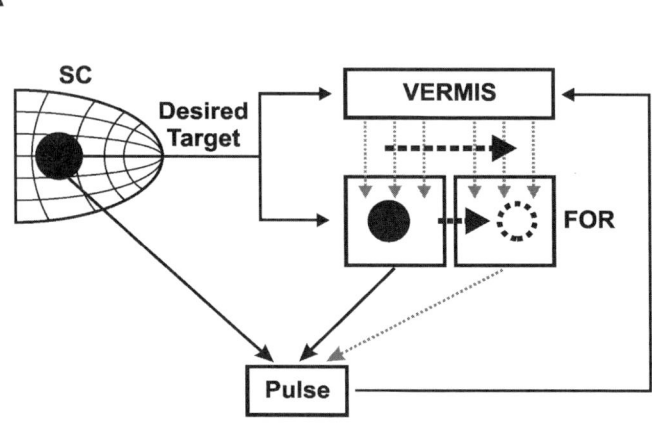

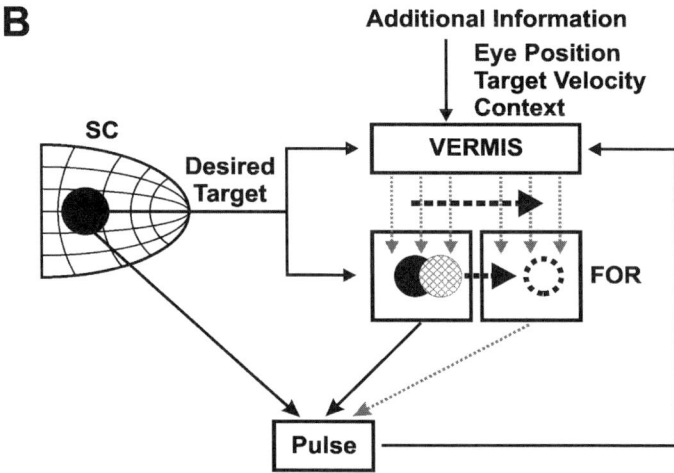

FIGURE 2. Effect of additional information on CB activity. (**A**) Loci of activity on the SC and CB maps. Desired target information excites both the vermis and the fastigial nucleus (FOR) over a broad range. Inhibition from the vermis (*dotted gray arrows*) prevent the FOR neurons from firing, except at the locus appropriate for the movement. As the saccade proceeds, a wave of inhibition sweeps across the vermis, allowing a wave of excitation to sweep across the FOR. The saccade ends when the choke overcomes the drive contributing to the Pulse. The *solid disks* indicate initial loci of activity, *dashed circle* shows final locus of activity. (**B**) When additional information is available (e.g., initial eye position or target velocity), the CB uses learned associations to cause the initial locus of activity on the pilot map to shift. In this example, the initial locus shifts from the *solid disk* to the *hatched disk*, which would make the movement smaller.

get in retinal coordinates to both the vermis and the fastigial nucleus (more precisely, to the caudal subdivision of this nucleus, the so-called fastigial oculomotor region or FOR[30,31]). We posit that the FOR acts as a topographically organized map, sufficient for all directions of movement. (Recent evidence suggests that the ventral posterior interpositus nucleus might be involved in vertical saccades,[32] but this nucleus is not yet included in our model.)

Different NRTP neurons project, in a topographically organized manner, to different cerebellar neurons. We propose that the projection spreads widely across this map, so that for each target location a large fraction of the neurons receives excitatory inputs. The same divergence occurs in both the vermis and FOR. However, we assume that the inhibition from the vermis (FIG. 2A, gray dashed lines) overcomes the excitation to the FOR, so that those cells are initially silent.

Just before the beginning of the movement (e.g., a 20° rightward saccade), the vermis releases inhibition at a site in the contralateral FOR whose location is a function of the desired displacement of the eyes (FIG. 2A, black disk in left FOR). In the first phase of the movement this activity will excite the contralateral MLBNs, thus driving the eyes in the appropriate direction. As the movement progresses, feedback from the brain stem about saccade velocity causes a wave of inhibition to sweep across the vermis (FIG. 2A, black dashed arrow). This wave in the vermis sequentially disinhibits neighboring parts of the FOR, so that the FOR activity appears to spread towards the opposite side. If the trajectory of the saccade is in error, say too far upward, the feedback signal will push the activity on the pilot map upward, which will drive the eye down, thus compensating, at least partially, for the error.[20,33]

Eventually the ipsilateral side of the FOR is activated (FIG. 2A, dashed circle). However, the projections from the ipsilateral side go to the inhibitory MLBNs on the contralateral side, and thus this activity reduces the pulse. As the inhibition grows, the pulse shrinks, until it gets to zero and the saccade ends. At this point the activity in the FOR stops spreading and slowly decays toward zero. Because the speed of this spread is directly proportional to the pulse (and thus to the speed of the movement), if the movement is fast, the activity will reach the other side quickly, whereas if it is slow it will reach there more slowly. Thus, the duration of the movement is inversely proportional to its speed, keeping the amplitude constant.

The Desired Displacement Signal

For this scheme to work, the CB must know, or must be able to infer, the desired displacement of the eyes. However, in our model this signal is not readily available, as the SC and the other cortical structures encode the location of the selected target in retinal coordinates, not the amplitude of the movement required to foveate it. Under the simplest behavioral paradigm (i.e., saccades directed to a stable target when the eyes are in primary position) these two signals coincide, but simple (and realistic) paradigm changes can break that equality.

Consider, for example, saccades to smoothly moving targets, or saccades to a target that is suddenly jumped, in a predictable (but not necessarily perceivable) fashion, during a saccade. It is well known that, in both cases, subjects can make a saccade straight to the final location of the target. It has also been shown that, as we posit in our model, under those conditions the collicular site activated encodes the location of the target and not the desired movement.[24,34,35] Thus, how does the brain know how to move the eye?

In our model, the CB receives inputs from the cerebral cortex and superior colliculus that only encode the location of the target in retinal coordinates. To guarantee that the appropriate movement is produced, other information, such as the speed of the target, the initial position of the eyes, and perhaps some information about the required behavior (i.e., the movement task) must also be provided. FIGURE 2B shows such additional information coming into the CB. In our model, the CB learns to use the target, context, and feedback information to initialize and update the state of the pilot map. The way in which the CB uses this information is fairly simple, and it can be easily explained through an example.

Suppose that a target appears in the periphery and moves toward the fovea with a predictable velocity. In this case, the SC represents the desired target by activity at a locus corresponding to the initial target location in retinotopic coordinates (e.g., the black disk in FIGURE 2B). We propose that the additional information about the target speed, conveyed to the CB by the pontine nuclei,[36] is used by the vermis to cause the initial locus of activity in the contralateral CB to shift toward the midline (e.g., the hatched disk in FIGURE 2B). The faster the target, the larger the shift. As the eye starts moving, the velocity feedback causes the activity to spread; however, because the site initially activated is more medial, the activity reaches the ipsilateral CB earlier, ending the saccade sooner and making it smaller than expected from the active site in the SC. The movement will thus have a shorter amplitude than required to fixate a stationary target, because the additional information about target motion was used to move the cerebellar site initially activated closer to the midline. It is the role of the cerebellar cortex to learn to associate different inputs and contexts with different initial sites of CB activity, thus determining saccade amplitude and direction.

Distributed Displacement Signals

One might be tempted to conclude that, in our model, the initially active cerebellar site encodes the desired displacement of the eyes. However, a closer inspection reveals that such a conclusion is unwarranted. Consider, for example, a 15° and a 20° saccade to stable targets (FIG. 3A, black solid and open symbols: + is fixation point, * is target). The SC and CB sites initially active would be the ones indicated in FIGURE 3C (dark gray and hatched disks). Now, consider a saccade directed to a target that starts 20° away from the fovea but moves towards the fovea at 50°/s (FIG. 3A, gray symbols). Suppose the time to acquire the target is 100 ms. Then, only a 15° saccade is needed, because the target will move 5° closer before the saccade gets there.

In this case, the SC site activated is the same as for the stable 20° target, because the SC only encodes which target to acquire, and at the beginning of the decision process that target is at 20° (FIG. 3C, dashed circle). However, the CB site initially activated (FIG. 3C, light gray disk) needs to be much closer to the midline than it was for a stable 15° target (FIG. 3C, hatched disk), even though the movement to be generated is the same, because the collicular drive coming from the 20° site will be stronger than that from the 15° SC site. Thus, the amplitude of the eye movement is a weighted average of the initial loci in the SC and the CB. As this example shows, in our model there is no explicit representation of the desired displacement signal.

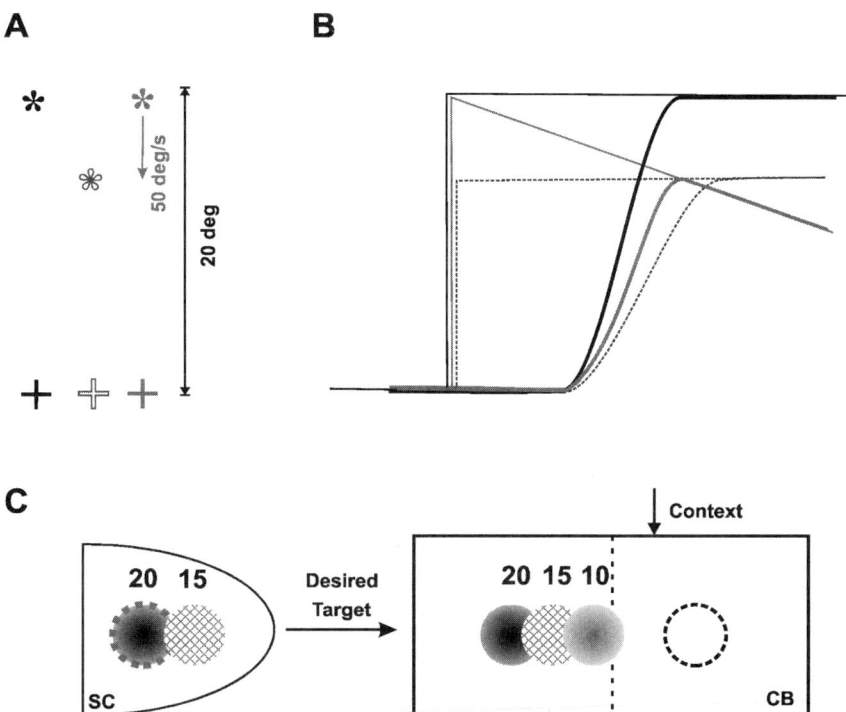

FIGURE 3. Distributed representation of desired saccade vector. (**A**) Saccades to stationary targets at 15° (*open symbols*) and 20° (*solid symbols*), and to a moving target appearing at 20° and moving centripetally at 50°/s (*gray symbols*). (**B**) Schematic drawing of saccades to the three targets in part **A**. (**C**) Activity in SC and CB for saccades in **A**. Activity in the SC appears, and remains, at the site corresponding to the initial visual location of the target. Activity in the CB for stationary targets is initially at the site on the pilot map corresponding to the site in the SC (*dark gray and hatched disks*). By the end of the movement, activity will spread to the ipsilateral side (*dashed circle*). For a target at 20° approaching at 50°/s, only a 15° saccade is required (B, *gray traces*). The SC activity reflects only the initial position of the target (20°, *dashed circle*). The initial locus in the CB needs to be closer to the midline (10°, *light gray disk*) than for a 15° saccade to a stable target (*hatched disk*), because the output of the SC at the 20° site is larger than the output at the 15° site. Note that none of the initial loci of activity correspond to the desired eye movement when the target is moving. Thus, the representation of desired ocular displacement is in neither the SC nor the CB, but is implicit in the distribution of activity across both the SC and the CB.

This signal is only implicit, and it is distributed across both the SC and the CB. No single area carries enough information to determine the desired eye movement.

An example of the hypothetical movements in this case are shown in FIGURE 3B. Note that we expect the initial speed of the 15° saccade to the moving target to be faster than the initial speed of the 15° saccade to the stationary target. This is because the drive from the SC comes from the 20° locus, which has a stronger projection to the brain stem than does the 15° site. The current implementation of our model does not contain a pursuit system. However, we can simulate the effect of dis-

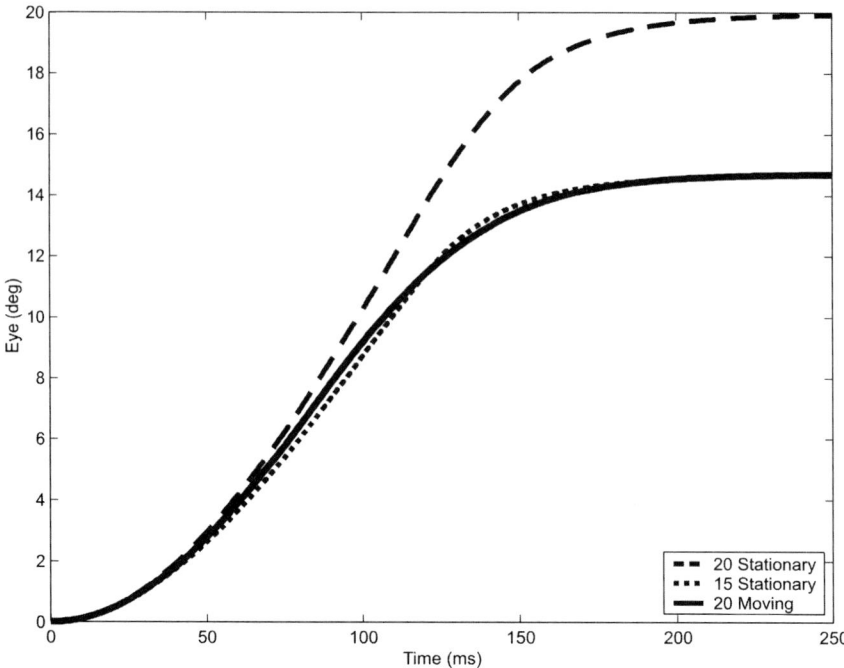

FIGURE 4. Simulated saccades. *Dashed and dotted lines* are saccades to stationary targets at 20° and 15°, respectively. *Solid line* shows saccade to a target at 20° approaching at 50°/s. (Note: there is no pursuit system in this model, so the saccade is not followed by a pursuit movement, as predicted from FIGURE 3B.) The shorter movement resulted from shifting the initial locus of activity on the pilot map in the CB closer to the midline.

placing the initial locus of activity on the CB toward the midline. FIGURE 4 shows three simulated saccades. Note that the initial speed of the 15° saccade to the moving target is slightly faster than the initial speed of the 15° saccade to the stationary target.

The Spatial-to-Temporal Transformation

The superior colliculus is commonly assumed to encode a desired target with a population code, such as the center-of-gravity,[37–39] or more precisely, the center-of-activity (COA). To determine the COA of a population, a calculation is required that includes a normalization step. In one dimension, if u is the distance from rostral to caudal on the SC, and if f(u) is the activity of a neuron at u, then the COA is given by:

$$\frac{\sum uf(u)}{\sum f(u)}$$

The numerator of this equation is the weighted sum of all the activity on the SC map, whereas the denominator is just the sum of all the activity on the map. The denominator is a normalization factor that requires a division operation. As a consequence of this normalization, the drive from the STT to the premotor neurons would be independent of the activity level in the SC. In fact, the activity of SC neurons is not constant, but depends upon experimental conditions. For example, SC activity is larger for a visually guided than for a memory guided saccade, and speeds are faster.[39–42] If the brain performed a normalizing operation in the STT, saccades would always have the same speed for a given amplitude, independent of the level of SC activation. However, it is well known that saccade speed does vary with the level of SC activation.[43] In our model, no division operation is needed, because there is no normalization step. The SC signal to the brain stem provides a drive signal, and a stronger level of SC activity will indeed lead to a faster saccade. However, the SC signal to the CB is a projection from one map to another. In this case, the level of activity is irrelevant. Thus, in our model the eye gets on target because the velocity feedback from the brain stem to the CB shifts the activity on the pilot map. (Note that in earlier models a spread of the map on the SC was proposed,[4,16,18,19,44] but such a spread is inconsistent with lesion experiments in the SC.[45]) In our model, no explicit STT is required, because the signal encoded spatially on the SC (i.e., the retinal location of the target) does not shift and is never converted into a temporal code. An STT is needed if, and only if, the same signal (e.g., retinal location of the target or desired displacement) is encoded in both the spatial and temporal domains. In our model, the pilot map itself is updated dynamically, so no signal appears in both spatial and temporal domains.

DISCUSSION

Our neuromimetic model is based on what is known about the physiology and anatomy of the saccadic system, and we propose that the superior colliculus represents a *target map* and the cerebellum a *pilot map*. In this model, the signals are somewhat unorthodox. Instead of a desired eye displacement motor command, there is only a desired, or selected, target signal (in retinal coordinates) and a distributed motor command. The neurons in the SC and the CB give rise to multiple signals, because their outputs go to many different structures (SC: *veto, what,* and *retinal drive*; CB: *progress, choke,* and *pilot drive*). The most common elements of classical models, the dynamic motor error, the comparator that computes it (as the difference between desired eye displacement and current displacement), and the displacement integrator are not even represented in our model.[12] Instead, their roles are played by a *dynamic pilot map*, which uses associative learning of target inputs, movement context and velocity feedback to initialize and update its state during the movement. The spread of activity in the CB replaces the displacement integrator in classical models of the saccadic system, which was necessary to keep track of saccade's progress and provide the current displacement signal to the comparator.

Benefits

The distributed model we proposed has many advantages over classical approaches to the control of movement. A lumped scheme relies on the computation of a desired displacement signal, and then on comparing that signal with an estimate of the current displacement. Our model is less complex, because it does not require computation of either of those signals, and thus does not require a comparator to produce a motor error signal. Instead, the CB simply learns to associate a certain pattern of inputs with a specific site of initial activation. The task of computing the desired displacement is thus reduced to a simpler, pattern recognition task.

Classic schemes are also very sensitive to noise, because some temporally coded signals (e.g., desired displacement) are outside the feedback loop. Thus, noise on those signals propagates to the output. Our model makes extensive use of spatial codes, so fluctuations in single cell discharges have very little effect because the outputs of both the SC and the CB are the sum of the activity of a huge number of neurons. The only element sensitive to noise is the pulse generator (FIG. 1, MLBN), where the discharge level is very high and the noise across neurons is likely to be correlated. However, that signal is under direct feedback control, and so such fluctuations are automatically compensated.

Finally, a common characteristic of all classic models is that they are made up of computationally complex modules, and thus have almost no resistance to failure. If any one element goes, the ability to make saccades is lost. More importantly, each module is so complex that, if it were to fail, it would be almost impossible for another structure to take over its functions. Even lumped models with adaptive controllers can only compensate for simple changes in the gain of the circuit; they cannot overcome structural changes (e.g., caused by lesions). In contrast, in our scheme the situation is very different; virtually any block (except the pulse generator) could be destroyed without causing massive deficits. For example, if the SC were damaged, the FEF could still provide the target location to the CB. Then, all that would be needed to evoke saccades would be another mechanism to shut off the OPNs; reinforcing the existing connections between FEF and the OPNs might be sufficient. If the cerebellum were lesioned, things would be slightly worse, as the complexity of its function would make it impossible for another structure to replace it. However, the effects would not be devastating for vision. The system would become purely feed-forward, so the accuracy, and consistency, of saccades would be lost. But at least saccades could still be made, and if the lesion were limited to only the vermis, some recovery, at least in the average behavior, would still be possible. The ability to compensate for the motion of a target would also be lost, but that is not a big loss, as saccades wouldn't be accurate anyway. Only combined damages to at least two structures would completely obliterate saccades; but, as the FEF, LIP, SC and CB are far apart, it is highly unlikely that a lesion could do so without being fatal to the individual. Of course, destroying the pulse generator in our model creates a complete failure of saccades, but this is consistent with experimental results; lesions of the paramedian pontine reticular formation (PPRF) do eliminate saccades.[46,47]

Thus, our model's resistance to failure can be ascribed to three factors: first, no single block is vital to the functioning of the circuit. Second, functionality is distributed across areas that are far apart in the brain. And third, the structure is such that one pathway (through the superior colliculus) provides a motor drive in retinal coordinates that is good enough for survival, while the other pathway (cerebellar) improves the movement's accuracy and consistency.

Generalization

The roles of the SC and CB proposed here may not be unique to saccadic eye movements, but may be generalized to other motor systems. We hypothesize that the role of sensory cortical areas is to select what to do, which implicitly triggers action unless it is vetoed.[48] The role of the cerebellar cortex is to learn to associate different sets of inputs (sensory information, state of the organism, desired behavior) with different initial starting points on a pilot map. The role of the pilot map, formed by the cerebellar cortex and deep nuclei, is to guide the ongoing behavior, using feedback to cause an evolution of the behavior from its initial to its final state. Therefore, the transformation from sensory to motor reference frames occurs implicitly, during the movement, in the cerebellum. There need be no explicit spatial-to-temporal transform, as proposed by Robinson.[13]

CONCLUSION

This neuromimetic model is significantly different from earlier models of movement control. Not only is it more successful at simulating behavioral and neuronal events, it suggests new hypotheses about the function of brain areas in the control of movement. First, it de-emphasizes the role of the superior colliculus in the control of saccades. The SC functions to select targets and initiate movements, but it neither steers nor stops them. Second, it suggests a new role for the cerebellum: it controls the movement by setting the initial locus of activity on the pilot map, and then using feedback to guide the movement. This role is non-computational, in the sense that it does not use explicit signals for physical variables, such as desired saccade vector and current displacement, from which an error signal must be computed. Nor does it use explicit kinematic models (inverse or forward). Instead, learning and pattern recognition replace these classical computations with mass action, or network properties, of neurons. Furthermore, no module explicitly performs a normalizing spatial-to-temporal transformation to obtain an explicit temporally encoded signal of the desired eye displacement. The STT is an emergent property of velocity feedback onto a dynamic spatial map.

REFERENCES

1. ROBINSON, D.A. 1975. Oculomotor control signals. *In* Basic Mechanisms of Ocular Motility and Their Clinical Implications. G. Lennerstrand & P. Bach-y-Rita, Eds.: 337–374. Pergamon Press. Oxford.
2. ZEE, D.S., *et al.* 1976. Slow saccades in spinocerebellar degeneration. Arch. Neurol. **33**: 243–251.
3. JÜRGENS, R., W. BECKER & H.H. KORNHUBER. 1981. Natural and drug-induced variations of velocity and duration of human saccadic eye movements: evidence for a control of the neural pulse generator by local feedback. Biol. Cybern. **39**: 87–96.
4. GROSSBERG, S., K. ROBERTS, M. AGUILAR & D. BULLOCK. 1997. A neural model of multimodal adaptive saccadic eye movement control by superior colliculus. J. Neurosci. **17**: 9706–9725.
5. DEAN, P. 1995. Modelling the role of the cerebellar fastigial nuclei in producing accurate saccades: the importance of burst timing. Neuroscience **68**: 1059–1077.

6. SCUDDER, C.A., A.F. FUCHS & T.P. LANGER. 1988. Characteristics and functional identification of saccadic inhibitory burst neurons in the alert monkey. J. Neurophysiol. **59**: 1430–1454.
7. VAN GISBERGEN, J.A.M., A.J. VAN OPSTAL & J.J.M. SCHOENMAKERS. 1985. Experimental test of two models for the generation of oblique saccades. Exp. Brain Res. **57**: 321–336.
8. CRAWFORD, J.D. & D. GUITTON. 1997. Visual-motor transformations required for accurate and kinematically correct saccades. J. Neurophysiol. **78**: 1447–1467.
9. KAWATO, M. 1999. Internal models for motor control and trajectory planning. Curr. Opin. Neurobiol. **9**: 718–727.
10. BHUSHAN, N. & R. SHADMEHR. 1999. Computational nature of human adaptive control during learning of reaching movements in force fields. Biol. Cybern. **81**: 39–60.
11. MIALL, R.C. & D.M. WOLPERT. 1996. Forward models for physiological motor control. Neural Networks **9**: 1265–1279.
12. OPTICAN, L.M. & Q. QUAIA. 2001. From sensory space to motor commands: lessons from saccades. Proc. 23rd IEEE EMBS, Istanbul, Turkey. In press.
13. ROBINSON, D.A. 1973. Models of the saccadic eye movement control system. Kybernetik **14**: 71–83.
14. DEUBEL, H., W. WOLF & G. HAUSKE. 1984. The evaluation of the oculomotor error signal. *In* Theoretical and Applied Aspects of Eye Movement Research. A.G. Gale & F. Johnson, Ed.: 55–62. Elsevier-North-Holland. Amsterdam.
15. HOUK, J.C., H.L. GALIANA & D. GUITTON. 1992. Cooperative control of gaze by the superior colliculus, brainstem and cerebellum. *In* Tutorials in Motor Behavior II. G.E. Stelmach & J. Requin, Eds. :443–474. Elsevier. Amsterdam.
16. LEFÈVRE, P. & H.L. GALIANA. 1992. Dynamic feedback to the superior colliculus in a neural network model of the gaze control system. Neural Networks **5**: 871–890.
17. ARAI, K., E.L. KELLER & J.A. EDELMAN. 1994. Two-dimensional neural network model of the primate saccadic system. Neural Networks **7**: 1115–1135.
18. DROULEZ, J. & A. BERTHOZ. 1991. A neural network model of sensoritopic maps with predictive short-term memory properties. Proc. Natl. Acad. Sci. USA **88**: 9653–9657.
19. VAN OPSTAL, A.J. & H. KAPPEN. 1993. A two-dimensional ensemble coding model for spatial-temporal transformation of saccades in monkey superior colliculus. Network **4**: 19–38.
20. LEFÈVRE, P., C. QUAIA & L.M. OPTICAN. 1998. Distributed model of control of sacades by superior colliculus and cerebellum. Neural Networks **11**: 1175–1190.
21. QUAIA, C., P. LEFÈVRE & L.M. OPTICAN. 1999. Model of the control of saccades by superior colliculus and cerebellum. J. Neurophysiol. **82**: 999–1018.
22. KELLER, E.L., R.M. MCPEEK & T. SALZ. 2000. Evidence against direct connections to PPRF EBNs from SC in the monkey. J. Neurophysiol. **84**: 1303–1313.
23. KLIER, E.M., H. WANG & J.D. CRAWFORD. 2001. The superior colliculus encodes gaze commands in retinal coordinates. Nat. Neurosci. **4**: 627–632.
24. GOLDBERG, M.E. *et al.* 1993. The role of the cerebellum in the control of saccadic eye movements. *In* Cerebellum and Basal Ganglia in the Control of Movement, N. Mano, Ed. Elsevier. Amsterdam.
25. GLIMCHER, P.W. & D.L. SPARKS. 1992. Movement selection in advance of action in the superior colliculus. Nature **355**: 542–545.
26. BASSO, M.A. 1998. Cognitive set and oculomotor control. Neuron **21**: 665–668.
27. BASSO, M.A. & R.H. WURTZ. 1998. Modulation of neuronal activity in superior colliculus by changes in target probability. J. Neurosci. **18**: 7519–7534.
28. QUAIA, C. & L.M. OPTICAN. 1998. Commutative saccadic generator is sufficient to control a 3-D ocular plant with pulleys. J. Neurophysiol. **79**: 3197–3215.
29. THIER, P., P.W. DICKE, R. HAAS & S. BARASH. 2000. Encoding of movement time by populations of cerebellar Purkinje cells. Nature **405**: 72–76.
30. NODA, H. *et al.* 1988. Saccadic eye movements evoked by microstimulation of the fastigial nucleus of macaque monkeys. J. Neurophysiol. **60**: 1036–1052.
31. ROBINSON, F.R., A. STRAUBE & A.F. FUCHS. 1993. Role of the caudal fastigial nucleus in saccade generation. II. Effects of muscimol inactivation. J. Neurophysiol. **70**: 1741–1758.

32. ROBINSON, F.R. 2000. Role of the cerebellar posterior interpositus nucleus in saccades I. Effect of temporary lesions. J. Neurophysiol. **84:** 1289–1302.
33. QUAIA, C., M. PARE, R.H. WURTZ & L.M. OPTICAN. 2000. Extent of compensation for variations in monkey saccadic eye movements. Exp. Brain Res. **132:** 39–51.
34. FRENS, M.A. & A.J. VAN OPSTAL. 1997. Monkey superior colliculus activity during short-term saccadic adaptation. Brain Res. Bull. **43:** 473–483.
35. KELLER, E.L., N.J. GANDHI & J.M. SHIEH. 1996. Endpoint accuracy in saccades interrupted by stimulation in the omnipause region in monkey. Vis. Neurosci. **13:** 1059–1067.
36. SUZUKI, D.A. & E.L. KELLER. 1984. Visual signals in the dorsolateral pontine nucleus of the alert monkey: their relationship to smooth-pursuit eye movements. Exp. Brain Res. **53:** 473–478.
37. ROBINSON, D.A. 1972. Eye movements evoked by collicular stimulation in the alert monkey. Vis. Res. **12:** 1795–1808.
38. SPARKS, D.L., C. LEE & W.H. ROHRER. 1990. Population coding of the direction, amplitude, and velocity of saccadic eye movements by neurons in the superior colliculus. Cold Spring Harbor Symp. Quant. Biol. **55:** 805–811.
39. LEE, C., W.H. ROHRER & D.L. SPARKS. 1988. Population coding of saccadic eye movements by neurons in the superior colliculus. Nature **332:** 357–360.
40. HIKOSAKA, O. & R.H. WURTZ. 1985. Modification of saccadic eye movements by GABA-related substances. I. Effect of muscimol and bicuculline in monkey superior colliculus. J. Neurophysiol. **53:** 266–291.
41. HIKOSAKA, O. & R.H. WURTZ. 1986. Saccadic eye movements following injection of lidocaine into the superior colliculus. Exp. Brain Res. **61:** 531–539.
42. SPARKS, D.L. & L.E. MAYS. 1990. Signal transformations required for the generation of saccadic eye movements. Annu. Rev. Neurosci. **13:** 309–336.
43. STANFORD, T.R., E.G. FREEDMAN & D.L. SPARKS. 1996. Site and parameters of microstimulation: evidence for independent effects on the properties of saccades evoked from the primate superior colliculus. J. Neurophysiol. **76:** 3360–3381.
44. WAITZMAN, D.M., T.P. MA, L.M. OPTICAN & R.H. WURTZ. 1991. Superior colliculus neurons mediate the dynamic characteristics of saccades. J. Neurophysiol. **66:** 1716–1737.
45. QUAIA, C., H. AIZAWA, L.M. OPTICAN & R.H. WURTZ. 1998. Reversible inactivation of monkey superior colliculus: II. Maps of saccadic deficits. J. Neurophysiol. **79:** 2097–2110.
46. COHEN, B., A. KOMATSUZAKI & M.B. BENDER. 1968. Electrooculographic syndrome in monkeys after pontine reticular formation lesions. Arch. Neurol. **18:** 78–92.
47. HENN, V., W. LANG, K. HEPP & H. REISINE. 1984. Experimental gaze palsies in monkeys and their relation to human pathology. Brain **107:** 619–636.
48. QUAIA, C. & L.M. OPTICAN. 1999. No "when" without "where." Behav. Brain Sci. **22:** 696–697.

The Cerebellar Contribution to Eye Movements Based upon Lesions

Binocular Three-Axis Control and the Translational Vestibulo-Ocular Reflex

DAVID S. ZEE,[a] MARK F. WALKER,[a] AND STEFANO RAMAT[a,b]

[a]*Departments of Neurology and Ophthalmology, Johns Hopkins University School of Medicine, Baltimore, Maryland 21287, USA*

[b]*Dipartimento di Informatica e Sistemistica, Università di Pavia, Italy*

ABSTRACT: The study of human cerebellar patients and monkeys with experimental cerebellar lesions has taught us much about the role of the cerebellum in normal ocular motor control. Here we emphasize recent findings that point to a role for the cerebellum in (1) the control of the three-dimensional axis about which the eye rotates in response to visual and vestibular stimuli, and (2) the generation of the translational VOR. Findings in cerebellar patients include abnormalities of eye torsion during attempted fixation that suggest a cerebellar role in the control of torsion so that Listing's law is obeyed. Abnormal torsion during vertical pursuit suggests that central processing of information for smooth pursuit may be based upon a phylogenetically old, semicircular canal coordinate scheme. Inappropriate and disconjugate vertical and torsional eye movements ("cross-coupling") occur during brief, high-acceleration rotations of the head. This suggests a role for the cerebellum in the binocular control of the rotation axis of the VOR. Finally, abnormalities of the modulation of the translational VOR with near viewing in cerebellar patients, but with sparing of the very initial 25–30 msec of response, suggests an important role for the cerebellum in the translational VOR.

KEYWORDS: cerebellum; vestibulo-ocular reflex; torsion; ocular misalignment

INTRODUCTION

The cerebellum plays a pivotal role in the generation of eye movements of all types, both in their immediate, on-line control and in their long-term adaptive calibration. Much of this information has been gleaned from the study of effects of lesions (reviewed in Robinson and Fuchs[1]). Here, we discuss the results from studies of the effects of cerebellar lesions—both those that occur naturally in patients and experimentally in monkeys—that bear on three relatively neglected aspects of cerebellar influences on eye movements: control of torsion, control of eye conjugacy, and control of the response to head translation. We will emphasize that an important

Address for correspondence: David S. Zee, M.D., Path 2-210, Johns Hopkins Hospital, Baltimore, MD 21287. Voice: 410-955-3319; fax: 410-614-1746.
dzee@dizzy.med.jhu.edu

Ann. N.Y. Acad. Sci. 956: 178–189 (2002). © 2002 New York Academy of Sciences.

role for the cerebellum is the binocular, three-axis (horizontal, vertical, torsion) control of eye movements and, more specifically, the ensuring that the globe rotates around the correct axis in response to a particular visual or vestibular stimulus. We will also present preliminary information about the role of the cerebellum in the control of the translational vestibulo-ocular reflex (tVOR). Some of this material has been reviewed previously.[2]

Abnormalities of Listing's Law in Cerebellar Patients

One of the more controversial issues about the control of eye torsion is related to the mechanisms that underlie Donders' and Listing's laws. Donders' law states that during steady fixation, the torsional orientation of the eyes is fixed for a given eccentric eye position. Listing's law describes the way in which Donders' law is implemented. Listing's law specifies a primary position of the eyes, and if this position is chosen as the reference position, all other positions can be reached from this position by rotation around an axis that lies in a plane (Listing's plane) that is perpendicular to primary position. This arrangement satisfies Donders' law. An important mathematical relationship between *angular velocity* and *eye position* for Listing's law to be obeyed is the "half-angle rule". For horizontal or vertical eye trajectories that do not pass through primary position, the angular *velocity* vector tilts by half the angle of the orthogonal eye *position* eccentricity.[3]

We recently reported that patients with degenerative cerebellar disease show abnormal torsion during fixation that is suggestive of a violation of Listing's law.[4] Further evidence for a role for the cerebellum in the control of Listing's law comes from the study of horizontal pursuit and optokinetic responses in cerebellar patients. Because of the half-angle rule, the horizontal angular velocity vector should tilt during pursuit according to the degree of vertical eccentricity. The results of this type of analysis in a group of cerebellar patients and in normal subjects are summarized in FIGURE 1. We calculated the tilt of the angular velocity vector (arctangent of the ratio of torsional to horizontal velocity) at the various vertical eccentricities and then computed the relationship of the tilt to vertical eye position (calling this number the "tilt slope"). For Listing's behavior, the tilt slope should be 0.5. Note that the values for the normal subjects cluster tightly around 0.65, while there was considerable spread in the values of the patients, which ranged from about 0.3 to 1.2. Thus, using an angular velocity analysis, many cerebellar patients show non-Listing's law behavior during smooth pursuit.

We also performed a similar analysis on the slow phases of optokinetic nystagmus (OKN) in response to a full-field stimulus. To induce OKN, subjects were rotated in a vestibular chair in darkness at a constant velocity, and when the vestibular response died away, the lights were turned on as the subjects continued to rotate. Subjects were then asked to change their vertical eye position at various times while still rotating. As shown in FIGURE 1, in normal subjects the tilt slope was about 0.25, a value close to the tilt expected from a rotational vestibular stimulus.[5] Note that this value differs from that during smooth pursuit, and is another piece of evidence that smooth pursuit and OKN have, at least in part, separate premotor circuitry. This distinction between smooth pursuit and OKN parallels the distinction between the translational and rotational VOR.[6] Like pursuit, the translational VOR roughly obeys Listing's law, and like OKN, the rotational VOR does not.[2,7]

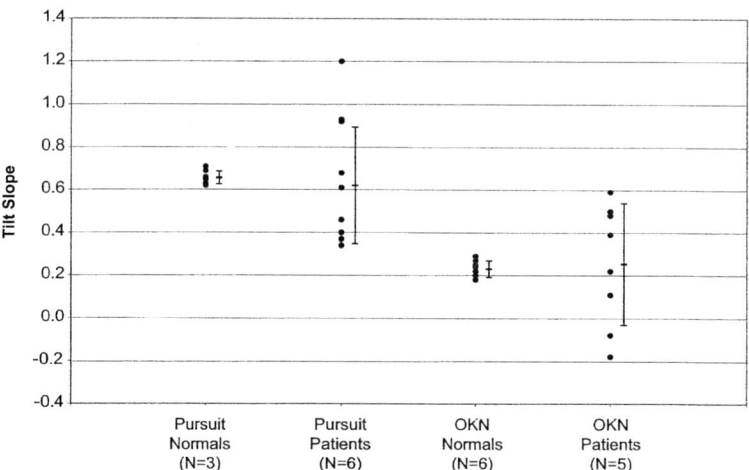

FIGURE 1. Tilt slopes for all normal subjects and cerebellar patients for smooth pursuit and for OKN. The values for the normal subjects cluster tightly; those for the patients scatter over a much wider range. (From Walker et al.[2] Reproduced by permission of Springer-Verlag.)

In our cerebellar patients we also found that the tilt slopes for OKN had a much wider spread of values than that for normals, ranging from about –0.2 to 0.6. Furthermore, the tilt slopes for pursuit and OKN were not well correlated in many patients. This dissociation suggests—but does not prove—that separate structures within the cerebellum control the axis around which the eye rotates for pursuit and for OKN.

The Role of the Cerebellum in the Control of Listing's Law

What might be the role of the cerebellum, then, in the elaboration of Listing's law? Recently, Demer and colleagues have shown the importance of orbital "pulleys"—connective tissue, smooth muscle sleeves through which the eye muscle tendons pass—in determining eye torsion[8] (Demer, this volume). The position of these pulleys, like the trochlea itself, determines the pulling direction of the eye muscles on the globe, and hence specifies the degree of torsion associated with a given horizontal and vertical gaze position during eye fixations as well as the axis around which the eye rotates during movements. In other words, the pulleys provide the orbital substrate for implementation of Listing's law.[9] Furthermore, the positions of the pulleys can be influenced by contraction of extraocular muscles fibers within the orbital layer, which are innervated by the ocular motor nerves themselves.[10] Hence, the position of the pulleys could be modified by changes in the innervation to either the smooth muscles that surround them or to the orbital layer of the extraocular muscles.

The potential for modification of pulley position provides a possible mechanism by which Listing's law could be implemented to its high degree of specificity in the long-term. Consequently, we predict that there must be a central mechanism that monitors torsional eye position as a function of gaze eccentricity, and adjusts innervation either to muscle fibers that can directly alter the position of the pulleys, or to the muscle fibers that insert directly on the globe, or to both, to ensure that Listing's law is obeyed over time. The cerebellum may be the structure that performs this function. This specific hypothesis is similar to other ideas about cerebellar ocular motor function, only extended to the third axis of rotation. For example, the cerebellum assures that the pulse, slide, and step of innervation to the ocular muscles is correctly matched according to the mechanical properties of orbital tissues so that there is no unwanted drift immediately following each saccade.[11] The suppression of horizontal and vertical postsaccadic drift allows for stabilization of images on the fovea. The suppression of torsional postsaccadic drift may relate to maintaining the retinal meridians aligned for optimal depth perception. Indeed, one aspect of the implementation of Listing's law may be the control of postsaccadic torsional disparity.

These considerations, of course, imply that control of eye torsion as a function of gaze position (i.e., Listing's law) is mutable. By manipulating disparity with optical means to require sustained fusion to maintain eye alignment, both Schor et al. (this volume) and ourselves[12] have shown that this is the case in normal subjects. Whether or not the cerebellum mediates this change remains to be discovered.

A Labyrinthine Coordinate Scheme for Smooth Pursuit: Torsion during Vertical Pursuit

Thus far we have considered the role of the cerebellum in Listing's law. One can also ask whether the cerebellum makes other contributions to the control of torsional velocity in the slow-phase response to visual and vestibular stimuli. We studied three patients with isolated cerebellar lesions (cavernous angiomas) in the region of the middle cerebellar peduncle close to the fourth ventricle.[13] Each showed a direction-changing torsional nystagmus during vertical smooth pursuit and during vertical VOR cancellation when fixing upon a target moving with the head, but not during vertical saccades or the vertical VOR in darkness. During upward smooth tracking, the upper poles of the eyes rotated toward the side of the lesion and during downward smooth tracking, toward the side opposite the lesion. The slow-phase velocity of the torsional eye movement was proportional to that of the vertical component.

How do we explain this unusual pattern of torsion? One hypothesis is that signal processing normally associated with pursuit uses a substrate based upon a vestibular, labyrinthine coordinate system. Consider, for example, that during pitch of the head, both anterior canals are excited, and a purely vertical slow phase of vestibular nystagmus is produced because oppositely directed torsional components cancel. One can then interpret our patients' inappropriate torsion as arising from inadequate cancellation of oppositely directed torsional signals. In our case, a loss of anterior canal encoded information on one side could account for the pattern of torsion during up and downward tracking. Because of the consistent location of the lesions in these patients we suggest that the middle cerebellar peduncle carries visual information (probably relaying information from the pontine nuclei to the vestibulocerebellum and dorsal vermis) encoded in "anterior SCC coordinates" and that interruption of this pathway leads to torsional nystagmus during vertical pursuit. Whether the lack of cancellation of torsional signals occurs in the cerebellum per se, or in more downstream, outflow pursuit pathways such as the vestibular nuclei, is not settled.

There are other examples in the cerebellum in which the influences upon anterior and upon posterior semicircular canal pathways are disparate. The flocculus, for example, has inhibitory projections to the vestibular nuclei that mediate anterior but not posterior canal responses. This dichotomy has been suggested as the cause for the frequent finding of downbeating nystagmus in cerebellar patients[14–16] and, as we will describe below, for inappropriate vertical slow phases during horizontal head rotation.

Inappropriate Torsional Responses to Vestibular Stimulation: "Cross-Coupling" in the VOR

We have previously reported that cerebellar patients frequently show inappropriate vertical slow-phase eye movements in response to yaw-axis (horizontal) head rotation.[17] This can occur during responses to brief, high-acceleration, head "thrusts," which are presumably composed of relatively high-frequency components, as well as with more sustained, relatively low-frequency vestibular responses, such as during constant-velocity head rotations in the dark. Recently, we have looked at the torsional as well as the vertical responses to horizontal head thrusts in cerebellar patients, and some have shown a response compatible with a disturbance in the control of anterior semicircular canal pathways.[2] For example, FIGURE 2 shows the response to horizontal head thrusts from a patient with cerebellar degeneration. The main features include (1) vertical cross-coupling that was always upwards; (2) torsional cross-coupling that changed direction (clockwise for rightward head thrusts and counterclockwise for leftward head thrusts); and (3) disconjugate cross-coupling (torsion was greater in the ipsilateral eye and vertical greater in the contralateral eye). This last response is a "dynamic skewing" of the eyes during head rotation.

A hypothesis to explain this abnormal pattern of vestibular response is that the cerebellar lesion has led to a "dynamic" release of inhibition upon vestibular pathways that carry information from the anterior semicircular canal contralateral to the direction of horizontal head rotation. A second assumption is that excitation of a semicircular canal is a more effective stimulus than inhibition (Ewald's second law), so that the excited anterior semicircular canal (which is in the opposite labyrinth relative to the direction of head rotation) contributes more to the response than the inhibited anterior semicircular canal on the other side. In this way, one might expect an upward slow phase to be associated with horizontal head rotation in either direction. In addition, the disconjugate pattern of torsion and vertical motion could reflect the fact that the primary excitatory connections of the anterior semicircular canal pathways are to the ipsilateral superior rectus and contralateral inferior oblique muscles, hence leading to relatively more vertical rotation in one eye and torsion in the other. There are, of course, other explanations for this pattern of abnormal vestibular response, but our hypothesis is an example—the explanations for primary position downbeat nystagmus and torsional nystagmus during vertical pursuit described above are others—in which an alteration of activity in anterior semicircular pathways is suggested as a cause of a vestibular disturbance in cerebellar disease. Finally, the disconjugacy of the abnormal responses implicates the cer-

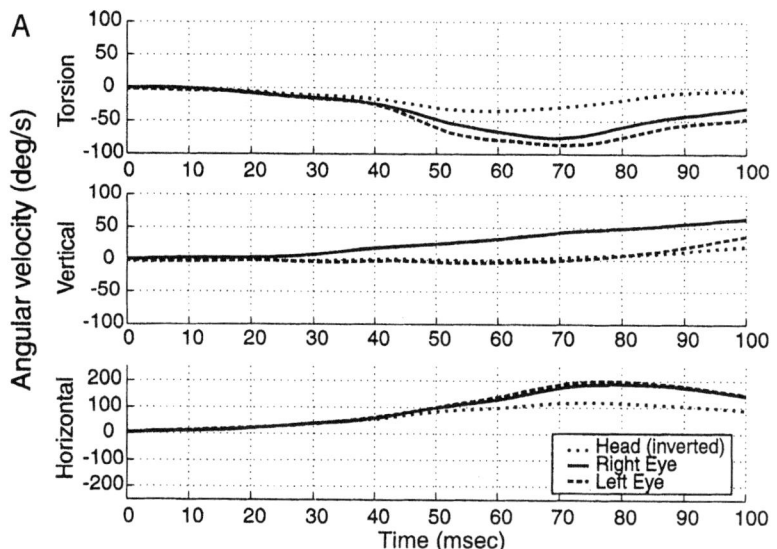
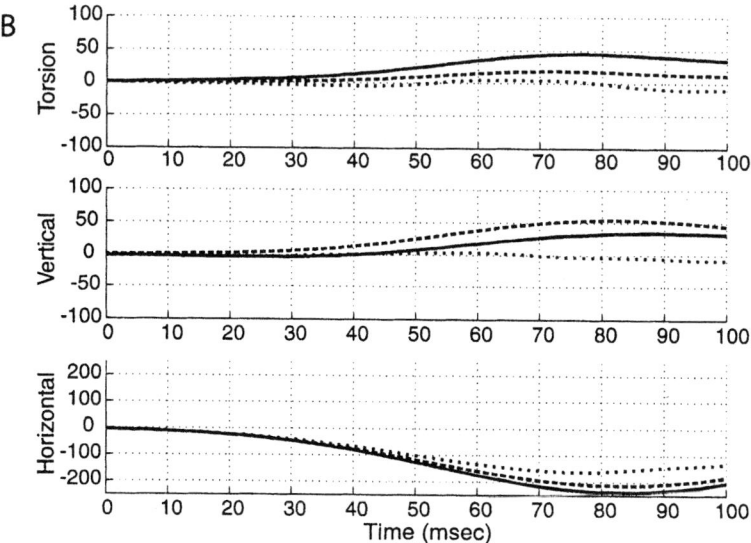

FIGURE 2. Binocular dual-axis scleral search coil recordings in a patient with cerebellar degeneration. Angular velocity components for both eyes and the head (inverted) are shown for head "impulses" in both horizontal directions. Note the increased horizontal gain, the upward vertical cross-coupling with horizontal head rotations in either direction, and the torsional cross-coupling that changes direction with the direction of horizontal head rotation. The responses are also disconjugate. (**A**) Rightward head rotation. (**B**) Leftward head rotation. Right, up, and clockwise are positive here. (From Walker et al.[2] Reproduced by permission of Springer-Verlag.)

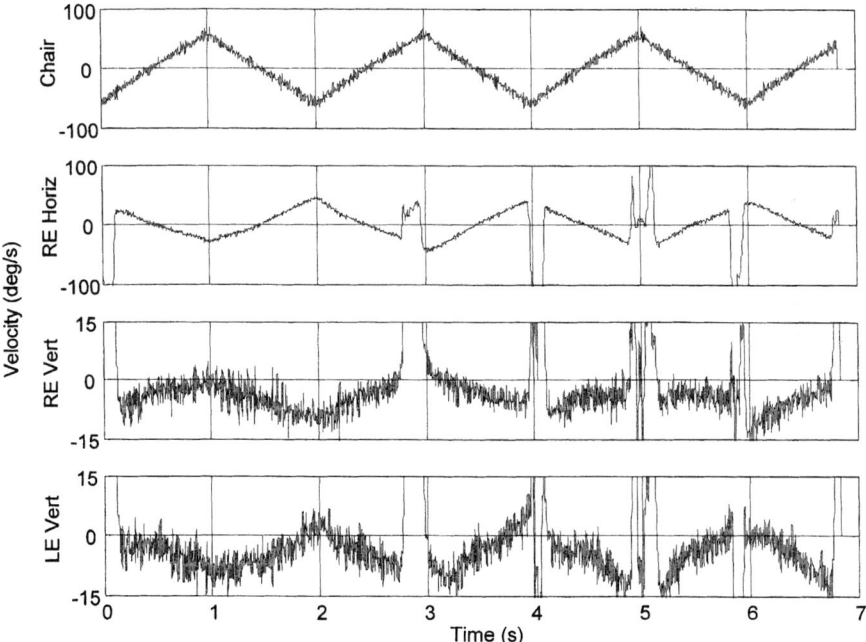

FIGURE 3. Yaw-axis chair rotation (0.5 Hz) in the dark in a rhesus monkey following bilateral lesions of the flocculus and paraflocculus, showing chair velocity and horizontal (one eye) and vertical (both eyes) components of angular eye velocity. The horizontal components were conjugate. Signs follow the right-hand rule: positive velocities are leftward, upward, and clockwise (from the animal's perspective). Note the abnormal "cross-coupling" of vertical eye velocity with horizontal chair velocity. The "cross-coupling" is disconjugate.

ebellum in the maintenance of dynamic eye alignment during high-acceleration, high-frequency head rotation.

Although the exact structures within the cerebellum responsible for control of the axis of eye rotation in response to head thrusts are not known, we have recent evidence from a monkey that the vestibulocerebellum might be involved. FIGURE 3 shows the response to rotation of the head of an animal after an ablation of some of the paraflocculus and flocculus. Pre-lesion there was virtually no vertical eye motion during yaw-axis rotation. Post lesion, however, there is an inappropriate upward component that is much greater in the right eye during rightward head rotation and much greater in the left eye during leftward head rotation. FIGURE 4 shows the response to a step of velocity in the same animal and shows that the torsion is also dissociated between the two eyes. Note that this pattern of vertical and torsional disconjugacy is in the opposite sense for the cerebellar patient described above. A simple interpretation based upon a loss of inhibition of information processed in anterior canal pathways from the labyrinth contralateral to the direction of rotation does not suffice here. Nevertheless, these data suggest that abnormally directed vertical and torsional eye motion during horizontal head rotation points to a lesion in the vestibulocerebellum.

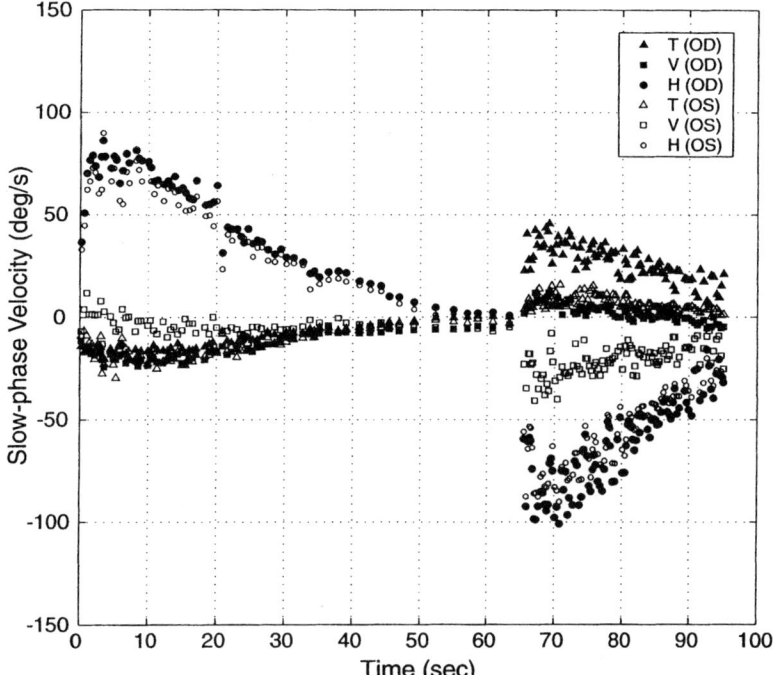

FIGURE 4. Per- and post-rotatory responses to rightward yaw-axis velocity step (120 deg/sec) in the dark for the same lesioned rhesus monkey described in Figure 3. The three components (T = torsion, V = vertical, H = horizontal) of median angular velocity for each slow phase are plotted. *Solid symbols* represent right-eye (OD) velocity and *open symbols*, left-eye (OS) velocity. Signs follow the right-hand rule, as in the previous figure. Note the inappropriate vertical and horizontal components: there is an upward velocity in the right eye associated with leftward slow phases, and an upward velocity in the left eye with rightward slow phases. The torsional component is also disconjugate, particularly in the post-rotatory phase.

The Cerebellum and the Translational VOR

Finally, we will discuss the translational VOR (tVOR), another area of relative neglect in our knowledge of cerebellar control of eye movements. First, in the cerebellar-lesioned monkey described above, we also found "cross-coupling"; there was an inappropriately directed vertical component of the slow-phase response to an interaural translation of the head. The pattern of vertical "cross-coupling," however, was not precisely the same as that associated with yaw-axis head rotation. Nevertheless, the results implicate the vestibulocerebellum in the control of the axis of eye rotation during translation as well as rotation of the head.

We also present preliminary data from two patients (aged 42 and 56 years) with isolated signs of vestibulocerebellar involvement, including downbeat nystagmus. The tVOR was measured in response to abrupt, high-acceleration head translations, either using a "head sled" apparatus or manual head heaves[18] (Ramat and Zee, this volume).

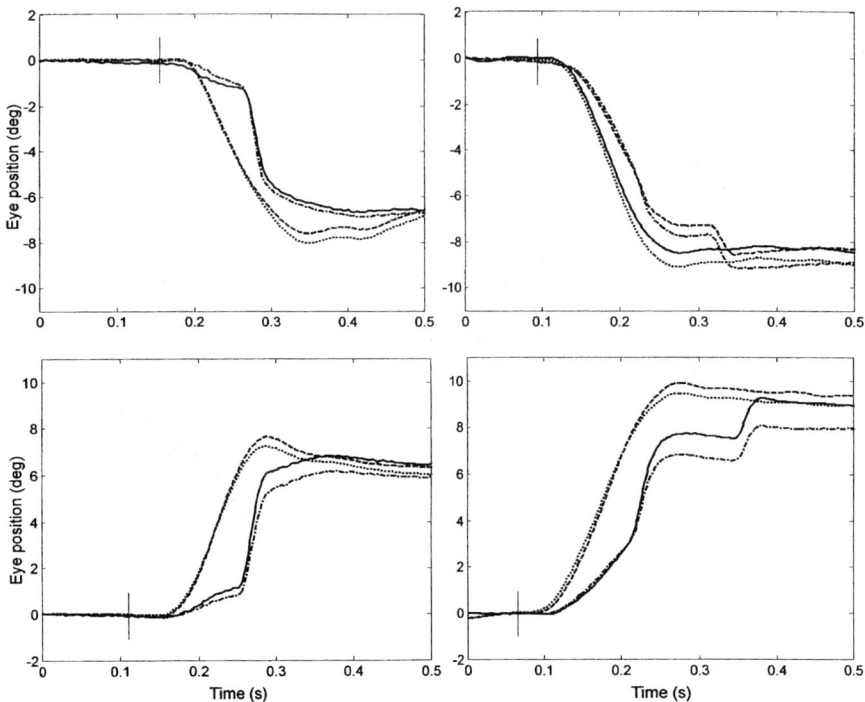

FIGURE 5. Responses to high-acceleration interaural head translations in patient P1 (*left column*) and in a representative normal subject (*right column*). *Continuous trace*: recorded right eye; *dash-dotted trace*: recorded left eye; *dashed trace*: ideal right eye; *dotted trace*: ideal left eye. *Vertical line*: head movement onset.

For both patients the "gain" was measured as the ratio of recorded/ideal eye velocity. The ideal response was based on geometrical requirements that took into account target location (15 cm in front of the subject) and interpupillary distance. Eye velocities were averaged over a 20-msec period centered at the time of peak head velocity (about 90 msec after the onset of head translation). The gain in the two patients was low (0.16 and 0.22), with the normal range in our laboratory being 0.33 to 0.54. FIGURE 5 shows responses from one patient (P1) compared with responses from a typical normal subject. Note that both P1 and the normal subject generated catch-up saccades quite early in the response, often with latencies in the 100–110 msec range.

FIGURE 6 compares average position traces (10 trials) from P1 and from the normal subject, aligned on the onset of head motion. The inset shows that the average pattern of translation of the head was virtually identical for P1 and the normal subject. The eye position traces of P1 and the normal subject did not diverge until 25 to 35 msec (vertical lines for right and left directions) after the onset of head movement (dashed line), implying that the initial tVOR was unaffected by the cerebellar lesion. Perhaps related to this finding is that we have found that first 25–30 msec of the tVOR response to the head translation stimulus in normal subjects is also unaffected

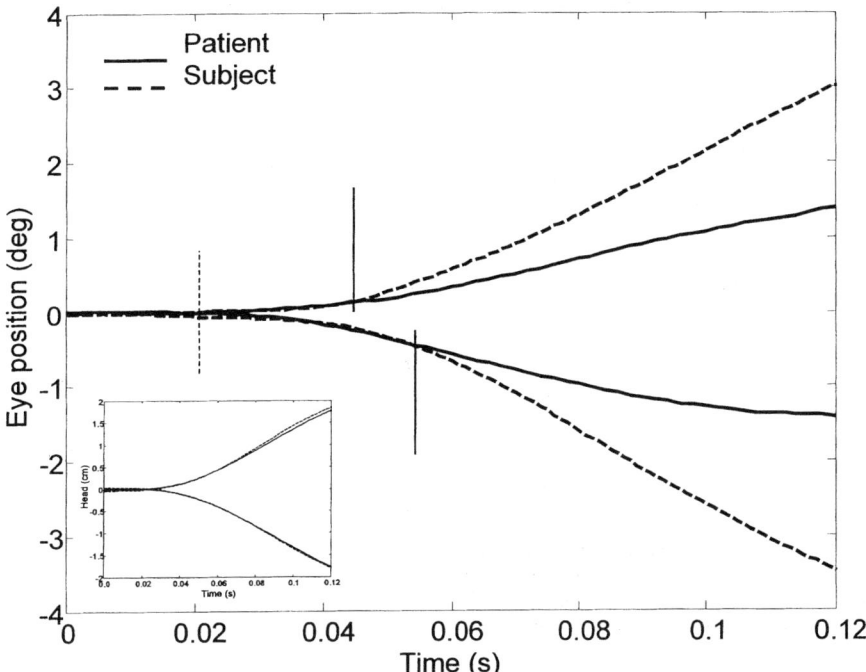

FIGURE 6. Comparison of the first 100 msec of the near viewing response between patient P1 (*continuous lines*) and the normal subject (*dashed lines*). The patient response diverges (*vertical continuous lines*) from that of the subject about 25 to 30 msec after the onset of head movement (*dashed vertical line*). The *inset* shows that the stimuli for patient P1 (*continuous line*) and for the normal subject (*dashed line*) are almost indistinguishable.

by target distance. Taken together, an interpretation of these findings is that there is a relatively immutable early tVOR response that is not modulated by viewing distance (as shown by normals), and that the cerebellum plays a role in the modulation of the later part of the response with viewing distance (as shown by patients).

In the second patient (P2) we recorded the response to head translations at three viewing distances (15, 30 and 45 cm). Unlike normal subjects (Ramat and Zee, this volume) there was little modulation of the response, although the patient was able to adequately converge and maintain binocular fixation of the targets at each distance. FIGURE 7 shows gains and sensitivities for the three distances. The gain (recorded/ideal) decreased and sensitivity (degree/[cm of head translation]) did not change as the target became closer. The sensitivity measured at the time of peak head velocity was appropriate for fixation on a target at about 80 cm. This patient's baseline phoria, measured relative to that at a target at 124 cm was about 2 degrees eso (eyes turned inward), which corresponds to a near point of about 71 cm. The closeness of the near point based on phoria and that based on tVOR sensitivity may be coincidental, but is certainly worthy of additional investigation.

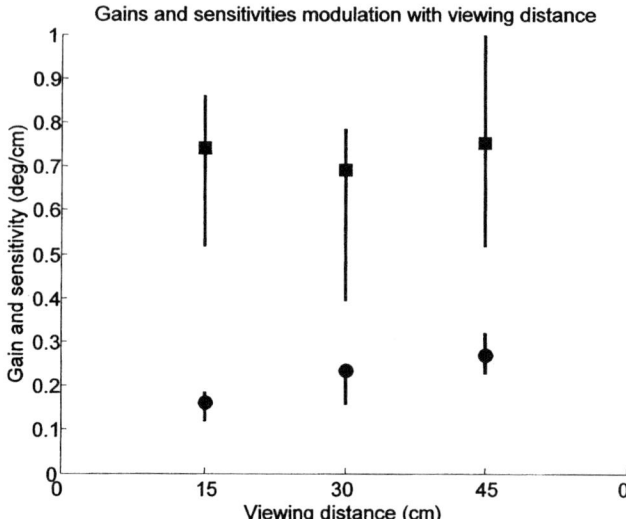

FIGURE 7. Gains and sensitivities for patient P2 measured at the time of peak head velocity in responses to head heaves with three target distances. *Round symbols*: median gain values. *Square symbols*: median sensitivity (degrees/cm) values. *Vertical bars* extend between the 25th and the 75th percentile of the recorded data.

In P1 we were only able to study the effect of viewing distance on the rotational VOR (targets at 15 and 124 cm); the VOR gain (eye velocity/head velocity) only slightly increased (from 0.79 to 0.83) for the near target (normal subjects show a considerably greater increase with near viewing).

To conclude: these preliminary studies in patients suggest that the cerebellum plays an important role in the generation of the tVOR, and that modulation of the response with target distance may be a major component. This idea is compatible with previous clinical studies[19,20] as well as with neurophysiological recordings of activity within Purkinje cells in the cerebellar flocculus during the VOR at different viewing distances.[21] Furthermore, the patients could generate corrective saccades quite early in the response, as do normal subjects. Because of their low latencies, they are unlikely to be triggered simply by visual error signals. Rather, these corrective saccades may reflect a preprogrammed response, perhaps driven by internal estimates of where the eye is relative to where it should be.[18,22] The preservation of these responses in our patients raises the issue of whether or not the cerebellum is involved with generating these early corrective saccades and, if so, which parts of the cerebellum.

ACKNOWLEDGMENTS

This work was supported by NIH Grants RO1-EY01849 and K23-EY00400, the Arnold-Chiari Foundation, and the Robert M. and Annetta J. Coffelt Endowment for PSP Research.

REFERENCES

1. ROBINSON, F.R. & A.F. FUCHS. 2001. The role of the cerebellum in voluntary eye movements. Annu. Rev. Neurosci. **24:** 981–1004.
2. WALKER, M.F., H. STEFFEN & D.S. ZEE. 2002. Three-axis approaches to ocular motor control: a role for the cerebellum. *In* Levels of Perception. L.R. Harris & M. Jenkin, Eds. Springer-Verlag. Heidelberg.
3. HASLWANTER, T. 1995. Mathematics of three-dimensional eye rotations. Vision Res. **35:** 1727–1739.
4. STRAUMANN, D., D.S. ZEE & D. SOLOMON. 2000. Three-dimensional kinematics of ocular drift in humans with cerebellar atrophy. J. Neurophysiol. **83:** 1125–1140.
5. FETTER, M., D. TWEED, H. MISSLISCH & E. KOENIG. 1994. Three-dimensional human eye movements are organized differently for the different oculomotor subsystems. Neuro-ophthalmology **14:** 147–152.
6. MILES, F.A. 1997. Visual stabilization of the eyes in primates. Curr. Opin. Neurobiol. **7:** 867–871.
7. WALKER, M.F., *et al.* 2000. Variation of eye velocity axis with vertical eye position during horizontal pursuit, interaural translation, and yaw rotation in normal humans. Soc. Neurosci. Abstr. **26:** 1718.
8. CLARK, R.A., J.M. MILLER & J.L. DEMER. 2000. Three-dimensional location of human rectus pulleys by path inflections in secondary gaze positions. Invest. Ophthalmol. Visual Sci. **41:** 3787–3797.
9. MILLER, J.M. & J.L. DEMER. 1997. New orbital constraints on eye rotations. *In* Three-Dimensional Kinematics of Eye, Head, and Limb Movements. M.T. Fetter *et al.*. Eds. Harwood Academic Publishing. The Netherlands.
10. DEMER, J.L., S.Y. OH & V. POUKENS. 2000. Evidence for active control of rectus extraocular muscle pulleys. Invest. Ophthalmol. Visual Sci. **41:** 1280–1290.
11. OPTICAN, L.M., D.S. ZEE & F.A. MILES. 1986. Floccular lesions abolish adaptive control of post-saccadic drift in primates. Exp. Brain Res. **64:** 596–598.
12. STEFFEN, H., M. WALKER & D.S. ZEE. 2002. Changes in Listing's plane following sustained vertical fusion. Invest. Ophthalmol. Visual Sci. In press.
13. FITZGIBBON, E.J., *et al.* 1996. Torsional nystagmus during vertical pursuit. J. Neuroophthalmol. **16:** 79-90.
14. ZEE, D.S., A. YAMAZAKI, P.H. BUTLER & G. GÜCER. 1981. Effects of ablation of the flocculus and paraflocculus on eye movements in primate. J. Neurophysiol. **46:** 878–899.
15. BALOH, R.W. & R.D. YEE. 1989. Spontaneous vertical nystagmus. Rev. Neurol. (Paris) **145:** 527–532.
16. BÖHMER, A. & D. STRAUMANN. 1998. Pathomechanism of downbeat nystagmus: a simple hypothesis. Neurosci. Lett. **250:** 127–130.
17. WALKER, M.F. & D.S. ZEE. 1999. Directional abnormalities of vestibular and optokinetic responses in cerebellar disease. Ann. N.Y. Acad. Sci. **871:** 205–220.
18. RAMAT, S., D.S. ZEE & L.B. MINOR. 2001. Translational vestibuloocular reflex evoked by a "head heave" stimulus Ann. N.Y. Acad. Sci. **942:** 95–113.
19. CRANE, B.T., J.-R. TIAN & J.L. DEMER. 2000. Initial vestibulo-ocular reflex during transient angular and linear acceleration in human cerebellar dysfunction. Exp. Brain Res. **130:** 486-496.
20. BALOH, R.W., Q. YUE & J.L. DEMER. 1995. The linear vestibulo-ocular reflex in normal subjects and patients with vestibular and cerebellar lesions. J. Vestib. Res. **5:** 349-361.
21. SNYDER, L.H. & W.M. KING. 1996. Behavior and physiology of the macaque vestibulo-ocular reflex response to sudden off-axis rotation: computing eye translation. Brain Res. Bull. **40:** 293–301.
22. TIAN, J.-R., B.T. CRANE & J.L. DEMER. 2000. Vestibular catch-up saccades in labyrinthine deficiency. Exp. Brain Res. **131:** 448–457.

Spatial Orientation of Caloric Nystagmus

YASUKO ARAI,[a] SERGEI B. YAKUSHIN,[b] MINGJIA DAI,[b] MIKHAIL KUNIN,[c] THEODORE RAPHAN,[c] JUN-ICHI SUZUKI,[d] AND BERNARD COHEN[b]

[a]*Department of Otolaryngology, Tokyo Women's Medical University, Daini Hospital, Tokyo, Japan*

[b]*Departments of Neurology and Physiology/Biophysics, Mount Sinai School of Medicine, New York, New York, USA*

[c]*Department of Computer and Information Science, Brooklyn College, Brooklyn, New York, USA*

[d]*Department of Otolaryngology, Teikyo University School of Medicine, Tokyo, Japan*

ABSTRACT: The spatial orientation of the slow-phase eye velocity of caloric nystagmus was investigated in cynomolgus monkeys after all six semicircular canals had been plugged. Normal animals generate responses that have dominant convective components produced by movement of the endolymph in the lateral canal toward or away from gravity. As a result, the direction of horizontal slow-phase velocity induced by cold-water irrigation changes direction with changes in head position with regard to gravity. Plugging produced a dense overgrowth of bone that blocked the flow of endolymph, but the end organs were intact. Robust caloric nystagmus was elicited after recovery, but the horizontal (yaw) component was now always toward the stimulated (ipsilateral) side, regardless of head position re gravity. The induced caloric nystagmus had strong spatial orientation properties after canal plugging. With animals upright, the three-dimensional velocity vector of the caloric nystagmus was close to the yaw axis with small vertical and roll components. Roll components became stronger in supine and prone positions and vertical components were enhanced in the right- and left-side down positions. In each instance, the addition of the roll and vertical components moved the velocity vector of the nystagmus closer to the spatial vertical. Modeling supported the postulate that the caloric nystagmus after canal plugging is influenced by three factors: (1) a reduction in neural activity in the ampullary nerves on the stimulated side due to cooling of the nerves; (2) contraction of the endolymph in the closed space between the cupula and the plug due to cooling, which resulted in deflection of the cupula and hair cells toward the plug (ampullofugal deflection); and (3) alignment of eye velocity to gravity due to the orientation properties of velocity storage. Although convection is the most prominent factor in producing caloric responses in the normal state, our results suggest that alteration of nerve activity due to thermal effects, endolymph contraction or expansion, and velocity storage are also likely to contribute to the total response.

KEYWORDS: caloric nystagmus; semicircular canals; plugging; convection; velocity storage; vestibulo-ocular reflex; eye movements; monkey

Address for correspondence: Bernard Cohen, M.D., Department of Neurology, Mount Sinai School of Medicine, Box 1135, 1 East 100th Street, New York, NY 10029. Voice: 212-241-7068; fax: 212-831-1610.
bernard.cohen@mssm.edu

INTRODUCTION

Cold-water irrigation of the external auditory canal with the head tilted up 30° from supine, which aligns the lateral canals with the spatial vertical, induces horizontal nystagmus with ipsilateral slow phases. Irrigation with warm water produces an oppositely directed nystagmus. If the head is placed 30° down from the prone position, inverting the orientation of the lateral canals with regard to the space vertical, the nystagmus produced by cold and hot stimuli reverses direction. Bárány inferred that the induced nystagmus was the result of convection currents in the lateral canal produced by local cooling or warming of the endolymph. He postulated that the thermal stimuli caused the endolymph to contract or expand locally, changing its specific gravity. Thus, when the lateral canals are tilted with regard to the spatial horizontal, the cooled or heated endolymph moves toward or away from gravity, causing deflection of the hair cells.[1] In turn, this deflection activates eye movements over the vestibulo-ocular reflex arc (VOR).

Studies of caloric stimulation following canal plugging and during space flight have identified other factors that contribute to the caloric response. Direct cooling or warming of the vestibular nerves changes the frequency of firing,[2] simulating excitation with heating and inhibition with cooling. It is believed that this is responsible for the bias in eye velocity of caloric nystagmus, so that the induced nystagmus is more intense when subjects are supine rather than prone.[3–6] It has been estimated that about 75% of the caloric response is the result of convection currents, and about 25–30% of the response to warming or cooling of the nerve.[7] Additionally, it is postulated that the thermally induced changes in endolymph volume can produce pressure effects that result in deflection of the cupula and hair cells.[8–10] This could account, at least in part, for the nystagmus induced by caloric stimulation in microgravity, in the absence of convection currents, although this is controversial.[7]

Central circuits in the vestibular system, termed "velocity storage,"[11–13] that are activated by semicircular canal, otolith, and visual stimulation, could also contribute to the caloric response.[5,14–19] These circuits contribute importantly to per- and post-rotatory nystagmus, as well as to optokinetic nystagmus (OKN), and they are primarily responsible for optokinetic after-nystagmus (OKAN). A striking feature of velocity storage is its spatial orientation, which tends to align eye velocity with the spatial vertical.[20–24] Since velocity storage responds to canal stimulation, it should be activated by caloric stimuli[16–19] and affect the spatial orientation of the caloric responses. As yet, however, there is little information about whether this occurs, primarily because the convection currents associated with caloric stimulation are also oriented to gravity and overwhelm all other responses.

In this paper, we describe the effects of plugging all six semicircular canals on the three-dimensional caloric response of monkeys. Plugging interrupts the flow of endolymph in the semicircular canals, leaving the end organs histologically intact[25–31] and the resting discharge in the afferent fibers unaltered.[30,32] The time constant of the cupula is reduced after plugging, from ≈4 sec to ≈0.07 msec.[29] This makes the canals responsive only to angular head movements above 1 Hz, and unresponsive to convection currents, which have a much lower frequency content. A consequence of this is that after plugging, caloric stimulation should produce the same activation or inhibition of the cupula, hair cells and nerves, regardless of the position of the head with regard to gravity. We took this as a unique opportunity to study the contribution of central circuits to the orientation of caloric nystagmus to gravity. A structural model of canal activation could then be developed to help understand the contribution of the various peripheral and central mechanisms that contribute to the caloric response. A preliminary report was presented earlier.[33]

METHODS

Techniques of Animal Preparation

Experiments were performed on cynomolgus monkeys. The experiments conformed to the Guide for the Care and Use of Laboratory Animals[34] and were approved by the Institutional Animal Care and Use Committee. Surgical procedures are described in detail elsewhere.[29,35] Eye movements were recorded in three dimensions (yaw, pitch, and roll) with scleral search coils.[35–37] The semicircular canals were plugged by grinding across them with a fine diamond burr.[27,29,35,38–43] After recovery, the bone fused to provide an impenetrable block to the flow of endolymph, but the hair cells of the canals and otolith organs were intact. Histological sections showing the plugged canals and the intact sensory apparatus from the right labyrinth of M9308 (one of the monkeys in this series) are shown in FIGURE 1. Data presented in this study were obtained one to two years after canal plugging, when the animals had fully recovered from the acute effects of operation. After canal plugging, there was a characteristic loss of the vestibulo-ocular response to sinusoidal rotation at 0.2 Hz (60°/sec peak velocity), but gains and phases normalized as the animals were rotated at frequencies above 1 Hz.[29] This has led to the understanding that canal plugging does not inactivate the canals; rather, it changes their dominant time constants from 4–5 sec to 0.07 sec (70 msec). [29]

Experimental Protocol

During testing, animals sat in a primate chair in a multi-axis vestibular stimulator (Neurokinetics) positioned upright, right-side down (RSD), left-side down (LSD), supine, and prone. Only data obtained in the upright, right-side down and left-side down positions will be presented in this report. Caloric stimuli consisted of irrigation of one ear with 10 ml of cold water at 20°C over 15 sec through a 22-gauge plastic tube introduced into the external canal. A plastic cuff prevented insertion of the plastic tube too far into the external canal, insuring that the tube did not puncture the drum.

Computation of Orientation Vectors

Orientation vectors, i.e., the eigenvectors of velocity storage during the caloric response, shown in FIGURES 3 and 4, were calculated using the same technique as for calculating the spatial orientation of OKAN and post-rotatory nystagmus.[20,44] The eigenvector is the best-fitting tangent line to the three-dimensional eye velocity trajectory as it approaches zero in state space using a minimum mean square error criterion.[20,44] To estimate a vector along this line, the data of the caloric response were windowed over the last 37% of the response. At this time, the activity in the vestibular nerve is presumably close-to or back to its resting level, and the predominant contribution to the response is from velocity storage in the central vestibular system.[45] To

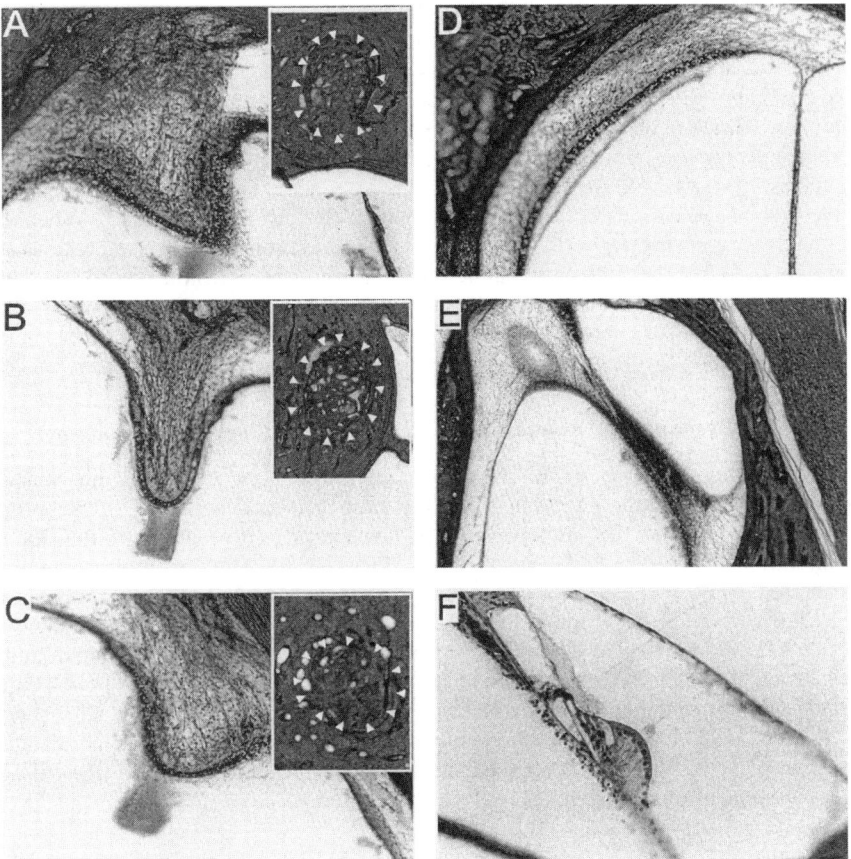

FIGURE 1. Histological sections from right labyrinth of M9308. **A–C:** *Insets* show that each canal (**A**, anterior canal; **B**, lateral canal; and **C**, posterior canal) was completely occluded by a dense bone overgrowth after plugging. *Arrowheads* show the borders of the canals. The hair cells of the crista in each of the canals (**A–C**), the utricular macula (**D**), the saccular macula (**E**), and the cochlea (**F**) were intact.

compare the orientation vectors for each head position, the vectors were normalized with respect to the length of vector at the culmination of the caloric response.

RESULTS

Caloric Responses of Normal Animals

In the upright position, cold caloric stimulation of the right ear induced nystagmus with horizontal (yaw) slow-phase velocity to the right along the −Z direction (FIG. 2A, bottom trace). There were also clockwise roll relative to the animal (+X; FIG. 2A, top trace) and downward components (+Y; FIG. 2A, middle trace). Yaw slow-phase eye velocity rose to a peak value, culminating at approximately 180°/sec about 40 sec after the start of irrigation and then declined to zero. The duration of the response was approximately 130 sec. This is about 50 sec beyond the end of the effective cooling of the bone,[45] which suggests that the duration of the effective stimulus was approximately 80 sec. With *the left side down* (LSD) (FIG. 2C), irrigation of the right ear produced a strong leftward (+Z) yaw component of eye velocity with a small counter-clockwise (−X) torsional component. An upward (−Y) pitch component relative to the head was also present in this position (FIG. 2C, middle trace). When the animal was *right side down* (RSD) (FIG. 2E), the yaw component reversed and was to the right (−Z), but the pitch component remained upward (−Y) in the head frame. The peak pitch component of the nystagmus culminated about 30 sec after the culmination of the yaw response. There was also a small clockwise (+X) roll, which had a vector component along the positive naso-occipital axis, out the front of the head.

Caloric Responses of Animals with All Six Canals Plugged (NC Animals)

Despite interruption of endolymph flow after plugging, robust nystagmus was induced by caloric stimuli in the canal-plugged animals (FIG. 2B, D, and F). In contrast to the normal animal, the yaw eye velocity, however, was now always to the side of stimulation, independent of head position (FIG. 2B–F). Thus, right-ear irrigation generated a yaw component of eye velocity to the right (−Z) in the upright position (FIG. 2B), when left-side down (FIG. 2D) and right-side down (Fig. 2F). Pitch components developed in side-down positions that were direction-specific relative to the horizontal eye velocity. For left-side down, a strong positive (down) (+Y) pitch component in head coordinates appeared (FIG. 2D, second trace). When the animal was right-side down (FIG. 2F), the pitch component was negative (up) (−Y) in head coordinates (FIG. 2F, second trace). Roll components were negligible in upright and side-down positions after plugging (FIG. 2B, D, and F).

Orientation of Eye Velocity Vectors in 3-D

The orientation vectors for the caloric responses in three dimensions were consistent across the data sets in all animals (FIGS. 3 and 4). When upright, the average vectors induced by right-ear (FIG. 3, black) and left-ear (FIG. 3, gray) stimulation were close to the yaw axis for both the normal (FIG. 3A and B) and the canal-plugged animals (FIG. 3D and E) and there was little roll (FIG. 3A and D) or pitch (FIG. 3B and E). The direction of the yaw component was (+Z) for left-ear irrigation and (−Z) for right ear irrigation. Confirming this, the average vectors had yaw components ranging from 0.96 to 0.99 (FIG. 3C and F).

When the animals were in side-down positions, the results of caloric stimulation in the normal and canal-plugged animals were quite different (FIG. 4). For left-side down in the normal animals (FIG. 4A and B), the velocity vectors were to the left (+Z) for both right- and left-ear stimulation (FIG. 4A, B), and were to the right (−Z) when the animals were right-side down (FIG. 4D and E). In the normal monkeys, the roll components were inconsistent (FIG. 4A and D), but the pitch components were negative (up) (−Y) in head coordinates for both left- and right-side down independent of which ear was irrigated (FIG. 4B and E). The spatial directions of the pitch components were in accordance with the direction of the induced yaw com-

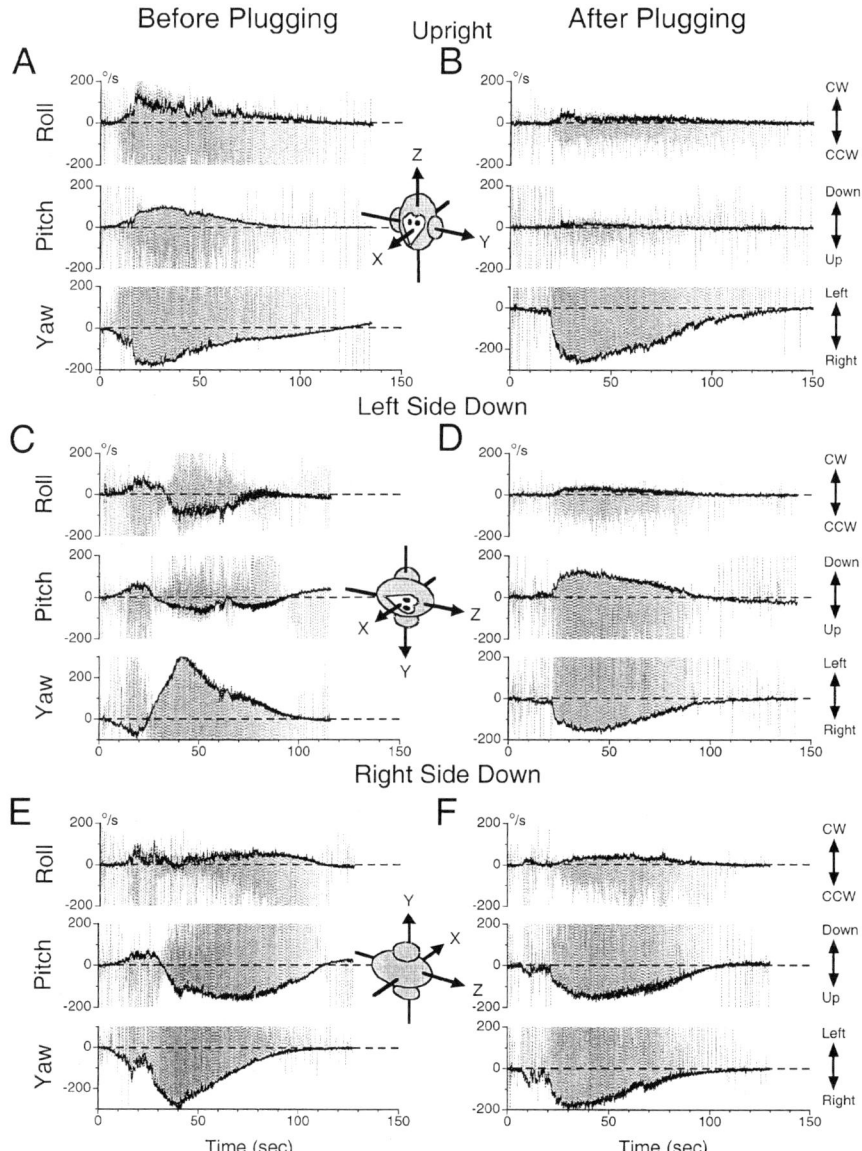

FIGURE 2. Caloric nystagmus in a monkey (M9308) before (**A, C, E**) and after plugging (**B, D, F**) of the six semicircular canals in the upright (**A, B**), left-side down (**C, D**), and right-side down (**E, F**) positions. Nystagmus was induced by irrigation with cold water (20° C) into the right ear for 15 sec in light in the upright position. At the end of stimulation, the animal was brought to the test position and the lights were extinguished. Positive (upward) directions for each of the components was **clockwise (CW)** for roll, **down** for pitch and to the **left** for yaw, from the animal's point of view. The *insets* show the orientation of the monkey's head in each of the positions used for stimulation and the head-fixed coordinate axes (**+X, +Y, +Z**).

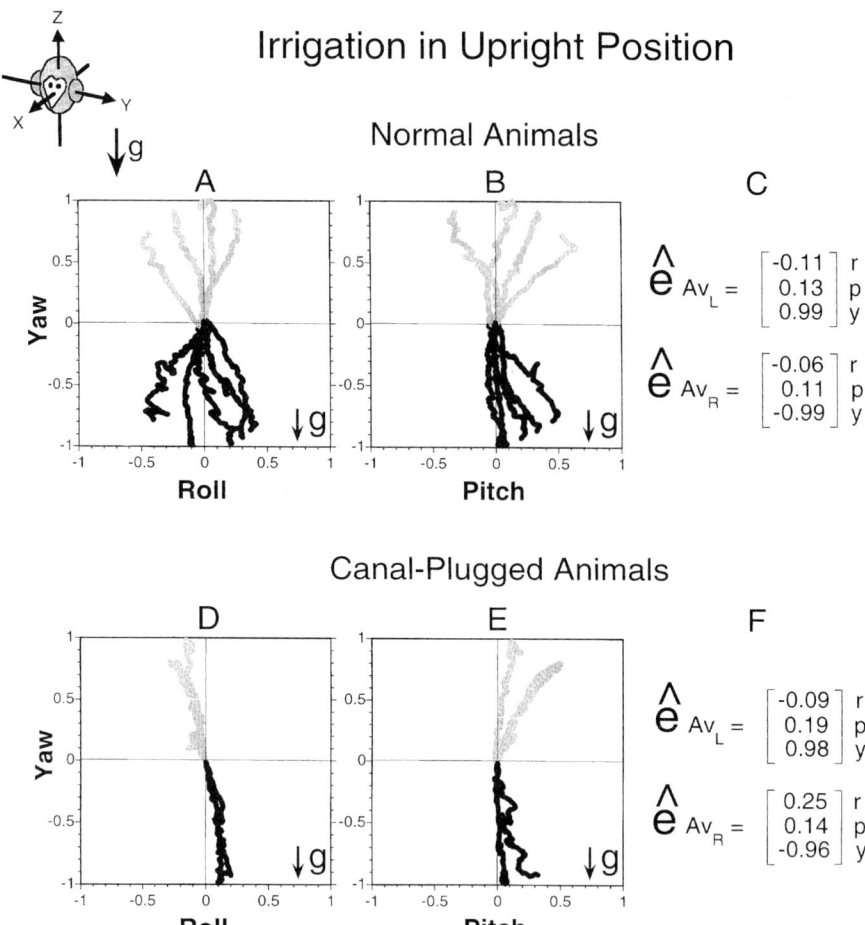

FIGURE 3. Two-dimensional projections (**A, B, D, E**) and the orientation vectors (**C, F**) of caloric nystagmus in five normal monkeys (**A–C**) and in two six-canal-plugged animals (**D-F**) while upright. The trajectories of the decaying phases of the caloric vectors in the yaw-roll plane (**A, D**) and yaw-pitch plane (**B, E**) were plotted in gray for left-ear stimulation and in black for right-ear stimulation. g in the graphs indicates the direction of gravitational force. The caloric vectors were normalized by the length of the vector at culmination as ±1.0. **C, F**: The 3 x 1 matrices of the averaged orientation vector for the left ear ($\hat{e}_{avL}$) and the right ear ($\hat{e}_{avR}$) are shown in row 3. The responses from left-and right-ear stimulation were symmetrical, and the predominant velocity was in yaw, along the body and the earth-vertical axis. *Abbreviations*: r, roll; p, pitch; y, yaw.

ponent. For left-side down, up in space corresponds to (−Y) in the head. Correspondingly, an upward yaw (+Z) in head coordinates was associated with an upward spatial pitch component (−Y) (FIG. 4C). When right side down, down in space corresponds to (−Y). Consistently, the vector of eye velocity rotated down in spatial coordinates (−Y) in accordance with the downward yaw component (−Z) (FIG. 4F).

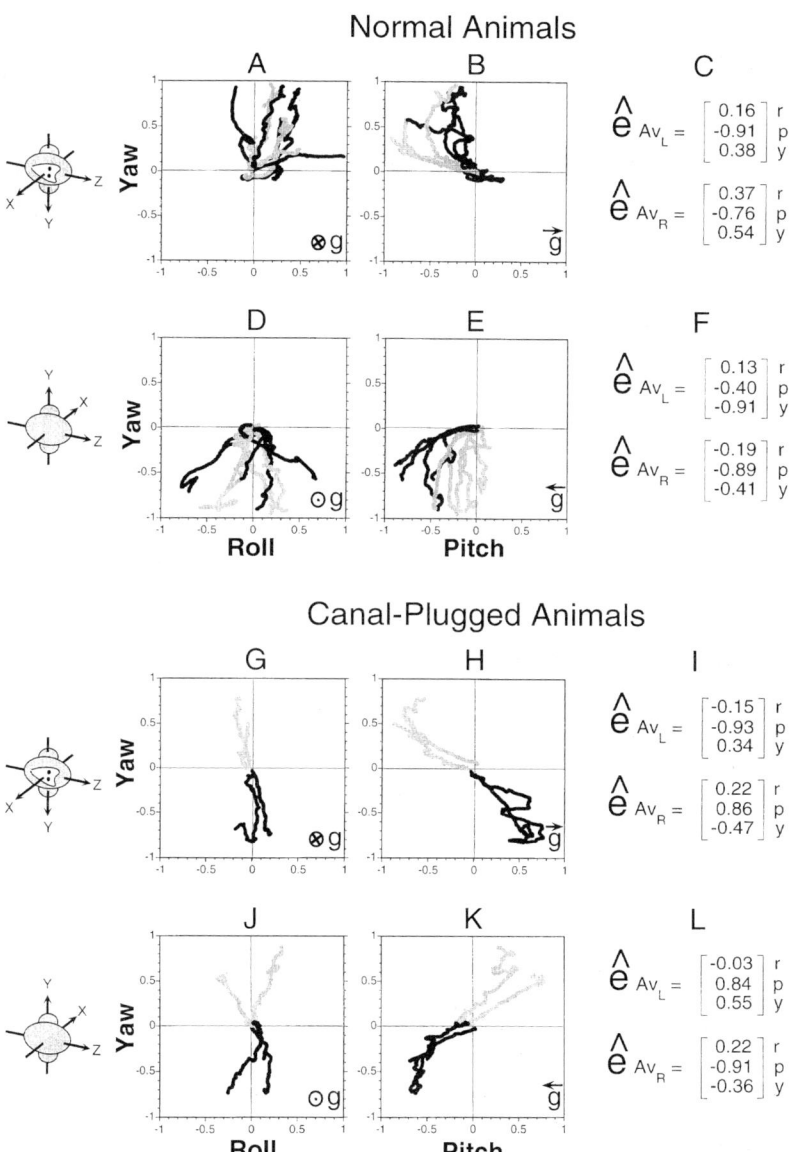

FIGURE 4. Two-dimensional projections and the orientation vectors of caloric nystagmus in the left-side down (**A–C, G–I**), and right-side down position (**D–F, J–L**) in normal (**A–F**) and six canal-plugged animals (**G–L**). The arrangement of the figures is the same as in FIGURE 3. The **g** vector is shown as a *circle with an embedded X* when the vector is pointed away from the reader, into the page, and as a *circle with an embedded point* when the vector is pointed toward the reader, out of the page. See text for details.

Thus, the direction of the dominant pitch component was in the same direction in space as the yaw component, relative to the head frame (+Z, −Y for LSD and −Z, −Y when RSD), consistent with the orientation properties of velocity storage.[20,21] It should be noted that the delays in timing of the culmination of the pitch relative to the yaw components (FIG. 2C, E) were also consistent with delays due to the development of cross-coupling through velocity storage.[20,21]

Irrigation of canal-plugged animals in side-down positions (FIG. 4G–K) produced an ipsilateral yaw component that was positive (+Z) from left-ear stimulation (gray) and negative (−Z) from right-ear stimulation (black). In each instance, a pitch component was induced along the spatial vertical in the same direction as the yaw component in the head frame. Thus, for left-ear down (FIG. 4H), the pitch was negative (−Y) for left-ear irrigation (gray) and positive (+Y) for right-ear irrigation (black), corresponding to the negative (−Z) and positive (+Z) yaw components induced. These vectors were reversed when the right ear was down (FIG. 4K). The average roll components were insignificant in the side-down positions (FIG. 4G and J), as in the upright position (FIG. 3D). Thus, as in the normal animal the dominant effect of unilateral caloric stimulation was to induce spatial components of slow-phase eye velocity components whose direction in space was the same as the direction of the yaw component in the head coordinate frame, consistent with the orientation of velocity storage. Again, the delay in timing of the culmination of the pitch relative to the yaw components (FIG. 2D and F) was consistent with the development of cross-coupling through velocity storage.[20,21]

Modeling Canal Activations as a Function of Head Orientation

In order to gain insight into the factors contributing to the caloric response after plugging, we utilized the model of 3-D eye-velocity generation developed in Yakushin *et al.*[29,35] to determine whether activations related to specific canal planes could induce the observed eye velocities. It was assumed that excitation of the canal nerves corresponded to positive directions of activation of the canal planes, which are shown as vectors normal to the canal planes on the right side according to right-hand rule (FIG. 5A). Compensatory eye rotations would be opposite to these excitations, similar to the eye movements produced by electrical stimulation of the ampullary nerves.[41,46]

We first assumed that cooling of the nerve would produce about 30% reduction in activity in the canal nerves.[7] The components of the canal vector were chosen as (−0.3, −0.3, −0.3) to reflect the negative equal activation of the canal planes. When this vector was inserted into the model, the computed eye velocity (FIG. 5B, "Eye Velocity 1") did not correspond to the observed Actual Eye Velocity vector (FIG. 5B), which was directed predominantly along the −Z axis (0.25, 0.14, −0.96). Beyond that, there was no set of all negative canal activations that could generate an eye-velocity vector outside the quadrant defined by the canal vectors, to predict the actual direction of eye velocity. Therefore inhibition of the canal nerves alone could not have produced the observed eye velocity.

To correct for this, the canal activations were modified on the assumption that the cooling had caused contraction of the endolymph between the plug and the cupula. This would cause deviation of the hair cells in the direction of the plug in all three canals, similar to the utriculofugal (ampullofugal) deflection produced by convec-

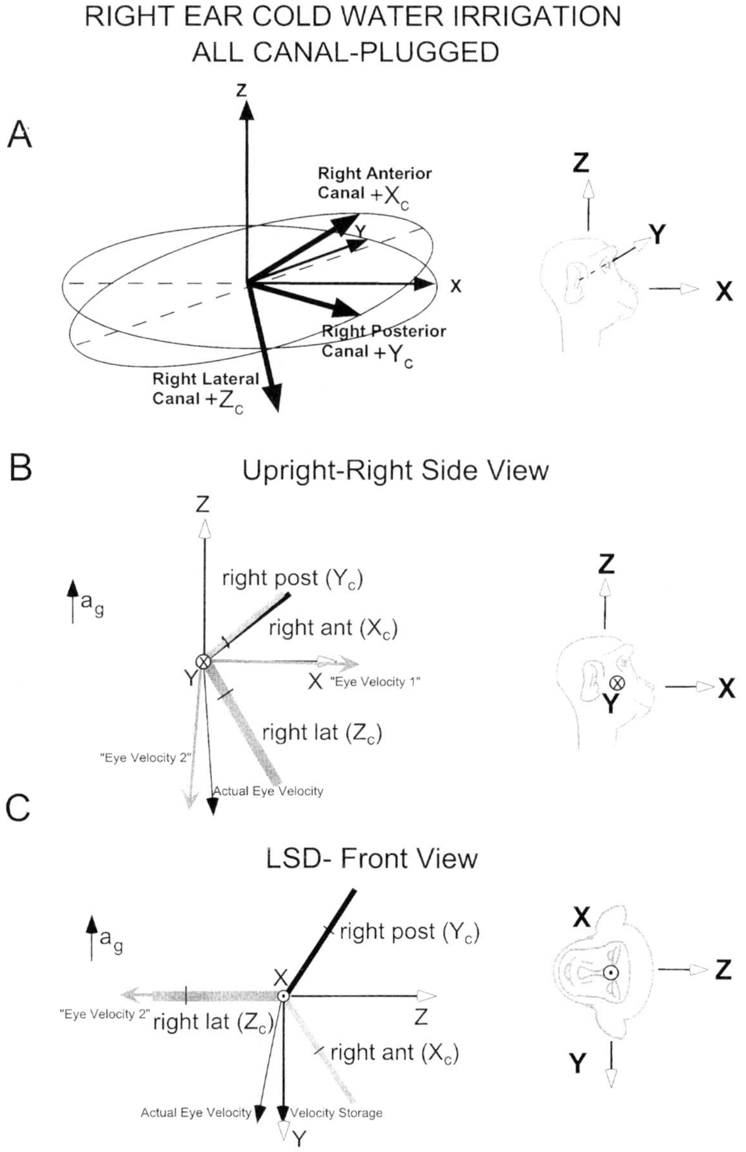

FIGURE 5. Coordinate frame (**A**) and model predictions of direction of the eye-velocity orientation vector induced by cold caloric stimulation of the left ear in the upright (**B**) and left-side down (**C**) positions. The coordinate frames show the relationship between the head frame (*open arrows*, +X,+Y, +Z), and the canal frame for the right labyrinth, which are the direction of the normals to the canal planes (right lat [Z_c]; right ant [X_c]; right post [Y_c]). Eye-velocity orientation vectors can be directly related to the position of the animal's head shown by the *insets* on the right of each vector plot. The basis for the simulation is the model developed in Ref. 29. *Circles* and *dots* show axes that project directly out of the picture (+X

tion, which results in inhibition of the lateral canal nerves and excitation of the anterior and posterior canal nerves.[25] Therefore, the canal activation vector was modified to increase the inhibition of the lateral canal afferent activity and to generate excitation of the anterior and posterior canal nerves. The vector was chosen as (0.5, 0.5, −0.75). This produced a resultant eye-velocity direction that closely approximated the eye-velocity vector in the canal-plugged animal when upright (FIG. 5B, "Eye Velocity 2" vs. Actual Eye Velocity). Although the explicit values used for the canals were somewhat arbitrary, the important factor was that the anterior and posterior canal values should be positive and be the same, and that the value for the lateral canal should be negative in the absence of other factors.

Another possibility we considered was that the observed response may have arisen due to lateral canal inhibition alone, and that there was no contribution from the vertical canals. Velocity storage, which is oriented toward the spatial vertical, would have provided the additional rotation toward the spatial vertical observed in the data. However, this would still have required the proposed contraction mechanism to counter the inhibition of the nerve cooling of the vertical canal nerves and to increase the inhibition of the lateral canal.

We tested the hypothesis that velocity storage contributed to the orientation of the caloric response by considering the model predictions for tilted head positions. Because the nature of the cooling was the same in every head orientation, it would be expected that the direction of eye velocity would be the same in all positions, if only nerve cooling and no other factors were involved. The Actual Eye Velocity vectors were significantly different from the direction of "Eye Velocity 2" for side-down positions (FIG. 5C). Instead, the Actual Eye Velocity vector had a large pitch component that brought it close to the spatial vertical, in the direction of velocity storage. Because of the absence of convection in the canal-plugged animals, the most likely cause for this substantial shift in the direction of the eye velocity vector was the spatial orientation of velocity storage. Therefore, velocity storage must also have contributed an additional factor in adjusting the orientation of the eye velocity in the upright position to achieve spatial verticality (FIG. 5B).

---

in **C**), while axes that project into the page are shown by Xs enclosed in circles (−Y in **B**). The *upward arrows* on the left of **B** and **C** show the acceleration of gravity (a_g), which is also the direction of orientation of velocity storage. The *perpendicular dashes* across the normals of the canals are the postulated relative activations of each canal by the caloric stimulus. Viewed from the side (**B**), the anterior posterior canal normals overlay each other, while in C, the anterior canals project 45° into the page and the posterior canals 45° out of the page. In **B**, these normals have been offset slightly to show the normals of the anterior (*gray*) and posterior (*black*) canals. Actual Eye Velocity is the average eye-velocity vector obtained from the matrices for right-ear stimulation shown in FIG. 3C and F and FIG. 4C, F, I, and L. "Eye velocity 1" and "Eye Velocity 2" are predicted eye velocities for canal nerve inhibition and for canal nerve inhibition plus contraction of the endolymph due to cooling, respectively. A contribution from velocity storage is necessary in the side-down position in **C** to shift the vector from "Eye Velocity 2" to the Actual Eye Velocity. See text for details.

DISCUSSION

This study shows that caloric nystagmus is readily induced in animals after all six canals have been plugged. The caloric responses were robust and comparable to those in the normal monkey. Thus, blockage of the flow of endolymph in the canal did not prevent the production of caloric nystagmus, despite the absence of convection currents. The data also show that the induced caloric nystagmus exhibited significant orientation towards the acceleration of gravity in canal-plugged animals. Caloric nystagmus in canal-plugged animals was consistent with the assumptions that plugging had eliminated convection currents and that there was approximately the same stimulus to the canals in each head position in the plugged animals. After plugging, there was no longer a reversal of the yaw component of eye velocity when the head position was reversed, but the magnitude of the eye-velocity vector and the duration of the nystagmus were not affected by changes in head position in the plugged animals. This supports the contention that the activation of the nerve of the canal was the same for all head positions. Consequently, the orientation of the caloric nystagmus could be directly compared in each head position in the plugged animals. This analysis indicated that the spatial orientation of caloric nystagmus could be predicted from orientation properties of velocity storage in both the canal-plugged and the normal animals.

Two lines of evidence support this idea. When caloric nystagmus is suppressed by exposure to a stationary visual surround during its early phases, slow-phase velocity recovers slowly, suggesting reactivation of central circuits by afferent input.[14,47] Additionally, if caloric nystagmus is suppressed by light in its terminal stages, eye velocity never recovers, presumably because cupula deflection has ceased and the terminal portions of the caloric response are produced solely by central activity from velocity storage. Both findings have been modeled by a central integrator that adds to the activity produced by cupula deflection in its early stages, but which outlasts cupula deflection.[14] From experiments on the recovery of slow-phase velocity after visual suppression of varying durations, it has been estimated that the last 20 sec of the caloric response is produced solely by activation of velocity storage.[45] This would predict about 80 sec of cooling in the bone, which is consistent with other estimates.[48]

On the basis of the histological demonstration that the membranous and bony canals were obliterated after plugging, we assumed that convection currents were permanently eliminated in these animals. Therefore, the response to caloric stimulation was due either to cooling of the canal nerves or to cooling of the nerves plus other factors. A likely possibility is that cooling of the endolymph in the space between the plug and the cupula produced constriction of the endolymph, causing the hair cells to deflect toward the plug (i.e., in the ampullofugal direction). Scherer and Clarke and colleagues have shown that caloric nystagmus is present in microgravity where there are no convection currents and no orientation of velocity storage toward gravity.[10] They postulate that this activation is due to unequal pressure effects on both sides of the cupula, a postulate that was rejected by Minor and Goldberg[15] on the grounds that the pressure would quickly equilibrate on both sides of the labyrinth if there were no impediment to flow. However, as shown by the projection of the eye velocity vector along the naso-occipital axis in FIG. 5B ("Eye Velocity 1"), inhibition alone could not have produced yaw axis nystagmus. Therefore, the possibility that caloric nystagmus in space was produced by pressure effects is supported by our findings. These conclusions are also supported by studies of Rabbitt et al.,[30] who have shown that distension and contraction takes place in the compliant utricular vestibule. It has been suggested that this could cause an inhibitory movement of the cupula in the lateral canal.[30] Our studies suggest further that such endolymph contraction due to cooling could additionally cause excitatory movement of the cupulae in the vertical canals and contribute to the spatial properties of the caloric response.

It is difficult to determine pressure effects in the caloric nystagmus of normal subjects. Nevertheless, we postulate that there are four factors that produce caloric nystagmus in normal subjects: convection currents, cooling of the canal nerves, pressure effects, and orientation of velocity storage to gravity. The findings in this paper give a reasonable explanation of the caloric nystagmus in canal-plugged animals. The use of animals with various combinations of canal-plugging in conjunction with a model-based approach now offers the possibility for a more complete understanding of this response approximately 100 years after its discovery.

ACKNOWLEDGMENTS

This study was funded in part by the Japanese National Science and Research Fund 11671707 and by NIH Grants EY11812, EY04148, and EY01867, DC 03787, and DC03284.

REFERENCES

1. BÁRÁNY, R. 1906. Untersuchungen ueber den vom Vestibularaparat des Ohres reflektorisch ausgeloesten Nystagmus und seine Begleiterscheinungen. Monatsschr. Ohrenheilk. **4:** 193–297.
2. KLINKE, R. 1992. Thermal effects on the vestibular hair cell synapse have to be considered for the explanation of caloric nystagmus [letter]. Acta Otolaryngol. (Stockh.) **112:** 574–578.
3. COATS, A.C. & S.Y. SMITH. 1967. Body position and the intensity of caloric nystagmus. Acta Otolaryngol. (Stockh.) **63:** 515–532.
4. PAIGE, G.D. 1985. Caloric responses after horizontal canal inactivation. Acta Otolaryngol. **100:** 321–327.
5. CLARKE, A.H., H. SCHERER & J. SCHLEIBINGER. 1988. Body position and caloric nystagmus response. Acta Otolaryngol. **106:** 339–347.
6. STAHLE, J. 1990. Controversies on the caloric response: from Barany's theory to studies in microgravity. Acta Otolaryngol. **109:** 162–167.
7. MINOR, L.B. & J.M. GOLDBERG. 1990. Influence of static head position on the horizontal nystagmus evoked by caloric, rotational and optokinetic stimulation in the squirrel monkey. Exp. Brain Res. **82:** 1–13.
8. CLARKE, A.H., W. TEIWES, P. OELHAFEN & H. SCHERER. 1993. Three-dimensional aspects of caloric nystagmus in humans: I. The influence of increased gravitoinertial force. Acta Otolaryngol. **113:** 687–692.
9. CLARKE, A.H., W. TEIWES & H. SCHERER. 1993. Vestibulo-oculomotor testing during the course of a spaceflight mission. Clin. Invest. **71:** 740–748.
10. SCHERER, H. & A.H. CLARKE. 1985. The caloric vestibular reaction in space: physiological considerations. Acta Otolaryngol. **100:** 328–336.
11. COHEN, B., V. MATSUO & T. RAPHAN. 1977. Quantitative analysis of the velocity characteristics of optokinetic nystagmus and optokinetic after-nystagmus. J. Physiol. **270:** 321–344.

12. RAPHAN, T., V. MATSUO & B. COHEN. 1979. Velocity storage in the vestibulo-ocular reflex arc (VOR). Exp. Brain Res. **35:** 229–248.
13. RAPHAN, T. & B. COHEN. 1996. How does the vestibulo-ocular reflex work? *In* Disorders of the Vestibular System. R. Baloh & G.M. Halmagyi, Eds.: 20–47. Oxford University Press. New York.
14. RAPHAN, T. & B. COHEN. 1981. The role of integration in oculomotor control. *In* Models of oculomotor behavior and control. B.L. Zuber, Ed.: 91–109. CRC Press. Boca Raton, FL.
15. MINOR, L.B. & J.M. GOLDBERG. 1990. Vestibular-nerve inputs to the vestibulo-ocular reflex: a functional-ablation study in the squirrel monkey. J. Neurosci. **11:** 1636–1648.
16. ARAI, Y., V. HENN, A. BÖEHMER & J. SUZUKI. 1989. How could canal-pluggings result in intensive direction changing type of positional nystagmus? Acta Otolaryngol. (Stockh.) Suppl **468:** 159–164.
17. KAWACHI, N. 1992. [The vertical component in a caloric nystagmus and the existence of a second phase of the nystagmus—the possibility of canal otolithic interaction in normal subjects]. Nippon Jibiinkoka Gakkai Kaiho **95:** 1409–1420.
18. TSHUCHIYA, Y. 1995. Caloric nystagmus in the lateral recumbent position in normal subjects—the possibility of a participation of velocity storage and other non-convection factors. J. Otolaryngol. Jpn. **98:** 1006–1020.
19. ARAI, Y., S.B. YAKUSHIN, M. DAI, *et al.* 1998. Position dependency of caloric nystagmus in monkeys with all semicircular canals plugged. Int. Tinnitus J. **4:** 1–2.
20. RAPHAN, T. & D. STURM. 1991. Modelling the spatio-temporal organization of velocity storage in the vestibulo-ocular reflex (VOR) by optokinetic studies. J. Neurophysiol. **66:** 1410–1421.
21. DAI, M.J., T. RAPHAN & B. COHEN. 1991. Spatial orientation of the vestibular system: dependence of optokinetic after-nystagmus on gravity. J. Neurophysiol. **66:** 1422–1439.
22. WEARNE, S., T. RAPHAN & B. COHEN. 1996. Nodulo-uvular control of the central vestibular dynamics determines spatial orientation of the angular vestibulo-ocular reflex. *In* New Directions in Vestibular Research. S.M. Highstein, B. Cohen & J. A. Buttner-Ennever, Eds. Ann. N.Y. Acad. Sci. **781:** 364–384.
23. WEARNE, S., T. RAPHAN & B. COHEN. 1997. Contribution of vestibular commissural pathways to spatial orientation of the angular vestibuloocular reflex. J. Neurophysiol. **78:** 1193–1197.
24. WEARNE, S., T. RAPHAN & B. COHEN. 1998. Control of spatial orientation of the angular vestibulo-ocular reflex by the nodulus and uvula. J. Neurophysiol. **79:** 2690–2715.
25. EWALD, J.R. 1892. Physiologische Untersuchungen uber das Endorgan des Nervus Octavus. Weisbaden.
26. CAMIS, M. 1930. The Physiology of the Vestibular Apparatus. Clarendon. Oxford, UK.
27. SUZUKI, J.-I., A. KODAMA, B. COHEN & V. HENN. 1991. Canal-plugging in the rhesus monkey: a tool to study the contribution of individual canals to nystagmus generation. Acta Otolaryngol. (Suppl). **481:** 91–93.
28. ANGELAKI, D.E., B.J.M. HESS, Y. ARAI & J.-I. SUZUKI. 1996. Adaptation of primate vestibular reflex to altered peripheral vestibular inputs. I. Frequency-specific recovery of horizontal VOR after inactivation of the lateral semicircular canals. J. Neurophysiol. **76:** 2941–2953.
29. YAKUSHIN, S.B. *et al.* 1998. Dynamics and kinematics of the angular vestibulo-ocular reflex in monkey: effects of canal plugging. J. Neurophysiol. **80:** 3077–3099.
30. RABBITT, R. D., R. BOYLE & S. M. HIGHSTEIN. 1999. Influence of surgical plugging on horizontal semicircular canal mechanics and afferent response dynamics. J. Neurophysiol. **82:** 1033–1053.
31. ARAI, Y., *et al.* 1996. Canal plugging did not abolish caloric nystagmus in monkey. J. Vestib. Res. **6:** S41 (CAL5).
32. GOLDBERG, J.M. & C. FERNANDES. 1975. Responses of peripheral vestibular neurons to angular and linear accelerations in the squirrel monkey. Acta Otolaryngol. (Stockh.) **80:** 101–110.

33. ARAI, Y., J. SUZUKI, B.J. HESS & V. HENN. 1991. Caloric nystagmus in three dimensions under otolithic control in rhesus monkeys. A preliminary report. ORL J. Otorhinololaryngol Relat. Spec. **52:** 218–225.
34. NATIONAL RESEARCH COUNCIL. 1996. Guide for the care and use of laboratory animals. National Academy Press. Washington, DC.
35. YAKUSHIN, S.B., *et al.* 1995. Semicircular canal contributions to the three-dimensional vestibuloocular reflex: a model-based approach. J. Neurophysiol. **74:** 2722–2738.
36. COHEN, H., B. COHEN, T. RAPHAN & W. WAESPE. 1992. Habituation and adaptation of the vestibulo-ocular reflex: a model of differential control by the vestibulocerebellum. Exp. Brain Res. **90:** 526–538.
37. DAI, M., *et al.* 1994. Effects of spaceflight on ocular counterrolling and the spatial orientation of the vestibular system. Exp. Brain Res. **102:** 45–56.
38. COHEN, B., J. SUZUKI, S. SHANZER & M.B. BENDER. 1964. Semicircular canal control of eye movements. *In* The Oculomotor System. Chapt. 6. M.B. Bender, Ed.: 163–172. Hoeber Medical Division, Harper & Row. New York.
39. SUZUKI, J.I. & B. COHEN. 1964. Head, eye, body and limb movements from semicircular canal nerve. Exp. Neurol. **10:** 393–405.
40. MONEY, K.E. & J.W. SCOTT. 1962. Functions of separate sensory receptors of nonauditory labyrinth of the cat. Am. J. Physiol. **202:** 1211–1220.
41. SUZUKI, J., B. COHEN & M.B. BENDER. 1964. Compensatory eye movements induced by vertical semicircular canal stimulation. Exp. Neurol. **9:** 137–160.
42. SUZUKI, J.-I. & B. COHEN. 1966. Integration of semicircular canal activity. J. Neurophysiol. **29:** 981–995.
43. SUZUKI, J.-I. *et al.* 1997. Paired and unpaired plugging of the semicircular canals in monkeys related with physiological and pathological findings. *In* Three-Dimensional Kinematic Principles of Eye, Head, and Limb Movements in Health and Disease. M. Fetter *et al.*, Eds.: 275–279. Harwood Academic Publishers. Amsterdam.
44. RAPHAN, T., M. DAI & B. COHEN. 1992. Spatial orientation of the vestibular system. Ann. N.Y. Acad. Sci. **656:** 140–157.
45. ARAI, Y., S.B. YAKUSHIN, J.-I. SUZUKI & B. COHEN. 2000. Duration of effective thermal stimulation in caloric test. Equilibrium Res. **59:** 206–214.
46. COHEN, B., J. SUZUKI & M.B. BENDER. 1964. Eye movements from semicircular canal nerve stimulation in the cat. Ann. Otol. Rhinol. Laryngol. **73:** 153–169.
47. TAKEMORI, S. & B. COHEN. 1974. Visual suppression of vestibular nystagmus in rhesus monkeys. Brain Res. **72:** 203–212.
48. STEER, R.W., Y.T. LI, L.R. YOUNG & J.L. MEIRY. 1968. Physical properties of the labyrinthine fluids and quantification of the phenomenon of caloric stimulation, Vol. SP-152.: 409–420. Third Symposium on the Role of the Vestibular Organs in Space Exploration. U.S. Aeronautic and Space Administration. Washington, DC.

The Role of the Lateral Intraparietal Area of the Monkey in the Generation of Saccades and Visuospatial Attention

MICHAEL E. GOLDBERG,[a,c] JAMES BISLEY,[a,c] KEITH D. POWELL,[a] JACQUELINE GOTTLIEB,[a] AND MAKOTO KUSUNOKI [a,b]

[a]*The Laboratory of Sensorimotor Research, National Eye Institute, Bethesda, Maryland 20892-4435, USA*

[b]*University College, London, United Kingdom*

[c]*David Mahoney Center for Mind and Brain, Center for Neurobiology and Behavior, Departments of Neurology and Psychiatry, Columbia University, and the New York State Psychiatric Institute, New York, New York, USA*

ABSTRACT: The brain cannot monitor or react towards the entire world at a given time. Instead, using the process of attention, it selects objects in the world for further analysis. Neuronal activity in the monkey intraparietal area has the properties appropriate for a neuronal substrate of attention: instead of all objects being represented in the parietal cortex, only salient objects are. Such objects can be salient because of their physical properties (recently flashed objects or moving objects) or because they can be made important to the animal by virtue of a task. Although lateral intraparietal area (LIP) neurons respond through the delay period of a memory-guided saccade, they also respond in an enhanced manner to distractors flashed during the delay period of a memory-guided saccade being generated to a position outside the receptive field. This activity parallels the monkey's psychophysical attentional process: attention is ordinarily pinned at the goal of a memory-guided saccade, but it shifts briefly to the locus of a task-irrelevant distractor flashed briefly during the delay period and then returns to the goal. Although neurons in LIP have been implicated as being directly involved in the generation of saccadic eye movements, their activity does not predict where, when, or if a saccade will occur. The ensemble of activity in LIP, however, does accurately describe the locus of attention.

KEYWORDS: parietal; saccade; monkey; attention; salience

INTRODUCTION

We live in a world of sensory overload. Sights, sounds, smells, and touches bombard our sensory apparatus constantly, and the primate brain cannot possibly deal with all of them simultaneously. Instead, it chooses among this intense sensory

Address for correspondence: Michael E. Goldberg, The Laboratory of Sensorimotor Research, National Eye Institute, 49 Convent Drive, Room 2A-50, Bethesda, MD 20892-4435, USA. Voice: 301-496-1060; fax: 301-402-0511.

meg@lsr.nei.nih.gov

world the objects most relevant to its behavior for further processing. This act of selection is called attention. James[1] described attention as "the taking possession by the mind in clear and vivid form, of one out of what seem several simultaneously possible objects or trains of thought....It implies withdrawal from some things in order to deal effectively with others." He then described two different kinds of attention: "It is either passive, reflex, non-voluntary, effortless or active and voluntary. In passive immediate sensorial attention the stimulus is a sense-impression, either very intense, voluminous, or sudden...big things, bright things, moving things...blood." More recently, these two kinds of attention have been described as exogenous and endogenous. Primates usually look at the objects of visual attention, and investigators have described two different kinds of attention along a different axis, depending on whether the subject actually looks at the object (covert attention) or responds without looking (overt attention).[2]

The parietal cortex has long been thought to be important in the neural mechanisms underlying spatial attention. One area in particular, the lateral intraparietal area (LIP), has been implicated in attentional and oculomotor processes. Although it is clear that LIP has a visual representation, it is not clear if this visual representation is dedicated to processing saccadic eye movements or if it has a more general attentional function independent of the generation of any specific movement. In this review, we describe three different experiments that examine the role of attention in LIP and its relation to the generation of saccadic eye movements. The first deals with the nature of the visual representation in LIP, the second with the independence of LIP activity from saccade planning, and the third with the nature and determinants of visual attention in the monkey. We begin with a description of the methods common to all three experiments.

GENERAL METHODS

Six rhesus monkeys were trained to perform various visual tasks for liquid reward and then were prepared, under sterile surgical procedures, for neurophysiological and eye position recording. All animal protocols were approved as conforming to the National Institutes of Health guidelines for animal care and use by the National Eye Institute Animal Care and Use Committee. We recently described our physiological, anatomical, and data analytical methods elsewhere.[3,4]

The monkeys were trained to fixate on a red laser spot that appeared on a tangent screen 86 cm in front of them. They were rewarded for maintaining their eye within a fixation window (2 degrees) in width. When the fixation point moved, the monkeys followed it with a saccade.[5] They also quickly learned a memory-guided delayed saccade task: while the monkey looked at the central fixation point, a peripheral stimulus was flashed for 200 ms. After a delay of 500–1,000 ms, the fixation point disappeared and the monkey made saccades to the remembered spatial location of the now vanished target.[6] Having learned these standard tasks, the monkeys were ready to learn tasks that were more complicated, that is, the stable array, the distractor, and the go–no go saccade tasks.

NATURE OF VISUAL REPRESENTATION IN THE LATERAL INTRAPARIETAL AREA

Since the development of the fixation task by Wurtz,[7] the standard method for determining a visual response of a neuron has been the response of the neuron to a stimulus that appears suddenly in its receptive field. This definition has a problem, however. Abruptly appearing stimuli are not only associated with photons exciting rods and cones; as James[8] noted, they are also attentional attractors. Stimuli can enter receptive fields in several ways: one is when a light appears suddenly in the receptive field; a second is when a saccade brings a stable object into the receptive field. Since activity in the parietal cortex is associated with attention as well as with vision, the question arises as to whether the 'visual responses' of parietal neurons are visual, that is, responding to photons on the retina, like a retinal ganglion cell, or attentional.

To distinguish between these alternatives we devised a number of tasks in which the stimulus, rather than appearing *de novo* in the receptive field, entered the receptive field by virtue of a saccadic eye movement. This enabled us to stimulate the receptive field using stimuli that did not have the attentional tag of abrupt onset. In these stable array tasks,[9] the monkeys were presented with an array of eight stimuli arranged uniformly in a circular array. These stimuli did not appear or disappear from trial to trial. Instead, they were constant for a block of trials. The stimuli were roughly 2 degrees in diameter and varied in shape and color. They were not equated for luminance. They were positioned so that when the monkey fixated on the center of the array, at least one stimulus appeared in the receptive field of the neuron under study. In the simplest of these tasks the monkey fixated at a position outside the array so that no stimulus was in the receptive field of the neuron being studied, and then, when the red fixation point jumped, the monkey made a saccade to the center of the array (FIG. 1). This saccade brought one of the stable stimuli into the receptive field.

The typical neuron had a brisk response to the sudden appearance of a stimulus in its receptive field during a fixation task (FIG. 2A), and a much smaller response when the same stimulus as a member of the stable array entered the receptive field (FIG. 2B). The decrement of response could have been related to the behavioral irrelevance of the stable target or it could have been due to a series of other confounds. For example, the movement of the stimulus into the receptive field by the saccade is not exactly the same as its appearance from the flash; the other members of the array might exert some purely visual local inhibition that suppresses the response. To test if these other factors could be responsible for the diminished response to the stable target, we developed the recently flashed stimulus task. In this task, the stable array contained only seven stimuli, but not the one that would be brought into the receptive field by the saccade. This eighth stimulus appeared while the monkey was fixating at the initial position, and it remained on throughout the trial. The monkey then made a saccade that brought this recently appeared stimulus into the receptive field. The neuron responded almost as briskly in that case as it did to the abrupt appearance of the stimulus in the receptive field (FIG. 2C; compare with FIG. 2A). Therefore, the difference between the fixation case and the stable target case was due not to the visual or oculomotor differences between the tasks, but to the lack of salience of a stable component of the visual environment. Note that the neuron began to respond at or before the end of the saccade. This was a much lower latency than that when the

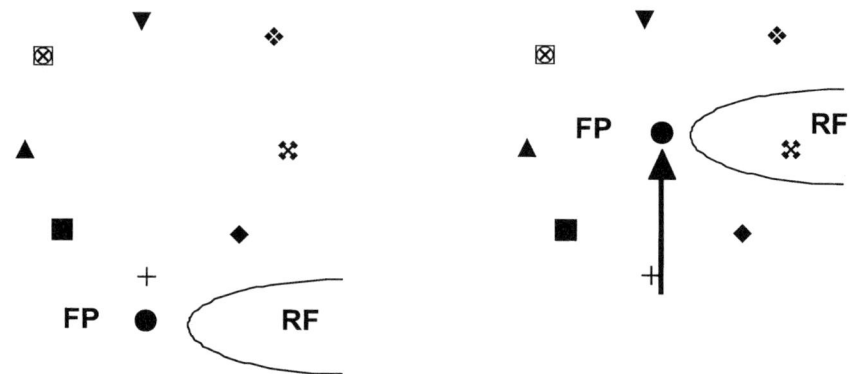

FIGURE 1. Stable array task. An array of symbols remains on the screen unchanging throughout the task. (**Left**) The monkey looks at a fixation point (*black dot*, marked FP) situated so no member of the array is in the receptive field (*parabolic solid line*, RF) of the neuron. (**Right**) The fixation point jumps and monkey makes a saccade (*arrow*) to follow it, bringing the receptive field onto the spatial location of a symbol (in this case the X). Adapted from Gottlieb et al.[9]

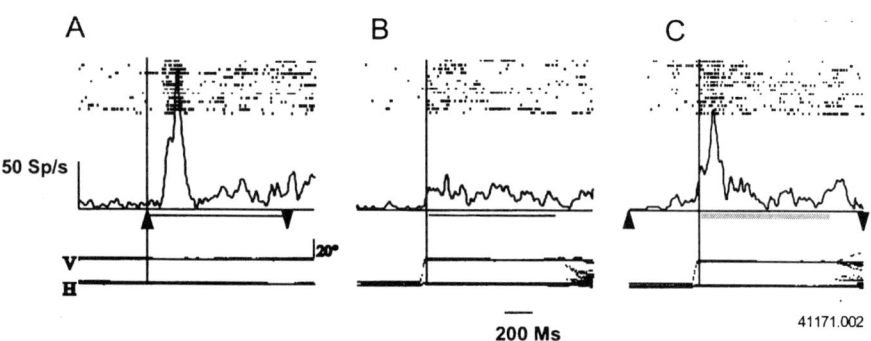

FIGURE 2. Effect of recent flash on stable array response. Each diagram is a raster diagram. Each dot is a cell discharge. Each line represents cell activity for one trial. Successive lines are synchronized on an event that occurs at the vertical line. Spike density histograms are shown beneath each raster. The *gray bar* at the bottom of the spike density histogram shows when, during the trial, the stimulus is in the receptive field of the neurons. *Up arrows* represent the onset of the flashed stimulus; *down arrows* represent its disappearance. Horizontal (H) and vertical (V) eye position traces for each raster line are shown superimposed beneath the spike density diagram. (**A**) Stimulus flashes in receptive field during fixation task; activity synchronized on stimulus appearance. (**B**) Stable array task: monkey makes saccade that brings stable stimulus into receptive field; activity synchronized on saccade end. (**C**) Recent stimulus task: monkey makes saccade that brings recently flashed stimulus into receptive field. Stimulus appears at *up arrow*, roughly 500 ms before saccade; activity synchronized on saccade end. Adapted from Gottlieb et al.[9]

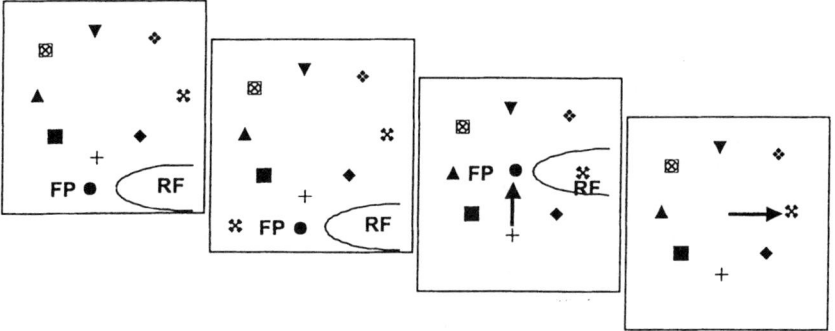

FIGURE 3. Stable target task. (**First panel**) The monkey fixates so that all symbols in the array are outside the receptive field. (**Second panel**) A cue appears, also outside the receptive field. (**Third panel**) The fixation point jumps, and the monkey makes a saccade that brings a symbol into the receptive field. In this example the symbol in the receptive field matches the cue. (**Fourth panel**) The fixation point disappears, and the monkey makes a saccade to the symbol that matches the cue. Adapted from Gottlieb et al.[9]

stimulus appeared in the receptive field abruptly (compare FIG. 2A with FIG. 2C). Presumably this occurred because of the predictive response described previously[10]: neurons in LIP may respond to stimuli that will be brought into their receptive field by saccades earlier than they respond to the abrupt appearance of the same stimulus in their receptive fields. The recently appeared stimulus evoked a greater response across the population than did the stable stimulus (using an average response in an interval 200 ms after the end of the saccade: $p < 0.001$ by Wilcoxon signed rank test; 31 neurons), and it evoked a statistically significantly greater response in a majority (23/31, $p < 0.05$ by two-tailed t test) of single neurons.

Salience does not only arise from intrinsic properties of the stimulus. Stable objects can become important by virtue of their relevance to current behavior, and under those circumstances a member of a stable array can evoke a response from a neuron in LIP. We can show this using the stable target task, a more complicated version of the stable array task (FIG. 3). In this task, the monkey fixated so that the stimulus was not in the receptive field, and a cue appeared during the first fixation. This cue matched one of the symbols in the stable array. The fixation point then jumped to the center of the array and the monkey tracked it with a saccade. Finally, when the fixation point disappeared, the monkey made a saccade to the member of the array that had matched the cue. The target indicated by the cue was randomly chosen on each trial among the members of the array. Neurons responded strongly to stable stimuli brought into their receptive field if these were designated as the target of the next saccade (FIG. 4A). The neuron discharged from the first saccade to the second. By contrast, if the identical stable stimulus entered the receptive field but was not designated as the saccade target (the monkey was instructed to saccade elsewhere), neurons responded minimally (FIG. 4B).

It is possible that LIP neurons respond in the stable target task because the monkey is planning a purposive saccade,[11] and the activity is less related to the salience of the stable target than it is to the processes underlying saccade planning. To see

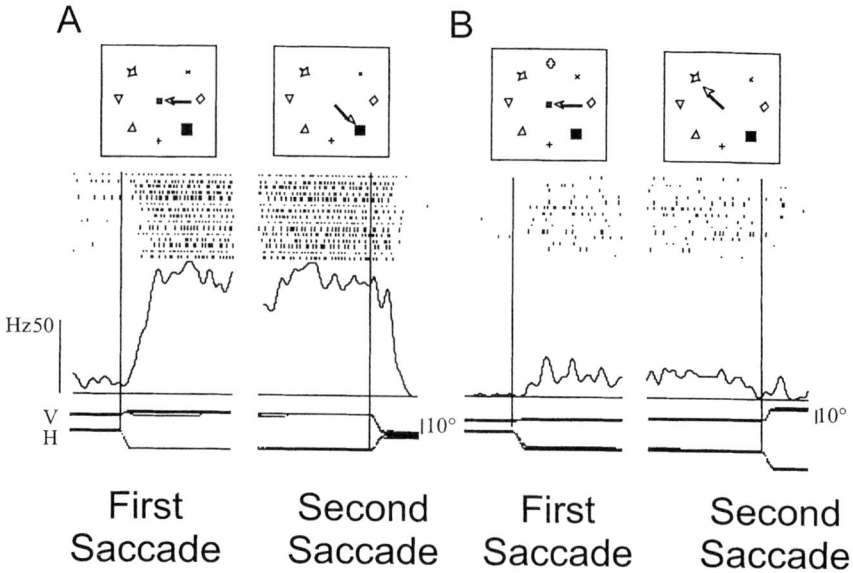

FIGURE 4. Response of the neuron in the stable target task. Each pair of rasters shows the response of the neuron in the same trials synchronized on first saccade beginning (*left*) and second saccade beginning (*right*). (**A**) Second saccade to receptive field. Cue appears before the first saccade and matches stable symbol in the receptive field. (**B**) Second saccade to receptive field. Cue matches target 180 degrees away from receptive field. Adapted from Kusunoki *et al.*[3]

how much activity in LIP can be allocated to the planning and generation of a saccade itself we used a task in which the monkey had to make a saccade to a spatial location that had no visual stimulus (the "black hole" task). LIP neurons active in the stable target task frequently failed to respond in this task, and no neurons responded more in this task than in the stable target task when the cue, and not the target, was in the receptive field.[9]

EFFECT OF SACCADE PLANNING ON VISUAL RESPONSES OF NEURONS IN THE LATERAL INTRAPARIETAL AREA

It could be, however, that LIP is only dedicated to the planning of visually guided saccades, and learned saccades such as those in the black hole task are irrelevant to LIP. If LIP were exclusively involved in planning saccades, then once the monkey was committed to a given saccade, a stimulus appearing far away from the saccade goal should evoke a weaker response than a stimulus at the saccade goal.[4] To test this we first studied neurons in the delayed saccade task and only chose those that had delay period and/or presaccadic activity (FIG. 5A and B). The neuron illustrated in FIGURE 5 has a visual response, a striking delay period response but a lesser presaccadic increase.

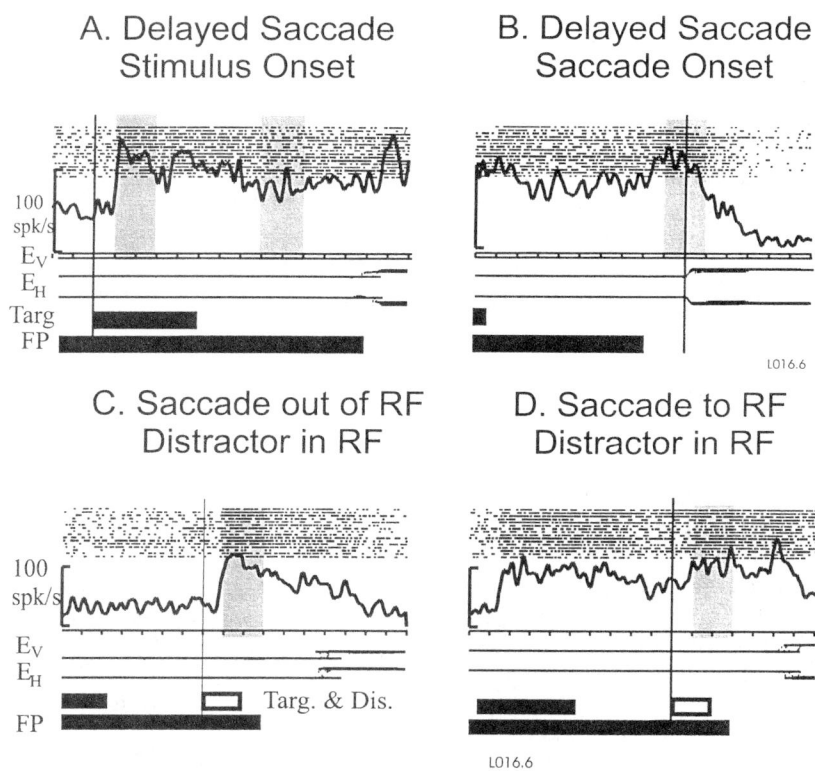

FIGURE 5. Response of a neuron to a distractor flashed in its receptive field. Each panel shows a raster, spike density histogram, and stimulus traces for fixation point, saccade target, and distractor. *Shaded areas* in the raster show stimulus, delay, and distractor activity periods used for quantitative analyses. (**A**) Response in the delayed saccade task synchronized on the stimulus appearance. (**B**) Same activity synchronized on the saccade beginning. (**C**) Response of the same neuron to the stimulus flashed in the receptive field (RF) during the delay period of a saccade made outside the receptive field. (**D**) Response of the same neuron during the delay period of a saccade made to the receptive field. Adapted from Powell and Goldberg.[4]

We then studied 27 neurons with visual and delay and/or presaccadic activity in the distractor task. In this task the monkey performs a delayed saccade task, but 300 ms before the fixation disappeared we flashed a distractor for 200 ms that was identical to the saccade target. We randomly interleaved six types of trials: simple delayed saccade tasks to the receptive field or to a second point that was outside the receptive field; trials with saccades to the receptive field and distractor either in the receptive field or away; and trials with saccades away from the receptive field and the distractor either in the receptive field or at the saccade goal.

We found that when a stimulus flashed in the receptive field of an LIP neuron while the monkey was planning a memory-guided saccade away from the receptive field, not only was the response of the neuron not suppressed, but also it was slightly enhanced relative to the case when the stimulus appeared at the saccade goal. FIGURE 5C and D illustrates this for the same neuron whose activity in the memory-guided saccade task was shown in FIGURE 5A and B. This was true across the population of neurons and for individual neurons as well. If one looked at the activity of the neurons in the 50 ms before the beginning of the saccade, no difference was noted between the case when the monkey made a saccade to the receptive field and the distractor was elsewhere and the case when the monkey made a saccade away from the receptive field and the distractor was in the receptive field. Although the distractor evoked significant activity before the saccade, it had no effect on any measure of saccadic performance: velocity, reaction time, accuracy, or even the early trajectory.[4]

DYNAMICS OF ATTENTION IN THE MONKEY

Neurons in LIP clearly do not have information about where or when a saccade will occur, although they do respond to saccade targets. It is possible, however, that neurons in LIP specify the locus of attention. If this were true, the activity in the distractor task enables two predictions about attention in the monkey: (1) the default location of attention during the delay period of the delayed saccade task is the saccade goal; and (2) a distractor appearing during a delayed saccade will transiently pull attention away from the saccade goal even though it will not disrupt the saccade plan. It is known in humans that attention is ordinarily pinned at a saccade goal[12,13] and is also drawn to the locus of an abruptly appearing stimulus,[14] but the interaction of these effects has not been studied in humans or monkeys.

The locus of attention has been determined in two different ways. It is the area of space associated with the shortest reaction times[2] or the lowest perceptual threshold.[15] Because using perceptual thresholds would enable us to study the temporal as well as the spatial aspects of attention, we trained two rhesus monkeys on a perceptual task.[16] The task began with a monkey fixating on a spot for 1,000–1,200 ms after which a saccade target appeared and disappeared at one of four loci equidistant from the fovea and reflected across the horizontal or vertical meridia. Then 500 to 1,500 ms later a cue appeared for one video frame (roughly 17 ms). The cue consisted of a stimulus at each possible saccade goal: three circles and a Landolt ring with its gap on either the right or the left. If the gap were on the left, the monkey had to make the saccade when, 500 ms later, the fixation point disappeared. If the gap were on the right, the monkey had to continue fixating. Then we were able to measure a contrast sensitivity curve for the cue, which easily fit a classic Weibull function. We defined the threshold contrast sensitivity as that contrast at which the monkey perceived the cue 75% of the time. When the cue appeared at the saccade goal, the threshold was at 1% contrast above background. When the cue appeared at any of the other sites (the other possible saccade goals used during that experimental day), the threshold was at 2% contrast. There were two levels of contrast sensitivity, the advantaged one at the saccade goa, and the unadvantaged level elsewhere. The saccade goal was perceptually advantaged throughout the delay period, even 1,700 ms after the appearance of the saccade target. Thus, the default locus of attention rests at the goal of a memory-guided saccade task.

In the next version of the task we inserted a task-irrelevant distractor between the target and the cue to see if the distractor could draw the locus of attention away from the cue. Five hundred milliseconds after the saccade target appeared the distractor, identical to the saccade target, appeared for 100 ms at one of the four possible cue locations. After a variable period, the cue appeared for one video frame, and the fixation point disappeared 500 ms later. Two hundred milliseconds after the distractor appeared, the contrast sensitivity at the distractor site was at the advantaged level, and the contrast sensitivity at the saccade goal was at the unadvantaged level. Six hundred milliseconds after the distractor appeared, the saccade goal was again advantaged, as if the locus of attention had moved from the distractor site back to the saccade goal.

Neurons in LIP during this task had activity paralleling the movement of the attentional focus. The saccade target evoked a large burst of activity, which then fell to a delay period level significantly above background, but lower than the peak of the burst. The distractor evoked a large response, which on average returned to a level below the delay period activity between 340 ms and 455 ms depending on the monkey. If the cue flashed at the time when the activity evoked by the distractor was identical to the delay period activity for a given monkey, there was no attentional advantage at either site. The delay period activity was clearly greater than the distractor activity for the first monkey at the time curves crossed for the second monkey. There was a clear attentional advantage at the saccade goal for the first monkey at that time.

DISCUSSION

In these experiments we have demonstrated two different aspects of the representation of the salient visual world in LIP. These different aspects correspond to the objects of James' different kinds of attention. Recently appeared objects evoke James' passive, reflex, or involuntary attention and they also evoke responses in LIP even when they are irrelevant to the ongoing task, whether they appear in the receptive field directly or enter the receptive field by a saccade shortly after their appearance. Sudden motion also evokes activity in LIP, in a directionally nonselective fashion.[3]

Attended stable objects evoke James' active or voluntary attention. Such objects are also represented in LIP. They are not salient by themselves, but become so by virtue of their relevance to the task. Whether they have just entered the receptive field or have been in the receptive field already, they evoke a response much greater than that evoked by stable objects that are irrelevant to the animal's behavior.

Neurons in LIP discharge throughout the interval of a memory-guided delayed saccade task. They respond to the abrupt onset of the stimulus, continue to discharge during the delay period, and burst slightly before the saccade. This pattern of activity has led a number of investigators to assume that LIP is providing a motor intention signal for the saccade.[17,18] However, there are many examples that demonstrate instances for which the signal in LIP contradicts a saccade plan: LIP neurons describe stimuli that describe saccades elsewhere[9] and respond in an enhanced manner to stimuli that flash in the receptive field of a neuron when a monkey is planning a saccade elsewhere. The distractor does not interfere with the performance of the saccade, even though immediately before the saccade one cannot distinguish between activity evoked by a memory-guided saccade to the receptive field when a distractor appeared elsewhere and a distractor in the receptive field when the saccade goal is elsewhere. Neurons in LIP respond to saccade targets, but discharge much less before saccades not made to visual targets.[9] This is in distinction to movement neurons in the frontal eye field which are defined by their equivalent discharge in association with visually guided and learned saccades.[19] Neurons in LIP do not predict where, when, or if a saccade will occur.

Instead, the activity of neurons in LIP describes salient spatial locations, and the area of the visual field associated with the greatest activity in LIP corresponds to the locus of visual attention. This is a winner-take-all mechanism rather than a graded one. The intensity of discharge in LIP does not correlate with the amount of attention; instead, it correlates with the probability that a given location will win.[20] Ordinarily the locus of visual attention is pinned at the saccade goal throughout the delay period of a memory-guided saccade, and this is consistent with the delay period activity of the parietal neurons as well as with a saccade plan. However, when a distractor appears, the locus of attention moves to the site of the distractor, and this too is consistent with the activity of neurons in LIP, but not with a saccade plan.

ACKNOWLEDGMENTS

This work was supported by the National Eye Institute and the Human Frontiers Science Program. We are grateful to the staff of the National Eye Institute for assistance in all aspects of this work: Drs. James Raber and Ginger Tansey for veterinary care; Dr. John McClurkin for display programming; Thomas Ruffner and Altah Nichols for machining; Lee Jensen for electronics; Mitchell Smith for histology; Art Hays for computer systems; Brian Keegan for technical assistance; and Becky Harvey and Jean Steinberg for facilitating everything. The Laboratory of Diagnostic Radiology of the Clinical Center provided MRI services.

REFERENCES

1. JAMES, W. 1890. The Principles of Psychology. Holt. New York.
2. POSNER, M.I. 1980. Orienting of attention. Q. J. Exp. Psychol. **32:** 3–25.
3. KUSUNOKI, M., J. GOTTLIEB & M.E. GOLDBERG. 2000. The lateral intraparietal area as a salience map: the representation of abrupt onset, stimulus motion, and task relevance. Vision Res. **40:** 1459–1468.
4. POWELL, K.D. & M.E. GOLDBERG. 2000. Response of neurons in the lateral intraparietal area to a distractor flashed during the delay period of a memory-guided saccade. J. Neurophysiol. **84:** 301–310.
5. SPARKS, D.L. 1975. Response properties of eye movement-related neurons in the monkey superior colliculus. Brain Res. **90:** 147–152.
6. HIKOSAKA, O. & R.H. WURTZ. 1983. Visual and oculomotor functions of monkey substantia nigra pars reticulata. III. Memory-contingent visual and saccade responses. J. Neurophysiol. **49:** 1268–1284.
7. WURTZ, R.H. 1969. Visual receptive fields of striate cortex neurons in awake monkeys. J. Neurophysiol. **32:** 727–742.
8. EGETH, H.E. & S. YANTIS. 1997. Visual attention: control, representation, and time course. Annu. Rev. Psychol. **48:** 269–297.

9. GOTTLIEB, J., M. KUSUNOKI & M.E. GOLDBERG. 1998. The representation of visual salience in monkey parietal cortex. Nature **391**: 481–484.
10. DUHAMEL, J.-R., C.L. COLBY & M.E. GOLDBERG. 1992. The updating of the representation of visual space in parietal cortex by intended eye movements. Science **255**: 90–92.
11. SNYDER, L.H., A.P. BATISTA & R.A. ANDERSEN. 1997. Coding of intention in the posterior parietal cortex. Nature **VI-386**: 167–170.
12. DEUBEL, H. & W.X. SCHNEIDER. 1996. Saccade target selection and object recognition: evidence for a common attentional mechanism. Vision Res. **36**: 1827–1837.
13. KOWLER, E., E. ANDERSON, B. DOSHER & E. BLASER. 1995. The role of attention in the programming of saccades. Vision Res. **35**: 1897–1916.
14. YANTIS, S. & J. JONIDES. 1984. Abrupt visual onsets and selective attention: evidence from visual search. J. Exp. Psychol. Hum. Percept. Perform. **10**: 601–621.
15. BASHINSKI, H.S. & V.R. BACHARACH. 1980. Enhancement of perceptual sensitivity as the result of selectively attending to spatial locations. Percept. & Psychophys. **28**: 241–248.
16. BISLEY, J.W. & M.E. GOLDBERG. 2001. The locus of visual attention can be predicted by the responses of neurons in the lateral intraparietal area. Soc. Neurosci. Abstr. **27**: 384.
17. ANDERSEN, R.A., L.H. SNYDER, D.C. BRADLEY & J. XING. 1997. Multimodal representation of space in the posterior parietal cortex and its use in planning movements. Annu. Rev. Neurosci. **20**: 303–330.
18. PLATT, M.L. & P.W. GLIMCHER. 1997. Responses of intraparietal neurons to saccadic targets and visual distractors. J. Neurophysiol. **78**: 1574–1589.
19. BRUCE, C.J. & M.E. GOLDBERG. 1985. Primate frontal eye fields: I. Single neurons discharging before saccades. J. Neurophysiol. **53**: 603–635.
20. DESIMONE, R. & J. DUNCAN. 1995. Neural mechanisms of selective visual attention. Annu. Rev. Neurosci. **18**: 183–222.

Effects of Cortical Lesions on Saccadic Eye Movements in Humans

CH. PIERROT-DESEILLIGNY,[a] C.J. PLONER,[b] R.M. MÜRI,[c] B. GAYMARD,[a] AND S. RIVAUD-PÉCHOUX[a]

[a]*INSERM 289 et Service de Neurologie 1, Hôpital de la Salpêtrière, Paris, France*

[b]*Klinik für Neurologie, Charité, Berlin, Germany*

[c]*Eye Movement Research Laboratory and Department of Neurology, Inselspital, Bern, Switzerland*

ABSTRACT: Our knowledge of the cortical control of saccadic eye movements (saccades) in humans has recently progressed mainly because of lesion and transcranial magnetic stimulation (TMS) studies, but also because of functional imaging. It is now well known that the frontal eye field is involved in the control of intentional saccades, the parietal eye field in that of reflexive saccades, the supplementary eye field (SEF) in the initiation of motor programs comprising saccades, the pre-SEF in the learning of these programs, and the dorsolateral prefrontal cortex (DLPFC) in saccade inhibition, prediction and spatial working memory. Saccades may also be used as a convenient model of motricity to study general cognitive processes such as motivation and spatial memory. Thus, it has been shown that the posterior part of the anterior cingulate cortex, called the cingulate eye field, is involved in motivation and the preparation of all intentional saccades, but not in reflexive saccades. Recently, our understanding of the cortical control of spatial memory has noticeably progressed by using the simple visuo-oculomotor model represented by the memory-guide saccade paradigm, in which a single saccade is made to the remembered position of a unique visual item presented a while before. Transcranial magnetic stimulation studies have determined that after a brief stage of spatial integration in the posterior parietal cortex (inferior to 300 ms), short-term spatial memory (i.e., up to 15–20 seconds) is controlled by the DLPFC. Behavioral and lesion studies have shown that medium-term spatial memory (between 15 and 20 seconds and a few minutes) is specifically controlled by the parahippocampal cortex, before long-term memorization (i.e., after a few minutes) in the hippocampal formation. These different but complementary study methods used in humans have thus contributed to a better understanding of both eye movement physiology and general cognitive processes preparing motricity as whole.

KEYWORDS: cingulate cortex; eye movements; frontal lobe; medial temporal lobe; motivation; saccade; spatial memory; working memory

Address for correspondence: Ch. Pierrot-Deseilligny, Service de Neurologie 1, Hôpital de la Salpêtrière, 47 Bd de l'Hôpital, 75651 Paris cedex 13, France. Voice: (+33)14216 1828; fax: (+33) 14424 5247.

cp.deseilligny@psl.ap-hop-paris.fr

INTRODUCTION

Saccadic eye movements (saccades) are generated in the brain stem, but are triggered by the cerebral hemispheres. In humans, several methods may be used to study the cortical control of saccades. Lesion studies are required to determine which cortical areas are crucial to performing a saccade paradigm. However, lesion studies cannot tell us at what specific time in the execution of this paradigm each area plays a significant role. Transcranial magnetic stimulation (TMS), recently used in research to study different cerebral functions, can help to answer such questions. This technique, which inhibits or inactivates the stimulated areas for a few milliseconds, that is, with the same effect as a brief functional lesion, has relatively good temporal resolution, but poor spatial resolution. Conversely, functional imagery, that is, PET-scan and functional magnetic resonance imaging (fMRI), has the best spatial resolution but relatively poor temporal resolution and frequent false-positive or false-negative activity. Functional imagery is therefore mainly useful to determine the precise locations of the ocular motor areas in humans. Finally, all these methods provide complementary information in the study of any human cerebral function, including the control of the different types of saccades.

Saccades may be reflexive, externally triggered by the sudden appearance of a visual target (reflexive, visually guided saccade) (FIG. 1A) or intentional, internally triggered towards a target either already present (intentional visually guided saccade), not yet present (predictive saccade), or no longer visible (memory-guided saccade) (FIG. 1B). Antisaccades, made in the opposite direction of a suddenly appearing visual target, are also intentional saccades (FIG. 1A). Lastly, there are spontaneous saccades, made, for example, at rest in darkness. These different saccades may be studied *per se* to improve our knowledge of eye movement physiology. However, saccades may also be used as a convenient model of motricity to understand more complex neuropsychological processes such as motivation, prediction, spatial integration, and spatial memory. Therefore, after a brief review of our current knowledge of the cortical areas triggering or inhibiting saccades in humans, this paper focuses on the study of some of the neuropsychological processes preparing saccades as well as general motricity.

CORTICAL AREAS TRIGGERING SACCADES

Experimental studies have shown that three cerebral areas are capable of triggering saccades[1,2]: the frontal eye field (FEF), the supplementary eye field (SEF), and the parietal eye field (PEF). In humans, the FEF is located in the precentral gyrus and sulcus,[3–6] mainly around the intersection with the superior frontal sulcus (FIG. 2A). Lesions studies have shown that the FEF is involved in the control of intentional saccades, such as intentional, visually guided saccades, correct antisaccades (FIG. 1A), memory-guided saccades (FIG. 1B), and predictive saccades.[5,7–11] The increase in latency of such saccades observed in these studies and the eye displacements elicited by electrical stimulation of the FEF[6] suggest that this area plays a significant role in the triggering process. TMS studies of antisaccades[12] or memory-guided saccades[13] have confirmed that the FEF is more particularly involved in the triggering of such saccades. By contrast, the FEF does not appear to be crucial for the trig-

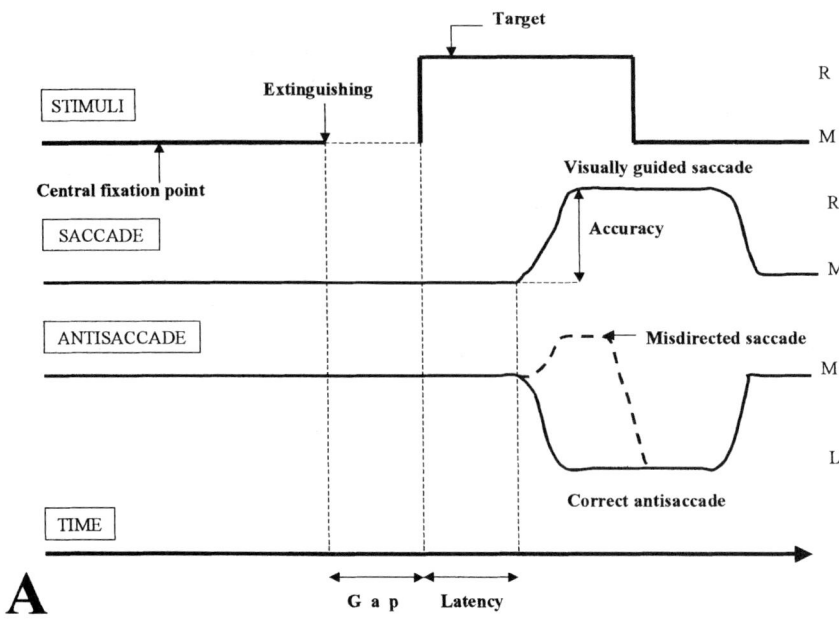

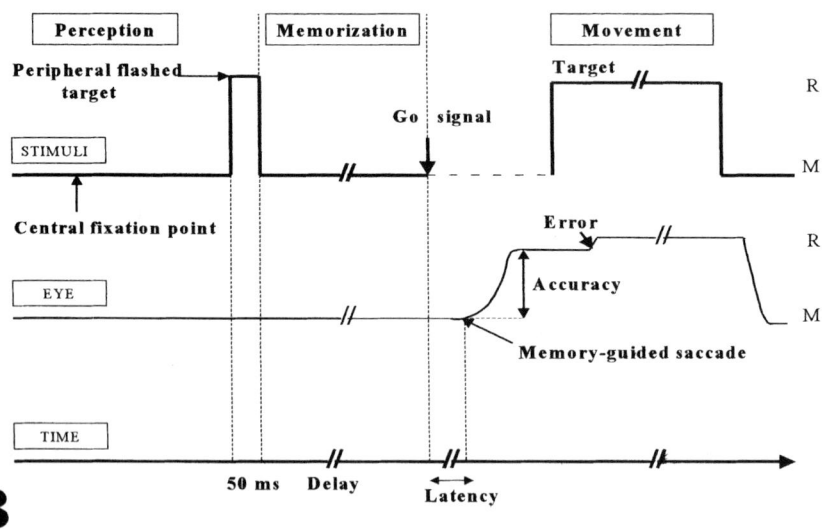

FIGURE 1. *Legend on following page.*

gering of reflexive, visually guided saccades.[7] The human SEF is located anterior to the supplementary motor area (SMA) (FIG. 2A), namely, in the upper part of the paracentral sulcus.[4,14] The SEF, in particular on the left side, appears to be involved in the control of motor programs comprising either several successive saccades,[15] that is, a sequence of saccades, or a saccade combined with a body movement.[10,16] During sequences of saccades, TMS studies have shown that the SEF could be crucial at two distinct times,[2] that is, during the learning phase (presentation of the visual targets)[17] and just after the go signal,[18] that is, when the subject has to initiate the sequence of saccades. More recently, results of fMRI studies have suggested that the learning phase of a sequence of saccades could be controlled by a pre-SEF area, located just anterior to the SEF.[19,20] Similarly, a pre-SMA exists anterior to the SMA for general motricity. Thus, the learning phase and initiation of a sequence of saccades could be controlled by two close but distinctive areas in the SEF region. The human PEF, which corresponds to the lateral intraparietal area (LIP) in the monkey, is located in the intraparietal sulcus (FIG. 2A).[21–23] This area appears to be involved mainly in the triggering of reflexive, visually guided saccades, because after a PEF lesion (but not after an FEF or an SEF lesion), latency of these saccades is significantly increased,[6,24] particularly when damage involves the right cerebral hemisphere. TMS over the posterior parietal cortex may also increase the latency of visually guided saccades.[25] Therefore, three cortical areas contribute to the triggering of saccades with apparently different roles depending on the type of saccade to be performed; the PEF could mainly control reflexive saccades, whereas the FEF appears to be more involved in controlling different modalities of intentional saccades and the SEF in controlling relatively more complex motor programs, comprising either several successive saccades or saccades combined with body movements. However, although the control exerted by the PEF and FEF, respectively, is relatively specialized in terms of the distinct types of saccades to be performed, these two areas also act complementarily in saccade triggering in general, because only bilateral lesions affecting both the PEF and the FEF result in a severe and long-lasting disturbance of this triggering.[26]

FIGURE 1. Main saccade paradigms used. (**A**) Reflexive visually guided saccade and antisaccade paradigms. In the saccade paradigm, the subject, fixating on a central fixation point, has to make a reflexive visually guided saccade as soon as the peripheral visual target occurs. Latency is a reflection of the triggering mechanism of this reflexive saccade. Note that a gap is used, that is, the central point is extinguished 200 ms before the appearance of the visual target in order to facilitate the disengagement of fixation. In the antisaccade paradigm, the subject is instructed to make a saccade in the opposite direction to the visual target. A reflexive misdirected saccade to the target is an error, and the percentage of errors is a reflection of inhibition mechanisms controlling saccades. Latency of correct antisaccades is a reflection of the triggering mechanisms of an intentional saccade. (**B**) Memory-guided saccade paradigm. A peripheral visual target is flashed while the subject is fixating on a central fixation point. After a delay (memorization), the central point is switched-off ("go signal"), and the subject then has to make a saccade to the remembered position of the flash. Accuracy of this memory-guided saccade (or of the final eye position if there are several saccades) is a reflection of spatial memory. Latency is a reflection of the triggering mechanisms of an intentional saccade. L, left; M, midline; R, right.

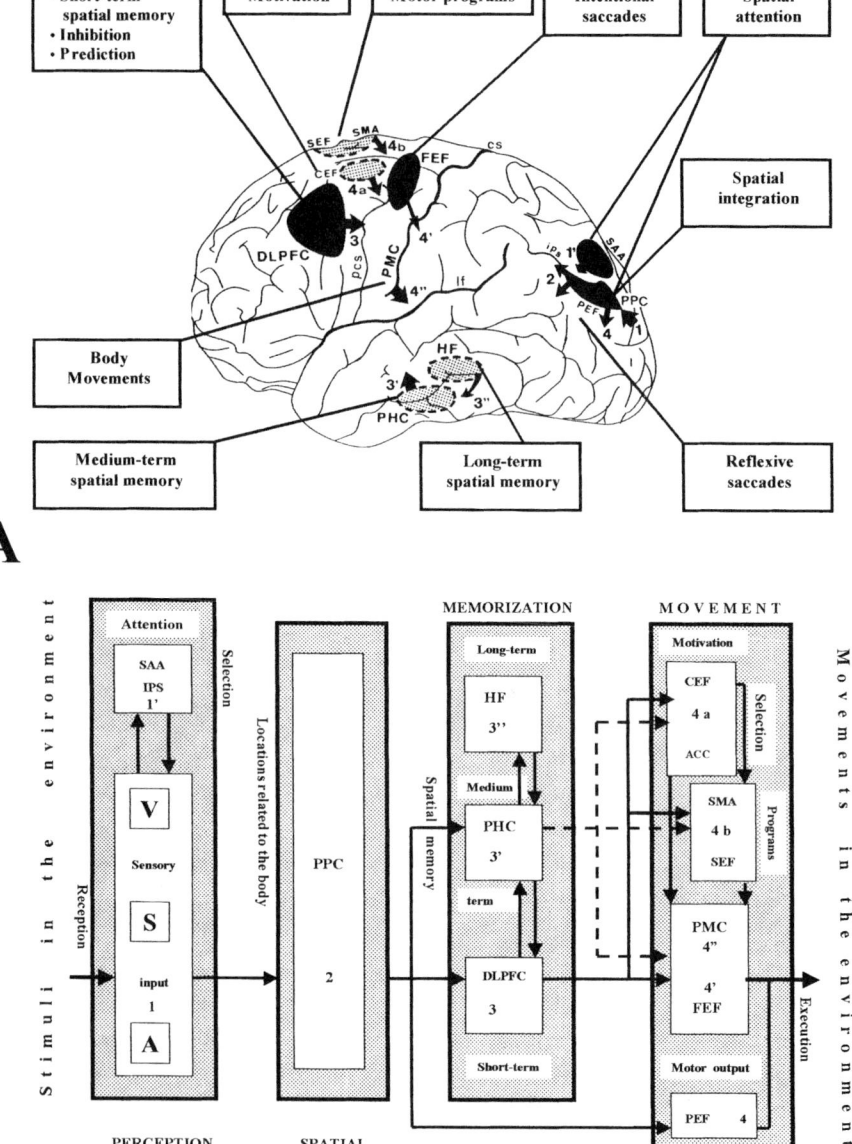

FIGURE 2. *Legend on following page.*

CORTICAL AREA INHIBITING SACCADES

Inhibition of saccades is useful when these eye movements may disturb another essential task, such as relatively prolonged fixation. The antisaccade paradigm tests saccade inhibition, because before performing a correct intentional antisaccade in the opposite direction of a suddenly appearing visual target, the subject has first to inhibit a reflexive, visually guided saccade to the target (FIG. 1A). The percentage of errors (misdirected reflexive saccades) is a reflection of saccade inhibition. We have shown that at the cortical level, only patients with lesions affecting the dorsolateral prefrontal cortex (DLPFC) (FIG. 2A) have an increased percentage of errors in this paradigm.[7] In humans, the DLPFC is located in area 46 and the adjacent area 9 of Brodmann, namely, in the middle frontal gyrus anterior to the FEF (FIG. 2A).[8,22,27–29] Our patients with an FEF lesion had a normal percentage of errors,[5,7,9] but, as just noted herein, an increase in the latency of intentionally correct antisaccades. The role of the DLPFC in the antisaccade paradigm has also been shown using fMRI,[30] and its more particular implication in saccade inhibition has been confirmed in a recent TMS study.[31] Therefore, the first part of execution of the antisaccade paradigm, that is, inhibition of reflexive misdirected saccades, is under the control of the DLPFC, whereas the second part, that is, the triggering of the intentionally correct antisaccade, depends on FEF control.

FIGURE 2. Cortical areas and pathways. Schematized representation of the cortical areas (**A**) and pathways (**B**) preparing movements in response to environmental stimuli. (1) At the perceptive stage, several areas (1) are involved in the reception of information (sensory input), depending on the sensory modalities, and different attentional areas (1') contribute to the selection of salient information at this stage. (2) At the spatial integration stage, salient information is integrated in the PPC (2) to localize further the stimuli in relation to the body. (3) At the memorization stage, that is, when information has not immediately been used, there is a storage in the DLPFC (3), PHC (3'), and HF (3''), successively, for short-term, medium-term, and long-term memorization, respectively. (4) At the movement stage, the motor areas were previously prepared for possible intentional forthcoming movements by motivation, that is, by the ACC (body movements) and/or CEF (intentional saccades), with a selection of the useful corresponding areas by enhancing their basis activity (4a); if several movements are planed, their sequences were previously learned by the pre-SEF (saccades) and the pre-SMA (body movements) (not shown, but located just anterior to the SEF and SMA), before initiation of these sequences by the SEF (several successive saccades and/or a saccade with body movements) and/or SMA (body movements) (4b); movements (motor output) are executed either immediately after spatial integration by the PEF (reflexive saccades) (4), or, after a delay, by the FEF (intentional saccades) (4') and/or the PMC (body movements) (4''). A, auditory areas; ACC, anterior cingulate cortex; CEF, cingulate eye field; cs, central sulcus; DLPFC, dorsolateral prefrontal cortex; FEF, frontal eye field; HF, hippocampal formation; ips and IPS, intraparietal sulcus; lf, lateral fissure; pcs, precentral sulcus; PEF, parietal eye field; PHC, parahippocampal cortex; PMC, primary motor cortex; PPC, posterior parietal cortex; S, somesthesic areas; SAA, spatial attentional area; SEF, supplementary eye field; SMA, supplementary motor area; V, visual areas.

CORTICAL AREAS PREPARING SACCADES

A saccade is a rather simple motor act, relatively easy to record and interpret. This is why saccades have frequently been used as a convenient model of motricity to study complex neuropsychological processes preparing these eye movements as well as other body movements. We mainly develop here the examples of motivation and spatial memory.

Motivation

Motivation is a neuropsychological process preparing motor areas to perform forthcoming intentional motor acts. The anterior cingulate cortex could be involved in this process.[32,33] In a PET-scan study,[34] it was shown that the posterior part of the anterior cingulate cortex (FIG. 2A), which may be called the "cingulate eye field" (CEF),[32] is involved in voluntary saccades made in darkness, that is, a type of intentional saccades. We have shown that small lesions affecting the same area result in deficits (latency increase and accuracy decrease) of different types of intentional saccades, whereas reflexive saccades remain normal.[32] Because the anterior cingulate cortex appears to be involved very early before a motor act[35] and is connected with the motor areas,[33] whereas the CEF is strongly connected with all frontal ocular motor areas and the brain stem,[33] the role of the anterior cingulate cortex (in general motricity) and the CEF (in eye movements) could be to enhance the basic activity of the motor areas involved in the control of the forthcoming intentional motor acts, in order to facilitate their triggering when the right time of execution has come. It should also be noted that the anterior cingulate cortex is reciprocally connected with the DLPFC and the parahippocampal cortex (see below)[32] and could therefore contribute to both maintaining the salience of memorized information and facilitating its transmission to the motor areas. This could represent the physiological mechanisms by which motivation is expressed at the neuronal level for eye movements in the CEF as well as for general motricity in the remainder of the anterior cingulate cortex.

Spatial Memory

Spatial memory is another essential cognitive function preparing motricity, namely, eye movements as well as other body movements. In animals and humans, spatial memory has been extensively studied for either short delays (of a few seconds) or long delays (superior to a few minutes), but little is known about intermediate delays, that is, between a few seconds and a few minutes. To study spatial memory, the visuo-oculomotor model, represented by the memory-guided saccade (MGS) paradigm (FIG. 1B), has frequently been used because of its simplicity; a simple visual item is first flashed and memorized; then, after a delay, a single MGS is made to the remembered location of the visual item; the accuracy of this saccade is a reflection of spatial memory, provided that perceptive and oculomotor deficits have been ruled out. The results relating to short-term, medium-term, and long-term spatial memory are reviewed successively.

Short-Term Spatial Memory. Short-term spatial memory is the working memory used for current, on-going behavior. In monkeys, single-neuron recordings have shown that the DLPFC is involved in MGS.[36] In humans, our various lesion

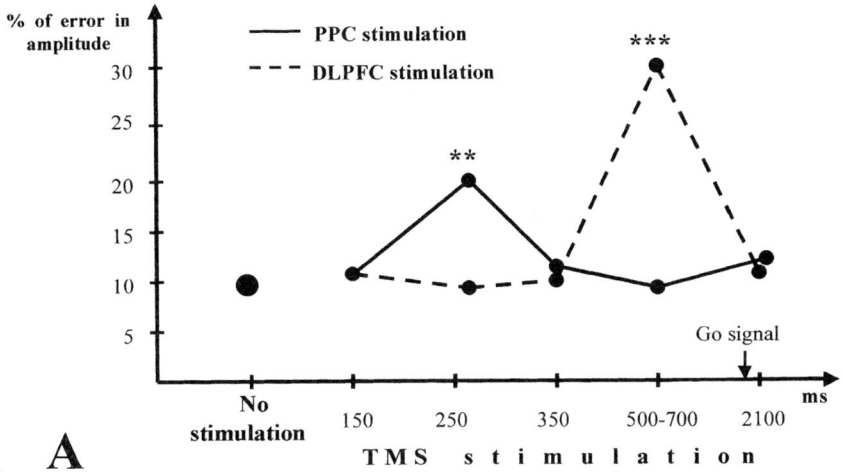

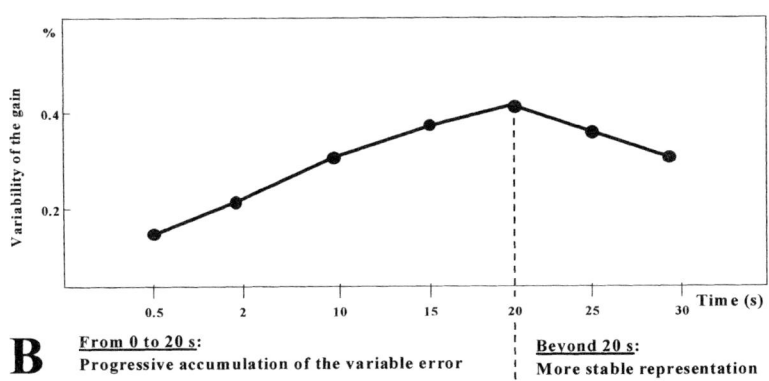

FIGURE 3. Spatial memory results. (**A**) Results of a transcranial magnetic stimulation (TMS) study made on the right hemisphere in eight normal subjects performing memory-guided saccades (with a 2-second delay)[37]: results of TMS are compared to those without stimulation; note that 1) after PPC stimulation, saccade accuracy was significantly impaired when stimulation was performed 260 ms after the flashed peripheral target ($p < 0.01$), (2) after DLPFC stimulation, saccade accuracy was impaired when stimulation was performed during the memorization period (i.e., beyond 500 ms) ($p < 0.001$), and (3), after the go signal (2 seconds), accuracy was normal in both cases. DLPFC, dorsolateral prefrontal cortex; PPC, posterior parietal cortex. (**B**) Delay-dependency of variable targeting errors of memory-guided saccades in 16 normal subjects.[44] The delay varied from 0.5 to 30 seconds. Only

studies[5,8-10,15,16] have shown that several cortical areas, such as the DLPFC, the FEF, and the posterior parietal cortex, but not the SEF, control MGS. The posterior parietal cortex, that is, the equivalent of area 7a in the monkey, is located in the vicinity of the PEF and is probably involved in spatial integration; this process allows the subject to know the locations of stimuli related to the body and not simply to the eyes. In fact, after the visual target presentation, the MGS paradigm comprises three successive physiological phases: spatial integration, memorization, and saccade triggering. To determine the specific role(s) played by the different cortical areas involved in the control of MGS, a method other than lesion studies (or functional imaging) is needed. Several TMS studies have allowed us to resolve this question. TMS applied over the posterior parietal cortex or DLPFC showed that: (1) the right posterior parietal cortex (but not the left posterior parietal cortex) is involved at 260 ms after the target presentation,[37,38] but not later in the paradigm, whereas (2) both DLPFC are involved during the memorization phase (around 1 second after the target presentation), but not earlier in the paradigm[37-39] (FIG. 3A). Furthermore, other TMS studies have suggested that the FEF is involved in MGS triggering,[12] with a possible contribution of the parietal lobe at this stage, namely, probably the PEF.[37,38]

Overall, these different results suggest that: (1) the right posterior parietal cortex is involved in the control of saccade accuracy during the brief initial visuospatial integration stage, immediately after the stimulus presentation, but not later during the memorization phase of the paradigm, and is therefore not crucial in the control of short-term spatial memory; (2) the left posterior parietal cortex is either uninvolved or less involved in the control of saccade accuracy, a result that is probably related to the well-known dominance in humans of the right cerebral hemisphere for visuospatial functions; (3) by contrast, both the right and the left DLPFC control saccade accuracy during the memorization phase of this paradigm, that is, short-term spatial memory; (4) lastly, the FEF is mainly involved in MGS triggering, to which the posterior parietal cortex may also contribute. Thus, these TMS studies succeeded in determining the chronology of the cortical control of MGS (FIG. 2). In the monkey, the DLPFC is reciprocally connected with both the posterior parietal cortex and the FEF[40] with, therefore, the existence of an anatomical substrate for such a cortical control. Lastly, it should also be noted that according to the results of our recent lesion studies, the FEF may be prepared by short-term memorized information maintained in the DLPFC, in order to perform possible forthcoming ocular saccades corresponding to such salient information.[5,41]

Medium-Term Spatial Memory. How long the transient DLPFC control of spatial memory lasts was the next question to resolve. Accuracy of MGS decreases during the first seconds of the memory delay.[42] This is mainly reflected in increased variable targeting errors, that is, scatter of MGS endpoints around the to-be-remembered target position, and it probably represents information loss in short-term spatial memory.[42,43] In a recent study, we used this delay dependency of targeting errors to

---

group results are shown. Note that the variability of the gain increased up to a 20-second delay, but, surprisingly, significantly improved (i.e., decreased) after 20 seconds. This means that after 20 seconds a more stable representation is involved. Therefore, probably two different structures, working in parallel, control spatial memory before and after 20 seconds, respectively.

investigate possible temporal limits for short-term spatial memory.[44] Normal subjects performed an MGS task with unpredictable varied memory delays of 0.5–30 seconds. The results confirmed that variable targeting errors initially increased linearly with delay length (FIG. 3B). Surprisingly, after a maximum error around delays of 20 seconds, this process did not stabilize, but reverted partially, and MGS accuracy improved significantly with longer delays. This time-limited decay of spatial information strongly suggests involvement of a more stable memory system at longer delays and thus allows for the hypothesis of an upper temporal limit of about 20 seconds for short-term spatial memory in humans. This temporal limit fits remarkably well with the limited temporal stability of spatially selective neuronal activity in the monkey DLPFC and the known temporal properties of short-term plasticity in prefrontal neurons.[36,43]

This new finding of improvement of spatial memory at delays longer than 20 seconds then led us to determine which cerebral structure could be involved in such medium-term spatial memory. Previous lesion studies in humans have suggested that the medial temporal lobe is the region containing the neuronal substrate of the putative medium-term memory, because spatial memory was deficient at delays of about 20 seconds but normal at shorter delays.[45,46] However, the medial temporal lobe is a complex anatomical region consisting of the hippocampal formation (i.e., hippocampus and entorhinal cortex), the perirhinal cortex, and the parahippocampal cortex.[47] The hippocampal formation is directly connected with both cortices, but this is mainly the parahippocampal cortex, which is reciprocally connected with neocortical areas involved in spatial cognition, that is, the posterior parietal cortex and the DLPFC (FIG. 2),[48,49] and motor areas such as the FEF.[50] In two recent studies, we investigated the question of possible differential contributions of medial temporal lobe subregions to spatial memory in humans with postsurgical lesions of the right medial temporal lobe.[51,52] The results showed that only patients with parahippocampal cortex involvement exhibited significant inaccuracy of MGS at delays longer than 20 seconds contralateral to the lesion side. MGS with short delays (i.e., those less than 20 seconds) were normal, a result confirming the findings of our previous study of patients with large lesions of the right medial temporal lobe.[53] Thus, these results show that the right parahippocampal cortex contributes to spatial memory at delays beyond DLPFC-based short-term spatial memory and therefore that the parahippocampal cortex, and not the hippocampal formation or the perirhinal cortex, is a likely neuronal substrate for medium-term spatial memory of single visual items in humans.

Long-Term Spatial Memory. What is the next structure involved in spatial memory, that is, in long-term spatial memory, and at which critical time? There is ample evidence, both from single-neuron recordings and lesion studies in monkeys, that the hippocampal formation is involved in spatial memory processes.[54] The hippocampal formation, by virtue of its extensive direct and indirect connections with areas involved in spatial processing and object recognition,[47,49] is situated at the highest level of integration of perceptual information and is thus in a privileged position to bind together information from different sensory modalities and spatial coordinate frames into a spatial memory map of the environment.

Furthermore, previous studies with patients with autoptically verified lesions restricted to the hippocampal formation have shown that memory for locations of a simple spatial array of objects is impaired with delays of 5 minutes.[55,56] Thus, it appears that at delays of some minutes the hippocampus may be actively involved in spatial memory, at least for largely nonrelational spatial information. This interpretation would parallel the known delay-dependent memory deficits for nonrelational object information seen in monkeys with lesions to the hippocampal formation, where memory is normal at delays of 60 seconds, but impaired at delays of 10 minutes.[57,58] Thus, the hippocampal formation is a likely candidate region for a long-term spatial memory system, subserving spatial memory formation beyond parahippocampal cortex-based medium-term spatial memory.

In conclusion of this review of spatial memory, the results of the different studies using such a simple visuo-oculomotor model allow us to suggest that after a brief initial visuospatial integration stage under the control mainly of the right posterior parietal cortex (up to about 300 ms), spatial memory is controlled, successively, by the DLPFC for its short-term component (up to 15–20 seconds), the parahippocampal cortex for its medium-term component (between 15 and 20 seconds and probably a few minutes) and, lastly, the hippocampal formation for its long-term component (after a few minutes) (FIG. 2). It should be noted that such an organization of spatial memory involves a new concept of an intermediary link between short-term and long-term memorizations, with involvement of the parahippocampal cortex in medium-term memorization. It remains to be determined, however, how memorized information is coordinated between the different structures successively involved in spatial memory, that is, in parallel, according to certain behavioral results[44] (see above), or serially, according to other preliminary TMS results (Müri et al., unpublished data), or, as therefore seems likely, by a combination of the two. Furthermore, is this simple visuo-oculomotor model of spatial memory applicable to the memorization of other types of stimuli (i.e., more complex visual, auditory, and somesthesic stimuli) and the rest of body motricity? Although this appears probable, inasmuch as the DLPFC, posterior parietal cortex, and hippocampal formation are involved in other types of stimuli and the preparation of other movements,[46,59,60] the demonstration of such general mechanisms and of the systematic involvement of the parahippocampal cortex requires other studies.

GENERAL CONCLUDING REMARKS

The brain is continuously organizing possible multiple accurate motor responses to sensory stimuli occurring in the environment. A visuo-oculomotor model comprising a simple visual item as the stimulus and a single saccade as the motor response appears to be convenient to understand the multiplicity of possible movements and the relatively complex neuropsychological processes preparing them. As early as the perceptive stage, different attentional processes select salient information[61–63] (FIG. 2). This information is then spatially integrated in the posterior parietal cortex in order to know the location of the stimulus related to the whole body and not simply to a part of it. Just after this stage, a reflexive (saccade) response without any delay may be triggered in the motor part of the parietal lobe (PEF) towards a suddenly (possibly interesting and/or threatening) occurring stimulus. By contrast, if reflexive saccades have to be inhibited, the dorsolateral prefrontal cortex (DLPFC, but not the frontal eye field, FEF) plays a major role. When the motor response is planned but not immediately made, salient memorized information is stored, successively, in the DLPFC[63] (short-term memory), parahippocampal cortex (medium-term memory), and hippocampal formation (long-term memory) depending on the time of the delayed response. When the response is predicted (predictive saccade) to a specific location before the appearance of the target, the DLPFC also appears to be involved in such prediction.[64] At the memorization stage, the anterior cingulate cortex and cingulate eye field could already select and prepare the motor areas that might have to respond. If several movements are planned, the SMA and the supplementary eye field (SEF) are involved in the preparation of motor programs, previously learned thanks to the pre-SMA and pre-SEF. Finally, the responses are executed by the primary motor cortex and/or the FEF if body movements and/or intentional saccades, respectively, are required.

REFERENCES

1. LEIGH, R.J. & D.S. ZEE. 1999. The Neurology of Eye Movements. Oxford University Press. New York.
2. PIERROT-DESEILLIGNY, C., S. RIVAUD, B. GAYMARD, et al. 1995. Cortical control of saccades. Ann. Neurol. **37:** 557–567.
3. PAUS, T. 1996. Location and function of the human frontal eye field: a selective review. Neuropsychology **734:** 475–483.
4. CARTER, N. & D.S. ZEE. 1997. The anatomical localization of saccades using functional imaging studies and transcranial magnetic stimulation. Curr. Opin. Neurol. **10:** 10–17.
5. GAYMARD, B., C.J. PLONER, S. RIVAUD-PECHOUX & C. PIERROT-DESEILLIGNY. 1999. The frontal eye field is involved in spatial short-term memory but not in reflexive saccade inhibition. Exp. Brain Res. **129:** 288–301.
6. LOBEL, E., P. KAHANE, U. LEONARDS, et al. 2001. Localization of the human frontal eye fields: anatomical and functional findings from fMRI and intracerebral electrical stimulation. J. Neurosurg. In press.
7. PIERROT-DESEILLIGNY, C., S. RIVAUD & B. GAYMARD. 1991. Cortical control of reflexive visually guided saccades in man. Brain **114:** 1473–1485.
8. PIERROT-DESEILLIGNY, C., S. RIVAUD, B. GAYMARD & Y. AGID. 1991. Cortical control of memory-guided saccades in man. Exp. Brain Res. **83:** 607–617.
9. RIVAUD, S., R.M. MÜRI, B. GAYMARD, et al. 1994. Eye movement disorders after frontal eye field lesions in humans. Exp. Brain Res. **102:** 110–120.
10. ISRAËL, I., S. RIVAUD, B. GAYMARD, et al. 1995. Cortical control of vestibular-guided saccades in man. Brain **118:** 1169–1183.
11. HEIDE, W. & D. KÖMPF. 1998. Combined deficits of saccades and visuo-spatial orientation after cortical lesions. Exp. Brain Res. **123:** 164–171.
12. MÜRI, R.M., C.W. HESS & O. MEIENBERG. 1991. Transcranial stimulation of the human frontal eye field by magnetic pulses. Exp. Brain Res. **86:** 219–223.
13. WIPFLI, M., J. FELBLINGER, U.P. MOSIMANN, et al. 2001. Double pulse transcranial magnetic stimulation over the frontal eye field facilitates triggering of memory-guided saccades. Eur. J. Neurosci. **14:** 571–575.
14. GROBRAS, M.H., E. LOBEL, P.F. VAN DE MOORTELE, et al. 1999. An anatomical landmark for the supplementary eye fields in human revealed with functional magnetic resonance imaging. Cereb. Cortex **9:** 705–711.
15. GAYMARD, B., S. RIVAUD & C. PIERROT-DESEILLIGNY. 1993. Role of the left and right supplementary motor areas in memory-saccade sequences. Ann. Neurol. **34:** 404–406.
16. PIERROT-DESEILLIGNY, C., I. ISRAËL, A. BERTHOZ, et al. 1993. Role of the different frontal lobe areas in the control of the horizontal component of memory-guided saccades in man. Exp. Brain Res. **95:** 166–171.
17. MÜRI, R.M., S. RIVAUD, A.I. VERMERSCH, et al. 1995. Effects of transcranial magnetic stimulation on the supplementary motor area region during sequences of memory-guided saccades. Exp. Brain Res. **104:** 163–166.

18. Müri, R.M., K.M. Roesler & C.W. Hess. 1994. Influence of transcranial magnetic stimulation on the execution of memorized sequences of saccades in man. Exp. Brain Res. **101:** 521–524.
19. Kawashima, R., J. Tanji, K. Okada, et al. 1998. Oculomotor sequence learning: a positron emission tomography study. Exp. Brain Res. **122:** 1–8.
20. Grosbras, M.H., U. Leonards, E. Lobel, et al. 2001. Human cortical networks for new and familiar sequences of saccades. Cereb. Cortex **11:** 606–618.
21. Müri, R.M., M.T. Iba-Zizen, C. Derosier, et al. 1996. Location of the human posterior eye field with functional magnetic resonance imaging. J. Neurol. Neurosurg. Psychiatry **6:** 445–448.
22. Heide, W., F. Binkofski, R.J. Seitz, et al. 2001. Activation of frontoparietal cortices during memorized triple-step sequences of saccadic eye movements: an fMRI study. Eur. J. Neurosci. **13:** 1177–1189.
23. Milea, D., E. Lobel, S. Lehericy, et al. 2001. Functional MRI mapping of parietal and cingular activity during voluntary saccades. Soc. Neurosci. Abstr. **27.** In press.
24. Braun, D., H. Weber, T.H. Mergner, et al. 1992. Saccadic reaction times in patients with frontal and parietal lesions. Brain **115:** 1359–1386.
25. Elkington, P.T., G.K. Ker & J.S. Stein. 1992. The effect of electromagnetic stimulation of the posterior parietal cortex on eye movements. Eye **6:** 510–514.
26. Pierrot-Deseilligny, C., J.C. Gautier & J. Loron. 1988. Acquired ocular motor apraxia due to bilateral frontoparietal infarcts. Ann. Neurol. **23:** 199–202.
27. Sweeney, J.A., M.A. Mintun, M. Kwee, et al. 1996. Positron emission tomography study of voluntary saccadic eye movements and spatial working memory. J. Neurophysiol. **75:** 454–468.
28. Rajkowska, G. & P.S. Goldman-Rakic. 1995. Cytoarchitectonic definition of prefrontal areas in the normal human cortex. II. Variability in locations of area 9 and 46 and relationship to the Talairach coordinate system. Cereb. Cortex **5:** 323–337.
29. Petrides, M. & D.N. Pandya. 1999. Dorsolateral prefrontal cortex: comparative cytoarchitectonic analysis in the human and the macaque brain and corticocortical connection patterns. Eur. J. Neurosci. **11:** 1011–1036.
30. Müri, R.M., O. Heid, A.C. Nirkko, et al. 1998. Functional organisation of saccades and antisaccades in the frontal lobe in humans: a study with echo planar functional magnetic resonance imaging. J. Neurol. Neurosurg. Psychiatry **65:** 374–377.
31. Müri, R.M., J. Felblinger, Y. Ottiger, et al. 2000. Transcranial magnetic stimulation of the dorsolateral prefrontal cortex reduces antisaccade performance. Soc. Neurosci. Abstr. **26:** 1076.
32. Gaymard, B., S. Rivaud, J.F. Cassarini, et al. 1998. Effects of anterior cingulate cortex lesions on ocular saccades in humans. Exp. Brain Res. **120:** 173–183.
33. Paus, T. 2001. Primate anterior cingulate cortex: where motor control, drive and cognition interface. Natl. Rev. Neurosci. **2:** 417–424.
34. Petit, L., C. Orsaud, N. Tzourio, et al. 1993. PET study of voluntary saccadic eye movements in humans: basal ganglia-thalamocortical system and cingulate cortex involvement. J. Neurophysiol. **69:** 1009–1017.
35. Shima, K., K. Aya, H. Mushiake, et al. 1991. Two movement-related foci in the primate cingulate cortex observed in signal-triggered and self-paced forelimb movements. J. Neurophysiol. **65:** 188–202.
36. Goldman-Rakic, P.S. 1996. Regional and cellular fractionation of working memory. Proc. Natl. Acad. Sci. USA **93:** 13473–13480.
37. Müri, R.M., A.I. Vermersch, S. Rivaud, et al. 1996. Effects of single-pulse transcranial magnetic stimulation over the prefrontal and posterior parietal cortices during memory-guided saccades in humans. J. Neurophysiol. **76:** 2102–2106.
38. Müri, R.M., B. Gaymard, S. Rivaud, et al. 2000. Hemispheric asymmetry in cortical control of memory-guided saccades. A transcranial magnetic stimulation study. Neuropsychology **38:** 1105–1111.
39. Brandt, S.A., C.J. Ploner, B.U. Meyer, et al. 1998. Effects of repetitive transcranial magnetic stimulation over dorsolateral prefrontal and posterior parietal cortex on memory-guided saccades. Exp. Brain Res. **118:** 197–204.

40. CHAFEE, M.V. & P. GOLDMAN-RAKIC. 2000. Inactivation of parietal and prefrontal cortex reveals interdependence of neural activity during memory-guided saccades. J. Neurophysiol. **83:** 1550–1566.
41. PLONER, C.J., S. RIVAUD-PECHOUX, B. GAYMARD, et al. 1999. Errors of memory-guided saccades in humans with lesions of the frontal eye field and the dorsolateral prefrontal cortex. J. Neurophysiol. **82:** 1086–1090.
42. WHITE, J.M., D.L. SPARKS & T.R. STANFORD. 1994. Saccades to remembered target locations: an analysis of systematic and variable errors. Vision Res. **34:** 79–92.
43. WANG, X.J. 2001. Synaptic reverberation underlying mnemonic persistent activity. Trends Neurosci. **24:** 455–463.
44. PLONER, C.J., B. GAYMARD, S. RIVAUD, et al. 1998. Temporal limits of spatial working memory in humans. Eur. J. Neurosci. **10:** 794–797.
45. SIDMAN, M., L.T. STODDARD & J.P. MOHR. 1968. Some additional observations of immediate memory in a patient with bilateral hippocampal lesions. Neuropsychology **6:** 245–254.
46. RAINS, G.D. & B. MILNER. 1994. Right-hippocampal contralateral-hand effect in the recall of spatial location in the tactual modality. Neuropsychology **32:** 1233–1242.
47. AMARAL, D.G. & R. INSAUSTI. 1990. The hippocampal formation. *In* The Human Nervous System. G. Paxinos, Hrsg. Academic Press. San Diego.
48. GOLDMAN-RAKIC, P.S., L.D. SELEMON & M.L. SCHWARTZ. 1984. Dual pathways connecting the dorsolateral prefrontal cortex with the hippocampal formation and parahippocampal cortex in the rhesus monkey. Neuroscience **12:** 719–743.
49. SUZUKI, W.A. & D.G. AMARAL. 1994. Perirhinal and parahippocampal cortices of the macaque monkey: cortical afferents. J. Comp. Neurol. **350:** 497–533.
50. BARBAS, H. & M.M. MESULAM. 1981. Organization of afferent input to subdivisions of area 8 in the rhesus monkey. J. Comp. Neurol. **200:** 407–431.
51. PLONER, C.J., B. GAYMARD, N. EHRLÉ, et al. 1999. Spatial memory deficits in patients with lesions affecting the medial temporal neocortex. Ann. Neurol. **45:** 312–319.
52. PLONER, C.J., B. GAYMARD, S. RIVAUD-PECHOUX, et al. 2000. Lesions affecting the parahippocampal cortex yield spatial memory deficits in humans. Cereb. Cortex **10:** 1211–1216.
53. MÜRI, R.M., S. RIVAUD, S. TIMSIT, et al. 1994. The role of the right medial temporal lobe in the control of memory-guided saccades. Exp. Brain Res. **101:** 165–168.
54. ROLLS, E.T. 1999. Spatial view cells and the representation of place in the primate hippocampus. Hippocampus **9:** 467–480.
55. CAVE, C.B. & L.R. SQUIRE. 1991. Equivalent impairment of spatial and nonspatial memory following damage to the human hippocampus. Hippocampus **1:** 329–340.
56. REMPEL-CLOWER, N.L., S.M. ZOLA, L.R. SQUIRE & D.G. AMARAL. 1996. Three cases of enduring memory impairment after bilateral damage limited to the hippocampal formation. J. Neurosci. **16:** 5233–5255.
57. ALVAREZ, A., S. ZOLA-MORGAN & L.R. SQUIRE. 1995. Damage limited to the hippocampal region produces long-lasting memory impairment in monkeys. J. Neurosci. **15:** 5637–5807.
58. LEONARD, B.W., D.G. AMARAL, L.R. SQUIRE & S. ZOLA-MORGEN. 1995. Transient memory impairment in monkeys with bilateral lesions of the entorhinal cortex. J. Neurosci. **15:** 5637–5659.
59. GOLDMAN-RAKIC, P.S., J.F. BATES & M.V. CHAFEE. 1992. The prefrontal cortex and internally generated motor acts. Curr. Opin. Neurobiol. **2:** 830–835.
60. CULHAM, J.C. & N.G. KANWISHER. 2001. Neuroimaging of cognitive functions in human parietal cortex. Curr. Opin. Neurobiol. **11:** 157–163.
61. COLBY, C.L. & M.E. GOLDBERG. 1999. Space and attention in parietal cortex. Annu. Rev. Neurosci. **22:** 319–349.
62. PLONER, C.J., F. OSTENDORF, S.A. BRAND, et al. 2001. Behavioral relevance modulates access to spatial working memory in humans. Eur. J. Neurosci. **13:** 357–363.
63. KASTNER, S. & L.G. UNGERLEIDER. 2000. Mechanisms of visual attention in the human cortex. Annu. Rev. Neurosci. **23:** 315–341.
64. PIERROT-DESEILLIGNY, C., R.M. MÜRI, S. RIVAUD & B. GAYMARD. 1995. Eye movement disorders after prefrontal cortex lesions in humans. Soc. Neurosci. Abstr. **21:** 1270.

Visual-Vestibular and Visuovisual Cortical Interaction

New Insights from fMRI and PET

THOMAS BRANDT, STEFAN GLASAUER, THOMAS STEPHAN, SANDRA BENSE, TAREK A. YOUSRY, ANGELA DEUTSCHLÄNDER, AND MARIANNE DIETERICH

Department of Neurology, Klinikum Grosshadern, Ludwig-Maximilians University, 81366 Munich, Germany

ABSTRACT: PET and fMRI studies have revealed that excitation of the vestibular system by caloric or galvanic stimulation not only activates the parietoinsular vestibular cortex but also bilaterally deactivates the occipital visual cortex. Likewise, visual motion stimulation not only activates the visual cortex but also deactivates the parietoinsular vestibular cortex. These findings are functionally consistent with the hypothesis of an inhibitory reciprocal visual-vestibular interaction for spatial orientation and motion perception. Transcallosal visuovisual interaction between the two hemispheres was found by using half-field visual motion stimulation: activation of motion-sensitive areas hMT/V5 and deactivations of the primary visual cortex contralateral to the stimulated hemisphere. The functional significance of these inter- and intrasensory interactions could be that they (A) allow a shift of the sensorial weight between two incongruent sensory inputs and (B) ensure a correspondence of the two hemispheres during evaluation of contradictory motion stimulation of the right and left hemifields. In terms of mathematical modeling, these findings may reflect the concepts of a sensory conflict mechanism or a mismatch between expected and actual sensory input.

KEYWORDS: vestibular system; visual system; spatial orientation; motion perception; intrasensory interactions; sensory input; fMRI, PET

INTRODUCTION

Most human brain activation studies describe the processing of either the vestibular or the visual system (e.g., with respect to retinotopic mapping of early visual areas in the human cortex or functional specialization for motion, color, and face perception). We focus on the interaction between the visual and the vestibular systems which is necessary for adequate perception of verticality and motion and for distinguishing between self-motion and the motion of surroundings. The individual who observes moving visual scenes may perceive himself/herself as either a station-

---

Address for correspondence: Dr. Thomas Brandt, Department of Neurology, Klinikum Grosshadern, Ludwig-Maximillians University, Marchioninistr. 15, 81366 Munich, Germany. Voice: 49-89-7095-2570; fax: 49-89-7095-8883.

tbrandt@brain.nefo.med.uni-muenchen.de

Ann. N.Y. Acad. Sci. 956: 230–241 (2002). © 2002 New York Academy of Sciences.

ary observer of movement in the surroundings or a moving observer of stationary surroundings (i.e., vection).

Visual perception of motion is possible when the eyes move (pursuit eye movements, optokinetic nystagmus [OKN]) or do not move (fixation of a stationary target). The largely distributed cortical ocular motor network is always involved during dynamic spatial orientation and perception of motion. To complicate matters further, apart from these specific intersensory and sensorimotor processes, attention,[1] cognition, and timing must be considered. All of these factors contribute to the sometimes confusing experience that even simple paradigms such as to-and-fro saccades or OKN evoke highly complex patterns of activation in the cortical, basal ganglia, brain stem, and cerebellar regions.[2] Consequently, experimental paradigms in brain activation studies should try to compare stimulus conditions that are as similar as possible (differing in only one property) rather than to compare a complex stimulus condition with rest.

In this paper, we discuss cortical brain activation studies of the differential effects of visual motion stimulation, vestibular stimulation, and their reciprocal interaction.

VISUAL-VESTIBULAR INTERACTION

It is becoming increasingly apparent that relative deactivations may be as important for brain function as are activations. A chance observation[3] revealed that stimulation of the vestibular system by caloric irrigation not only activated the parietoinsular vestibular cortex (PIVC), but also at the same time bilaterally deactivated the occipital visual cortex, which covers Brodmann areas (BA) 17, 18, and 19. Likewise, large-field visual motion stimulation, which induces an apparent self-motion (vection), not only caused activation of the visual cortex (a medial parieto-occipital visual area separate from motion-sensitive areas MT/MST), but also simultaneously caused deactivation of the PIVC.[4] In a recent PET study the patterns of rCBF increases and decreases were compared for unimodal vestibular, unimodal visual, and simultaneous vestibular and visual stimulation.[5] Healthy volunteers were exposed to (a) caloric vestibular stimulation, (b) small-field visual motion stimulation in roll, and (c) simultaneous caloric vestibular and visual pattern stimulation. Unimodal vestibular stimulation led to activations of vestibular cortex areas, particularly the PIVC, and concurrent deactivations of visual cortical areas (BA 17-19). Unimodal visual motion stimulation led to activations of the striate visual cortex and the motion-sensitive area in the middle temporal/middle occipital gyri (BA 19/37) with concurrent deactivations of the PIVC. Simultaneous bimodal stimulation resulted in activation of the cortical representation of both sensory modalities (FIG. 1). Activations were significantly smaller in bimodal stimulation than in unimodal stimulation. These findings are consistent with the concept of an inhibitory sensori-sensory interaction in all three stimuli conditions.

This finding led to the functional hypothesis that inhibitory reciprocal visual-vestibular interaction is a form of sensori-sensory interaction for orientation in space and perception of motion. This hypothesis is supported by earlier findings[6] that thresholds for detecting vestibular body accelerations (vestibular system) are significantly increased during optokinetically induced vection (visual system) in a combined rotatory chair-drum system. A similar interaction seems to exist between other

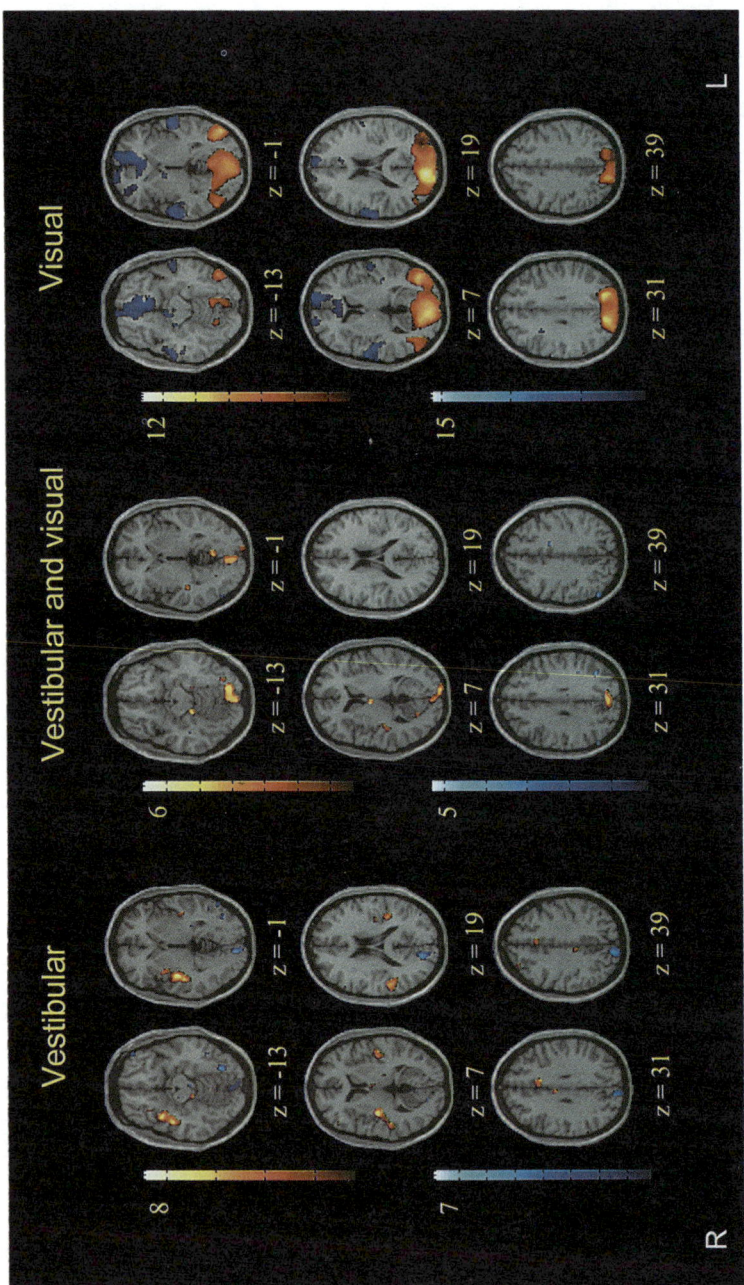

FIGURE 1. Activation maps (rCBF increases and decreases in PET) showing activations (*red*) and deactivations (*blue*) during unimodal vestibular (**A**), simultaneous (**B**), and unimodal visual (**C**) stimulation versus rest condition (**D**) (group analysis: $n = 13$; $p < 0.001$), superimposed onto the MNI standard brain template. Caloric vestibular stimulation activates the parietoinsular vestibular cortex and deactivates the visual cortex. Visual motion stimulation activates the visual cortex and deactivates the vestibular cortex. During simultaneous visual-vestibular stimulation, activations and deactivations are smaller.

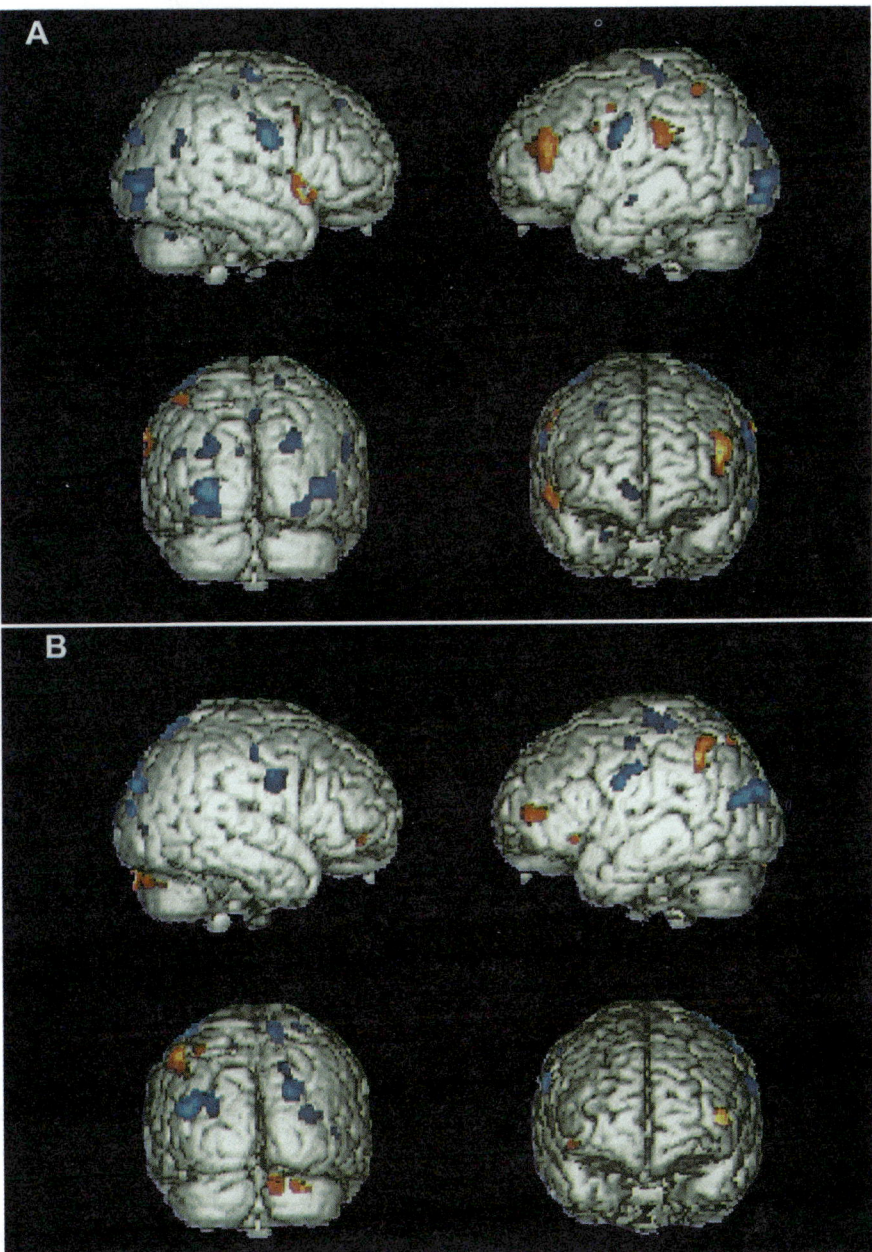

FIGURE 2. Blood oxygen-level dependent functional magnetic resonance imaging during galvanic vestibular and galvanic pain stimulation. Color-coded maps for blood oxygen-level dependent functional magnetic resonance imaging (MRI) signal increases (*activation in red-yellow*) and signal decreases (*deactivation in blue-white*) of a group of six healthy individuals **(A)** during galvanic vestibular stimulation (GVS) at mastoid level and **(B)** during

sensory systems (e.g., the nociceptive and the somatosensory systems; FIG. 2). In an fMRI study galvanic cutaneous pain stimulation not only activated nociceptive areas in the frontal lobe and the anterior/median insula, but also simultaneously deactivated the somatosensory cortex bilaterally (BA 2/3/4).[7]

The functional significance of inhibitory sensori-sensory interactions is obvious; it allows a potential mismatch between two incongruent or misleading sensory inputs to be suppressed by shifting the sensorial weight to the dominant or more reliable modality. The perception of self-motion can be dominated either by vestibular input, provided mainly by head acceleration, or by visual input, during passive locomotion at a constant velocity with head still, as in a train, depending on the prevailing mode of stimulation. This mechanism makes it possible to relate the dominant perception of self-motion or orientation to the actual input of one of the two modalities and thus avoid perceptual ambiguity.[8]

Such interactions have functional implications not only for perception but also for sensorimotor control. During locomotion, for example, destabilizing vestibular input,[9,10] monosynaptic stretch reflex responses in the human leg, ascending spinal group I afferents,[11] and muscular sense[12] are significantly attenuated. The obvious reason for this attenuation is to prevent interference with a highly automatic spinal locomotor pattern.

VISUOVISUAL INTERACTION

Orientation and perception of motion rely not only on multisensory interaction or fusion but also on "intrasensory" interaction, especially between the two hemispheres. Visuovisual interaction may serve as an example.

In an fMRI study, we stimulated patients with complete homonymous hemianopia due to acute infarctions of the posterior cerebral artery with coherent visual pattern motion[13] and observed bilateral activation of occipitotemporal areas, the human homologue (MT/V5)[14] of the motion-sensitive middle temporal (MT) and middle superior temporal (MST) areas in the monkey. This finding indicated activation of hMT/V5 on the infarcted hemisphere in the absence of input from the ipsilateral primary visual cortex. Two possible explanations were suggested: first, involvement

---

galvanic pain stimulation (GPS) at neck C5/6 level superimposed onto a three-dimensional rendered template brain. For purposes of illustration, a significance level of $p \leq 0.001$ (uncorrected) was used. GVS caused activations of the precentral gyrus (Brodmann area [BA] 6; frontal eye field [FEF]), prefrontal cortex (middle frontal gyrus, BA 46/9), middle temporal gyrus (BA 37; middle temporal/medial superior temporal), and inferior parietal lobule (BA 40). Activations in the anterior, medial, and posterior parts of the insula are not mapped. GPS mainly activated the prefrontal cortex (middle/inferior frontal gyrus; BA 10/46/9; prefrontal cortex), the precentral gyrus (BA 6/44; FEF), and the inferior/superior parietal gyrus (BA 40/7). Under both conditions, significant signal decreases were seen in the central sulcus region, predominantly in the postcentral gyrus, analogous to the primary somatosensory cortex (BA 2/3/4 and BA 3/4/6). GVS showed further decreases of the visual cortex bilaterally (fusiform/inferior occipital gyrus; BA 18 and 19) and in the right precuneus (BA 7). By contrast, GPS led to signal decreases in temporo-occipital cortex areas (middle temporal/occipital gyrus; BA 18, 19, and 39).[7]

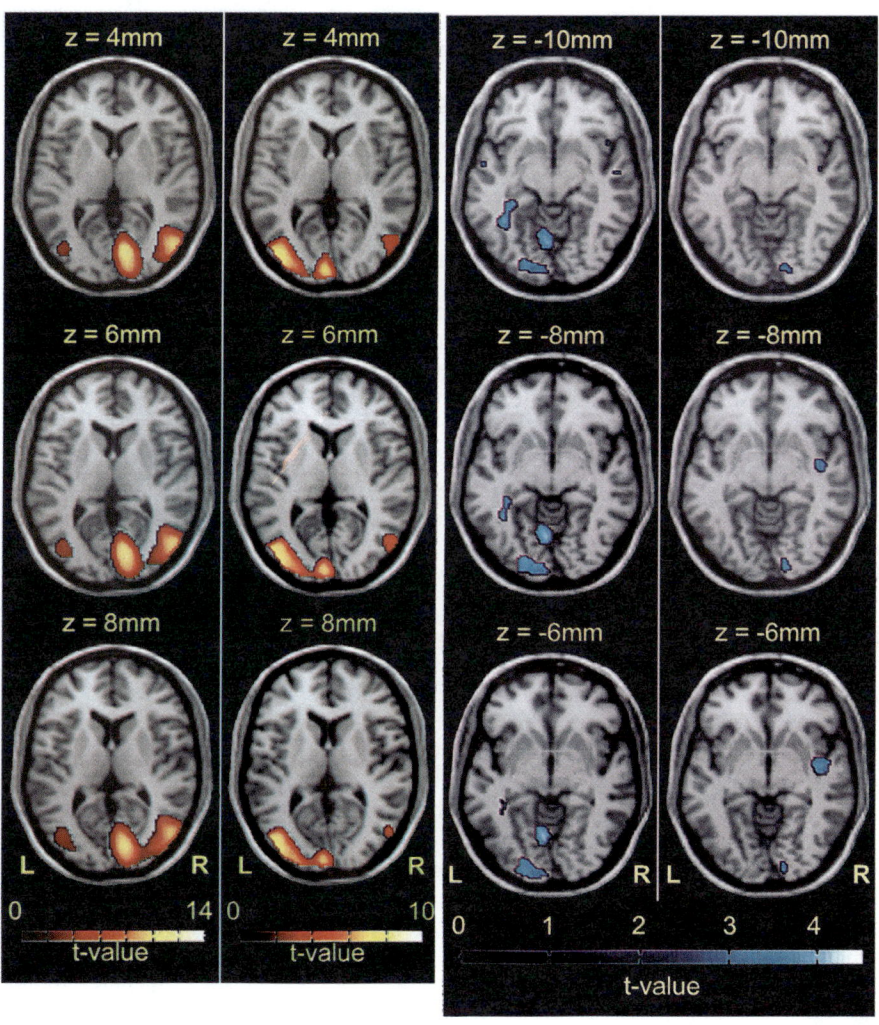

FIGURE 3. (A) Areas with signal increases during visual motion stimulation of the left (*left column*) and right (*right column*) hemifields obtained by statistical group analysis ($n = 9$; $p \leq 0.001$). Activation maps are superimposed onto selected transverse sections of a standard brain template 4–8 mm above the anteroposterior commissure. Activations were found in the primary visual cortex contralateral to the stimulated hemifield and in the medial occipital gyrus covering V5 on both the stimulated hemisphere as well as the hemisphere without input from ipsilateral primary visual cortex. (B) Signal decreases were found in the primary visual cortex and the optic radiation in the lower white matter of the medial temporal gyrus ipsilateral to the stimulated hemifield (no direct visual input), cerebellar vermis, and posterior insula.[19]

of ipsilateral direct extrastriatal visual pathways; and second, interhemispheric callosal transfer between right and left hemispheres. Finley and co-workers[15] described a patient with residual visual function in his right 'blind' hemifield who, during movement stimulation in the normal left hemifield, reported a sensation of movement in both hemifields. They also discussed direct subcortical extrastriate projections or transcallosal pathways as the possible source of fMRI responses found in the dorsal extrastriate areas of the damaged hemisphere.[16]

To examine the existence of transcallosal visuovisual connections for motion perception, as earlier described anatomically in animal experiments,[17,18] we conducted an fMRI study in normal subjects using half-field visual motion stimulation.[19] The vertical edge of the motion pattern field was located 8 degrees from the fixation point to avoid stimulating the vertical meridian, which is represented retinotopically in both hemispheres. Bilateral activation was significant in the middle occipital gyrus (motion-sensitive middle temporal/middle superior temporal areas; BA 19/37). A negative signal change was found in the primary visual cortex including the lingual and fusiform gyri (BA 18/17) contralateral to the stimulated hemisphere (FIG. 3). These data are compatible to an interhemispheric transfer of visual motion information, most likely through the corpus callosum. Transcallosal transfer of visual motion information was evident as increases (BA 19/37) and decreases (BA 18/17) of the fMRI signals.

The question arises as to the possible functional significance of an interhemispheric visuovisual interaction. Both hemispheres have to interact in stimulus situations in which both visual hemifields have contradictory information about motion, as when sitting in a train reading. Here information about constant velocity self-motion is provided by the optic flow pattern in one hemifield, while the other hemifield is filled with stationary contrasts from inside the train. Because we cannot perceive two different states of body motion at the same time, the two hemispheres have to correspond to each other to be able to determine an actual and unique perception of self-motion or absence of motion.[8,19] Prestriatal visual areas have transcallosal interhemispheric connections for this kind of bilateral interaction for perception and initiation of motor response.[17,20] MT/V5 provides us with intermediate information about visual motion in extraretinal coordinates, which can be used for motor control of not only the eyes but also the body. As animal experiments have shown, interhemispheric interaction requires neural and interhemispheric synchronization that is mediated by corticocortical connections.[21,22]

The inhibitory intrasensory visuovisual interaction found in the fMRI study would theoretically allow attention to be shifted between the two hemispheres (two visual hemifields) by raising the perceptual thresholds in the visual hemifield with the currently less relevant input. In this context, visual test results are striking, as in the famous hemianopic patient, GY, who exhibited some residual visual function in his "blind hemifield." In their initial studies on GY, Barbur and co-workers[23] demonstrated that uniform illumination of the normal hemifield exerted a weak inhibitory influence on GY's sensitivity to transient light presented within the blind hemifield. Later, Finlay and co-workers[15] showed in the same patient that movement in the normal hemifield generated a motion perception localized in the blind hemifield which could not be distinguished qualitatively from that associated with direct stimulation of the blind hemifield. The two different findings of induction or sup-

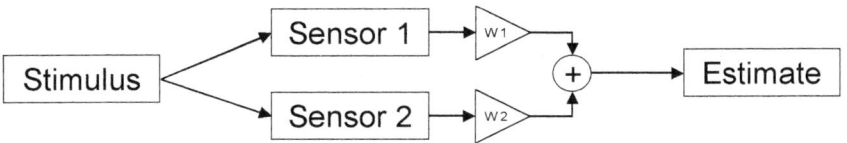

FIGURE 4. Weighted averaging for two sensors with equal characteristics. Weights W1 and W2 were adjusted to obtain an optimal estimate of the stimulus.

pression of a perception in the visual cortex contralateral to the stimulated hemisphere agree with our findings with visual hemifield stimulation.

Thus, the simple paradigm of visual motion stimulation restricted to one hemifield provided two observations of a significant visuovisual interhemispheric interaction: activation of motion-sensitive areas hMT/V5 and "deactivation" of the primary visual cortex contralateral to the stimulated hemisphere.

MODELING VISUAL-VESTIBULAR INTERACTION

PET and fMRI brain activation studies have provided fascinating insights into the neuronal network that mediates motion perception, orientation, and ocular motor control. At the same time, however, they have revealed a dilemma involving the correlation of a particular locus to a particular function. System analysis and mathematical modeling are necessary to understand sensorisensory and sensorimotor brain functions. We are, however, still struggling with basic questions on fusion mechanisms and are far from being able to model visual-vestibular interaction at the cortex level.

Principles of Sensor Fusion

One of the simplest and most common methods of fusing information from two different sensors is weighted averaging (FIG. 4). This method yields optimal results if both sensors, which have equal sensor characteristics, report a noisy estimate of the variable to be measured. Weights are adjusted to the inverse variances of each sensor.

Weighted averaging is, by itself, limited when dealing with different sensor types such as vestibular and visual sensors. Consider, for example, a situation in which a constant rotation around an earth vertical axis is applied. Both visual and vestibular sensors provide a good estimate of self-rotation in the beginning; thus, weighted averaging could be applied initially. However, vestibular afferent information decays after a certain time, whereas visual information remains available. Weighted averaging of both cues at this time would lead to an inaccurate estimate, because zero vestibular and full visual information would be averaged. A similar problem occurs during rotation in darkness or when only visual information is available (circular vection).

It is necessary, therefore, to pre-process sensory information to provide signals that are suitable for weighted averaging, as follows: (1) signals from sensors that measure different physical variables have to be transformed into a common representation by internal models of physical laws; and (2) signals from sensors with different temporal characteristics have to be filtered by, for example, internal models of sensors, to ensure that the temporal information is comparable (frequency completion or frequency segregation). For example, during rapid tilt relative to gravity, the semicircular canals provide information about angular velocity, whereas the otoliths measure the changing orientation of gravity with respect to the head. Therefore, current models of otolith-canal interaction assume that signals from the otoliths are transformed into an angular velocity signal.[24,25] Conversely, the canal signal can be used to predict the change of the gravity vector[24,26,27] or to compute linear translation[28] and thus help to distinguish gravitational acceleration from linear translation. Both operations require computational mechanisms that mimic physical laws called internal models.[29]

If the signals to be fused have the same physical meaning, frequency completion/segregation alone can be used as a sensor fusion mechanism.[30] The vestibulo-ocular and the optokinetic reflexes, which cooperate at the brain stem level, provide an example of such signals. Under the eyes-open condition, the high-pass characteristics of the vestibulo-ocular reflex are enhanced by visual input via the optokinetic reflex, which has complementary low-pass characteristics. This approach, however, may result in a loss of information: the visual system may, for example, provide high-frequency cues that are not used if the vestibular input is missing. Frequency completion is often modeled by means of internal models of sensor dynamics.

Other models propose the use of weighted averaging after frequency completion and transformation of each type of sensory information. To achieve this, frequency completion is performed on internal estimates rather than on primary sensory information. Thus, such an approach involves feedback loops, but it may result in better estimation procedures. A similar approach is optimal filtering, which is formalized for linear systems as so-called Kalman filters. Optimal filters try to minimize the errors between an afferent sensory input and a predicted sensory input, the latter resulting from the estimated variable (FIG. 5).

However, these mechanisms alone cannot explain several experimental findings, for example, in visual-vestibular interaction for self-motion perception, such as the long latency of onset of circular vection. Therefore, most model-based considerations of visual-vestibular interaction propose threshold or conflict mechanisms, which determine whether a sensory cue is to be used. This can, for example, be achieved by having the conflict mechanism adjust the averaging weights[31] or by suppressing the visual signals, depending on the amount of conflict between two sensory inputs.[32] Conflict mechanisms can thus be viewed as basic decision-making: if one sensory system is considered inaccurate or unreliable, it is eliminated from the averaging process.

The inhibitory reciprocal visual-vestibular interaction found in brain-activation studies can be interpreted as a manifestation of such a conflict mechanism. If one sensory system receives a strong input, which is missing in the other, it is likely that the sensory system without input is less reliable and has to be suppressed. Even the simple diagram in FIGURE 5 provides a model for interpreting activations and deactivations.

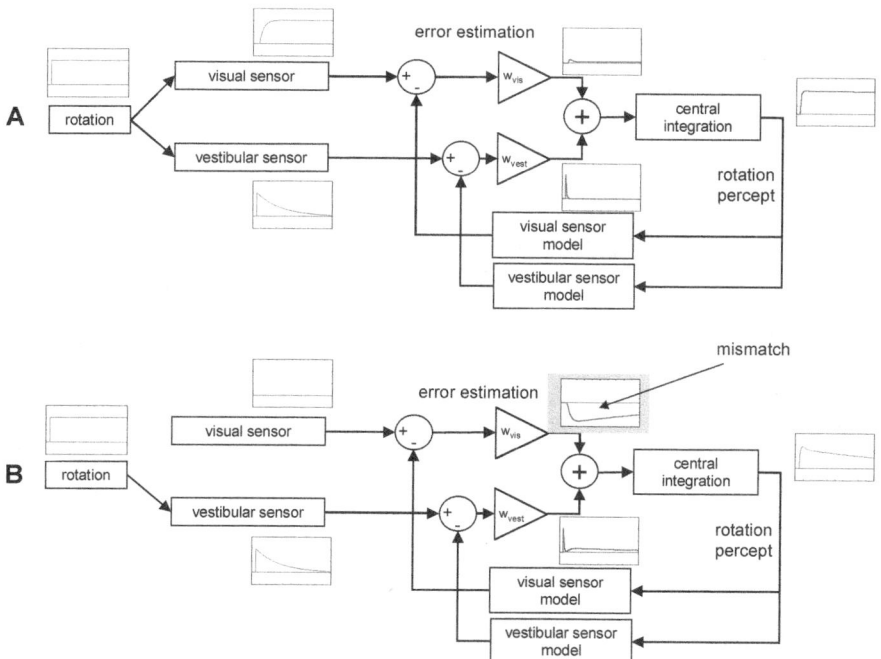

FIGURE 5. Two-sensor optimal filter for estimation of rotation from visual and vestibular cues. The estimate of rotation (rotation percept) is fed through internal models of sensor characteristics to predict the sensory input. The errors between input and predicted input are weighted and summed. The central integration minimizes the weighted sum of errors. In contrast to simple weighted averaging, as shown in FIGURE 4, where the sensory signals are averaged, the sensory errors are averaged and minimized here by the feedback loop. *Insets* show schematic time courses of variables following a step of rotational velocity (*left inset*) as stimulus to yield rotation estimate (*rightmost inset*). (**A**) Rotation in light. Although both sensory outputs considerably differ, estimated errors are very small (except for an initial peak, which indicates that the onset of vestibular input constitutes an unexpected event) and the rotation estimate is accurate. (**B**) Rotation in darkness. The visual sensor is expected to deliver a signal that corresponds to the vestibular stimulus. Therefore, a sensory mismatch occurs which leads to a large negative visual error (*gray inset*). The corresponding vestibular error is positive, because the vestibular stimulus is larger than expected from the visual sensory input. Note that the rotation percept decays with a longer time constant than the vestibular signal (experimental finding). Cortical activation and deactivation, as found with fMRI and PET, may correspond to a similar scheme in which activation signals more sensory input than expected, and deactivation less input than expected. Interestingly, this scheme also predicts that matching simultaneous vestibular and visual sensory input causes less activation than does sensory mismatch (see **A**).

Whereas approaches such as weighted averaging or optimal filtering can be understood as methods that yield the best possible estimate (as a Bayesian maximum likelihood estimate), conflict mechanisms have so far not been related to any optimality criterion. However, there are several theoretical approaches to sensor fusion that can include conflict mechanisms.

Sensory conflicts may be treated as transitions from one sensory state to another; this has been formalized mathematically by McCollum and co-workers[33] for postural control. Nonlinear attractor dynamics, proposed for sensor fusion in robotics,[34] provides a unification of weighted averaging, conflict mechanisms, and even thresholds, while allowing for the aforementioned mechanisms of internal models of physics and sensors. Another method of sensor fusion, which may be suited to formalizing perceptual sensor fusion mechanisms, is based on Dempster-Shafer theory[35] rather than on Bayesian statistics (like weighted averaging). In Dempster-Shafer theory, evidence is represented as Shafer belief function rather than as probability density function. Evidence is combined according to Dempster's rule of combination, which has the important property of being able to generate a measure of conflict between belief functions. This can be used to determine consistency between observations and to reject unreliable sensory information.

The PET and fMRI data on visual-vestibular and visuovisual interactions in the form of activations and deactivations serve as a source of inspiration for the modeling of cortical mechanisms of orientation and motion perception (FIG. 5). The construction of a mathematical model to simulate brain function, however, requires further quantification of psychophysical variables for defined sensory stimuli and verification of basic theoretical principles, as just discussed.

ACKNOWLEDGMENT

The authors thank J. Benson for copyediting the manuscript.

REFERENCES

1. CORBETTA, M. *et al.* 1998. A common network for functional areas for attention and eye movements. Neuron **21:** 761–773.
2. DIETERICH, M. & TH. BRANDT. 2000. Brain activation studies on visual-vestibular and ocular motor interaction. Curr. Opin. Neurol. **13:** 13–18.
3. WENZEL, R. *et al.* 1996. Deactivation of human visual cortex during involuntary ocular oscillations: a PET activation study. Brain **119:** 101–110.
4. BRANDT, T. *et al.* 1998. Reciprocal inhibitory visual-vestibular interaction: visual motion stimulation deactivates the parieto-insular vestibular cortex. Brain **121:** 1749–1758.
5. DEUTSCHLÄNDER, A. *et al.* 2001. Sensory system interactions during simultaneous vestibular and visual stimulation in PET. Hum. Brain Map. In press.
6. PROBST, T., A. STRAUBE & W. BLES. 1985. Differential effects of ambivalent visual-vestibular-somatosensory stimulation on the perception of self motion. Behav. Brain Res. **16:** 71–79.
7. BENSE, S. *et al.* 2001. Multisensory cortical signal increases and decreases during vestibular galvanic stimulation (fMRI). J. Neurophysiol. **85:** 886–899.
8. BRANDT, TH. & M. DIETERICH. 1999. The vestibular cortex: its locations, functions, and disorders. Ann. N.Y. Acad. Sci. **871:** 293–312.
9. BRANDT, T., M. STRUPP & J. BENSON. 1999. You are better off running than walking with acute vestibulopathy. Lancet **354:** 746.
10. JAHN, K. *et al.* 2000. Differential effects of vestibular stimulation on walking and running. NeuroReport **11:** 1745–1748.
11. BROOKE, J.D. *et al.* 1997. Sensori-sensory afferent conditioning with leg movement: gain control in spinal reflex and ascending paths. Prog. Neurobiol. **51:** 393–421.

12. COLLINS, D.F. et al. 1998. Muscular sense is attenuated when humans move. J. Physiol. **508:** 635–643.
13. BRANDT, T. et al. 1998. Bilateral functional MRI activation of the basal ganglia and middle temporal/medial superior temporal motion-sensitive areas: optokinetic stimulation in homonymous hemianopia. Arch. Neurol. **55:** 1126–1131.
14. ZEKI, S. & S. SHIPP. 1988. The functional logic of cortical connections. Nature **335:** 311–317.
15. FINLAY, A.L. et al. 1997. Movement in the normal visual hemifield induces a percept in the 'blind' hemifield of a human hemianope. Proc. Roy. Soc. Lond. B **264:** 267–275.
16. BASELER, H.A., A.B. MORLAND & B.A. WANDELL. 1999. Topographic organization of human visual areas in the absence of input from primary cortex. J. Neurosci. **19:** 2619–2627.
17. MAUNSELL, J.H. & D.C. VAN ESSEN. 1987. Topographic organization of the middle temporal visual area in the Macaque monkey: representational biases and the relationship to callosal connections and myeloarchitectonic boundaries. J. Comp. Neurol. **266:** 535–555.
18. WELLER, R.E., J.T. WALL & J.H. KAAS. 1984. Cortical connections of the middle temporal visual area (MT) and the superior temporal cortex in owl monkeys. J. Comp. Neurol. **228:** 81–104.
19. BRANDT, TH. et al. 2000. Hemifield visual motion stimulation: an example of interhemispheric crosstalk. NeuroReport **11:** 2803–2809.
20. VAN ESSEN, D.C., W.T. NEWSOME & J.L. BIXBY. 1982. The pattern of interhemispheric connections and its relationship to extrastriate visual areas in the macaque monkey. J. Neurosci. **2:** 265–283.
21. NOWAK, L.G. et al. 1995. Structural basis of cortical synchronization. I. Three types of interhemispheric coupling. J. Neurophysiol. **74:** 2379–2400.
22. MUNK, M.H. et al. 1995. Structural basis of cortical synchronization. II. Effects of cortical lesions. J. Neurophysiol. **74:** 2401–2414.
23. BARBUR, J.L., K.H. RUDDOCK & V.A. WATERFIELD. 1980. Human visual responses in the absence of the geniculo-calcarine projection. Brain **103:** 905–928.
24. DROULEZ, J. & C. DARLOT. 1989. The geometric and dynamic implications of the coherence constraints in three-dimensional sensorimotor interactions. In Attention and Performance. M. Jeannerod, ed. :495–562. Lawrence Erlbaum. New Jersey.
25. MERGNER, T. & S. GLASAUER. 1999. A simple model of vestibular canal-otolith signal fusion. Ann. N.Y. Acad. Sci. **871:** 430–434.
26. GLASAUER, S. 1992. Interaction of semicircular canals and otoliths in the processing structure of the subjective zenith. Ann. N.Y. Acad. Sci. **656:** 847–849.
27. MERFELD, D.M. et al. 1993. A multi-dimensional model of the effect of gravity on the spatial orientation of the monkey. J. Vestib. Res. **3:** 141–161.
28. HESS, B.J.M. & D.E. ANGELAKI. 1999. Inertial processing of vestibulo-ocular signals. Ann. N.Y. Acad. Sci. **871:** 148–161.
29. MERFELD, D.M., L. ZUPAN & R.J. PETERKA. 1999. Humans use internal models to estimate gravity and linear acceleration. Nature **398:** 615–618.
30. MERGNER, T. & T. ROSEMEIER. 1998. Interaction of vestibular, somatosensory and visual signals for postural control and motion perception under terrestrial and microgravity conditions: a conceptual model. Brain Res. Rev. **28:** 118–135.
31. ZACHARIAS, G.L. & L.R. YOUNG. 1981. Influence of combined visual and vestibular cues on human perception and control of horizontal rotation. Exp. Brain Res. **41:** 159–171.
32. MERGNER, T. et al. 2000. Visual contributions to human self-motion perception during horizontal rotation. Arch. Ital. Biol. **138:** 139–166.
33. MCCOLLUM, G., C.L. SHUPERT & L.M. NASHNER. 1996. Organizing sensory information for postural control in altered sensory environments. J. Theor. Biol. **180:** 257–270.
34. STEINHAGE, A. 1999. Nonlinear attractor dynamics: a new approach to sensor fusion. In Sensor Fusion and Decentralized Control in Robotic Systems. II. Proceedings of SPIE Vol. 3839. P.S. Schenker & G.T. McKee, eds. :31–42.
35. MURPHY, R.R. 1996. Biological and cognitive foundations of intelligent sensor fusion. IEEE Trans. Sys., Man, Cybern. **26:** 42–51.

Scanpaths: The Path to Understanding Abnormal Cognitive Processing in Neurological Disease

CHRISTOPHER KENNARD

Division of Neuroscience and Psychological Medicine, Faculty of Medicine, Imperial College of Science, Technology and Medicine, Charing Cross Campus, London, W6 8RP, United Kingdom

ABSTRACT: Scanpaths, the sequence of saccades and fixations during visual search, provide the means to study a range of cognitive activities: planning, visuospatial attention, and spatial working memory. By measuring scanpaths, we have been able to identify impairment of working memory in patients with spatial hemineglect. Scanpaths during cognitively demanding tasks indicate a defect of working memory in patients with Parkinson's disease. Studies of scanpaths in patients with homonymous hemianopia have provided the means to develop a new strategy for visual rehabilitation.

KEYWORDS: scanpaths; saccades; fixations; memory; Parkinson's disease; homonymous hemianopia

INTRODUCTION

During natural behavior we move our gaze around a complex visual environment actively searching for information relevant to current motivations and goals. This gives rise to a sequence of saccades and fixations, which have been termed scanpaths. Detailed analysis of eye position during such scanpaths can provide a rich description of task performance. Studies that have analyzed scanpaths during cognitively demanding tasks have revealed that particular activities are associated with specialized gaze-shifting strategies indicative of specific cognitive processes, which could not be identified by the use of gross measures such as reaction times and error rates.[1–4]

Among the various cognitive processes required for normal visual search are planning, visuospatial attention, a system that is able to select and focus visual processing on certain locations in visual space, and spatial working memory, a system that allows an individual to hold "on-line" and manipulate spatial memory. In this review, our studies analyzing scanpaths during visual search in specific paradigms are reported in patient's with spatial hemineglect, Parkinson's disease, and homonymous hemianopia.

Address for correspondence: Christopher Kennard, Division of Neuroscience and Psychological Medicine, Faculty of Medicine, Imperial College of Science, Technology and Medicine, Charing Cross Campus, St Dunstan's Road, London, W6 8RP, United Kingdom. Voice: 44 (0)20 8846 7598; fax: 44 (0)20 8846 7715.

c.kennard@ic.ac.uk

A TRANS-SACCADIC DEFICIT OF SPATIAL WORKING MEMORY IN SPATIAL HEMINEGLECT

Spatial hemineglect refers to the difficulties of patients with unilateral brain damage, usually involving the right inferior parietal lobe, in exploring the side of space contralateral to the lesion and reporting stimuli presented in that region of space. These patients, therefore, appear to show a bias to orientate towards objects or events located to their right, while ignoring those to the left. The standard clinical bedside tests such as line bisection, target cancellation, copying, and drawing show a bias of attention to the right. The left-hand side of space is ignored, despite the fact that the subject is free to move his head and eyes and is given unlimited time to complete the task. Over the last 20–30 years many hypotheses have been proposed to account for this phenomenon.[5,6] While it is recognized that multiple components underlie the neglect syndrome, such as a lateral bias in attention towards the right and/or difficulty in disengaging attention from objects on the right side, the relative contribution of each may vary from patient to patient.

One such component that has received little attention until recently is a possible deficit of memory for visual locations across saccades during visual search. When trying to identify targets in a visual scene of multiple distractors, for example, it is necessary to retain the location of those previously identified to prevent searching the same locations repeatedly. This task therefore requires, in addition to cognitive processes such as planning and the ability to shift attention, intact spatial working memory. In a series of experiments we combined analysis of scanpaths in hemineglect patients and normal controls with the performance of a visual search paradigm with a measure of whether observers were fixating on a particular target for the first time or were judged to have examined it previously.[7]

Our tests involved variations on a search task resembling a standard cancellation test, in which patients and controls had to look for specified targets (letter Ts) embedded in randomly located distractors (letter Ls) displayed on a computer monitor. However, it differed from the bedside test as the subject was never required to leave a visible mark on the targets visited (as occurs during visual search in real life). Instead, the eye position was monitored measuring whether particular targets were fixated and re-fixated (FIG. 1). Re-fixations by themselves would not prove a deficit in retaining searched locations across saccades, because they could arise as a result of checking rather than forgetting which locations had already been searched. Subjects, therefore, had to click a response button only when they looked at what they considered was a "new" target and not one they had previously located. Re-clicking on a target indicated failure to retain its location as already searched, across intervening saccades.

By appropriate manipulations of the number and location of both targets and distractors in the examination of a 68-year-old patient, G.K., with infarction of the right inferior parietal lobe, a deficit in the trans-saccadic working memory was identified. This patient consistently neglected left targets but made saccades equally leftwards and rightwards and frequently re-fixated targets, whereas age-matched normal control subjects rarely did. The patient also showed an abnormally high re-click rate. In particular, the re-click rate increased systematically with an increase in the number of targets, and the degree of neglect was significantly correlated with the rate of re-clicking on targets across search tasks.

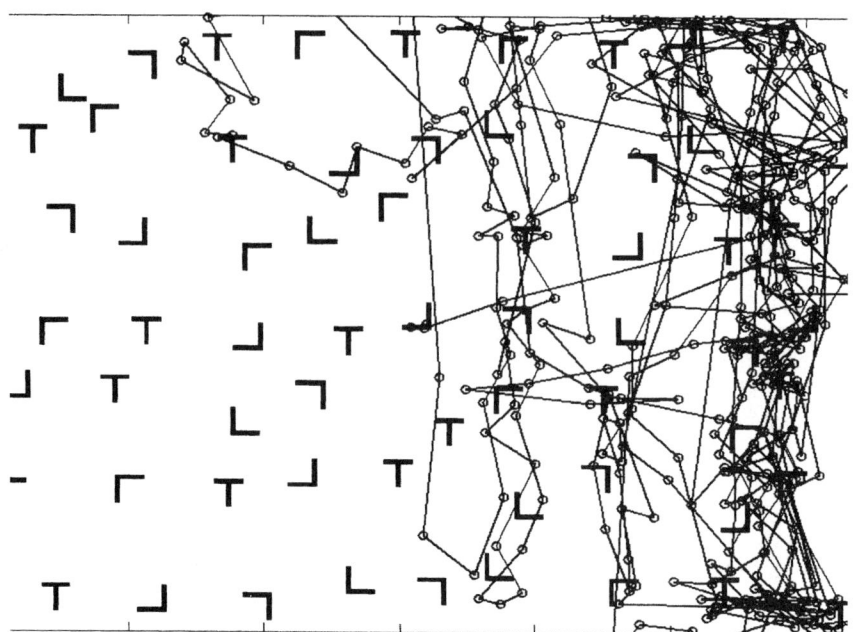

FIGURE 1. Example of typical search array (19 Ts + 44 Ls) on which are superimposed fixations (*circles*) and interpolated scanpaths (*black lines*) of a patient with visual hemineglect. Note the profound neglect of the left side of the array and the many re-fixations of stimuli on the right.

To determine the prevalence of a trans-saccadic spatial working memory deficit in spatial hemineglect, an additional 16 patients were studied.[8] The re-fixation rate was significantly higher in the neglect group (mean = 142%) than in a group of 15 age-matched normal controls (mean = 38%) ($p < 0.001$). Eleven of sixteen patients had re-fixation rates beyond the upper 95% confidence limits for normal controls. Neglect patients had a higher re-click rate (mean = 30%) than did normal controls (mean = 6%) ($p < 0.02$). Of the 11 patients who showed abnormally high re-fixation rates, 10 also had re-click rates beyond the 95% confidence interval for normal subjects.

When the location of the lesions was evaluated in this group of hemineglect patients with high re-click rates, we found that although most had lesions in the right frontal lobe, others had lesions in the inferior right parietal lobe. These studies therefore provide compelling evidence that a deficit of trans-saccadic spatial working memory may be a relatively common component of neglect and that when this is present in patients with spatial hemineglect, the lesion may be in either the right frontal or the inferior parietal lobe. It is important to emphasize that this working memory deficit is not proposed as the only cause for neglect. Rather, when combined with the well established lateral attentional bias to the right, it may lead to recursive searching of the right side with continued failure to explore left space. While the prefrontal cortex is generally considered to have an important role in the processing of spatial working memory[9,10] and lesions in the area can result in spatial hemineglect,[11] several studies indicate involvement of the inferior parietal lobe in spatial working memory in addition to attention. Electrophysiological recordings in monkey posterior parietal lobe indicate that neurons may be involved in representing visual location across saccades and in maintaining a memory trace for the location of saccadic targets across delays.[12,13] In addition, functional imaging studies in humans have revealed a network of brain areas involved in maintaining information about visual location in a variety of spatial working memory tasks including saccades to remembered locations.[14–16] This network includes damaged areas in patients with spatial hemineglect who show a deficit in spatial working memory.

STRATEGIC CONTROL OF VISUAL SEARCH IN A PLANNING TASK: EVIDENCE OF A DEFICIT OF SPATIAL WORKING MEMORY IN PARKINSON'S DISEASE

An exemplar of the insights into cognitive processing that may be revealed by analysis of scanpaths during the execution of problem solving and "executive" function is described using the Tower-of-London (TOL) task in normal subjects and in patients with Parkinson's disease (PD). The TOL task was developed to test for the subtle deficits in behavior that are observed following frontal lobe damage in humans.[17] The test involves the presentation of two different arrangements of colored balls. The subject's task is to rearrange the first array of balls (referred to as the Workspace) so that it matches the second array of balls (the Goalspace) using the minimum number of moves possible. The positioning of the balls is constrained to the location of three pegs or pockets in each half of the display. Because of this, complex problems demand that the sequence of moves is carefully planned in advance before attempting the first move. Failure to engage in advanced planning of the sequence will result in initial moves blocking subsequent ball moves. Owen et al.[18] have described a "one-touch" version of the TOL task in which the incentive for individuals to plan solutions in advance is enhanced further. In this variant of the task, subjects are required to inspect the problems visually and then make a single motor response to indicate how many moves would be required for the ideal solution. In this way, the one-touch task isolates the cognitive planning component of the test by requiring the internal planning of solutions without actually executing the appropriate moves.

In normal subjects the time taken to plan successful solutions as well as the number of errors made in executing the final sequence increases with the minimum number of moves to solve the TOL task. Many neuropsychological and neuroimaging studies have identified the brain areas involved in performing the TOL task, which include the frontal lobes[17,19,20] and basal ganglia.[18] We chose the one-touch TOL task combined with analysis of the scanpaths undertaken during its execution in a group of normal subjects[21] and a group of patients with PD.[22] It was of particular interest to determine if there are limiting factors restricting problem-solving performance, such as the processing capacity of the working memory system, or difficulties in selecting between competing behavioral goals, and if subjects inspect the

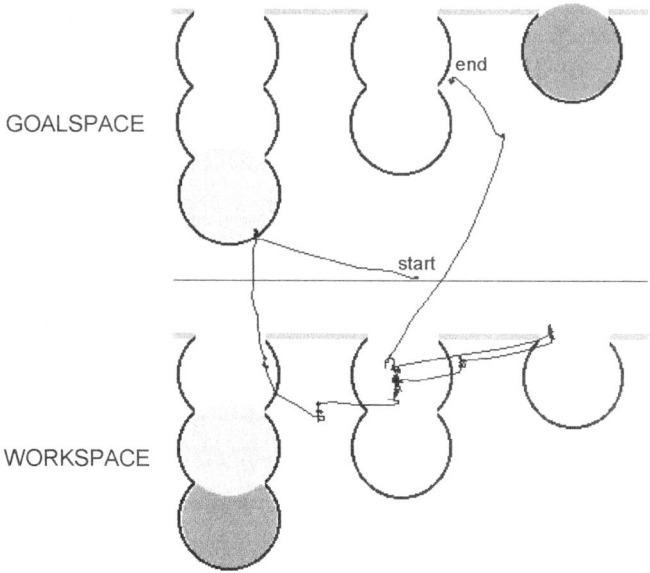

FIGURE 2. Example of scanpaths superimposed over relevant Tower-of-London 3-move display. Subject had to plan moves required to rearrange lower balls in the Workspace to match upper balls in the Goalspace. Fixations tended to land on Goalspace towards the beginning and end of the trial and on the Workspace in the middle of the trial.

problem in a strategic manner, that is, if specific components of gaze relate to acquisition of stimulus configuration and elaboration of problem solutions.

In a series of experiments we found that the duration of the fixation time spent in the Workspace increased with the task complexity, that is, the number of moves required to rearrange the balls, while fixation time in the Goalspace remained constant.[21] Scanpaths revealed gaze-shifting strategies, which consisted of three phases in the performance of this planning task. In an initial assessment phase, fixation was mainly in the Goalspace and the problem was assessed (Phase 1). Next, fixations were directed to the Workspace during a solution elaboration phase during which the different operations were rehearsed and assessed (Phase 2). Finally, gaze was directed again to the Goalspace in a final verification phase in which internal representation of the planned solution was compared with the goal configuration (Phase 3) (FIG. 2). This pattern was found regardless of whether the subjects solved problems by rearranging balls in the lower or upper visual fields, demonstrating that this strategy correlates with discrete phases in problem solving. A second experiment showed that efficient planners direct their gaze selectively towards the problem critical balls in the Workspace. By contrast, individuals who make errors spend more time looking at irrelevant items and are strongly influenced by the movement strategy needed to solve the preceding problem.

Study of a group of patients with idiopathic PD revealed interesting alterations in the scanpath just described.[22] Whereas normal subjects spent proportionally more time fixating in the Workspace region as the problem complexity increased, patients with PD spent equal amounts of time fixating in the Workspace and Goalspace. In addition, patients with PD made an increased number of errors, which correlated with disease severity, although they did not show any increase in total response time. We had originally reasoned that eye movement analysis would allow us to discriminate between two potential explanations for abnormal problem solving in PD: a general slowing of mental imagery (bradyphrenia) and defective encoding/maintenance of current goals on-line (goal-processing deficit). The data clearly favor the goal-processing hypothesis, because failure to find a dissociation between Goalspace and Workspace fixation times suggested that patients with PD have difficulty encoding and/or maintaining task goals, leading to increased response errors. The simplest explanation of these findings is that the trans-saccadic working memory for the color and location of balls is impaired in PD. Put simply, patients keep forgetting the arrangement of balls every time they look away. Therefore, they spend proportionately less time fixating at the Workspace and more time fixating at the Goalspace. Alternatively, it is also possible that normal subjects use covert attention to continuously monitor the goals in peripheral vision, while fixating on the Workspace. If this were the case, then the present results imply an impairment in the control of covert attention in PD. Although we cannot discriminate between these two accounts using the results of our experiments, recent studies of a closely related block-copying task indicate that covert attention is not used to monitor goals in this manner. Instead, the color and location of relevant items are acquired during separate fixations and are held in working memory across gaze shifts between different regions of the display.[3]

VISUAL SEARCH IN HOMONYMOUS HEMIANOPIA DUE TO LESIONS IN THE POSTERIOR VISUAL PATHWAYS

Patients with lesions of the posterior visual pathways resulting in a visual hemifield defect, called homonymous hemianopia, have considerable disabilities. They often have particular difficulties with reading and visual exploration, often failing to note relevant objects or objects on their affected side, resulting in collisions. Since 30% of patients presenting with stroke have a homonymous hemianopia that usually persists, this is a frequently encountered problem, yet most patients are still not offered any systematic form of therapy.

Patients with homonymous hemianopia often intrinsically try to compensate for their visual loss. They tend to paradoxically concentrate their gaze towards their blind hemifield, because this brings more of the visual scene into their seeing hemifield.[23,24] Similarly, they adopt compensatory strategies when performing repetitive saccades to predictable targets.[25] These patients also show difficulties scanning a visual scene of targets embedded among distractors. They repeat saccades and fixations to the same target, resulting in longer search times and longer unsystematic scanpaths.[26,27]

We studied eight patients with homonymous hemianopia without neglect to determine the effect of homonymous hemianopia on scanpaths generated when viewing naturalistic visual scenes.[28] Subjects viewed 22 images of real scenes for 3 seconds, initially in a spatially filtered form in which much of the semantic content had been removed and then in the unfiltered, original form. Hemianopic patients fixated on different spatial positions than did normal controls, made more fixations of shorter duration, and spent a greater proportion of their total fixation time in the area corresponding to their blind hemifield. Patients made more saccades into their blind hemifield and had longer scanpaths than did controls. Filtering out the high spatial content of the visual scene accentuated these differences. A correlation was noted between the degree of difference between patients and controls of their scanpaths and the duration of the hemianopia, reflecting the evolution of a spontaneous compensatory eye movement strategy.

Several groups have tested the efficacy of various saccadic training techniques in the rehabilitation of patients with homonymous hemianopia on the basis that training serves to systematically reinforce such compensatory oculomotor strategies, thereby enlarging the field of search.[29–32] Varying degrees of success have been reported, but most techniques require the use of complex equipment and have been hospital based. We recently devised a novel rehabilitation tool using portable, low-cost equipment that is easy to deploy with patients.[33] The method used a visual search paradigm of multiple presentations of a randomly positioned target among distractor elements that differed by a single feature (size and orientation), displayed on a 21-inch television monitor. We assessed response time and error rates during visual search as well as tests representing activities of daily living in a group of 29 hemianopic patients before and after 20 daily sessions lasting 40 minutes over 1 month. Patients showed a significantly shorter response time confined to the training period with no significant practice effect or spontaneous improvement. They also performed activities of daily living tasks significantly faster after training and reported subjective improvements. Importantly, the improvements were maintained beyond the treatment phase of the study.

CONCLUSIONS

Analysis of scanpaths in visual search and planning paradigms can provide useful insights into cognitive processing and can increase our understanding of the deficits that occur in neurological conditions due to brain damage.

ACKNOWLEDGMENTS

We gratefully acknowledge the support of the Stoke Association and the Wellcome Trust in funding this research.

REFERENCES

1. BUSWELL, G.T. 1935. How People Look at Pictures. University of Chicago Press. Chicago, IL.
2. YARBUS, A. 1967. Movements of the Eyes. Plenum Press. New York.
3. HAYHOE, M.H. *et al.* 1997. Task constraints in visual working memory. Vision Res. **38:** 125–137.
4. LAND, M.F. & S. FURNEAX. 1997. The knowledge base of the oculomotor system. Phil. Trans. Roy. Soc. Lond. Series B **352:** 1231–1239.
5. VALLAR, G. 1998. Spatial hemineglect in humans. Trends Cognit. Sci. **2:** 87–97.

6. MESULAM, M.-M. 1999. Spatial attention and neglect: parietal, frontal and cingulate contributions to the mental representation and attentional targeting of salient extrapersonal events. Phil. Trans. Roy. Soc. Lond. Series B **354:** 1325–1346.
7. HUSAIN, M. et al. 2001. Impaired spatial working memory across saccades contributes to abnormal search in parietal neglect. Brain **124:** 941–952.
8. MANNAN, S. et al. 2001 Prevalence of trans-saccadic spatial working memory deficit in hemispatial neglect. Submitted.
9. GOLDMAN-RAKIC, P.S. 1996. The prefrontal landscape: implications of functional architecture for understanding human mentation and the central executive. Phil. Trans. Roy. Soc. Lond. Series B **351:** 1445–1453.
10. DUNCAN, J. & A.M. OWEN. 2000. Common regions of the human frontal lobe recruited by diverse cognitive demands. Trends Neurosci. **23:** 473–483.
11. HUSAIN, M. & C. KENNARD. 1998. Personal communication.
12. MAZZONI, P. et al. 1996. Spatially tuned auditory responses in area LIP of macaques performing delayed memory saccades to acoustic targets. J. Neurophysiol. **75:** 1233–1241.
13. COLBY, C.L. & M.E. GOLDBERG. 1999. Space and attention in parietal cortex. Ann. Rev. Neurosci. **22:** 319–349.
14. OWEN, A.M. et al. 1966. Evidence for a two-stage model of spatial working memory processing within the lateral frontal cortex. A positron emission tomography study. Cereb. Cortex **6:** 31–38.
15. LABAR, K.S. 1999. Neuroanatomic overlap of working memory and spatial attention networks: a functional MRI comparison within subjects. Neuroimage **10:** 695–704.
16. ANDERSON, T.J. et al. 1994. Cortical control of saccades and fixation in man. Brain **117:** 1073–1084.
17. SHALLICE, T. 1982. Specific impairments of planning. Phil. Trans. Roy. Soc. Lond. Series B **298:** 199–209.
18. OWEN, A.M. et al. 1995. Dopamine-dependent fronto-striatal planning deficits in early Parkinson's disease. Neuropsychology **9:** 126–140.
19. OWEN, A.M. et al. 1990. Planning and spatial working memory deficits following frontal lobe lesions in man. Neuropsychologia **28:** 1021–1034.
20. OWEN, A.M. et al. 1996. Planning and spatial working memory examined with positron emission tomography (PET). Eur. J. Neurosci. **8:** 353–364.
21. HODGSON, T.L. et al. 2000. The strategic control of gaze direction in the Tower of London task. J. Cognit. Neurosci. **12:** 894–907.
22. HODGSON, T.L. et al. 2002. Abnormal gaze strategies during problem solving in Parkinson's disease. Neuropsychologia **40:** 411–422.
23. ISHIAI, S. et al. 1987. Eye-fixation patterns in homonymous hemianopia and unilateral spatial neglect. Neuropsychologia **25:** 675–679.
24. GASSEL, M.M. & D. WILLIAMS. 1963. Visual function in patients with homonymous hemianopia. Part II. Oculomotor mechanisms. Brain **86:** 1–36.
25. MEIENBERG, O. et al. 1981. Saccadic eye movement strategies in patients with homonymous hemianopia. Ann. Neurol. **9:** 537–544.
26. CHEDRU, F. et al. 1973. Visual searching in normal and brain-damaged subjects (contribution to the study of unilateral inattention). Cortex **9:** 94–111.
27. ZIHL, J. 1995. Visual scanning behaviour in patients with homonymous hemianopia. Neuropsychologia **33:** 287–303.
28. PAMBAKIAN, A.L.M. et al. 2000. Scanning the visual world: a study of patients with homonymous hemianopia. J. Neurol. Neurosurg. Psychiatry **69:** 751–759.
29. ZIHL, J. 2000. Rehabilitation of Visual Disorders after Brain Injury. Psychology Press. Hove, UK.
30. ZIHL, J. 1995. Eye movement patterns in hemianopic dyslexia. Brain **118:** 891–912.
31. KERKHOFF, G. et al. 1992. Rehabilitation of homonymous scotomata in patients with postgeniculate damage of the visual system: saccadic compensation training. Restor. Neurol. Neurosci. **4:** 245–254.
32. KARSTEN, E. et al. 1998. Computor based training for the treatment of partial blindness. Nature Med. **4:** 1083–1087.
33. PAMBAKIAN, A.L.M. 2000. Saccadic visual search training: a successful treatment for patients with homonymous hemianopia. Submitted.

Antisaccades and Task Switching

Studies of Control Processes in Saccadic Function in Normal Subjects and Schizophrenic Patients

JASON J. S. BARTON,[a,b,c] MARIYA V. CHERKASOVA,[a] KRISTEN LINDGREN,[a] DONALD C. GOFF,[d] JAMES M. INTRILIGATOR,[a] AND DARA S. MANOACH[a,d]

Departments of [a]Neurology and [b]Ophthalmology, Beth Israel Deaconess Medical Center, Harvard Medical School, Boston, Massachusetts, USA

[c]*Department of Bioengineering, Boston University, Boston, Massachusetts, USA*

[d]*Department of Psychiatry, Massachusetts General Hospital, Boston, Massachusetts, USA*

>ABSTRACT: Executive functions allow us to respond flexibly rather than stereotypically to the environment. We examined two such functions, task switching and inhibition in the antisaccade paradigm, in two studies. One study involved 18 normal subjects; the other, 21 schizophrenic patients and 16 age-matched controls. Subjects performed blocks of randomly mixed prosaccades and antisaccades. Repeated trials were preceded by the same type of trial (i.e., an antisaccade following an antisaccade), and switched trials were preceded by a trial of the opposite type. We measured accuracy rate and latency as indices of processing costs. Whereas schizophrenic patients had a threefold increase in error rate for antisaccades compared to normals, the effect of task switching on their accuracy did not differ from that in normal subjects. Moreover, the accuracy rate of trials combining antisaccade and task switching was equivalent to a multiplication of the accuracy rates from trials in which each was done alone. Schizophrenic latencies were disproportionately increased for antisaccades, but again they were no different from normal subjects in the effect of task switching. In both groups the effect of task switching on antisaccades was a paradoxical latency reduction. We conclude that the executive dysfunction in schizophrenia is not generalized but selective, sparing task switching from exogenous cues, in which the switch is limited to a stimulus-response remapping. The accuracy data in both groups support independence of antisaccade and task-switching functions. The paradoxical task-switching benefit in antisaccadic latency effects challenges current models of task switching. It suggests either carryover inhibition by antisaccadic performance in the prior trial or facilitation of antisaccades by simultaneous performance of other cognitive operations.
>
>KEYWORDS: antisaccades; task switching; executive; attention; latency; schizophrenia

Address for correspondence: Jason J.S. Barton, Neurology, KS-454, Beth Israel Deaconess Medical Center, 330 Brookline Avenue, Boston, MA 02215. Voice: 617-667-1243; fax: 617-975-5322.

jbarton@caregroup.harvard.edu

INTRODUCTION

Antisaccades[1] are an example of controlled processing in which a habitual act (looking towards a suddenly appearing target with a prosaccade) must be overruled by a highly novel response (looking away from the target). The antisaccade/prosaccade relation is an example of a response pair with a dominance asymmetry. Dominance arises when one response gains an advantage over the other through prior experience, intraexperimental practice, or stimulus-response compatibility.[2] Thus, in the Stroop test,[3] reading the name of a color is dominant over stating the color of the ink when the two conflict. Dominance asymmetries vary along a spectrum.[4] The antisaccade/prosaccade pairing lies on the extreme of this continuum.

Dominance asymmetries have gained interest in studies of task switching. Task switching, another example of controlled processing, usually incurs added costs in prolonged latencies and increased error compared to task repetition (e.g., Refs. 5–7). It is not clear what generates these costs. Some studies report that the increase in latency induced by switching from a dominant to a nondominant task is less than the increase generated by switching in the reverse direction.[5] This has been interpreted as evidence of either negative stimulus-response priming[8] or carryover of inhibition of the current response from the prior trial, when it was an inappropriate response.[5] Thus, a nondominant response requires strong inhibition of the dominant response, which must be overcome to switch back to a dominant response in the next trial. In the reverse direction, a dominant trial does not need much inhibition of the nondominant alternative, and hence little inhibition carries over in the switch to a nondominant trial. The resulting asymmetry of switch costs with dominance asymmetry is hard to account for with other explanations of task-switching costs.[7]

Others note, however, that not all response pairings with dominance asymmetries engender asymmetric switch costs.[2] Monsell et al.[2] hypothesized that asymmetric switch costs may only occur with pairings that are highly asymmetric in dominance. Other factors may play a role also. Task switching can be a complex of many changing cognitive processes. Switches might require a shift in the stimulus dimension attended (word versus ink color in the Stroop test), the stimulus location attended, the classification of the stimulus needed, the response mode to use (verbal versus manual), and the stimulus-response mappings made, among others. The contributions of each of these factors to switch costs are relatively unknown.

The antisaccade/prosaccade pairing has some advantageous features for exploring these issues in task switching. First, the dominance asymmetry is extreme: most naive subjects have never performed an antisaccade, even though they perform prosaccades many times a minute while awake, every day of their lives. Second, the switch between prosaccades and antisaccades minimizes the number of changing task features. The stimulus for both prosaccades and antisaccades is a small peripheral light, with the same locations, the same relevant attribute (spatial location), and the same classification (right or left). Both tasks require the same response mode (saccade) with only two possible values (right or left). The key difference remaining is the stimulus-response mapping, which is reversed for antisaccades. Hence, if asymmetries between switching to antisaccades and switching to prosaccades are found, this would be strong evidence that the carryover of inhibitory influences from the prior trial are generated at the level of stimulus-response mapping.

Another issue of note in the interaction of task dominance and task switching is that of independence. Models of task switching tend to treat the settings from current and prior trials as independent effects.[6] If these processes are indeed independent, the accuracy cost of a response that requires both functions (the switched antisaccade) should equal the product of the accuracy rates of each function in isolation (a switched prosaccade and a repeated antisaccade).[9] The possibility that these are not independent is raised by the fact that damage to similar prefrontal areas impairs both task switching[10] and antisaccade performance.[11]

We explored these issues in both a normal population and a patient group to determine if pathologic effects on antisaccade and task-switching costs were correlated or independent. We chose to study patients with schizophrenia. These patients consistently show deficits on nondominant tasks such as the Stroop test and the antisaccade task[12] and have dysfunctional task switching based on instruments such as the Wisconsin card sorting test.[13,14]

METHODS

Participants

Normal Study. Eighteen subjects (6 male) participated, with ages ranging from 13 to 54 years (mean 30.8 years, SD = 9.5). None had previously performed an antisaccadic task.

Schizophrenic Study. This included 21 outpatients maintained on stable doses of antipsychotic drugs for at least 6 weeks. Diagnoses were confirmed with the Structured Clinical Interviews for DSM-IV.[15] Clinical status was characterized with the Positive and Negative Syndrome Scale (PANSS)[16] and the Brief Psychiatric Rating Scale (BPRS).[17] Sixteen normal subjects matched for age, sex, handedness, and parental socioeconomic status[18] served as the controls.

Participants also completed two manual tests of sustained attention on a computer, the Vigil Continuous Performance Test (The Psychological Corporation, Harcourt Brace & Company, 1998) and an abbreviated version of the California Computerized Assessment Package (CalCAP).[19] Twenty of the 21 schizophrenic patients also completed a computerized version of the Wisconsin Card Sorting test (WCST, CyberMetrics Testing Services) with these results classified by published age and education-matched normative data.[20]

Apparatus and Eye Movement Protocol

We recorded eye movements with a magnetic search coil technique (Crist Instruments, Bethesda, MD). Displays were generated by a Power Macintosh 9600/233, using programs written in C++ on the Vision Shell programming platform (www.kagi.com/visionshell) and back-projected with an Eiki LC-7000U projector. Eye position was digitized at 500 samples/s and a five-point central difference algorithm[21] derived velocity from eye position.

The initial display had a dark background with a white, 1-degree fixation ring at the center (FIG. 1). The fixation ring was flanked by two 0.7-degree white dots at right and left 20 degrees. Trials started when the subject's eye was within 3 degrees

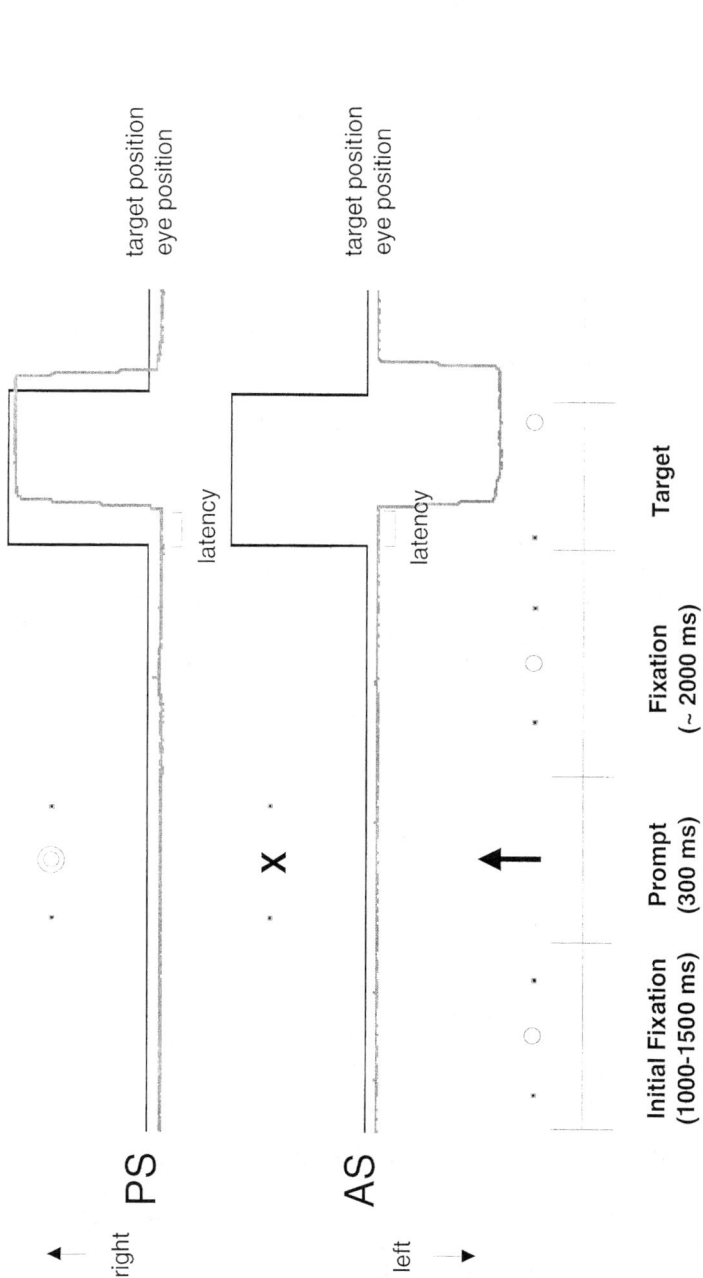

FIGURE 1. Trial illustration. Progress over time is from *left to right*. *Top lines* show horizontal position traces of targets (*black smooth lines*) and eyes (*grey irregular lines*) for a correct prosaccade (PS, *top*) and antisaccade (AS, *below*). Rightward motion is shown as up, by convention. *Bottom diagrams* show what the screen shows at each interval. The trial begins with a fixation period, with the eyes and target (ring) at zero position, or mid-screen. Two small dots mark the possible right and left locations of the target at all times. At the prompt, different screens are shown for prosaccades and antisaccades. The former are cued by a yellow double ring, the latter by a blue cross. The fixation screen then returns, followed by the appearance of the target, which triggers an eye movement response, either toward (prosaccade) or away from (antisaccade) the target. The trial is terminated when the eye enters a zone surrounding the desired eye location.

TABLE 1. Definition of relative latency effects

	Mixed-task			Residual switch costs
	Repeat	Switch		
Prosaccade (PS)	A	B	for PS:	B − A
Antisaccade (AS)	C	D	for AS:	D − C
Antisaccade (AS) costs:	C − A	D − B		
	for repeat	for switch		

of center. After 1–1.5 seconds, the fixation point was replaced by one of two prompts, a yellow "O" 4.5 degrees in diameter for prosaccade trials or a blue "X" spanning 4.5 degrees for antisaccade trials. Prompts were replaced after 300 ms by the white fixation ring. After a mean interval of 2 seconds the ring target shifted to one of the two peripheral dots.

Single-task blocks had 26 trials, either all prosaccades or all antisaccades. Mixed-task blocks had 52 trials, a random mix of prosaccades and antisaccades. Each block was repeated four times, generating 104 trials of each type. In the mixed-task blocks, about half required similar (repeated) and half required different (switched) responses from the previous trial. Blocks were given in a counterbalanced order to militate against the effects of learning and fatigue. In total there were 12 blocks between which short rests were provided. All subjects performed a practice session of 20 trials of each of the three different blocks.

Analyses

Trials from mixed-task blocks could be either "repeated trials," preceded by a trial requesting the same response (e.g., antisaccade preceded by an antisaccade), or "switched trials," preceded by a trial requesting a different response (e.g., an antisaccade preceded by a prosaccade). Consequently, there were three *conditions* — blocked (from single-task blocks), repeated, and switched — for both saccadic *tasks*, prosaccades and antisaccades, yielding six different saccade groups.

The first trial of each block was eliminated from analysis. Accuracy was calculated for each subject on each of the six saccade groups. Means and standard deviations for latencies of correct trials were calculated for each subject.

This analysis focuses on the comparison of switched and repeated responses, to identify the "residual switch costs" for both prosaccades and antisaccades ("residual" because this reflects the cost that cannot be eliminated by advance preparation during the 2-second period between the prompt and the stimulus[7]). Cost is identified as the subtraction of repeated from switched trial results. A similar subtraction of prosaccade latencies from antisaccade latencies within each condition yields the estimate of antisaccade latency costs (TABLE 1). *A priori* paired *t* tests were used for specifically identified costs in the normal study. In the schizophrenia study, ANOVA was used to compare effects between the two subject groups.

RESULTS

Normal Study

Task switching lowered the accuracy of prosaccades from 98.7 to 91.9% (paired t test, $p < 0.002$). Switching reduced antisaccade accuracy from 90.2 to 84.3% (paired t test, $p < 0.01$) (FIG. 2A).

A correct response on a switched antisaccade (ASs) trial requires both a correctly performed task switch and a correctly performed antisaccade. If these two functions are independent, the proportion of correct ASs responses should be equivalent to the proportion correct for antisaccades without switching (repeated antisaccades, ASr) multiplied by the proportion correct for task switching without antisaccades (switched prosaccades, PSs).[9] Thus:

$$ASs = ASr \cdot PSs \qquad (1)$$

A paired t test comparing ASs to ASr•PSs showed no significant difference ($p = 0.76$). A more stringent test across individual subjects used an error rate linear regression of ASs versus ASr • PSs. This yielded a significant correlation ($r = 0.62$) with a slope of 0.65 and an intercept of 6.2 (FIG. 2B), not differing significantly from a slope of 1 and an intercept of zero ($p = 0.14$). These results are consistent with the hypothesis that these are independent effects.

Task switching increased the latency of prosaccades by 14 ms (SD = 22, $p < 0.02$). However, antisaccades showed the reverse relation: switching reduced latencies by 16 ms (SD = 20, $p < 0.004$). Thus, rather than a switch cost, there was a switch benefit for antisaccades. This paradoxical reduction occurred in 14 of 18 subjects (FIG. 3B). The result was to reduce the antisaccade cost from 64 ms (SD = 32) in repeated trials to only 34 ms (SD = 35) in switched trials (FIG. 3A).

How could such an unexpected reduction arise? We first considered whether it reflected a speed/accuracy trade-off. Certain participants may have been primed to make more rapid responses to the antisaccade prompt when it followed prosaccade trials than when it followed antisaccade trials. If so, latency and accuracy switch effects for antisaccades should be positively correlated; this was not found ($r = 0.14$). Another possibility is that rather than a switch cost from the prior trial, there was an "antisaccade cost," that is, that an antisaccade in the prior trial increased the latency of the next response, whether prosaccade or antisaccade. If so, the switch effect on antisaccades should be negatively correlated with the switch effect on prosaccades; again, this was not found ($r = 0.27$) (FIG. 3B). Rather, the only correlations noted were those from comparisons with the manual reaction time measures of attention from the VIGIL and CalCAP tests. The paradoxical switch effect for antisaccades (but not prosaccades) significantly correlated with three separate measures on these tests; the shorter the reaction times (i.e., the more attentive the subject), the smaller the paradoxical task-switch effect on antisaccadic latency.

Schizophrenic Study

Antisaccades were significantly less accurate than prosaccades (task main effect: $F(1,35) = 58.49$, $p < 0.001$). There was a significant group-by-task interaction ($F(1,35) = 11.06$, $p = 0.002$), with schizophrenic subjects similar to normal subjects

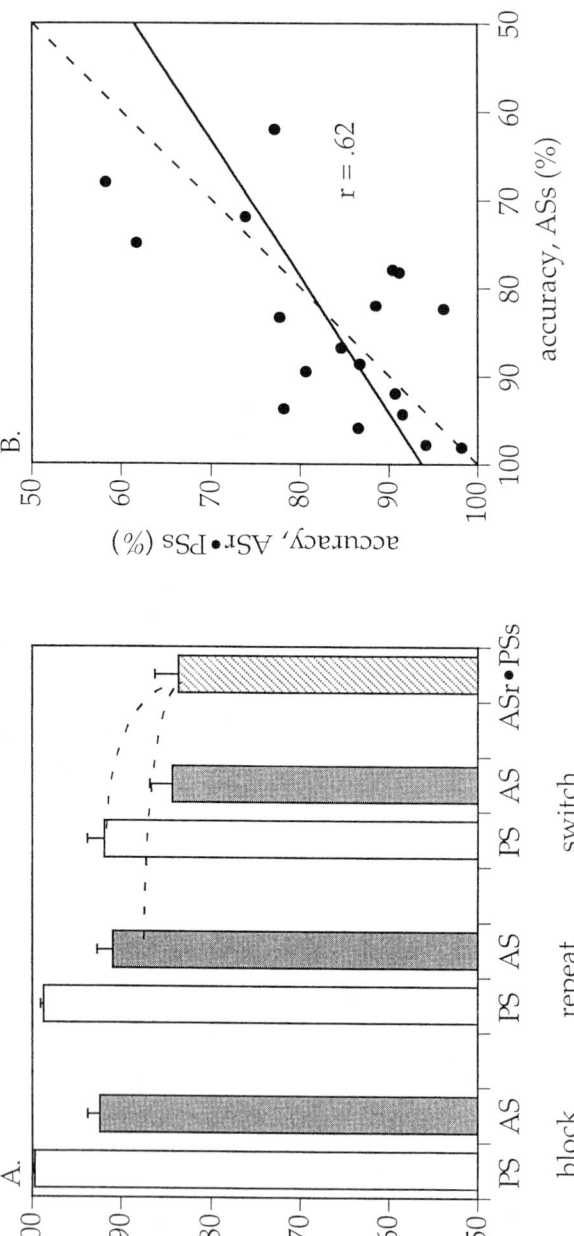

FIGURE 2. Error costs, normal study. (**A**) Mean accuracy for prosaccades (PS) and antisaccades (AS) under the three different conditions. "Block" data are from single-task blocks; "repeat" and "switched" data from mixed-task blocks; repeated trials are those with the same type of response requested in the prior trial (e.g., antisaccades preceded by an antisaccade), and switched trials are those with the other response requested in the prior trial. The "ASr • PSs" column is the mean of the product of the accuracy rates of switched PS and repeated AS. If task switching and antisaccades are independent, this should equal the switched AS cost in the adjacent column. Error bars are 1 standard error. (**B**) Linear regression of the switched AS cost (ASs) with the product ASr • PSs across all subjects, indicated as a *solid line*. Independence predicts a line with a slope of 1 and an intercept of zero (*dashed line*).

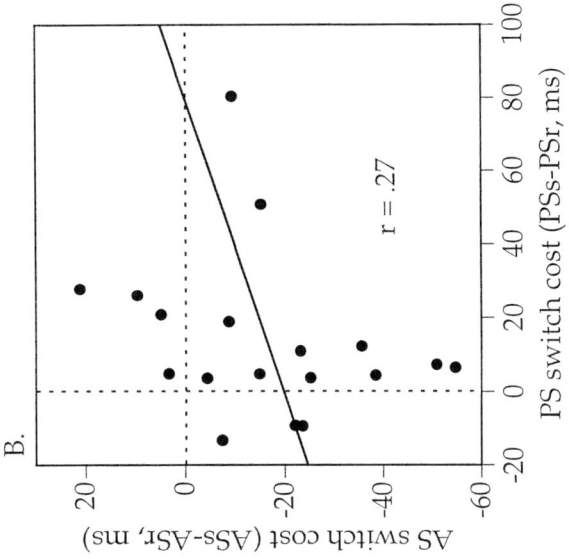
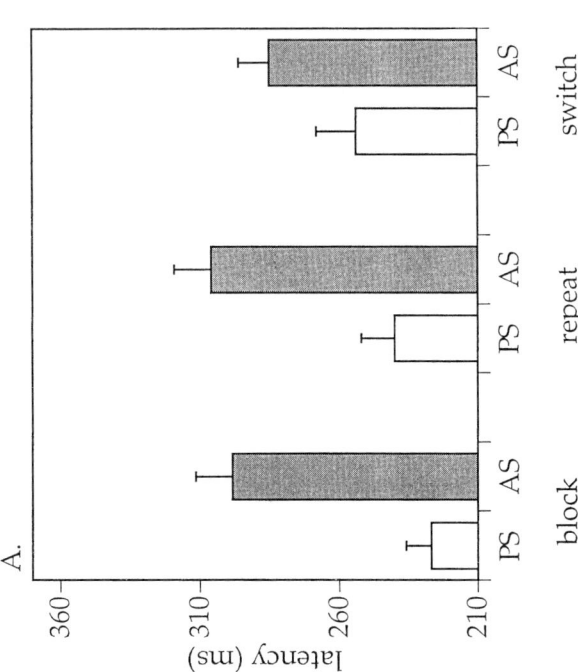

FIGURE 3. Latency costs, normal study. **(A)** Mean latencies for prosaccades (PS) and antisaccades (AS) under the three different conditions. Error bars indicate 1 standard error. Note that the mean latency of switched antisaccades is paradoxically shorter than that of repeated antisaccades. **(B)** Correlation of the switch cost for task switching (mean difference of switched minus repeated trials) for prosaccades versus antisaccades. No significant relation is demonstrated.

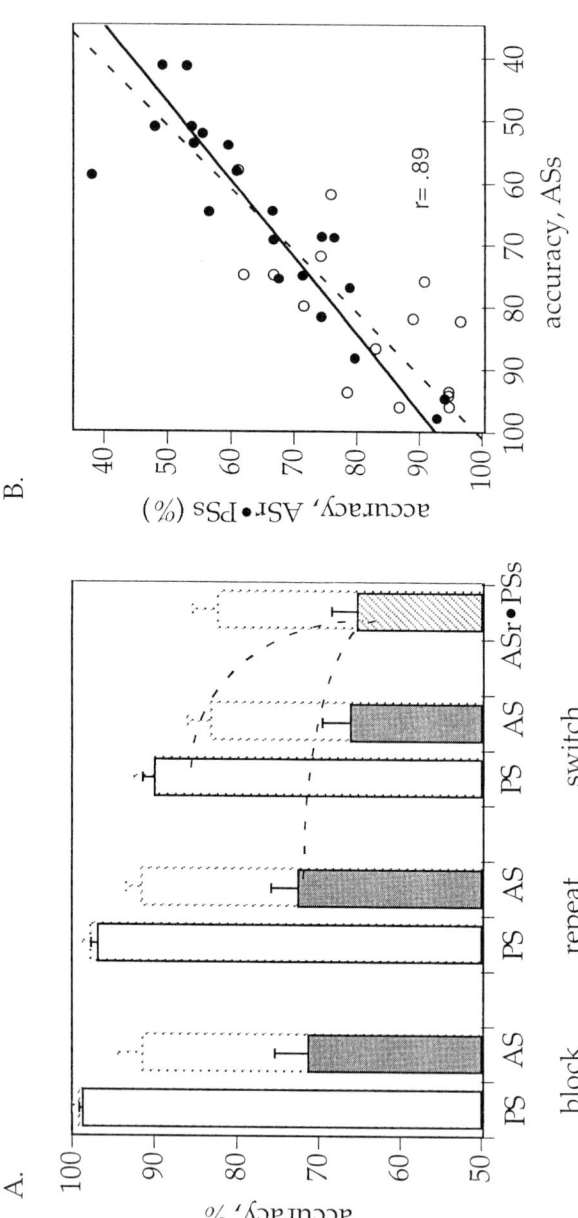

FIGURE 4. Error costs, schizophrenic patients. (**A**) Mean accuracy for prosaccades (PS) and antisaccades (AS) under the three different conditions. Conventions as in FIGURE 2. Dotted bars show data for normal age-matched controls. Error bars are 1 standard error. There is a significant reduction of accuracy for antisaccades, but the effect of switching on either prosaccades or antisaccades is no different from that in normal subjects (FIG. 2). (**B**) Linear regression of the switched AS cost (ASs) with the product ASr • PSs across all subjects (schizophrenia, *black circles*; control, *white circles*), indicated as a *solid line*. Independence predicts a line with slope of one and intercept of zero (*dashed line*). The schizophrenic data conform closely to this relationship.

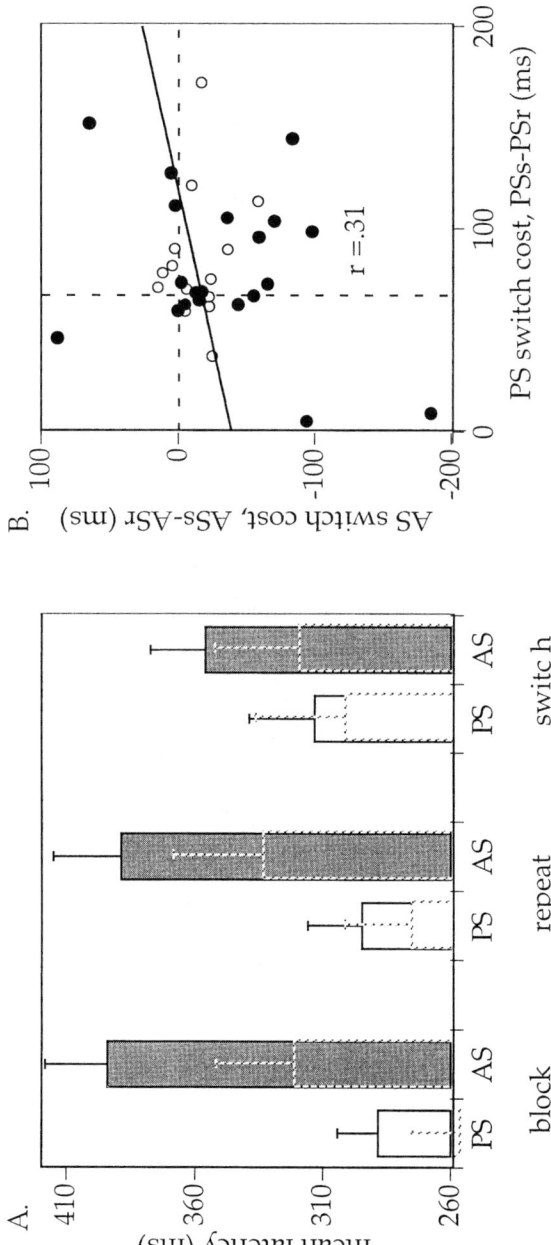

FIGURE 5. Latency costs, schizophrenic patients. **(A)** Mean latencies for prosaccades (PS) and antisaccades (AS) under the three different conditions. Dotted bars show data for normal age-matched controls. Error bars indicate 1 standard error. Compared to those in normal subjects, the antisaccade costs are increased, but the task-switching costs are not. Again, the mean latency of switched antisaccades is paradoxically shorter than that of repeated antisaccades. **(B)** Correlation of the switch cost for task switching (mean difference of switched minus repeated trials) for prosaccades versus antisaccades. Schizophrenia, *black circles*; control, *white circles*. No significant relation is demonstrated.

in prosaccade accuracy (t(35) = 0.59, p = 0.56), but less accurate on antisaccades (t(35) = 4.57, p ≤0.0001). Switched trials were less accurate than repeated trials (switch main effect: $F(1,35)$ = 34.81, p <0.0001), but there were no significant interactions with group or task (FIG. 4A). Whereas in the normal group the effects on accuracy of antisaccades and task switching were approximately equal (t(15) = 0.47, p = 0.65), the accuracy costs for schizophrenia were much greater for antisaccades than for task switching (t(20) = 4.88, p < 0.0001).

The tests for independence of switching and antisaccade function (ASs = ASr • PSs) again showed that ASs did not differ from ASr • PSs (t(20) = 0.79, p = 0.44) and that there was a strong correlation of ASs with ASr • PSs, with r of 0.89, ($F(1,19)$ = 33.04, p <0.0001), a slope of 0.81, and an intercept of 7.4, again not significantly different from a slope of 1 and an intercept of zero (FIG. 4B).

The latency data (FIG. 5) showed that antisaccades were more delayed than prosaccades (task main effect: $F(1,35)$ = 170.53, p <0.0001). There was a group-by-task interaction ($F(1,35)$ = 9.25, p = 0.002); although schizophrenic patients had longer latencies than normal subjects on both tasks (prosaccade t(35) = 3.33, p = 0.0009; antisaccade t(35) = 7.09, p ≪.0001), they were much slower to initiate antisaccades. Switching affected the latency of prosaccades and antisaccades differently (switch-by-task interaction: $F(1,35)$ = 22.08, p <0.0001). Switching prolonged prosaccade latency (t(35) = 3.52, p = 0.0004) but reduced antisaccade latency (t(35) = 3.15, p = 0.002). Group did not interact with switch ($F(1,35)$ = 1.28, p = 0.26) or with switch by task ($F(1,35)$ = 1.58, p = 0.21). Thus, schizophrenic patients had similar task-switching effects to those of normal subjects, and the paradoxical task-switching reduction for antisaccade latency was reproduced, being present in 15 of 21 patients. Again, an explanation based on a speed–accuracy trade-off was not supported by correlation analyses of accuracy and latency switch effects in either group (normal: r = –0.22, p = 0.41; schizophrenia: r = 0.05, p = 0.82).

Of great interest was the comparison in our schizophrenic patients of task-switching costs with performance on the WCST, a standard clinical instrument purported to measure switching behavior. The group means were in the mildly impaired range for total errors (μ = 82 ± 16) and perseverative errors (μ = 83 ± 18). However, there was no correlation between either total or perseverative WCST errors and our task-switching costs in accuracy or latency. Even schizophrenic patients with abnormal WCST performance could have normal task-switching costs.

DISCUSSION

Our estimates of the antisaccade effects in normal subjects accord with prior results, particularly those of the largest study to date (168 subjects), which documented latency costs of 50–80 ms and accuracy rates of 90%.[22] The finding of much reduced antisaccade accuracy and increased latency costs in schizophrenia is also consistent with prior reports[23–25] (although some studies did not find increased latency costs [26,27]).

We found that the error rates of task switching to prosaccades and of antisaccade performance without task switching (i.e., repeated antisaccades) in normal subjects were similar, about 9%. This was not true in schizophrenic patients, where the antisaccade error rate tripled that of normal subjects but task switching to prosaccades was as accurate as in controls. Furthermore, the fact that the interaction of accuracy costs of task switching and of antisaccade performance fit a multiplicative interaction (equation 1) is consistent with independence of current-trial dominance effects from prior-trial–switching effects.[9]

The latency data showed that antisaccade costs were nearly four times the costs of task switching in normal subjects. Again, whereas antisaccade latency costs were elevated by schizophrenia, task-switching costs were not. Schizophrenic patients showed the same pattern of effects of task switching on prosaccades and antisaccades that was seen in controls. Thus, both the accuracy and latency data support the hypotheses that task switching and antisaccade performance are independent and selectively vulnerable to pathology. This has obvious implications for the debate about whether all executive control processes that govern volitional behavior are mediated by a single attentional system or are distributed among distinct prefrontal networks.[28,29]

The fact that our measures of saccadic task switching did not correlate with the WCST results deserves comment. Although instruments such as the WCST are thought to measure task switching, they are multidimensional, requiring several cognitive processes for successful performance. Poor performance on the WCST, for example, can reflect problems in sustained attention, concept formation, or working memory as well as task switching.[30,31] Our results suggest caution in drawing conclusions from these multidimensional tests. On the other hand, it must be stressed that our saccadic task switch involves a fairly pure stimulus-response remapping. Paradigms with additional switches between stimulus dimension, location, response mode, value, or even other factors such as sequence predictability may reveal differences attributable to schizophrenia. At the least, our results place some boundary limits on where any hypothetical task-switching deficit must lie in this condition.

The latency results confirm that asymmetric switch costs are indeed found with a highly asymmetric dominance task pair like the antisaccade/prosaccade relation. This is true even though the switch between these tasks is limited to stimulus-response remapping. However, not only is the cost reduced for our (nondominant) antisaccade task, but also it is reversed, to give a switch *benefit* to antisaccades. Whereas an asymmetry might be construed as consistent with task-set inhibition [5] or stimulus-cued negative priming[8] hypotheses of task switching, no current model of task-switching processes can account for the paradoxical reduction of antisaccade (nondominant) latencies by task switching, which we believe to be a novel finding. Although small, the paradoxical reduction was consistent across subjects and was found in both the normal and the schizophrenic studies. This reduction is not due to a speed/accuracy trade-off, but it is less in subjects who are more attentive, with shorter manual reaction times on tests of vigilance.

How could a paradoxical reduction in latency arise? There are at least two possible explanations. First is that rather than a task-switching cost, there may be a "nondominant stimulus-response mapping cost" carried over from the prior trial, affecting both prosaccades and antisaccades. Thus, an antisaccade stimulus-response mapping in the prior trial may inhibit the saccade system in general in the current trial. Although we could not demonstrate a correlation between task switch costs of prosaccades and switch costs for antisaccades, this does not entirely exclude this possibility, given the magnitude of the within-subject variance in saccadic latencies.

Second, rather than a general detrimental antisaccade effect carrying over from the prior trial, it may be that the operation of a second cognitive function, such as task switching, facilitates the execution of nondominant responses such as antisaccades specifically and yet delays habitual responses such as prosaccades. Some support for this can been found in a recent study of antisaccades performed simultaneously with an attentionally demanding perceptual discrimination task.[32] These investigators found that simultaneous perfor.nance of other attentional tasks may interfere with the programming of reflexive responses, both delaying them and also facilitating nondominant responses. In our study, the possibility of an attentional basis to this facilitatory effect on the nondominant antisaccade response is indicated by significant correlations with manual reaction time measures of vigilance. These showed that the paradoxical effect is smallest in those subjects who are most attentive. Thus, those subjects who are most adept at deploying attention may actually need to devote less resources to the secondary cognitive operation of task switching, resulting in less facilitation of the primary operation of antisaccade generation. Which of these two fairly different accounts is responsible for this interesting effect of task switching on antisaccade latency requires further investigation.

ACKNOWLEDGMENTS

This work was supported by a grant from the NINDS (to J.B.) and a NARSAD grant and NIMH Grant K23MH01829-01 (to D.M.).

REFERENCES

1. HALLETT, P. 1978. Primary and secondary saccades to goals defined by instructions. Vision Res. **18:** 1279–1296.
2. MONSELL, S., N. YEUNG & R. AZUMA. 2000. Reconfiguration of task-set: is it easier to switch to the weaker task? Psychol. Res. **63:** 250–264.
3. STROOP, J. 1935. Studies of interference in serial verbal reactions. J. Exp. Psychol. **18:** 643–662.
4. COHEN, J., K. DUNBAR & J. MCCLELLAND. 1990. On the control of automatic processes: a parallel distributed processing account of the Stroop effect. Psychol. Rev. **97:** 332–361.
5. ALLPORT, A., E. STYLES & S. HSIEH. 1994. Shifting intentional set: exploring the dynamic control of tasks. *In* Attention and Performance XV. C. Umiltà & M. Moscovitch, eds. :421–452. Erlbaum. Hillsdale, NJ.
6. MEIRAN, N. 2000. Modeling cognitive control of task-switching. Psychol. Res. **63:** 234–249.
7. ROGERS, R.D. & S. MONSELL. 1995. Costs of a predictable switch between simple cognitive tasks. J. Exp. Psychol. Gen. **124:** 207–231.
8. WYLIE, G. & A. ALLPORT. 2000. Task switching and the measurement of "switch costs." Psychol. Res. **63:** 212–233.
9. SCHWEICKERT, R. 1985. Separable effects of factors on speed and accuracy: memory scanning, lexical decision, and choice tasks. Psychol. Bull. **97:** 530–546.
10. STUSS, D. & D. BENSON. 1984. Neuropsychological studies of the frontal lobe. Psychol. Bull. **95:** 3–28.
11. GUITTON, D., H. BUCHTEL & R. DOUGLAS. 1985. Frontal lobe lesions in man cause difficulties in suppressing reflexive glances and in generating goal-directed saccades. Exp. Brain Res. **58:** 455–472.

12. LEVY, D., N. MENDELL, C. LAVANCHER, et al. 1998. Disinhibition in antisaccade performance in schizophrenia. *In* Origins and Development of Schizophrenia. M. Lenzenweger & R. Dworkin, eds. :185–210. American Psychological Association. Washington, DC.
13. BRAFF, D., R. HEATON, J. KUCK, et al. 1991. The generalized pattern of neuropsychological deficits in outpatients with chronic schizophrenia with heterogeneous Wisconsin Card Sorting Test results. Arch. Gen. Psychiatry **48:** 891–898.
14. PERRY, W. & D. BRAFF. 1998. A multimethod approach to assessing perseverations in schizophrenia patients. Schizophrenia Res. **33:** 69–77.
15. FIRST, M., R. SPITZER, M. GIBBON, et al. 1997. Biometrics Research. New York State Psychiatric Institute. New York.
16. KAY, S., A. FISZBEIN & L. OPLER. 1987. The positive and negative syndrome scale (PANSS) for schizophrenia. Schizophren. Bull. **13:** 261–276.
17. OVERALL, J. & D. GORHAM. 1962. The brief psychiatric rating scale. Psychol. Rep. **10:** 799–812.
18. HOLLINGSHEAD, A. 1965. Two Factor Index of Social Position. Yale University Press. New Haven, CT.
19. MILLER, E., P. SATZ & B. VISSCHER. 1991. Computerized and conventional neuropsychological assessment of HIV-1-infected homosexual men. Neurology **41:** 1608–1616.
20. HEATON, R., G. CHELUNE, J. TALLEY, et al. 1993. Psychological Assessment Resources, Inc. Odessa, FL.
21. BAHILL, T. & J. MCDONALD. 1983. Frequency limitations and optimal step size for the two-point central difference derivative algorithm with applications to human eye movement data. IEEE Trans. Biomed. Eng. **30:** 191–194.
22. MUNOZ, D., J. BROUGHTON & J. GOLDRING. 1998. Age-related performance of human subjects on saccadic eye movement tasks. Exp. Brain Res. **121:** 391–400.
23. MÜLLER, N., M. RIEDEL, T. EGGERT, et al. 1999. Internally and externally guided voluntary saccades in unmedicated and medicated schizophrenic patients. Part II. Saccadic latency, gain, and fixation suppression errors. Eur. Arch. Psychiatry Clin. Neurosci. **249:** 7–14.
24. FUKUSHIMA, J., K. FUKUSHIMA, N. MORITA, et al. 1990. Further analysis of the control of voluntary saccadic eye movements in schizophrenic patients. Biol. Psychiatry **28:** 943–958.
25. FUKUSHIMA, J., K. FUKUSHIMA, K. MIYASAKA, et al. 1994. Voluntary control of saccadic eye movement in patients with frontal cortical lesions and parkinsonian patients in comparison with that in schizophrenics. Biol. Psychiatry **36:** 21–30.
26. MARUFF, P., J. DANCKERT, C. PANTELIS, et al. 1998. Saccadic and attentional abnormalities in patients with schizophrenia. Psychol. Med. **28:** 1091–1100.
27. CLEMENTZ, B., J. MCDOWELL & S. ZISOOK. 1994. Saccadic system functioning among schizophrenia patients and their first-degree biological relatives. J. Abnormal Psychol. **103:** 277–287.
28. NORMAN, D. & T. SHALLICE. 1986. Attention to action. Willed and automatic control of behaviour. *In* Consciousness and Self-Regulation. R. Davidson, G. Schwartz & D. Shapiro, Eds.: 1–18. Plenum. New York.
29. STUSS, D., T. SHALLICE, M. ALEXANDER, et al. 1995. A multidisciplinary approach to anterior attentional functions. Ann. N.Y. Acad. Sci. **769:** 191–211.
30. SULLIVAN, E., D. MATHALON, R. ZIPURSKY, et al. 1993. Factors of the Wisconsin Card Sorting Test as measures of frontal-lobe function in schizophrenia and in chronic alcoholism. Psychiatry Res. **46:** 175–199.
31. GOLD, J., C. CARPENTER, C. RANDOLPH, et al. 1997. Auditory working memory and Wisconsin Card Sorting Test performance in schizophrenia. Arch. Gen. Psychiatry **54:** 159–165.
32. KRISTJANSSON, A., Y. CHEN & K. NAKAYAMA. 2000. Less attention is more, in preparation of antisaccades. Invest. Ophthalmol. Visual Sci. **41:** s315.

Neural Mechanisms for the Control of Vergence Eye Movements

PAUL D.R. GAMLIN

Vision Science Research Center, University of Alabama at Birmingham, Birmingham, Alabama 35294, USA

ABSTRACT: Our recent studies in non-human primates have identified and characterized cerebro-ponto-cerebellar pathways involved in the control of vergence eye movements. Specifically, within the deep cerebellar nuclei and nucleus reticularis tegmenti pontis, we have identified neurons that are related to either the near response (convergence and increased ocular accommodation) or the far response (divergence and decreased ocular accommodation). In addition, within the prearcuate region (area 8a), we have characterized neurons related not only to either the far response or the near response, but also to the sensorimotor transformations underlying these eye movements. Because the vergence-related prearcuate region abuts the frontal eye fields, we suggest that the extent of the frontal eye fields be expanded to include this region. We further suggest that with inclusion of this vergence-related region, the frontal eye fields are important for all voluntary eye movements.

KEYWORDS: cerebellum; fastigial nucleus; frontal eye fields; frontal cortex; prearcuate; interpositus nucleus; pontine nuclei; nucleus reticularis tegmenti pontis

INTRODUCTION

Whenever eye movements are made between objects located at different distances, appropriate changes in accommodation and vergence are required. For example, when looking from a far to a near object, appropriate convergence and increases in accommodation ensure that the object is clearly focused on the fovea of each eye. Vergence eye movements can be elicited by changes in both disparity and blur-driven accommodation.[1–5] Our research has concentrated on identifying the neural substrates for such vergence eye movements. Specifically, we have conducted anatomical and electrophysiological experiments in alert rhesus monkeys (*Macaca mulatta*) trained to perform vergence eye movements and other oculomotor tasks. Based on these and other studies, FIGURE 1 provides an overview of our current knowledge of the pathways underlying the neural control of vergence eye movements. Characteristics of the cerebro-ponto-cerebellar pathways shown in FIGURE 1 are described.

Address for correspondence: Dr. Paul Gamlin, Vision Science Research Center, 658 Worrell Building, University of Alabama at Birmingham, Birmingham, AL 35294. Voice: 205-934-0322; fax: 205-934-5725.
pgamlin@uab.edu

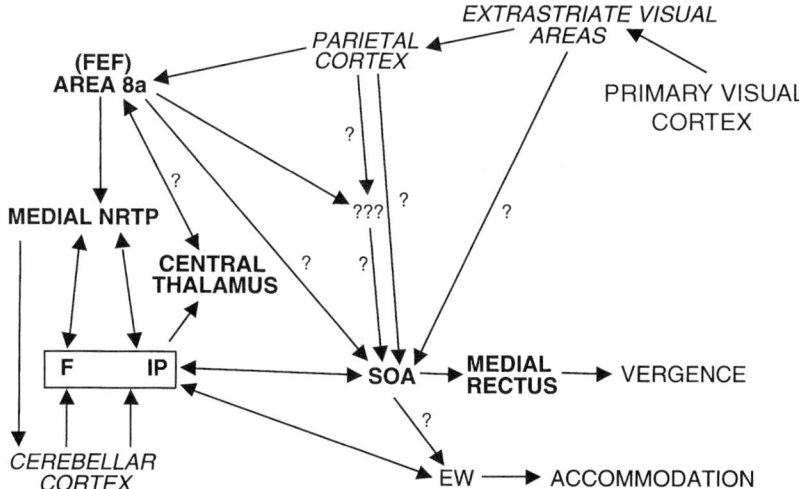

FIGURE 1. Neural pathways involved in the control of vergence eye movements. Areas that are known to contain cells related to vergence eye movements are shown with *bold lettering*. The areas that, based on anatomical studies or electrophysiological studies, appear to contain cells related to vergence eye movements are *italicized*, and the areas that remain to be identified are indicated by *question marks*. Abbreviations: EW, nucleus of Edinger-Westphal; F, fastigial nucleus; FEF, frontal eye fields; IP, posterior interposed nucleus; NRTP, nucleus reticularis tegmenti pontis; SOA, supraoculomotor area.

CEREBELLAR INFLUENCES ON VERGENCE

The supraoculomotor area (SOA) and adjacent reticular formation around the oculomotor nucleus contain premotor neurons related to vergence eye movements.[6–8] We have shown connections between these regions and the deep cerebellar nuclei.[9] Specifically, we found that neurons in the posterior interposed nucleus (IP) and fastigial nuclei projected to the SOA. Because these anatomical experiments revealed a clear pattern of connections that could underlie the cerebellar modulation of vergence eye movements, we investigated these cerebellar nuclei electrophysiologically. To date, the most complete information is available for neurons in the IP. Studies have revealed phasic-tonic cells in a localized region of the IP whose activity is modulated as a function of divergence and far accommodation (FIG. 2).[10] Microstimulation in this region of the IP produced divergence eye movements and matching decreases in accommodation.[10] These studies suggest that this region of the IP is involved in the control of the far response possibly by way of its projection to the midbrain near-response region.

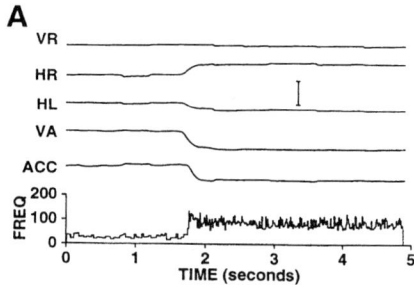
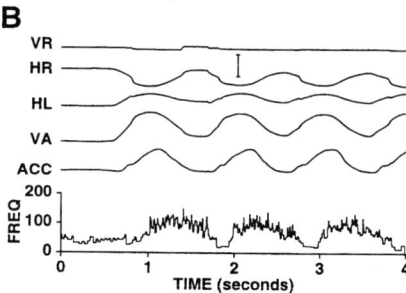
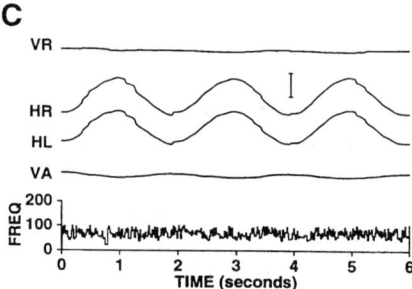

FIGURE 2. (**A**) Behavior of a posterior interposed nucleus (IP) far-response cell for a symmetrical far-response movement in response to a step change in vergence and accommodative demand during normal binocular viewing. In this and subsequent figures, convergence, accommodation for near, and rightward eye movements are represented as upward movements of the traces. (**B**) Response of this cell during sine-wave tracking of a target moving in depth. (**C**) Activity of this cell is unaffected by smooth pursuit of a horizontally moving target at optical infinity. (Scale bar = 4 meter angles [MA] in **A** and **B**; 4 degrees in **C**.) Abbreviations: ACC, accommodation; HL, horizontal left eye position; HR, horizontal right eye position; VA, vergence angle; VR, vertical right eye position. (Modified from Zhang & Gamlin.[10])

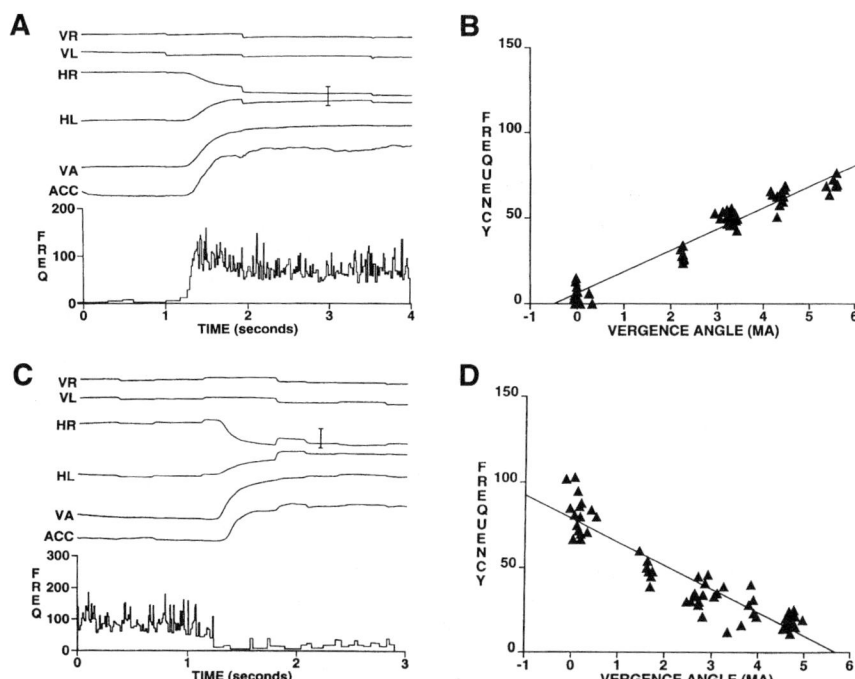

FIGURE 3. (**A**) Behavior of a nucleus reticularis tegmenti pontis (NRTP) near-response cell during an increase in vergence angle and accommodation that occurred in response to the target being optically stepped towards the animal by 4 meter angles (MA). (**B**) A summary plot of firing rate as a function of vergence angle for this cell. (**C**) Behavior of an NRTP far-response cell during an increase in vergence angle and accommodation that occurred in response to the target being optically stepped towards the animal by 3 MA. (**D**) A summary plot of firing rate as a function of vergence angle for this cell. (Scale bar in **A** and **C** = 2 MA). Abbreviations: VL, vertical left eye position. (Modified from Gamlin and Clarke, 1995.)

BEHAVIOR OF NRTP FAR-RESPONSE AND NEAR-RESPONSE NEURONS DURING VERGENCE

Given the vergence-related activity recorded in the posterior interposed nucleus, we decided to identify those precerebellar nuclei that might be providing the requisite input signals. Single-unit recording from one precerebellar nucleus, the nucleus reticularis tegmenti pontis (NRTP), identified phasic-tonic cells that exhibited increases in their firing rate during the near response.[11] Many of these cells had a tonic firing rate that increased as a function of increases in convergence and accommodation. An example of one of these cells is shown in FIGURE 3A and B. Approximately equal numbers of phasic-tonic cells were encountered in the NRTP that showed increased activity during the far response.[11] With far viewing, many of these cells displayed tonic activity which declined to zero with effective viewing distances of 1–0.3 meters. An example of one of these cells is shown in FIGURE 3C and D. We found that many of these recorded cells were close to neurons displaying saccade-related activity, and marking lesions confirmed that the recording sites were in the medial NRTP in a location similar to that reported for saccade-related neurons.[12]

ROLE OF FRONTAL EYE FIELDS IN VERGENCE EYE MOVEMENTS

The frontal eye fields are known to be involved in saccadic[13–15] and smooth pursuit eye movements[16–18] and to project to the NRTP.[19–21] Therefore, based on our recordings from neurons in the NRTP, we hypothesized that the frontal eye fields or an adjacent region within the frontal cortex might play a role in controlling vergence eye movements. As hypothesized, in the prearcuate cortex anterior to the saccade-related region of the frontal eye fields, we located neurons displaying predominantly phasic activity that was significantly correlated with vergence eye movements.[22] Twenty-five of these neurons increased their activity for near viewing and nine increased their activity for far viewing. Electrical microstimulation at the site of these recorded neurons often produced vergence eye movement.[22] FIGURE 4 is an example of the behavior of one of the convergence-related prearcuate neurons during vergence eye movements.

The possibility arose that the neural activity observed during vergence trials, as shown in FIGURE 4A, was related not to the vergence movement but to the retinal disparity of the target. To investigate this possibility, the behavior of neurons was examined during vergence movements elicited by targets ramped, not stepped, in depth. As can be seen for the responses shown in FIGURE 4B, the relation between neural activity and vergence velocity is comparable to that seen in FIGURE 4A and thus is good evidence that the behavior of this neuron is more closely related to the movement than to the retinal disparity of the target that elicited it.

To examine the possibility that neurons displayed visual responses, their behavior was examined during binocular vergence tracking of a target that moved sinusoidally in depth and was briefly extinguished (250 ms) during the tracking period (FIG. 4C). The responses in this figure show continued activity during this blanking period despite the absence of visual stimulation. This result was similar for 20 neurons tested.

The behavior of most neurons was investigated during trials in which a target was presented, but a motor response was either delayed or not required. During these trials, most neurons, including the one shown in FIGURE 4, did not respond to the appearance of the visual target or during the time between its appearance and the movement. However, some neurons showed clear modulation of their activity under these conditions. An example of one of these visuovergence neurons is shown in

---

FIGURE 4. Response properties of a prearcuate vergence-related neuron. In **A**, the raster showing unit activity is aligned on eye movement onset. In **B** and **C**, they are aligned on target motion. Peristimulus time histograms at the base of each plot represent the average of the rasters. (**A**) Neuronal response during vergence eye movements to a target stepped in depth. (**B**) Neuronal response during vergence eye movements to a target ramped in depth. (**C**) Response of the neuron during sinusoidally modulated vergence responses in which the target was briefly extinguished (blank in TARG trace). AD, average disparity; AVA, average vergence angle; AVV, average vergence velocity; TARG, target position. (Scale bar = 4 degrees and 20 degrees.s^{-1}). (Modified from Gamlin & Yoon.[22])

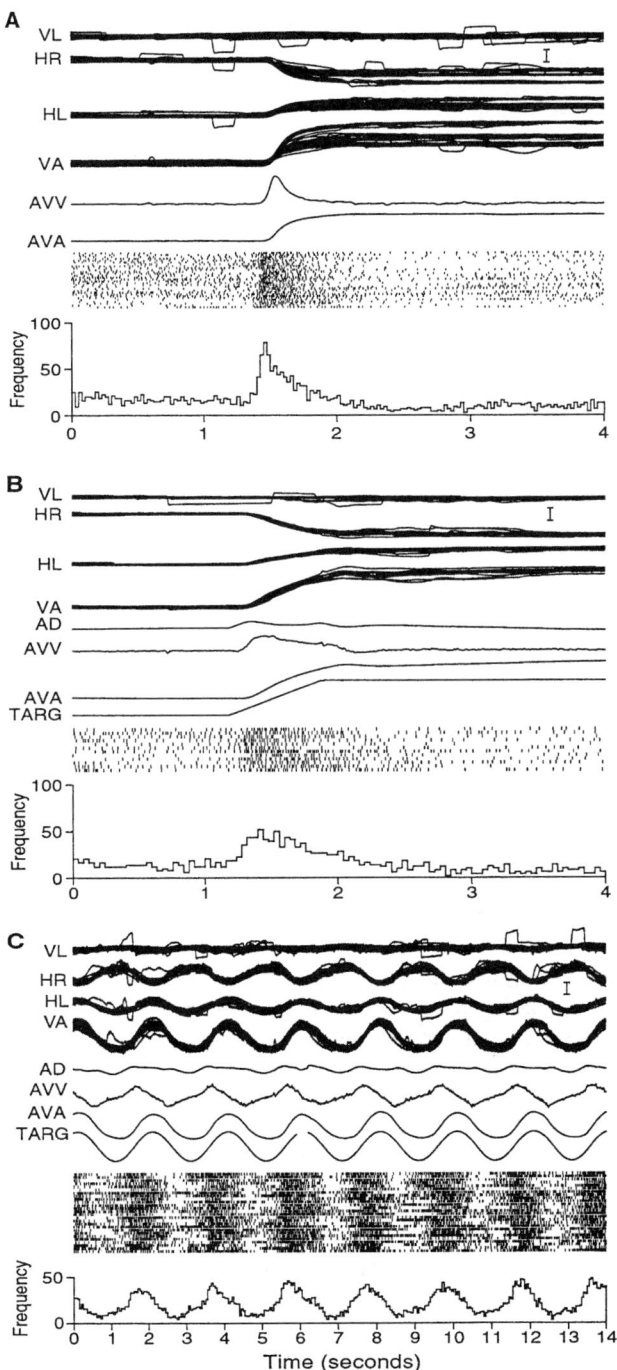

FIGURE 4. *Legend on previous page.*

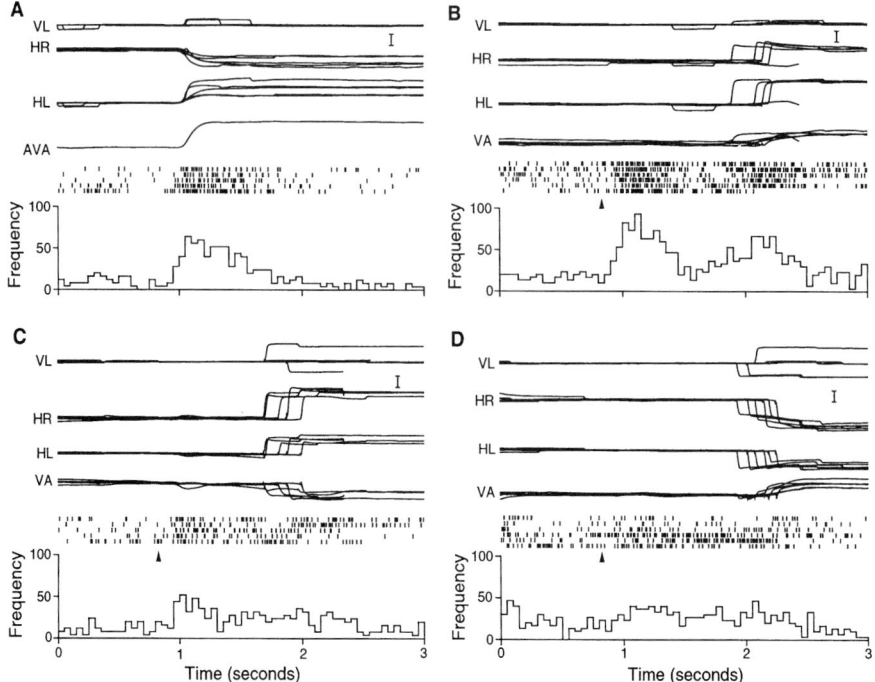

FIGURE 5. Response properties of a prearcuate visuovergence neuron. In **A**, the raster showing unit activity is aligned on eye movement onset. In **B, C,** and **D,** the rasters are aligned on target appearance (indicated by the *arrowhead*). (**A**) Neural response during convergence eye movements elicited by stepped increases in vergence demand. (**B**) Response to the appearance of a near target (+4° disparity) in the contralateral hemifield. (**C**) Response to the appearance of a far target (−4° disparity) in the contralateral hemifield. (**D**) Response to the appearance of a near target (+4° disparity) in the ipsilateral hemifield. (Scale bar = 4 degrees.) (Modified from Gamlin & Yoon.[22])

FIGURE 5. This neuron was initially identified as being vergence related based on its response during convergence eye movements elicited by stepped vergence targets (FIG. 5A). However, we found that when the animal was required to make delayed movements to near targets in the contralateral hemifield, the neuron significantly increased its activity in response to the appearance of the target well in advance of the movement (FIG. 5B). By contrast, the activity of this neuron was modulated only weakly during the delay period for movements to far targets within the contralateral hemifield (FIG. 5C) or near targets within the ipsilateral hemifield (FIG. 5D). Thus, this neuron behaved as if it had both movement and visual receptive fields tuned for near space within the contralateral hemifield. Overall, the neurons that responded during these trials behaved as if they had both movement and visual receptive fields tuned in depth within the contralateral hemifield.

Based on the localization of our recording sites, we generated a composite diagram (FIG. 6) showing the approximate extent of the region of frontal cortex related to vergence and accommodation. This diagram demonstrates that this prearcuate

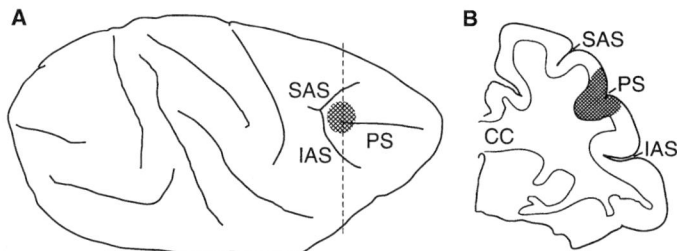

FIGURE 6. Location of near-response and far-response neurons. The location of these neurons is shown as the *shaded area* in a side view of the rhesus cortical surface (**A**) and a coronal section (**B**) at the level indicated by the *dashed line* in **A**. Abbreviations: CC, corpus callosum; IAS, inferior arcuate sulcus; PS, principal sulcus; SAS, superior arcuate sulcus. (From Gamlin & Yoon.[22] Reprinted with permission from *Nature*.)

region lies anterior to the saccade-related region of the frontal eye fields and includes the surface cortex between the arcuate sulcus and the posterior pole of the principal sulcus.

These results demonstrate that neurons within a specific region of the frontal cortex play a role in vergence eye movements. Thus, the arcuate region of frontal cortex contains areas specialized for the control of all classes of voluntary eye movements, including saccades, smooth pursuit, and vergence. This finding suggests that the current boundaries of primate frontal eye fields should be expanded to include this vergence-related region that lies anterior to the saccade-related region.

SUMMARY

The cerebro-ponto-cerebellar pathways that are presumed to control or modulate vergence eye movements in primates are reviewed. Clearly, a significant number of questions on the neural control of these eye movements remain to be answered. For example, despite our recent report of involvement of the frontal eye fields in vergence eye movement control, little is currently known about the location and role of other cortical areas involved in these eye movements or the way in which the sensory signals controlling them are processed and transformed into the appropriate motor output. Also, we know little of the neural substrates underlying the ability to adapt the static and dynamic characteristics of vergence eye movements. However, it is likely that the cerebellar circuits described in this review may play a role in this plasticity. Future studies to investigate these important issues are clearly needed.

ACKNOWLEDGMENTS

I thank Sam Hayley for computer programming. This research was supported by NEI Grant EY07558 and NEI CORE Grant P30 EY03039.

REFERENCES

1. FINCHAM, E.F. & J. WALTON. 1957. The reciprocal actions of accommodation and vergence. J. Physiol. Lond. **137:** 488–508.
2. RASHBASS, C. & G. WESTHEIMER. 1961. Disjunctive eye movements. J. Physiol. **159:** 339–360.
3. KRISHNAN, V.V., S. PHILLIPS & L.A. STARK. 1973. Frequency analysis of accommodation, accommodative vergence, and disparity vergence. Vision Res. **13:** 1545–1554.
4. KRISHNAN, V.V. & L. STARK. 1977. A heuristic model for the human vergence eye movement system. IEEE Trans. Biomed. Eng. **24:** 44–49.
5. SCHOR, C.M. 1992. A dynamic model of cross-coupling between accommodation and convergence: simulations of step and frequency responses. Optom. Vision Sci. **69:** 258–269.
6. MAYS, L.E. 1984. Neural control of vergence eye movements: convergence and divergence neurons in the midbrain. J. Neurophysiol. **51:** 1091–1108.
7. JUDGE, S.J. & B.G. CUMMING. 1986. Neurons in the monkey midbrain with activity related to vergence eye movement and accommodation. J. Neurophysiol. **55:** 915–930.
8. ZHANG, Y., L.E. MAYS & P.D.R. GAMLIN. 1992. Characteristics of near response cells projecting to the oculomotor nucleus. J. Neurophysiol. **67:** 944–960.
9. MAY, P.J., J.D. PORTER & P.D.R GAMLIN. 1992. Interconnections between the cerebellum and midbrain near response regions. J. Comp. Neurol. **315:** 98–116.
10. ZHANG, H.Y. & P.D.R GAMLIN. 1998. Neurons in the posterior interposed nucleus of the cerebellum related to vergence and accommodation. I. Steady-state characteristics. J. Neurophysiol. **79:** 1255–1269.
11. GAMLIN, P.D.R. & R.J. CLARKE. 1995. Single-unit activity in the nucleus reticularis tegmenti pontis related to vergence and ocular accommodation. J. Neurophysiol. **73:** 2115–2119.
12. CRANDALL, W.F. & E.L. KELLER. 1985. Visual and oculomotor signals in nucleus reticularis tegmenti pontis in alert monkey. J. Neurophysiol. **54:** 1326–1345.
13. BIZZI, E. 1968. Discharge of frontal eye field neurons during saccadic and following eye movements in unanesthetized monkeys. Exp. Brain Res. **6:** 69–80.
14. BRUCE, C.J. & M.E. GOLDBERG. 1985. Primate frontal eye fields. I. Single neurons discharging before saccades. J. Neurophysiol. **53:** 603–635.
15. DIAS, E.C., M. KIESAU & M.A. SEGRAVES. 1995. Acute activation and inactivation of macaque frontal eye field with GABA-related drugs. J. Neurophysiol. **74:** 2744–2748.
16. LYNCH, J.C. 1987 Frontal eye field lesions in monkeys disrupt visual pursuit. Exp. Brain Res. **68:** 437–441.
17. MACAVOY, M.G., J.P. GOTTLIEB & C.J. BRUCE. 1991. Smooth-pursuit eye movement representation in the primate frontal eye field. Cereb. Cortex **1:** 95–102.
18. GOTTLIEB, J.P., M.G. MACAVOY & C.J. BRUCE. 1994. Neural responses related to smooth-pursuit eye movements and their correspondence with electrically elicited smooth eye movements in the primate frontal eye field. J. Neurophysiol. **72:** 1634–1653.
19. LEICHNETZ, G.R., D.J. SMITH & R.F. SPENCER. 1984. Cortical projections to the paramedian tegmental and basilar pons in the monkey. J. Comp. Neurol. **228:** 388–408.
20. HUERTA, M.F., L.H. KRUBITZER & J.H. KAAS. 1986. Frontal eye field as defined by intracortical microstimulation in squirrel monkeys, owl monkeys, and macaque monkeys. I. Subcortical connections. J. Comp. Neurol. **253:** 415–439.
21. STANTON, G.B. M.E. GOLDBERG & C.J. BRUCE. 1988. Frontal eye field efferents in the macaque monkey. I. Subcortical pathways and topography of striatal and thalamic terminal fields. J. Comp. Neurol. **271:** 473–492.
22. GAMLIN, P.D. & K. YOON. 2000. An area for vergence eye movement in primate frontal cortex. Nature **407:** 1003–1007.

Neural Basis of Disjunctive Eye Movements

W.M. KING[a] AND WU ZHOU[b]

Departments of Neurology,[a,b] Anatomy,[a,b] and Surgery,[b] University of Mississippi Medical Center, Jackson, Mississippi 39216-4505, USA

ABSTRACT: New evidence has challenged a widely accepted interpretation of Hering's law of equal innervation, which states that disjunctive saccades are produced by the linear addition of conjugate and vergence innervation commands produced by independent oculomotor subsystems. We hypothesize, instead, that saccades are produced by a monocular premotor control network. A model, based on this hypothesis and consistent with known brain-stem anatomy, simulates realistic disjunctive saccades including initial and late slow vergence movements.

KEYWORDS: Hering's law; vergence eye movements; binocular vision; saccades; neural models

INTRODUCTION

Disjunctive horizontal eye movements occur naturally in a variety of species either to view different regions of visual space with each eye or to binocularly position the two eyes to facilitate stereopsis. Any disjunctive eye movement can be constructed from components that are equal in amplitude and direction in both eyes (conjugate eye movements) and components that are equal in amplitude but opposite in direction (symmetric vergence eye movements). These components are a linear sum of the monocular movements of each eye (*Eq. 1*).

$$\text{Conjugate} = (\text{Eye}_{\text{right}} + \text{Eye}_{\text{left}})/2 \tag{1a}$$

$$\text{Vergence} = \text{Eye}_{\text{left}} - \text{Eye}_{\text{right}} \tag{1b}$$

Based on these relationships, Hering[1] proposed that the each eye's innervation was equal in magnitude but could specify movement in the same or opposite direction. Hering's "law of equal innervation" has been interpreted to mean that binocular coordination is the result of each eye receiving similar conjugate and vergence motor innervations (for a discussion of this point, see Howard and Rogers[2]). Several lines of anatomical and physiological evidence have been interpreted as support for the notion that separate conjugate and vergence subsystems are responsible for premotor control of saccades. For example, the abducens nucleus contains internuclear as well as motor neurons.[3,4] Abducens internuclear neurons (AINs) provide an excitatory

---

Address for correspondence: W.M. King, Ph.D., Department of Neurology, University of Mississippi Medical Center, 2500 North State St., Jackson, MI 39216-4505. Voice: 601-984-5491; fax: 601-815-4115.

mike@vor.umsmed.edu

input to contralateral medial rectus motoneurons.[5] AIN discharge activity is similar to abducens motoneuron activity, suggesting that the innervation conveyed by AINs to contralateral medial rectus motoneurons is at least qualitatively similar to the innervation of ipsilateral abducens motoneurons.[6,7] Thus, the internuclear pathway could provide a basis for conjugate innervation of both eyes. Near response cells, located in the supraoculomotor region, encode a pure vergence signal and project to medial rectus motoneurons providing support for the notion of a separate subsystem for vergence innervation of oculomotoneurons.[8–12] However, other observations suggest the possibility of independent monocular control of each eye. An early finding was based on the much slower time course of symmetric vergence eye movements (hundreds of ms) compared to conjugate saccades (10s of ms).[13,14] During disjunctive saccades, intrasaccadic vergence components have a rapid time course only slightly slower than that of conjugate saccades, an observation that cannot be reconciled with linear addition of rapid conjugate components and much slower symmetric vergence components (Eq. 1). Furthermore, during rapid eye movement (REM) sleep, binocular alignment is lost and the lines of sight of the two eyes fail to intersect.[15] Rapid eye movements during sleep are disjunctive and occasionally even monocular, suggesting an underlying monocular organization.[15] Finally, single unit recordings of premotor cells in the brain stem strongly suggest that eye position and speed are encoded monocularly.[16] For example, although the paramedian pontine reticular formation is regarded classically as the "conjugate gaze center," excitatory and inhibitory burst neurons (EBNs and IBNs) encode neural commands for saccades in a monocular coordinate system and can be identified as left or right eye burst neurons. Although it is conceivable that Hering's law could be implemented at a premotor level, it is more parsimonious to interpret these data as evidence for monocular control of each eye (for a review, see King and Zhou[17]). Thus, we hypothesize that the neural system for generating conjugate and disjunctive saccades is inherently monocular.

METHODS

Binocular eye movements were measured using a magnetic search coil method.[18] Monkeys were trained to fixate on small visible targets projected by lasers onto a far screen (located 1.8 m from the monkey's eyes) or onto a near screen in a horizontal plane extending 8–25 cm from the animal's nose. The targets could be positioned to elicit conjugate saccades using the far screen or disjunctive saccades using the near screen. For some trials, the target's distance from the monkey was varied along a trajectory that was aligned with one or the other eye in order to elicit a "monocular" saccade by the unaligned eye or along a trajectory equidistant between the two eyes in order to elicit a symmetric vergence response. All animal procedures were consistent with the National Institutes of Health "Guide for the Care and Use of Laboratory Animals" and were approved by the University of Mississippi Medical Center Animal Care Committee (IACUC).

Target and eye position signals were digitized at either 1 or 2 kHz sampling rates per channel with 16-bit resolution using a CED (Cambridge Electronic Design Limited, Cambridge, UK), data acquisition system. The data were subsequently filtered,

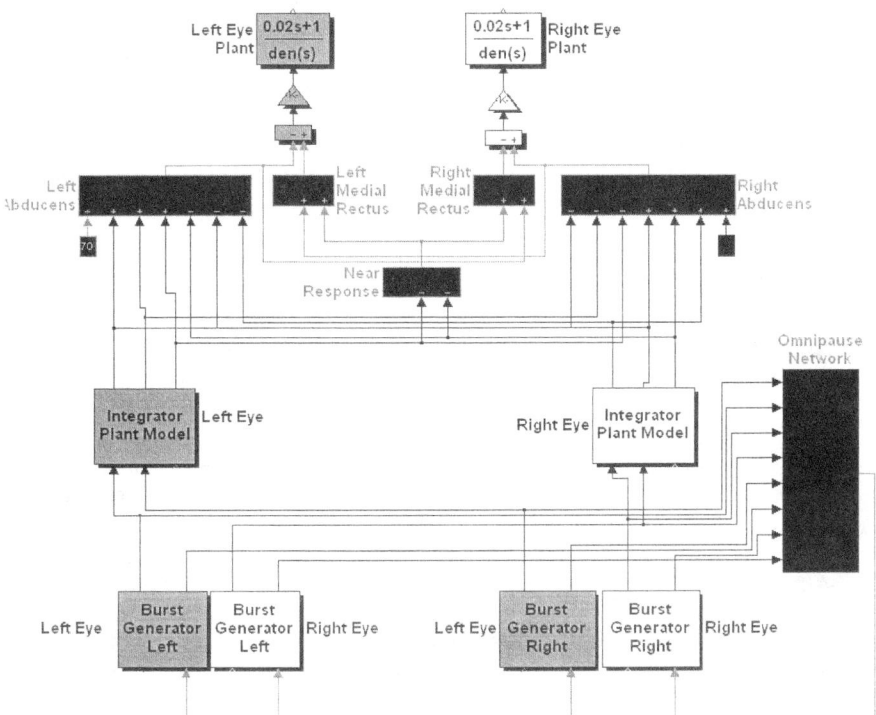

FIGURE 1. Simulink model of the saccade generator based on monocular premotor command signals.

and eye velocity was computed using Spike4 software provided by CED. Eye movement data from individual trials were displayed by the analysis routine, and occasional trials were rejected when the monkey failed to properly perform the task or fixate on the targets. The remaining trials were grouped by saccade amplitude and direction and analyzed for latency, amplitude, peak velocity, and duration of movement in each eye. Saccades were presumed to begin and end when eye velocity crossed a threshold of 8 degrees/s. Saccades were aligned on their onsets and averaged to illustrate typical waveforms.

DISJUNCTIVE EYE MOVEMENT MODEL

A simplified block diagram of the disjunctive controller model is presented in FIGURE 1. This model differs from previous models of saccade generators in three ways: burst generators are monocular, motoneurons receive binocular inputs, and monocular neural integrator circuits provide an input to near response vergence cells. Each of these components is described in detail below.

To generate saccades, the burst generator employs a local feedback model.[19,20] However, the motor error signal is hypothesized to be monocular. In the model, there are four monocular burst generators, one for each eye and for each direction. Consider the right eye burst generators (FIG. 1, white background). For rightward saccades, the output is connected to the integrator plant model as an excitatory input; for leftward saccades, the output is connected as an inhibitory input. A similar pattern is followed for the left eye burst generators (FIG. 1, gray background). The omnipause network (shown in black), which acts as a latch, is assumed to be common to all burst generators. The response of burst neurons to a motor error signal is nonlinear. The shape of the nonlinearity is similar to that employed by Zee *et al.*,[21] but parameters have been adjusted to reproduce the characteristics of monkey saccades.

The extraocular plants are modeled as second-order linear systems with slide components.[22,23] We assume that the eye velocity signals produced by burst generators are matched to the plants by brain stem circuits consisting of direct pathways ("pulse"), slide components ("slide"), and neural integrators ("step") for each eye. Up to this point, the premotor circuit is entirely monocular except for the omnipause network. In the final premotor and motoneuron stages these signals are combined to produce binocular innervation of the eyes. First, the monocular outputs of the pulse, step, and slide elements are summed binocularly at the level of the abducens nucleus. Although there may be subtle differences, we assume that abducens motoneurons and AINs encode the same signal. Thus, AINs provide a binocular control signal to contralateral medial rectus motoneurons appropriate for conjugate eye movements. These connections are consistent with known anatomy and physiology. Second, we hypothesize that the monocular "step" outputs of the neural integrators are subtracted at the level of the mesencephalic near response cells to generate a pure vergence position signal. During disjunctive eye movements, these neurons provide binocular vergence innervation to medial rectus motoneurons. Although connections between the eye position integrators and near response cells have not been demonstrated, they are not excluded by any evidence. We also assume that the "pulse" and "slide" components reach motoneurons via the abducens and internuclear pathways but not via near response neurons.

SIMULATION OF DISJUNCTIVE SACCADES

FIGURE 2 illustrates averages of eye position (upper panels) and velocity (lower panels) for three exemplary saccades. Left hand panels show a rightward conjugate saccade of about 12 degrees. The left eye position trace is displaced with respect to the right eye position trace (upper panel) because the monkey's eyes are slightly converged (~1.2 degrees for the target at 1.8 m). Eye velocity is conjugate and attains a peak of nearly 600 degrees/s. The middle panels show a disjunctive saccade where the amplitude of the right eye component is about 50% of the left eye component (total vergence about 6 degrees). The right hand panel shows a monocular saccade (total vergence about 12 degrees). Three characteristics of disjunctive saccades are readily observed[24–26]: peak velocity is lower than that for conjugate saccades of similar amplitude, and there are initial and late slow vergence components. Notice also the movement of the aligned eye during the monocular saccade. Further details of these findings are illustrated in component amplitude-duration and amplitude-velocity

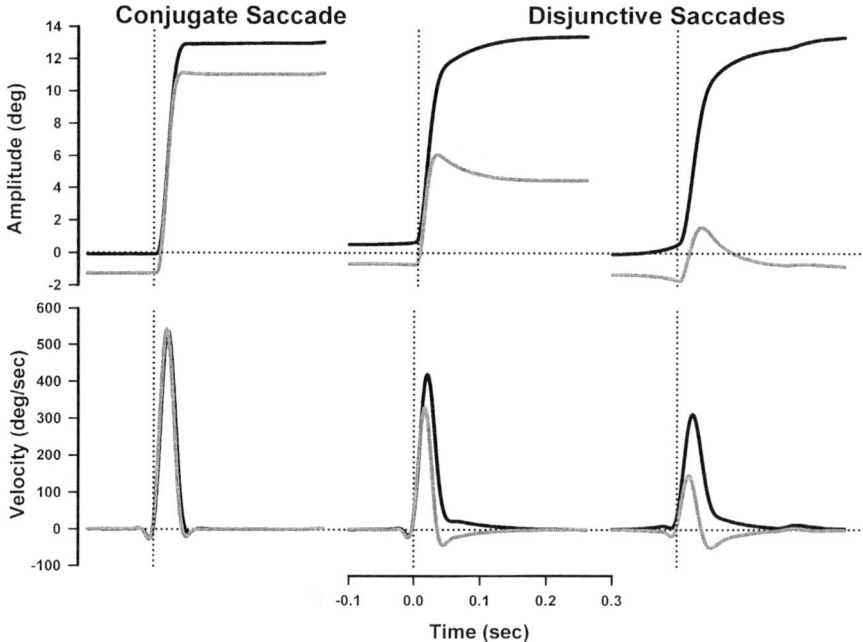

FIGURE 2. Examples of conjugate (*left panels*), disjunctive (*middle panels*), and monocular saccades (*right panels*). Each trace is an average of 25 saccades aligned on saccade onset. *Black traces*, left eye position and velocity; *gray traces*, right eye position and velocity.

plots (FIG. 3) for the individual saccades that comprise these averages. The conjugate saccades are illustrated as circles (FIG. 3). These saccades have the shortest durations (upper panels) and highest peak velocities (lower panels). By contrast, the monocular saccades (squares) have the longest durations and smallest peak velocities. Despite the fact that the amplitudes of the left eye's saccades are nearly equal for the conjugate and disjunctive saccades, their durations and velocities are quite different.

FIGURE 4 illustrates how effectively the model replicates these characteristics for a similar series of simulated saccades. The burst generator parameters and plant models were adjusted to reproduce the main sequence characteristics of conjugate monkey saccades and were left unchanged for the simulation of disjunctive saccades.[27] The model realistically reproduces the characteristic features of conjugate and disjunctive saccades (compare FIGS. 2 and 4). First, peak component velocity declines systematically with the fastest velocities associated with the conjugate saccade (left hand panels) and the slowest with the monocular saccade (right hand panels). Second, there is a long period of slow post-saccadic vergence associated with disjunctive saccades (middle and right hand panels). Third, there is an early interval of slow symmetric vergence that precedes the actual saccade. In the model, the early symmetric vergence is produced by assuming that the near-response cells receive

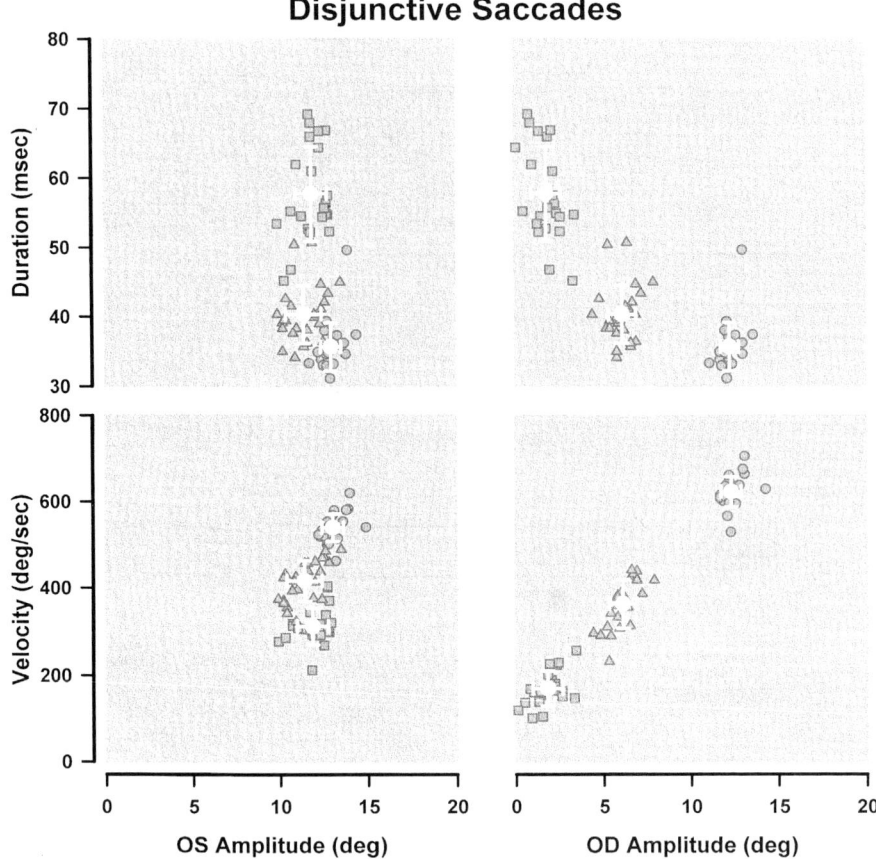

FIGURE 3. Main sequence plots of the individual saccades used to compute the averages shown in FIGURE 2. *Large white circles* and *bars* show mean values plus and minus standard error. Conjugate saccades are plotted as *circles,* disjunctive saccades as *triangles,* and monocular saccades as *square symbols.*

motor error inputs from the monocular saccade generators. For disjunctive saccades, the motor error signals will be unequal, and it is their difference that drives near-response cells. Since near-response cells innervate both eyes equally, the resultant vergence is symmetric.

DISCUSSION

Previous models of the saccade generator either have been limited to simulation of conjugate eye movements[19,20] or have employed nonlinear coupling of parallel conjugate and vergence controllers.[21] The most successful model[21] assumes that EBNs encode the conjugate component of saccades and that a separate population of

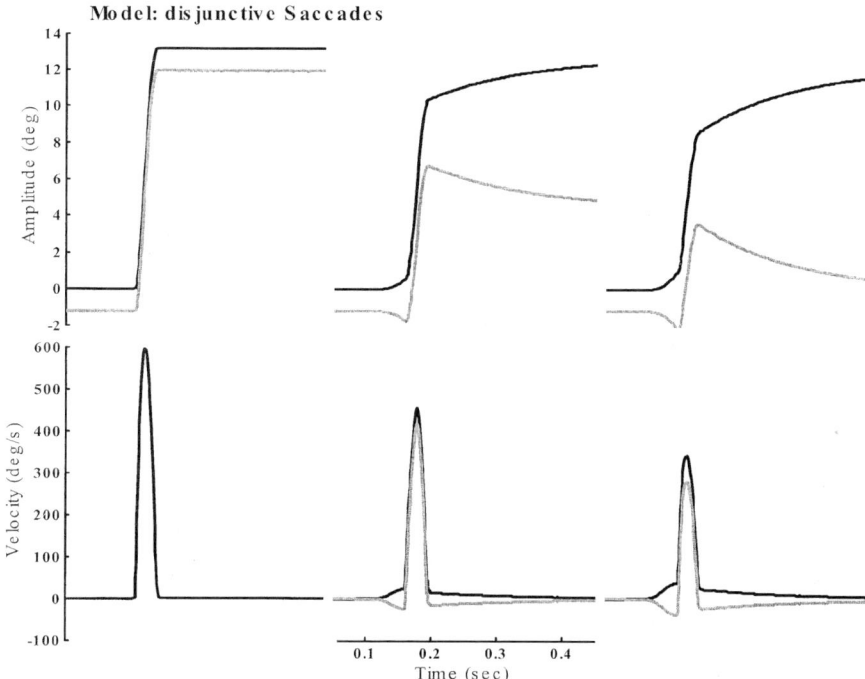

FIGURE 4. Examples of simulated conjugate (*left panels*), disjunctive (*middle panels*), and monocular saccades (*right panels*). Traces as in FIGURE 2.

burst neurons[28] encode the vergence component of saccades. Similarly, there must be separate neural integrators for the conjugate and vergence eye movement components. A recent study of EBNs, however, showed that they encode a monocular eye movement command.[16] This finding is inconsistent with the Zee *et al.* model unless one assumes that the conjugate and vergence command signals are summed at or occur prior to the EBNs. We believe it is more parsimonious to assume that the premotor signals for saccades are computed separately for each eye. The model presented here demonstrates that this arrangement can produce realistic saccade trajectories.

Two aspects of this model merit further discussion. First, we hypothesize that discharge of the near-response cells reflects the difference in the eye position signals encoded by the monocular integrators. These signals could reach near-response cells via direct projections from nucleus prepositus and vestibular neurons or via more indirect pathways. This idea does not exclude a traditional role for near-response neurons to mediate components of the near triad related to disparity vergence and accommodation.[9,29] In fact, we believe that disparity-driven vergence plays an important role in maintaining ocular alignment during fixation.

The simulations suggest that the postsaccadic slow vergence characteristic of disjunctive saccades is caused by a pulse-step mismatch. To understand why this mismatch occurs, consider the examples of FIGURE 4. During a conjugate saccade, two

TABLE 1. Conjugate step

	Left eye	Right eye
Integrator	−2.5	2.5
Near response	0	0
Abducens internuclear neurons	−5	−5
Abducens	−5	5
Media rectus	5	−5
Eye command	10	10

TABLE 2. Monocular step

	Left eye	Right eye
Integrator	−2.5	0
Near response	5	5
Abducens internuclear neurons	−2.5	2.5
Abducens	−2.5	2.5
Media rectus	7.5	2.5
Eye command	10	0

of the four burst generators are active ("left eye" and "right eye" burst generator right). The bursts are integrated by the monocular plant models to produce innervation "steps." Quantitatively, the step produced by each plant model is only part of what is required by the lateral or medial recti to actually hold either eye at the new desired position (TABLE 1, first row). In the tables, the units are arbitrary; it is their ratios that are important. For example, a value of −2.5 for the integrator may be interpreted as the decrement in firing needed to generate an eye command of 10). For conjugate saccades the appropriate step amplitude is produced by the "push-pull" arrangement of the circuit. Left abducens neurons are inhibited by the sum of the "step" inputs from the left and right eye plant models. This summation produces the correct step amplitude for the lateral recti and, via the internuclear pathway, the correct step amplitude for the medial recti muscles (TABLE 1, gray cells). Note that these pathways treat each eye symmetrically; thus, they cannot produce the correct "step" innervation for disjunctive saccade components when the "step" is different in each eye. For the monocular saccade, only one of the four burst generators is active (left eye burst generator right), so the step amplitude is 50% less than that for a comparable conjugate saccade (compare TABLES 1 and 2). Because the left and right eye integrator outputs are different, a vergence signal is conveyed to near-response cells (TABLE 2, row 2). For the left eye, the size of the eye command is the same as that for the conjugate saccade, but more of the step is generated by medial rectus motoneurons than by lateral rectus motoneurons (TABLE 2, gray cells). For the right eye, the net eye command is zero because the medial and lateral recti are equally coactivated.

The "pulse" and "slide" components inhibit the left abducens and excite the right abducens nucleus (TABLE 3, rows 3 and 4). We assume that the pulse and slide com-

TABLE 3. Monocular pulse

	Left eye	Right eye
Integrator	−2.5	0
Near response	0	0
Abducens internuclear neurons	−2.5	2.5
Abducens	−2.5	2.5
Media rectus	2.5	−2.5
Eye command	5	5

ponents are not input to near-response cells, so medial recti activity simply mirrors abducens activity bilaterally (TABLE 3, gray cells). Because of this symmetry, the left and right eye pulse and slide components must be equal even though the saccadic step is disjunctive. In the left eye, the pulse and slide components are too small and eye position at the end of the "saccade" is less than that needed to match the step. The left eye drifts slowly to the right until the step innervation is matched. A similar explanation accounts for the intrasaccadic movement of the aligned right eye. In this case, a pulse and slide are present, but the step is zero. Thus, the right eye moves transiently to the right and then drifts back to zero position. The reduced pulse and slide amplitudes also account for the smaller peak velocities attained during disjunctive saccades (FIG. 3).

In a pioneering study, Gamlin et al.[7] reported that abducens motoneurons and AINs appeared to carry inappropriate signals during vergence. Specifically, they found that during convergence, the firing rates of abducens neurons failed to decrease as much as they would have had the eye reached the same position with a conjugate movement. Comparison of the data in TABLES 1 and 2 confirms that the model replicates this finding. For both the monocular and conjugate saccades, the left eye command is 10. For the conjugate saccade, abducens activity is reduced 5 units (TABLE 1, rows 3 and 4), but for the monocular saccade, the reduction is only 2.5 units. As a result, there will be coactivation of lateral and medial rectus motoneurons. A recent study by Miller (J. Miller, personal communication), however, failed to find evidence of co-contraction of these muscles during vergence. Any interpretation is complicated, however, by the actions of muscle pulleys which alter their effective insertion points during eye movements and hence the relations between muscle length, tension, and eye position.[30]

These data demonstrate that a saccade generator model based on monocular eye movement control signals can accurately replicate conjugate and disjunctive saccade trajectories. Furthermore, unlike models based on conjugate and vergence premotor controllers, this model is consistent with the monocular discharge characteristics of EBNs involved in saccade generation, with the monocular discharge characteristics of prepositus hypoglossi and vestibular neurons that are part of the eye position integrator, and with the binocular discharge characteristics of oculomotoneurons. A crucial assumption of the model, that near-response cells receive monocular eye position inputs, requires further data for verification. Similarly, the hypothesized role of the motor error signals in producing the initial early vergence is hypothetical.

ACKNOWLEDGMENTS

This work was supported by Public Health Service Grant EY04045 (to W.M.K.).

REFERENCES

1. HERING, E. 1977. The Theory of Binocular Vision. Plenum Press. New York.
2. HOWARD, I.P. & B.J. ROGERS. 1996. Binocular Vision and Stereopsis. Oxford University Press. New York.
3. BAKER, R. & S.M. HIGHSTEIN. 1975. Physiological identification of interneurons and motoneurons in the abducens nucleus. Brain Res. **91:** 292–298.
4. BUETTNER-ENNEVER, J.P. & K. AKERT. 1981. Medial rectus subgroups of the oculomotor nucleus and their abducens internuclear input in the monkey. J. Comp. Neurol. **197:** 17–27.
5. HIGHSTEIN, S.M. & R. BAKER. 1978. Excitatory termination of abducens internuclear neurons on medial rectus motoneurons: relationship to syndrome of internuclear ophthalmoplegia. J. Neurophysiol. **41:** 1647–1661.
6. FUCHS, A.F., C.A. SCUDDER & C.R. KANEKO. 1988. Discharge patterns and recruitment order of identified motoneurons and internuclear neurons in the monkey abducens nucleus. J. Neurophysiol. **60:** 1874–1895.
7. GAMLIN, P.D., J.W. GNADT & L.E. MAYS. 1989. Abducens internuclear neurons carry an inappropriate signal for ocular convergence. J. Neurophysiol. **62:** 70–81.
8. CLENDANIEL, R.A. & L.E. MAYS. 1994. Characteristics of antidromically identified oculomotor internuclear neurons during vergence and versional eye movements. J. Neurophysiol. **71:** 1111–1127.
9. ZHANG, Y., L.E. MAYS & P.D.R. GAMLIN. 1992. Characteristics of near response cells projecting to the oculomotor nucleus. J. Neurophysiol. **67:** 944–960.
10. ZHANG, Y., P.D.R. GAMLIN & L.E. MAYS. 1991. Antidromic identification of midbrain near response cells projecting to the oculomotor nucleus. Exp. Brain. Res. **84:** 525–528.
11. MAYS, L.E. 1984. Neural control of vergence eye movements: convergence and divergence neurons in midbrain. J. Neurophysiol. **51:** 1091–1108.
12. JUDGE, S.J. & B.G. CUMMING. 1986. Neurons in the monkey midbrain with activity related to vergence eye movement and accommodation. J. Neurophysiol. **55:** 915–930.
13. BAHILL, A.T., K.J. CIUFFREDA, R. KENYON & L. STARK. 1976. Dynamic and static violations of Hering's law of equal innervation. Am. J. Optom. Physiol. Opt. **53:** 786–796.
14. KENYON, R.V., K.J. CIUFFREDA & L. STARK. 1980. Unequal saccades during vergence. Am. J. Optom. Physiol. Opt. **57:** 586–594.
15. ZHOU, W. & W.M. KING. 1997. Binocular eye movements not coordinated during REM sleep. Exp. Brain. Res. **117:** 153–160.
16. ZHOU, W. & W.M. KING. 1998. Premotor commands encode monocular eye movements. Nature **393:** 692–695.
17. KING, W.M. & W. ZHOU. 2000. New ideas about binocular coordination of eye movements: is there a chameleon in the primate family tree? Anat. Rec. (New Anat.) **261:** 153–161.
18. ROBINSON, D.A. 1963. A method of measuring eye movement using a scleral search coil in a magnetic field. IEEE Trans. Biomed. Eng. **10:** 137–145.
19. SCUDDER, C.A. 1988. A new local feedback model of the saccadic burst generator. J. Neurophysiol. **59:** 1455–1475.
20. JUERGENS, R., W. BECKER & H.H. KORNHUBER. 1981. Natural and drug induced variations of velocity and duration of human saccadic eye movements: evidence for a control of the pulse generator by local feedback. Biol. Cybern. **39:** 87–96.
21. ZEE, D.S., E.J. FITZGIBBON & L.M. OPTICAN. 1992. Saccade-vergence interaction in humans. J. Neurophysiol. **68:** 1624–1641.

22. OPTICAN, L.M. & F.A. MILES. 1985. Visually induced adaptive changes in primate saccadic oculomotor control signals. J. Neurophysiol. **54:** 940–958.
23. GOLDSTEIN, H.P. & D.A. ROBINSON. 1986. Hysteresis and slow drift in abducens unit activity. J. Neurophysiol. **55:** 1044–1056.
24. MAXWELL, J.S. & W.M. KING. 1992. Dynamics and efficacy of saccade-facilitated vergence eye movements in monkeys. J. Neurophysiol. **68:** 1248–1260.
25. VAN LEEUWEN, A.F., H. COLLEWIJN & C.J. ERKELENS. 1998. Dynamics of horizontal vergence movements: interaction with horizontal and vertical saccades and relation with monocular preferences. Vision Res. **38:** 3943–3954.
26. ERKELENS, C.J., R.M. STEINMAN & H. COLLEWIJN. 1989. Ocular vergence under natural conditions. II. Gaze shifts between real targets differing in distance and direction. Proc. Roy. Soc. Lond. B Biol. Sci. **236:** 441–465.
27. BAHILL, A.T., M.R. CLARK & L. STARK. 1975. The main sequence: a tool for studying human eye movements. Math. Biosci. **24:** 191–204.
28. MAYS, L.E., J.D. PORTER, P.D. GAMLIN & C.A. TELLO. 1986. Neural control of vergence eye movements: neurons encoding vergence velocity. J. Neurophysiol. **56:** 1007–1021.
29. MAYS, L.E. & P.D. GAMLIN. 1995. Neuronal circuitry controlling the near response. Curr. Opin. Neurobiol. **5:** 763–768.
30. CLARK, R.A., J.M. MILLER & J.L. DEMER. 2000. Three-dimensional location of human rectus pulleys by path inflections in secondary gaze positions. Invest. Ophthalmol. & Visual Sci. **41:** 3787–3797.

Population Coding in Cortical Area MST

A. TAKEMURA,[a] K. KAWANO,[a] C. QUAIA,[b,c] AND F.A. MILES[b]

[a]*Neuroscience Research Institute, National Institute of Advanced Industrial Science and Technology (AIST), Tsukubashi, Ibaraki 305-8568, Japan*

[b]*Laboratory of Sensorimotor Research, The National Eye Institute, National Institutes of Health, Bethesda, Maryland 20892, USA*

[c]*Dipartimento di Elettronica, Elettrotecnica ed Informatica, Università degli Studi di Trieste, 34100 Trieste, Italy*

ABSTRACT: Disparity steps applied to large patterns elicit vergence eye movements at ultrashort latencies. Disparity tuning curves, describing the dependence of the amplitude of the initial vergence responses on the amplitude of the disparity steps, resemble the derivative of a gaussian and indicate that appropriate servo-like behavior occurs only with small disparity steps (<1 degree). Lesion data from monkeys suggest that these vergence responses are mediated, at least in part, by neurons in the medial superior temporal area of the cerebral cortex, and we here review a recent study of the associated single unit activity in that area. Few medial superior temporal neurons have disparity tuning curves whose shapes resemble the tuning curve for vergence. Yet, when the disparity tuning curves for all of the disparity-sensitive cells recorded from a given monkey are summed together, they match the tuning curves for the vergence responses of that monkey very closely, even reproducing that animal's idiosyncracies. When all of the spike trains elicited by a given disparity step are summed together to give an average discharge profile for the whole population of recorded cells, many are noisy, but others that are less so match the temporal profile of the motor response, vergence velocity, quite well. We conclude that the discharges of the disparity-sensitive cells in the medial superior temporal area each represent only a very limited aspect of the sensory stimulus (and/or associated motor response?), but when pooled together, they provide a complete description of the vergence velocity motor response: population coding.

KEYWORDS: vergence eye movements; disparity-selective neurons; population coding

INTRODUCTION

The coding of information by the activity of populations of neurons has received considerable attention in recent years. In some instances this coding relates to sensory events, such as the motion of a visual stimulus[1–3] or the orientation of the head,[4] and in others to motor responses, such as the magnitude and direction of a saccadic eye movement[5–17] or the direction of a hand movement.[18–28] The population coding in these cases generally centered on the mechanisms whereby the aggregate activity

---

Address for correspondence: F.A. Miles, Laboratory of Sensorimotor Research, The National Eye Institute, National Institutes of Health, Bethesda, MD 20892. Voice: 301-496-2455; fax: 301-402-0511.

fam@lsr.nei.nih.gov

of neurons with broad, overlapping tuning functions could achieve a finer representation of a particular sensory or motor function. Our interest, however, is in the possibility that the population activity conveys information that is not available at the level of the individual neurons, in the same way that words convey information that is not evident at the level of the individual letters. In the present paper we review a recent study of ours[29] in which the activity of neurons in the medial superior temporal (MST) area of cortex was described in association with a simple sensorimotor paradigm. This area of cortex has traditionally been treated as a pure sensory (i.e., visual) area.[d] We found that the individual neurons in MST area each discharge in relation to only a very limited aspect of the sensory stimulus (and/or associated motor response?), but the combined activity of the whole population of such cells provides a complete description of the motor response.

The sensory input in our study was a disparity step applied to a large stationary random-dot pattern, and the motor output was a vergence eye movement that was elicited at ultrashort latency (<60 ms). The neurons were recorded in the MST cortical area, because lesions of this area have caused major deficits in these eye movements.[30–32] We first review *spatial* aspects of the stimulus-response relationships and their associated single unit discharges in the MST area, concentrating initially on the disparity tuning of the individual cells and then on the disparity tuning of the summed (population) activity. We then review *temporal* aspects of the stimulus-response relationships and their associated single unit discharges in the MST area, again starting with the individual cells before considering the population activity. Finally, we discuss some consequences of population coding and some features of our study that we think allowed us to uncover it.

SHORT-LATENCY, DISPARITY-VERGENCE EYE MOVEMENTS

Before we can discuss the single unit data, we must first describe the sensorimotor paradigm that was used to probe the activity in the MST area. The paradigm that we used was first described by Busettini *et al.*,[33] who employed a dichoptic viewing arrangement to permit independent control of the images seen by each of the two eyes. Busettini and coworkers showed that brief horizontal disparity steps (≤200 ms in duration) applied to large correlated random-dot patterns (in which the two eyes saw identical patterns of dots) elicited vergence eye movements at ultrashort latencies in both monkeys[33,34] and humans.[34,35] Some sample responses taken from the study of Masson *et al.*[34] are shown in FIGURE 1A; see the continuous traces. These studies plotted disparity tuning curves describing the dependence of the amplitude of the initial vergence response on the amplitude of the disparity step and showed that these curves were well-fit by Gabor functions, resembling the derivative of a gaussian. The effective stimulus range was always small, so that appropriate servo-like behavior, in which increases in the disparity input resulted in roughly proportional increases in the vergence output (in the compensatory direction), was seen only for steps of less than 1 degree or so: see the closed symbols in

---

[d]An exception is the suggestion of Sakata, Shibutani & Kawano[51] and Newsome, Wurtz & Komatsu[52] that some MST neurons carry efference copy signals.

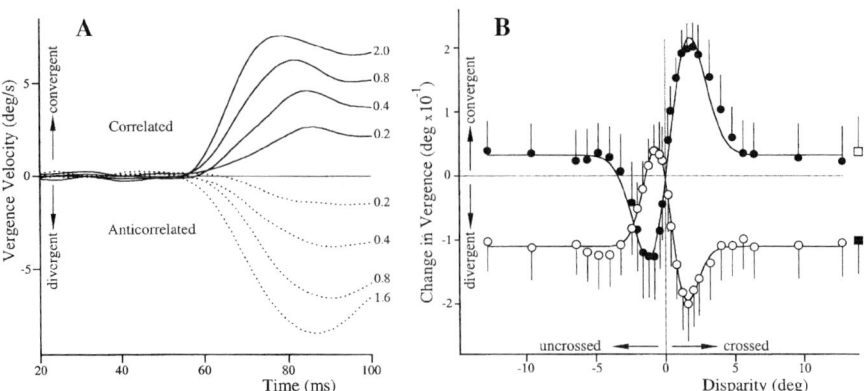

FIGURE 1. Vergence eye movements: dependence on the amplitude and direction of the disparity step. (**A**) Mean vergence velocity over time in response to crossed disparity steps applied to correlated (*continuous line*) and anticorrelated (*dotted line*) random-dot patterns. Upward deflections denote increased vergence angle, and the numbers at the ends of the traces indicate the magnitudes of the steps in degrees. (**B**) The mean change in vergence position in degrees (measured over the 33-ms period starting 60 ms after the disparity step) is plotted against the magnitude of the step in degrees: disparity tuning curves. *Closed circles*, correlated pattern. *Open circles*, anticorrelated pattern. *Continuous lines* are least-square, best-fit Gabor functions. Data from one monkey. Error bars, 1 SD. From Masson, Busettini, and Miles.[34]

FIGURE 1B. Similar disparity steps applied to anticorrelated random-dot patterns, in which the dots seen by the two eyes were of opposite contrast so that each black dot seen by one eye was matched to a white dot seen by the other eye, elicited similar vergence eye movements except that they had the opposite sign and hence were said to be "anticompensatory"[34] (dotted traces in FIG. 1A and open symbols in FIG. 1B). Interestingly, when the disparity steps were applied to the correlated patterns, subjects perceived clear changes in the depth of those patterns, provided that the steps were not too large, whereas this was *never* the case when the disparities were applied to the anticorrelated patterns regardless of the size of the steps.[34,36,37] These findings with anticorrelated patterns were used to argue that the short-latency vergence eye movements are generated independent of perception.[34]

NEURONAL RESPONSES IN THE MEDIAL SUPERIOR TEMPORAL AREA

Spatial Coding by Individual Cells

In our recent study,[29] which is the major focus of the rest of the present article, about 20% of the neurons recorded in the MST area were sensitive to disparity steps applied to large correlated random-dot patterns, and disparity tuning curves describing the dependence of their initial (open-loop) discharges on the magnitude of the disparity step were constructed. Based on the *shape* of their disparity tuning curves, MST cells were sorted into four groups using objective criteria and the fuzzy c-

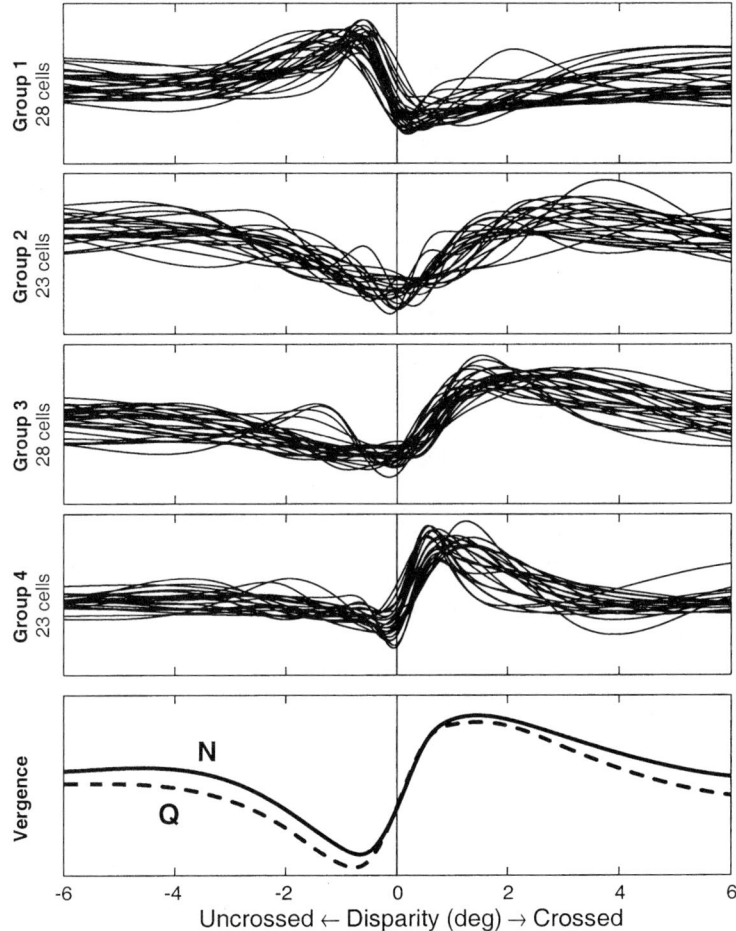

FIGURE 2. Disparity tuning curves for individual MST cells (correlated stimuli). *Upper 4 graphs*: mean change in discharge rate (measured over the 60-ms period starting 40 ms after the disparity step) is plotted against the magnitude of the disparity step in degrees; curves are normalized (all have same overall mean and same peak-to-peak amplitude) and are arranged in four groups based on the outcome of the fuzzy c-means clustering algorithm of Bezdek.[38] *Bottom*: The disparity tuning curves for the vergence responses of the two monkeys that yielded most of the data (N and Q). All traces are spline interpolations. From Takemura et al.[29]

means clustering algorithm of Bezdek[38] (FIG. 2). These four groups had features in common with four of the classes of disparity-selective neurons that others have described in striate cortex,[39] although groups 2, 3, and 4 appeared to be part of a continuum. About half the cells that responded to disparity steps applied to correlated patterns were also tested with disparity steps applied to anticorrelated random-dot patterns and all modulated significantly ($p < 0.005$, 1-way ANOVA). Thus, the dis-

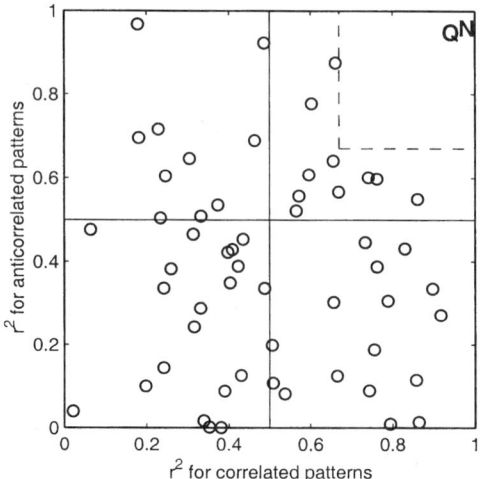

FIGURE 3. Coding of vergence by individual cells: correlated *vs.* anticorrelated stimuli. The disparity tuning curves of the individual cells were fitted to the disparity tuning curves of the associated vergence responses and r^2 values for the least-square, best fits were computed. This graph plots the r^2 values for the data obtained with anticorrelated patterns against those obtained with correlated patterns. No cell had r^2 values that exceeded 0.67 for both stimuli (indicated by the *dashed lines*). Also shown (indicated by their identifying letters) are the r^2 values for the fits between the summed activity and the vergence responses for the two monkeys, N and Q (from which all the unit data plotted here were obtained). From Takemura *et al.*[29]

parity-sensitive cells in the MST area responded even when the disparity stimuli were not perceived in depth.

A few MST neurons had disparity tuning curves whose shapes closely resembled the shapes of the tuning curve for vergence when the stimuli were *either* correlated *or* anticorrelated stimuli, but no neuron had curves that showed a close match to the vergence responses with *both* types of stimuli. This was examined quantitatively by fitting each neuron's disparity tuning curves to the disparity tuning curves for vergence (gain and offset free parameters), treating the data from each animal separately. The goodness of these fits was then assessed by computing the fraction of the disparity-induced variation in vergence accounted for by the fits (r^2). FIGURE 3 plots the r^2 values obtained with correlated stimuli against the r^2 values obtained with anticorrelated stimuli for each of the cells examined with both types of disparity stimuli (circular symbols). It is evident that there is considerable scatter in these data, the goodness of the fit with the correlated stimulus clearly having no relation to the goodness of the fit with the anticorrelated stimulus. The upper right corner of the plot, which is where pure vergence encoding cells would be expected, is devoid of any cells. Thus, there were a few cells that had high r^2 values with *either* correlated *or* anticorrelated stimuli but none that had high r^2 values with *both* types of stimuli: the goodness of the fit clearly depended on the *type of stimulus* used to generate the vergence, indicating that there were no pure vergence-encoding cells.

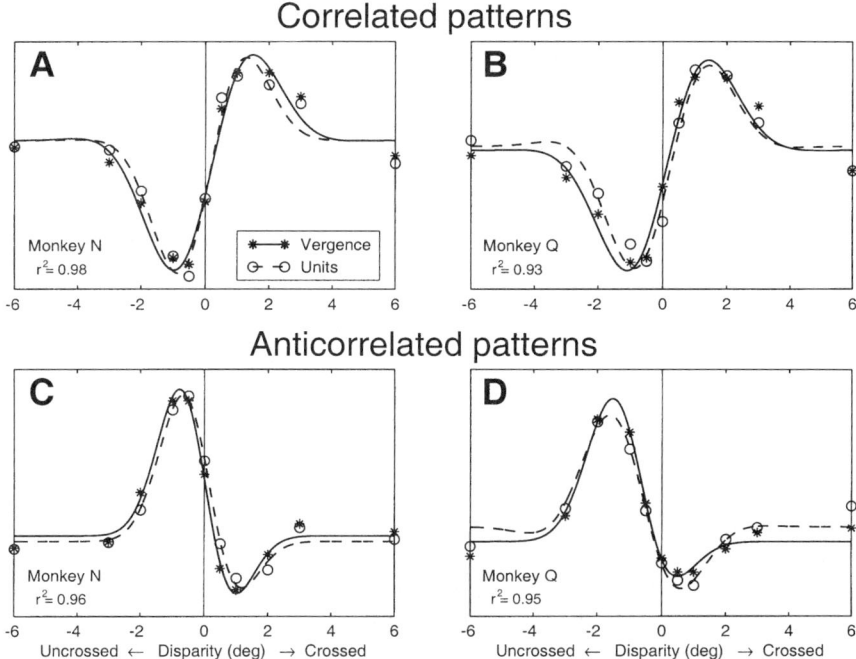

FIGURE 4. Spatial coding of vergence by populations of cells. (**A,B**) correlated stimuli. (**C,D**) anticorrelated stimuli. Disparity tuning data for vergence (*stars*) and for the least-square, best-fit summed activity (*circles*) for monkey N (**A,C**) and monkey Q (**B,D**). Note that before summing, the individual unit curves with negative slopes over the servo range (i.e., for disparities over the range ±1 degree) were inverted. Curves are the least-square, best-fit Gabor functions for the vergence data (*continuous line*) and the summed activity data (*dashed line*). The four disparity tuning curves for the summed activity were fitted to the disparity tuning curves for the associated vergence responses, and the r^2 values for the least-square, best fits are shown at the bottom left in each panel. From Takemura et al.[29]

Spatial Coding by the Population of Cells

Most of the unit data were recorded from two monkeys (designated N and Q), and when the tuning curves of all cells obtained from a given monkey were summed together (using a simple, as opposed to weighted, sum), they fitted the tuning curves for the vergence responses of that same monkey very well, always accounting for at least 93% of the disparity-induced variability.[e] This was true for the unit data obtained with both correlated and anticorrelated stimuli: see FIG. 4, and the symbols N and Q plotted in FIGURE 3 (the latter represents the r^2 values for the *summed ac-*

---

[e]One manipulation of the data was critical to achieve the good fits: the sign of those tuning curves that had a negative slope in the important servo range, ±1 degree, was reversed; this involved all cells in Group 1 and one in Group 2. Given that the sign of the vergence responses was positive over the range in question—solely by convention—meant that all cells in all groups would make a positive net contribution to the population vergence signal. Of course, sign differences like this can be achieved by appropriate excitatory/inhibitory connections.

tivity of each of the two monkeys from which all of the data in this figure were recorded).

The shapes of the disparity tuning curves for the vergence eye-movement data obtained with correlated stimuli were similar for the two animals (compare FIG. 4A and 4B), but the shapes of the disparity tuning curves for the vergence data obtained with anticorrelated stimuli differed significantly for the two animals (compare FIG. 4C and 4D). These differences in the vergence data obtained with anticorrelated stimuli were such that the summed neural activity from monkey N gave a poor fit to the vergence data obtained from monkey Q ($r^2 = 0.29$) and vice versa ($r^2 = 0.35$). Thus, the summed neural activity for a given animal matched only the vergence data obtained from that same animal, that is, the population data reproduced the idiosyncratic differences between the vergence responses of the two monkeys, a finding that encourages the belief that the ensemble coding of vergence in the MST area has biological significance.

We were also interested in the relative contributions of each of the four groups of cells that were identified by fuzzy cluster analysis to the population coding of vergence: it was possibe that one or more groups actually made the fits between aggregate activity and vergence responses worse than they would otherwise have been. Sufficient data to examine this question were available only for correlated stimuli, and these suggested that all four groups of cells were necessary for a good fit between summed activity and vergence, especially for monkey N. Thus, excluding all of the cells in any one of the four groups always decreased the goodness of fit in monkey N. Furthermore, randomly excluding equivalent numbers of cells indicated that the probability of achieving these results by chance was always <0.008: bootstrap statistic.[40] The data from monkey Q were less compelling in that only exclusion of Group 1 resulted in significant worsening of the fit. Using a genetic algorithm made it possible to obtain an estimate of the subsets of neurons whose tuning curves, when summed together, gave the best fit to the vergence tuning curves. These subsets invariably included cells from all four groups. These findings indicate that the encoding of vergence depended on contributions from across the entire spectrum of tuning curves found among disparity-selective MST cells. Thus, the population code depends crucially on the aggregate activity of a heterogeneous collection of cells.

Temporal Coding

Additional analyses of the spike trains elicited by a given disparity step revealed considerable variation across cells in the latency, amplitude, and time course of the changes in discharge rate: see FIGURE 5, which shows the discharge frequency profiles of 20 cells recorded from monkey N in response to 2-degree crossed-disparity steps. When all of the spike trains recorded from a given monkey in response to a given disparity step were summed together, providing a population discharge profile for that animal's response to that step, many were rather noisy, ruling out any possibility that they might match the profile of the associated vergence responses. However, other summed discharge profiles were much cleaner and matched the temporal profile of the vergence *velocity* response quite well (free parameters: gain, y-offset, and x-offset). An example is seen at the bottom of FIGURE 5; compare the dashed trace, depicting the summed neural activity, with the continuous trace, depicting the vergence velocity profile. In the example shown, the x-offset that gave the best fit de-

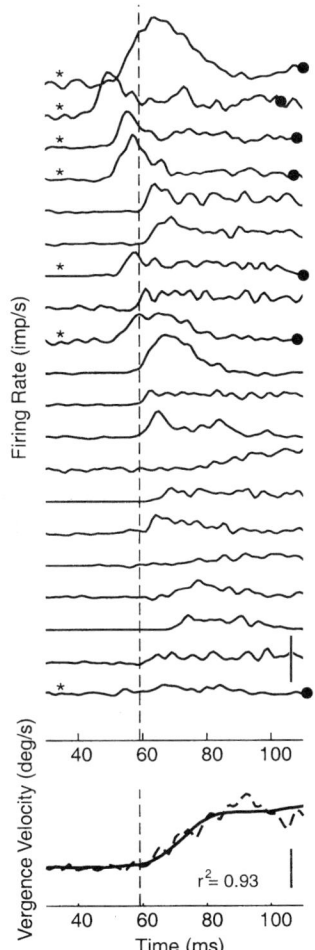

FIGURE 5. Time course of the neuronal responses (correlated stimuli). *Upper traces* show the changes in mean discharge rate over time in response to 2-degree crossed-disparity steps for each of the 20 units that modulated most with this stimulus (ranked in descending order of their mean discharge rates over the period 30–110 ms after the step). *Vertical dashed line* is the estimated latency of the vergence response (59 ms). Seven of the units, indicated by *asterisks*, had response latencies <51 ms, and the dots on the traces indicate the estimated times at which the closing of the disparity feedback loop could first influence the discharges of these units. (The response latencies of the remaining units were too long for the disparity-feedback loop to close during the 110-ms time window shown.) *Bottom traces* show the changes over time in vergence velocity (*continuous line*) and in least-square, best-fit summed activity (*dashed line*, with a time lag of 18 ms). Calibration bars: 500 imps/s, 5°/s. Data from monkey N. From Takemura et al.[29]

layed the neural response by 18 ms, and the summed activity accounted for 93% of the disparity-induced variation in vergence. In fact, r^2 values were greater than 0.9 for 40% of the fits (8/20, there being temporal data for 10 disparity steps for both monkeys). In view of the noise problems inherent in spike trains and the fact that many discharge profiles of the individual cells showed a strong initial transient, we were surprised that so many summed-activity profiles approximated the vergence velocity profile, which generally showed a monotonic rise with no hint of an initial transient overshoot (FIG. 5). Thus, whereas initial phasic components were common in the averaged discharge profiles of the individual cells, they were rare in the population average profiles, presumably because of the latency jitter: smoothing by temporal summation.

DISCUSSION

Our findings show that the summed activity of the disparity-sensitive cells that we recorded in the MST area effectively describes the magnitude, direction, and time course of the initial (open-loop) vergence velocity responses elicited by disparity steps applied to large textured patterns. Thus, the population activity provides a complete description of the vergence motor output. On the other hand, there was little or no hint of this vergence information at the level of the individual cells, implying that the representation of vergence velocity in the MST area is an *emergent property* of the population activity. Our latency data, as well as the x-offsets that gave the best fits to the temporal profiles, suggest that this activity in the MST area occurs early enough to play an active role in the generation of even the earliest vergence eye movements, consistent with other findings that lesions of the MST area result in significant deficits in such eye movements.[30–32] It is possible that the individual cells carry pure sensory signals (binocular disparity), in which case the pure motor signal conveyed by the population would be nothing more than the arithmetic sum of a set of sensory signals, and the sensory-motor transformation could be accomplished by a simple summing junction. This means that discharges in the MST area could be viewed as both sensory and motor depending on the level of scrutiny (individual cells or population) and would represent a clear example of parallel (rather than serial or hierarchical) signal processing. It has been argued previously that parallel processing is necessary to achieve the ultrashort latency that characterizes the motor responses in our study.[41]

Subsequent Signal Processing?

Various detailed schemes have been proposed for decoding the information embedded in the activity of populations of neurons: for recent reviews see Abbott and Sejnowski.[42] Based on the effects of excluding any one of the four groups of cells from the population sum and on the outcome of the genetic algorithm, we concluded that the encoding of vergence motor responses in the MST area requires contributions from across the entire spectrum of disparity-selective cells that we recorded. This raises the possibility that a random selection of these cells would suffice to generate the vergence responses and that the projection from the MST area to the next stage in the processing of the vergence drive signal need not involve complex computations or connectivity rules. That the aggregate activity of the disparity-selective cells in the MST area relates to vergence *velocity* (rather than vergence *position*) is consistent with the commonly held view that the MST area is exclusively involved in the processing of motion signals; see Eifuku and Wurtz[43,44] for recent references.

There are a number of anatomical routes by which the vergence signals in the MST area might gain access to the vergence premotor centers: see Takemura *et al.*[29] for a recent review. Contrary to earlier notions, it is now clear that medial rectus motoneurons carry both position and velocity signals for the control of vergence[45] and that both types of signals are evident in the midbrain neurons thought to carry the command signals for vergence eye movements.[46–48] This means that the vergence velocity signal conveyed by the population activity in the MST area would need to be supplemented with a position signal in order to provide the complete command signal to the midbrain neurons generating vergence eye movements. This could be achieved by a single integration in the projection pathways, similar to that which has long been postulated to occur in the pathways from the semicircular canals to the oculomotor motoneurons.[49,50] It is interesting that the initial phasic component that is often seen in the discharge profiles of individual MST cells (FIG. 5)—and might be thought to represent a vergence acceleration signal, albeit rather crude—is filtered out of the population response by the latency jitter and temporal summation.

Multiplexing?

Population coding raises the possibility that these same MST cells can participate in other functions unrelated to vergence. This might involve subgroupings of the cell population that we have recorded operating through other output pathways to achieve some other function necessitating disparity information. Another possibility is that these same cells also carry signals unrelated to disparity and belong to other groupings/populations of cells that combine to achieve other purposes through their shared connections. In such a distributed network, the functionality depends critically on the pooling achieved by the shared connectivity. Clearly, such multiplexing would involve some delicate balancing of inputs to minimize cross-talk and thereby render the different functions of the individual cells orthogonal at the population level. Of course, failure to do this at one level of the system might be corrected by adding appropriate compensatory signals at subsequent levels.

Concerning the Stimulus Sets Required to Uncover Population Codes

We think that the stimulus set in our study has a number of features, some novel, that, in retrospect, were crucial for revealing the population coding in the MST area. Firstly, we sought only to identify the possible contributions of the cells, both individually and collectively, to one particular well-defined behavior: disparity vergence. Thus, we ignored the possibility that these cells discharge in relation to other stimuli and/or motor responses. In particular, we did *not* use an extensive set of stimulus parameters as others have often done in hopes of identifying the stimuli "preferred" by the individual cells. Aside from the problem that one can *never* be sure if there is some untried stimulus that would be "preferred" over all others, we now think it is possible that there is orthogonal multiplexing as discussed earlier, that is, the cells discharge in relation to more than one stimulus dimension. Secondly, we think it is also significant that our stimuli extended well beyond what we presume to be the biologically useful (servo) range. It is the responses to the outlying stimuli that give the tuning curves their *individual shapes*, that is, their identity. Thirdly, we included stimuli—anticorrelated patterns—that are rarely, if ever, encountered in the real world and that elicit vergence responses with idiosyncratic features.[f] That these subject-specific features of the motor responses were mirrored by the aggregate neu-

---

[f]The relatively modest intersubject variability in the vergence responses to correlated stimuli is presumably due to the operation of adaptive mechanisms that function to optimize performance and, given the common biological matrix and survival imperative, converge on similar solutions in different subjects. In contrast, anticorrelated stimuli are outside normal experience, so that the system's anticompensatory responses to them have no biological significance; perceptually, the brain appears to deal with *anti*correlated stimuli the same way it deals with *un*correlated stimuli.

ronal responses helps to persuade us that the population coding is not simply a fortuitous epiphenomenon but has biological significance.

REFERENCES

1. MAUNSELL, J.H.R. & D.C. VAN ESSEN. 1983. Functional properties of neurons in middle temporal visual area of the macaque monkey. I. Selectivity for stimulus direction, speed, and orientation. J. Neurophysiol. **49:** 1127–1147.
2. PRIEBE, N.J., M.M. CHURCHLAND & S.G. LISBERGER. 2001. Reconstruction of target speed for the guidance of pursuit eye movements. J. Neurosci. **21:** 3196–3206.
3. STEINMETZ, M.A. et al. 1987. Functional properties of parietal visual neurons: radial organization of directionalities within the visual field. J. Neurosci. **7:** 177–191.
4. SCHOR, R.H., A.D. MILLER & D.L. TOMKO. 1984. Responses to head tilt in cat central vestibular neurons. I. Direction of maximum sensitivity. J. Neurophysiol. **51:** 136–146.
5. HENN, V. & B. COHEN. 1976. Coding of information about rapid eye movements in the pontine reticular formation of alert monkeys. Brain Res. **108:** 307–325.
6. BÜTTNER, U., J.A. BÜTTNER-ENNEVER & V. HENN. 1977. Vertical eye movement related unit activity in the rostral mesencephalic reticular formation of the alert monkey. Brain Res. **130:** 239–252.
7. SCHLAG-REY, M. & J. SCHLAG. 1977. Visual and presaccadic neuronal activity in thalamic internal medullary lamina of cat: a study of targeting. J. Neurophysiol. **40:** 156–173.
8. THIER, P. et al. 2000. Encoding of movement time by populations of cerebellar Purkinje cells. Nature **405:** 72–76.
9. VAN GISBERGEN, J.A.M., A.J. VAN OPSTAL & A.A.M. TAX. 1987. Collicular ensemble coding of saccades based on vector summation. Neuroscience **21:** 541–555.
10. SPARKS, D.L., C. LEE & W.H. ROHRER. 1990. Population coding of the direction, amplitude, and velocity of saccadic eye movements by neurons in the superior colliculus. Cold Spring Harbor Symp. Quant. Biol. **55:** 805–811.
11. LEE, C., W.H. ROHRER & D.L. SPARKS. 1988. Population coding of saccadic eye movements by neurons in the superior colliculus. Nature **332:** 357–360.
12. ANDERSON, R.W. et al. 1998. Two-dimensional saccade-related population activity in superior colliculus in monkey. J. Neurophysiol. **80:** 798–817.
13. MUNOZ, D.P. & R.H. WURTZ. 1995. Saccade-related activity in monkey superior colliculus. II. Spread of activity during saccades. J. Neurophysiol. **73:** 2334–2348.
14. OPTICAN, L.M. 1995. A field theory of saccade generation: temporal-to-spatial transform in the superior colliculus. Vision Res. **35:** 3313–3320.
15. WURTZ, R.H. 1996. Vision for the control of movement. The Friedenwald Lecture. Invest. Ophthalmol. & Visual Sci. **37:** 2131–2145.
16. QUAIA, C., P. LEFÈVRE & L.M. OPTICAN. 1999. Model of the control of saccades by superior colliculus and cerebellum. J. Neurophysiol. **82:** 999–1018.
17. VAN OPSTAL, A.J. & J.A.M. VAN GISBERGEN. 1989. A nonlinear model for collicular spatial interactions underlying the metrical properties of electrically elicited saccades. Biol. Cybern. **60:** 171–183.
18. GEORGOPOULOS, A.P. et al. 1982. On the relations between the direction of two-dimensional arm movements and cell discharge in primate motor cortex. J. Neurosci. **2:** 1527–1537.
19. GEORGOPOULOS, A.P., A.B. SCHWARTZ & R.E. KETTNER. 1986. Neuronal population coding of movement direction. Science **233:** 1357–1460.
20. GEORGOPOULOS, A.P., R.E. KETTNER & A.B. SCHWARTZ. 1988. Primate motor cortex and free arm movements to visual targets in three-dimensional space. II. Coding of the direction of movement by a neuronal population. J. Neurosci. **8:** 2928–2937.
21. KALASKA, J.F., R. CAMINITI & A.P. GEORGOPOULOS. 1983. Cortical mechanisms related to the direction of two-dimensional arm movements: relations in parietal area 5 and comparison with motor cortex. Exp. Brain Res. **51:** 247–260.

22. SCHWARTZ, A.B. 1993. Motor cortical activity during drawing movements: population representation during sinusoid tracing. J. Neurophysiol. **70:** 28–36.
23. SCHWARTZ, A.B. 1994. Direct cortical representation of drawing. Science **265:** 540–542.
24. MORAN, D.W. & A.B. SCHWARTZ. 1999. Motor cortical representation of speed and direction during reaching. J. Neurophysiol. **82:** 2676–2692.
25. MORAN, D.W. & A.B. SCHWARTZ. 1999. Motor cortical activity during drawing movements: population representation during spiral tracing. J. Neurophysiol. **82:** 2693–2704.
26. SCHWARTZ, A.B. & D.W. MORAN. 1999. Motor cortical activity during drawing movements: population representation during lemniscate tracing. J. Neurophysiol. **82:** 2705–2718.
27. MAYNARD, E.M. et al. 1999. Neuronal interactions improve cortical population coding of movement direction. J. Neurosci. **19:** 8083–8093.
28. SCHWARTZ, A.B. & D.W. MORAN. 2000. Arm trajectory and representation of movement processing in motor cortical activity. Eur. J. Neurosci. **12:** 1851–1856.
29. TAKEMURA, A. et al. 2001. Single unit activity in cortical areas MST associated with disparity-vergence eye movements: evidence for population coding. J. Neurophysiol. **85:** 2245–2266.
30. TAKEMURA, A. et al. 1999. Evidence that disparity-sensitive cells in medial superior temporal area contribute to short-latency vergence eye movements. Soc. Neurosci. Abstr. **25:** 1400.
31. TAKEMURA, A., Y. INOUE & K. KAWANO. 2000. The role of MST neurons in short-latency visual tracking eye movements. Soc. Neurosci. Abstr. **26:** 1715.
32. TAKEMURA, A., Y. INOUE & K. KAWANO. 2002. Visually driven eye movements elicited at ultra-short latency are severely impaired by MST lesions. Ann. N.Y. Acad. Sci. This volume.
33. BUSETTINI, C., F.A. MILES & R.J. KRAUZLIS. 1996. Short-latency disparity vergence responses and their dependence on a prior saccadic eye movement. J. Neurophysiol. **75:** 1392–1410.
34. MASSON, G.S., C. BUSETTINI & F.A. MILES. 1997. Vergence eye movements in response to binocular disparity without depth perception. Nature **389:** 283–286.
35. BUSETTINI, C., E.J. FITZGIBBON & F.A. MILES. 2001. Short-latency disparity vergence in humans. J. Neurophysiol. **85:** 1129–1152.
36. CUMMING, B.G. & A.J. PARKER. 1997. Responses of primary visual cortical neurons to binocular disparity without depth perception. Nature **389:** 280–283.
37. COGAN, A.I., A.J. LOMAKIN & A.F. ROSSI. 1993. Depth in anticorrelated stereograms: Effects of spatial density and interocular delay. Vision Res. **33:** 1959–1975.
38. BEZDEK, J.C. 1981. Pattern Recognition with Fuzzy Objective Function Algorithms. Plenum. New York.
39. POGGIO, G.F., F. GONZALEZ & F. KRAUSE. 1988. Stereoscopic mechanisms in monkey visual cortex: binocular correlation and disparity selectivity. J. Neurosci. **8:** 4531–4550.
40. EFRON, B. & R. TIBSHIRANI. 1991. Statistical data analysis in the computer age. Science **253:** 390–395.
41. MILES, F.A. 1998. The neural processing of 3-D visual information: evidence from eye movements. Eur. J. Neurosci. **10:** 811–822.
42. ABBOTT, L. & T.J. SEJNOWSKI. 1999. Neural Codes and Distributed Representations: Foundations of Neural Computation. The MIT Press. Cambridge, MA.
43. EIFUKU, S. & R.H. WURTZ. 1998. Response to motion in extrastriate area MSTl: center-surround interactions. J. Neurophysiol. **80:** 282–296.
44. EIFUKU, S. & R.H. WURTZ. 1999. Response to motion in extrastriate area MSTl: disparity sensitivity. J. Neurophysiol. **82:** 2462–2475.
45. GAMLIN, P.D.R. & L.E. MAYS. 1992. Dynamic properties of medial rectus motoneurons during vergence eye movements. J. Neurophysiol. **67:** 64–74.
46. JUDGE, S.J. & B.G. CUMMING. 1986. Neurons in the monkey midbrain with activity related to vergence eye movement and accommodation. J. Neurophysiol. **55:** 915–930.

47. MAYS, L.E. et al. 1986. Neural control of vergence eye movements: neurons encoding vergence velocity. J. Neurophysiol. **56:** 1007–1021.
48. ZHANG, Y., P.D. GAMLIN & L.E. MAYS. 1991. Antidromic identification of midbrain near response cells projecting to the oculomotor nucleus. Exp. Brain Res. **84:** 525–528.
49. SKAVENSKI, A. & D.A. ROBINSON. 1973. Role of abducens motoneurons in the vestibuloocular reflex. J. Neurophysiol. **36:** 724–738.
50. GALIANA, H.L. & J.S. OUTERBRIDGE. 1984. A bilateral model for central neural pathways in vestibuloocular reflex. J. Neurophysiol. **51:** 210–241.
51. SAKATA, H., H. SHIBUTANI & K. KAWANO. 1983. Functional properties of visual tracking neurons in posterior parietal association cortex of the monkey. J. Neurophysiol. **49:** 1364–1380.
52. NEWSOME, W.T., R.H. WURTZ & H. KOMATSU. 1988. Relation of cortical areas MT and MST to pursuit eye movements. II. Differentiation of retinal from extraretinal inputs. J. Neurophysiol. **60:** 604–620.

Adaptive Control of Vergence in Humans

CLIFTON M. SCHOR, JAMES S. MAXWELL, JEFREY McCANDLESS, AND ERICH GRAF

University of California at Berkeley, School of Optometry, Berkeley, California 94720, USA

ABSTRACT: Vergence eye alignment minimizes horizontal, vertical, and cyclodisparities to optimize stereo-depth perception. Only the horizontal component of vergence is under voluntary control. Couplings with voluntary version and horizontal vergence guide vertical vergence and cyclovergence. Can these couplings be modified in response to sensory demands on binocular vision? We have modified vertical vergence and cyclovergence in response to optical changes in disparity. Vertical vergence was stimulated with aniseikonic lenses that exaggerated vertical disparity in tertiary gaze. Vertical vergence adapted in an hour to produce nonconcomitant changes in vertical phoria that varied with vertical eye position in tertiary gaze. Cyclovergence was stimulated with cyclodisparities that varied with gaze elevation and convergence angle. Cyclovergence adapted within 2 hours to produce nonconcomitant changes in cyclophoria that varied with gaze elevation and convergence. The adaptive couplings for vertical vergence and cyclovergence are modeled as a combination of passive orbital mechanics and active gain control of the vertical recti and obliques. Vergence adaptation is a calibration process that adjusts the innervation for horizontal, vertical, and torsion components of vergence to the physical constraints set by the extraocular muscles and orbital connective tissues. Passive orbital mechanics simplify the neural control for precise vertical vergence and cyclovergence that are needed to achieve binocular alignment under open-loop conditions in response to perceived spatial location.

KEYWORDS: cyclovergence; Listing's law; orbital mechanics

INTRODUCTION

The primary goal of binocular alignment is to optimize disparity stimuli for stereo-depth perception and binocular sensory fusion. Vergence movements minimize horizontal, vertical, and torsional disparities at the fovea. Residual disparities elsewhere in the visual field are cues to steropsis, which is one of our primary cues to spatial layout. Stereo is used to perceive relative depth, surface slant, and object volume. All three components of disparity are needed to interpret depth and orientation of objects relative to the head. Horizontal disparity by itself is an ambiguous spatial cue. For example, the gap between the front and back surface of a cup subtends different horizontal disparities at different distances. If we are to grasp the cup,

Address for correspondence: Clifton M. Schor, University of California at Berkeley, School of Optometry, Berkeley, CA 94720. Voice: 510-642-1130; fax: 510-643-5109.
schor@socrates.berkeley.edu

we need to know its three-dimensional volume. Similarly, several fronto-parallel surfaces viewed in symmetrical and asymmetrical convergence subtend different patterns of horizontal retinal disparity even thought they have the same orientation relative to the head. Horizonal disparity information is disambiguated with information about target distance and direction. Information about target location can be obtained from extraretinal cues, that is, sensed eye position, or from retinal cues in the form of vertical and cyclodisparity.[1,2] To perceive space accurately from these retinal cues we need to distinguish between disparities produced by the optical geometry of target location relative to the head and errors of binocular eye alignment. This distinction is made possible by precise vergence control of three-dimensional eye alignment.

Two classes of stimuli are used for fine and coarse stages of vergence control. Retinal image disparity provides a feedback signal to refine and maintain binocular eye alignment. Disparity vergence and sensory fusion are sensitive to horizontal retinal disparities ranging from several seconds of arc to 1 degree from the plane of fixation. Larger disparities appear diplopic and only support stereopsis for brief periods of time (transient stereopsis).[3–5] They have less influence on static disparity vergence than do small disparities, so that when one target is fixated binocularly, another target at a different distance does not engage vergence and disrupt the current state of fixation. Volitional changes in convergence to newly attended targets are in response to spatiotopic cues or perceived distance.[6] Disparities associated with large gaze shifts are beyond the operating range of disparity vergence, so that retinal image feedback is not used to guide the initial gaze shift. The horizontal vergence system makes an open-loop response that reduces the initial large disparity of the attended target until it falls within the operating range of the disparity vergence system. Then feedback from retinal image disparity becomes available to refine the response in a coarse to fine strategy.[6]

CROSS-COUPLING OF VOLUNTARY AND INVOLUNTARY COMPONENTS OF THE NEAR RESPONSE

Horizontal, vertical, and torsional components of vergence respond during both coarse and fine stages of a gaze shift; however, only the horizontal component is under voluntary control. Vertical vergence and cyclovergence are not under voluntary control, yet they still accompany the voluntary gaze shift. These involuntary systems are guided by low-level couplings with convergence.[7,8] The association of cyclovergence with convergence is described as part of the near response.[9] We expand the definition of the near response to include changes of vertical vergence associated with convergence. The coupling between convergence and vertical vergence can be demonstrated by occluding one eye and fixating a target in tertiary gaze. When viewing a near target in symmetrical convergence, tertiary targets seen in the periphery lie at different distances from the two eyes. These tertiary targets subtend different retinal image sizes that result in vertical disparities as described in Fick coordinates. Yet the two eyes are precisely aligned with the target, even without feedback from retinal image disparity.[10–12] The magnitude of vertical vergence, described as the ratio of left/right eye position in Fick coordinates, normally nulls the vertical disparity associated with a given tertiary target location, even when one eye is occluded. The coupling between vertical vergence and the combination of horizontal version (H) and convergence (C) components of eye position is described by the following relationship:

$$V_l/V_r = K_v \tan^{-1}[\cos(H_l - C/2)]/\tan^{-1}[\cos(H_r + C/2)]$$

where K_v quantifies the magnitude of the coupling. The coupling results in an accurate response when $K_v = 1.0$, and the eyes are aligned with the tertiary target under open-loop conditions. Empirical open-loop measures of vertical vergence while viewing tertiary targets illustrate that K_v normally equals 1.0.[10] Note that $V_l/V_r = K_v$ when vertical eye position is described in Helmholtz coordinates. We have investigated whether the precision of vertical eye alignment is a consequence of adjustments to Hering's law by an active neural coupling with convergence and horizontal eye position or the consequence of passive orbital mechanics.

The coupling between convergence and cyclovergence can be demonstrated by observing cyclovergence while tracking a spot that has no cyclodisparity cues. Like vertical and horizontal eye position, torsion is described with respect to a frame of reference or primary position of gaze. Listing's law provides a convenient way to empirically measure primary position.[13] Listing's law describes zero torsion in any direction of gaze for eye orientations that would result from rotating the eye from primary position to another position about an axis lying in a single plane. That plane is approximately fronto-parallel and was referred to by Helmholtz as Listing's plane.[14] Primary position is orthogonal to Listing's plane, which can be measured empirically by a plane fit to measures of eye position expressed as rotation vectors.[15] Hering noted that when the eyes were converged, cyclotorsion varied systematically with vertical gaze from the orientation predicted by Listing's law.[16] The eyes intorted during elevation and extorted during depression. Until recently these were considered exceptions to Listing's law.[9,17] However, recent studies reveal that when ocular torsion is measured while the eyes are converged, eye orientation is still consistent with Listing's law. A single plane describes the orientation of the eye in any position; however, the orientation of this plane is different from Listing's plane when convergence is zero. When the eyes are converged, Listing's planes for the two eyes are diverged by approximately the amount the eyes are converged.[18–20]

The divergence or yaw orientation of Listing's planes during convergence benefits stereoscopic depth perception. If Listing's plane did not diverge during convergence, then the horopter would cease to exist when gaze was elevated. Horizontal lines in space would subtend incyclo-torted retinal image disparities when gaze was elevated and excyclo-torted retinal image disparities when gaze was depressed. If Listing's planes diverged by the full amount of convergence, horizontal lines would be imaged precisely on corresponding horizontal meridians of the retina, and the horopter would exist when gaze was elevated. The cyclovergence necessary for coplanar alignment of the planes of regard depends on convergence and eye elevation. In Helmholtz coordinates, it is described by the following relationship:

$$T = K_t * 4(\tan^{-1}[\tan C/4 * \tan V/2])^{14}$$

When expressed in radians, $T = K_t * V * C/2$.[21] Cyclovergence equals right–left eye position, and sign convention follows a right-hand rule. Horizontal and vertical eye position is described in Helmholtz coordinates, so that foveal alignment is achieved by equal elevation of the two eyes. When $K_t = 1.0$, cyclodisparity of horizontal lines in space remains zero during gaze elevation in symmetrical convergence. Empirical measures of cyclovergence during convergence indicate that K_t ranges from 0.5–0.8 rather than the ideal 1.0. Because K_t is less than 1, there are residual retinal cyclodisparities that vary from excyclo-disparity in down gaze to incyclo-disparity in up gaze.[22] The lower value is interpreted as a compromise between obtaining an economy of movement and an optimization of retinal image disparity for stereoscopic depth perception.[13] We investigated whether the change in orientation of Listing's plane with horizontal vergence is controlled actively by neural control or passively by orbital mechanics.

PLASTICITY OF VERTICAL VERGENCE COUPLING

The couplings between convergence and the two involuntary components of vergence are a benefit to binocular vision if they have values that keep the eyes aligned even if binocular retinal feedback is temporarily interrupted. The precision of vertical vergence in tertiary gaze and the change in orientation of Listing's planes for the two eyes during convergence suggest that the couplings respond to the sensory demands of binocular vision. If so, can these couplings be modified in response to novel demands placed on binocular vision? Normally targets in the right field subtend larger images in the right eye than left eye, and targets in the left field subtend larger images in the left eye than the right eye. This image size difference (aniseikonia) produces right hyperdisparities in the upper field and right hypodisparities in the lower field. Schor and McCandless[7] used magnifying lenses (6% magnification) during asymmetrical convergence to test if subjects could alter their vertical eye alignment in relation to an exaggerated version of this relationship. When targets were presented to the right (14 degrees), a larger than normal right hyperdisparity was seen in the upper field (up 14 degrees) and a larger than normal right hypodisparity was seen in the lower field (down 14 degrees). When targets were presented to the left (14 degrees), the situation was reversed, that is, an increased left hyperdisparity was seen in the upper field and a hypodisparity in the lower. Subjects attempted to fuse the vertical disparities presented at these four tertiary positions for 1 hour. In a second experiment,[7] vertical disparities were presented as a function of both vertical eye position and horizontal vergence. So, for example, at far distance a right hyperdisparity was presented in the upper field and a right hypodisparity in the lower, and at a near distance a right hypodisparity was presented in the upper field and a right hyperdisparity in the lower. In both experiments the vertical phoria changed by approximately half the optically induced vertical disparity. The nonconcomitant vertical phoria varied in magnitude with version and vergence eye position. The results demonstrate plasticity in the coupling between open-loop vertical vergence and specific combinations of vertical and horizontal eye position and horizontal convergence. Clearly the gain of the couplings with vertical vergence responded to sensory demands placed on binocular vision.

TABLE 1. Pre-adjusted and postadapted gains in K values

	Subject K-value				
	CMS	JSM	EWG	MH	Average
Reversed condition					
Pre-	0.74	1.04	0.66	0.85	0.82
Post-	0.33	0.26	0.00	0.57	0.29
Exaggerated condition					
Pre-	0.84	0.94	0.82	0.66	0.82
Post-	1.23	1.09	1.07	0.76	1.04

PLASTICITY OF CYCLOVERGENCE COUPLING

The normal variations of open-loop cyclovergence that occur during convergence with gaze elevation[18–20] also exhibit plasticity in response to sensory demands on binocular vision.[23,24] Because the coupling between cyclovergence and vertical eye position normally has a gain of less than 1 ($K_t < 1.0$), horizonal lines subtend excyclodisparities in downward gaze and incyclodisparities in upward gaze.[22] These disparities would be at least twice as large if there were no changes in the orientation of Listing's planes with convergence. In our experiments, sensory demands on cyclofusion were altered by either exaggerating or reversing the sign of this pattern of cyclodisparity.[23] In the exaggerated condition, 5-degree cyclodisparities were presented in the midsagittal plane with excyclodisparity in 10-degree upward gaze and incyclodisparity in 10-degree downward gaze during far fixation (zero convergence), and the opposite pattern of cyclodisparity was presented in relation to vertical eye position in near fixation (10-degree convergence). In the reversed condition, the pattern of cyclodisparities was opposite that in the exaggerated condition. Cyclovergence partially adapted within the 2-hour training period to produce nonconcomitant changes in cyclophoria that varied with gaze elevation and convergence. When plotted with rotation vectors,[15] Listing's planes diverged more than the pre-adapted state with convergence after adaptation to the exaggerated condition, and they diverged less than the pre-adapted state after adaptation to the reversed condition. The changes in cyclotorsion can be expressed by the coupling constant K_t. The pre-adapted and post-adapted gains are shown in TABLE 1. The pre-adapted gains averaged 0.82 for our four subjects, which is similar to values reported in earlier studies.[18,20,21,25] The reversed adaptation condition decreased these gains to an average value of 0.29, which is a 65% reduction. The exaggerated condition increased the gain to 1.04, which is a 27% elevation. Clearly the coupling between convergence and cyclovergence responded to the sensory demands placed on binocular vision.

MODEL SIMULATIONS

How might these changes in the vertical vergence and cyclovergence be implemented? Physiological studies indicate that simple gain changes of the obliques and

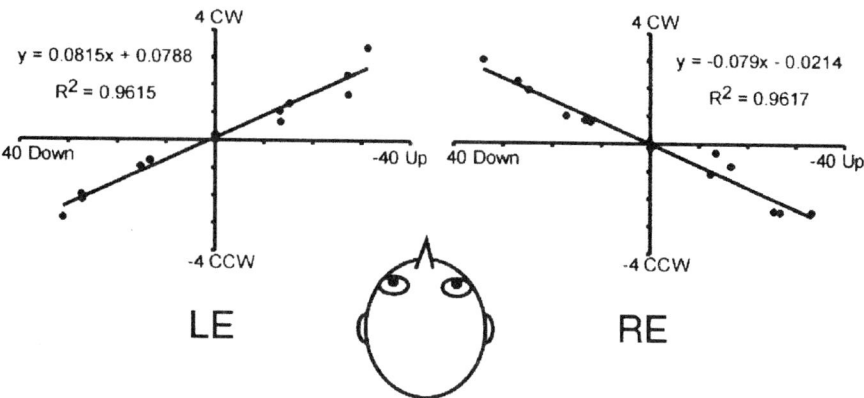

FIGURE 1. A top-down view of simulated displacement planes for a straight-ahead reference direction are shown of the right and left eye during 20 degrees of convergence. Planes were derived from Orbit[tm] simulations of three-dimensional eye position with modified gains of the obliques and vertical recti. Orbit simulations are for a 15% decreased gain of the obliques and a 15% increased gain of the vertical recti. Torsion, plotted in degrees, is simulated for vertical and horizontal changes in eye positions over a 60-degree horizontal and vertical range.

perhaps the vertical recti might be involved. Mays et al.[26] found that convergence-dependent changes of cyclovergence with gaze elevation were associated with a reduced discharge rate of trochlear motor neurons and an implied relaxation of the superior oblique muscle during convergence. The modulation of trochlear activity with convergence varied systematically with gaze elevation and was largest in downward gaze. Perhaps simple gain changes of the obliques and vertical recti account for changes in orientation of Listing's planes with convergence. This hypothesis was tested by simulating three-dimensional eye position with Orbit (Eidactics, San Francisco), a biomechanics model that simulates binocular eye position based on the relationships of the six extraocular muscles, their tendons and supportive connective tissues including muscle sheaths or pulleys, innervation level, and motor nucleus connection weights (innervation gain) according to equations given, in part, by Robinson[27] and Miller and Robinson.[28] Simulations were conducted with 15% gain reductions to the obliques and 15% gain increases to the vertical recti. The gain changes only depended on convergence angle and did not vary with vertical eye position. Eye positions were simulated during 20 degrees of convergence, while eccentric gaze was varied horizontally and vertically over ±30 degrees. Convergence was simulated by increasing the bilateral innervation to the medial recti and decreasing the innvervation to the lateral recti. Hering's law was simulated during convergence by finding the innervation to an assumed normal following the eye that would produce the same gaze direction as that of the fixating eye when the eyes were not converged. The simulated eye positions were converted from Fick coordinates to rotation vectors and fit to planes. Before any gain adjustments to the obliques and vertical recti muscles, the resulting orientation of Listing's plane at the near convergence distance was fronto-parallel. With gain adjustments, the primary positions diverged by 20 degrees for a simulated convergence of 20 degrees (FIG. 1). These gain changes might be modified to describe the adaptive plasticity of K_t demonstrated in our experiments. The adaptation results of the exaggerated condition could be simulated with greater gain changes, and the results of the reversed condition could be simulated with smaller gain changes. The simulation demonstrates that simple convergence-related gain changes of the vertical ocular muscles are sufficient to transform the innervation pattern appropriate for torsion at far viewing distances into ones consistent with the binocular extension of Listing's law at near viewing distances.

Vertical eye alignment in tertiary gaze was preserved during the gain adjustments to the vertical recti and obliques that produced the binocular extension of Listing's law. Vertical eye alignment is indicated when the amplitude ratio of left/right vertical eye position equals 1.0 when expressed in Helmholtz coordinates and when $K_v = 1.0$. Vertical eye alignment was accurate independent of the gain adjustments we made to the vertical ocular muscles, indicating that changes in cyclophoria that were consistent with Listing's extended law did not disrupt vertical eye alignment in tertiary gaze. The simulation suggests that binocular vertical eye alignment is maintained in tertiary gaze at near viewing distances, because the yoked innervation for vertical eye position to the following eye is the same as that during distance fixation with parallel lines of sight. Orbital mechanics produces a different elevation of the two eyes when horizontal position of the following eye is modified by asymmetrical convergence. As a consequence, the same innervation aligns the lines of sight with distant or near tertiary targets when the fixing eye is aimed in a direction common to both target distances. For normal eye alignment, it is not necessary to alter the innervation to vertical vergence to obtain binocular alignment of tertiary targets at near and far viewing distances.[10–12] The simulation suggests that normal binocular vertical eye alignment is primarily a consequence of Hering's law and the passive biomechanics of the oculomotor system.[29,30] However, our adaptation study of vertical vergence illustrates that it is possible to modify the innervation for eye alignment and change K_v as might be necessary during development and in response to disease or injury.

SUMMARY

Vergence adaptation is a calibration process that adjusts the innervation for horizontal, vertical, and torsion components of vergence to complement the physical constraints set by the extraocular muscles and orbital connective tissues. Our simulations illustrate that passive orbital mechanics simplify the neural control for precise vertical vergence and cyclovergence that are needed to achieve binocular alignment under open-loop conditions in response to perceived spatial location. Normally, cyclovergence varies with convergence and vertical eye position. These variations can be achieved with simple gain adjustments of the vertical muscles that are proportional to the amount of convergence. Orbital mechanics constrain cyclovergence with eye elevation to produce the cyclovergence necessary to reduce retinal image disparity. Similarly, in tertiary gaze, vertical vergence (in Fick coordinates) normally varies with both convergence and versional eye position to null vertical disparity. In natural viewing conditions, orbital mechanics constrains vertical vergence in tertiary gaze and achieves binocular alignment during convergence with the same innervation patterns used to align the eyes while viewing distant targets that do not subtend vertical disparities. In cases of pathology or optical distortion from spectacles, it is possible to modify the open-loop innervation to vertical vergence and cyclovergence to achieve binocular alignment as long as retinal image disparity can be sensed to guide the adaptation process. In cases of strabismus where binocular sensory systems are disrupted and do not code binocular disparity, the calibration process is disrupted without feedback from retinal image disparity, and the couplings between the three components of convergence are abnormal.[31,32]

ACKNOWLEDGMENT

This work was supported by National Institutes of Health Grant EYO-3532.

REFERENCES

1. GARDING, J., J. PORRILL, J.E.W. MAYHEW & J.P. FRISBY. 1995. Stereopsis, vertical disparity and relief transformations. Vision Res. **35:** 703–722.
2. BACKUS, B.T., M.S. BANKS, R. VAN EE & J.A. CROWELL. 1999. Horizontal and vertical disparity, eye position, and stereoscopic slant perception. Vision Res. **39:** 1143–1170.
3. OGLE, K.N. 1962. Spatial localization through binocular vision. In The Eye. Vol 4. H. Davson, ed.: 271–324. Academic Press. New York.
4. KUMAR, T. & D. GLASER. 1994. Some temporal aspects of stereoacuity. Vision Res. **34:** 913–925.
5. POPE, D., M.E. EDWARDS & C.M. SCHOR. 1999. Extraction of depth from opposite-contrast stimuli: transient system can, sustained system can't. Vision Res. **39:** 4010–4017.
6. SCHOR, C.M., J. ALEXANDER, L. CORMACK & S. STEVENSON. 1992. A negative feedback control model of proximal convergence and accommodation. Ophthal. Physiol. Opt. **12:** 307–318.
7. SCHOR, C.M. & J.W. MCCANDLESS. 1997. Context-specific adaptation of vertical vergence to multiple stimuli. Vision Res. **37:** 1929–1938.
8. BRADSHAW, M.F. & B.J. ROGERS. 1994. Is cyclovergence state affected by the inclination of stereoscopic surfaces? Invest. Ophthalmol. Visual Sci. **35:** 1316.
9. ALLEN, M.J. & J.H. CARTER. 1967. The torsion component of the near reflex. Am. J. Optom. **44:** 343–349.
10. SCHOR, C.M., J. MAXWELL & S.B. STEVENSON. 1994. Isovergence surfaces: the conjugacy of vertical eye movements in tertiary positions of gaze. Ophthal. Physiol. Opt. **14:** 279–286.
11. COLLEWIJN, H. 1994. Vertical conjugacy: what coordinate system is appropriate? In Contemporary Ocular Motor and Vestibular Research: A Tribute to David A. Robinson. A.F. Fuchs, T. Brandt, Z.U. Buttner & D.S. Zee, Eds.: 296–303. Thieme. Stuttgart.
12. YGGE, J. & D.S. ZEE. 1995. Control of vertical eye alignment in three-dimensional space. Vision Res. **35:** 3169–3181.
13. TWEED, D. 1997. Visual-motor optimization in binocular control. Vision Res. **37:** 1939–1951.
14. HELMHOLTZ, H. 1910. Treatise on Physiological Optics. J.P.C. Southal, Ed. Dover. New York.

15. HAUSTEIN, W. 1989. Considerations on Listing's law and the primary positions by means of a matrix description of eye position control. Biol. Cybern. **60:** 411–420.
16. HERING, E. 1868. The Theory of Binocular Vision. B. Bridgeman & L. Stark, Eds. & translators. 1977. Plenum. New York.
17. NAKAYAMA, K. 1983. Kinematics of normal and strabismic eyes. Chap. 16. *In* Vergence Eye Movements: Basic and Clinical Aspects. C.M. Schor & K. Ciuffreda, Eds.: 543–564. Butterworths. Boston.
18. MOK, D., A. CADERA, W. RO, *et al.* 1992. Rotation of Listing's plane during vergence. Vision Res. **32:** 2055–2064.
19. VAN RIJN, L.J. & A.V. VAN DEN BERG. 1993. Binocular eye orientation during fixations: Listing's law extended to include eye vergence. Vision Res. **33:** 691–708.
20. MINKEN, A.W.H. & J.A.M. VAN GISBERGEN. 1994. A three-dimensional analysis of vergence movements at various levels of elevation. Exp. Brain Res. **101:** 331–345.
21. SOMANI, R.A.B., J.F.X. DESOUZA, D. TWEED & T. VILIS. 1998. Visual test of Listing's law during vergence. Vision Res. **38:** 911–923.
22. SCHREIBER, K., J.D. CRAWFORD, M. FETTER & D. TWEED. 2001. The motor side of depth vision. Nature **410:** 819–822.
23. SCHOR, C.M., J. MAXWELL & E. GRAF. 2001. Plasticity of convergence-dependent variations of cyclovergence with vertical gaze. Vision Res. **41:** 3351–3367.
24. MAXWELL, J., E. GRAF & C.M. SCHOR. 2001. Adaptation of torsional eye alignment in relation to smooth pursuit and saccades. Vision Res. **4l:** 3735–3749.
25. MIKHAEL, S., D. NICOLLE & T. VILIS. 1995. Rotation of Listing's plane by horizontal, vertical and oblique prism-induced vergence. Vision Res. **35:** 3243–3254.
26. MAYS, L.E., Y. ZHANG, M.H. THORSDTAD & P.D.R. GAMLIN. 1991. Trochlear unit activity during ocular convergence. J. Neurophysiol. **65:** 1484–1491.
27. ROBINSON, D.A. 1975. A quantitative analysis of extraocular muscle cooperation and squint. Invest. Ophthalmol. Visual Sci. **14:** 801–825.
28. MILLER, J.M. & D.A. ROBINSON. 1984. A model of the mechanics of binocular alignment. Comput. Biomed. Res. **17:** 436–470.
29. MILLER, J.M. & J.L. DEMER. 1992. Biomechanical analysis of strabismus. Binocular Vision Eye Muscle Surg. Q. **7:** 233–248.
30. PORRILL, J., P.A. WARREN & P. DEAN. 2000. A simple control law generates Listing's positions in a detailed model of the extraocular muscle system. Vision Res. **40:** 3743–3758.
31. VAN DEN BERG, A.V., L.J. VAN RIJN & J.T.H. DE FABER. 1995. Excess cyclovergence in patients with intermittent exotropia. Vision Res. **35:** 3265–3278.
32. MELIS, B., J. CRUYSBERG & J. VAN GISBERGEN. 1996. Listing's plane dependence on alternating fixation in a strabismus patient. Vision Res. **37:** 1355–1366.

Inferior Vestibular Neuritis

G.M. HALMAGYI, S.T. AW, M. KARLBERG, I.S. CURTHOYS, AND M.J. TODD

Neurology Department, Royal Prince Alfred Hospital, Camperdown, NSW 2050, Sydney, Australia

ABSTRACT: Sudden, spontaneous, unilateral loss of vestibular function without simultaneous hearing loss or brain stem signs is generally attributed to a viral infection involving the vestibular nerve and is called acute vestibular neuritis. The clinical hallmarks of acute vestibular neuritis are vertigo, spontaneous nystagmus, and unilateral loss of lateral semicircular function as shown by impulsive and caloric testing. In some patients with vestibular neuritis the process appears to involve only anterior and lateral semicircular function, and these patients are considered to have selective *superior* vestibular neuritis. Here we report on two patients with acute vertigo, normal lateral semicircular canal function as shown by both impulsive and caloric testing, but selective loss of posterior semicircular canal function as shown by impulsive testing and of saccular function as shown by vestibular evoked myogenic potential testing. We suggest that these patients had selective *inferior* vestibular neuritis and that contrary to conventional teaching, in a patient with acute spontaneous vertigo, unilateral loss of lateral semicircular canal function is not essential for a diagnosis of acute vestibular neuritis.

KEYWORDS: vestibular neuritis; labyrinthitis; vestibulo-ocular reflex; head impulse; canal paresis

INTRODUCTION

Sudden spontaneous unilateral loss of vestibular function with preserved hearing and no signs of brain stem dysfunction is generally attributed to viral infection and is called acute vestibular neuritis (aVN).[1,2] It has been suggested that reactivation of herpes simplex type 1 virus[3] could cause aVN in a manner resembling facial palsy (Bell's palsy) and sudden unilateral hearing loss. In patients with aVN, lateral semicircular canal (SCC) function as shown by impulsive or caloric testing is, by definition, abolished or severely reduced on the affected side.

Normally the superior vestibular nerve innervates the SCC, the anterior SCC, and the utricle, whereas the inferior vestibular nerve innervates the posterior SCC and the saccule. In some patients with aVN the process appears to affect predominantly the superior vestibular nerve. In these patients: the axis of the spontaneous nystagmus is aligned with either the lateral SCC alone or the resultant from the lateral and anterior SCCs;[4] the vestibulo-ocular reflex (VOR) shows asymmetry, suggesting involvement of only the lateral SCC or the lateral and anterior SCCs, sparing the pos-

---

Address for correspondence: Dr. G.M. Halmagyi, Neurology Department, Royal Prince Alfred Hospital, Missenden Road, Camperdown, NSW 2050, Sydney, Australia. Voice: +61 2 515 8300; fax: +61 2 9515 8345.
michael@icn.usyd.edu.au

terior SCC;[4] about one patient in three with aVN will develop posterior SCC benign paroxysmal positional vertigo, implying preservation of posterior SCC function;[5] vestibular evoked myogenic potentials (VEMPs) are preserved in two of three patients with aVN, implying preservation of saccular function.[6]

Does aVN sometimes affect only the inferior vestibular nerve and if so what vestibular deficits should such patients have and how would one recognize them? Such patients might have spontaneous nystagmus but it shouldn't be horizontal but should be downbeating and torsional; they will have normal caloric tests since these only measure lateral SCC function. On impulsive testing lateral SCC function should be normal but posterior SCC function should be impaired on one side. The inferior vestibular nerve also innervates the saccule so that VEMPs should also be impaired on same side. Here we report on two patients, each with a single attack of acute spontaneous vertigo, who had just such deficits of vestibular function and suggest that they did indeed have a selective inferior vestibular neuritis.

CASE REPORTS

CASE 1. The patient, A.T., a 26-year-old IT executive, presented with mild persistent imbalance following a 20-minute attack of acute spontaneous vertigo with nausea 2 months previously. He had never had headaches or difficulty with hearing, and his health had been unremarkable. On examination there was no spontaneous, gaze-evoked, head-shaking, or positional nystagmus and no positional vertigo; the Romberg test on a mat and the Unterberg test both gave negative results.

Results of routine investigations were as follows: electronystagmogram, normal; caloric test: minimal directional preponderance (30%) to the left; vestibular evoked myogenic potentials: absent to clicks and taps from the left ear; subjective visual horizontal: 0 degrees (normal); audiogram: normal; MRI brain scan with contrast, normal.

CASE 2. The patient, P.C., a 61-year-old previously well male business executive, developed sudden intense vertigo and nausea while driving home from work. He had to stop his car, vomited, and called for help. He was taken by ambulance to a hospital emergency room. On admission he was in distress with vertigo, retching, and vomiting and was unable to stand. There was no spontaneous or positional nystagmus with or without visual fixation. A CT brain scan gave normal results. He was admitted to the hospital with the provisional diagnosis of cerebellar infarct. The following day he felt better and could stand without support. However, he was unable to hear from his left ear. Over the next 3 days his balance continued to improve but his hearing did not do so.

Routine investigations at that time revealed the following: electronystagmogram: minimal left beating gaze-evoked nystagmus in the dark; caloric tests: normal (R30 = 27 degrees/s; R44 = 28 degrees/s; L30 = 24 degrees/s; L44 = 20 degrees/s); vestibular evoked myogenic potentials: absent to clicks and to taps from left ear; subjective visual horizontal: 4 degrees to the left (just above our control range of up to 3 degrees): audiogram: severe (90 dB) flat sensorineural hearing loss left ear; slight conductive loss right ear; MRI brain scan with contrast showed no abnormality, particularly no cerebellar infarction, and no contrast enhancement of the inner ear.

His balance continued to improve, but 2 weeks later he still rotated to the left on the Unterberger (Fukuda) test and fell on the matted Romberg test. Within 1 month his hearing also improved, but was not back to normal, whereas the VEMPs had returned to normal.

METHODS

Head Impulses

The Recording System

The scleral search coil technique was used to record head and left eye positions using a method previously described.[7–10] Head and left eye positions were recorded with dual-search coils (Skalar, Delft, The Netherlands) and were precalibrated before each recording. The recording system has a 16-bit resolution and minimum resolution of 0.1 minute of arc with maximum errors and cross-coupling at 2%.

Experimental Protocols

A head impulse is a passive, unpredictable, low-amplitude (10–20 degree), high acceleration (3,000–4,000 degrees/s^2) head rotation approximately in the plane of corresponding SCC pairs in yaw, left anterior and right posterior (LARP), or right anterior and left posterior (RALP) directions.[10] The terms yaw, LARP, and RALP refer to the head impulse directions defined with reference to the subject. Yaw-left, left anterior, and right anterior rotations are positive, whereas yaw-right, right posterior, and left posterior rotations are negative. Counterclockwise direction means that the upper pole of the head or eye is rotated towards the patient's left, and clockwise direction means rotation towards the patient's right.

Data Analysis

We have expressed three-dimensional head, gaze, and eye positions as rotation vectors.[11] "Head" and "gaze" positions are the orientations in space-fixed coordinates of head and eye, respectively. Eye position is the orientation of eye in head-fixed coordinates. Head, gaze, and eye velocities are calculated from their respective positions. Impulsive canal paresis (iCP) is defined as the ratio of gaze velocity to head velocity in SCC coordinates at close to peak head velocity, in response to a head rotation towards the on-direction of a SCC, along its SCC plane. The rationale is that the VOR mediated by SCC receptor hair cells stabilizes gaze during any angular head rotation, but gaze instability only occurs in the absence of SCC function. Impulsive canal paresis measures the gaze instability when the head is rotated towards a paretic SCC. Gaze and head velocities were normalized by dividing each velocity by the magnitude of peak head velocity in each trial. We then determined the gaze and head velocity in SCC coordinates.[12,13] For each head impulse direction, we determined the mean head and gaze velocity in SCC coordinates from 10 trials. Impulsive canal paresis is determined at 100 ms from the onset of the head impulse and expressed as a percentage, as follows:

$$\text{iCP} = \frac{\hat{g}}{h} * 100$$

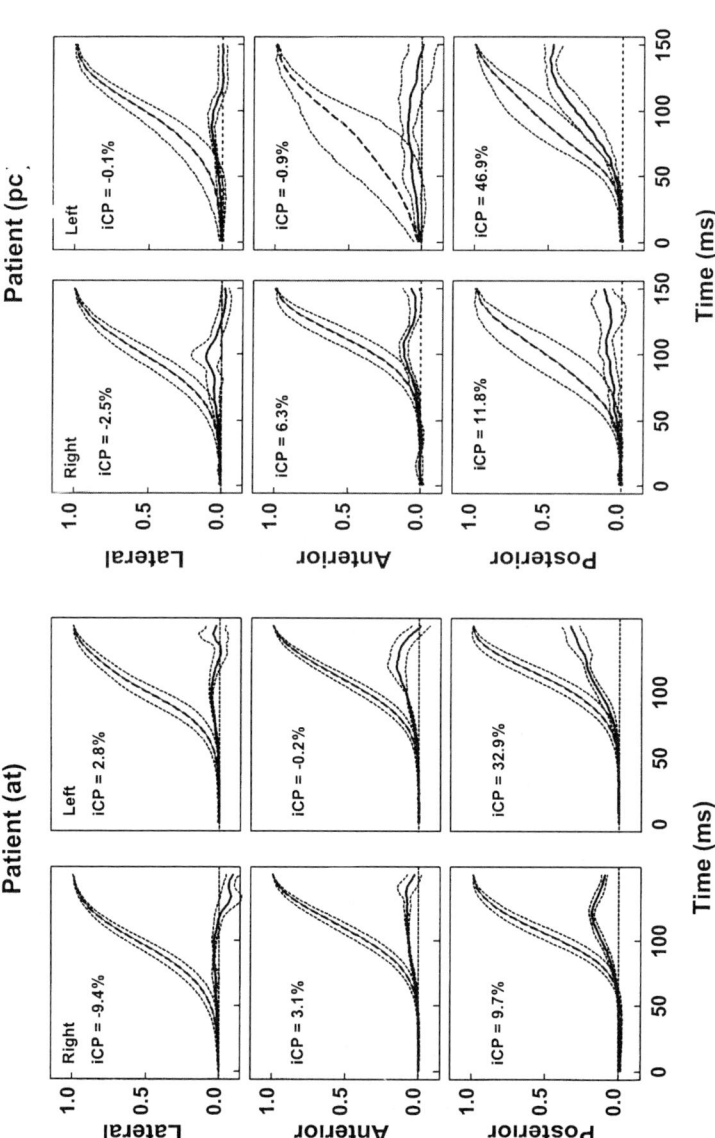

FIGURE 1. Impulsive testing in acute inferior vestibular neuritis. The two columns on the *left* show the results of impulses from patient A.T. and the two on the right from patient P.C. For each patient, impulses from the right lateral, anterior, and posterior SCCs are shown in one column and from the left lateral, anterior, and posterior SCC in the other column. Normalized head velocity is shown on the Y axis and time in milliseconds on the X axis. Mean 1 ± SD head velocity is shown with *broken lines*; mean ± 1 SD gaze velocity is shown with *continuous lines*. Note that both patients had an isolated deficit of posterior SCC function on impulsive testing; patient A.T. had a 33% impulsive canal paresis, and patient P.C. had a 47% impulsive canal paresis.

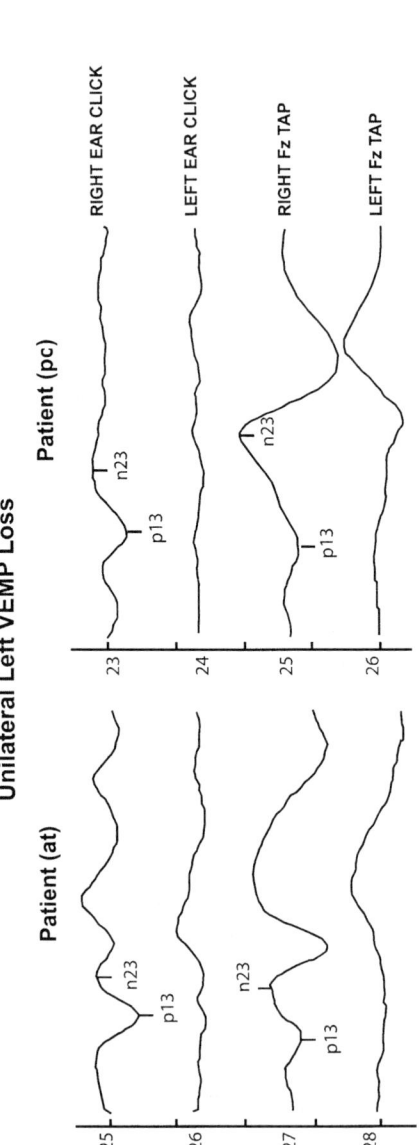

FIGURE 2. Vestibular evoked myogenic potentials in acute inferior vestibular neuritis. Neither patient had p13–n23 potentials from the left side in response to click or in response to tap stimulation. On the normal side, click-evoked p13–n23 potentials were 47 µV (A.T.) and 96 µV (PC); the tap-evoked potentials were 90 µV (A.T.) and 847 µV (P.C).

where iCP is the canal paresis, $\hat{h}$ is head velocity in SCC coordinates, and $\hat{g}$ is gaze velocity in SCC coordinates. We excluded the contribution of smooth pursuit by restricting our analysis to an interval of 100 ms from the onset of the impulse, because smooth pursuit have latencies ≥100 ms.[14]

Vestibular Evoked Myogenic Potentials

Vestibular evoked myogenic potentials were recorded using our standard methodology previously reported.[15,16] In brief, averaged surface electromyographic potentials were recorded from each sternomastoid muscle, during active contraction of each muscle, in response to loud, brief clicks to each ear and then in response to taps to the forehead. The normal response to such stimuli is a positive potential at around 13 ms followed by a negative potential at 23 ms, the VEMP. In response to a click the potential appears only in the ipsilateral sternomastoid muscle; in response to a tap the response appears in both sternomastoid muscles. The VEMP is abolished by vestibular nerve section but not in cases of total sensorineural hearing loss; it is progressively attenuated and then abolished by conductive hearing loss. Animal experiments indicate that the VEMP arises from stimulation of the saccule.[17–20]

RESULTS

FIGURE 1 shows the results of impulsive testing. Head velocities stimulating the affected posterior SCC produce a significant gaze perturbation, so that the value of iCP is 33% in patient 1 and 47% in patient 2 (normal 12 ± 7%). By contrast, head velocities stimulating the anterior and lateral SCCs on the affected side and in the planes of all three SCCs on the normal side produced only slight perturbutations of gaze, so that the values for iCP were within or just above the control range (lateral SCC = 0 ± 5%; anterior 7 ± 7%; posterior 12 ± 7%).[21] FIGURE 2 shows that VEMPs were absent from the affected side in both patients, both in response to clicks and in response to taps.

DISCUSSION

Three-dimensional measurement and vector analysis of the angular VOR in response to the head impulse test are an established method of evaluating individual SCC function.[8–10,21,22] The SCC planes do not lie along traditional head planes, yaw (i.e., horizontal), pitch (i.e., sagittal), and roll (i.e., coronal or frontal). For example, the plane of lateral SCC is oriented about 25° above the yaw plane, as defined by Reid's line.[12] Interpretation of impulsive tests requires familiarity with the reference planes and directions for a three-dimensional description of the VOR. Therefore, VOR gain and canal paresis are best understood when expressed in SCC planes.

We define iCP as the ratio of gaze velocity to head velocity in SCC planes at close to peak head velocity, and use iCP to quantify individual SCC deficits. The VOR mediated by a normal SCC stabilizes gaze when the head is rotated towards the SCC, so that gaze velocity would be close to zero.[8] Our data show that iCP from a normal SCC is ≤12%. When the head is rotated towards an abnormal SCC, gaze is unstable, so that gaze velocity is high.[9,10] Mean iCP values from an affected SCC after aVN and after surgical deafferentation is ≥65%.

The iCP values from normal vertical SCCs are slightly higher due to a lower (0.7) torsional VOR gain in response to head impulses in the roll plane even in normal subjects.[8] As the vertical SCCs are positioned approximately halfway between the pitch and roll planes, a smaller roll-torsional gain will result in the higher iCP value when gaze velocity is resolved into SCC planes.

A head impulse stimulates both the on- and off-directions of a pair of SCCs, that is, left and right lateral SCCs (yaw plane), right anterior and left posterior SCCs (RALP plane), and left anterior and right posterior SCCs (LARP plane). The iCP value from in-plane SCC on the side opposite an abnormal SCC would be slightly higher because of the "push-pull" behavior of the SCCs; the contribution from disinhibition is absent;[23] therefore, the iCP value is higher than from normal subjects.

The two patients we report on here both presented with acute spontaneous vertigo and had little or no spontaneous nystagmus, normal lateral SCC function but unilateral loss of posterior SCC function and saccular function, features that we propose are the hallmarks in inferior vestibular neuritis.

The likely reason that patients with inferior vestibular neuritis are difficult to recognize clinically is that with conventional testing they have little spontaneous nystagmus during the acute phase of the illness and have normal caloric responses. Without these clinical signs, a patient with inferior vestibular neuritis would not fulfill the conventional clinical criteria of aVN and therefore would be considered to have a central vestibular lesion. This is particularly so, because it has been reported that one in five patients with acute pontine lesions present with signs that mimic acute unilateral peripheral vestibulopathy.[24] Therefore, until now, patients with inferior vestibular neuritis would have needed some other clinical feature to indicate that the cause of the acute vertigo was acute unilateral peripheral vestibulopathy. In one of our cases, that additional feature was simultaneous ipsilateral hearing loss. Our patients had normal lateral SCC function on caloric and impulsive testing but a posterior SCC deficit on impulsive testing, indicating that they had selective involvement of the inferior vestibular nerve; however, it was simultaneous ipsilateral hearing loss in one patient that indicated to the attending clinicians that a unilateral inner ear disturbance was the likely cause of the problem.

ACKNOWLEDGMENTS

This work was supported by the Garnett Passe and Rodney Williams Memorial Foundation, the Australian National Health and Medical Research Council, the Swedish Medical Research Council, the Wenner-Gren Foundation, The Maggie Stephens Foundation, and the Neurology Department Trustees, Royal Prince Alfred Hospital.

REFERENCES

1. DIX, M.R. & C.S. HALLPIKE. 1952. The pathology, symptomatology and diagnosis of certain common disorders of the vestibular system. Ann. Otol. Rhinol. Laryngol. **61:** 987–1016.

2. STRUPP, M. & T. BRANDT. 1999. Vestibular neuritis. *In* Vestibular Dysfunction and Its Therapy. U. Buttner, ed. Adv. Otorhinolaryngol. **55:** 111–136.
3. SCHULZ, P., V. ARBUSOW, M. STRUPP, *et al.* 1998. Highly variable distribution of HSV-1-specific DNA in human geniculate, vestibular and spiral ganglion. Neurosci. Lett. **252:** 139–142.
4. FETTER, M. & J. DICHGANS. 1996. Vestibular neuritis spares the inferior division of the vestibular nerve. Brain **119:** 755–763.
5. BÜCHELE, W. & T. BRANDT. 1988. Vestibular neuritis: a horizontal semicircular canal paresis? Adv. Otorhinolaryngol. **42:** 157–161.
6. MUROFUSHI, T., G.M. HALMAGYI, R.A. YAVOR & J.G. COLEBATCH. 1996. Absent vestibular evoked myogenic potentials in vestibular neurolabyrinthitis. Arch. Otolaryngol. Head Neck Surg. **122:** 845–848.
7. ROBINSON, D.A. 1963. A method of measuring eye movement using a scleral search coil in a magnetic field. IEEE Trans. Biomed. Eng. **10:** 137–145.
8. AW, S.T., T. HASLWANTER, G.M. HALMAGYI, *et al.* 1996. Three-dimensional vector analysis of the human vestibuloocular reflex in response to high-acceleration head rotations. I. Responses in normal subjects. J. Neurophysiol. **76:** 4009–4020.
9. AW, S.T., G.M. HALMAGYI, T. HASLWANTER, *et al.* 1996. Three-dimensional vector analysis of the human vestibuloocular reflex in response to high-acceleration head rotations. II. Responses in subjects with unilateral vestibular loss and selective semicircular canal occlusion. J. Neurophysiol. **76:** 4021–4030.
10. CREMER, P.D., G.M. HALMAGYI, S.T. AW, *et al.* 1998. Semircircular canal plane head impulses detect absent function of individual semicircular canals. Brain **121:** 699–716.
11. HAUSTEIN, W. 1989. Considerations on Listing's law and the primary position by means of a matrix description of eye position control. Biol. Cybern. **60:** 411–420.
12. BLANKS, R.H.I., I.S. CURTHOYS & C.H. MARKHAM. 1975. Planar relationships of the semicircular canals in man. Acta Otolaryngol. **80:** 185–196.
13. AW, S.T., T. HASLWANTER, M. FETTER, *et al.* 1998. Contribution of the vertical semicircular canals to the caloric nystagmus. Acta Otolaryngol. **118:** 618–627.
14. CARL, J.R. & R.S. GELLMAN. 1987. Human smooth pursuit: stimulus-dependent responses. J. Neurophysiol. **57:** 1446–1463.
15. COLEBATCH, J.G., G.M. HALMAGYI & N.F. SKUSE. 1994. Myogenic potentials generated by a click-evoked vestibulocollic reflex. J. Neurol. Neurosurg. Psychiatry **57:** 190–197.
16. HALMAGYI, G.M., R.A. YAVOR & J.G. COLEBATCH. 1995. Tapping the head activates the vestibular system: a new use for the clinical reflex hammer. Neurology **45:** 1927–1929.
17. MUROFUSHI, T. & I.S. CURTHOYS. 1997. Physiological and anatomical study of click-sensitive primary vestibular afferents in the guinea-pig. Acta Otolaryngol. (Stockh.) **117:** 66–72.
18. MUROFUSHI, T., I.S. CURTHOYS & D.P. GILCHRIST. 1996. Response of guinea pig vestibular nucleus neurons to clicks. Exp. Brain Res. **111:** 149–152.
19. MUROFUSHI, T., I.S. CURTHOYS, A.N. TOPPLE, *et al.* 1995. Responses of guinea pig primary vestibular neurons to clicks. Exp. Brain Res. **103:** 174–178.
20. KUSHIRO, K., M. ZAKIR, Y. OGAWA, *et al.* 1999. Saccular and utricular inputs to sternocleidomastoid motoneurons of decerebrate cat. Exp. Brain Res. **126:** 410–416.
21. AW, S.T., M. FETTER, P.D. CREMER, *et al.* 2001. Individual semicircular function in superior and inferior vestibular neuritis. Neurology **57:** 768–774.
22. SCHMID-PRISCOVEANU, A., D. STRAUMANN, A. BOHMER & H. OBZINA. 1999. Vestibulo-ocular responses during static head roll and three-dimensional head impulses after vestibular neuritis. Acta Otolaryngol. **119:** 750–757.
23. SHIMAZU, H. & W. PRECHT. 1966. Inhibition of central vestibular neurons from the contra-lateral labyrinth and its mediating pathway. J. Neurophysiol. **29:** 467–492.
24. THOMKE, F. & H.C. HOPF. 1999. Pontine lesions mimicking acute peripheral vestibulopathy. J. Neurol. Neurosurg. Psychiatry **66:** 340–349.

Otolith Function: Basis for Modern Testing

GARY D. PAIGE

Department of Neurobiology and Anatomy, and the Center for Visual Science, University of Rochester, Rochester, New York 14642, USA

ABSTRACT: The major challenge in developing a robust test of otolith function, particularly with regard to linear vestibulo-ocular reflex (LVOR) and perceptual measures, is to find a way in which graded lesions are reflected in graded response properties and abnormalities. The ability of the vestibulo-ocular reflex (VOR) to compensate and adapt to dysfunction and pathology presents formidable challenges for registering localizing clinical findings, whether in the angular vestibulo-ocular reflex (AVOR), the LVOR, or both. Based on a variety of considerations, various forms of eccentric rotation seem to provide the most convenient, and potentially the most useful, means to generate motion profiles from which otolith function can be directly assessed. Both translational and tilt responses can be recorded depending on the stimulus profile. The near-centric version is particularly enticing because of the ability to study one labyrinth at a time, much like calorics. In that case and in others in which the tilt-LVOR is prominent, measures of the perceived visual vertical are useful and by all accounts similar to ocular torsion. The latter does hold the important advantage of being an objective measure, requiring no intervention on the part of the patient. The translational-LVOR can be derived from eccentric rotation responses with the head displaced forward as well as backward, while viewing near targets in hopes of generating a large addition or subtraction (even inversion) of an otherwise AVOR-driven reflex. These considerations provide an impetus to pursue improved methods of quantifying otolith function in a clinical population. The sobering caveat is that the diagnosis of total unilateral vestibular loss presents little challenge either clinically or by classic testing (e.g., calorics), and yet most of our efforts in developing quantifiable measures of dysfunction over the years have yielded results that are modest and hardly compelling.

KEYWORDS: otolith; vestibulo-ocular reflex; eccentric rotation; calorics

OVERVIEW

The goals of any useful clinical test are to identify the existence, progression, and potential recovery of a pathologic process. There are two major paths by which individuals arrive at clinical testing. The first is based on symptoms expressed by patients to health care providers who then request specific tests to quantify relevant structural or functional variables. The second entails the screening of a population to identify those with pathology. We focus here on the first class of testing.

Address for correspondence: Gary D. Paige, Department of Neurobiology and Anatomy, Box 603, University of Rochester, 601 Elmwood Avenue, Rochester, NY 14642, USA. Voice: 716-275-2591; fax: 716-442-8766.
gary_paige@urmc.rochester.edu

The ideal clinical test should be sensitive to a particular dysfunction and specific for it. Findings should be both reproducible and graded, reflecting the severity of pathology or intensity of dysfunction. Finally, in a pragmatic sense, a test should be cost effective, simple to perform, and comfortable to endure for patients. There are nuances that arise that rightfully modify the ideal. Magnetic resonance imaging (MRI) is an example that might be considered cost effective when judged in light of the potential impact of test outcome.

In applying ideals to vestibular function testing, several limitations are apparent. The vestibular system is ancient, following hundreds of millions of years of natural selection to generate a highly robust and well protected system. Part of this protection includes resilience against pathology and mechanisms that compensate and adapt to it. Furthermore, the vestibular system is tightly integrated with other sensory systems that drive common behaviors, including eye movements and the various somatic behaviors known to be influenced by vestibular input. The vestibular endorgans are remarkably protected by virtue of their position near the center of the skull, encased among the densest bones of the body. This attribute makes them inaccessible to the kind of clinical investigation that characterizes, for example, ophthalmologic examination of the eye. Finally, the methods employed to test vestibular function, like all tests, mimic direct challenges to the system under study. In the case of vestibular assessment, this means cumbersome methods to control movement and orientation of the head and body. The challenge poses technological limitations as well as concerns related to patient comfort.

Despite limitations, a variety of vestibular testing strategies have been developed. The most successful ones to date have addressed the function of the semicircular canals. Arguably, the closest to ideal in the testing armamentarium remains the caloric test,[1] now roughly a century old and still standing. It has the undeniable advantage of allowing the investigator to specifically stimulate the endorgan one side at a time, and it yields at least coarsely graded responses and therefore a graded test result. Few alternatives over the years have produced tests that approach these characteristics. A promising useful example exploits high-acceleration transient head rotations both in the laboratory[2] and at the bedside.[3]

The development of otolith function tests has been a venerable goal for years. However, progress has been impeded in part because our understanding of otolith physiology is surprisingly modern and ongoing. Three approaches have been taken that are relevant to the development of otolith function tests. These include otolith influences on eye movements (the linear vestibulo-ocular reflex, LVOR), perception (measures of orientation or motion), and head or postural orientation (spinal reflexes). This discussion focuses primarily on otolith influences on eye movements and perception. This is not to belittle posture and head orientation, but recognizes that we know less about the specific connections and effects of otolith inputs related to head and postural movements. An exception is the vestibular-evoked myogenic potential (VEMP), thought to reflect a direct saccular influence on the sternocliedomastoid muscle activated by high-intensity clicks delivered to each ear.[4-6] A clinical test of saccular function is gathering momentum and raises a promising and evolving approach. In the sections below, attention is focused on methods by which the LVOR might be exploited in search of otolith function testing, with emphasis on practical implementations.

LINEAR VESTIBULO-OCULAR REFLEX

The overall goal of the vestibulo-ocular reflex (VOR) is to maintain binocular fixation on external targets of interest. This defining statement applies strictly to frontal eyed bi-foveate animals such as primates and humans. If binocular fixation is indeed maintained during head movements, then the visual image will remain fixed on the fovea of both eyes. It is often stated that the goal of the VOR is to maintain gaze stability. In fact, this is not generally true in foveate species and applies only when the eyes fixate on targets that are far in the distance.[7] We now know that the VOR generally requires (and usually delivers) modification based on viewing distance,[7-12] and this is why a revised definition with reference to binocular fixation is required.

The VOR can be conveniently parsed into two fundamental classes, the angular VOR (AVOR), driven by the semicircular canals in response to head rotation, and the LVOR, driven by the otolith organs in response to head linear acceleration. The AVOR responds to a signal closely related to instantaneous head angular velocity, with the endorgan itself performing an initial integration of its required rotational acceleration input.[13] The otolith organs perform no such integration and carry linear acceleration signals directly to the brain.[14] All else must be derived from central processing. Further, otolith input suffers from a fundamental ambiguity not shared by the semicircular canals. That is, the otolith organs respond equally to linear acceleration due to translational motion of the head as well as to head tilt with respect to gravity, a manifestation of Einstein's equivalency principle. Thus, a given otolith afferent in the left eighth nerve that is activated by a head roll-tilt towards the left ear is also activated equivalently by a translational acceleration towards the right ear.[15] We do not witness prolonged translational accelerations during natural behavior. These tend to be brief, as entailed in the rapid linear oscillations produced by walking and running. However, head tilt relative to gravity tends to be prolonged, as when lying down. Furthermore, head tilt entails head rotation and therefore activates the AVOR in complement to tilt-related LVOR responses.[16-18]

The difference in the way natural behavior encounters tilt and translation provides a useful clue in understanding how the brain has evolved to distinguish these two forms of movement. On the basis of physics, the brain can never do a perfect job of distinguishing the two under all circumstances for all time.[19,20] Instead, it must perform a reasonable job most of the time and with relation to natural challenges. Based on theoretical, psychophysical, and physiologic evidence, otolith input is distributed in large part to translation and tilt-related functions based on the frequency content of the linear acceleration input.[16,21] A frequency parsing mechanism resembles the crossover network of a two-way loudspeaker. Low-frequency sound is preferentially directed towards the woofer, while high-frequency sound is directed towards the tweeter, although with considerable overlap in the mid-range. Similarly, low-frequency and prolonged linear acceleration is directed towards perceptions and functions related to head tilt, whereas high-frequency linear accelerations seem directed towards pathways and functions that convey translation. This scheme strictly applies to the processing of otolith input and in no way constrains the brain from utilizing canal, visual, somatosensory, cognitive, or other cues to develop an overall sense of orientation and motion in space. Simply put, otolith input alone is parsed and processed in this way. The process is clearly imperfect, as there is a broad range of overlap at middle frequencies. Even a modicum of introspection reveals some essential elements of tilt and translation parsing. When we lie down (even in the dark), we perceive that we are lying on our side and not hurtling into space. By contrast, when we walk briskly with our head level, the side to side motion of the head is not perceived as large rolling oscillations, but instead as a translation back and forth, as is indeed the case. Interestingly, at modest frequencies (e.g., 0.25 Hz) in the laboratory, we sense a little bit of tilt and translation together, such that the overall perception is a combination of rocking and translating termed "the hilltop illusion."[22] These coarse perceptual phenomena can be quantified more rigorously and reflexively in the LVOR.

Translational-Linear Vestibulo-ocular Reflex

Background. One important attribute of the translational-LVOR, but not the tilt-LVOR, is that the former is strongly modulated by viewing distance,[23] as one would expect in order for the LVOR to maintain binocular fixation on a target during head motion. For a target that is far in the distance, translation of the head requires little compensatory eye movement to maintain fixation. However, as the target approaches closer to the head, a progressively augmented response is required. This has been demonstrated in a variety of primate species and constitutes a general property of the translational-LVOR.[7,8,10,24] This fact has implications in the development of clinically useful tests of the translational-LVOR, because a change in response amplitude can reflect a natural change in viewing distance as opposed to a pathologic change in endorgan function. The same caveat holds for studies of adaptive plasticity in the LVOR, underlying the need to control and measure the context of viewing distance in studies of the LVOR. This variable is most readily acquired by recording eye movements binocularly and taking the angle between the two eyes (vergence) as a close reflection of viewing distance. Accommodation might influence the LVOR, as reported in monkeys,[10] but it seems less important in humans.[25] This is useful given the numbers of aging presbyopic individuals in our population, because over the age of 55 we have little remaining accommodative function.

Clinical Applications. Several attempts have been made to exploit the translational-LVOR in the development of a useful clinical test. Typically this entails interaural (IA, or side to side) translation in the form of either rapid transient accelerations or sinusoidal oscillation. A useful example is a study by Lampert *et al.*[26] that utilized transient motion to the left and right before and after surgical destruction of one ear. Roughly symmetric response amplitudes were measured prior to surgery, perhaps due to partial to minimal preexisting destruction of endorgan input in the patients studied and an obvious postsurgical reduction of the LVOR response limited to translation towards the lesioned ear. However, response symmetry was restored when next measured 10 weeks after surgery. Thus, the utility of the method as a clinical test is diminished due to the adaptive restorative properties of the reflex. A single otolith organ seems capable of generating a roughly symmetrical response after a few weeks of recovery at most. This has been confirmed in monkeys during transient as well as sinusoidal motion.[27,28] Combined with the cumbersome challenge of generating translational motion of the head and body and based on today's knowledge, there would be little reason for a typical vestibular testing laboratory to invest in this class of test. However, there is the potential for enhancing the test given knowledge that combining high linear acceleration with near fixation enhances response asymmetry long after endorgan destruction.[27] Another provocative finding is a tilt of the LVOR during IA motion by virtue of the development of a vertical response component after unilateral vestibular loss.[27,29] This likely reflects the fact that the saccular organ is not parallel to the sagittal plane and in fact leans medially.[15] When the head is translated laterally, an asymmetry between the sacculi results in an unbalanced vertical drive to the LVOR, resulting in a response tilt opposite the lesioned side. However, this phenomenon is subject to adaptive recovery like so many other attributes of the VOR.

Tilt-Linear Vestibulo-ocular Reflex

The tilt-LVOR has typically been quantified during prolonged IA force (e.g., on a centrifuge) or during slow roll-tilt of the head so as to avoid excessive canal activation. Either will produce torsion of the eyes,[30,31] commonly referred to as ocular counter-rolling. A simpler method of assessing ocular counter-rolling, particularly in patients, is to measure ocular torsion with the subject sitting still and upright.[32] Any abnormal and persistent ocular torsion presumably reflects an abnormality in the balance of activity between the two sets of otolith organs which, in the case of the tilt-LVOR, produces a static DC torsional response (12 o'clock position of the eye towards the lesioned side). Ocular counter-rolling is greatest acutely after unilateral vestibulopathy and declines through the process of vestibular compensation to a level that, on average, remains abnormal, although with some normal overlap.[31,32] There is usually no horizontal nystagmus due to an otolith-based imbalance, because the translational-LVOR operates with high-pass dynamics and produces no DC response.[28] Thus, a normal individual who lies down on one side demonstrates ocular counter-rolling, but not horizontal nystagmus despite the strong 1G IA force being applied to the head. Of course, positional nystagmus is a well known pathologic entity,[33] and at least some such cases likely reflect a failure of high-pass processing in CNS pathways related to the translational-LVOR. Regardless, a spontaneous horizontal nystagmus of vestibulopathic origin is usually taken to reflect a central asymmetry between inputs from coplanar canals from the two sides and the resulting AVOR manifestation.

A phenomenon associated with ocular counter-rolling is a perceived change in the visual vertical (SVV), or the equivalent perceptual change of the visual horizontal.[34–36] Both ocular counter-rolling and SVV have been recorded in patients, and given their rough similarity of outcome (roll towards the lesioned side),[32] there is considerable impetus to pursue the rather promising and simple clinical option of a clinical test of SVV. All a subject must do is align a vertical bar in darkness and provide the examiner with an endpoint that can be recorded in degrees of roll in space. In a useful study by Tabak et al.,[36] both postsurgical patients devoid of one labyrinth as well as those with partial lesions were studied. Enduring abnormalities in SVV were observed in which subjects persistently tilted the 12 o'clock position of the bar towards the pathologic side. Evidence hinted at a graded relation between response abnormality and lesion, with correlation between SVV and at least some measures of canal function (post headshake nystagmus and AVOR during transient head turns), although conclusions must await a more definitive data set.

ECCENTRIC ROTATION

One means of producing a robust otolith stimulus is to rotate the subject with the head and/or body displaced from the axis of rotation. This exploits the presence of centripetal and tangential accelerations, which can be further manipulated depending on the characteristics of the motion profile. Eccentric rotation (ER) always entails activation of the canals and therefore the AVOR, at least somewhere in the motion profile. Long duration trapezoidal velocity profiles have frequently been employed to study both perception and reflex eye movements in response to what is effectively a low-speed centrifugation.[37,38] These tend to emphasize tilt-LVOR responses with some translational-LVOR components added. For example, centrifugation with the head oriented with the IA line radially, thereby producing centripetal acceleration along the IA axis, generates a transient horizontal response that adds to the AVOR but also produces more substantial perceptions and measures of head-tilt during this prolonged IA force profile. Studies utilizing this form of eccentric rotation have been performed in patients, and indeed, differences between patients and normals have been reported.[37] However, there is a fair amount of overlap with normals as well as the need for the substantial challenges of generating angular motion on a 1- or 2-meter centrifuge arm.

Eccentric rotation can be used to exploit the translational-LVOR as opposed to the tilt-LVOR. This is done by emphasizing transient or high-frequency motion with the subject eccentrically displaced.[39–42] For example, with the head facing radially and nose out, the LVOR and AVOR act synergistically to generate large horizontal eye movements that are highly modulated by viewing distance. The AVOR and translational-LVOR interact linearly during this type of eccentric rotation. One fascinating attribute arises when the subject is turned to face nose in (towards the axis of rotation). In this case, the LVOR now subtracts from the AVOR. When the subject fixates far away and little LVOR response is present, the AVOR dominates and generates eye movements that are out of phase with the rotational direction of the head. However, when a subject fixates on a point close to the head and nearer than the axis of fixation, the ideal eye movement must now reverse in order to maintain fixation, implying that the LVOR must rise to a greater amplitude than the AVOR to invert the direction of eye movements and produce the appropriate compensatory response. Early attempts[43] to demonstrate this phenomenon failed largely because the LVOR was not activated at sufficiently high stimulus frequency, head eccentricity, or vergence to force the LVOR larger than the AVOR. Recent demonstrations of the inversion phenomenon[40,44] have revealed a robust illustration of a strong LVOR overwhelming the AVOR and therefore constitutes an interesting version of an otolith response. Whether this could be exploited as an unequivocal otolith test is promising, but remains to be demonstrated.

Eccentric rotation with the nose out was originally employed in the testing of patients by Bronstein and colleagues.[41] A clever attribute was the use of a "garden variety" rotatory chair with the subject simply leaning forward and with the head held eccentrically. Thus, with modest modifications of existing technology, a rotatory laboratory can be converted to include an eccentric rotation test. However, the clinical yield at this point remains questionable, as for other measures of the translational-LVOR. Asymmetries and abnormalities tend to be subtle and/or impermanent.

Nevertheless, borrowing from parallel comments above on the translational-LVOR in general, the test might be improved by exploiting high stimulus acceleration and frequency with near-viewing distance.[42] To implement eccentric rotation in a typical clinical rotatory laboratory might benefit by driving the rotatory stimulus not by the motorized chair itself, but by an eccentrically placed actuator (a solid chair with a passive axis will do as well). This serves to exploit an eccentric lever arm to produce high-frequency or transient tangential accelerations of sufficient magnitude but over short excursions. An alternative to eccentric rotation in the head's horizontal plane (i.e., head upright and nose out or nose in) is to tilt the subject back in the chair so that the head sits eccentric to the rotation axis but now faces upwards. Earth-horizontal rotation in this case activates the AVOR in response to head roll, thereby producing ocular torsion, whereas the LVOR drives horizontal responses to IA acceleration, as in the upright case. This alternative method isolates the LVOR from the AVOR and is therefore arguably more selective in quantifying the LVOR, as demonstrated in the monkey.[23] In all cases, controlling fixation distance is essential and can be conveniently accomplished by providing brief visual presentations of LEDs at two or three distances to help fix vergence.

A fascinating variation of eccentric rotation[45] might be termed "near-centric rotation." This technique exploits the fact that the labyrinth itself is slightly eccentric within the head. By rotating the subject at a very high constant rate (e.g., 300 degrees/s) and allowing the AVOR to extinguish, the tilt- LVOR (OCR) generated from each labyrinth can be studied by shifting head position a few centimeters to the right and left, thereby centering the left and right endorgan, respectively. This trick provides considerable clinical promise. In one important sense, the ability to test one side at a time resembles calorics, and if this test can also prove permanent and graded, a promising test of otolith function will have been devised. Like eccentric rotation with the larger head eccentricities described above, a modified rotatory chair could be used. Small lateral head displacements could be built into a bite bar, for example. An important caveat is that ocular counter-rolling is not readily measured using classic eye movement recording techniques and represents a challenge using most video-based systems. However, borrowing from the relation between eye movements and perception, ocular counter-rolling is closely related to perceived visual roll and therefore SVV techniques. It is a simple matter to instrument a rotatory chair to accomplish near-centric rotation while recording SVV.

OFF VERTICAL AXIS ROTATION

Constant velocity rotation around an axis that is tilted from vertical generates an unusual condition in which a fixed amplitude linear acceleration rotates in a direction opposite the head rotation and therefore activates the LVOR.[46–48] In this case, because the acceleration vector rotates around the head, one can imagine a variety of LVOR components arising together. Thus, a combination of translational and tilt LVORs is buried in the response. Classically, off vertical axis rotation is performed with the head rotating in yaw (in the head-horizontal plane). The induced horizontal response includes a "modulation" component (translational-LVOR), which is an oscillatory waveform that matches the frequency of the acceleration cycle around the head (e.g., 360 degrees/s = 1 Hz), and a persistent "bias" (or DC) component that is directed opposite the head rotation. Findings from off vertical axis rotation responses in patients[49,50] have been somewhat limited and relegated largely to asymmetries in the bias component. It remains doubtful that the yield overcomes the often nauseagenic and uncomfortable process of performing off vertical axis rotation routinely in patients.

REFERENCES

1. BALOH, R.W., A.W. SILLS & V. HONRUBIA. 1977. Caloric testing. 3. Patients with peripheral and central vestibular lesions. Ann. Otol. Rhinol. Laryngol. **86** (Suppl. 44): 24–30.
2. AW, S.T., G.M. HALMAGYI, R.A. BLACK, et al. 1999. Head impulses reveal loss of individual semicircular canal function. J. Vest. Res. **9**: 173–180.
3. HALMAGYI, G.M. & I.S. CURTHOYS. 1988. A clinical sign of canal paresis. Arch. Neurol. **45**: 737–739.
4. MUROFUSHI, T. & I.S. CURTHOYS. 1997. Physiological and anatomical study of click-sensitive primary vestibular afferents in the guinea pig. Acta Otolaryngol. (Stockh.) **117**: 66–72.
5. COLEBATCH, J.G. 2001. Vestibular evoked potentials. Curr. Opin. Neurol. **14**: 21–26.
6. DE WAELE, C., H.P. TRANBA, J.P. DIARD, et al. 1999. Saccular dysfunction in Meniere's patients. A vestibular-evoked myogenic potential study. Ann. N.Y. Acad. Sci. **871**: 392–397.
7. PAIGE, G.D. & D.L. TOMKO. 1991. Eye movement responses to linear head motion in the squirrel monkey. II. Visual-vestibular interactions and kinematic considerations. J. Neurophysiol. **65**: 1183–1196.
8. PAIGE, G.D., L. TELFORD, S.H. SEIDMAN & G.R. BARNES. 1998. Human vestibuloocular reflex and its interactions with vision and fixation distance during linear and angular head motion. J. Neurophysiol. **80**: 2391–2404.
9. VIIRRE, E., D. TWEED, K. MILNER & T. VILIS. 1986. A reexamination of the gain of the vestibuloocular reflex. J. Neurophysiol. **56**: 439–450.
10. SCHWARZ, U. & F.A. MILES. 1991. Ocular responses to translation and their dependence on viewing distance. I. Motion of the observer. J. Neurophysiol. **66**: 851–863.
11. HINE, T. & F. THORN. 1987. Compensatory eye movements during active head rotation for near targets: effects of imagination, rapid head oscillation and vergence. Vision Res. **27**: 1639–1657.
12. SNYDER, L.H. & W.M. KING. 1992. The effect of viewing distance and location of the axis of head rotation on the monkey's vestibulo-ocular reflex: I. Eye movement responses. J. Neurophysiol. **67**: 861–874.
13. GOLDBERG, J.M. & C. FERNÁNDEZ. 1971. Physiology of peripheral neurons innervating semicircular canals of the squirrel monkey. I. Resting discharge and response to constant angular accelerations. J. Neurophysiol. **34**: 635–660.
14. FERNÁNDEZ, C. & J.M. GOLDBERG. 1976. Physiology of peripheral neurons innervating otolith organs of the squirrel monkey. III. Response dynamics. J. Neurophysiol. **39**: 996–1008.
15. FERNÁNDEZ, C. & J.M. GOLDBERG. 1976. Physiology of peripheral neurons innervating otolith organs of the squirrel monkey. I. Response to static tilts and to long-duration centrifugal force. J. Neurophysiol. **39**: 970–984.
16. PAIGE, G.D. & D.L. TOMKO. 1991. Eye movement responses to linear head motion in the squirrel monkey. I. Basic characteristics. J. Neurophysiol. **65**: 1170–1182.
17. RUDE, S.A. & J.F. BAKER. 1988. Dynamic otolith stimulation improves the low frequency horizontal vestibulo-ocular reflex. Exp. Brain Res. **73**: 357–363.
18. TOMKO, D.L., C. WALL, III, F.R. ROBINSON & J.P.S TAAB. 1987. Gain and phase of cat vertical eye movements generated by sinusoidal pitch rotations with and without head tilt. Aviat. Space Environ. Med. **58**: 186–188.

19. HOLLY, J.E. & G. MCCOLLUM. 1996. The shape of self-motion perception I. Equivalence of classification for sustained motions. Neuroscience **70:** 461–486.
20. HOLLY, J.E. & G. MCCOLLUM. 1996. The shape of self-motion perception. 2. Framework and principles for simple and complex motion. Neuroscience **70:** 487–513.
21. MAYNE, R. 1974. A systems concept of the vestibular organs. In Handbook of Sensory Physiology, Vol.VI/2: Vestibular System. H.H. Kornhuber, ed. :493–580. Springer-Verlag. Berlin, Heidelberg, New York.
22. GLASAUER, S. 1995. Linear acceleration perception: frequency dependence of the hilltop illusion. Acta Otolaryngol. (Stockh.) **520:** 37–40.
23. TELFORD, L., S.H. SEIDMAN & G.D. PAIGE. 1997. Dynamics of squirrel monkey linear vestibuloocular reflex and interactions with fixations distance. J. Neurophysiol. **78:** 1775–1790.
24. BUSETTINI, C., F.A. MILES, U. SCHWARZ & J.R. CARL. 1994. Human ocular responses to translation of the observer and of the scene: dependence on viewing distance. Exp. Brain Res. **100:** 484–494.
25. PAIGE, G.D. 1991. Linear vestibulo-ocular reflex (LVOR) and modulation by vergence. Acta Otolaryngol. **481** (Suppl): 282–286.
26. LEMPERT, T., M.A. GRESTY & A.M. BRONSTEIN. 1999. Horizontal linear vestibulo-ocular reflex testing in patients with peripheral vestibular disorders. Ann. N.Y. Acad. Sci. **871:** 232–247.
27. ANGELAKI, D.E., S.D. NEWLANDS & J.D. DICKMAN. 2000. Primate translational vestibuloocular reflexes. IV. Changes after unilateral labyrinthectomy. J. Neurophysiol. **83:** 3005–3018.
28. PAIGE, G.D. & S.H. SEIDMAN. 1999. Characteristics of the VOR in response to linear acceleration. Ann. N.Y. Acad. Sci. **871:** 123–135.
29. PAIGE, G.D., L. TELFORD, S.H. SEIDMAN & P. BOULOS. 1996. The linear vestibulo-ocular reflex (LVOR) following labyrinthectomy [abstr]. Neurosci. Abstr. **22:** 661.
30. DAI, M.J., I.S. CURTHOYS & G.M. HALMAGYI. 1989. A model of otolith stimulation. Biol. Cybern. **60:** 185–194.
31. DIAMOND, S.G. & C.H. MARKHAM. 1983. Ocular counterrolling as an indicator of vestibular otolith function. Neurology **33:** 1460–1469.
32. CURTHOYS, I.S., M.J. DAI & G.M. HALMAGYI. 1991. Human ocular torsional position before and after unilateral vestibular neurectomy. Exp. Brain Res. **85:** 218–225.
33. BARBER, H.O. 1984. Positional nystagmus. Otolaryngol. Head Neck Surg. **92:** 649–655.
34. DAI, M.J., I.S. CURTHOYS & G.M. HALMAGYI. 1989. Linear acceleration perception in the roll plane before and after unilateral vestibular neurectomy. Exp. Brain Res. **77:** 315–328.
35. BOHMER, A. & F. MAST. 1999. Assessing otolith function by the subjective visual vertical. Ann. N.Y. Acad. Sci. **871:** 221–231.
36. TABAK, S., H. COLLEWIJN & L.J. BOUMANS. 1997. Deviation of the subjective vertical in long-standing unilateral vestibular loss. Acta Otolaryngol. (Stockh.) **117:** 1–6.
37. CURTHOYS, I.S., T. HASLWANTER, R.A. BLACK, et al. 1998. Off-center yaw rotation: effect of naso-occipital linear acceleration on the nystagmus response of normal human subjects and patients after unilateral vestibular loss. Exp. Brain Res. **123:** 425–438.
38. WEARNE, S., T. RAPHAN & B. COHEN. 1999. Effects of tilt of the gravito-inertial acceleration vector on the angular vestibuloocular reflex during centrifugation. J. Neurophysiol. **81:** 2175–2190.
39. TELFORD, L., S.H. SEIDMAN & G.D. PAIGE. 1996. Canal-otolith interactions driving vertical and horizontal eye movements in the squirrel monkey. Exp. Brain Res. **109:** 407–418.
40. TELFORD, L., S.H. SEIDMAN & G.D. PAIGE. 1998. Canal-otolith interactions in the squirrel monkey vestibulo-ocular reflex and the influence of fixation distance. Exp. Brain Res. **118:** 115–125.
41. BARRATT, H.J., A.M. BRONSTEIN & M.A. GRESTY. 1987. Testing the vestibular-ocular reflexes: abnormalities of the otolith contribution in patients with neuro-otological disease. J. Neurol. Neurosurg. Psychiatry **50:** 1029–1035.

42. CRANE, B.T. & J.L. DEMER. 1998. Human horizontal vestibulo-ocular reflex initiation: effects of acceleration, target distance, and unilateral deafferentation. J. Neurophysiol. **80:** 1151–1166.
43. BRONSTEIN, A.M. & M.A. GRESTY. 1991. Compenstory eye movements in the presence of conflicting canal and otolith signals. Exp. Brain Res. **85:** 697–700.
44. SEIDMAN, S.H., R.D. TOMLINSON, G.D. PAIGE, *et al.* 1999. Linearity of canal-otolith interactions in humans. Neurosci. Abstr. **25:** 659.
45. CLARKE, A.H. & A. ENGELHORN. 1998. Unilateral testing of utricular function. Exp. Brain Res. **121:** 457–464.
46. RAPHAN, T. & B. COHEN. 1981. Effects of gravity on rotatory nystagmus in monkeys. Ann. N.Y. Acad. Sci. **374:** 44–55.
47. RAPHAN, T. & B. COHEN. 1985. Velocity storage and the ocular response to multidimensional vestibular stimuli. *In* Reviews of Oculomotor Research. Vol. I. Adaptive Mechanisms in Gaze Control. A. Berthoz & G. Melvill Jones, eds. :123–143. Elsevier. Amsterdam, New York, Oxford.
48. ANGELAKI, D.E. & B.J.M. HESS. 1996. Three-dimensional organization of otolith-ocular reflexes in rhesus monkeys. 1. Linear acceleration responses during off-vertical axis rotation. J. Neurophysiol. **75:** 2405–2424.
49. PAYMAN, R.N., C. WALL, III & R. ASH-BERNAL. 1995. Otolith function tests in patients with unilateral vestibular lesions. Acta Otolaryngol. (Stockh.) **115**.
50. FURMAN, J.M., C. WALL, III & D.B. KAMERER. 1989. Earth horizontal axis rotational responses in patients with unilateral peripheral vestibular deficits. Ann. Otol. Rhinol. Laryngol. **98:** 551–555.

Vergence-Mediated Modulation of the Human Horizontal Angular VOR Provides Evidence of Pathway-Specific Changes in VOR Dynamics

DAVID M. LASKER,[a] STEFANO RAMAT,[b,c] JOHN P. CAREY,[a] AND LLOYD B. MINOR[a]

[a]*Department of Otolaryngology—Head & Neck Surgery, The Johns Hopkins University School of Medicine, Baltimore, Maryland 21287, USA*

[b]*Department of Neurology, The Johns Hopkins University School of Medicine, Baltimore, Maryland 21287, USA*

[c]*Dipartimento di Informatica e Sistemistica, Università di Pavia, Pavia, Italy*

ABSTRACT: The horizontal vestibulo-ocular reflex (VOR) evoked by passive, high-acceleration, head-on-body rotations (head thrusts) while viewing a far (124-cm) or near (15-cm) target was recorded (scleral search coil) in four subjects with normal vestibular function and in one subject with unilateral vestibular hypofunction. For responses in the subjects with normal vestibular function, the latency of responses relative to the onset of head movement was 7.5 ± 1.5 ms for the VOR and 21.6 ± 1.2 ms for the vergence-mediated increase in VOR gain. The gain of the VOR at the peak of the velocity response while viewing a far target was 1.01 ± 0.06; while viewing a near target, it was 1.25 ± 0.08 ($p < 0.003$). The responses were modeled with two pathways based on the different latencies. The "far-viewing" pathway was represented by a constant gain term. The "near-viewing" pathway was represented by a first-order lead term, a gain that was dependent on viewing distance, and a delay. Analysis of the responses revealed that the lead term was greater for the adducting than the abducting eye. In the subject with unilateral vestibular hypofunction, ipsilesional responses showed no change in VOR gain with respect to viewing distance. Contralesional responses retained the vergence-dependent increase in gain. A bilateral model was developed based on the data from the subjects with normal vestibular function. Simulations of this model when inputs were eliminated from one side predict the changes observed in the subject with unilateral vestibular hypofunction. The response asymmetries arise because the near-viewing pathway is more susceptible to inhibitory cutoff than is the far-viewing pathway.

KEYWORDS: vestibulo-ocular reflex; vergence; modulation; vestibular function

Address for correspondence: Dr. Lloyd Minor, Department of Otolaryngology—Head & Neck Surgery, Johns Hopkins University School of Medicine, 601 N. Caroline St., Rm. 6253, Baltimore, MD 21287-0910. Voice: 410-955-3403; fax: 410-955-0035.
lminor@jhmi.edu

INTRODUCTION

In studies of the horizontal angular vestibulo-ocular reflex (VOR) evoked by high-frequency, high-acceleration rotations in squirrel monkeys, we identified two pathways involved in control of the reflex.[1] A linear pathway is responsible for the constant gain of the reflex across the range of stimulus frequencies of 0.5–15 Hz for peak stimulus velocities of 20 degrees/s. The linear pathway has a phase lead that rises with frequency. This phase lead compensates for the phase lag that would be expected based on the 7-ms delay in the reflex, thereby keeping the VOR relatively in phase, but opposite in direction, to head velocity.

Our studies in squirrel monkeys also provide evidence of a nonlinear pathway that augments the responses provided by the linear pathway. There are two components of the nonlinearity: (1) a frequency- and velocity-dependent rise in gain and (2) a rectification due to inhibitory cutoff. The phasic dynamics of the nonlinear pathway match those that have been recorded in irregularly discharging afferents.

The gain of the nonlinear pathway is more modifiable than is the gain of the linear pathway. A selectively greater increase in the gain of the nonlinear than the linear pathway on the intact side after unilateral plugging of the three semicircular canals[2] or unilateral labyrinthectomy[3] is responsible for the recovery of contralesional gain after these lesions. The VOR following spectacle-induced adaptation provides compelling evidence in support of these two pathways.[4] In responses to steps of acceleration (3,000 degrees/s^2 acceleration reaching a peak velocity of 150 degrees/s) following long-term adaptation to 2.2 × magnifying spectacles, the VOR gain during the step of acceleration rises to almost equal the power of the lenses. By contrast, the VOR gain at the velocity plateau of the stimulus averaged 1.5, a value commonly observed in previous studies of the VOR that employed stimuli that were of lower frequency and acceleration.[5,6]

We sought to determine if there were similar pathways in the human VOR. We hypothesized that responses for viewing near in comparison to far targets for rapid, unpredictable, head-on-body rotations (head thrusts) might reveal similar pathway-specific effects. Our findings indicate that responses when viewing the near target manifest a phase lead. This lead is greater in the adducting than in the abducting eye. These changes in VOR dynamics with viewing distance were absent in the ipsilesional responses for a patient with unilateral vestibular hypofunction. Future studies are needed to evaluate the effect of changes in the axis of rotation with respect to the otoliths on response dynamics.

METHODS

Subjects

Three-dimensional eye movements were recorded from four subjects (aged 20–50 years) with normal vestibular function and one patient with left vestibular hypofunction. This patient, a 48-year-old woman, had intractable vertigo from left Meniere's disease and was treated with intratympanic gentamicin at an outside institution. A wick was inserted in her left tympanic membrane, and she placed gentamicin drops in the ear twice per day for 3 weeks. She developed signs of left vestibular hypofunction: right beating spontaneous nystagmus, right beating nystagmus after horizontal head shaking, and hypometric horizontal VOR evoked by rapid head thrusts to the left. She was tested 3 months after receiving the gentamicin. All subjects gave informed consent for eye movement recording through a protocol approved by the Joint Committee on Clinical Investigation of the Johns Hopkins University School of Medicine.

Recording of Eye and Head Movements

The movements of both eyes were recorded around all three axes of rotation (horizontal, vertical, and torsional) with the magnetic search coil technique. The field coil system consisted of a cubic frame with a side length of 1.02 m producing three orthogonal magnetic fields (frequencies: 55.5, 83.3, and 42.6 kHz; intensity: 0.088 G). The dual search coils (manufactured by Skalar, Delft, The Netherlands) yielded two sensitivity vectors each characterized by voltages induced in one of the two coils by the three orthogonal fields. An *in vitro* calibration was performed before each experiment in which voltage offsets were nullified by placing the coils into a metallic tube that completely shielded the coil from the magnetic fields. The coil was then placed on a gimbal system located in the center of the magnetic coil frame. Coil gains were determined by aligning the sensitivity vectors of each coil with each of the three magnetic fields. The output signals of the coils were filtered with a bandwidth of 0–90 Hz and sampled at 1,000 Hz with 12-bit resolution. The annuli were placed on each eye after administration of a topical anesthetic (proparacaine HCl 0.5%). Head rotations were sensed with a dual search coil embedded in a bite bar that was molded to the subject's dental occlusion. Further details of the calibration and recording procedures can be found in Straumann *et al.*[7]

Head Rotation Paradigms

The subject, seated with the head centered in the magnetic field, was instructed to gaze at a light-emitting diode (LED) located either 124 cm or 15 cm directly in front of the eyes. The examiner stood behind the subject and grasped the head over the temporoparietal area. The axis of rotation was approximately centered with respect to the otoliths. The head was kept stationary in a comfortable, "upright" position prior to each head thrust. This position placed Reid's stereotactic line (inferior orbital rim to superior external auditory canal) 7 degrees nose-up from the earth-horizontal plane. From this position, the examiner rapidly rotated the head by 10–20 degrees in the horizontal plane. Peak acceleration for these stimuli ranged from 800–3,000 degrees/s^2 and peak velocity was from 80–150 degrees/s. Head thrusts were randomized in terms of direction.

Analysis of Eye and Head Movements

Successive head rotations in each direction and for each viewing condition were aligned based on the onset of the head rotation. The onset of the head and eye movements was determined to be the points at which the velocity signals deviated from the mean value measured before onset of the stimulus for head and eye velocity, respectively, by >2 SD. The difference in these two onset points measured for each trial was defined as response latency. Latency for the vergence-induced increase in VOR

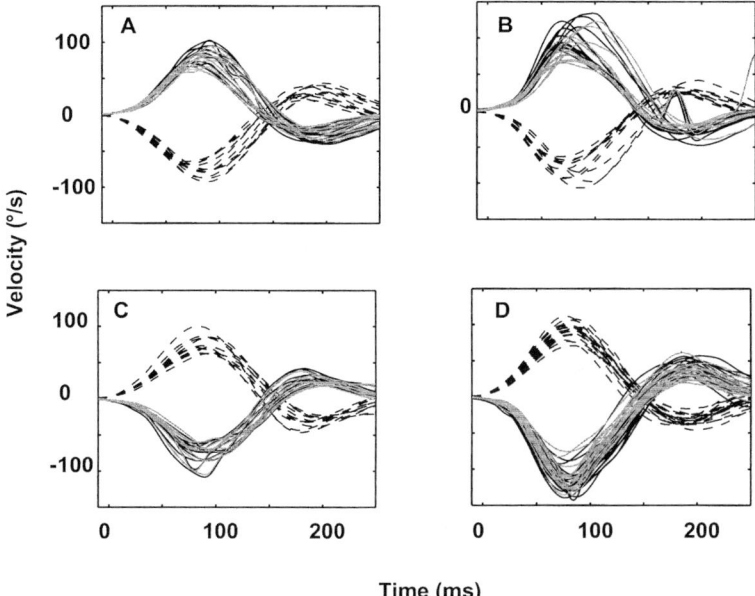

FIGURE 1. Head velocity (*dashed traces*) and eye velocity (left eye, *black*; right eye, *gray*) for head thrusts while viewing far (124 cm; **A,C**) or near (15 cm; **B,D**) target in Subject 3. (**A,B**) Responses for head thrusts to the left. (**C,D**) Responses for head thrusts to the right. On the velocity axis, positive values represent eye or head movement to the right and negative values represent eye or head movement to the left.

gain was measured as the time at which the eye velocity while viewing a near target differed from the eye velocity while viewing a far target by 1 SD.

Gain was determined as the ratio of eye to head velocity at the peak of head velocity. Comparisons of gain based on viewing condition and direction of the head thrust were made with t tests.

Modeling of Responses

Mathematical models of the VOR were formulated in Simulink® (The MathWorks). The Dormand-Prince method with a fixed step size of 0.001 s was used for simulations of the ordinary differential equations.

RESULTS

Responses to rightward and leftward head thrusts under near- and far-viewing conditions in one normal subject are shown in FIGURE 1. The responses were symmetric for both directions of rotation and for each viewing condition. Averaged responses for each viewing distance and ideal responses for each eye are shown in FIGURE 2A and B. The response in the far-viewing condition (FIG. 2A) more closely

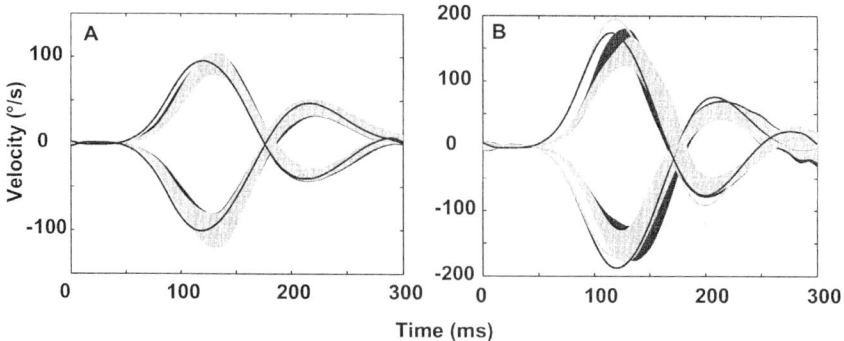

FIGURE 2. Averaged responses for head thrusts while viewing far (**A**) or near (**B**) target in Subject 2. Rightward eye velocity (resulting from leftward head thrust) is positive, and leftward eye velocity (resulting from rightward head thrust) is negative. Data are displayed as a shaded window depicting the region of 1 SD surrounding the mean. Responses shown by *gray shading* are for the right eye and those with *black shading* are for the left eye. The *solid gray and black traces* show ideal responses for the right and left eyes, respectively. Note that the ideal responses for the left and right eye are the same for far target viewing and therefore only the *black trace* is noted on the plot. For near-target viewing, the ideal response of the abducting eyes has a larger gain than does that of the adducting eye.[8] This difference arises because of the location of the axis of rotation posterior to the eyes. As described in TABLE 2, the gain values for the two eyes were similar in these subjects. The trajectory of the responses for the adducting eye shows phasic dynamics that are more prominent than those recorded from the abducting eye.

matches the ideal response than does the response in the near-viewing condition (FIG. 2B).

Our analysis of these responses involved calculation of the latency of the VOR as well as the latency of the vergence-induced change in the VOR. We also calculated the gain (defined as peak eye velocity/peak head velocity) for each viewing condition. From these parameters of latency and gain, we developed mathematical models to fit and analyze our data.

Latency of the Responses

The responses to head thrusts under near- and far-viewing conditions from the four normal subjects are shown in FIGURE 3. For the display and analysis of these data, head thrusts were divided into two groups based on the mean acceleration from 10–40 ms after the onset of the stimulus. The latency of the VOR during the far-target viewing condition (124 cm, FIG. 1A) measured 7.7 ± 1.2 ms for rotations in the leftward direction and 7.4 ± 2.0 ms in the rightward direction ($p > 0.7$). The latency measured during the near-target viewing condition measured 7.2 ± 1.7 ms for rotations in the leftward direction and 7.8 ± 1.0 for rotations in the rightward direction ($p > 0.6$). There was no difference in latency of the onset of the VOR at the two different target distances ($p > 0.5$).

The average latency, measured from the onset of the head movement, for the vergence-induced change in gain measured 21.6 ± 1.2 ms in the four subjects.

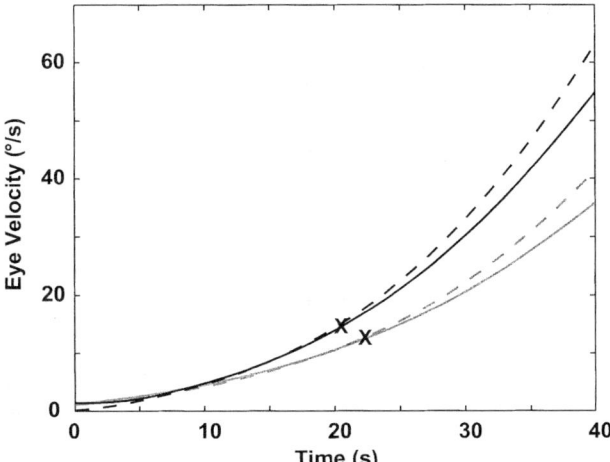

FIGURE 3. Latency of the vergence-mediated change in gain of the VOR. Data from the four subjects with normal vestibular function were pooled. Leftward eye velocity (resulting from rightward head movements) was inverted so that traces in both directions could be combined. The responses were then divided into two groups according to the peak acceleration of the head during the stimuli. The group of responses to lower acceleration stimuli had peak head acceleration of 800–1,200 degrees/s^2 (*lower traces*), whereas those in the higher acceleration group had peak acceleration of 1,300–2,000 degrees/s^2 (*upper traces*). Average responses for far targets are shown by the *solid traces,* and average responses for near targets are shown by the *dashed traces*. Each trace represents an average of 60–80 stimulus repetitions. The X represents the time at which the near-target eye velocity differs by 1 SD from the far-target velocity.

Dynamics of the Vestibulo-Ocular Reflex

The VOR in response to high acceleration head rotations is shown in FIGURE 1. Peak gains were measured by dividing the peak eye velocity by the peak head velocity. The results for the four subjects during the far-target viewing condition are shown in TABLE 1. The gain for leftward rotations measured 1.02 ± 0.06 for the left eye and 1.02 ± 0.07 for the right eye ($p > 0.5$). The gain for rightward rotations measured 1.0 ± 0.07 for left eye and 0.99 ± 0.03 for the right eye ($p > 0.7$). After pooling the data for the two eyes, the gain for head thrusts in the leftward direction was 1.01 ± 0.06 and the gain for head thrusts in the rightward direction was 1.01 ± 0.05 ($p > 0.9$).

The results for the four subjects during the near-target viewing condition are shown in TABLE 2. The gain for leftward rotations when measured from the left eye was 1.24 ± 0.09 and when measured from the right eye was 1.26 ± 0.09 ($p > 0.5$). The gain for rightward rotations when measured from the left eye was 1.28 ± 0.06 and when measured from the right eye was 1.19 ± 0.08 ($p > 0.3$). After pooling the data for the left and right eyes, the gain measured 1.25 ± 0.08 for the leftward direction and 1.24 ± 0.08 for the rightward direction ($p > 0.8$). There was a significant difference between the gains for the far-target viewing condition and the near-target viewing condition ($p < 0.003$).

TABLE 1. Response gain for far-target viewing conditions

Subject No.	Left eye		Right eye	
	Leftward	Rightward	Leftward	Rightward
1	0.95	1.00	0.92	0.95
2	1.09	1.06	1.07	1.03
3	1.05	1.10	1.04	1.00
4	1.00	0.93	0.95	0.99

TABLE 2. Response gain for near-target viewing condition

Subject No.	Left eye		Right eye	
	Leftward	Rightward	Leftward	Rightward
1	1.21	1.36	1.24	1.24
2	1.35	1.26	1.38	1.25
3	1.27	1.30	1.22	1.08
4	1.14	1.21	1.19	1.20

As observed for the responses during far-target viewing (FIG. 2A), there were differences in the trajectories of the eye velocity for the adducting in comparison with the abducing eye. The adducting eye had a velocity profile with a larger lead than that of the abducting eye (FIG. 2B). This effect was more prominent in the near-target than in the far-target condition.

In the subject tested following intratympanic gentamicin (FIG. 4), gains were measured while viewing a far and a near target for head thrusts delivered towards (ipsilesional) and away from (contralesional) the treated side. The gain in the contralesional direction during the far-viewing condition measured 0.69 ± 0.04 and was 0.87 ± 0.08 during the near-viewing condition ($p < 0.0001$). As noted in responses from normal subjects, the adducting eye for the responses to contralesional head thrusts showed more phasic dynamics than the abducting eye. The gain measured in the ipsilesional direction during the far-viewing condition was 0.38 ± 0.05 and was 0.36 ± 0.06 during the near-viewing condition ($p > 0.15$). Responses for the adducting and abducting eye were similar for ipsilesional rotations.

Mathematical Models of the Data

We used mathematical models with two pathways to analyze the responses for near and far viewing (FIG. 5A). The pathways differ in their dynamic properties and in latency relative to the onset of the head movement. The far-viewing pathway has a constant gain. The near-viewing pathway has a gain that is inversely proportional to viewing distance (higher gain for near targets) and a time delay that is determined by the onset of the viewing distance-dependent increase in gain. The response predicted from this model for the near-viewing condition used in this study is shown in FIGURE 5C (dotted trace). The predicted responses from this model lag behind the recorded data during the head movement.

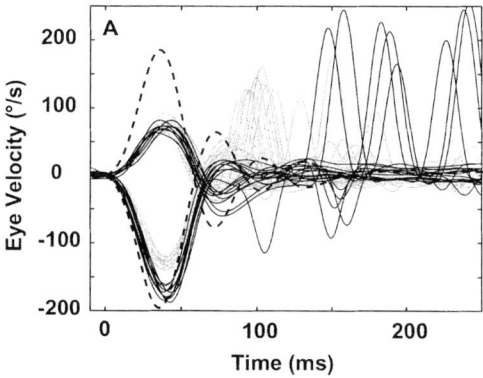

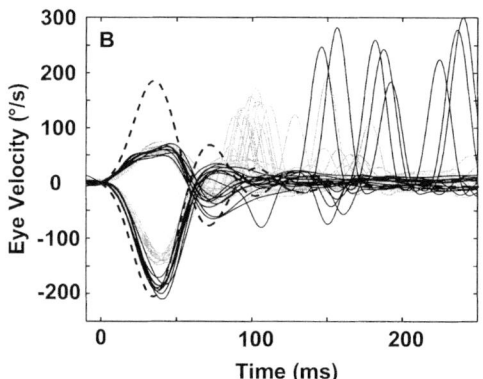

FIGURE 4. Head and eye velocity recorded from head thrusts delivered with a far or near target in a 48-year-old woman tested 3 months after intratympanic gentamycin treatment in the left ear. Positive values for head and eye velocity represent rightward movement and negative values depict leftward movement. *Dashed traces* show average head velocity. Responses from the left eye are shown in **A** and responses from the right eye in **B**. *Gray traces* are eye velocity while viewing a far target. *Black traces* are eye velocity while viewing a near target.

We added a phasic component to the near-viewing pathway to provide better representation of the data (FIG. 5B). The dashed trace in FIGURE 5C shows the response from the model for the adducting eye when $T_N = 0.16$ and $g_N = 0.11$, values obtained for fits to the data from Subject 3. The model provides a good fit to the data obtained from the adducting eye for the VOR evoked while viewing a near target in a normal subject (FIG. 5C; solid gray trace). We used a least-squares fit technique to optimize the model parameters for the adducting and abducting eye in each subject. The values obtained for the adducting eye in the four subjects were as follows (mean ± SD): $g_F = 0.99 \pm 0.05$, $g_N = 0.11 \pm 0.08$, and $T_N = 0.16 \pm 0.08$ s. The following values were obtained for the abducting eye: $g_F = 1.0 \pm 0.03$, $g_N = 0.16 \pm 0.09$, and $T_N = 0.05$

A

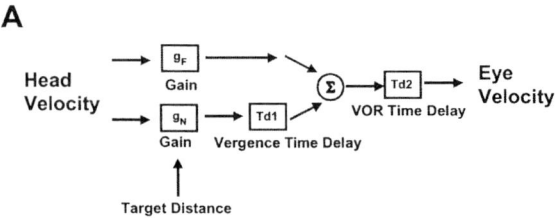

B

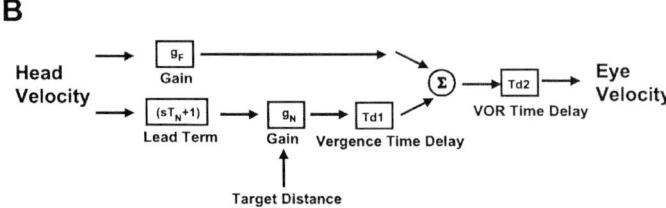

C

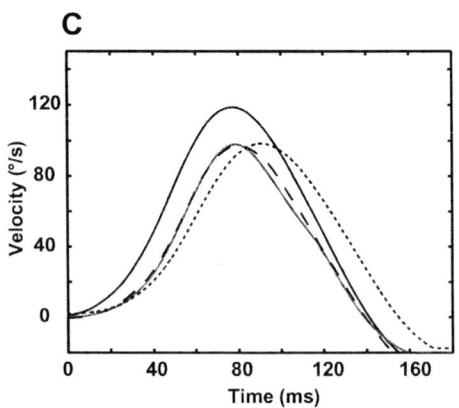

FIGURE 5. Mathematical models used to fit responses for viewing a near target. (**A**) Two pathways were formulated based on the difference in the latency of the VOR and the latency of the vergence-mediated change in the gain of the VOR. The first pathway is modeled by a single gain term (g_F). The second pathway is modeled by a gain term (g_N) and a time delay (T_{d1}) that represents the time after the onset of the VOR at which the vergence-mediated change in gain is noted. Both pathways sum together and go through a second time delay (T_{d2}) that represents the latency of the VOR. For the simulations, $g_F = 0.99$, $g_N = 0.2$, $T_{d1} = 8$ ms, and $T_{d2} = 13$ ms. (**B**) The model in **A** was modified to include a lead term ($sT_N + 1$) to obtain a better fit to the data. Simulations were performed separately for the adducting and abducting eye. The data show that more phasic responses were noted for the adducting eye. Simulations for the responses of the adducting eyes were obtained with $T_N = 0.16$ s and $g_N = 0.11$. All other coefficients in the model remained the same as those in **A**. (**C**) Simulations of eye velocity in the adducting eye for responses while viewing the near target in Subject 3. The *black line* (upper trace) is the ideal eye velocity. The *gray line* is the average eye velocity response. The *dotted line* is the simulated eye velocity based on the model in **A**. The *dashed line* is the simulated eye velocity based on the model in **B**.

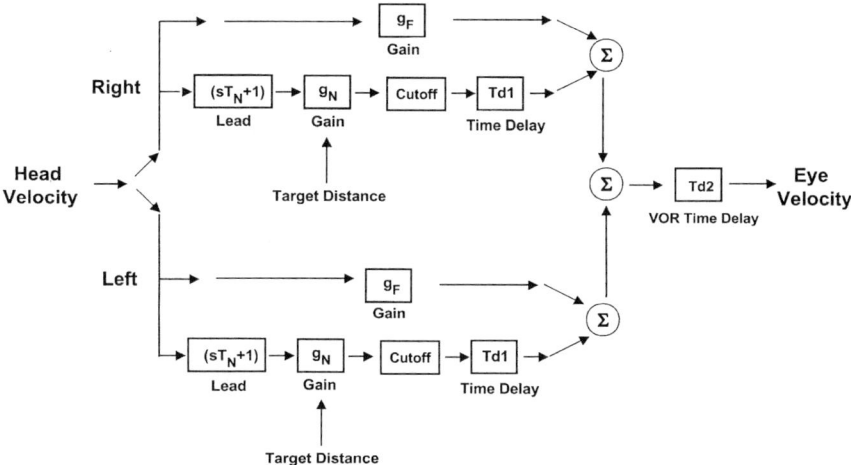

FIGURE 6. Bilateral model of vergence-mediated changes in VOR. The phasic pathway has a lead term that is comparable to the lead noted in the responses of irregularly discharging vestibular-nerve afferents. A cutoff block in this phasic pathway indicates that it makes little contribution to inhibitory responses. The following parameters were used in the simulations: $g_F = 0.43$, $T_N = 50$ ms, $T_{d1} = 13$ ms, and $T_{d2} = 8$ ms. For the far-viewing condition, $g_N = 0.15$. For the near-viewing condition, $g_N = 0.25$.

± 0.03 s. The principal difference between these parameters is the larger value of T_N for the adducting than for the abducting eye.

We developed a bilateral model to understand the changes in the vergence-dependent responses that would be expected in cases of unilateral vestibular hypofunction (FIG. 6). The pathways in this model are similar to those that we used to describe changes in the VOR of the squirrel monkey after unilateral plugging of the three semicircular canals, unilateral labyrinthectomy, and spectacle-induced adaptation. To model these responses after unilateral vestibular hypofunction, we removed the input to the left side. One key feature of the model is that the phasic pathway is rapidly driven into inhibitory cutoff. The dynamics of the far-viewing pathway are similar to those of regularly discharging afferents, and the dynamics of the near-viewing pathway show an additional component that is similar to the dynamics of irregularly discharging afferents.

FIGURE 7 shows a comparison of responses for ipsilesional (leftward) and contralesional (rightward) head thrusts for near and far viewing in the patient with left vestibular hypofunction. The eye movements displayed in this figure were recorded from the patient's left eye. Ipsilesional responses did not change with viewing distance. The model predicts this observation (dashed traces, light and dark gray). This loss of the increase in gain with near viewing arises because the near-viewing pathway is driven into inhibitory cutoff during the ipsilesional rotations. Contralesional responses continue to show vergence-dependent changes in gain.

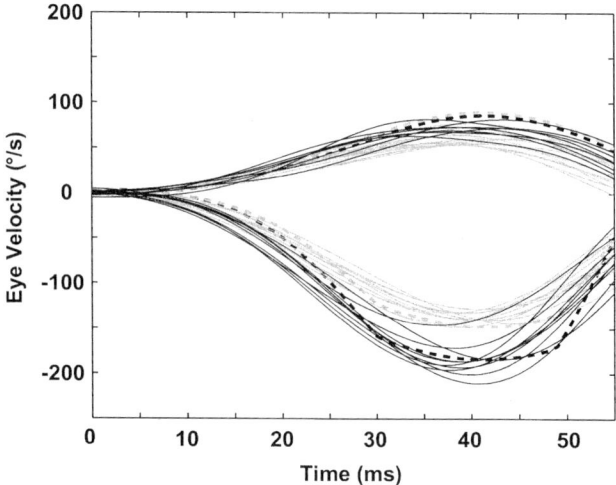

FIGURE 7. Responses recorded from the left eye in the patient with left vestibular hypofunction for ipsilesional and contralesional head thrusts. *Gray traces* are responses while viewing a far target, and *black traces* are responses while viewing a near target. Simulations from the model in FIGURE 6 are shown in the *heavy dashed traces* for far (*gray*) and near (*black*) targets. Positive values represent eye movement to the right, and negative values values represent eye movement to the left.

DISCUSSION

Dynamics of Pathways Mediating the Horizontal Angular Vestibulo-Ocular Reflex

The requirements for maintaining binocular stability of images on the fovea during horizontal head rotations change with viewing distance and with the location of the axis of rotation relative to the eyes and the otoliths.[8] The VOR gain increases with near-target viewing and with axes of rotation that are located posterior to the otoliths. A vergence-mediated increase in VOR gain in humans is also noted for rotations about an axis that is centered with respect to the otoliths when subjects are viewing a near in comparison to a far target.[9]

Highly modifiable neural processes are required for the moment-to-moment modifications in VOR gain that occur with changes in target distance. Chen-Huang and McCrea[10] studied viewing distance-related changes in the horizontal VOR in squirrel monkeys during far- and near-target viewing. The effects of linear translation produced by positioning the axis of rotation anterior or posterior to the otoliths were also assessed. The anodal polarization paradigm for functionally ablating the discharge of irregular afferents[11] was used to evaluate the role of these afferents in the changes in gain that occurred with viewing distance. These investigators found that although viewing distance-related changes in the horizontal VOR could not always be evoked and were often small in amplitude, the galvanic anodal polarizations reduced the increase in gain associated with near targets by an average of 64%.

The findings of Chen-Huang and McCrea have particular significance in terms of the putative pathways for processing of signals arising from regularly and irregularly discharging afferents. In studies using intrasomatic recordings from central vestibular neurons in squirrel monkeys, it was found that 30% of the monosynaptic excitatory postsynaptic potential to neurons projecting to the extraocular motor nuclei arise from irregularly discharging afferents.[12,13] In contrast, studies of the VOR using a galvanic polarization paradigm to selectively and reversibly ablate the discharge of irregular afferents did not result in any change in the horizontal VOR for stimuli at 0.5 Hz, ± 60 degrees/s; 4 Hz, ± 10 degrees/s; and steps of acceleration (600 degrees/s^2).[11] To explain the apparent discrepancy between the findings in the intrasomatic recording experiments and the functional ablation paradigm used in alert monkeys, it was hypothesized that there was a convergence of monosynaptic excitatory and polysynaptic inhibitory inputs from irregularly discharging afferents onto the central neurons that project to the extraocular motor nuclei.[11]

Evidence in favor of the notion that there are convergent excitatory and inhibitory inputs from irregularly discharging afferents onto central vestibular neurons was provided by extracellular recordings from central vestibular neurons in alert squirrel monkeys.[14] Galvanic anodal polarization led to an increase in response gain in some horizontal position-vestibular-pause neurons (PVPs) (the interneurons mainly responsible for signals mediating the horizontal VOR) and to a decrease in response gain in others. When the entire sample of PVPs was considered together, the excitatory and inhibitory effects almost completely cancelled one another so that there was no change in the overall signal from PVPs as a consequence of silencing irregular afferents.

Recent studies of the horizontal angular VOR evoked by high-frequency, high-acceleration head rotations in the squirrel monkey have provided evidence of linear and nonlinear (phasic) pathways mediating the reflex.[1] The rotations used in these studies were delivered in darkness, and the vergence angle was less than 1.5 meter-angle (reciprocal of the fixation distance), the position of the eyes that is observed when the animal is not attending to a specific target. The linear pathway is responsible for the relatively constant gain and phase of the VOR with respect to frequency over the range of 0.5–15 Hz for peak head velocity up to 20 degrees/s. The dynamics of regularly discharging afferents have been shown to closely match those of the linear pathway.[15] The nonlinear pathway is phasic (gain and phase lead that rise with frequency), is highly modifiable, and has two nonlinear properties: inhibitory cutoff and gain that increases with the cube of velocity for frequencies greater than 2 Hz. Irregularly discharging afferents have the phasic dynamics and inhibitory cutoff that are noted in the nonlinear pathway. The nonlinear pathway became most important for large changes in VOR gain (brought about through processes of vestibular compensation or adaptation) during rapid head rotations. The frequencies and velocities of the rotational stimuli used in the study of Minor and Goldberg[11] would have elicited little contributionm from the nonlinear pathway.

The findings in the present study show that there is an increase in the phasic contribution with near-target viewing. This increase could be related to the changes that are seen in the squirrel monkey VOR after compensation to unilateral vestibular injury. The increase in VOR gain with near viewing was absent for ipsilesional rotations in the subject with unilateral vestibular hypofunction. Similar results were noted for the near-viewing responses to passive, whole-body rotations in subjects with unilateral vestibular hypofunction.[7] An increase in gain for responses to the near target than to the far target was noted for contralesional rotations. These findings suggest that the vergence-mediated increase in gain arises from a rectified or ipsilateral pathway that makes a substantial excitatory contribution to the response but that is rapidly driven into inhibitory cutoff. The model in FIGURE 6 accounts for these effects.

Vergence-Mediated Changes in the Dynamics of the Human Vestibulo-Ocular Reflex

The four subjects with normal vestibular function in this study showed a robust increase in VOR gain for responses to head thrusts when viewing near in comparison with far targets. The rotational stimuli in these studies were head-on-body rotations delivered passively (head thrusts) with the axis of rotation approximately centered with respect to the otoliths. The axis of rotation for these stimuli is located behind the eyes, and compensatory eye movements in this condition are expected to be larger for the abducting than for the adducting eye. Instead, we observed comparable gains for both eyes (when measured as peak eye velocity/peak head velocity); however, the response for the adducting eye showed a larger phase lead than did the abducting eye.

The accentuated phasic response for the adducting eye suggests a role for the ascending tract of Deiter's (ATD) in the signals that mediate these effects. The ATD projects ipsilaterally from secondary vestibular neurons receiving inputs from horizontal canal afferents to the medial rectus subdivision of the oculomotor nucleus. The sensitivity of ATD neurons to angular and linear head velocity has been shown to increase for fixation of a near target than a far target.[16]

These findings are provocative, but additional studies are needed to define the origin of the effects we have observed. Experiments in which the axis of rotation is varied will help to characterize the asymmetries between adducting and abducting eye movements. Larger changes in VOR gain with respect to viewing distance would be expected from responses to rotations about axes that are displaced posterior to the otoliths. The eye movements in the VOR for viewing a near target involve different profiles of activation for the medial and lateral rectus in each eye. Both medial rectus muscles are contracted to achieve vergence prior to the head thrust. During the VOR, the response in the abducting eye involves contraction of the lateral rectus with concomitant relaxation of the medial rectus. By contrast, the response in the adducting eye involves further contraction of the medial rectus and further relaxation of the lateral rectus. Experiments in which the vergence angle and eccentricity of the target are varied are needed to evaluate the role of the dynamics of the oculomotor plant in these effects.

ACKNOWLEDGMENTS

This work was supported by National Institutes of Health grants R01 DC02390 and K23 DC00196.

REFERENCES

1. MINOR, L.B., D.M. LASKER, D.D. BACKOUS & T.E. HULLAR. 1999. Horizontal vestibuloocular reflex evoked by high-acceleration rotations in the squirrel monkey. I. Normal responses. J. Neurophysiol. **82:** 1254–1270.
2. LASKER, D.M., D.D. BACKOUS, A. LYSAKOWSKI, et al. 1999. Horizontal vestibuloocular reflex evoked by high-acceleration rotations in the squirrel monkey. II. Responses after canal plugging. J. Neurophysiol. **82:** 1271–1285.
3. LASKER, D.M., T.E. HULLAR & L.B. MINOR. 2000. Horizontal vestibuloocular reflex evoked by high-acceleration rotations in the squirrel monkey. III. Responses after labyrinthectomy. J. Neurophysiol. **83:** 2482–2496.
4. CLENDANIEL, R.A., D.M. LASKER & L.B. MINOR. 2001. Horizontal vestibuloocular reflex evoked by high-acceleration rotations in the squirrel monkey. IV. Responses after spectacle-induced adaptation. J. Neurophysiol. **86:** 1594–1611.
5. MILES, F.A. & B.B. EIGHMY. 1980. Long-term adaptive change in primate vestibuloocular reflex. I. Behavioral observations. J. Neurophysiol. **43:** 1406–1425.
6. LISBERGER, S.G. & T.A. PAVELKO. 1986. Vestibular signals carried by pathways subserving plasticity of the vestibulo-ocular reflex in monkeys. J. Neurosci. **6:** 346–354.
7. STRAUMANN, D., D.S. ZEE, D. SOLOMON, et al. 1995. Transient torsion during and after saccades. Vision Res. **35(23/24):** 3321–3334.
8. VIRRE, E., D. TWEED, K. MILNER & T. VILIS. 1986. A reexamination of the gain of the vestibuloocular reflex. J. Neurophysiol. **56:** 439–450.
9. CRANE, B.T. & J.L. DEMER. 1998. Human horizontal vestibulo-ocular reflex initiation: Effects of acceleration, target distance, and unilateral deafferentation. J. Neurophysiol. **80:** 1151–1166.
10. CHEN-HUANG, C. & R.A. MCCREA. 1998. Contribution of vestibular nerve irregular afferents to viewing distance-related changes in the vestibulo-ocular reflex. Exp. Brain Res. **119:** 116–130.
11. MINOR, L.B. & J.M. GOLDBERG. 1991. Vestibular-nerve inputs to the vestibulo-ocular reflex: a functional-ablation study in the squirrel monkey. J. Neurosci. **11:** 1636–1648.
12. HIGHSTEIN, S.M., J.M. GOLDBERG, A.K. MOSCHOVAKIS & C. FERNANDEZ. 1987. Inputs from regularly and irregularly discharging vestibular nerve afferents to secondary neurons in the vestibular nuclei of the squirrel monkey. II. Correlation with output pathways of secondary neurons. J. Neurophysiol. **58:** 719–738.
13. BOYLE, R., J.M. GOLDBERG & S.M. HIGHSTEIN. 1992. Inputs from regularly and irregularly discharging vestibular nerve afferents to secondary neurons in the squirrel monkey vestibular nuclei. III. Correlation with vestibulospinal and vestibuloocular output pathways. J. Neurophysiol. **68:** 471–484.
14. CHEN-HUANG, C., R.A. MCCREA & J.M. GOLDBERG. 1997. Contributions of regularly and irregularly discharging vestibular-nerve inputs to the discharge of central vestibular neurons in the alert squirrel monkey. Exp. Brain Res. **114:** 405–422.
15. HULLAR, T.E. & L.B. MINOR. 1999. High-frequency dynamics of regularly discharging canal afferents provide a linear signal for angular vestibuloocular reflexes. J. Neurophysiol. **82:** 2000–2005.
16. CHEN-HUANG, C. & R.A. MCCREA. 1998. Viewing distance related sensory processing in the ascending tract of Deiters vestibulo-ocular reflex pathway. J. Vestib. Res. **8:** 175–184.

Genetics of Familial Episodic Vertigo and Ataxia

ROBERT W. BALOH[a,b] AND JOANNA C. JEN[a]

[a]*Department of Neurology and* [b]*Division of Surgery (Head and Neck), UCLA School of Medicine, Los Angeles, California 90095-1769, USA*

ABSTRACT: The familial episodic ataxias are prototypical inherited channelopathies that result in episodes of vertigo and ataxia triggered by stress and exercise. Episodic ataxia type 1 (EA-1) is caused by missense mutations in the potassium channel gene KCNA1, whereas episodic ataxia type 2 (EA-2) is caused by missense and nonsense mutations in the calcium channel gene CACNA1A. These ion channels are crucial for both central and peripheral neurotransmission. Within the last few years, the genetic mechanisms underlying these relatively rare familial episodic ataxia syndromes have been worked out. They provide a model for understanding the mechanisms of more common recurrent vertigo and ataxia syndromes, particularly those associated with migraine. Migraine affects as many as 15-20% of the general population, and it has been estimated that about 25% of patients with migraine experience spontaneous attacks of vertigo and ataxia. We identified 24 families with migraine and benign recurrent vertigo inherited in an autosomal dominant fashion. These families have numerous features in common with EA-1 and EA-2 (particularly EA-2), suggesting that benign recurrent vertigo may be an inherited channelopathy. An ion channel mutation shared by brain and inner ear could explain the combined central and peripheral features of the syndrome.

KEYWORDS: calcium channel; migraine; mutations; channelopathy

INTRODUCTION

Several different familial episodic vertigo and ataxia syndromes have been described, and recently there has been a rapid advance in our understanding of the genetics of these disorders. Episodic ataxia type 1 (EA-1) is characterized by brief episodes of ataxia and interictal myokymia, whereas episodic ataxia type 2 (EA-2) is manifest by longer episodes of ataxia with interictal nystagmus. Litt and colleagues[1] mapped EA-1 to chromosome 12q13 near a cluster of three potassium channel genes. Browne *et al.*[2] discovered four different missense mutations in KCNA1 in four unrelated EA-1 pedigrees. This was the first report of a mutation in a human potassium channel gene and the first known ion channel mutation involving brain. We and others localized the disease locus for EA-2 to a region on chromosome 19p previously shown to be the disease locus for familial hemiplegic migraine (FHM).[3,4] FHM is a rare inherited type of migraine with aura characterized by re-

Address for correspondence: Robert W. Baloh, M.D., UCLA Department of Neurology, Box 951769, Los Angeles, CA 90095-1769. Voice: 310-825-5910; fax: 310-206-1513.
rwbaloh@ucla.edu

TABLE 1. Comparison of three dominantly inherited episodic vertigo and ataxia syndromes

	EA-1	EA-2	BRV
Age of onset (yr)	2-20	2-30	2-50
Male/female ratio	1/1	1/1	1/2
Duration of attacks	seconds-minutes	hours	minutes-hours
Triggers	Stress, exercise startle	Stress, exercise, caffeine	Stress, sleep deprivation, hormones
Migraine headaches	Unknown	~50%	~75%
Interictal signs	Myokymia	Nystagmus, mild ataxia	Vestibulopathy
Acetazolamide response	~50%	~90%	~70%
Gene	KCNA1	CACNA1A	Unknown

ABBREVIATIONS: EA-1, episodic ataxia type 1; EA-2, episodic ataxia type 2; BRV, benign recurrent vertigo.

current episodes of headache and hemiparesis. Recovery between attacks is usually complete, but in some families there is interictal nystagmus and ataxia similar to that seen with EA-2. A calcium channel gene mapped to this locus on chromosome 19p, and Ophoff and colleagues[5] analyzed the exons and flanking introns of CACNA1A, identifying point mutations that resulted in a premature stop codon or interfered with splicing in two families with EA-2 and missense mutations in four families with FHM. These findings confirmed that EA-2 and FHM were allelic disorders.

In 1964, Basser[6] described an episodic disorder in children that he called benign paroxysmal vertigo. The spells typically lasted for several minutes, and afterwards the child was usually able to return to play without any untoward effects. These recurrent vertigo spells typically began before the age of 4 and would recur throughout childhood, either spontaneously remitting or persisting into adulthood. In 1979, Slater[7] described a series of patients who experienced recurrent episodes of vertigo, nausea, and vomiting, usually beginning in adulthood, that he called benign recurrent vertigo (BRV). The attacks often occurred on awakening in the morning, being particularly prominent in women around the time of their menstrual period. Duration varied from a few minutes to as long as 3 or 4 days, and patients were asymptomatic between spells. Most of Slater's patients had either migraine themselves or a strong family history of migraine, and the episodes of vertigo had several features in common with migraine including precipitation by alcohol, lack of sleep, and stress and an increased prevalence in women. Some of Slater's patients had vertigo onset in childhood, and subsequent studies in patients with benign paroxysmal vertigo of childhood showed that most of these patients eventually developed migraine. We recently studied 24 families with BRV and concluded that BRV is a migraine syndrome inherited in an autosomal dominant fashion with decreased penetrance in men.[8] It shares numerous features in common with EA-1 and EA-2 (particularly EA-2), suggesting that it may also represent an inherited channelopathy (TABLE 1).

CLINICAL FEATURES

EA-1

Episodic ataxia type 1 (EA-1) is characterized by sudden episodes of ataxia, typically triggered by exercise, startle, or emotional upset, lasting from seconds to a few minutes.[2,9,10] Aura-like symptoms, including a feeling of weakness or falling, dizziness, and blurring of vision, are common. During the episode, the ataxia involves the trunk and extremities and speech is slurred. Between episodes of ataxia, there is typically a continuous myokymia (muscle rippling), which may be either clinically evident or only detectable with EMG. Myokymia is most easily observed in the periorbital region and fingers. The episodes of ataxia usually diminish as the child becomes older and may completely disappear in later life.

EA-2

Episodic ataxia type 2 (EA-2) is characterized by episodes of ataxia lasting hours along with interictal nystagmus.[3,4,11] The episodes vary from pure ataxia to combinations of symptoms, suggesting involvement of the cerebellum and brain stem, and more than 50% of patients will have vertigo, nausea, and vomiting with the attacks. As with EA-1, episodes are triggered by exercise and emotional stress and often relieved by acetazolamide.[12] About half the patients with EA-2 have migraine headaches.[11] On examination between episodes, patients typically have gaze-evoked nystagmus with features typical of rebound nystagmus. Spontaneous vertical nystagmus, particularly downbeat nystagmus, occurs in about a third of cases.

BRV

Episodes of BRV typically last minutes to hours, although occasionally they can last a few days.[8] Attacks are commonly triggered by emotional stress, sleep deprivation, and, in women, their menses. Unlike EA-1 and EA-2, there is a female preponderance of about 2 to 1 in families with BRV. Most patients with BRV also have migraine headaches, but the headaches usually occur independently of the episodic vertigo. Examination during an episode of vertigo typically shows spontaneous nystagmus with features of a peripheral vestibular lesion (i.e., horizontal and torsion that does not change direction with gaze). Patients with BRV can develop a bilateral progressive vestibulopathy after having episodic vertigo for many years.[13]

MOLECULAR PATHOGENESIS

EA-1

The KCNA1 gene codes for six transmembrane segments (S-1 to S-6) of the Kv1.1 potassium channel subunit (FIG. 1A). Four of these subunits join together along with other auxiliary units to form the Kv1.1 potassium channel. The four subunits are believed to be arranged in a ring, like the staves of a barrel, around a central pore (FIG. 1B). The four pore loops (between S-5 and S-6) reach into the barrel and confer the ion conduction properties. The KCNA1 protein is localized in a variety of

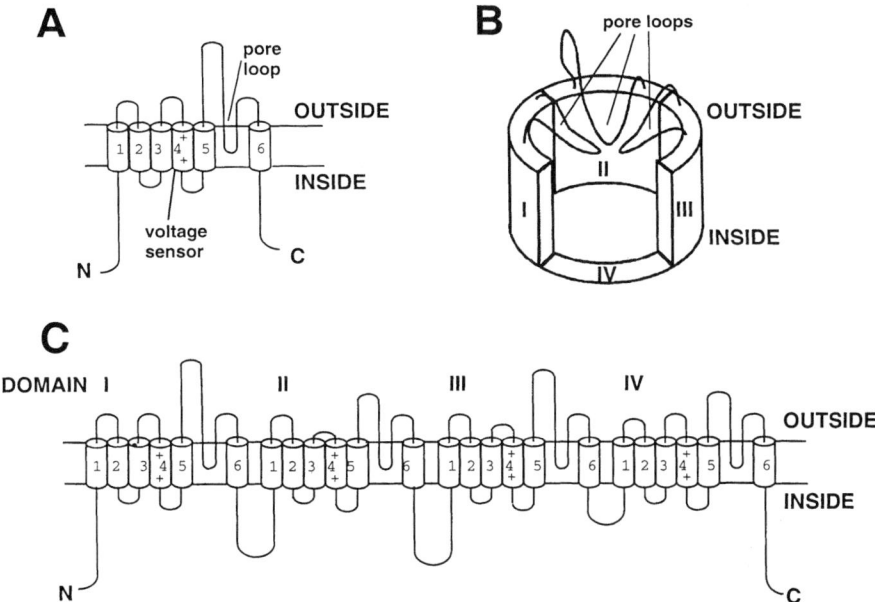

FIGURE 1. (**A**) Schematic drawing of the transmembrane subunit coded for by KCNA1. The pore-forming region (pore loop) is between segments 5 and 6. Segment 4 is the voltage-sensing region. (**B**) Illustration of how four of the subunits in **A** join together to form a potassium channel. The central pore is formed by the pore loops from each of the four subunits. (**C**) Schematic drawing of the transmembrane subunit coded for by CACNA1A. Each domain is equivalent to one of the subunits coded for by KCNA1.

brain and peripheral nerve regions. It is heavily expressed in Purkinje cells and basket and granular cells of the cerebellum and in the juxtaparanodal regions of nodes of Ranvier of peripheral nerves.[14,15] A range of missense mutations in KCNA1 have been identified in families with EA-1, most involving either the transmembrane segments or the intracellular linkers affecting highly conserved amino acids.[2,16]

EA-2

CACNA1A encodes for the α1A subunit of the P-Q calcium channel. Each of the four homologous domains of the α1A calcium channel subunit has six putative alpha helical membrane-spanning segments (S-1 to S-6) similar to the six transmembrane segments of the potassium channel Kv1.1 (FIG. 1C). CACNA1A is expressed throughout the brain but is particularly prominent in the cerebellum.[17] It is also heavily expressed in the neuromuscular junction where it tightly couples with neurotransmitter release.[18] Ophoff et al.[5] identified two mutations disrupting the reading frame in CACNA1A, thus predicting truncated α1A subunits in two families with EA-2. We identified a missense mutation in a family with severe progressive ataxia in some members and superimposed episodes of vertigo and ataxia in others.[19] This mutation predicted a glycine to arginine substitution at codon 293, a

TABLE 2. List of known mutations in CACNA1A[a]

exon	nt Change	aa Change	Domain	Phenotype	Exon	nt Change	aa Change	Domain	phenotype
4	811G>A	R192Q	I S4	FHM	25	4248A>G	K1338E	III S3-S4	FHM
4	820G>A	R195K	I S4	FHM	26	4396A>C	Y1387C	III S5	FHM/PA
5	889C>T	S218L	I S4-S5	coma	26*	4453T>G	F1406C	III p	EA2
6*	1096 G>A	C287Y	I S5-S6	EA2	27	4611G>T	V1459L	III S5-S6	FHM
6*	1113G>A	G293R	I S5-S6	PA	29*	4881C>T	R1549X	IV S1	EA2
11*	1718-9delCA	S495; 553X	II S1	EA2	30	4745-47delCTT	Y1597/A1596D	IV S1-S2	EA2
12*	1854G>A	G540R	II S2	EA2	32	5233G>A	R1666H	IV S2	EA2
13	1984G>A	R583Q	IIIp	FHM/PA	32*	5241C>T	R1669X	IV S4	EA2
16	2233 C>T	T666M	II S5-S6	FHM/PA	32	5250C>T	R1672W	IV S4	FHM/PA
16*	2278-9delAG	Q681;783X	II S5-S6	EA2	32	5298T>C	W1688R	IV S4-S5	FHM/PA
17	2377 T>C	V714A	II S6	FHM	33	5334G>A	V1700I	IV S5	FHM
17	2381C>G	D715E	II S6	FHM/PA	35	5517G>A	E1761K	IV-p loop	EA2
22	4040delC	1296X	III S1	EA2	36	5679A>C	I 1815L	IV S6	FHM
23*	4077C>T	R1281X	III S1-S2	EA2	42*	6351C>T	Q2039X	c-term	EA2
In24	4239+1 G>A	novel seq	III S3	EA2	47	7191(CAG)n	2319 polyQ	c-term	SCA6

[a]*Asterisk* indicates those mutations discovered in our laboratory. FHM, familial hemiplegic migraine; PA, progressive ataxia; EA-2, episodic ataxia; SCA6, spinocerebellar ataxia type 6; nt, nucleotide; aa, amino acid; in, intron; del, deletion.

highly conserved amino acid in the critical pore loop region of the α1A transmembrane subunit. We also identified a patient with EA-2 and no family history who showed a de novo mutation in exon 23 that predicted a premature stop code in a truncated protein.[20] Subsequently, several other families with mutations in CACNA1A and with the EA-2 phenotype have been reported, most with nonsense mutations, although a few have critical missense mutations.[21,22]

To date, we have studied 18 unrelated pedigrees with EA-2 large enough to perform linkage analysis. Of the 18 pedigrees, 13 showed linkage to chromosome 19p, whereas 5 did not. This clearly documents genetic heterogeneity with EA-2. We have been able to identify the causative mutation in CACNA1A in 8 of the 13 kindreds linked to chromosome 19p (TABLE 2). The difficulty in identifying mutations in linked families raises the possibility of nucleotide changes in noncoding regulatory regions affecting the expression and/or splicing of this complex gene.

BRV

Because of clinical similarities with EA-2, we tested several large families with BRV for linkage to the EA-2 locus on chromosome 19p. In each case, linkage to this region was ruled out. We then screened several of the probands in families with BRV but could identify no mutations in the calcium channel gene on chromosome 19p.[23] We therefore feel that the CACNA1A gene is not the BRV gene. BRV appears to be inherited in an autosomal dominant fashion, although the approximate 2 to 1 female preponderance in families needs to be explained. Possibly as with migraine, the female preponderance is the result of hormonal effects on an abnormal protein, specifically an abnormal ion channel.

DIAGNOSIS

Magnetic resonance imaging (MRI) of the brain shows typically normal results in patients with EA-1 and BRV, but it may show cerebellar atrophy, particularly of the midline, in patients with EA-2. Nonspecific paroxysmal slowing on electroencephalography (EEG) can be seen with EA-1 and EA-2.[24,25] Continuous muscle rippling (myokymia) can be detected in most patients with EA-1, particularly after provocation by limb ischemia. Eye movement recordings demonstrate a unique consistent oculomotor pattern in patients with EA-2.[11] Saccade velocity remains normal, but saccade dysmetria occurs. Smooth pursuit and optokinetic responses and suppression of vestibular-induced nystagmus are impaired. On the other hand, the vestibulo-ocular reflex gain is either high normal or increased. This pattern of findings is highly localized to the caudal midline cerebellum. Some patients with BRV show bilateral vestibulopathy with a decreased vestibulo-ocular reflex gain starting at low frequencies of rotation.[12]

Identification of the mutations in KCNA1 with EA-1 and CACNA1A with EA-2 is currently available only in a small number of research laboratories, because no single mutation occurs in the majority of families. We are currently developing an automated technique to screen CACNA1A for the more than 25 mutations currently identified (TABLE 2).

TREATMENT

Acetazolamide can be dramatic in controlling episodes of vertigo and ataxia with EA-2 and BRV.[12,26] It can also be beneficial with EA-1, although the results have been less impressive. There can be a variable response to acetazolamide, even within a single family with a known EA-2 mutation. Acetazolamide presumably works by altering the pH within the cerebellum, thus stabilizing the mutated calcium channel.[28] As noted earlier, the response to acetazolamide in families with BRV suggests the likelihood of a mutation in an ion channel. An ion channel expressed in both the inner ear and the brain could explain the combination of ear and brain symptoms reported in these patients. Whether the vertigo originates from the brain or inner ear is not clear, although the fact that some patients with BRV develop progressive bilateral vestibulopathy suggests a likely inner ear origin for the vertigo.

One typically begins with a low dose of acetazolamide (125 mg/day) and then works up to an average effective dose of between 500 and 750 mg/day. Most patients will experience paresthesias of the extremities after taking the drug, but these symptoms usually decrease over time. The main long-term side effect is the development of kidney stones, which it can be decreased if the patient regularly drinks citrus juices. As far as other drugs that might be effective, two children with EA-2 responded to the centrally active calcium channel blocker flunarizine.[29] We have tried another centrally active calcium channel blocker, nimodipine, and other peripheral calcium channel blockers including verapamil in patients with EA-2, without success. However, we have found that calcium channel blockers and beta blockers are effective in patients with BRV, but to date there have been no controlled studies to document this clinical impression.

REFERENCES

1. LITT, M., P. KRAMER, D. BROWNE, et al. 1994. A gene for episodic ataxia/myokymia maps to chromosome 12p13. Am. J. Hum. Genet. **55:** 702–709.
2. BROWNE, D.L., S.T. GANCHER, J.G. NUTT, et al. 1994. Episodic ataxia/myokymia syndrome is associated with point mutations in the human potassium channel gene, KCNA1. Nat. Genet. **8:** 136–140.
3. KRAMER, P.L., Q. YUE, S.T. GANCHER, et al. 1995. A locus for the nystagmus-associated form of episodic ataxia maps to an 11-cM region on chromosome 19p. Am. J. Hum. Genet. **57:** 182–185.
4. Vahedi, K., A. Joutel, P. Van Bogaert, et al. 1995. A gene for hereditary paroxysmal cerebellar ataxia maps to chromosome 19p. Ann. Neurol. **37:** 289–293.
5. OPHOFF, R.A., G.M. TERWINDT, M.N. VERGOUWE, et al. 1996. Familial hemiplegic migraine and episodic ataxia type-2 are caused by mutations in the Ca^{2+} channel gene CACNA1A. Cell **87:** 543–552.
6. BASSER, L.S. 1964. Benign paroxysmal vertigo of childhood: a variety of vestibular neuronitis. Brain **87:** 141–152.
7. SLATER, R. 1979. Benign recurrent vertigo. J. Neurol. Neurosurg. Psychiatry **42:** 363–367.
8. OH, A.K., H. LEE, J.C. JEN, et al. 2001. Familial benign recurrent vertigo. Am. J. Med. Genet. **100:** 287–291.
9. VAN DYKE, D.H., R.C. GRIGGS, M.J. MURPHY & M.N. GOLDSTEIN. 1975. Hereditary myokymia and periodic ataxia. J. Neurol. Sci. **25:** 109–118.
10. BRUNT, E.R. & T.W. VAN WEERDEN. 1990. Familial paroxysmal kinesigenic ataxia and continuous myokymia. Brain **113:** 1361–1382.
11. BALOH, R.W., Q. YUE, J.M. FURMAN & S.F. NELSON. 1997. Familial episodic ataxia: clinical heterogeneity in four families linked to chromosome 19p. Ann. Neurol. **41:** 8–16.
12. GRIGGS, R.C., R.T. MOXLEY, R.A. LAFRANCE & J. MCQUILLEN. 1978. Hereditary paroxysmal ataxia response to acetazolamide. Neurology **28:** 1259–1264.
13. BALOH, R.W., K. JACOBSON & T. FIFE. 1994. Familial vestibulopathy: a new dominantly inherited syndrome. Neurology **44:** 20–25.
14. RHODES, K.J., B.W. STRASSLE, M.M. MONAGHAN, et al. 1997. Association and colocalization of the $Kv\beta_1^-$ and $Kv\beta_2^-$ subunits with Kv1 α-subunits in mammalian K^+ channel complexes. J. Neurosci. **17:** 8246–8258.
15. ZHOU, L., C.-L. ZHANG, A. MESSING & S.Y. CHIU. 1998. Temperature-sensitive neuromuscular transmission in Kv1.1 null mice: role of potassium channels under the myelin sheath in young nerves. J. Neurosci. **18:** 7200–7215.
16. BRETSCHNEIDER, F., A. WRISCH, F. LEHMANN-HORN & S. GRISSMAR. 1999. Expression in mammalian cells and electrophysiological characterization of two mutant Kv1.1 channels causing episodic ataxia type 1 (EA-1). Eur. J. Neurosci. **11:** 2403–2412.
17. MORI, Y., I. FRIEDRICH, M.S. KIM, et al. 1991. Primary structure and functional expression from complementary DNA of a brain calcium channel. Nature **350:** 398–402.
18. JEN, J. 1999. Calcium channelopathies in the central nervous system. Curr. Opin. Neurobiol. **9:** 274–280.
19. YUE, Q., J.C. JEN, S.F. NELSON & R.W. BALOH. 1997. Progressive ataxia due to a missense mutation in a calcium-channel gene. Am. J. Hum. Genet. **61:** 1078–1087.
20. YUE, Q., J.C. JEN, M.M. THWE, et al. 1998. De novo mutation in CACNA1A caused acetazolamide-responsive episodic ataxia. Am. J. Med. Genet. **77:** 298–301.
21. DENIER, C., A. DUCROS, K. VAHEDI, et al. 1999. High prevalence of CACNA1A truncations and broader clinical spectrum in episodic ataxia type 2. Neurology **52:** 1816–1821.
22. JEN, J., Q. YUE, S.F. NELSON, et al. 1999. A novel nonsense mutation in CACNA1A causes episodic ataxia and hemiplegia. Neurology **53:** 34–37.
23. KIM, J.-S., Q. YUE, J.C. JEN, et al. 1998. Familial migraine with vertigo: no mutations found in CACNA1A. Am. J. Med. Genet. **79:** 148–151.
24. ZASORIN, N.L., R.W. BALOH & L.B. MYERS. 1983. Acetazolamide-responsive episodic ataxia syndrome. Neurology **33:** 1212–1214.

25. ZUBERI, S.M., L.H. EUNSON, A. SPAUSCHUS, *et al.* 1999. A novel mutation in the human voltage-gated potassium channel gene (Kv1.1) associates with episodic ataxia type 1 and sometimes with partial epilepsy. Brain **122:** 817–825.
26. GRIGGS, R.C., R.T. MOXLEY, R.A. LAFRANCE & J. MCQUILLEN. 1978. Hereditary paroxysmal ataxia: response to acetazolamide. Neurology **28:** 1259–1264.
27. LUBBERS, W.J., E.R. BRUNT, H. SCHEFFER, *et al.* 1995. Hereditary myokymia and paroxysmal ataxia linked to chromosome 12 is responsive to acetazolamide. J. Neurol. Neurosurg. Psychiatry **59:** 400–405.
28. BAIN, P.G., M.D. O'BRIEN, S.F. KEEVIL & D.A. PORTER. 1992. Familial periodic cerebellar ataxia: a problem of molecular pH homeostasis. Ann. Neurol. **31:** 146–154.
29. BOEL, M. & P. CASAER. 1988. Familial periodic ataxia responsive to flunarizine. Neuropediatrics **19:** 218–220.

Animal Models for Visual Deprivation-Induced Strabismus and Nystagmus

RONALD J. TUSA, MICHAEL J. MUSTARI, VALLABH E. DAS, AND RONALD G. BOOTHE

Yerkes Research Institute and Emory University, Atlanta, Georgia 30322, USA

ABSTRACT: The development of gaze-stabilizing systems depends on normal vision during infancy. Monkeys reared with binocular lid suture (BLS) for the first 25-40 days of life have strabismus, optokinetic nystagmus deficits, latent nystagmus, and decreased binocular cells in the visual cortex and nucleus of the optic tract. When BLS is extended to 55 days, pendular and congenital nystagmus also occurs. Eyelids in infant monkeys are hairless and thin, but BLS still degrades sensory fusion, motion, and form perception. To determine to what extent these visual properties are critical in the development of normal gaze stabilization, we examined infant monkeys reared with one opaque contact lens over one eye, alternated to the fellow eye every other day (AMO); and monkeys reared in a 3-Hz strobe environment. Monkeys reared with AMO develop strabismus, but have normal optokinetic nystagmus and no spontaneous nystagmus. Area 17 is monocular, but the medial temporal area and the nucleus of the optic tract are binocular. Monkeys reared in strobe light develop pendular nystagmus but not strabismus. We were puzzled by the results of the AMO monkeys until we examined infant monkeys with BLS that were prevented from seeing form through the lids. This was done by leaving the tarsal plate intact behind the eyelid. They developed similar to the AMO monkeys. These results suggest that disruption of sensory fusion during infancy (BLS, AMO) causes strabismus. If strabismus occurs while the monkeys have some form vision from each eye (BLS without tarsal plate), then the nucleus of the optic tract becomes monocular, which causes optokinetic nystagmus deficits and latent nystagmus. Infant monkeys reared without visual motion develop pendular nystagmus.

KEYWORDS: strabismus; nystagmus; congenital nystagmus; latent nystagmus; amblyopia; monkey; development; stereovision; vision

INTRODUCTION

The normal development of eye growth and vision depends on normal vision during infancy.[1] Rearing monkeys with their eyelids sutured closed during the first year of life has been a useful model for deprivation-induced eye growth problems and amblyopia.[1] We found that lid suture is also useful in understanding the normal development and capacity to calibrate gaze-stabilizing systems. Monkeys reared with binocular lid suture (BLS) for the first 25–40 days of life have permanent strabismus,

Address for correspondence: Ronald J. Tusa, M.D., Ph.D., Yerkes Research Center, 954 Gatewood Rd. NE, Emory University, Atlanta, GA 30322. Voice: 404-712-0996; fax: 404-712-1927.
rtusa@emory.edu

opotokinetic nystagmus deficits during monocular viewing, and latent nystagmus, and when lid suture is extended for 55 days, pendular and congenital nystagmus also occurs.[2–4] Rearing monkeys with BLS is not equivalent to dark rearing. The eyelids of infant monkeys are hairless and very thin. When eyelids are sutured closed, the iris can still be seen behind the eyelid. An optokinetic nystagmus response can be generated to a very slowly moving stimulus in infant monkeys with sutured eyelids. Despite the thinness of the eyelid, BLS still degrades motion perception, sensory fusion, and form vision. To determine to what extent these visual parameters are critical in the development of normal gaze-stabilization, we evaluated infant *Macaca mulatta* monkeys reared in a variety of visual environments at Yerkes Research Institute. It is hoped that these studies will generate animal models for infantile strabismus and nystagmus, two common disorders in infants often caused by structural problems in the visual system.

METHODS

Subjects

All *M. mulatta* monkeys came from the breeding colony at Yerkes Primate Research Center. Each infant was separated from its mother at birth and raised in an incubator until 2 months of age. The monkeys were then moved to a single cage in the nursery room with other animals and periodically allowed contact with other infants to facilitate social development. Monkeys were reared either with a contact opaque occlusion lens over both eyes, which eliminated all form and light; with occluder lens over one eye alternated to fellow eye daily, which disrupted sensory fusion; in a strobe environment at 3 Hz, which disrupted motion perception; or with lid suture. For strobe rearing, a Grass Photostimulator PS22e was used, which had a flash duration of 10 µs. Deprivation started within hours to 1 day of birth. The light cycle for both strobe light and continuous light was 16 hours on and 8 hours off per 24-hour period. The animals were inspected at least every 3–6 hours during the rearing period.

Behavioral Measurements

Ocular alignment was determined by corneal reflex photography.[4,5] A 35-mm camera equipped with a 90-mm macro lens with a ring flash was placed 30 cm in front of the monkey and multiple photographs on slide film were made with the animal fixating on a toy. Slides were then projected, and the ocular deviations in millimeters were determined using a Hirshberg ratio of 14 degrees per millimeter of light reflex displacement. Visual acuity was determined by preferential looking using sinusoidal grating (Teller acuity cards). Stereopsis was determined by using Landolt C's that varied in depth using a dichoptic random dot display designed by VRG (Vision Research Graphics, Durham, NH).

Eye Movement Recordings

Eye movements were recorded with search coil implanted around one or both eyes. The head was stabilized by a lightweight, removable aluminum halo of the type used on human beings, to fix the head to the chair. The monkeys were trained to fixate and follow a small target light rear projected onto a tangent screen located 75 cm in front of the animal. Spontaneous eye movements were measured while the monkey viewed a stationary optokinetic nystagmus drum or a Ganzfeld featureless screen (one half a clean ping-pong ball placed over the eye). Fixation was measured while the monkey viewed an 0.5-degree target light that was located in the primary and eccentric positions of gaze. A full-field grating background always surrounded the target light. Optokinetic nystagmus was measured by rotating a field drum around the animal. The drum contained a random pattern of black circles, each subtending a visual angle of 8–16 degrees on a white background.

Single Unit Electrophysiologic Recordings

Extracellular single-unit potentials were recorded with tungsten microelectrodes by conventional methods[3] in the nucleus of the optic tract and cerebral cortical areas 17 and medial temporal area when monkeys reached 2 years of age. We also examined ocular motor behavior during pharmacological inactivation and activation using GABA agonists and antagonists. Target- and eye-position signals, unit discharge, and other relevant signals were saved on analog or digital-audio tape for subsequent computer digitization and quantitative off-line data analysis. The eye- and target-positions were digitized at 1 kHz.

RESULTS

TABLE 1 summarizes the rearing conditions and the behavioral and electrophysiologic results at 1–2 years of age for monkeys reared with alternate monocular occlusion (AMO), bilateral continuous occlusion, BLS, and strobe light. In some monkeys with BLS, the tarsal plate was removed (BLSwoTP). The tarsal plate is thick cartilage attached to the underside of the upper lid. When the tarsal plate is left in place during BLS, it will allow transmission of light but prevent form vision. When it is removed during BLS, the lids are thin enough to allow some form vision. The strobe-reared monkeys are still being evaluated, so TABLE 1 shows data from 2 months of age. The monkeys with BLS without tarsal plate (BLSwoTP) reared for 25, 40, and 55 days have been presented before,[2,3] but they will be included here for comparison. We focus on strabismus and nystagmus in this paper.

Sensory-Induced Strabismus

Permanent strabismus occurred in four of four monkeys reared with AMO and seven of eight monkeys with BLS (TABLE 1, align; FIG. 1). One BLSwoTP (25-day duration) was orthotropic at 1 year, but had up to 10-degree esotropia for the first 30 days after the eyelids were opened. These results suggest that disruption of sensory fusion during infancy results in strabismus. Monkeys reared in strobe light or with bilateral continuous occlusion did not develop strabismus. Strobe rearing preserves sensory fusion depending on the frequency of the strobe light and pulse duration.[6] Bilateral continuous occlusion is equivalent to dark rearing, which is considered disuse deprivation. Jampolsky[7] has argued that disuse deprivation leaves the visual

TABLE 1. Summary of rearing conditions and behavioral and physiologic results at 1–2 years of age

Rearing type	n	Duration (days)	Align	Stereo	VA (Logmar)	Nys	Mono OKN	17	MT	NOT
AMO	4	122–182	22et-28xt	None	0–.15	None	Normal	Mono (3/3)	Bino (3/3)	Bino (2/2)
BCO	4	21	Ortho	33–73″	.22–.53	None	?	?	?	?
	1		Ortho	None	.55	None				
BLS woTP	1	25	Ortho	602″	.47	LN	T-N bias	?	Mono bias (1/1)	Mono bias (2/2)
	3	21–40	9–18xt		.30–.40	LN	T-N bias			
	1	55	27xt		.60	LN,PN,CN	T-N bias			
BLS wTP	3	21	9–10xt	None	.43	None	?	?	?	?
Strobe*	2	60	Ortho	?	?	PN,CN?	?	?	?	?

AMO, alternate monocular occlusion; BCO, bilateral continuous occlusion; BLSwoTP, bilateral lid suture without tarsal plate; BLSwTP, bilateral lid suture with tarsal plate; *Behavior data measured at 2 months of age; n, number of animals in group; Dur, duration of deprivation; Align, alignment; et, esotropia; xt, exotropia; ortho, orthotropic; Stereo, stereoacuity VA, visual acuity; nys, nystagmus; LN, latent nystagmus; PN, pendular nystagmus; CN, congenital nystagmus; Mono OKN, monocular OKN; T–N, temporal to nasal bias; 17, striate cortex, MT, Middle temporal cortex; NOT, nucleus optic tract; Mono, monocular bias; Bino, normal binocularity.

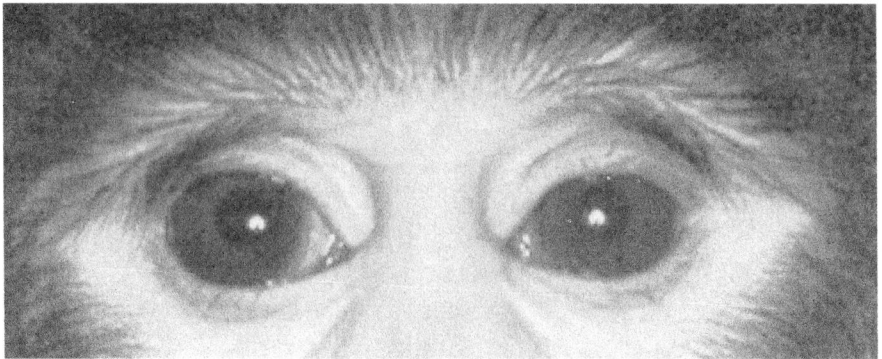

FIGURE 1. Sixteen-degree right exotropia in a monkey reared with AMO.

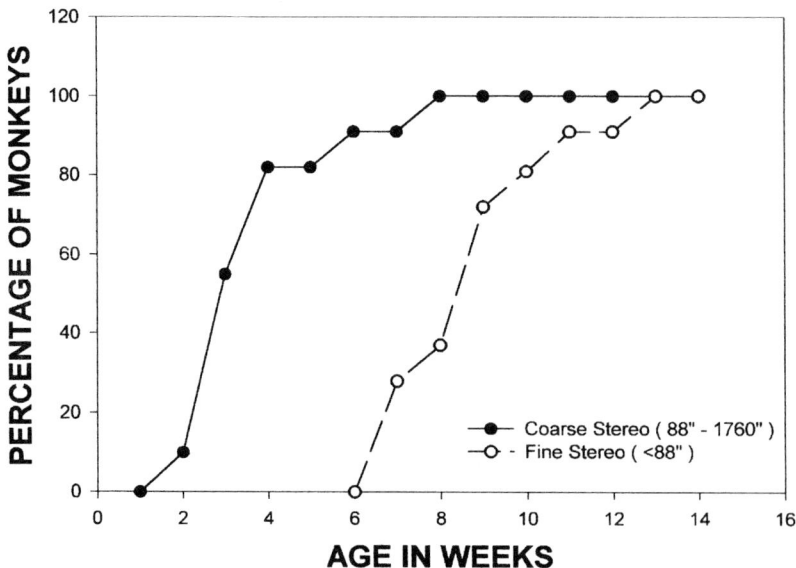

FIGURE 2. Longitudinal study of the development of stereopsis in 11 normal *M. mulatta* monkeys from birth. *Filled circles* represent the percentage of monkeys with measurable coarse stereopsis (88–1,760 seconds of arc) as a function of postnatal age. *Open circles* are the percentage of monkeys who demonstrate fine stereopsis (less than 88 seconds of arc). (Modified from O'Dell & Boothe[11].)

brain in an immature state and may not lead to any detrimental effects. In contrast, BLS and AMO are forms of misuse deprivation that lead to in visual degradation.

The critical period in which normal sensory fusion must be present in infancy to prevent permanent strabismus appears to be the first 5 weeks after birth (TABLE 2, column 2). If sensory fusion is disrupted for 3 weeks or longer during this time, strabismus will likely occur (TABLE 2, columns 3). This critical period overlaps with the development of coarse stereopsis in monkeys (FIG. 2.)[11] Coarse stereopsis begins to

TABLE 2. Percentage of Strabismus in visually-deprived monkeys

Deprivation Type	Start (wks)	Duration (wks)	No.	Percent Strabismus	Ref.
Occlusion (monocular) 50-100% of the day	1	>20	4	100%	5
Occlusion (monocular) AMO	1	>20	4	100%	8
Lid suture (reverse)	1	6	3	100%	4
Lid suture (monocular, or binocular)	1	3-8	10	100%	1,4,8
Aphakia or defocus (monocular)	1-2	>20	6	67%	5
Mild diffuser contact lens (monocular)	1	>20	5	60%	9
Lid suture (monocular & binocular)	5	2-16	7	0%	10

develop in normal monkeys by 2 weeks after birth and is complete in most by 1 month. Fine stereopsis begins to develop by 1½ months after birth and is complete in most by 3 months. Therefore, it appears that sensory-induced strabismus occurs if sensory fusion is disrupted during the development of coarse stereopsis.

We found that monkeys with strabismus also have significant saccade disconjugacy (present in three of three AMO monkeys.[12] In these studies, one eye was patched to avoid sensory confusion. FIGURE 3 shows 10-, 20-, and 30-degree amplitude saccades in one AMO monkey. The viewing eye was on target (top panel); the nonviewing eye always fell short (middle panel). This saccade disconjugacy can also be plotted as a change in ocular misalignment. The bottom panel of FIGURE 3 shows how the misalignment between the two eyes changes during the saccade. Before the saccades, there is a 12-degree esotropia, but immediately after the saccade the misalignment varied from 20-25-degree esotropia. Based on examining saccades of different amplitude made from different orbital positions, the degree of saccade disconjugacy depends on the orbital position of the eye (FIG. 4). The amount of disconjugacy in these monkeys is significantly more than that reported in human subjects with congenital strabismus.[13]

We have begun to measure behavior stereoacuity and the physiologic properties of visual cortex and nucleus of the optic tract in some of our monkeys (TABLE 1). Monkeys with strabismus have no stereoacuity up to 2,640 seconds of arc (BLSwTP) and have very few binocular cells in area 17 (AMO). The binocularity of the medial temporal area and the nucleus of the optic tract appears to have no bearing on the development of strabismus (AMO).

In summary, sensory-induced strabismus occurs if sensory fusion is disrupted at birth by misuse deprivation (BLS, AMO) but not by disuse deprivation (bilateral continuous occlusion, strobe rearing) during the first 5 weeks of birth, which overlaps with the development of coarse stereopsis. Strabismus results in a significant saccade disconjugacy that is orbital-position dependent. In monkeys with strabismus, striate cortex is monocular, but the medial temporal area and the nucleus of the optic tract may be binocular or monocular. To determine if the saccade disconjugacy is neurally mediated or due to changes in the ocular motor plant, we are examining the firing rate position curves in the ocularmotor nuclei in these strabismic monkeys.

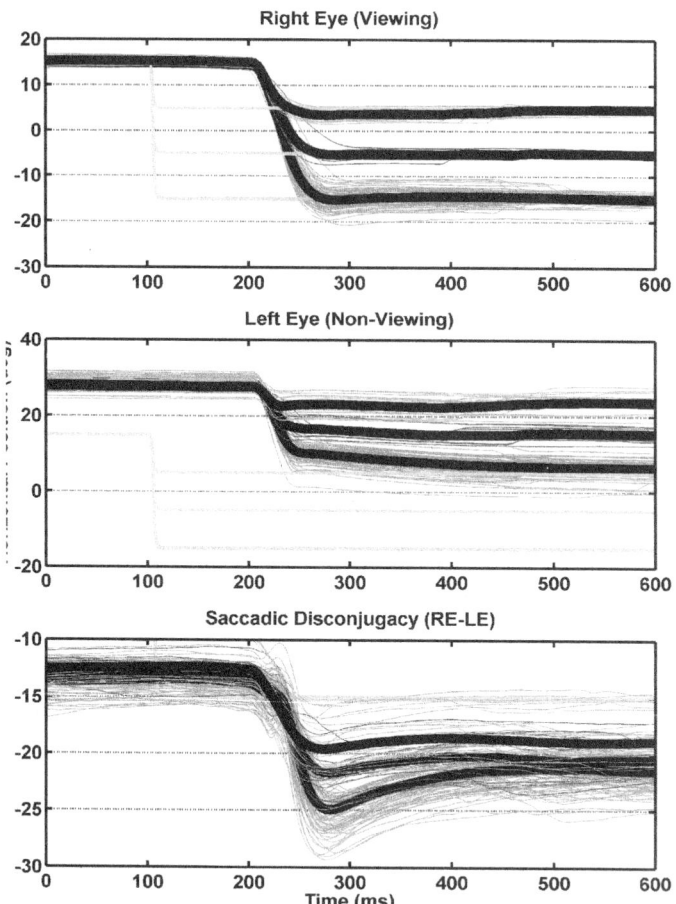

FIGURE 3. Disconjugacy of saccades in AMO4. This figure shows 10-, 20-, and 30-degree amplitude saccades to the left in one AMO monkey that had 12 degrees of esotropia (means of the three different amplitudes is a *dark line*). The left eye was patched to avoid sensory confusion. (*Top*) The viewing eye was on target. (*Middle*) The nonviewing eye always fell short of the target. (*Bottom*) Degree of ocular misalignment (right eye position minus left eye position). Before the saccades, there is a 12–degree esotropia, but immediately after the saccade the misalignment varied from 18- to 25-degree esotropia, which depended on the location of the eye at the end of the saccade.

SENSORY-INDUCED LATENT NYSTAGMUS

All monkeys reared with BLS without the tarsal plate (BLSwoTP) had a jerk nystagmus that persisted through the duration of the study (TABLE 1, latent nystagmus in the Nys column). When the right eye was covered, there was a conjugate, left-beating nystagmus, and when the left eye was covered, there was a right-beating nystagmus (FIG. 5). The term left and right beating is used here based on the direction

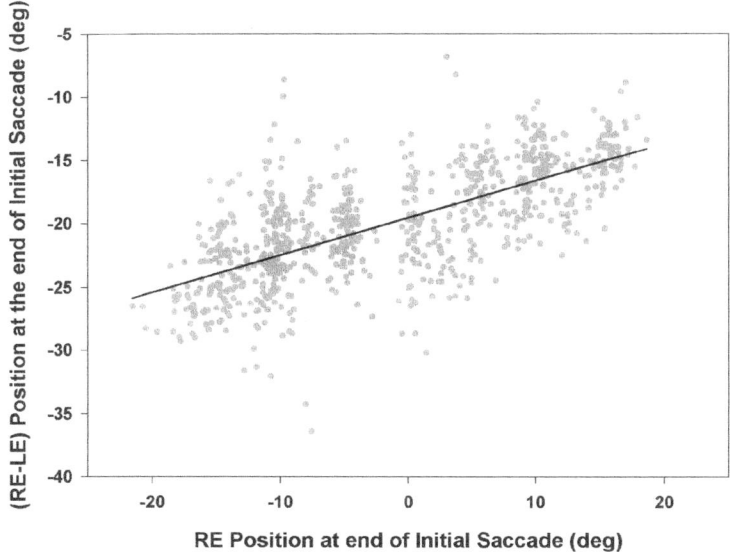

FIGURE 4. Relation of disconjugacy of the saccadic pulse with orbital position in AMO4. Each dot represents one saccade generated by the AMO4 monkey to a visual target viewed by the right eye. Right eye position immediately at the end of the saccade (saccadic pulse) is plotted on the abscissa. The ocular misalignment immediately at the end of the saccade (right eye position minus left eye position) is plotted on the ordinate. The degree of ocular misalignment decreases as the orbital position of the right eye moves from left to right.

of the quick phases of the nystagmus. When both eyes were open, all animals had a low-amplitude, horizontal nystagmus. This combination of a low-amplitude horizontal nystagmus with both eyes viewing and a strong horizontal nystagmus when only one eye was viewing (with slow phases always toward the nose in the viewing eye) resembles latent nystagmus or "manifest latent nystagmus" found in human subjects with infantile strabismus.

Latent nystagmus can also be induced in monkeys if surgical strabismus is artificially induced within the first 11 days after birth.[14] We believe monkeys with lid suture develop latent nystagmus because they developed strabismus behind the closed eyelids and were able to see some form from each eye. All monkeys with BLS in our study developed strabismus, but only those monkeys that had form vision through the sutured eyelids of both eyes developed latent nystagmus (TABLE 1, BLSwoTP). Those monkeys with BLS with no form vision (BLSwTP, tarsal plate left intact) did not develop latent nystagmus. Similarly, monkeys that developed strabismus but were not allowed to view out of each eye simultaneously (AMO) did not develop latent nystagmus.

The neural basis for the development of latent nystagmus and a defect in monocular optokinetic nystagmus in monkeys appears to be loss of binocularity in the nucleus of the optic tract, with most cells driven by the contralateral eye (FIG. 6).[2] The area that normally provides binocular eye input to the nucleus of the optic tract is the

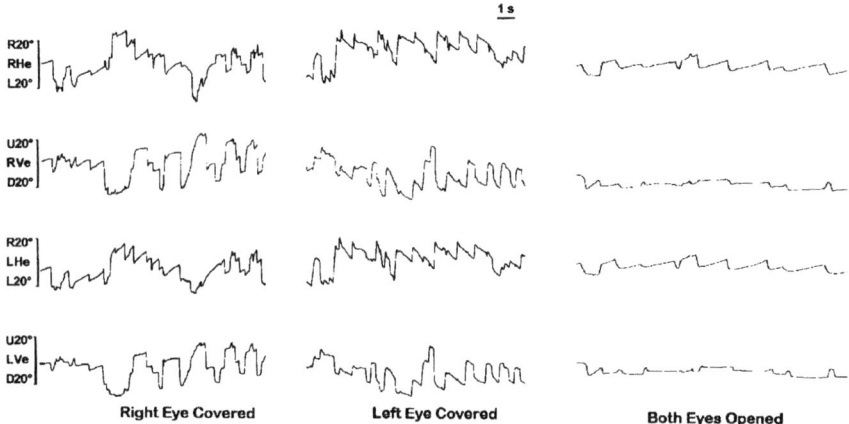

FIGURE 5. Eye movement trace showing latent nystagmus (LN) in the monkey reared with BLSwoTP for 55 days. RHe, right eye horizontal eye position; RVe, right eye vertical eye position; LHe, left eye horizontal eye position; LVe, left eye vertical eye position. One-second tick mark is shown above the trace. Upward deflections of traces indicate rightward (or upward) eye motion. (Modified from Tusa et al.[2])

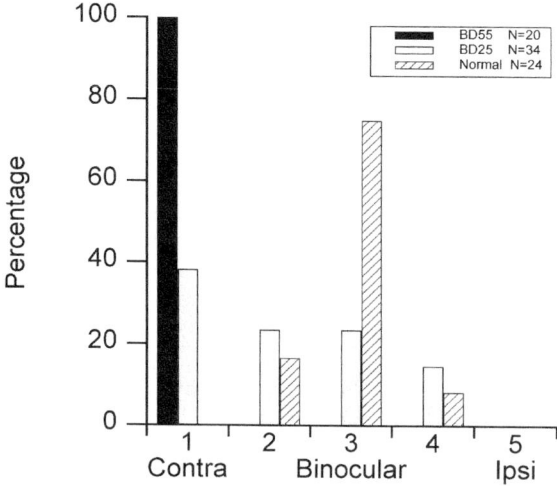

FIGURE 6. Nucleus of the optic tract unit ocular dominance preference. Units were first tested during binocular viewing. The optimal stimulus was determined and then used to test the unit's response during monocular viewing. Assignment to an ocular dominance group was determined using a ratio of response obtained during visual stimulation presented to each eye. Units that responded only during stimulation of one eye were placed in group 1 (contralateral eye) or group 5 (ipsilateral eye). Units with equal responses during stimulation presented to either eye were placed in group 3. Units with a bias towards the contralateral or ipsilateral eyes were placed in group 2 or group 4, respectively. *Open histogram bars*, normal; *shaded bars*, 25-day BLSwoTP; *dark bars*, 55-day BLSwoTP. (From Mustari et al.[3] Reprinted with permission of the *Journal of Neurophysiology*.)

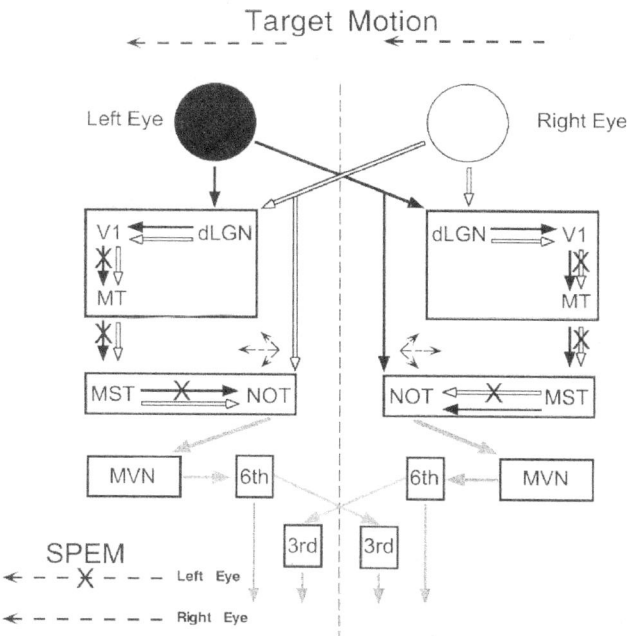

FIGURE 7. Schematic diagram of circuit mediating optokinetic nystagmus in monkeys with BLSwoTP. *Dashed lines* indicate target motion and slow-phase eye movement (SPEM) directions. *Continuous lines* indicate anatomical projections. A cortical afferent system includes a projection from the dorsolateral geniculate nucleus (dLGN) to striate cortex (V1) to the medial temporal area (MT). This system processes information about target motion from each eye. The cortical efferent system includes a projection from the medial superior temporal (MST) area to the nucleus of the optic tract (NOT). NOT also receives a direct retinal projection from the contralateral eye. Each NOT generates ipsilateral slow-phase eye movements via a projection to the medial vestibular nucleus (MVN), 6th nerve nucleus (6th) and 3rd nerve nucleus (3rd). In the schematic, *black* and *open arrows* indicate the visual pathways from the left and right eyes, respectively. *Gray arrows* indicate pathways downstream from NOT. Most neurons in NOT are normally activated only by target motion towards the ipsilateral side (indicated by the three small *dashed lines* above NOT). (From Tusa et al.[2] Reprinted with permission of the *Journal of Neurophysiology*.)

medial temporal area, and this structure also shows a loss of binocularity in monkeys with surgical strabismus-induced latent nystagmus[14] and in one monkey we examined that had BLSwoTP-induced latent nystagmus (TABLE 1).

Based on these animals studies, we propose that the presence of latent nystagmus and a defect in monocular optokinetic nystagmus indicates a loss of binocularity in the medial temporal area and the nucleus of the optic tract. In the BLSwoTP monkey, the nucleus of the optic tract loses its input from the ipsilateral eye (cortical input) and responds only to T-N motion viewed from the contralateral eye (FIG. 7). Consequently, optokinetic nystagmus to N-T stimuli during monocular viewing is not generated. In addition, when one eye is patched, visual input will only occur in the nucleus of the optic tract contralateral to the unpatched eye. Because nucleus of the

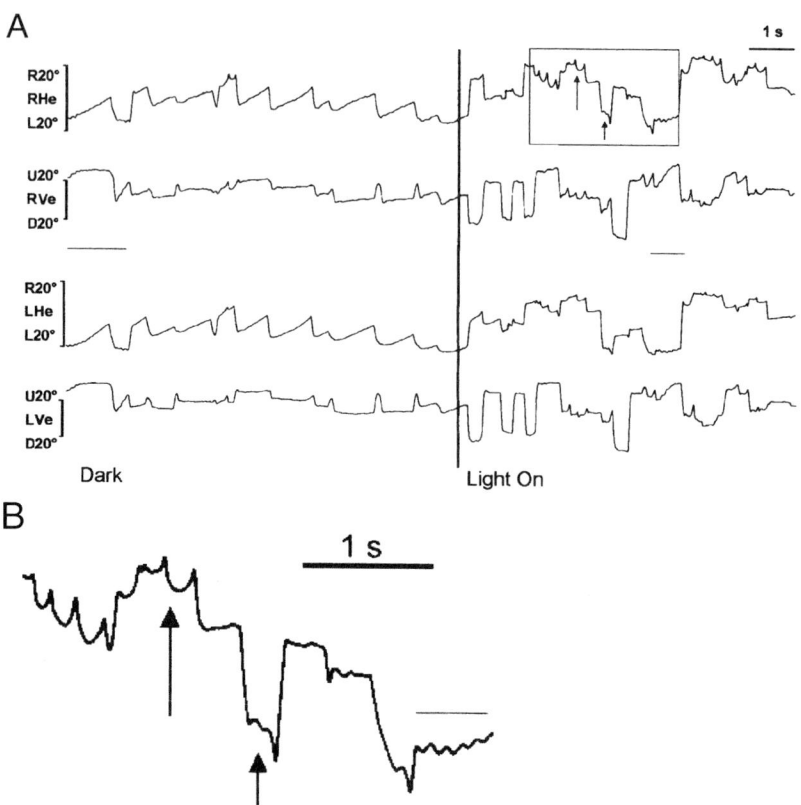

FIGURE 8. Spontaneous, jerk nystagmus with velocity-increasing slow phases and pendular nystagmus in the BLSwoTP monkey deprived 55 days. (**A**) Recordings were made while the monkey had both eyes open while sitting in the dark or while viewing an 0.5-degree light. In the dark there was a left-beating nystagmus with linear and velocity-increasing slow phases. In the light, this animal had marked velocity-increasing slow phases in eccentric gaze (*arrows*). Transient bursts of pendular nystagmus were also present (*horizontal bars*). (**B**) Magnified view of a segment of right eye horizontal eye position. RHe, right eye horizontal eye position; RVe, right eye vertical eye position; LHe, left eye horizontal eye position; LVe, left eye vertical eye position. One-second tick mark is shown above the trace. (Modified from Tusa *et al.*[2])

optic tract cells are partially driven by light, this increased activity on one side will potentially cause slow phases to occur in the T-N direction with respect to the unpatched eye, which is the same direction as latent nystagmus. We verified that this change in activity of the nucleus of the optic tract on one side is the source of latent nystagmus by the use of GABA agonist injections into the nucleus of the optic tract. Neurons in the nucleus of the optic tract have GABA receptors and when we injected a GABA agonist into both nucleus of the optic tract structures, latent nystagmus was abolished. We believe that this loss of binocularity in MT and in the nucleus of the optic tract will only occur if strabismus occurs during the first 2 weeks of life and

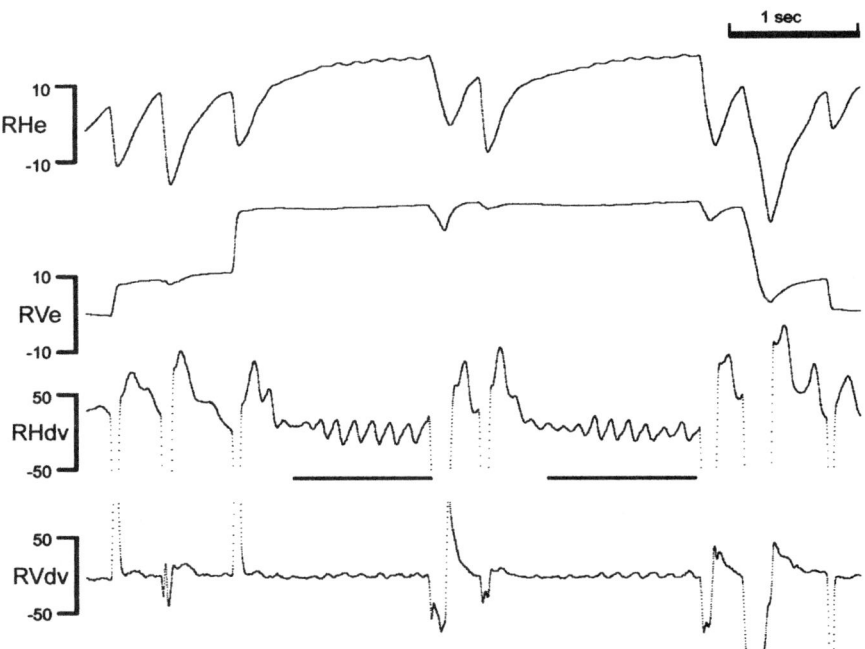

FIGURE 9. Low-amplitude, high-frequency pendular nystagmus in the BLSwoTP monkey deprived for 25 days. RHe, right horizontal eye position (in degrees); RVe, right vertical eye position (in degrees); RHdv, right horizontal eye velocity (in degrees/s); RVdv, right vertical eye velocity (in degrees/s). Transient bursts of pendular nystagmus were present (*horizontal bar*). One-second tick mark is shown above the trace. (Modified from Tusa et al.[2])

only if form vision is allowed in both eyes simultaneously. If one eye is always occluded, as in AMO, or if light is allowed to reach both eyes but no form vision is allowed, as in BLSwTP, then striate cortex still loses binocularity, but areas MT and the nucleus of the optic tract retain binocularity and latent nystagmus fails to develop (TABLE 1). The neural mechanism for this preservation of MT and the nucleus of the optic tract binocularity in monkeys with strabismus despite a monocular 17 is still unclear, but some insight may come from past studies in cats. In naturally strabismic cats and in cats made strabismic at birth, few binocular cells are found in area 17, yet the lateral suprasylvian area (homolog to area MT) still contains a near normal percentage of binocular cells depending on the degree of strabismus.[15] In the lateral suprasylvian area, the receptive fields from each eye overlap in a congruent way despite the motor misalignment.[15] This overlap of receptive fields may be the physiologic basis for anomalous retina correspondence described in humans with strabismus. Anomalous retina correspondence is a neural adaptation to eye misalignment in which noncorresponding retinal points project to single elements in visual cortex to provide binocular vision. We are currently examining monkeys with strabismus to determine if there are overlapping receptive fields in area MT that could

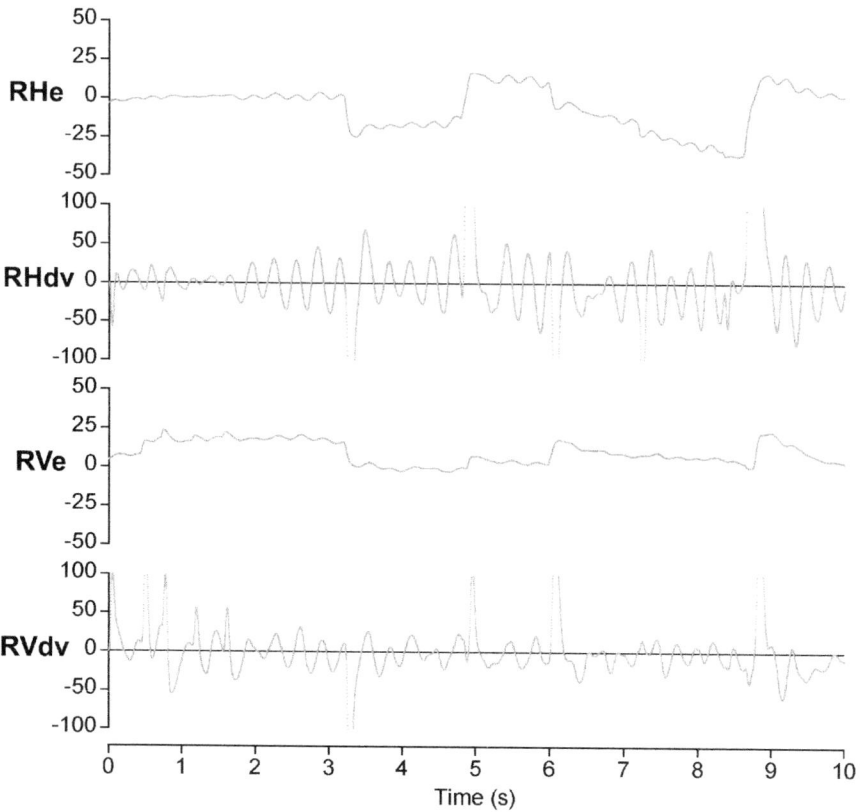

FIGURE 10. Low-amplitude, high-frequency pendular nystagmus in the monkey reared in strobe light for 2 months. RHe, right horizontal eye position (degrees); RHdv, right horizontal eye velocity (degrees/s); RVe, right vertical eye position (degrees); RVdv, right vertical eye velocity (degrees/s).

mediate anomalous retina correspondence and binocularity in AMO monkeys, which are absent in monkeys reared with BLSwoTP.

SENSORY-INDUCED PENDULAR NYSTAGMUS AND CONGENITAL NYSTAGMUS

All monkeys reared with BLSwoTP had some form of low-amplitude, conjugate pendular nystagmus (TABLE 1). This was found in the eye position trace in the monkey deprived for 55 days (FIG. 8, horizontal bars), but only in the eye velocity trace in the monkey deprived for 25 days (FIG. 9). The peak-to-peak amplitude was 0.5–2.0 degrees, and the frequency was 7–10 Hz. These values for pendular nystagmus were constant regardless if one eye was patched, if the monkey fixated on a target,

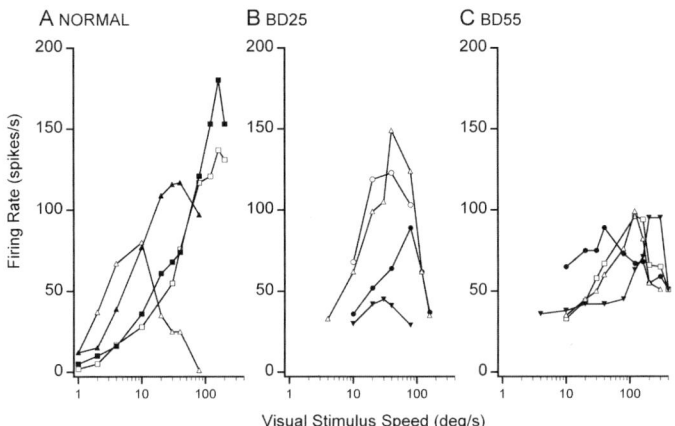

FIGURE 11. Examples of speed tuning for nucleus of optic tract units of normal (A), BLSwoTP 25-day deprived (B), and BLSwoTP 55-day BD deprived monkeys (C). Speed tuning was tested by moving a large-field visual stimulus in the preferred direction for each unit. The average firing rate over at least 10 cycles of such testing was used to plot the individual tuning curves shown. Data from different representative units are indicated by different symbols. (From Mustari et al.[3] Reprinted with permission of the *Journal of Neurophysiology*.)

or if the monkey was placed in the dark. The etiology for sensory-induced pendular nystagmus may be due to a lack of motion visual stimuli during a critical period of development. Cats reared for 1.5 years in a motion-deprived visual environment (8-Hz strobe environment) develop low-amplitude pendular nystagmus.[16] We have also noted persistent pendular nystagmus in two monkeys that were reared for 2 months in a 3.0-Hz strobe environment (TABLE 1, FIG. 10).

The monkey reared with BLSwoTP for 55 days also had a nystagmus that resembled congenital nystagmus in human subjects. Congenital nystagmus is a conjugate horizontal nystagmus with both pendular nystagmus and jerk nystagmus with velocity-increasing slow phases. This type of nystagmus was most apparent when the monkey fixated on a small target light (arrows in FIG. 8).

The neural basis for sensory-induced pendular and congenital nystagmus is unknown. Both pendular and congenital nystagmus can be simulated in control system models by making the neural integrator unstable.[17] Calibration of the neural integrator during infancy probably depends on retinal slip during eye movements. The nucleus of the optic tract is a major source of visual motion information to the neural integrator, and we have found the peak firing rate of nucleus of the optic tract cells to motion of appropriate direction and speed to be low in the BLSwoTP monkey deprived for 55 days compared to normal monkeys or the monkey deprived for 25 days (FIG. 11). The source of high-velocity sensitivity in nucleus of the optic tract comes from area MT. We have yet to examine velocity tuning in these areas in monkeys with pendular and congenital nystagmus, although cats reared in strobe light that develop pendular nystagmus have decreased motion sensitivity in striate cortex and the lateral suprasylvian area.[18,19]

REFERENCES

1. BOOTHE, R.G., V. DOBSON & D.Y. TELLER. 1985. Postnatal development of vision in human and nonhuman primates. Annu. Rev. Neurosci. **8:** 495–545.
2. TUSA, R.J. et al. 2001. Gaze-stabilizing deficits and latent nystagmus in monkeys with brief, early-onset visual deprivation: eye movement recordings. J. Neurophysiol. **86:** 651–661.
3. MUSTARI, M.J. et al. 2001. Gaze-stabilizing deficits and latent nystagmus in monkeys with early-onset visual deprivation: role of the pretectal not. J. Neurophysiol. **86:** 662–675.
4. TUSA, R.J. et al. 1991. Early visual deprivation results in persistent strabismus and nystagmus in monkeys. Invest. Ophthalmol. & Visual Sci. **32:** 134–141.
5. QUICK, M.W. & R.G. BOOTHE. 1989. Measurement of binocular alignment in normal monkeys and in monkeys with strabismus. Invest. Ophthalmol. & Visual Sci. **30:** 1159–1168.
6. CREMIEUX, J., P. BUISSERET & E. GARY-BOBO. Experimental evidence that rearing kittens in stroboscopic light retards maturation of the visual cortex. Vision Res. **32:** 41–34.
7. JAMPOLSKY, A. 1994. Consequences of retinal image clarity versus occlusion (absent) versus diffusion. Trans. Am. Ophthalmol. Soc. **17:** 349–376.
8. TUSA, R.J. et al. 2001. Unpublished data based on data at Yerkes Primate Research Institute.
9. TUSA, R.J., R.G. BOOTHE & D.V. BRADLEY. 2001. Mild diffuser contact lens induces exotropia in infant rhesus monkeys (abstr.). Invest. Ophthalmol. & Visual Sci. Suppl (ARVO) **42:** S168.
10. HARWERTH, R.S. et al. 1991. Functional effects of bilateral form deprivation in monkeys. Invest. Ophthalmol. & Visual Sci. **32:** 2311–2327.
11. O'DELL, C. & R.G. BOOTHE. 1997. The development of stereoacuity in infant rhesus monkeys. Vision Res. **19:** 2675–2684.
12. FU, L.N. et al. 2001. Saccadic disconjugacy in monkeys with strabismus (abstr.). Soc. Neurosci. In press.
13. KAPOULA, Z. et al. 1997. Impairment of the binocular coordination of saccades in strabismus. Vision Res. **37:** 2757–2766.
14. KIORPES, L. et al. 1996. Effects of early-onset artificial strabismus on pursuit eye movements and on neuronal responses in area MT of macaque monkeys. J. Neurosci. **16:** 6537–6553.
15. GRANT, S. & N.E.J. BERMAN. 1991. Mechanism of anomalous retinal correspondence: maintenance of binocularity with alteration of receptive-field position in the lateral suprasylvian (LS) visual area of strabismic cats. Visual Neurosci. **7:** 259–281.
16. MELVIL JONES, G. et al. 1981. Eye oscillations in strobe reared cats. Brain Res. **209:** 47–60.
17. DAS, V.E. et al. 2000. Experimental test of a neural-network model for ocular oscillations caused by disease of central myelin. Exp. Brain Res. **133:** 189–197.
18. OLSON, C.R. & J.D. PETTIGREW. 1974. Single units in visual cortex of kittens reared in stroboscopic illumination. Brain Res. **70:** 189–204.
19. SPEAR, P.D. et al. 1985. Developmentally induced loss of direction-selective neurons in the cat's lateral suprasylvian visual cortex. Dev. Brain Res. **20:** 281–285.

Development of New Treatments for Congenital Nystagmus

LOUIS F. DELL'OSSO

Ocular Motor Neurophysiology Laboratory, Veterans Affairs Medical Center; Department of Neurology, Case Western Reserve University and University Hospitals of Cleveland, Cleveland, Ohio 44106, USA

ABSTRACT: The use of ocular motor data as the basis for the development of both nonsurgical and surgical therapies for congenital nystagmus (CN) has been underway since the mid-1960s. This paper presents three nonsurgical therapies (composite prisms, soft contact lenses, and afferent stimulation) and a new surgical therapy (four-muscle tenotomy) hypothesized from analysis of ocular motor data. The expanded nystagmus acuity function test was developed to both predict and measure the effectiveness of CN therapies and for intersubject comparisons. Base-out prisms may be used to damp CN during distance fixation in patients whose CN damps during near fixation and who are binocular (i.e., they have no strabismus). Soft contact lenses may be used in those whose CN damps with afferent stimulation of the ophthalmic division of the trigeminal nerve. Cutaneous afferent stimulation (rubbing, vibration, or electricity) of the forehead or neck damps CN in some individuals. Finally, as first demonstrated in an achiasmic Belgian sheepdog and later in humans, tenotomy of the four horizontal rectus muscles and reattachment at their original sites may also damp CN. Taken together, these findings suggest the existence of one or more proprioceptive feedback loops acting to change the small-signal gain of the extraocular plant. Four-muscle tenotomy provides a needed therapeutic option for the many individuals with CN for whom other surgical therapies are contraindicated. Tenotomy may also prove useful in see-saw nystagmus (it abolished it in the aforementioned canine) or other types of nystagmus; further studies of the latter are required.

KEYWORDS: congenital nystagmus; composite prisms; soft contact lenses; afferent stimulation; tenotomy surgery

INTRODUCTION

Research into ocular motor dysfunction consists of studying "experiments of nature" where neither the experiment nor the methods are totally under the researcher's control, but the outputs, in the form of eye movements, are accurately accessible using sophisticated measurement techniques. Nature stresses the ocular motor system in ways that are difficult or impossible to duplicate in the laboratory on a normal subject and thereby reveals otherwise hidden capabilities of the control system. The

---

Address for correspondence: L.F. Dell'Osso, Ph.D., Ocular Motor Neurophysiology Laboratory, Veterans Affairs Medical Center (127A), 10701 East Boulevard, Cleveland, OH 44106. Voice: 216-421-3224; fax: 216-231-3461.

lfd@po.cwru.edu

potential of this research approach has been appreciated for more than 150 years. When I helped build and subsequently joined the new Ocular Motor Neurophysiology Laboratory of Robert B. Daroff at the Miami VA Hospital in the early 1970s, we combined this approach with control-system analysis and applied it to the study of congenital and neurological conditions that resulted in ocular motor dysfunction. It has proven to be productive in exposing underlying control-system architecture, defining diagnostic criteria, and suggesting possible therapies. Our studies have also benefited from the interdisciplinary collaboration of biomedical engineering and neurology, ophthalmology, and neuro-ophthalmology. The former has particularly influenced studies into the underlying mechanisms of congenital nystagmus (CN) and other oscillations. Although our main emphasis was on understanding the basics of ocular motor control instability, we remained alert to the possibility of clinically exploiting our findings, regardless of the level of our understanding of their theoretical basis. Fortunately, to design a therapy that exploits an observed beneficial effect, one need not know all the underlying details of how the responsible action caused the effect. The basic paradigm consists of: accurately documenting the eye movements of subjects with CN in response to normal stimuli (visual and otherwise); analyzing the data in the context of control systems; attempting to exploit any damping present in each individual; and, finally, incorporating all observations into an ocular motor system model capable of simulating normal and abnormal responses.

Obtaining accurate eye-movement data was not possible from the existing literature (circa 1965). Eye movements, and specifically nystagmus, had been described clinically based on visual observation alone with only occasional recordings made with primative equipment. To their credit, some early clinical observers were quite insightful, limited only by characteristics that could not be seen by the naked eye. One of the best of these pioneers was Alfred Kestenbaum, whose writing on the subject continues to reveal insights based on astute observation.[1] However, the absence of good data led to errors in interpretation and in the theories resulting from them. Despite the publication of ocular motility data from several investigators that contradicted some of those early theories, they continued to reappear in the literature, adding little but confusion.

Although the foregoing research approach has been fruitful in many ocular motor areas, this report will be limited to CN therapies that are based on: (1) exploiting inherent characteristics of the internally generated oscillation; (2) reducing the responsiveness of the ocular motor plant to that oscillation; or (3) interfering with the efferent CN signal. Some of the therapeutic suggestions that sprang from our data were straightforward and required little original thought. Others required that one ignore what was considered "common knowledge" and postulate an action unsupported (or even contradicted) by current theory. We hypothesized that the characteristics of responses from a stressed ocular motor system (e.g., one with CN) would provide insights into how the *normal* ocular motor system is organized (on a functional basis) and also suggest, via computer modeling, the functional source(s) of the CN waveforms. From these studies, possible therapeutic approaches might arise. Implicit in this approach was the presumption that the ocular motor system of an individual with CN was substantially the same as that of a normal person. We hypothesized that the complex waveforms that identify CN represented the otherwise normal ocular motor system's attempts to modify one or more simple, basic oscillations with the aim of maximizing target foveation and visual acuity. Indeed, as will be discussed, the observations made in the primordial systems study of CN[2] led to our first nonsurgical, acuity-improving therapy for this disorder, composite prisms.

The joining of the disciplines of biomedical engineering and neurology became a model for other investigators in this new and soon-to-be rapidly expanding area of research. Our approach has proven to be successful in elucidating the structure and function of the normal ocular motor system, in providing diagnostic criteria, and in suggesting new therapies (the subject of this report). In addition to defining our work for more than three decades, the combination of a biomedical engineering systems approach and a neuroanatomical and neurophysiological medical approach has also benefited other investigators and (most importantly) patients with congenital and acquired neurological conditions.

METHODS

Several ocular motility recording methods have been used in the Ocular Motor Neurophysiology Laboratory and new methods tried as they were developed.

Recording

Some eye movement recordings are made using infrared reflection. The infrared signal from each eye is calibrated with the other eye behind cover to obtain accurate position information and to document small tropias and phorias hidden by the nystagmus. Eye positions and velocities (obtained by analog differentiation of the position channels) are displayed on a strip chart recording system. Other data are recorded by means of a phase-detecting revolving magnetic field technique. Eye position data are digitized and stored in a computer.

Protocol

Written consent is obtained from subjects before testing. All test procedures are carefully explained to the subject before the experiment begins and are reinforced with verbal commands during the trials. Subjects are seated in a chair with headrest and either a bite board or a chin stabilizer, far enough from an arc of red LEDs to prevent convergence effects. The room light can be adjusted from dim down to blackout. Experiments consist of from 1 to 10 trials, each lasting under a minute with time allowed between trials for the subject to rest. Trials are kept this short to guard against boredom because CN intensity is known to decrease with inattention.

Analysis

Analysis is carried out in the MATLAB environment using software specifically written for each study as needed.

RESULTS

The nonsurgical and surgical therapies for CN that have resulted from ocular motility studies performed in the Ocular Motor Neurophysiology Laboratory will be discussed chronologically.

Nonsurgical Therapies

Optical Therapy— Composite Prisms

Individuals with CN who exhibit a gaze-angle null may benefit from the use of base-left or base-right version prisms that place their eyes at that angle. Similarly, those whose CN damps with convergence may benefit from base-out vergence prisms that induce convergence during fixation of a distant target. For individuals with both types of nulls, composite prisms combine the effects of shifting gaze and inducing convergence; they are base-out prisms of unequal powers. During my initial study of ocular motor control in a subject with CN, I observed that the intensity of the CN diminished with both gaze in the null region and near fixation.[2] When I presented these observations to Larry Stark while visiting his lab at Berkeley, we verified them with further recordings and converted the numbers into prism diopters that would provide both a gaze-angle shift to the null and induce convergence. Because the convergence induces unwanted accommodation, −1.00S O.U. was added to the refraction to ensure clear distance vision.[3] The prisms damped the CN significantly and improved visual acuity by two Snellen lines to 20/25.[4] Further explanations regarding the use of composite prisms appeared in subsequent reports.[4,5]

Optical Therapy—Soft Contact Lenses

To many ophthalmologists, the presence of CN was a contraindication for the use of contact lenses. The constant motion of the eyes was thought to cause lens slippage with possible irritation to the cornea. However, eye motion due to the CN is much less likely to cause slippage than the high eye accelerations of the many voluntary saccades performed daily by normal persons. We studied the effects of soft contact lenses on the CN waveform under normal conditions and with local anesthesia.[6] Contact lenses damped the CN over all gaze angles, but local anesthesia abolished the effect and returned the CN to baseline levels (i.e., with no contact lenses). Contact lenses with partial correction damped the CN but not as well as full correction.

Afferent-Stimulation Therapy—Vibration and Electrical

Discussions regarding possible mechanisms for the CN damping produced by contact lenses, led Bob Daroff to suggest we try excitation of the ophthalmic division of the trigeminal nerve by cutaneous stimulation above one eye.[7] It produced a 50% diminution of the CN amplitude. This led to studies of afferent stimulation using vibration and electrical stimuli of the forehead and the neck.[8,9] Both methods at both sites were successful. As FIGURE 1 shows, vibratory and electrical stimulation damped the CN and improved foveation periods. Vibratory stimulation had a positive effect in a larger number of subjects (9 of 13) than did electrical (3 of 10). Positive vibratory effects occurred in 19 of 30 (63%) trials, and positive electrical effects occurred in 5 of 11 (45%) trials. FIGURE 1b and c demonstrates that damping of CN is not necessary for improved acuity; lengthening foveation periods and reducing their position variation are the most important effects of therapy for functional improvement; amplitude reduction does result in cosmetic improvement.

The NAF(X). For this study, we further developed an objective measure of the CN waveform, the Nystagmus Foveation Function (NFF) used in earlier studies,[10–12] into the Nystagmus Acuity Function (NAF), named for its relation to predicted, best-

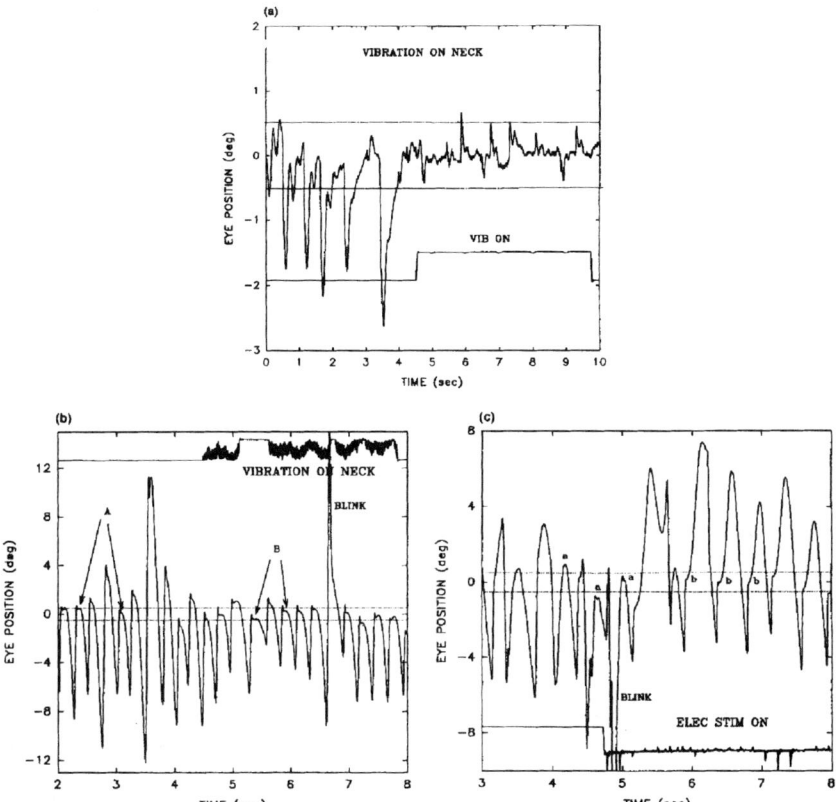

FIGURE 1. Positive effects of afferent stimulation showing (**a**) the reduction of amplitude of the nystagmus (from ~2-3 to ~0.2-0.5 degrees peak-to-peak) and longer foveation period durations during vibration on the neck and (**b**) an increase in the duration of the foveation periods during stimulation, but without much change in amplitude. Compare durations of foveation periods marked "A" (before stimulation) with those marked "B" (during stimulation). An effect of decreased variability in the positions of the foveation periods during stimulation is shown in **c**). Compare the alignment of the foveation periods marked "a" (before stimulation) with those marked "b" (during stimulation). The stimulus traces indicate only the intervals of stimulation and not the stimulus signals. Noise in the vibration traces resulted from the effects of the vibration on the resistive contact responsible for this signal. (From Sheth *et al.*, 1972.)

corrected visual acuity. Increases in this function when afferent stimuli were applied provided objective evidence of the beneficial effects. In the ensuing 8 years, the NAF has been used to assess subjects with CN and latent/manifest latent nystagmus (LMLN). Because some individuals cannot meet the stringent, well-developed foveation requirements of the NAF (i.e., repeated foveation of the target to within ±0.5 degrees and ±4 degrees/s simultaneously), the NAF's foveation window had to become expandable (based on an individual's foveation abilities) while maintaining the

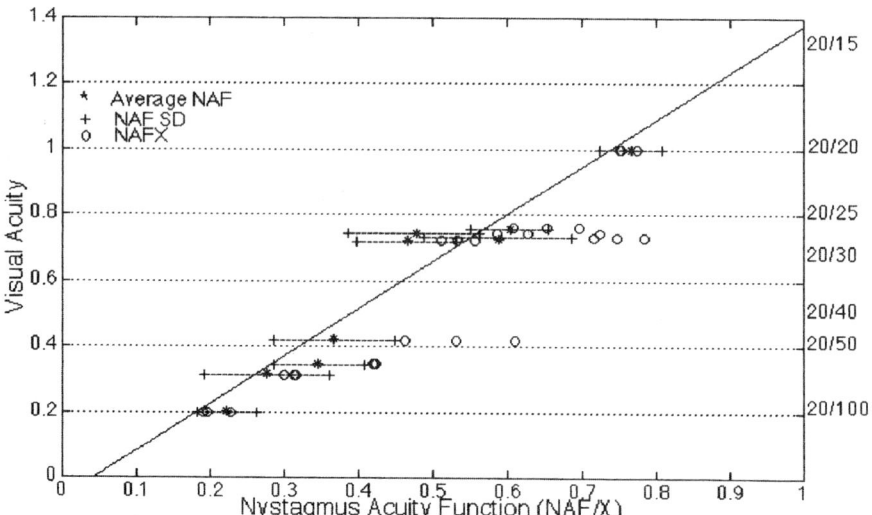

FIGURE 2. Nystagmus acuity function. Data from nine CN subjects with neither afferent deficits nor significant fixation-attempt effects on their waveforms, showing the linear relation between the NAF and potential, best-corrected visual acuity and the equivalent NAFX values calculated using arbitrary foveation-window sizes, determined by a protocol developed from our experience using the NAFX over the last 5 years. For clarity, the data points for two individuals (mean NAFs of 0.588 and 0.602) were slightly lowered or raised, respectively. Each individual had the same acuity (0.743 = 20/25-3) as the individual whose data are straddled (mean NAF of 0.478). (From Dell'Osso and Jacobs, 2001. In press.)

NAF's relationship to predicted, best-corrected visual acuity. The resulting eXpanded Nystagmus Acuity Function (NAFX) may be applied to any nystagmus waveform whose foveation variability (position and velocity) lies within foveation windows ranging from that of the original NAF up to the maximum expanded window of ±6 degrees and ±10 degrees/s.[13] Neither CN amplitude nor frequency is used to calculate any of these functions. FIGURE 2 demonstrates the relation between the NAFX and the predicted, best-corrected visual acuity of several subjects. This function has proven to be robust enough to allow analysis and acuity prediction of all subjects we have recorded thus far.

Surgical Therapies

The postulation and clinical trials of a new surgical therapy for CN with far-reaching theoretical implications had its roots in two areas of ocular motor research, documenting the effects of the Anderson-Kestenbaum recession-resection surgery on humans[14] and studying the eye movements of achiasmatic Belgian sheepdogs with CN and see-saw nystagmus (SSN).[15,16]

Anderson-Kestenbaum Surgery Results

In an early study of CN surgery (circa 1977), we noted a profound change in the shape of the nystagmus intensity versus gaze angle function. After Anderson-

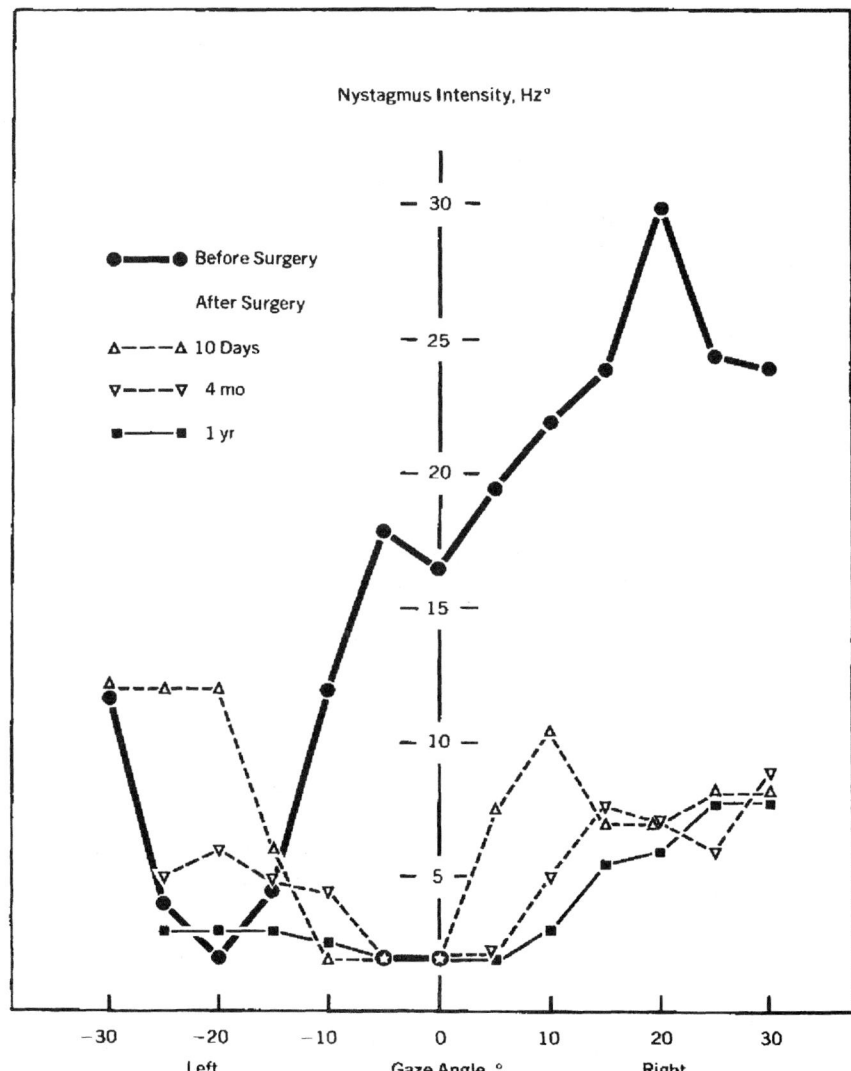

FIGURE 3. Pre- and postoperative plots of nystagmus intensity versus gaze angle for an individual, indicating the time course of improvement in primary and secondary effects of the Anderson-Kestenbaum surgery. (From Dell'Osso and Flynn, 1979.)

Kestenbaum surgery, in addition to the expected shift of this plot towards primary position, the breadth of the null region increased and the overall nystagmus intensity decreased.[14] FIGURE 3 shows these effects during the first year post-surgery; they remained unchanged after 5 years. We reasoned that there were two independent effects of this surgery due to different mechanisms. The expected null-shift towards primary position was mechanical due to the effective rotation of the globe opposite

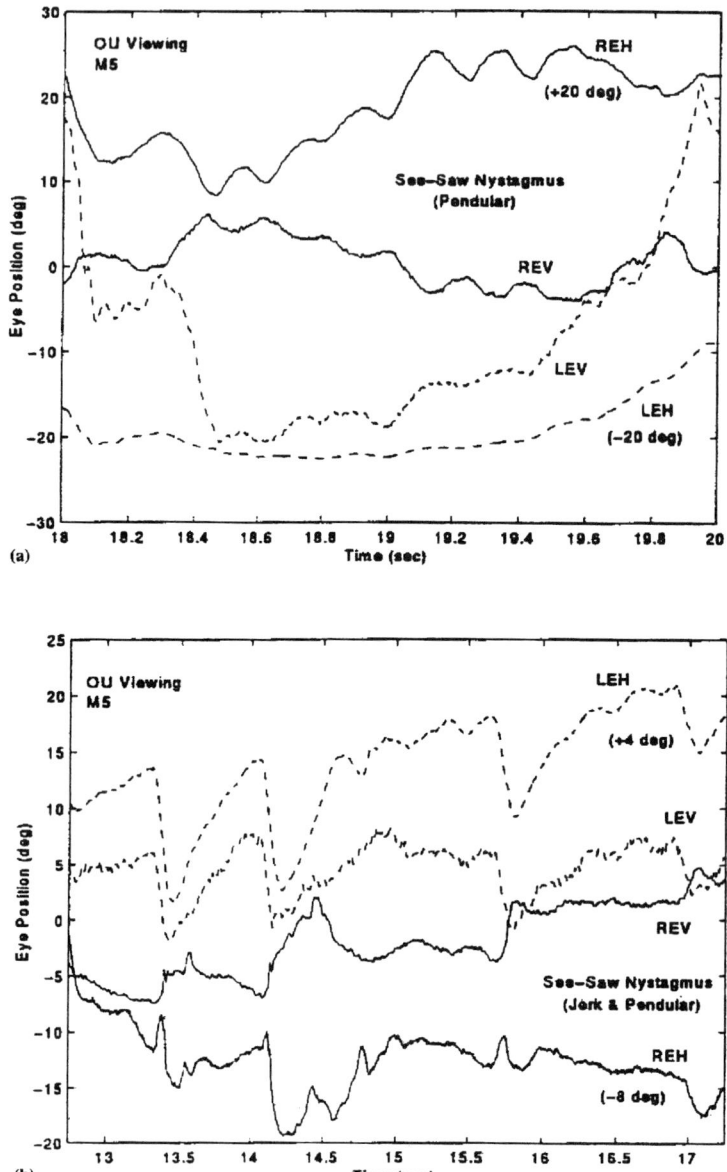

FIGURE 4. CN waveforms of an achiasmatic Belgian sheepdog. Right- and left-eye horizontal and vertical recordings during OU viewing. (**a**) Simultaneous pendular SSN and CN at the same frequency resulted in a diagonal pendular SSN, moving up-and-left (down-and-right) in the right eye and down-and-left (up-and-right) in the left eye. (**b**) Dual jerk (jerk plus pendular) SSN with simultaneous jerk-right CN resulted in a diagonal dual jerk SSN, beating up-and-left in the right eye and down-and-left in the left eye. During the slow phases, there was a superimposed diagonal pendular SSN, as in **a**. The horizontal traces in

the null angle and the subsequent innervation required to move the eyes back to primary position (i.e., the same innervation that placed the eyes in the null position preoperatively). Because the only surgery performed in addition to the resections and recessions was the accompanying tenotomies of the muscles, it was hypothesized that they alone were responsible for the CN damping at all gaze angles.[17] It was to take 20 years before we would have access to, and could test this hypothesis on, an animal model of naturally occurring CN (exhibiting the pathognomonic CN waveforms).

Achiasmatic Belgian Sheepdogs

In 1991, I was contacted by Robert W. Williams of the University of Tennessee, Memphis. He provided videos of the horizontal nystagmus of several achiasmatic Belgian sheepdogs, asking if they resembled human CN. I was struck by the similarity of the head posturing of the dogs and children with CN and arranged to document the waveforms to establish the diagnosis. The videos also revealed a see-saw nystagmus. In 1992, we documented horizontal CN waveforms and vertical-torsional see-saw nystagmus in the dogs.[15,16] This was the first time I had seen nystagmus waveforms (i.e., the horizontal components) from an animal that had the characteristics of human CN, including foveating and braking saccades. In subsequent recordings, we verified that all achiasmatic dogs had both CN and SSN,[18] including a dog with hemichiasma.[19] FIGURE 4 shows the pendular and jerk, horizontal and vertical waveforms of one of the achiasmatic dogs. The animal model for CN that we had been seeking to test our surgical hypothesis had been found. In addition, SSN was identified as a sign of chiasmal abnormalities in canines and humans.[20]

Four-Muscle Tenotomy Procedure

Canine. My colleague, Richard W. Hertle, performed the hypothetical tenotomy procedure in two stages: (1) all four horizontal recti were tenotomized and immediately reattached at their original insertion sites; and (2) 4 months later, the same procedure was performed on the four vertical recti and four obliques.[21] The first surgery tested the efficacy of tenotomy on horizontal CN, and the second, on SSN. The results were immediately obvious; there was a profound damping of the horizontal CN over a broad range of gaze angles (stage 1) and the SSN was abolished (stage 2). Eye movement recordings over the next 7 months verified the profound and persistent nystagmus damping. FIGURE 5 compares the pre- and postsurgical CN and SSN in this canine (note the expanded scales in B and D). The dog also exhibited immediate behavioral improvements that supported the prediction that the CN damping would increase his acuity.

Human. As a result of our hypothesis based on analysis of human eye movement data after Anderson-Kestenbaum procedures and demonstration that tenotomy damped the CN in a canine model of human CN, we began a clinical trial of this procedure under the auspices of the National Eye Institute with R.W. Hertle as the sur-

---

a and **b** were shifted by the indicated amounts for clarity. In this and FIGURE 5: RE, right eye; LE, left eye; OU or BE, both eyes; H, horizontal; V, vertical; upward (+) deflections indicate rightward (or upward) eye rotations; and viewing and fixation conditions are indicated. (From Dell'Osso *et al.*, 1998.)

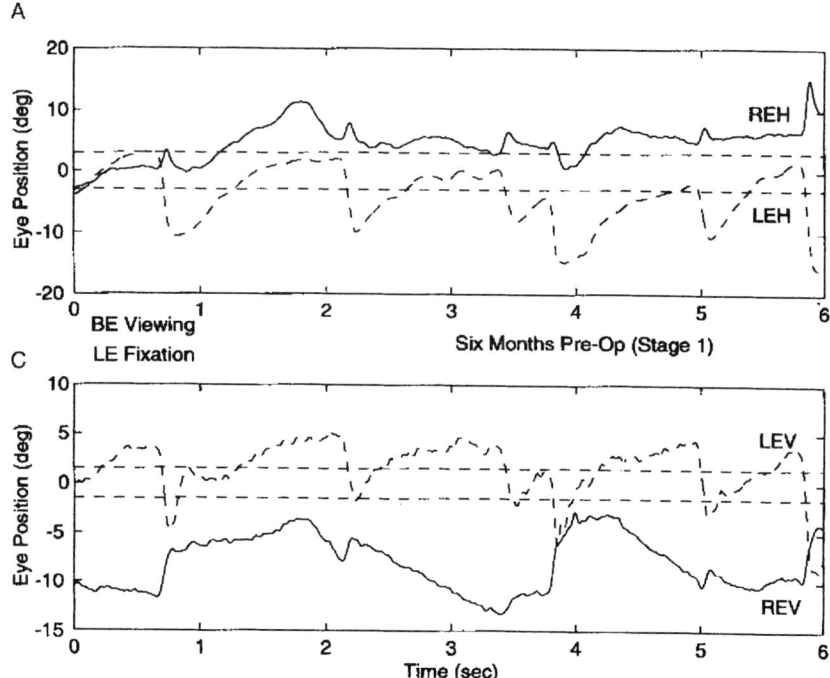

FIGURE 5. The effects of tenotomy on the CN and SSN in an achiasmatic Belgian sheepdog. (**A** and **C**) The horizontal CN and vertical SSN while fixating in primary position prior to stage 1 (horizontal recti) tenotomy. *Dashed lines* indicate the extent of the horizontal (±3 degrees) or vertical (±1.5 degrees) area centralis. (**B,D**) The horizontal and vertical CN, made during fixation in primary position 4 months postsurgical (stage 1) and 1 day postsurgical (stage 2). A 25-second interval of steady, right-eye fixation in both planes showing well developed centralization is shown. The CN was of low amplitude and no SSN was present. b, blink. (Modified from Dell'Osso *et al.*, 1999.)

geon.[21] Pre- and post-tenotomy eye movement data were recorded at the Laboratory of Sensorimotor Research, NEI, under the direction of E.J. Fitzgibbon. The data were blinded by D. Thompson and analyzed by this author at the Ocular Motor Neurophysiology Lab. The first phase included 10 adults and the second, 5 children. After the data up to 6 weeks post-tenotomy for the first five adult subjects were analyzed, they were unblinded to determine if phase 2 could be started. Preliminary results of these data showed improved NAFX scores and improved visual function.[22] FIGURE 6 shows the pre- and 6-weeks postoperative CN of one of the subjects (S2); tenotomy resulted in a 52% increase in the NAFX. As FIGURE 7 shows, the NAFX values for all patients improved (average improvement was 48%).

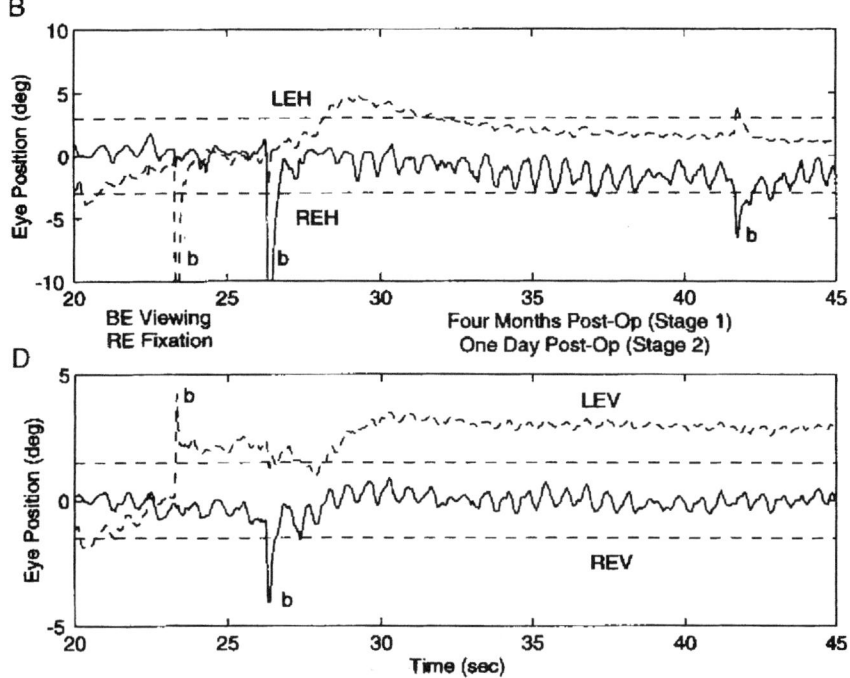

FIGURE 5. *Continued.*

DISCUSSION

Accurate diagnosis is the most important factor in the treatment of CN, or any medical condition. CN must be distinguished from a number of types of nystagmus that appear in infancy; their different waveforms imply different underlying mechanisms. Ocular motility data allow us to make these otherwise difficult differential diagnoses.[23–27] The variability in the success rates reported for CN surgery is, in great part, due to the absence of ocular motor data that would have produced more accurate diagnoses. Because of their different mechanisms and interrelationships with strabismus, CN, LMLN, the nystagmus blockage syndrome, and spasmus nutans do not respond to the same therapies; the lack of consistency in surgical results on patients diagnosed clinically without ocular motility data supports this view.

The first step, before considering therapies, was the establishment of definitive, differential diagnostic criteria for CN and other forms of nystagmus. Here, accurate ocular motor recordings quickly demonstrated that they were both essential and invaluable. The key waveform characteristics that are pathognomonic for CN became easily distinguishable.[23,28] No longer would "nystagmus" be an acceptable diagnosis. Eye movement recordings were to eventually be used to identify many different types of nystagmus and saccadic oscillations.[27,29] An integral part of waveform descriptions was identification of the portions of each cycle during which the image of

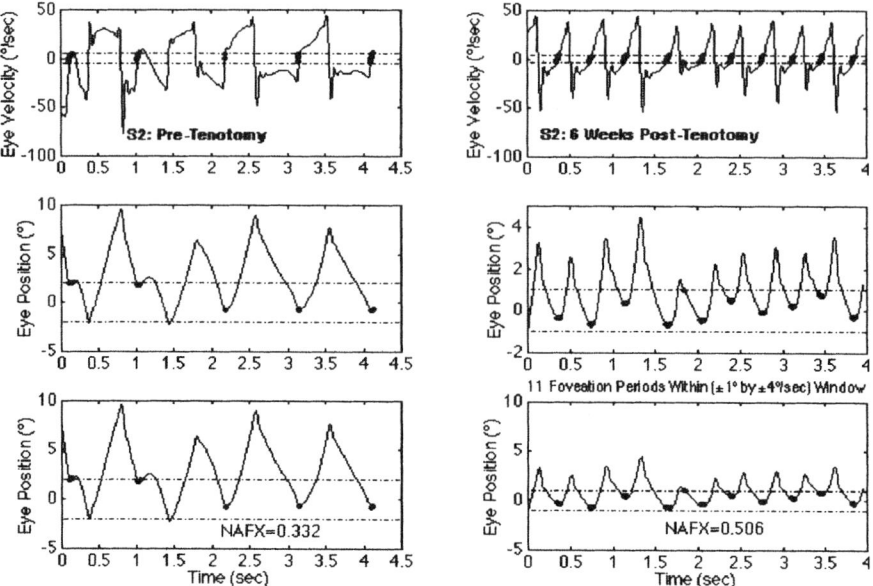

FIGURE 6. Effects of tenotomy on the CN in a human. Eye velocity (*top panels*) and position (*middle and bottom panels*) are shown with equivalent scales (*top and bottom panels*) for ease of comparison.

the target was on the fovea (foveation periods) during fixation[10,30] as well as during smooth pursuit and vestibulo-ocular movements.[11,12,31] Documentation of these important characteristics of CN waveforms by accurate ocular motor recordings provided the data necessary for both diagnosis and evaluation of the efficacy of specific therapies.

Evaluation and comparison of therapies require a quantitative measure of the change in the specific characteristic(s) that is the primary outcome of the therapy. Although the goals of CN therapy may be cosmetic and acuity improvements, they are not the primary outcomes. Because measured visual acuity is the result of several variables (e.g., stress, afferent deficits) whose relation to the CN waveform is idiosyncratic, it is not always a good measure of real-world acuity. CN amplitude is the characteristic most directly related to cosmetic appearance; however, amplitude is not a good predictor of acuity. A therapy that reduces amplitude may not improve acuity, and one that does improve acuity may not reduce amplitude. The waveform characteristics related to acuity are: foveation time and beat-to-beat foveation position and velocity variation. The NAFX is a quantitative function that includes all three of these primary characteristics. It predicts best possible visual acuity, which is ≤ *real-world* acuity which is ≤ measured acuity.

This paper has concentrated on the treatments stemming from studies conducted at the Ocular Motor Neurophysiology Lab. Other investigators have studied the effects on CN of acupuncture,[32,33] biofeedback,[34–42] and the injection of botulinum toxin.[43–48] These treatments have had variable success in both CN and acquired nys-

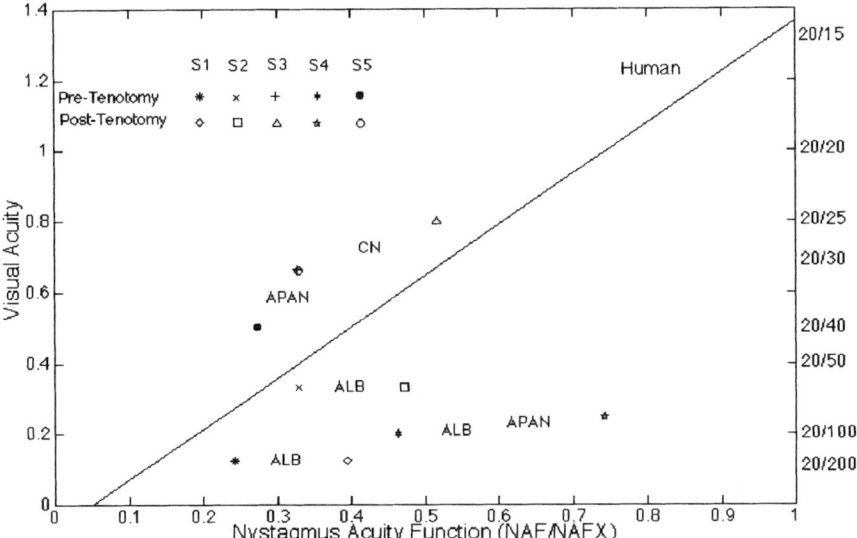

FIGURE 7. Expanded nystagmus acuity function. Effects of tenotomy on the NAFX of the CN in five human subjects (S1—S5). ALB, albino; APAN, asymmetric, (a) periodic alternating nystagmus.

tagmus. Others have also studied the effects on CN of prisms[49–51] and contact lenses.[34,52–56] The ocular motor effects of surgical procedures (Anderson-Kestenbaum and bimedial recession) have also received some attention.[57–59]

Composite Prisms

The mechanisms by which version and vergence prisms damp CN differ. The gaze-angle null exploited by version prisms is a region of reduced oscillation that increases as gaze is moved in either direction. By contrast, the vergence null exploited by base-out prisms appears to result from a reduction in the ocular motor plant's small-signal responsiveness (i.e., nonsaccadic gain) to the underlying oscillation. The mechanism is unknown, but it is probably related to repositioning of the extraocular muscle pulleys as a result of convergence. The resulting reduced gain is relatively unaffected by gaze angle, allowing the eye oscillation to remain damped as the latter changes.

Version prisms become impractical for null angles greater than 5 or 6 degrees due to the large prism size and the resulting chromatic aberration. Vergence prisms (with –1.00S O.U. added in prepresbyopic patients) do not have this limitation because the CN damping in most individuals occurs with less than 20D total convergence (i.e., 10D per eye). The use of composite prisms was an attempt to maximize the total damping by combining the required conjugate version shift to the gaze-angle null with the approximate vergence shift that damped the CN. Subsequent development of the NAFX allowed us to more accurately measure the visual-acuity-related effects at different vergence angles and identify the point of maximal damping. In an individual who had initially worn 16D of convergence prism (relaxed to 14D several years later), we found that 12D yielded the maximal NAFX value, allowing further reduction in the prisms. Once converged, CN remains damped at most gaze angles (especially those in the central ±20 degrees of gaze); therefore, we now recommend equal-value, base-out prisms rather than composite prisms. This minimizes the amount of prism in each eye and reduces chromatic distortion. Thus, for the aforementioned individual, the initial values of 12D and 4D base-out were replaced with two 6D base-out prisms, with the same improved acuity. Prisms can also be used to fine tune the effects of less-than-adequate surgery (i.e., failure to fully shift the null to primary position). The use of base-out prisms to damp nystagmus has also been extended to acquired nystagmus.[60,61]

Soft Contact Lenses

The damping effect in CN produced by soft contact lenses may be due to either reducing the ocular motor plant's gain or interfering with the CN signal at a higher level. Soft lenses provide a nonsurgical therapy for individuals with no CN nulls. They are useful in sports, where spectacles might be an impediment, as well as in social situations, where an individual might feel more comfortable with contact lenses. Soft lenses can also be used interchangeably with prism glasses without a period of adjustment.

Afferent Stimulation

Presumably, the same mechanisms described for contact lenses apply to cutaneous stimulation of the forehead. Vibratory stimulation caused positive effects in a larger percentage of subjects and caused them more frequently. Vibration may stimulate deeper tissue and muscle, altering afferent proprioceptive signals to the nuclei of the upper spinal cord and brain stem. The greater effect of vibration on the neck than on the forehead may be due to the latter's absence of deep tissue. The suprathreshold electrical stimulation may have only stimulated cutaneous afferents. Not only did simple scratching of the forehead or vibration on the neck produce transient CN damping and increased acuity, but also air blowing on the forehead had a similar effect. Thus, for someone with CN, playing the outfield on a windy day should be easier than on a calm day!

Although afferent stimulation has not yet been reduced to a practical therapeutic device, it is not hard to imagine a battery-operated neck collar or head band that can provide either a vibratory stimulus or a mild electric shock at the press of a button when higher acuity is desired (e.g., as a street or traffic-control sign approaches while driving). Such vibratory collars are currently available for stress relief and can be used to damp CN.

Four-Muscle Tenotomy Procedure

The initial successes of tenotomy in the achiasmic canine and in humans have prompted us to speculate on the possible mechanism involved. We initially postulated that the surgery altered a proprioceptive loop in the EOM.[21] Motor fibers in the proximal ends of the distal tendons have recently been discovered.[62] Evidence of neural fibers in the distal ends of these tendons has also been found (personal communication, R.W. Hertle); in patients with CN, there were abnormalities in some of these structures. Thus, in addition to the two proprioceptive gain-control loops (i.e., ophthalmic division of the trigeminal nerve and neck receptors) identified by contact lenses and afferent stimulation, there appear to be two others in the EOM itself. The efferent signals of one appear to be the motor fibers terminating in the EOM pulleys and of the other, motor fibers in the distal muscle and proximal tendon. The pulleys appear to provide gain-control mechanically by altering the point of action of the EOM. The second loop may provide gain-control by altering muscle tension. However, feedback loops require afferent as well as efferent information. We hypothesized that the fibers in the distal tendon provide that information. Others have provided evidence that proprioceptive inflow is used in ocular motor control.[63,64] Clearly, the demonstration of the efficacy of tenotomy has necessitated both a reevaluation of prevailing, simplistic concepts of the organization of the ocular motor plant and the differentiation between dynamic nystagmus procedures (aimed at reducing the oscillation) and static strabismus procedures (aimed at correcting a misalignment).

The use of tenotomy to damp CN in patients is now the subject of two clinical trials. Further analysis of the data taken in the first (described above) is currently under way. In the second study, patients from the Children's Hospital Medical Center of Akron are recorded pre- and postoperatively in the Ocular Motor Neurophysiology Lab; the surgery is accomplished by R.A. Burnstine. These data have not yet been analyzed but several of the patients have had improvements in their acuity, allowing them to pass the required tests to obtain a driver's license.

Combining Tenotomy with Other Procedures

If tenotomy proves to be an effective treatment for CN, it can also be combined with nystagmus procedures aimed at damping CN or with strabismus procedures in patients with CN. Thus, tenotomy of the lateral recti can be combined with bimedial recessions in binocular CN patients whose nystagmus damps with convergence;[21] both are nystagmus procedures. In cases of CN with no exploitable nulls plus strabismus, a one- or two-muscle strabismus procedure can be replaced by resections and recessions on both eyes, which will *de facto* include four-muscle tenotomies. Alternatively, the remaining muscles can be tenotomized to include the nystagmus-damping effects of this procedure. Note, these recommendations for combining tenotomy with strabismus procedures are for strabismus patients with CN, not LMLN.

FUTURE RESEARCH DIRECTIONS

The effectiveness of four-muscle tenotomy in LMLN or different types of acquired nystagmus (including SSN) needs to be assessed.

Chemical/Mechanical Tenotomy. Although surgical tenotomy has been shown to damp CN, it may be possible to achieve the same effect by microinjection of specific chemicals into the distal tendon (chemical tenotomy) or deform it with a hemostat (mechanical tenotomy). Alternatively, injection into the proximal tendon, where motor fibers have been identified, may prove beneficial. The NAFX provides an objective, quantitative measure of the effectiveness of various drugs, dosages, and injection sites; the studies remain to be done.

The Ocular Motor Plant. Prior to demonstrating that tenotomy of the extraocular tendons can damp CN, the plant was thought of and modeled as a passive low-pass filter (e.g., 2 poles or 2 poles + 1 zero). Clearly, although these representations may be useful in control-system models, the plant is a much more complex, actively controlled feedback system. The exact nature of the feedback loops and control signals (including those to the pulleys) needs to be studied. The functions of the neural fibers in the distal tendon and of the motor fibers in the proximal tendon need to be elucidated. Perhaps, with such knowledge, new and more precise therapeutic intervention will become possible for CN as well as other ocular motor disorders.

In summary, the foregoing therapies successfully damp CN by either exploiting the idiosyncratic null angle of the oscillation (version prisms or the Anderson-Kestenbaum procedure) or altering the responsiveness of the ocular motor plant to that oscillation (base-out prisms, bimedial recession, and four-muscle tenotomy). Contact lenses and afferent stimulation may use the latter mechanism or reduce the efferent CN signal. Ocular motility recordings and the NAFX have resulted in new therapies and have provided a means to objectively assess their efficacy.

ACKNOWLEDGMENTS

This work was supported in part by the Office of Research and Development, Medical Research Service, Department of Veterans Affairs. I am deeply grateful to Robert B. Daroff who, in 1972, provided me with the initial opportunity to apply the systems approach towards our mutual goals of diagnosing, understanding, and treating conditions with ocular motor symptoms and signs. I was trained as a biomedical engineer (then, a new and virtually unknown discipline) and had neither a medical nor a neurophysiological background. Throughout our collaborative research, Bob was instrumental in the design, execution, and analysis of the results of many studies into acquired and congenital ocular motor oscillations and dysfunctions.

REFERENCES

1. KESTENBAUM, A. 1961. Clinical Methods of Neuro-Ophthalmologic Examination, 2nd Ed. Grune & Stratton. New York.
2. DELL'OSSO, L.F. 1968. A dual-mode model for the normal eye tracking system and the system with nystagmus. (Ph.D. dissertation). Ph.D. thesis. University of Wyoming, Laramie.
3. DELL'OSSO, L.F., G. GAUTHIER, G. LIBERMAN, *et al.* 1972. Eye movement recordings as a diagnostic tool in a case of congenital nystagmus. Am. J. Optom. Arch. Am. Acad. Optom. **49:** 3–13.
4. DELL'OSSO, L.F. 1973. Improving visual acuity in congenital nystagmus. *In* Neuro-Ophthalmology Symposium of the University of Miami and the Bascom Palmer Eye Institute, Vol. VII. J.L. Smith & J.S. Glaser, Eds: 98–106. CV Mosby Co. St. Louis.
5. DELL'OSSO, L.F. 1976. Prism exploitation of gaze and fusional null angles in congenital nystagmus. *In* Orthoptics: Past, Present, Future. S. Moore, ed. :135–142. Symposia Specialists. New York.
6. DELL'OSSO, L.F., S. TRACCIS, L.A. ABEL, *et al.* 1988. Contact lenses and congenital nystagmus. Clin. Vision Sci. **3:** 229–232.
7. DELL'OSSO, L.F., R.J. LEIGH & R.B. DAROFF. 1991. Suppression of congenital nystagmus by cutaneous stimulation. Neuroophthalmology **11:** 173–175.

8. SHETH, N.V. 1994. The effects of afferent stimulation on congenital nystagmus. (M.S. thesis). Ph.D. thesis. Case Western Reserve University, Cleveland.
9. SHETH, N.V., L.F. DELL'OSSO, R.J. LEIGH, et al. 1995. The effects of afferent stimulation on congenital nystagmus foveation periods. Vision Res. **35:** 2371–2382.
10. DELL'OSSO, L.F., J. VAN DER STEEN, R.M. STEINMAN, et al. 1992. Foveation dynamics in congenital nystagmus. I. Fixation. Doc. Ophthalmol. **79:** 1–23.
11. DELL'OSSO, L.F., J. VAN DER STEEN, R.M. STEINMAN et al. 1992. Foveation dynamics in congenital nystagmus. II. Smooth pursuit. Doc. Ophthalmol. **79:** 25–49.
12. DELL'OSSO, L.F., J. VAN DER STEEN, R.M. STEINMAN, et al. 1992. Foveation dynamics in congenital nystagmus. III. Vestibulo-ocular reflex. Doc. Ophthalmol. **79:** 51–70.
13. DELL'OSSO, L.F. & J.B. JACOBS. 2001. An expanded nystagmus acuity function: intra- and intersubject prediction of best-corrected visual acuity. Doc. Ophthalmol. In press.
14. DELL'OSSO, L.F. & J.T. FLYNN. 1979. Congenital nystagmus surgery: a quantitative evaluation of the effects. Arch. Ophthalmol. **97:** 462–469.
15. DELL'OSSO, L.F. & R.W. WILLIAMS. 1995. Ocular motor abnormalities in achiasmatic mutant Belgian sheepdogs: unyoked eye movements in a mammal. Vision Res. **35:** 109–116.
16. DELL'OSSO, L.F. 1996. See-saw nystagmus in dogs and humans: an international, across-discipline, serendipitous collaboration. Neurology **47:** 1372–1374.
17. DELL'OSSO, L.F. 1998. Extraocular muscle tenotomy, dissection, and suture: a hypothetical therapy for congenital nystagmus. J. Pediatr. Ophthalmol. Strab. **35:** 232–233.
18. DELL'OSSO, L.F., R.W. WILLIAMS, J.B. JACOBS, et al. 1998. The congenital and see-saw nystagmus in the prototypical achiasma of canines: comparison to the human achiasmatic prototype. Vision Res. **38:** 1629–1641.
19. DELL'OSSO, L.F., D. HOGAN, J.B. JACOBS, et al. 1999. Eye movements in canine hemichiasma: does human hemichiasma exist? Neuroophthalmology **22:** 47–58.
20. DELL'OSSO, L.F. & R.B. DAROFF. 1998. Two additional scenarios for see-saw nystagmus: achiasma and hemichiasma. J. Neuro-Ophthalmol. **18:** 112–113.
21. DELL'OSSO, L.F., R.W. HERTLE, R.W. WILLIAMS, et al. 1999. A new surgery for congenital nystagmus: effects of tenotomy on an achiasmatic canine and the role of extraocular proprioception. J. Am. Assoc. Pediatr. Ophthalmol. Strab. **3:** 166–182.
22. DELL'OSSO, L.F., R.W. HERTLE, E.J. FITZGIBBON, et al. 2000. Preliminary results of performing the tenotomy procedure on adults with congenital nystagmus (CN): a gift from "man's best friend." In Neuro-ophthalmology at the Beginning of the New Millennium. J.A. Sharpe, ed. :101–105. Medimond Medical Publications. Englewood.
23. DELL'OSSO, L.F. & R.B. DAROFF. 1975. Congenital nystagmus waveforms and foveation strategy. Doc. Ophthalmol. **39:** 155-182.
24. DELL'OSSO, L.F., D. SCHMIDT & R.B. DAROFF. 1979. Latent, manifest latent and congenital nystagmus. Arch. Ophthalmol. **97:** 1877–1885.
25. DELL'OSSO, L.F., J.C. ELLENBERGER, L.A. ABEL, et al. 1983. The nystagmus blockage syndrome: congenital nystagmus, manifest latent nystagmus or both? Invest. Ophthalmol. Vision Sci. **24:** 1580–1587.
26. DELL'OSSO, L.F., S. TRACCIS & L.A. ABEL. 1983. Strabismus: a necessary condition for latent and manifest latent nystagmus. Neuroophthalmology **3:** 247–257.
27. DELL'OSSO, L.F. 1985. Congenital, latent and manifest latent nystagmus: similarities, differences and relation to strabismus. Jpn J. Ophthalmol. **29:** 351–368.
28. DELL'OSSO, L.F. & R.B. DAROFF. 1976. Braking saccade: a new fast eye movement. Aviat. Space Environ. Med. **47:** 435–437.
29. DELL'OSSO, L.F. 1991. Nystagmus, saccadic intrusions/oscillations and oscillopsia. In Current Neuro-Ophthalmology, Vol. 3. S. Lessell & J.T.W. Van Dalen, Eds: 153–191. Mosby Year Book. Chicago.
30. DELL'OSSO, L.F. 1973. Fixation characteristics in hereditary congenital nystagmus. Am. J. Optom. Arch. Am. Acad. Optom. **50:** 85–90.
31. DELL'OSSO, L.F. 1986. Evaluation of smooth pursuit in the presence of congenital nystagmus. Neuroophthalmology **6:** 383–406.
32. ISHIKAWA, S., H. OZAWA & Y. FUJIYAMA. 1987. Treatment of nystagmus by acupuncture. In Highlights in Neuro-Ophthalmology. Proceedings of the Sixth Meeting of the

International Neuro-Ophthalmology Society (INOS). B.F. Boyd, Ed.: 227–232. Aeolus Press. Amsterdam.
33. BLEKHER, T., T. YAMADA, R.D. YEE, *et al.* 1998. Effects of acupuncture on foveation characteristics in congenital nystagmus. Br. J. Ophthalmol. **82:** 115–120.
34. ABADI, R.V., D. CARDEN & J. SIMPSON. 1979. Controlling abnormal eye movements. Vision Res. **19:** 961–963.
35. ABADI, R.V., D. CARDEN & J. SIMPSON. 1980. A new treatment for congenital nystagmus. Br. J. Ophthalmol. **64:** 2–6.
36. CIUFFREDA, K.J., S.G. GOLDRICH & C. NEARY. 1982. Use of eye movement auditory biofeedback in the control of nystagmus. Am. J. Optom. Physiol. Opt. **59:** 396–409.
37. KIRSCHEN, D.G. 1983. Auditory feedback in the control of congenital nystagmus. Am. J. Optom. Physiol. Opt. **60:** 364–368.
38. ABPLANALP, P. & H. BEDELL. 1987. Visual improvement in an albinotic patient with an alteration of congenital nystagmus. Am. J. Optom. Physiol. Opt. **64:** 944–951.
39. MEZAWA, M., S. ISHIKAWA & K. UKAI. 1990. Changes in waveform of congenital nystagmus associated with biofeedback treatment. Br. J. Ophthalmol. **74:** 472–476.
40. LEUNG, V., B. WICK & H.E. BEDELL. 1996. Multifaceted treatment of congenital nystagmus: a report of 6 cases. Optom. Vision Sci. **73:** 114–124.
41. EVANS, B.J., B.V. EVANS, J. JORDAHL-MOROZ, *et al.* 1998. Randomised double-masked placebo-controlled trial of a treatment for congenital nystagmus. Vision Res. **38:** 2193–2202.
42. SHARMA, P., R. TANDON, S. KUMAR, *et al.* 2000. Reduction of congenital nystagmus amplitude with auditory biofeedback. J. Aapos. **4:** 287–290.
43. CRONE, R.A., P.T.V.M. DE JONG & G. NOTERMANS. 1984. Behandlung des Nystagmus durch Injektion von Botulinustoxin in die Augenmuskeln. Klin. Mbl. Augenheilk. **184:** 216–217.
44. HELVESTON, E.M. & A.E. POGREBNIAK. 1988. Treatment of acquired nystagmus with botulinum A toxin. Am. J. Ophthalmol. **106:** 584–586.
45. LEIGH, R.J., R.L. TOMSAK, M.P. GRANT, *et al.* 1992. Effectiveness of botulinum toxin administered to abolish acquired nystagmus. Ann. Neurol. **32:** 633–642.
46. SPIELMANN, A. 1994. Nystagmus. Curr. Opin. Ophthalmol. **5:** 20–24.
47. CARRUTHERS, J. 1995. The treatment of congenital nystagmus with Botox. J. Pediatr. Ophthalmol. Strab. **32:** 306–308.
48. TOMSAK, R.L., B.F. REMLER, L. AVERBUCH-HELLER, *et al.* 1995. Unsatisfactory treatment of acquired nystagmus with retrobulbar injection of botulinum toxin. Am. J. Ophthalmol. **119:** 489–496.
49. RADIAN, A.B. 1980. [Electronystagmography and correction of congenital nystagmus by means of prisms]. Rev. Chir. Oncol. Radiol. O R L Oftalmol. Stomatol. Ser. Oftalmol. **24:** 173–176.
50. DICKINSON, C.M. 1986. The elucidation and use of the effect of near fixation in congenital nystagmus. Ophthal. Physiol. Opt. **6:** 303–311.
51. SENDLER, S., J. SHALLO-HOFFMANN & H. MÜHLENDYCK. 1990. Die Artifizielle-Divergenz-Operation beim kongenitalen Nystagmus. Fortschr. Ophthalmol. **87:** 85–89.
52. ABADI, R.V. 1979. Visual performance with contact lenses and congenital idiopathic nystagmus. Br. J. Physiol. Opt. **33:** 32–37.
53. ALLEN, E.D. & P.D. DAVIES. 1983. Role of contact lenses in the management of congenital nystagmus. Br. J. Ophthalmol. **67:** 834–836.
54. GOLUBOVIC, S., S. MARJANOVIC, D. CVETKOVIC, *et al.* 1989. The application of hard contact lenses in patients with congenital nystagmus. Fortschr. Ophthalmol. **86:** 535–539.
55. MATSUBAYASHI, K., M. FUKUSHIMA & A. TABUCHI. 1992. Application of soft contact lenses for children with congenital nystagmus. Neuroophthalmology **12:** 47–52.
56. SAFRAN, A.B. & Y. GAMBAZZI. 1992. Congenital nystagmus: rebound phenomenon following removal of contact lenses. Br. J. Ophthalmol. **76:** 497–498.
57. ROBERTI, G., P. RUSSO & G. SEGR. 1987. Spectral analysis of electro-oculograms in the quantitative evaluation of nystagmus surgery. Med. Biol. Eng. Comp. **25:** 573–576.
58. ABADI, R.V. & J. WHITTLE. 1992. Surgery and compensatory head postures in congenital nystagmus. A longitudinal study. Arch. Ophthalmol. **110:** 632–635.

59. ZUBCOV, A.A., N. STÄRK, A. WEBER, et al. 1993. Improvement of visual acuity after surgery for nystagmus. Ophthalmology **100:** 1488–1497.
60. LAVIN, P.J.M., S. TRACCIS, L.F. DELL'OSSO, et al. 1983. Downbeat nystagmus with a pseudocycloid waveform: improvement with base-out prisms. Ann. Neurol. **13:** 621–624.
61. TRACCIS, S., G. ROSATI, M.F. MONACO, et al. 1990. Successful treatment of acquired pendular elliptical nystagmus in multiple sclerosis with isoniazid and base-out prisms. Neurology **40:** 492–494.
62. BÜTTNER-ENNEVER, J.A., A.K.E. HORN, H. SCHERBERGER, et al. 2001 Motoneurons of twitch and non-twitch extraocular fibres in the abducens, trochlear and oculomotor nuclei of monkeys. J. Comp. Neurol. In press.
63. STEINBACH, M.J. & D. SMITH. 1981. Spatial localization after strabismus surgery: evidence for inflow. Science **213:** 1407–1409.
64. STEINBACH, M.J., E.L. KIRSHNER & M.J. ARSTIKAITIS. 1987. Recession vs. marginal myotomy surgery for strabismus: effects on spatial localization. Invest. Ophthalmol. & Visual Sci. **28:** 1870–1872.

A Neurobiological Approach to Acquired Nystagmus

R. JOHN LEIGH,[a] VALLABH E. DAS,[b] AND SCOTT H. SEIDMAN[c]

[a]*Neurology Service, Veterans Affairs Medical Center and Case Western University, Cleveland, Ohio 44106, USA*

[b]*Emory University, Atlanta, Georgia 30322, USA*

[c]*University of Rochester Medical Center, Rochester, New York 14642, USA*

ABSTRACT: The development of animal and mathematical models for several forms of acquired nystagmus has led to more comprehensive knowledge of these disorders. In the best understood forms, such as periodic alternating nystagmus, our range of knowledge includes an animal model, the neurotransmitters involved, and effective treatment. For some other forms, such as downbeat nystagmus, we have an animal model, but reliable treatment is lacking. In other cases, exemplified by acquired pendular nystagmus, we have only a provisional hypothesis for pathogenesis to account for the oscillations, without an animal model, but effective treatment is possible in some patients. The present trend of studying all aspects of the neurobiology of nystagmus, from molecules to behavior, seems to be the best approach to extend our knowledge and to identify new treatments, but much remains to be done.

KEYWORDS: periodic alternating nystagmus; downbeat nystagmus; pendular nystagmus; multiple sclerosis

INTRODUCTION

Historically, nystagmus in patients with neurological disease was approached from a phenomenological viewpoint, some forms being recognized as valuable for topological diagnosis, but others being unexplained, clinical curiosities. An initial advance was to classify nystagmus based on clinical features and recorded waveforms. Accumulated information from basic studies over the last 30 years has identified anatomical, physiological, and pharmacological substrates for several forms of nystagmus and has led to the development of animal models.[1]

Many forms of nystagmus can now be attributed to abnormalities of three mechanisms that normally contribute to steady gaze and clear vision during natural activities.[2] These include visual fixation, the vestibulo-ocular reflex, and the neural mechanism that makes it possible to hold the eyes at eccentric positions in the orbits (the neural integrator). Thus, disturbances of vision, especially the magnocellular, motion-processing system, can cause nystagmus. For example, specific early binoc-

---

Address for correspondence: Dr. R. John Leigh, Department of Neurology, University Hospitals of Cleveland, 11100 Euclid Ave., Cleveland, Ohio 44106. Voice: 216-421-3224; fax: 216-231-3461.

rjl4@po.cwru.edu

ular deprivation of vision leads to latent nystagmus in rhesus monkeys. It is currently hypothesized that latent nystagmus is due to an imbalance or bias in the input to a defective nucleus of the optic tract.[3,4] Extensive loss of visual inputs eventually causes the ocular motor system to become uncalibrated, with continuous nystagmus and a "wandering null" point, such as occurs in complete blindness.[5] Vestibular nystagmus arises from either imbalance (peripheral or central) or instability within the vestibular system. Disorders of the velocity-to-position integrator of ocular motor signals commonly cause centripetal drifts of the eyes, leading to gaze-evoked nystagmus.[2] Rarely, the neural integrator appears to become unstable, causing drifts of the eyes away from a central position.[6,7]

Other forms of nystagmus are recognized that cannot easily be accounted for by disturbance of any of these mechanisms, but in most cases we are able to provide at least a working hypothesis that is subject to experimental testing. In many cases, we now have an animal model and some knowledge of the neural circuits and neurotransmitters involved. In this paper, we do not attempt to provide a comprehensive review of the neurobiology of acquired nystagmus (which can be found in other sources).[1,2] Instead, we discuss three specific examples of acquired nystagmus, to illustrate the range of progress in this field.

AN EXAMPLE OF NYSTAGMUS FOR WHICH THERE IS AN ANIMAL MODEL AND EFFECTIVE TREATMENT

Perhaps the best understood form of acquired nystagmus is periodic alternating nystagmus (PAN), a spontaneous horizontal nystagmus, present in central gaze, that reverses its direction about every 2 minutes.[2] Acquired PAN occurs most commonly with disease involving the midline cerebellum. (A congenital form of PAN has different characteristics and probably has a different pathogenesis.) Experimental ablation of the monkey nodulus and uvula causes PAN when the animals are put into darkness.[8] The nodulus and uvula are known to control the time course of rotationally induced nystagmus (i.e., velocity storage),[9] and removal of these structures causes an increase in the duration of the response by increasing the time constant of velocity storage. It has proven possible to develop a formal mathematical model for PAN (FIG. 1)[10] based on the assumptions that (1) the time constant of velocity storage of rotationally induced nystagmus is prolonged excessively due to nodulus-uvula lesions, and (2) a vestibular "repair mechanism" is then activated (modeled as a simple adaptation operator)[10] to reverse the direction of this nystagmus, so producing the oscillations of PAN. These oscillations would ordinarily be suppressed by visual fixation, but disease of the cerebellum that causes PAN usually also impairs visually mediated eye movements. This model predicts that rotational stimuli, applied during critical points in the PAN cycle, will transiently stop the nystagmus, and this has been experimentally confirmed.[10] At the molecular level, it was demonstrated that control of the velocity storage by the nodulus and uvula is achieved, at least in part, by inhibitory pathways that use GABA.[11] Furthermore, baclofen, a $GABA_B$ agonist, inhibits velocity storage and is usually effective in treating periodic alternating nystagmus due to either experimental or clinical lesions of the nodulus and uvula.[8,12]

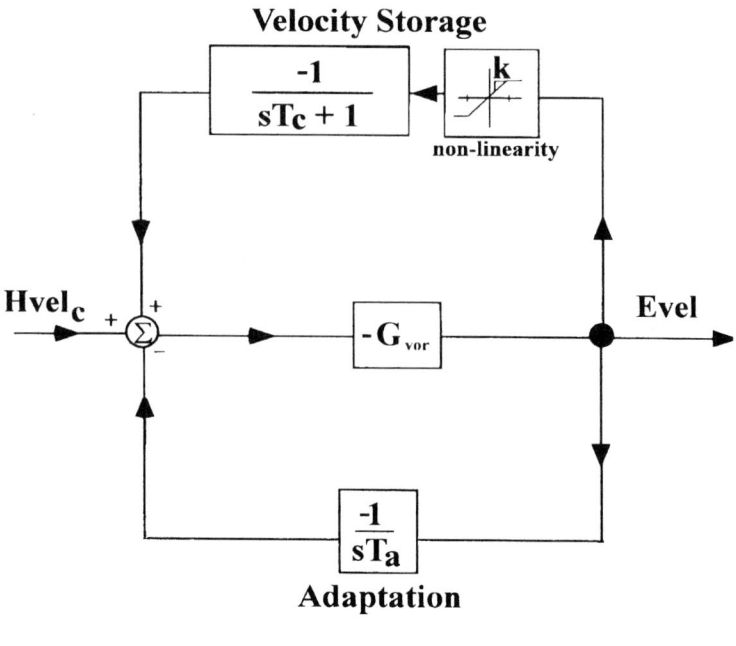

FIGURE 1. A mathematical model to account for periodic alternating nystagmus.[10] A head velocity signal from the semicircular canals (Hvel$_c$) is modulated by the gain of the "direct" vestibulo-ocular reflex pathway (G$_{vor}$) to produce an eye velocity command (Evel). The phenomenon of velocity storage (prolongation of the vestibulo-ocular time constant) is achieved by a positive feedback loop, of gain k. Current evidence suggests that the value of k is determined by the nodulus and that lesions of this structure cause velocity storage to become unstable. When this happens, an adaptation pathway periodically reverses the direction of the nystagmus. s, Laplace transform complex frequency; Ta, time constant of adaptation; Tc, time constant of cupula.

AN EXAMPLE OF NYSTAGMUS FOR WHICH THERE IS AN ANIMAL MODEL BUT NO EFFECTIVE TREATMENT

Downbeat nystagmus is clinically common and has been widely studied.[1,2] More than one mechanism may be responsible. Perhaps best understood is the hypothesis that vestibulocerebellar lesions (flocculectomy)[13] lead to downbeat nystagmus by causing a central imbalance of the inputs from the anterior and posterior semicircular canals of the vestibular labyrinth (FIG. 2A). This imbalance is apparently due to a quirk of evolution, in which the cerebellar flocculus sends inhibitory projections to

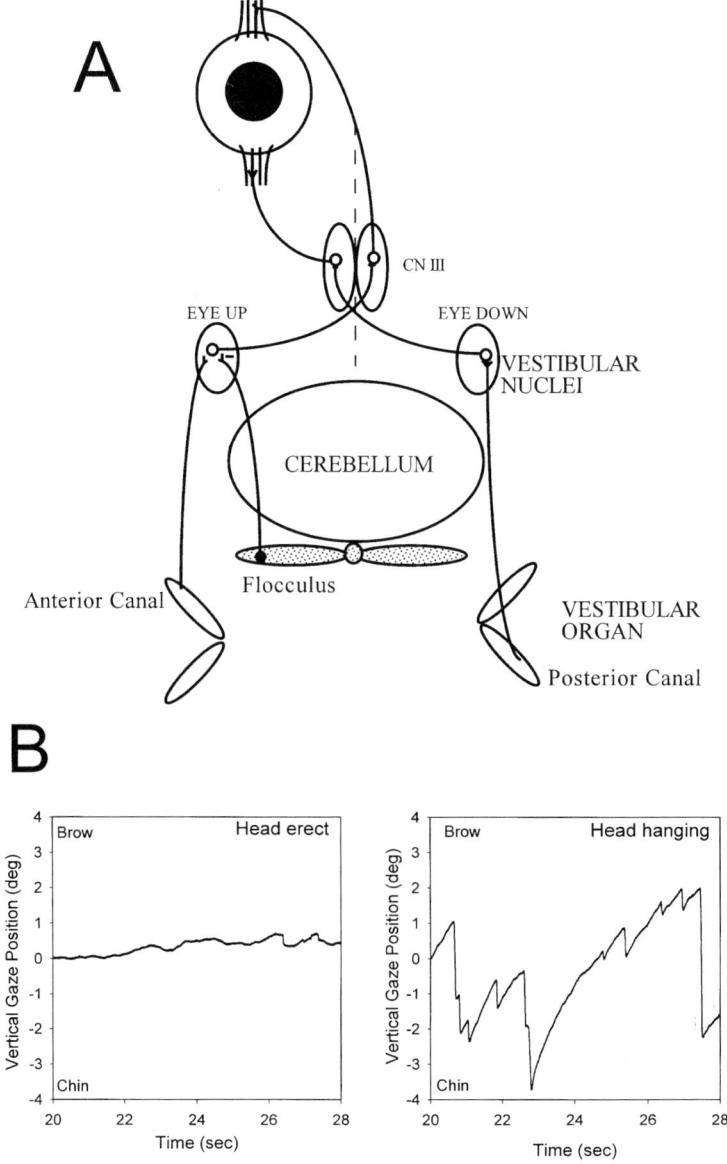

FIGURE 2. (**A**) Hypothesis for downbeat nystagmus based on the finding that the cerebellar flocculus governs the vertical vestibulo-ocular reflex by inhibiting inputs from the anterior, but not the posterior, semicircular canals.[14,15] Thus, floccular lesions should produce upward eye drifts (due to loss of inhibition of anterior canal inputs), leading to downbeat nystagmus. (**B**) Nystagmus in a normal subject with the head erect (*left*) and with the head upside down. Chin-beating nystagmus persists after several minutes and is therefore unlikely to be due to imbalance between semicircular canals; otolithic effects seem more likely.

govern projections from the anterior canals (which generate upward eye movements) but *not* from the posterior canals (which generate downward eye movements).[14,15] Consequently, after vestibulocerebellar lesions, the eyes drift upward and corrective quick phases cause downbeat nystagmus. Flocculectomy in the rhesus monkey causes downbeat nystagmus.[13]

It has been suggested that one reason for the inherent imbalance between inputs from the anterior and posterior canals is the geometric configuration of the labyrinthine semicircular canals.[16] Specifically, if one assumes that each semicircular canal elicits eye movements approximately in the direction of its anatomical plane,[17] then referring to the reported orientation of the canals in humans,[18] vector addition of the resting activity from all of the canals predicts an upward drift of the eyes.[16] Thus, there is a cancellation of the canal signals for horizontal and torsional eye movements, but an asymmetry of the vertical canals would be expected to lead to downbeat nystagmus. In health, the cerebellum compensates for this imbalance and prevents downbeat nystagmus.

A second hypothesis for downbeat nystagmus proposes an otolithic origin.[19,20] This hypothesis has received support from the observation that *static* positioning of normal human subjects' head upside-down causes "chin-beating" nystagmus (FIG. 2B).[21,22] Two other mechanisms have been proposed to account for downbeat nystagmus: vertical asymmetry of the neural signals governing smooth-pursuit eye movements[23] and instability of the neural integrator that normally guarantees gaze stability.[6] Each of these hypotheses has some experimental data to support it, so it is possible that one or several mechanisms may be responsible for downbeat nystagmus in patients.

Given this plethora of working hypotheses to account for downbeat nystagmus, it is somewhat disappointing that no reliable drug therapy exists for many patients with downbeat nystagmus. Based on the animal model for this disorder — flocculectomy in the monkey — it might be expected that loss of this cerebellar output would be corrected by administering drugs with GABAergic effects. However, clinical trials show inconsistent effects of drugs such as baclofen, clonazepam, and gabapentin,[24-26] although some patient without evidence of cerebellar degeneration do gain sustained improvement.[27] The possibility of other neurotransmitters has been investigated; intravenous scopolamine is reported to abolish downbeat nystagmus.[28] However, neither transdermal scopolamine[29,30] nor oral anticholinergic agents, such are trihexyphenidyl,[31] have proven reliable treatment for a range of forms of acquired nystagmus.

At this symposium,[32] it was demonstrated that mutant strains of mice (homozygous *rocker* mouse) bearing a point mutation in the gene encoding the pore protein of the P/Q calcium channel on the Purkinje cells, which impairs output from cerebellar cortex, show upward deviations of their eyes. This genetic disorder may have similarities with human mutations of the CACNA1A gene on chromosome 19p, which encodes the $\alpha 1A$ subunit of P/Q-type voltage-gated calcium channels.[33] Affected individuals manifest episodic ataxia type II, with downbeat nystagmus that does respond to acetazolamide. How acetazolamide works in this disorder is uncertain, although changes in blood pH may improve the functioning of calcium channels. In any case, the effectiveness of acetazolamide in downbeat nystagmus due to calcium channelopathy deserves further systematic investigation.

AN EXAMPLE OF NYSTAGMUS FOR WHICH THERE IS NO ANIMAL MODEL BUT SOME EFFECTIVE TREATMENT

Acquired pendular nystagmus (APN) consists of quasisinusoidal ocular oscillations typically ranging from 2–6 Hz. Several disorders may cause APN, including brain stem stroke (syndrome of oculopalatal myoclonus), Whipple's disease, and disorders of central myelin, most commonly multiple sclerosis.[1,2] It is possible that the pathogenesis of APN in each of these disorders is different. However, no animal model exists for APN. What formal hypotheses have been offered? First, it has been proposed that APN in multiple sclerosis is due to delayed visual feedback resulting from demyelination of the optic nerves.[34] However, neither darkness nor experimentally delaying the visual consequences of eye movements by electronic means changes the characteristics of pendular nystagmus, making it unlikely that these oscillations are caused by visual system dysfunction.[35] Furthermore, visual function is usually normal in patients with APN due to the syndrome of oculopalatal myoclonus or Whipple's disease. On the other hand, early-onset visual deprivation may induce a pendular form of nystagmus.[3]

Over 25 years ago, it was noted that APN in patients with multiple sclerosis often transiently stopped after a saccade.[36] This observation led us to address the question: could APN be due to an oscillation arising in the common neural integrator for conjugate eye movements and the temporary cessation of oscillations be due to the inhibitory effect of a large saccadic pulse on this integrator network?[37,38] More specifically, are the oscillations of pendular nystagmus generated within the network of neurons that constitute the neural integrator or from inputs to it? If pendular nystagmus did arise within the neural integrator, these oscillations would be susceptible to large inhibitory inputs, as occur during saccades. Specifically, oscillations arising after a saccade might be phase shifted compared with those occurring before one (FIG. 3A).[37] Alternatively, oscillations due to a premotor input to the neural integrator should not be phase shifted by saccades. Our first step towards answering these questions was to implement a neural network model of the ocular motor integrator, because recent studies support this conceptualization.[39] For this study, we selected the neural network described by Arnold and Robinson[40] to account for the role of the nucleus prepositus hypoglossi-medial vestibular nuclei region (NPH-MVN) in integrating vestibular inputs during horizontal head rotations. Next we looked for a plausible way to make this model oscillate. We first made the network unstable by increasing the weights of inhibitory connections across the midline commissure of the network. We then added negative feedback around the network. Such feedback might be provided by the cell groups of the paramedian tracts.[41] Paramedian tract cell groups receive inputs from essentially all premotor structures that are concerned with eye movements and may provide feedback control of gaze via their cerebellar projections. Experimental lesions of some of the paramedian tract cell groups disrupt neural integrator function.[42] Furthermore, in patients with APN due to multiple sclerosis, magnetic resonance imaging often indicates demyelinative lesions in the paramedian tegmentum of the brain stem that could involve the paramedian tract cell groups.[43] Thus, the biological substrate for our model might be summarized by the scheme shown in FIGURE 3C. When we increased negative feedback gain to a value of 2.0 around our unstable neural network, it oscillated with a frequency of 5 Hz

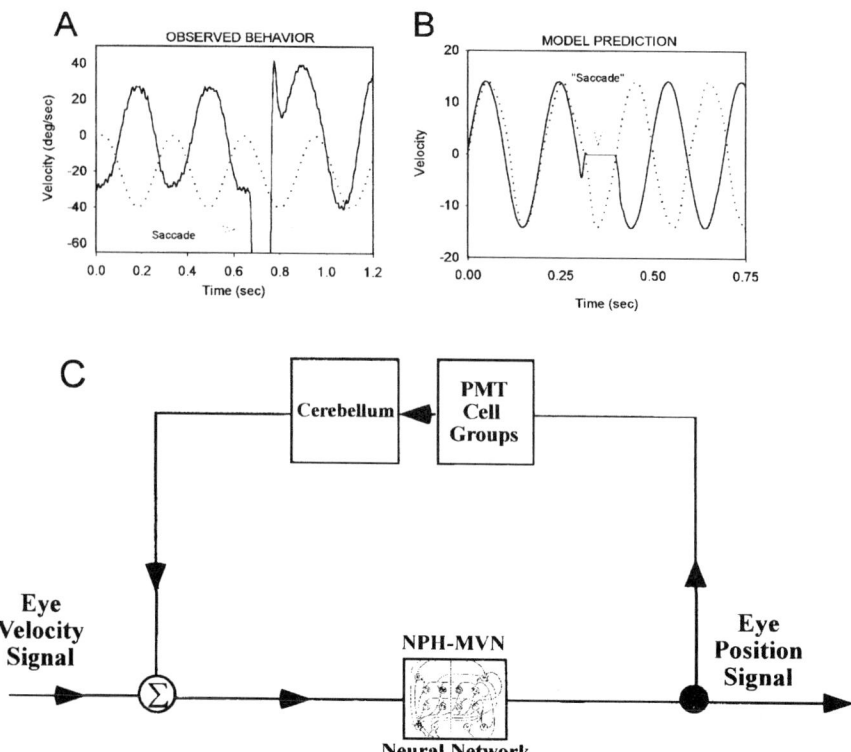

FIGURE 3. Acquired pendular nystagmus (APN) in multiple sclerosis. (**A**) Horizontal eye movement records (*solid line*) from a patient, showing how a large saccade caused a phase shift of the APN oscillations; the *dotted line* is a reference sine wave. (**B**) APN oscillations simulated by a network model of the neural integrator that was made unstable and subjected to increased-gain negative feedback.[37] The model simulation is the *solid line*; the *dotted line* is a reference sine wave. Note that after the oscillations are briefly silenced (by setting some of the network units to zero, which simulates the effect of a saccade), the oscillations are phase shifted, similar to behavior in the patient. (**C**) Possible biological substrate for the model used to simulate APN. A neural network of cells in the nucleus prepositus hypoglossi–medial vestibular nucleus (NPH-MVN) of the medulla, which contributes to the neural integration of eye movement signals, is governed by feedback control via paramedian tract (PMT) cell groups and the vestibular cerebellum. If the NPH-MVN network becomes unstable, and feedback loop is impaired due to involvement of PMT cell groups, oscillations of APN result.

(FIG. 3B). We tested this model experimentally by determining whether large saccades (which would be expected to transiently inhibit the neural network) would cause a phase shift of subsequent oscillations. If we set the activity of constituent units in the network to zero for a 100-ms period (simulating the effects of a large saccade), the oscillations momentarily stopped and subsequent oscillations showed a phase shift similar to the effects of large saccades on APN in patients (FIG. 3B). However, if oscillations were introduced on inputs into the stable neural network, a simulated saccade produced no phase shift. Our model made specific predictions about the effects of different-sized saccades on phase shifts: large saccades would be expected to produce large phase shifts and very small saccades would be predicted to cause a phase lead. These predictions were confirmed experimentally.[37] We also attempted to determine whether large saccades caused phase shifts of the APN in patients with oculopalatal tremor. However, their nystagmus is much less sinusoidal that the APN seen in multiple sclerosis, and it was not possible for us to make a reliable judgment.

Our model for APN in patients with multiple sclerosis suggested that the basic problem was abnormal cerebellar feedback of the brain stem integrator network. Although no animal model for APN exists, basic studies of neural integrator function might produce data to support or refute our hypothesis. Specifically, the technique of pharmacological inactivation of the neural integrator by microinjecting drugs with known effects on neurotransmitter systems in NPH-MVN has been studied.[44–48] Agents with agonist or antagonist actions at gamma-aminobutyric acid (GABA), glutamate, and kainate receptors all cause gaze-evoked nystagmus with centripetal eye drifts, implying that the neural integrator had been made "leaky" (decreased time constant of integration). Glycine and strychnine have no effect.[44] However, when the GABA-agonist muscimol was injected near the center of MVN, the eyes drifted away from the central position with increasing-velocity waveforms, implying an unstable neural integrator.[44] A similar effect was reported after injection of the GABA-antagonist bicuculline into the vestibular nuclei.[45] How can these results in which either agonist or antagonist of specific transmitter receptors had similar effects be explained? Pharmacological agents may prevent the constituent neurons of the neural network in NPH-MVN from modulating their discharge in response to inputs from other neurons; this would cause the integrator to become *leaky*. Such an effect would be expected both for agents such as muscimol, which hyperpolarizes cells (and drives them into cutoff), and drugs such as bicuculline, which depolarize cells (and drives their discharge rates into saturation). A suggested mechanism whereby the neural integrator could become unstable is that its constituent neurons remain functional, but that the feedback control of them, by the cerebellum, could be disrupted.[6,49] This concept is also schematized in FIGURE 3C. Experimentally, such instability is reported when either the $GABA_A$ agonist muscimol or the $GABA_A$ antagonist bicuculline is injected close to that portion of the MVN that receives floccular afferents and lies lateral to the marginal zone.[50] Floccular Purkinje cells are know to be GABAergic,[51] and so it seems possible that inactivating their projections to neurons in MVN might be responsible for instability of the neural integrator (by effectively increasing the value of feedback gain of the brain stem integrator network). Further support for this suggestion comes from reports of patients who show nystagmus with increasing velocity waveforms; such individuals have cerebellar, not brain stem, lesions.[6,7]

Although pharmacological inactivation has so far failed to produce an animal model of APN, the finding that an unstable integrator could be produced after injection of GABAergic agents into NPH-MVN suggested that drugs with GABAergic effects might be expected to suppress APN (because the increased gain of feedback might be equivalent to decreased inhibition by the cerebellum). A double-blind clinical trial of gabapentin and baclofen showed the prior drug to be more effective therapy,[26] and this was confirmed in subsequent studies.[52] However, gabapentin has effects other than those exerted on GABAergic mechanisms, and the drug vigabatrin, which is more purely GABAergic in its effects, has proven ineffective in treating APN.[52] Furthermore, the drug memantine, an agent with NMDA blocking, AMPA receptor modulation, and dopaminergic action, is reported to suppress APN in multiple sclerosis.[29] Thus, we still have only a partial understanding of the pathogenesis of the various forms of APN, but are fortunate in being able to treat some of these patients.

CONCLUSIONS

A true neurobiological understanding of acquired nystagmus, in which the neural circuits and their neurotransmitters are known to the point that a coherent quantitative hypothesis is achieved, is still lacking for many types of oscillations. Nonetheless, progress on several fronts, especially the development of animal models and the techniques of pharmacological inactivation, seem likely to advance us ultimately to the point that we will understand and be able to treat all forms of abnormal eye movements that disrupt clear vision.

ACKNOWLEDGMENTS

This work was supported by the Office of Research and Development, Medical Research Service, Department of Veterans Affairs; National Institutes of Health Grant EY06717; and the Evenor Armington Fund.

REFERENCES

1. STAHL, J.S., L. AVERBUCH-HELLER & R.J. LEIGH 2000. Acquired nystagmus. Arch. Ophthalmol. **118:** 544–548.
2. LEIGH RJ. & D.S. ZEE. 1999. The Neurology of Eye Movements, 3rd Ed. Oxford University Press. New York.
3. TUSA, R.J., M.J. MUSTARI, A.F. BURROWS, *et al.* 2001. Gaze-stabilizing deficits and latent nystagmus in monkeys with brief, early-onset visual deprivation: eye movement recordings. J. Neurophyiol. **86:** 651–661.
4. MUSTARI, M.J., R.J. TUSA, A.F. BURROWS, *et al.* 2001. Gaze-stabilizing deficits and latent nystagmus in monkeys with brief, early-onset visual deprivation: role of pretectal NOT. J. Neurophyiol. **86:** 662–675.
5. LEIGH, R.J., S.E. THURSTON, R.L. TOMSAK, *et al.* 1989. Effect of monocular visual loss upon stability of fixation. Invest. Ophthalmol. & Visual Sci. **30:** 288–292.
6. ZEE, D.S., R.J. LEIGH & F. MATHIEU-MILLAIRE. 1980. Cerebellar control of ocular gaze stability. Ann. Neurol. **7:** 37–40.
7. BARTON, J.J.S. & J.A. SHARPE. 1993. Oscillopsia and horizontal nystagmus with accelerating slow phases following lumbar puncture in the Arnold-Chiari malformation. Ann. Neurol. **33:** 418–421.
8. WAESPE, W., B. COHEN & T. RAPHAN 1985. Dynamic modification of the vestibuloocular reflex by the nodulus and uvula. Science **228:** 199–202.
9. COHEN, B., V. HENN, T. RAPHAN & D. DENNETT. 1981. Velocity storage, nystagmus, and visual-vestibular interactions in humans. Ann. N.Y. Acad. Sci. **374:** 421–433.

10. LEIGH, R.J., D.A. ROBINSON & D.S. ZEE. 1981. A hypothetical explanation for periodic alternating nystagmus: instability in the optokinetic-vestibular system. Ann. N.Y. Acad. Sci. **374:** 619–635
11. COHEN, B., D. HELWIG & T. RAPHAN. 1987. Baclofen and velocity storage: a model of the effects of the drug on the vestibulo-ocular reflex in the rhesus monkey. J. Physiol. (Lond.) **393:** 703–725.
12. HALMAGYI, G.M., P. RUDGE, M.A. GRESTY, et al. 1980. Treatment of periodic alternating nystagmus. Ann. Neurol. **8:** 609–611.
13. ZEE, D.S., A. YAMAZAKI, P.H. BUTLER & G. GUCER. 1981. Effects of ablation of flocculus and parafloccuIus on eye movements in primate. J. Neurophysiol. **46:** 878–899.
14. ITO, M., N. NISIMARU & M. YAMAMOTO. 1977. Specific patterns of neuronal connections involved in the control of the rabbit's vestibulo-ocular reflexes by the cerebellar flocculus. J. Physiol. (Lond.) **265:** 833–854.
15. BALOH, R.W. & J.W. SPOONER. 1981. Downbeat nystagmus: a type of central vestibular nystagmus. Neurology **31:** 304–310.
16. BÖHMER, A. & D. STRAUMANN. 1998. Pathomechanism of mammalian downbeat nystagmus due to cerebellar lesion: a simple hypothesis. Neurosci. Lett. **250:** 127–130.
17. SUZUKI, J. & B. COHEN. 1964. Head, eye, body and limb movements from semicircular canal nerves. Exp. Neurol. **10:** 333–405.
18. BLANKS, R.H.I., I. S. CURTHOYS & C.H. MARKHAM. 1975. Planar relationships of the semicircular canals in man. Acta Otolaryngol. **80:** 185–196.
19. CHAMBERS, B.R., J.J. ELL & M.A. GRESTY. 1983. Case of downbeat nystagmus influenced by otolith stimulation. Ann. Neurol. **13:** 204–207.
20. GRESTY, M., H. BARRATT, P. RUDGE & N. PAGE. 1986. Analysis of downbeat nystagmus: otolithic vs semicircular canal influences. Arch. Neurol. **43:** 52–55.
21. BISDORFF, A.R. et al. 2000. Positional nystagmus in the dark in normal subjects. Neuro-ophthalmology **24:** 283–290.
22. KIM, J.I. et al. 2001. Vertical nystagmus in normal subjects: effects of head position, nicotine and scopolamine. J.Vestib. Res. **10:** 291–300.
23. ZEE, D.S., A.R. FRIENDLICH, D.A. ROBINSON, et al. 1974. The mechanism of downbeat nystagmus. Arch. Neurol. **30:** 227–237.
24. DIETERICH, M., A. STRAUBE, T. BRANDT, et al. 1991. The effects of baclofen and cholinergic drugs on upbeat and downbeat nystagmus. J. Neurol. Neurosurg. Psychiatry **54:** 627–632.
25. CURRIE, J.N. & V. MATSUO. 1986. The use of clonazepam in the treatment of nystagmus-induced oscillopsia. Ophthalmology **93:** 924–932.
26. AVERBUCH-HELLER, L. et al. 1997. A double-blind controlled study of gabapentin and baclofen as treatment for acquired nystagmus. Ann. Neurol. **41:** 818–825.
27. YOUNG. Y.H. & T.W. HUANG. 2001. Role of clonazepam in the treatment of idiopathic downbeat nystagmus. Laryngoscope **111:** 1490–1493.
28. BARTON, J.J., A.G. HUAMAN & J.A SHARPE. 1994. Muscarinic antagonists in the treatment of acquired pendular and down beat nystagmus: a double blind, randomized trial of three intravenous drugs. Ann. Neurol. **35:** 319–325.
29. STARCK M. et al. 1997. Drug therapy for acquired pendular nystagmus in multiple sclerosis. J. Neurol. **244:** 9–16.
30. KIM, J.I., L. AVERBUCH-HELLER & R.J. LEIGH. 2001 Evaluation of transdermal scopolamine are treatment for acquired nystagmus. J. Neuro-ophthalmol. **21:** 188–192.
31. LEIGH, R.J., T.H. BURNSTINE, R.L. RUFF & R.J. KASMER. 1991. The effect of anticholinergic agents upon acquired nystagmus. A double-blind study of trihexyphenidyl and tridihexethyl chloride. Neurology **41:** 1737–1741.
32. STAHL, J.S. 2002. Calcium channelopathy mutants and their role in ocular motor research. Ann. N.Y. Acad. Sci. This volume.
33. BALOH, R.W. 2002. Genetics of familial episodic vertigo and ataxia. Ann. N.Y. Acad. Sci. This volume.
34. BARTON, J.J.S. & T.A. COX. 1993. Acquired pendular nystagmus in multiple sclerosis: clinical observations and the role of optic neuropathy. J. Neurol. Neurosurg. Psychiatry **56:** 262–267.

35. AVERBUCH-HELLER, L. et al. 1995. Investigations of the pathogenesis of acquired pendular nystagmus. Brain **188:** 369–378.
36. ASCHOFF, J.C., B. CONRAD & H.H. KORNHUBER. 1974 Acquired pendular nystagmus with oscillopsia in multiple sclerosis: a sign of cerebellar nuclei disease. J. Neurol. Neurosurg. Psychiatry **37:** 570–577.
37. DAS, V.E., P. ORUGANTI, P.D. KRAMER & R.J. LEIGH. 2000. Experimental tests of a neural-network model for ocular oscillations caused by disease of central myelin. Exp. Brain Res. **133:** 189–197.
38. MCFARLAND, J.L. & A.F. FUCHS. 1992. Discharge patterns in nucleus prepositus hypoglossi and adjacent medial vestibular nucleus during horizontal eye movement in behaving macaques. J. Neurophysiol. **68:** 319–332.
39. AKSAY, E., G. GAMKRELIDZE, H.S. SEUNG, et al. 2001. In vivo intracellular recording and perturbation of persistent activity in a neural integrator. Nature Neurosci. **4:** 184–193
40. ARNOLD, D.B. & D.A. ROBINSON. 1997. The oculomotor integrator: testing of a neural network model. Exp. Brain Res. **113:** 57–74.
41. BÜTTNER-ENNEVER, J.A. & A.K. HORN. 1996. Pathways from cell groups of the paramedian tracts to the floccular region. Ann. N.Y. Acad. Sci.**781:** 532–540.
42. NAKAMOGOE, K., Y. IWAMOTO & K. YOSHIDA. 2000. Evidence for brainstem structures participating in oculomotor integration. Science **288:** 857–859.
43. LOPEZ, L.I. et al. 1996 Clinical and MRI correlates in 27 patients with acquired pendular nystagmus. Brain **119:** 465–472.
44. ARNOLD, D.B., D.A. ROBINSON & R.J. LEIGH. 1999. Nystagmus induced by pharmacological inactivation of the brainstem ocular motor integrator in monkey. Vision Res. **39:** 4286–4295.
45. STRAUBE, A., R. KURZAN & U. BÜTTNER. 1991. Differential effects of bicuculline and muscimol microinjections into the vestibular nuclei on simian eye movements. Exp. Brain Res. **86:** 347–358.
46. METTENS, P.E. et al. 1994. Effect of muscimol microinjections into the prepositus hypoglossi and the medial vestibular nuclei on cat eye movements. J. Neurophysiol. **72:** 785–802.
47. METTENS, P., G. CHERON & E. GODAUX. 1994. NMDA receptors are involved in temporal integration in the oculomotor system of the cat. Neuroreport **5:** 1333–1336.
48. MORENO-LOPEZ, B., C. ESTRADA & M. ESCUDERO. 1998. Mechanism of action and targets of nitric oxide in the oculomotor system. J. Neurosci. **18:** 10672–10679.
49. KAMATH, B.Y. & E.L. KELLER. 1976. A neurological integrator for the oculomotor control system. Math. Biosci. **30:** 341–352.
50. BÜTTNER-ENNEVER, J.A. 1992. Patterns of connectivity in the vestibular nuclei. Ann. N.Y. Acad. Sci. **656:** 363–378.
51. MUGNAINI, E. & W.H. OERTEL. 1985. An atlas of the distribution of GABA-ergic neurons and terminals in the rat CNS as revealed by GAD-immunohistochemistry. In Handbook of Chemical Neuranatomy. Volume 4, part 1. GABA and Neuropeptides in the CNS. A. Björklund & T. Hökfelt, Eds.: 436–608. Elsevier. Amsterdam.
52. BANDINI, F., E. CASTELLO, L. MAZZELLA, et al. 2001. Gabapentin but not vigabatrin is effective in the treatment of acquired nystagmus in multiple sclerosis: how valid is the GABAergic hypothesis? J. Neurol. Neurosurg. Psychiatry **71:** 107–110.

Extraocular Muscle Gene Expression and Function after Dark Rearing

FRANCISCO H. ANDRADE,[a] ANITA P. MERRIAM,[b] AND JOHN D. PORTER[a,b,c]

[a]*Departments of* [a]*Neurology and* [b]*Ophthalmology, Case Western Reserve University and University Hospitals of Cleveland, Cleveland, Ohio 44106, USA*

[c]*Department of Neurosciences, Case Western Reserve University, Cleveland, Ohio 44106, USA*

KEYWORDS: extraocular muscle; critical period; cDNA microarray; contractile function; development

INTRODUCTION

The development of the visual system includes a postnatal time window, the "critical period" during which normal sensory input is needed to establish the adult phenotype. The final effectors of the ocular motor system, the extraocular muscles (EOMs), depend also on postnatal sensory experience for the induction of the adult myosin expression pattern and calcium handling systems.[1–3] The mechanisms regulating this period of postnatal EOM development, and the functional properties dependent on it, remain incompletely understood. For this study, we tested the hypothesis that absence of visual stimulation during the postnatal critical period would alter gene expression patterns and the contractile function of rat EOMs.

METHODS

Whole litters of Sprague-Dawley rats were reared in total darkness (DR) or in standard lighting conditions (12 h on–12 h off: control) for 45 days from birth. Total RNA was extracted from EOMs (recti and oblique muscles) pooled from animals taken from different litters, and used on Affymetrix rat cDNA microarrays to determine gene expression patterns. Analyses of microarray data from the two experimental groups generated an increase/no change/decrease difference call and a fold change for each probe set. Only difference calls $\geq$2-fold were considered significant. *In vitro* EOM contractile properties were determined using isolated superior rectus muscles as the "index" muscles. These were studied under isometric (fixed-length) conditions, and in a curarized saline solution (26 µM *d*-tubocurarine) to inactivate neuromuscular transmission. In consequence, the measured functional parameters

Address for correspondence: Francisco Andrade, Ph.D., Department of Neurology, University Hospitals of Cleveland, 11100 Euclid Avenue, Cleveland OH 44106-5040. Voice: 216-844-4793; fax: 216-844-4792.
fha@po.cwru.edu

TABLE 1. Changes in gene expression in extraocular muscles from DR rats

Increases	Decreases
Cathepsin E	CD44 (surface antigen)
Acyl-CoA synthetase	Early growth response 1 (egr1)
Soluble cytochrome $b5$	Ribosomal protein L5
Glutathione S-transferase	Calponin
Protein kinase C-η	
GTPase Rab8b	

TABLE 2. Contractile properties of extraocular muscles from control and DR rats[a]

	Control	DR
Time to peak force (msec)	14.9 ± 0.7	17.7 ± 0.3[b]
Half-relaxation time (msec)	17.0 ± 1.0	18.8 ± 0.4
Peak twitch force (N/cm^2)	0.55 ± 0.04	0.55 ± 0.03
Peak tetanic force (N/cm^2)	6.7 ± 0.2	5.6 ± 0.3[b]
Fatigue index (%)	61.4 ± 3.4	53.0 ± 2.1

[a]Data shown as means ± SEM.
[b]Significant difference between control and DR, $p<0.05$.

reflect intrinsic muscle properties. Absolute force measurements were normalized to muscle cross-sectional area.

RESULTS

The expression of 13 known genes and 26 expressed sequence tags (ESTs, or "candidate" genes) was altered in EOMs from DR rats. Several genes identified as involved in multiple signal transduction pathways (protein kinase C-η, GTPase Rab8b) and tissue remodeling (cathepsin, a calcium-dependent protease) were expressed at a higher level in DR EOMs (TABLE 1). Interestingly, DR decreased the expression of early growth response 1 (egr1), a marker of cellular de-differentiation. Isometric contractile function was also impaired by dark rearing (TABLE 2). Time to peak tension, an index of twitch kinetics, was significantly slower in DR EOMs, and twitch force was not altered. DR EOMs produced significantly less tetanic force, and their force/frequency relationship was shifted to the left. Curiously, DR EOMs showed evidence of excitation-contraction coupling failure; there was a significant drop in force at stimulation frequencies greater than 200 Hz. Dark rearing also rendered EOMs more fatigable, as indicated by the decrease in the fatigue index.

CONCLUSIONS

The changes in EOM gene expression and functional properties in response to dark rearing support our original hypothesis and are further evidence that visual experience is important to establish the normal EOM phenotype.

ACKNOWLEDGMENTS

This research was supported by the Evenor Armington Fund and by grants from the Muscular Dystrophy Association and NIH (EY12998 to F.H.A. and EY09834 to J.D.P.).

REFERENCES

1. BRUECKNER, J.K. & J.D. PORTER. 1998. Visual system maldevelopment disrupts extraocular muscle-specific myosin expression. J. Appl. Physiol. **85:** 584–592.
2. BRUECKNER, J.K. *et al.* 1999. Vestibulo-ocular pathways modulate extraocular muscle myosin expression patterns. Cell Tissue Res. **295:** 477–484.
3. PORTER, J.D. & P. KARATHANASIS. 1999. The development of extraocular muscle calcium homeostasis parallels visuomotor system maturation. Biochem. Biophys. Res. Commun. **257:** 678–683.

Conservation of Synapse-Signaling Pathways at the Extraocular Muscle Neuromuscular Junction

SANGEETA KHANNA[a] AND JOHN D. PORTER[a,b]

Departments of [a]Ophthalmology and [b]Neurology and Neurosciences, Case Western Reserve University, and University Hospitals of Cleveland, Cleveland, Ohio 44106, USA

KEYWORDS: extraocular muscle; neuromuscular junction; synapse; signal transduction

PURPOSE

The cell and molecular biology of extraocular muscle (EOM) is distinct from other skeletal muscles.[1,2] Novel aspects of the innervation pattern of EOM include: high motoneuron discharge rates, presence of multiply innervated muscle fiber types in the adult, retention of embryonic acetylcholine receptor isoforms at some neuromuscular junctions (NMJs), absence of significant postjunctional membrane foldings at most NMJs, and sensitivity of EOM to neuromuscular transmission disorders (myasthenia gravis) and distinct response to botulinum toxin. The formation and maturation of the NMJ require a series of inductive interactions between axon and muscle fiber. These interactions culminate in the juxtaposition of a highly specialized nerve terminal with an elaborate postsynaptic apparatus.[3,4] The sarcolemmal DGC and its associated proteins are central in this process of synapse formation and stabilization.[5,6] The central hypothesis of this work was that the unique phenotype, and functional properties, of EOM may require muscle group-specific adaptations to organize and maintain the NMJ. Therefore, the signaling and sarcolemmal organization at the NMJ may differ between EOM singly (SIF) and multiply innervated fiber types (MIF) and between EOM and other skeletal muscles.

METHODS

Here, we have examined cryostat sections of EOM of adult mice using immunohistochemistry for the neuregulin/erbB and agrin/MuSK pathway molecules and important components of the junctional DGC. The *en-plaque* (on SIFs) and the *en-grappe* junctions (on MIFs) were identified by their morphologic pattern using Tex-

Address for correspondence: Sangeeta Khanna, M.D., Research Institute, University Hospitals of Cleveland, 11100 Euclid Avenue, Cleveland, Ohio 44106. Voice: 216-844-1429; fax: 216-844-4792.
sxk128@po.cwru.edu

TABLE 1. Comparison of the cellular localization of the synaptic molecules

Protein	EOM SIFs		EOM MIFs		Skeletal muscle[a]	
	NMJ	Extra-synaptic	NMJ	Extra-synaptic	NMJ	Extra-synaptic
agrin	+	−	+	−	+	−
MuSK	+	−	+	−	+	−
rapsyn	+	−	+	−	+	−
neuregulin	+	−	+	−	+	−
erbB2	+	−	+	−	+	−
erbB3	+	−	+	−	+	−
erbB4	+	−	+	−	+	−
utrophin	+	−	+	−	+	−
α-syntrophin	+	+	+	+	+	+
β1syntrophin	+	−	+	+	+	+ (fast fibers)[7]
β2syntrophin	+	−	+	−	+	−
α-DB-1	+	−	+	+	+	−[6]
α-DB-2	+	+	+	+	+	+
s-laminin	+	−	+	−	+	−

ABBREVIATIONS: EOM, extraocular muscle; MIFs, multiply innervated fiber types; NMJ, neuromuscular junction; SIFs, singly innervated fiber types.
[a]Skeletal muscle data published elsewhere by several other groups.

as-red conjugated bungarotoxin (Molecular Probes; 4 µg/ml), which labels the AChRs. The other proteins were detected using monoclonal and polyclonal antibodies specific for: utrophin (NCL-DRP2, Novocastra Labs), α1-syntrophin (258), β1-syntrophin (37) and β2-syntrophin (28), rapsyn (1234), α-dystrobrevins DB-1 (670) and DB-2 (DB2) (all from Stan Froehner, University of Washington), s-laminin (D7; Joshua Sanes, Washington University), neuregulin (NDF; Jeffrey A Loeb, Wayne State University), erbB2 (Neu:sc-284; Santa Cruz Biotechnology, Inc. Santa Cruz), erbB3 and erbB4 (05-390, 06-572; Upstate Biotechnology, NY), agrin (AGR-550; StressGen Biotechnology Corp., Canada), and MuSK (Markus Ruegg, University of Basel). Junctions were examined for the cellular localization and colocalization of these proteins with AChRs using a scanning laser confocal microscope.

RESULTS

The components of the junctional DGC and the agrin/MuSK and the neuregulin/erbB nerve-muscle signaling pathways that we examined, colocalized with NMJ acetylcholine receptor aggregates in both the MIFs and the SIFs, similar to the typical skeletal muscles. There were some differences in the extrasynaptic localization of some DGC proteins, namely, β_1-syntrophin and α-dystrobrevin1 on EOM MIFs. The results are summarized in TABLE 1.

CONCLUSIONS

The EOM neuromuscular junctions differ in morphologic pattern from that of the typical NMJs, not only in the presence of MIFs in the adult but also in the SIF en-plaque junctions. However, we find here that these do not differ from the well-studied pattern of signaling mechanisms of typical skeletal muscle NMJ. The key components of the neuregulin/erbB and agrin/MuSK pathways and the junctional DGC proteins are present at EOM NMJs, establishing that synapse formation/maintenance pathways are conserved in all mammalian skeletal muscles, including the phenotypically novel EOMs. The inordinately high discharge rates of ocular motoneurons and the differential involvement of EOM in several neuromuscular diseases, however, suggest that there are other, as yet unidentified, features of NMJs that are unique to the extraocular muscle.

ACKNOWLEDGMENTS

We thank Denise Hatala for technical support. This work was supported by grants from Knights-Templar Eye Research Foundation, Research to Prevent Blindness, the Muscular Dystrophy Association USA, the Evenor Armington Fund, and by NIH Grants EY09834, EY12779, and P30 EY11373.

REFERENCES

1. PORTER, J.D., S. KHANNA, H.J. KAMINSKI, et al. 2001. Extraocular muscle is defined by a fundamentally distinct gene expression profile. Proc. Natl. Acad. Sci. USA. **98:** 12062–12067.
2. SPENCER, R.F. & J.D. PORTER. 1988. Structural organization of the extraocular muscles. Rev. Oculomot. Res. **2:** 33–79.
3. SANES, J.R. & J.W. LICHTMAN. 1999. Development of the vertebrate neuromuscular junction. Annu. Rev. Neurosci. **22:** 389–442.
4. SCHAEFFER, L., A. DE KERCHOVE D'EXAERDE & J.P. CHANGEUX. 2001. Targeting transcription to the neuromuscular synapse. Neuron **31:** 15–22.
5. RUEGG, M.A. 2001. Molecules involved in the formation of synaptic connections in muscle and brain. Matrix Biol. **20:** 3–12.
6. GRADY, R.M., H. ZHOU, J.M. CUNNINGHAM, et al. 2000. Maturation and maintenance of the neuromuscular synapse: genetic evidence for roles of the dystrophin–glycoprotein complex. Neuron **25:** 279–293.
7. PETERS, M.F., M.E. ADAMS & S.C. FROEHNER. 1997. Differential association of syntrophin pairs with the dystrophin complex. J. Cell Biol. **138:** 81–93.

Extraocular Muscle Fatigue

HENRY J. KAMINSKI[a,b] AND CHELLIAH R. RICHMONDS[a]

Departments of [a]Neurology and [b]Neurosciences, Case Western Reserve University School of Medicine, Louis Stokes Veterans Affairs Medical Center, University Hospitals of Cleveland, Cleveland, Ohio 44106, USA

KEYWORDS: extraocular muscles; fatigue

Extraocular muscles (EOM) are particularly resistant to fatigue as determined by functional measures in humans[1] and studies of isolated muscle preparations.[2] Fatigue resistance *in vivo* likely stems from EOM's high vascularity and mitochondrial content. Gene expression profiling also suggests that EOM rely on glucose uptake from blood rather than glycogen breakdown, as occurs in other skeletal muscle.[3] The high level of nitric oxide synthase and nitric oxide's effects on muscle contractility and glucose uptake may further contribute to EOM's fatigue resistance.[4]

To further evaluate fatigue properties of EOM, we compared the effects of fatigue protocols and temperature on *in vitro* tetanic contractions of lateral rectus (LR) and extensor digitorum longus (EDL) of mouse. Both these muscles share properties of rapid contractile speeds. At the end of a 5-minute fatigue regime, LR retained 58.5% of the original force (from 2.323 to 1.359 Newtons/cm^2), whereas EDL had only 16.93% (3.775 to 0.639 Newtons/cm^2). A significant decrease in force was already observed after 2 minutes. In temperature studies, LR showed a significant increase in force at 27°C compared to 37°C. In contrast, force generation of EDL was decreased at 27°C. Both the muscles exhibited a decrease in force at 20°C compared to higher temperatures. In contrast to EDL, LR maintains and, in fact, increases tetanic force levels after cooling.

The studies confirm work in other animals that show EOM to be particularly fatigue resistant.[2] The maintenance of tetanic force with cooling of LR suggests more efficient excitation-contraction coupling compared to EDL. The main effect of temperature on muscle function concerns the process of Ca^{2+} release and uptake. The response of the LR may be due to the small myofibrils, unique troponin isoform composition, abundant mitochondria and sarcoplasmic reticulum, high SR Ca^{2+} and Ca^{2+}-ATPase found in EOM. EOM may modulate intracellular free Ca^{2+} levels by both passive and active mechanisms since most EOM fibers are high in Ca^{2+} binding proteins and mainly express the fast isoform of Ca^{2+}-ATPase.[5] The differences in calcium characteristics may be particularly important in understanding the resistance of EOM to degeneration in muscular dystrophy, which may be in part mediated by calcium overload.[6]

Address for correspondence: Henry J. Kaminski, M.D., Department of Neurology, University Hospitals of Cleveland, 11100 Euclid Avenue, Cleveland, Ohio 44106. Voice: 216-368-0250; fax: 216-368-0249.

hjk3@po.cwru.edu

ACKNOWLEDGMENTS

This work was supported by the Office of Research and Development, Medical Research Service of the Department of Veterans Affairs and NIH Grant EY-11998 and Vision Core Grant EY-113373.

REFERENCES

1. FUCHS, A.F. & M.D. BINDER. 1983. Fatigue resistence of human extraocular muscles. J. Neurophysiol. **49:** 28–34.
2. FRUEH, B.R., A. HAYES, G.S. LYNCH & D.A. WILLIAMS. 1994. Contractile properties and temperature sensitivity of the extraocular muscles, the levator and superior rectus, of the rabbit. J. Physiol. **475:** 327–336.
3. PORTER, J.D. et al. Extraocular muscle is defined by a fundamentally distinct gene expression profile. Proc. Natl. Acad. Sci. USA **98:** 12062–12067.
4. RICHMONDS, C.R. & H.J. KAMINSKI. 2001 Nitric oxide synthase expression and nitric oxide effects on contractility in rat lateral rectus. FASEB J. **15:** 1764–1770.
5. AIDLEY, D. 1998. The Physiology of Excitable Cells. Cambridge University Press. Cambridge, England.
6. ANDRADE, F.H., J.D. PORTER & H.J. KAMINSKI. 2000. Eye muscle sparing by the muscular dystrophies: lessons to be learned? Microsc. Res. Tech. **48:** 192–203.

Nitric Oxide and cGMP Modulation of Extraocular Muscle Contraction

HENRY J. KAMINSKI[a,b] AND CHELLIAH R. RICHMONDS[a]

Departments of [a]Neurology and [b]Neurosciences, Case Western Reserve University School of Medicine, Louis Stokes Veterans Affairs Medical Center, University Hospitals of Cleveland, Cleveland, Ohio 44106, USA

Extraocular muscles (EOM) are understood to differ from other skeletal muscles in structural and physiological properties. In keeping with their specialized role in generation of eye movements, they demonstrate considerable fatigue resistance. Nitric oxide (NO) is a ubiquitous cell-signaling molecule involved in regulation of numerous homeostatic functions, and in skeletal muscle NO influences contraction, blood flow, and glucose uptake, making it likely to influence contractile properties of EOM.[1]

We evaluated the role of NO and a mediator of NO action, cGMP, on isometric contractile properties of EOM from Lewis rats. NO synthase inhibition (NG-nitro-L-arginine methyl ester) increased submaximal tetanic and peak twitch forces, and NO donors reduced submaximal tetanic and peak twitch forces.[2] The effect of NO on the contractile force of lateral rectus muscle is greater than previously observed on other skeletal muscle. A cGMP analogue (8-Br-cGMP) produced an increase of 19% in the peak isometric twitch tension and a 12% increase in tetanic tension, while cGMP synthase inhibition (LY83583) produced a 29% reduction in the isometric twitch tension and a tetanic force reduction of 23% (TABLE 1). The reduction produced by cGMP agents was roughly half of that produced by NO modulators, indicating that NO must influence EOM contractility by additional mechanisms.

NO's effect on contractility is greater in EOM compared to other skeletal muscles and this is only partially explained by cGMP modulation of isometric force. The cGMP analogue increased twitch tension and tetanic force of EOM, whereas LY83583 reduced force. Abraham *et al.* and Kobzik *et al.* found that cGMP could account for at most 50% of NO's reduction of tetanic force in diaphragm strips and intact leg muscles.[3,4] Despite the greater magnitude of the effect of NO on EOM, cGMP modulation contributes to a similar degree as in nonocular muscles. Maréchal and colleagues found that 8-Br-cGMP increased the maximum shortening velocity of mouse extensor digitorum longus, but, in contrast to the present study, isometric force was not changed.[5] However, our investigation used a higher concentration of 8-Br-cGMP (1mM vs. 0.1mM), as did Kobzik *et al.*, which may explain the differences. The group did find that LY83583 depressed maximum isometric force by 12 percent. Andrade *et al.* observed in single fibers that LY83583 reduced contractile force without influencing calcium sensitivity.[6]

Address for correspondence: Henry J. Kaminski, M.D., Department of Neurology, University Hospitals of Cleveland, 11100 Euclid Avenue, Cleveland, OH 44106. Voice: 216-368-0250; fax: 216-368-0249.

hjk3@po.cwru.edu

TABLE 1. cGMP effects on contractile force

cGMP agent	Tension (N/cm^2)	p value
8Br-cGMP	1.14 ± 0.06	<0.0001
Treated	1.26 ± 0.06	
LY83583	1.40 ± 0.07	<0.0001
Treated	1.06 ± 0.05	

We suspect that NO influences are particularly important for EOM by limiting muscle fatigue through dampening muscle contractile force, and increasing glucose uptake and muscle blood flow. Further, these findings demonstrate that second messenger systems important in force generation differ between EOM and other skeletal muscles. Such differences may be important in how these muscles respond to muscle disorders, in particular dystrophies and inflammatory myopathies, in which the EOM is spared.

ACKNOWLEDGMENTS

This work was supported by the Office of Research and Development, Medical Research Service of the Department of Veterans Affairs and by NIH Grant EY-11998 and Vision Core Grant EY-113373.

REFERENCES

1. KAMINSKI, H.J. & R.H. ANDRADE. 2001. Nitric oxide: biologic effects on muscle and role in muscle disease. Neuromusc. Disord. **11:** 517–524.
2. RICHMONDS, C.R. & H.J. KAMINSKI. 2001. Nitric oxide synthase expression and nitric oxide effects on contractility in rat lateral rectus. FASEB J. **15:** 1764–1770.
3. ABRAHAM, R.Z. et al. 1998. Cyclic GMP is a second messenger by which nitric oxide inhibits diaphragm contraction. Comp. Biochem. Physiol. **119A:** 177–183.
4. KOBZIK, L., M.B. REID, D.S. BREDT & J.S. STAMLER. 1994. Nitric oxide in skeletal muscle. Nature **372:** 546–548.
5. MARÉCHAL, G. & G. BECKERS-BLEUKX. 1998 Effect of nitric oxide on the maximal velocity of shortening of a mouse skeletal muscle. Pflüg. Arch. **436:** 906–913.
6. ANDRADE, F.H., M.B. REID, D.G. ALLEN & H. WESTERBLAD. 1998. Effect of nitric oxide on single skeletal muscle fibres from the mouse. J. Physiol. **509:** 577–586.

Ocular Motor Myotonic Phenomenon in Myotonic Dystrophy

MAURIZIO VERSINO,[a,b] SILVIA COLNAGHI,[b] GIORGIO SANDRINI,[a,b] AND VITTORIO COSI[a,b]

[a]*Fondazione Istituto Neurologico C. Mondino, IRCCS, Pavia, Italy*

[b]*Dipartimento di Scienze Neurologiche Università di Pavia, Pavia, Italy*

KEYWORDS: myotonic dystrophy; extraocular myotonia; warming-up phenomenon; ocular motor myotonic phenomenon

INTRODUCTION

Myotonic dystrophy (MyD) is an autosomal dominant multisystem disorder in which skeletal muscles may present dystrophic changes and a delay in muscle relaxation after contraction. This delay is the defining feature of the myotonic phenomenon, or myotonia.[1,2] Several and repetitive muscle contractions may momentarily reduce myotonia in the given muscle, and this has been called the warm-up phenomenon.[3] Although myotonic discharges are electromyographically detectable in extraocular muscles,[4] to our knowledge only one case report has addressed the question of the possible occurrence of extraocular myotonia.[5] The aim of the present report is to demonstrate that saccadic abnormalities may be related both to the extraocular myotonia and to the warming-up phenomenon.

MATERIALS AND METHODS

The study enlisted twenty MyD patients, 10 patients suffering from multiple sclerosis (MS) and 10 controls (Ctrl). All subjects showed visual acuity above 7/10 in both eyes, and none of them presented with clinically evident ocular motor signs. We included the MS group because their ocular motor abnormalities are attributable to a CNS rather than to an extraocular muscle dysfunction. Using scleral search coil technique (Skalar 3010 system), we recorded monocularly visually guided saccadic eye movements simultaneously from both eyes in binocular vision after the signal had been calibrated monocularly in monocular vision.

We used two different target displacements, that is, small (±25 degrees to primary position) and large (50 degrees, back and forth from +25 to −25 degrees), and two different interstimulus intervals (ISIs), that is, short (1 second) and long (5 seconds).

Address for correspondence: Maurizio Versino, M.D., Dipartimento Scienze Neurologiche Università di Pavia, Fondazione Istituto Neurologico C. Mondino, IRCCS, Via Palestro 3, 27100 Pavia, Italy. Voice: +39 0382 380340; fax: +39 0382 380286.

mversino@unipv.it

For each subject, for each eye, for each paradigm, and for each saccade direction, we computed the mean value, the duration, the peak velocity and skewness of the velocity profile, namely, the ratio of the duration of the acceleration period to the duration of the deceleration period. All the parameters were divided by the actual eye displacement (amplitude) values.

We hypothesized that short but not long ISIs would promote the warming-up phenomenon. Accordingly, for each subject and for each eye parameter, we computed the difference between short and long ISIs for corresponding parameters divided by the average of the two values. Finally, since the myotonic phenomenon is likely to involve the extraocular muscles on an individual basis, for each subject we considered the largest (as an absolute value) ISI difference with its sign among the four available (two eyes and two saccade directions) for each small- and large-amplitude paradigm.

RESULTS

The maximal short–long ISI difference mean values for saccade duration were invariably negative, that is, durations were greater for long than for short ISIs. The mean values for the three groups were significantly different both for small ($F_{2,37} = 3.33$, $p = 0.047$) and for large ($F_{2,37} = 7.32$, $p = 0.002$) amplitudes, and contrast analysis showed that MyD mean value was larger than those of the MS (contrast 1: $t_{37} = 2.13$ and $p = 0.04$ for small saccades; $t_{37} = 2.21$ and $p = 0.034$ for large saccades) and of the Ctrl (contrast 2: $t_{37} = 2.1$ and $p = 0.045$ for small saccades; $t_{37} = 3.68$ and $p = 0.001$ for large saccades) groups for both target displacements.

The maximal short–long ISI difference mean values for saccade skewness were negative. In other words, acceleration duration was greater than the deceleration duration for long ISIs than it was for short ISIs: in the MyD group, this was the case both for small and for large target displacements, while in the MS group, it was the case for large but not for small saccades; moreover the skewness maximal differences were positive in the control group. The mean values for the three groups differed significantly for small ($F_{2,37} = 6.94$, $p = 0.003$) but not for large ($F_{2,37} = 2.59$, $p = 0.089$) amplitudes. For small target displacements, contrast analysis showed that the MyD mean value was higher than those of the MS (contrast 1: $t_{37} = 3.19$ and $p = 0.03$) and of the Ctrl groups (contrast 2: $t_{37} = 2.88$ and $p = 0.007$).

The maximal short–long ISI difference mean values for saccade peak velocity were invariably positive, that is, peak velocities were usually higher for short than for long ISIs. The mean values for the three groups did not differ significantly either for small ($F_{2,37} = 0.68$, $p = 0.513$) or for large ($F_{2,37} = 2.45$, $p = 0.1$) amplitudes.

We compared the MyD and the MS groups for differences in abnormal individual maximal short–long ISIs. Any individual value was considered to be abnormal if it exceeded the mean ±2 standard deviations of the control groups. MyD patients showed a greater occurrence of abnormal duration (for small amplitudes only: 60% vs. 20%, $p = 0.045$) and of abnormal skewness (for small amplitudes only: 55% vs. 0%, $p = 0.003$), but not of abnormal peak velocity.

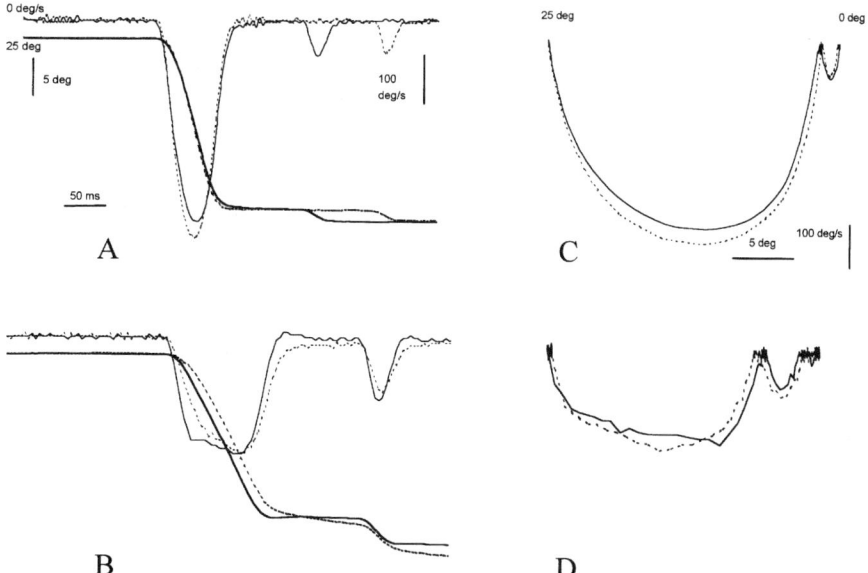

FIGURE 1. A and **B** show the velocity (*thinner lines*) and the position profile (*thicker lines*) of a leftward centripetal saccade to primary position after the 25 degrees position has been kept for 1 sec (short ISI: *dotted lines*) or for 5 sec (long ISI: *continuous lines*). **A** shows a control subject, whereas **B** shows a patient suffering from myotonic dystrophy. The patient makes a saccade which is slower than that of the control subject with both ISIs. In the control subject, both the saccadic duration and the velocity profile are not significantly modified by the difference in ISI, although the peak velocity is slightly lower for long than for short ISIs. In contrast, the patient's saccade reaches the same peak velocity with both ISIs, but with the long ISI the saccade lasts longer; this effect is due to a prolongation of acceleration period, which increases the skewness of the velocity profile. These same points are shown in **C** and **D**, where saccade velocity is plotted against eye position, starting from 25 degrees to primary position (0 degrees) during the same respective saccades shown in **A** and **B**. The control subject (**C**) shows the same profile for both ISIs. In contrast, the patient (**D**) shows a longer acceleration period for long (*continuous line*) than for short (*dotted line*) ISI. Note that this does not hold true for corrective saccades. (From Versino et al.[6] Reprinted with permission from the *Journal of Neurology, Neurosurgery & Psychiatry.*)

DISCUSSION

Our findings showed that saccade duration and skewness were influenced by the the inter-stimulus interval. In the MyD patients, prolongation of the stimulus interval led to an increase in saccade duration, without significant modification in saccade peak velocity, and to a reduction in saccade skewness; this latter derived from an increase in the acceleration duration with respect to the deceleration duration (FIG. 1)[6]. These modifications were more significant for small than for large target displacements, and were larger, and occurred more frequently, in the MyD than in the control or the MS groups. Taken together, these findings may be presumed to derive from a delay in de-contracting an extraocular muscle after fixation, in order to make a saccade in the off-direction for that muscle (for instance: a delay in de-contracting the right lateral rectus to make a leftward saccade starting from a rightward orbital position).

Moreover, the fact that the delay described here occurs more frequently the MyD than in the MS group suggests that this phenomenon likely derives from ocular muscle rather than from CNS dysfunction; we hypothesize that the delay may be considered an ocular motor myotonic phenomenon, although we cannot exclude a myopathic origin.

REFERENCES

1. MORGENLANDER, J.C & J.M. MASSEY. 1991. Myotonic dystrophy. Semin. Neurol. **11:** 236–243.
2. THORNTON, C. 1999. The myotonic dystrophies. Semin. Neurol. **19:** 25–33.
3. COOPER, R.G. et al. 1988. Physiological characterisation of the "warm up" effect of activity in patients with myotonic dystrophy. J. Neurol. Neurosurg. Psychiatry **51:**1134–1141.
4. DAVIDSON, S.I. 1961. The eye in dystrophia myotonica with a report on electromyography of the extra-ocular muscles. Br. J. Ophthalmol. **45:**183–196.
5. HANSEN, H.C. et al. 1993. Evidence for the occurrence of myotonia in the extraocular musculature in patients with dystrophia myotonica. Neuro-ophthalmology **13:**17–24.
6. VERSINO, M., B. ROSSI, G. BELTRAMI, et al. 2002. Ocular motor myotonic phenomenon in myotonic dystrophy. J. Neurol. Neurosurg. Psychiatry **72:** 236–240.

Midbrain Reticular Formation Circuitry Subserving Gaze in the Cat

PAUL J. MAY,[a] SUSAN WARREN,[a] BINGZHONG CHEN,[b] FRANCES J.R. RICHMOND,[c] AND ETIENNE OLIVIER[d]

[a]*Departments of Anatomy, Neurology and Ophthalmology, University of Mississippi Medical Center, Jackson, Missisippi 39216, USA*

[b]*Department of Anatomy and Neurobiology, University of Maryland at Baltimore, Baltimore, Maryland 21201, USA*

[c]*Department of Pharmacy and the Alfred E. Mann Institute of Biomedical Engineering, University of Southern California, Los Angeles, California 90033, USA*

[d]*Laboratory of Neurophysiology, University of Louvain School of Medicine, Brussels, B-1200 Belgium*

KEYWORDS: oculomotor; gaze; reticulospinal; reticulotectal

Physiological experiments in primates have identified a region of the midbrain tegmentum, the central mesencephalic reticular formation (cMRF), which plays a role in gaze control.[1] Specifically, stimulation of this region produces horizontal saccades; recordings from this region reveal neurons with saccade-related activity; and chemical inactivation of this region produces contralateral head tilt and spontaneous saccades.[1,2] More rostral regions of the midbrain reticular formation (MRF), just lateral to the interstitial nucleus of Cajal (InC), have neurons with vertical movement fields, and chemical inactivation of this region produces hypometric vertical saccades.[2] In the macaque, there is an extensive overlap of tectoreticular terminals and reticulotectal cells in the region corresponding to the cMRF.[3] The present study explored whether an area similar to the cMRF is present in the cat by analyzing the distribution of MRF neurons targeting the superior colliculus (SCol) and cervical spinal cord.

These studies were carried out in eight adult cats under approved protocols adhering to the NIH guidelines. In the single-injection animals, 0.1–0.5 µl of a 10% BDA was injected into either the SCol or cervical cord. The SCol was approached dorsally, by aspirating the overlying cortex, while the cervical cord (C1) was approached through the foramen magnum. In the dual-tracer animals, both approaches were used to inject 5% Fast Blue into the SCol and 5% Fluoro-Gold into C1 (0.5 µl each). Standard perfusion and processing techniques were used.[3]

Address for correspondence: Dr. Paul May, Department of Anatomy, University of Mississippi Medical Center, 2500 N. State Street, Jackson, MS 39216. Voice: 601-984-1662; fax: 601-984-1655.

pjm@anat.umsmed.edu

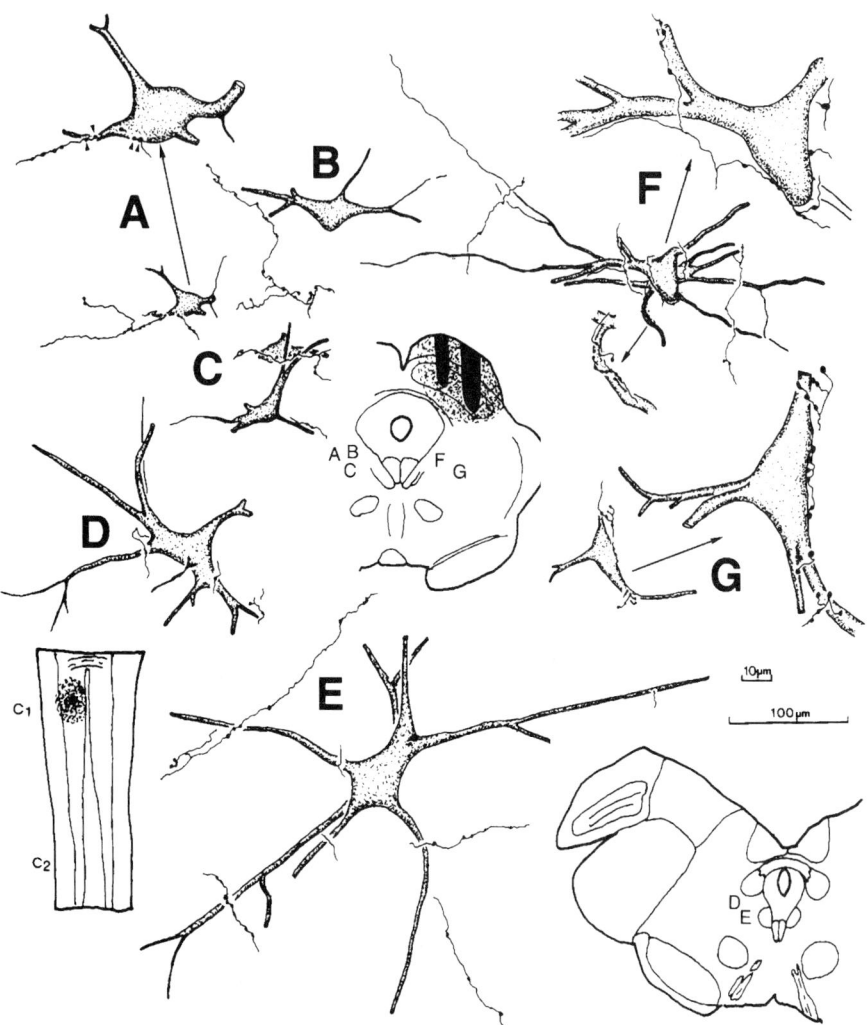

FIGURE 1. A–E show BDA-labeled reticulospinal neurons and spinoreticular axonal arbors in the rostral (**D–E**) and caudal (**A–C**) MRF. **F** and **G** show BDA-labeled reticulotectal neurons and tectoreticular axonal arbors.

Two populations of multipolar neurons were labeled in the MRF after an injection of BDA into the SCol. A few cells were found ipsilaterally, ventral to the nucleus of the posterior commissure (nPC), and appeared continuous with labeled cells within nPC. A second group was labeled bilaterally, caudal to the interstitial nucleus of Cajal (InC), in the MRF ventral to the caudal two-thirds of the SCol. These cells formed a mediolateral band across the MRF. The retrogradely labeled reticulotectal neurons in the caudal group were observed to lie among the labeled axons of the predorsal

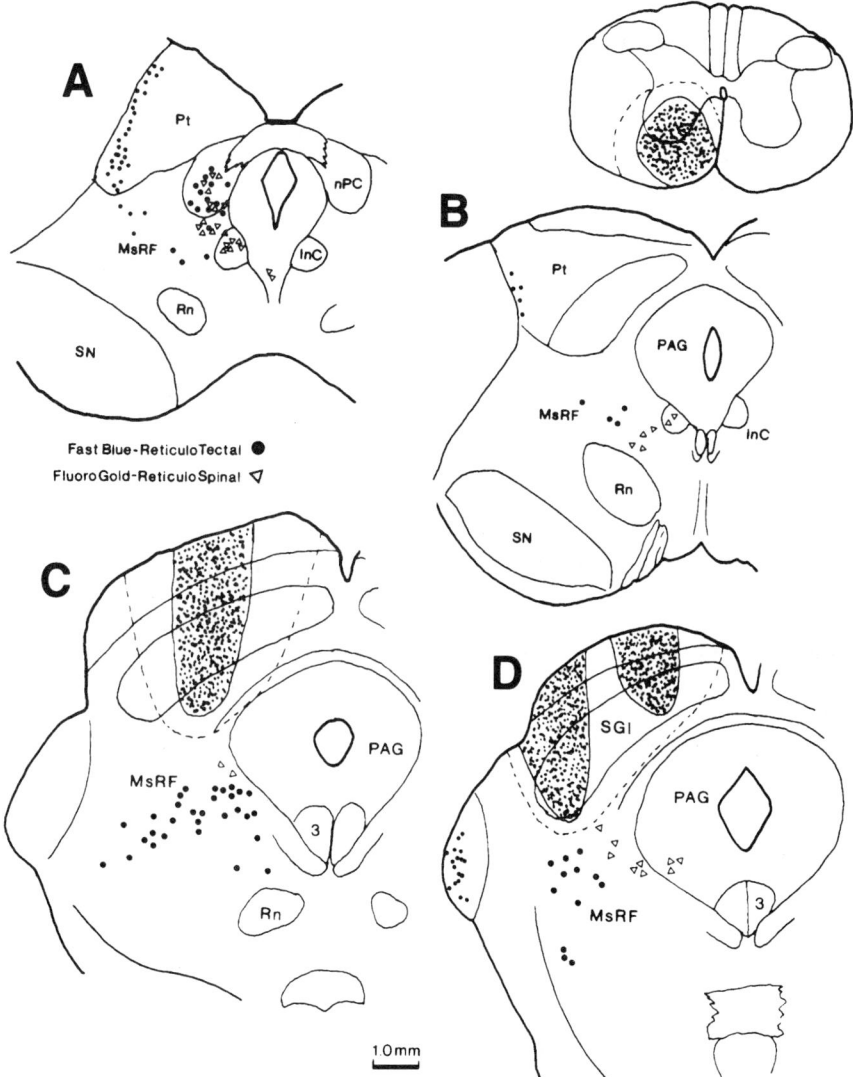

FIGURE 2. Distribution of fluorescent labeling of reticulotectal and reticulospinal MRF neurons.

bundle. Axonal boutons, which were anterogradely labeled with BDA, were often observed in close association with the reticulotectal somata and dendrites, suggesting monosynaptic input from the SCol (FIG.1 F and G).

An examination of the labeled MRF neurons projecting to the cervical spinal cord indicated that these multipolar reticulospinal neurons appear to be grouped into two cell clusters. The rostral cluster was located in the MRF between the nPC and

InC, and was continuous with labeled cells with these nuclei. Cells were present bilaterally, with an ipsilateral predominance. The caudal cluster was located in the MRF beneath the caudal two-thirds of the SCol. Most cells were located in the medial half of the ipsilateral MRF. BDA-labeled axonal boutons were often found in close association with the reticulospinal neurons in the caudal cluster, suggesting monosynaptic input from the cervical cord (FIG. 1A–C). BDA-labeled spinoreticular axons were also present rostrally, but close associations with reticulospinal cells were rarer (FIG. 1 D and E).

The distribution of fluorescently labeled neurons after injections of Fast Blue into the SCol and Fluoro-Gold into the cervical cord showed a similar pattern of label (FIG. 2). The distribution of the Fluoro-Gold-labeled reticulospinal neurons overlapped the more medial portion of the Fast Blue–labeled reticulotectal cell distribution, in both the rostral and caudal clusters. However, no double-labeled neurons were noted. Thus, while there is some overlap between the distributions of the two populations of MRF neurons, individual MRF neurons did not send branched axons to communicate directly with both the SCol and spinal cord.

In conclusion, these results indicate that there are two anatomically separate populations of neurons in the feline MRF. The caudal population may correspond to the physiologically defined cMRF[1,2] based on the fact that the reticulotectal cells are commonly found in association with numerous tectoreticular terminals. Thus, their reciprocal organization mimics that of the primate cMRF.[3] These neurons appear to be part of a monosynaptic circuit that sends a signal much like that carried by the predorsal bundle back to the SCol.[4] The fact that the cMRF reticulospinal neurons were not coextensive with the reticulotectal neurons may indicate that a portion of the cMRF is specialized for controlling gaze-related head movements. Our finding that cMRF reticulotectal cells do not also project to the spinal cord is in agreement with monkey data indicating that these cells lack non-tectal collaterals.[4] The rostral populations of reticulospinal and reticulotectal neurons may be related to vertical gaze, like the adjacent InC and nPC.

REFERENCES

1. COHEN, B., D.M. WAITZMAN, J.A. BÜTTNER-ENNEVER & V. MATSUO. 1986. Horizontal saccades and the central mesencephalic reticular formation. *In* Progress in Brain Research, Vol. 64. H.-J. Freund, U. Buttner, B. Cohen & J. Noth, Eds.: 243–256. Elsevier. Amsterdam.
2. WAITZMAN, D.M., V.L. SILAKOV, S. DEPALMA-BOWLES & A.S. AYERS. 2000. Effects of reversible inactivation of the primate mesencephalic reticular formation. II. Hypometric vertical saccades. J. Neurophysiol. **83:** 2285–2299.
3. CHEN, B.Z. & P.J. MAY. 2000. The feedback circuit connecting the superior colliculus and central mesencephalic reticular formation: a direct morphological demonstration. Exp. Brain Res. **131:** 10–21
4. MOSCHOVAKIS, A.K., A.B. KARABELAS & S.M. HIGHSTEIN. 1988. Structure-function relationships in the primate superior colliculus. II. Morphological identity of presaccadic neurons. J. Neurophysiol. **60:** 263–302.

Activity in the Primate Rostral Superior Colliculus during the "Gap Effect" for Pursuit and Saccades

RICHARD J. KRAUZLIS, NATALIE DILL, AND KRISTA KORNYLO

Systems Neurobiology Laboratory, Salk Institute for Biological Studies, La Jolla, California 92037, USA

KEYWORDS: superior colliculus; pursuit; saccade; gap effect; latency

The latencies of both pursuit and saccadic eye movements are reduced when a fixated visual target is extinguished several hundred milliseconds before a new target appears.[1–6] One possible neural substrate for this "gap effect" is the rostral superior colliculus (SC), which has recently been shown to participate in pursuit eye movements, in addition to its established role in the control of saccades. The discharge rate of buildup neurons in the rostral SC (rSC) is related to the retinal position of targets during pursuit, as well as during the preparation and execution of saccades.[7,8] In addition, microstimulation of the rSC disrupts the initiation of both types of eye movements.[9] To test whether these rostral buildup neurons could underlie the "gap effect," we have now studied their activity with this paradigm during both pursuit and saccades.

Monkeys tracked a target stimulus that appeared either immediately after the offset of the fixation spot (no-gap trials) or 200 msec after the offset of the fixation spot (gap trials). On saccade trials, the target stimulus was stationary and was presented at an eccentricity of ~3.5° along the horizontal meridian. On pursuit trials, the target stimulus appeared at an eccentricity of ~2° and moved toward the center of the display at 15°/sec. We adjusted the starting position of the stimulus to produce saccade-free pursuit and chose neurons with appropriate response fields. We identified the neurons in our sample ($n = 72$) as rostral buildup neurons on the basis of criteria described previously.[7]

The activity of most rostral buildup neurons showed a "gap effect" for both pursuit and saccades. For example, on no-gap trials, the firing rate of the neuron illustrated in FIGURE 1 remained nearly constant at 25 spikes/sec until 50–100 msec after target onset, at which point the firing rate increased to more than 100 spikes/sec. However, on gap trials, the firing rate of this neuron increased after the offset of the fixation spot, reaching a firing rate of 40–50 spikes/sec. These changes in activity

Address for correspondence: Richard J. Krauzlis, Ph.D., Salk Institute for Biological Studies, 10010 N. Torrey Pines Road, La Jolla, CA 92037-1099. Voice: 858-453-4100, ext. 1257; fax: 858-546-8526.

rich@salk.edu

Ann. N.Y. Acad. Sci. 956: 409–413 (2002). © 2002 New York Academy of Sciences.

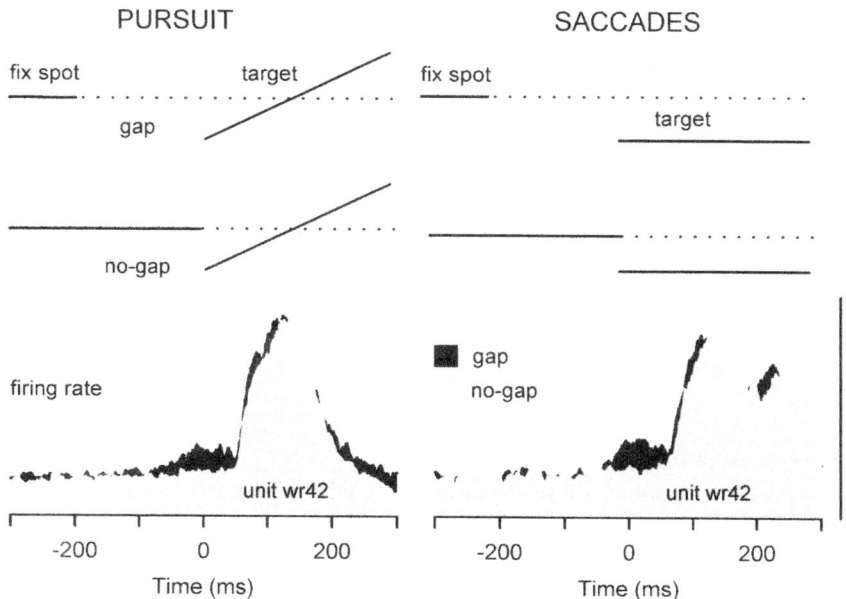

FIGURE 1. Activity of a sample rostral SC buildup neuron during pursuit and saccade trials. At top, traces indicate the timing of fixation spot offset and target onset. *Solid lines* indicate positions of visible stimuli; *dotted lines* indicate positions of stimuli no longer visible. At bottom, average firing rates from gap (*black*) and no-gap (*gray*) trials are shown superimposed. These represent the average activity from 40 gap and 64 non-gap trials during pursuit, and 35 gap and 34 non-gap trials during saccades. The *vertical scale bar* indicates 150 spikes/s.

were most notable around the time of target onset (defined as 0 msec), well before the onset of either pursuit or saccades. To quantify these changes in activity, we measured the activity in two epochs: (1) a "target onset" interval defined as 50 msec before target appearance to 50 msec after target appearance, and (2) a "fixation" interval defined as either the final 100 msec before the fixation spot was extinguished (gap trials) or a matching 100-msec interval starting 300 msec before the fixation spot was extinguished (no-gap trials).

As shown by the scatter plots in FIGURE 2, on gap trials most neurons in our sample had higher activity at target onset than during fixation. We found a significant difference in activity for 57 of 72 neurons prior to the onset of pursuit (solid symbols in FIG. 2A) and 53 of 72 neurons prior to the onset of saccades (FIG. 2B). In contrast, on no-gap trials, we found a significant difference for only 29 of 72 neurons for pursuit (FIG. 2C) and 23 of 72 neurons for saccades (FIG. 2D). This difference in the activity shows that rostral buildup neurons could represent the neural correlate of the behavioral "gap effect" observed previously for both pursuit and saccades.

To test whether these gap-related changes in activity were related to latency, we performed a linear regression analysis between the latency and firing rate measurements pooled across gap and no-gap trials. The results of this analysis for one sam-

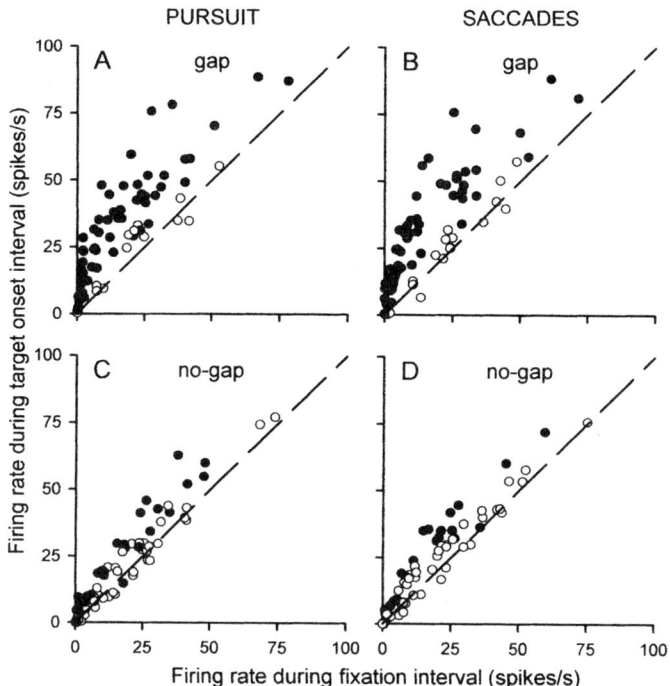

FIGURE 2. Summary of rostral buildup neuron activity during gap and non-gap trials. Average firing rate at target onset interval is plotted as a function of the average firing rate during fixation for gap (**A**) and no-gap (**C**) pursuit trials, and for gap (**B**) and no-gap (**D**) saccade trials. *Solid symbols* indicate significant differences between the two intervals (Wilcoxon rank sum test, $p<0.05$). Data from a total of 72 neurons are shown in each plot.

ple neuron (the same as in FIGURE 1) are shown in FIGURE 3. Pursuit and saccade latencies both exhibited significant negative correlations with firing rate ($r = -0.49$ and -0.53 for pursuit and saccades, respectively). Thus, increases in the activity of this rostral buildup neuron were correlated with the decreases in the latency of pursuit and saccades associated with the "gap effect." Across our sample of neurons, significant negative correlations were found for 25 of 72 neurons for pursuit and 44 of 72 neurons for saccades. Notably, 80% of the neurons that exhibited a significant negative correlation for pursuit also showed a significant negative correlation for saccades (20 of 25).

Our results show that most buildup neurons in the rostral SC increase their activity in the presence of a "gap" prior to either pursuit or saccades. These results indicate that a common process of motor preparation or release of fixation involving the rostral SC could be responsible for the previous findings of a behavioral "gap effect" for both pursuit and saccades.[1–6] The presence of such shared neural processing confirms the suggestion, based on behavioral evidence, that the two types of eye movements have shared inputs.[5] Although the function of these rostral buildup neurons remains unsettled, the increased activity we observed in the presence of a "gap"

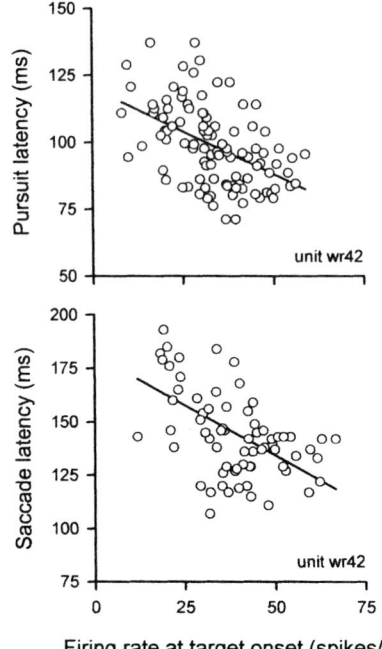

FIGURE 3. Correlation between latency and firing rate at target onset for pursuit (*top*, $n=104$) and saccades (*bottom*, $n = 69$) for the same neuron shown in FIGURE 1. Each symbol indicates a pair of measurements from one trial. The correlations for this neuron from pursuit (-0.49) and saccade (-0.53) trials were both significant ($p<0.05$, ANOVA).

could act to speed the initiation of both types of eye movements toward a common visual target.

ACKNOWLEDGMENTS

This work was supported by NIH Grant EY12212 and by the McKnight Foundation.

REFERENCES

1. SASLOW, M.G. 1967. Effects of components of displacement-step stimuli upon latency for saccadic eye movement. J. Opt. Soc. Am. **57:** 1024–1029.
2. FISCHER, B. & R. BOCH. 1983. Saccadic eye movements after extremely short reaction times in the monkey. Brain Res. **260:** 21–26.
3. FISCHER, B. & E. RAMSPERGER. 1984. Human express-saccades: extremely short reaction times of goal directed eye movements. Exp. Brain Res. **57:** 191–195.
4. KRAUZLIS, R.J. & F.A. MILES. 1996. Decreases in the latency of smooth pursuit and saccadic eye movements produced by the "gap paradigm" in the monkey. Vision Res. **36:** 1973–1985.
5. KRAUZLIS, R.J. & F.A. MILES. 1996. Release of fixation for pursuit and saccades in humans: evidence for shared inputs acting on different neural substrates. J. Neurophysiol. **76:** 2822–2833.
6. KNOX, P.C. 1996. The effect of the gap paradigm on the latency of human smooth pursuit of eye movement. Neuroreport **7:** 3027–3030.

7. KRAUZLIS, R.J., M.A. BASSO & R.H. WURTZ. 2000. Discharge properties of neurons in the rostral superior colliculus of the monkey during smooth-pursuit eye movements. J Neurophysiol. **84:** 876–891.
8. KRAUZLIS, R.J., M.A. BASSO & R.H. WURTZ. 1997. Shared motor error for multiple eye movements. Science **276:** 1693–1695.
9. BASSO, M.A., R.J. KRAUZLIS & R.H. WURTZ. 2000. Activation and inactivation of rostral superior colliculus neurons during smooth-pursuit eye movements in monkeys. J. Neurophysiol. **84:** 892–908.

Time-Optimality and the Spectral Overlap of Saccadic Eye Movements

M.R. HARWOOD[a] AND C.M. HARRIS[a,b]

[a]*Department of Ophthalmology, Great Ormond Street Hospital, London, UK*
[b]*Plymouth Institute of Neuroscience, University of Plymouth, Plymouth, UK*

KEYWORDS: saccade; saccade trajectories; optimization; Fourier transform; main sequence

Saccades exhibit remarkably stereotypical motor behavior, both in their bell-shaped velocity trajectories and in their peak velocity–amplitude and duration–amplitude relationships (the "main sequence"). It is sometimes assumed that this is the result of the system optimizing for movement duration, although no direct empirical evidence as such exists. However, formal time-optimal ("bang-bang") models based on linear control theory[1,2] do not produce realistic velocity profiles, as demonstrated recently in the Fourier domain.[3]

Here, we investigate high-frequency energy spectra for saccades of different amplitude and show that they overlap, indicating the existence of an absolute spectral limit. This provides the first clear empirical evidence that saccades are indeed optimal. The main sequence and trajectory shape reflect the minimum time with a spectral constraint.

METHODS

Reflexive saccades to symmetrical target steps of 2–40 degrees across the midline were recorded from 10 healthy adults using an infrared eye tracker (Skalar Medical). The manufacturer's Bode plot showed negligible attenuation before 100 Hz and a 3dB point at 185 Hz. The Fourier transform of each saccade was computed using padded, cosine windowed, fast Fourier transform of the *unfiltered* eye position.[3,4] The resulting spectra were multiplied by the square of frequency to give velocity energy spectra, and the first two minima (Mn1, Mn2) and maxima (Mx1, Mx2) were identified.

Address for correspondence: Mark Harwood, Department of Ophthalmology, Great Ormond Street Hospital, Great Ormond Street, London WC1N 3JH, UK. Voice: 44-207-405-9200, ext. 0283; fax: 44-207-829-8647.
    mharwood@ich.ucl.ac.uk
    cmharris@plymouth.ac.uk

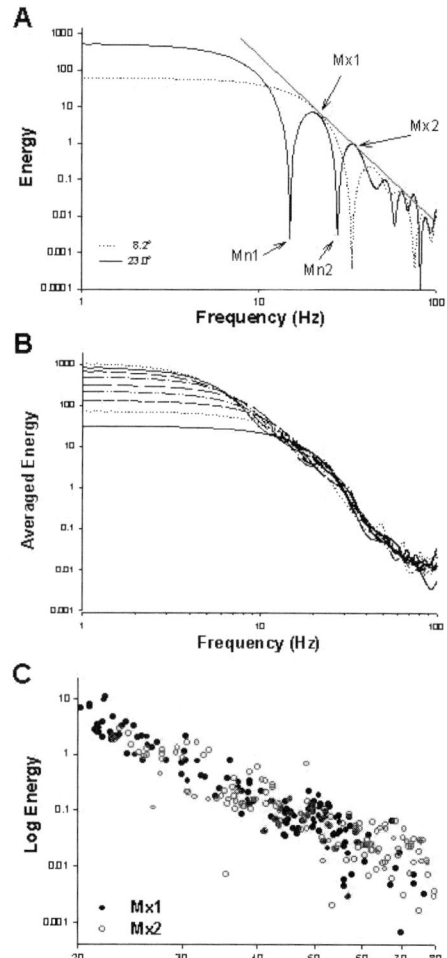

FIGURE 1. Spectral overlap in a typical subject. (**A**) The envelope for energy decline (*straight solid line*) coincides for 8 and 23× saccades; maxima (Mx1 and Mx2) and minima (Mn1 and Mn2) are also shown. (B) Averaged energy in amplitude bins collapses onto single line over 20–80 Hz. (C) Overlap of Mx1 and Mx2 across amplitude.

RESULTS

At low frequencies, energy spectra were flat and, as expected, proportional to amplitude squared. At high frequency individual spectra had maxima and minima at frequencies related to saccade duration (and so amplitude). Surprisingly the decline in the envelope of the energy spectra at high frequencies (20–80Hz) collapsed onto a single line regardless of amplitude. This overlap was seen from plotting maxima energy and frequency and in the smoothed-out averaged spectra. This overlap was approximately linear on log–log plots, until becoming submerged in recording noise above ~80Hz (FIG. 1).

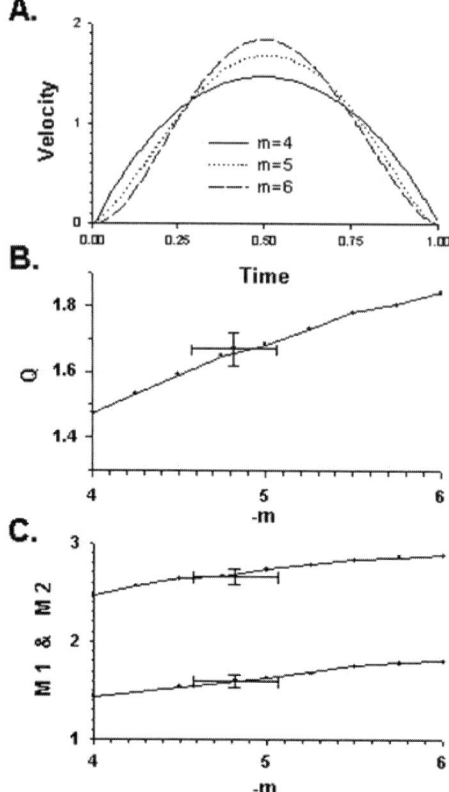

FIGURE 2. Optimal trajectories constrained by different overlap slopes. (**A**) Optimal velocity shapes that minimized time whilst not exceeding spectral slopes of $m = -4$ to -6. (**B**) The ratio of peak to mean velocity, Q, for optimal simulations and for data with 95% confidence limits, and (**C**) similarly for Mn1 (*lower curve*) and Mn2 (*upper curve*) vs. overlap slope, m.

There were no significant differences between intersubject slopes and intercepts of Mx1 or Mx2 over the linear 20–80 Hz range (MANOVA: $F_{2,17} = 0.67$; $p > 0.5$). Mx1 and Mx2 were pooled for each subject and the overall mean slope, m, was -4.82 (0.11) and intercept, k, 7.44 (0.15). Thus, for amplitude (A) and duration (T) the energy envelope (E) can be well described by $E_{A,T}(\omega) = 10^k \omega^m$. For $A < {\sim}15°$, velocity profiles were shown to be approximately self-similar (i.e., re-scaled versions of a template shape, see also Ref. 5). In the Fourier domain, re-scaling of a unit spectrum follows: $E_{A,T}(\omega) = A^2 E_{1,1}(\omega T)$. Hence, for self-similar overlap, $T = cA^{2/-m}$. For $A < {\sim}15°$, a power law fitted the T-A data well, and the mean exponent of 0.40 predicted an overlap slope of $m = -5$. The data fell within 95% confidence limits.

Numerical optimization found the normalized, parameterized polynomial (for order, $n = 5$–10) that minimized time given spectral overlap slopes of $m = -4$ to -6. The optimal trajectories were bell-shaped and symmetrical, and quantitative measures of their shape also fell within the data 95% confidence limits (FIG. 2).

DISCUSSION

The saccade system appears to be maximizing spectral energy up to some limit given by the overlap. As shorter durations for a given shape and amplitude would have spectra exceeding the limit, the implication is that saccades are time-optimal with this spectral constraint. An absolute spectral overlap requires the (non-linear) scaling of both duration and velocity, and predicts power law main sequences for self-similar saccades. This fits the data until symmetry breaks down for $A > 15°$, but the overlap remains, suggesting that it is the underlying constraint. If a spectral limit governs the time-optimal amplitude scaling of saccades, could the shape of their trajectory also be time-optimal given the same constraint? Numerical optimization confirms this suggestion (FIG. 2).

Thus, without any model-contingent assumptions, the current data strongly support the notion that saccades are time-optimal with a spectral constraint, and that temporal main sequences and trajectory shapes are interdependent. The reason for the spectral constraint is unknown. It could reflect a physiological limit in the neuromuscular pathway, although not that of "bang-bang" control.[3] Alternatively, it could reflect a further constraint such as accuracy in the presence of signal-dependent noise[4] or plant uncertainty (multiplicative noise).[6] Then, faster movements would incur greater output variance, and a speed-accuracy trade-off would tend to limit high-frequency energy.

REFERENCES

1. LEHMAN, S.W. & L. STARK. 1983. Multipulse controller signals. II Time optimality. Biol. Cybern. **48:** 5–8.
2. ENDERLE, J.D. & J.W. WOLFE. 1987. Time-optimal control of saccadic eye movements. IEEE Trans. Biomed. Eng. **34:** 43–55.
3. HARWOOD, M.R., L.E. MEZEY & C.M. HARRIS. 1999. The spectral main sequence of human saccades. J. Neurosci. **19:** 9098–9106.
4. HARRIS, C.M. 1998. The Fourier analysis of biological transients. J. Neurosci. Methofds **83:** 15–34.
5. ABRAMS, R.A, D.E. MEYER & S. KORNBLUM. 1989. Speed and accuracy of saccadic eye movements: characteristics of impulse variability in the oculomotor system. J. Exp. Psychol. Hum. Percept. Perf. **15:** 529–543.
6. HARRIS, C.M. 2001. Temporal uncertainty in reading the neural code (multiplicative noise). Presented at the 4th International Workshop: Neural Coding 2001, Plymouth, UK.

Knowledge of Future Target Position Influences Saccade-Associated Head Movements

JOHN S. STAHL

Departments of Neurology, Case Western Reserve University, and Cleveland Veterans Affairs Medical Center, Cleveland, Ohio 44106, USA

Keywords: eye-head; gaze saccades; visual scanning

The factors influencing the decision to move the head in association with a saccade remain poorly understood. We have explored the possibility that subjects are more likely to move the head if they expect that their visual attention will remain in the vicinity of the new target for some time, and less likely if they expect/intend to return gaze to the previous position following a brief glance at the new target. We used magnetic search coils to record eye and head angles in six normal subjects making saccades from a central illuminated target to targets appearing randomly along a semicircular array spanning ±90°. Two different blocks of stimuli were presented in each experimental session. In the short-dwell block, each test target was lit for 800–1000 msec and then the center LED was re-illuminated. In the long-dwell block, each test target was illuminated for 1400 msec and then was followed by a sequence of 3–7 "peri-test" targets (each lighting for 600 msec, located within ±2° of the test target), after which the center LED was re-illuminated. Subjects were instructed to fixate every lighted target as quickly and accurately as possible, moving the head or not as they desired. Subjects were explicitly instructed as to the difference between the two stimulus patterns, so as to encourage their forming expectations regarding the future direction/duration of gaze. The variation in stimulus timing affected head movement probability to different degrees in the six subjects. FIGURE 1 plots head movement amplitude versus target eccentricity for one subject who exhibited strong modifications in head movement probability. In the short-dwell condition (bottom panel) the range of target eccentricities over which the head movement was likely to be small or non-existent was appreciably broader than in the long-dwell condition (top panel). We quantified the tendency to move the head using the eye-only range (EOR), the span of target eccentricity with respect to the initial head position within which the probability of moving the head in conjunction with the saccade was less than 0.5.[1] Five of six subjects demonstrated a wider EOR (lesser tendency to move the head) in the short-dwell condition. Of the five, the increase in EOR ranged from 4.0–148% (median 13.4%). The one subject who showed the opposite effect exhib-

---

Address for correspondence: John S. Stahl, M.D., Ph.D., Department of Neurology, University Hospitals of Cleveland, 11100 Euclid Avenue, Cleveland, OH 44106-5040. Voice: 216-791-3800, ext. 5235; fax: 216-421-3040.

jss6@po.cwru.edu

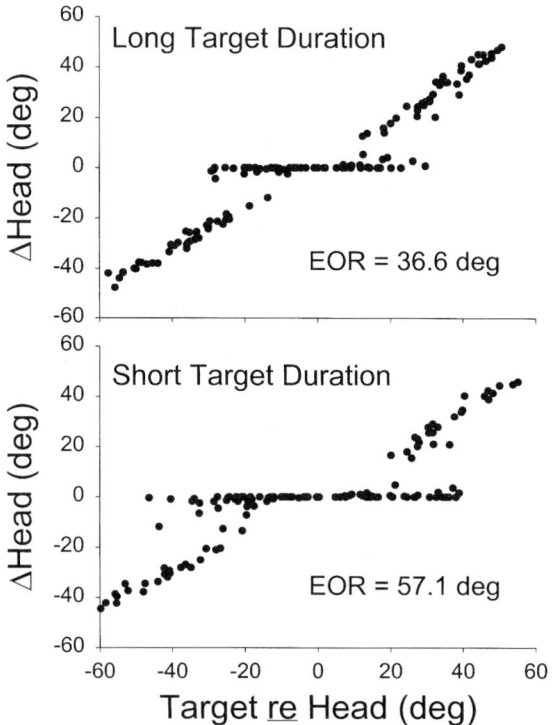

FIGURE 1. Plots of amplitude of saccade-associated head movement versus target eccentricity with respect to initial head position, taken from a single session of a subject who exhibited strong differences in head movement probability in the two stimulus blocks. *Top*: long-dwell stimulus block. *Bottom*: short-dwell stimulus block. EOR values are indicated in each plot.

ited a strong tendency to move the head in both long- and short-dwell conditions (EOR values of only 8.6 and 0.0°, respectively). Averaging all six subjects, EOR measured 44 ± 39° (mean ± SD) and 55 ± 38° for the long- and short-dwell stimuli, respectively. Thus, expectations of future gaze location can influence the decision to move the head, although the size of the effect varies widely. Innate variations in head movement tendencies may play a role in the wide variation of the effect size. Subjects who fall at the extremes of the spectrum of head movement tendencies (strong head movers or strong non-head movers) may be less affected by varying the expected duration of gaze. Studies of the task-related influences on head movement probability must accordingly either focus upon subjects with intermediate head movement tendencies, or modify other stimulus parameters so as to achieve an intermediate probability of moving the head in the context of the experiment.

ACKNOWLEDGMENT

This work was supported by the National Eye Institute, NIH grant EY-00356.

REFERENCE

1. STAHL, J. 1999. Amplitude of human head movements associated with horizontal saccades. Exp. Brain Res. **126:** 41–54.

Normal and Abnormal Slowing of Saccades

Are They One and the Same Phenomenon?

R.C. McGIVERN,[a] J.M. GIBSON,[b] D.W. JENNINGS,[a] K. LAVERY,[a] AND C. MONTGOMERY[b]

[a]*Medical Physics Agency, and* [b]*Department of Neurology and Neurophysiology, Royal Victoria Hospital, Belfast BT12 6BA, Northern Ireland*

KEYWORDS: internuclear ophthalmoplegia; saccades; infrared oculography

INTRODUCTION

Horizontal saccadic eye movements have been used to provide objective quantitative measurements of ocular motor function.[1–3] The most commonly used saccadic parameters are peak velocity, duration, and latency, a reduction in peak velocity with an associated increase in duration being accepted as potential evidence of pathology. However, similar changes are also observed in normal volitional (anti, remembered) saccades.[4] The aim of this work was to investigate the differences between saccade slowing in normal volitional saccades and that found in pathologically slowed saccades.

SUBJECTS AND METHODS

Infrared oculography was performed on 47 normal control subjects (aged 20–55 years; 29 female and 18 male). All were healthy with no past history of squint, head injury, or neurological problems, and none was taking any drugs known to affect eye movements. The patient group consisted of 10 patients (aged 32–67 years; 5 male and 5 female) clinically diagnosed as having incomplete right internuclear ophthalmoplegia (RINO). All subjects were seated in a hydraulic chair so that eyes were level with a tangential screen 1.5 m away. The head was rested on a neck pad and restrained with head tongs. The stimulus was produced by a series of light-emitting diodes (LEDs) positioned centrally and at positions on either side in the range left 10 degrees to right 10 degrees. Both reflexive and volitional saccadic stimuli were used. In the reflexive case, patients were instructed to track a randomly moving visible target. In the volitional case, two paradigms were employed, anti and remembered. For anti-saccades, subjects fixated a central LED and were then instructed to move their eyes in an equal and opposite direction to an eccentric target. For the remembered saccades, subjects were instructed to move to a target position after the

---

Address for correspondence: Dr. R.C. McGivern, Medical Physics Department, Royal Victoria Hospital, Grosvenor Road, Belfast BT12 6BA, Northern Ireland. Voice: 44-28-90263171; fax: 44-28-90322290.

Canice.mcgivern@royalhospitals.n-i.nhs.uk

target had been extinguished and on receipt of an audio stimulus. After calibration, typically 100 reflexive and volitional saccades were recorded for each control subject, while 100 reflexive saccades only were recorded for each member of the patient group. Eye-position data were sampled at 1 kHz after low-pass analogue filtering. Position and target data were displayed in real time, allowing the operator to monitor patient and equipment performance. After recording, individual saccades were selected for analysis. Data for latency, peak velocity, and duration were calculated for all selected saccades (saccade onset and offset were defined from the 20-deg s^{-1} velocity thresholds). In addition, the acceleration time τ_a, defined as the time from saccade onset to maximum velocity, and retarding time τ_r, defined as the time from maximum saccade velocity to saccade offset were evaluated for each saccade.

RESULTS

Results for the right eye only are presented here. In all cases saccade peak velocity, duration, τ_a and τ_r were pooled and averaged over 1-degree intervals between 0 and 10 degrees and 2-degree intervals from 10 degrees to 20 degrees.

Control Subject Group

FIGURE 1A gives a typical main sequence velocity plot for both reflexive and volitional saccades, clearly demonstrating the slowing of volitional with respect to the reflexive saccades. τ_a and τ_r values for the control group are shown in FIGURES 1B and 1C. From this it is apparent that the increase in duration observed in volitional saccades relative to reflexive saccades arises from an increase in saccade retardation time only.

Patient Group

FIGURE 2a displays the main sequence velocities for the control group and the RINO patient group, the asymmetry observed being typical of this condition. Unlike the case for the control subjects, FIGURES 2B and 2C indicate that in the presence of INO, both τ_a and τ_r are increased relative to those values for the control group. This indicates that pathological slowing of saccades is achieved through a mechanism that is different from that observed in normal volitional saccade slowing.

SUMMARY

Saccadic slowing may occur as a result of the testing paradigm or the presence of pathology. The standard peak velocity and duration measurements have difficulty in resolving the nature of this slowing. However, closer study of the saccade velocity profiles allows extraction of the acceleration and retardation times. The variation of these parameters shows distinct systematic differences between normal volitional and pathologically slowed saccades, suggesting that different neural mechanisms underpin the saccadic slowing that occurs with volitional rather than pathological reflexive saccades. We suggest that during volitional but not pathologically slowed reflexive saccades, vision may be preserved to enhance visual search.

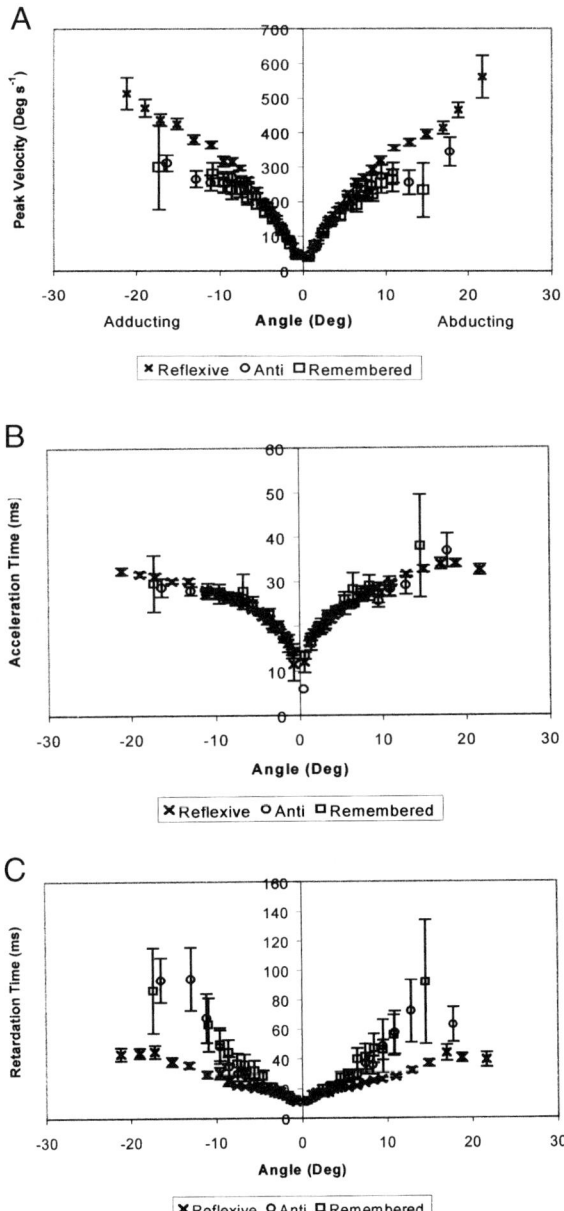

FIGURE 1. (A) Control subjects' peak velocities for reflexive, anti, and remembered saccades. Anti and remembered saccades have a reduced peak velocity relative to that for reflexive saccades. **(B)** Control subjects' acceleration times for reflexive, anti, and remembered saccades. Acceleration times for these saccades are the same. **(C)** Control subjects' retardation times for reflexive, anti, and remembered saccades. Slowing of volitional saccades is accomplished by an increase in retardation time only.

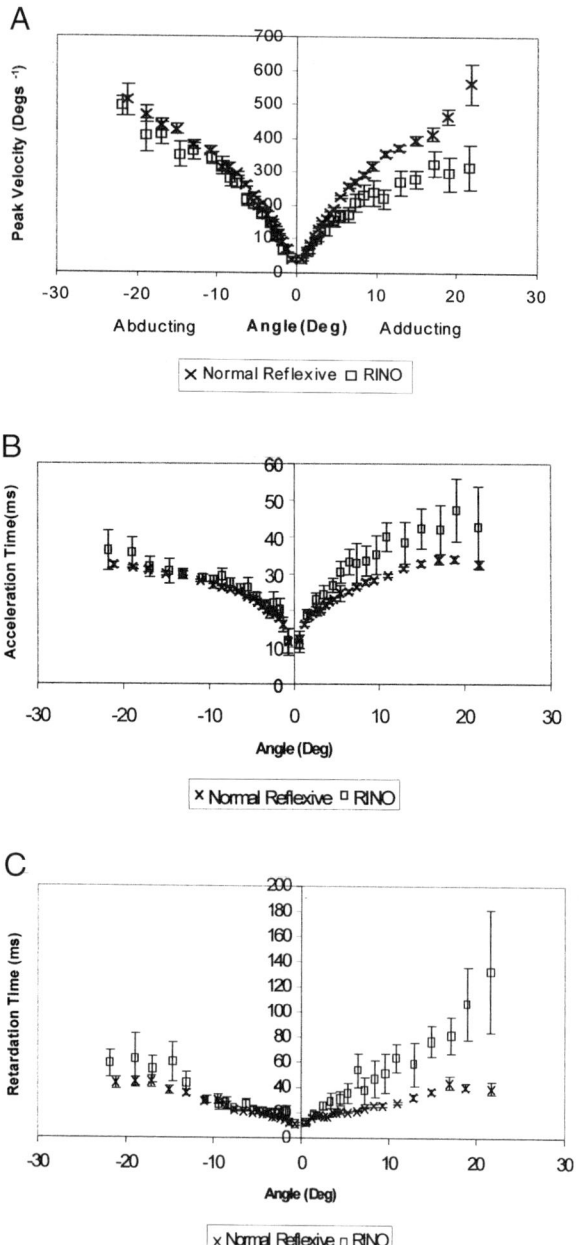

FIGURE 2. (A) Reflexive saccade peak velocities for control and RINO subjects. The RINO subjects display a characteristic reduction in adduction peak velocity. **(B)** Reflexive saccade acceleration times for control and RINO subjects. For RINO subjects, the acceleration time is increased relative to that for control subjects' reflexive saccades. **(C)** Reflexive saccade retardation times for control and RINO subjects. For RINO subjects, the retardation time is increased relative to that for control subjects' reflexive saccades.

REFERENCES

1. BALOH, R.W. *et al.* 1975. The saccade velocity test. Neurology **25:** 1071–1076.
2. BARTON, J.S. *et al.* 1995. Saccadic duration and intrasaccadic fatigue in myasthenia and nonmyasthenic ocular palsies. Neurology **45:** 2065–2072.
3. GIBSON, J.M. *et al.* 1993. Reflex horizontal saccades in multiple sclerosis: a clinically robust quantitative approach. Neuro-Ophthalmol. **13:** 289–295.
4. SMIT, A.C. *et al.* 1986. A parametric analysis of human saccades in different experimental paradigms. Vision Res. **27:** 1745–1762.

Smooth Pursuit Tracking

Saccade Amplitude Modulation during Exposure to Microgravity

J.T. SOMERS,[a] M.F. RESCHKE,[b] A. BERTHOZ,[c] AND L.C. TAYLOR[a]

[a]*Wyle Life Sciences, Houston, Texas 77058, USA*

[b]*Johnson Space Center (SK3), Houston, Texas 77058, USA*

[c]*Collège de France, Paris, France*

KEYWORDS: saccade; smooth pursuit; space flight; main sequence

INTRODUCTION

Russian investigators have reported changes in pursuit tracking of a vertically moving point stimulus during space flight.[1] Early in microgravity, changes were manifested by decreased eye movement amplitude (undershooting) and the appearance of correction saccades. As the flight progressed, pursuit of the moving point stimulus deteriorated while associated saccadic movements were unchanged. Immediately postflight there was an improved execution of active head movements, indicating that the deficiencies in pursuit function noted in microgravity may be of central origin.[1] In contrast, tests of two cosmonauts showed that horizontal and vertical smooth pursuit were unchanged inflight.[2] However, results of corresponding saccadic tasks showed a tendency toward the overshooting of a horizontal target early inflight, with high accuracy developing later inflight, accompanied by an increased saccade velocity and a trend toward decreased saccade latency. On the basis of these equivocal results, we have further investigated the effects of space flight (modified vestibular and sensory-motor function) on the smooth pursuit mechanism.

METHODS

On the first Life and Materials Spacelab flight (LMS-1) we measured smooth pursuit responses elicited by horizontal sinusoidal target movements in four male astronauts. Eye and head rotations were measured using EOG and angular rate sensors, respectively. Sinusoidal target movement was presented horizontally at frequencies of 0.33 and 1.0 Hz. Subjects were asked to perform two trials for each stimulus combination: (1) moving eyes-only (EO) and (2) moving eyes and head (EH) with the target motion. Peak amplitude was 30° for 0.33-Hz trials and 15° for the 1.0-Hz tri-

Address for correspondence: J.T. Somers, 1290 Hercules Drive, Suite 120, Houston, Texas 77058. Voice: 281-483-7485; fax: 281-244-5734.

jsomers@ems.jsc.nasa.gov

Ann. N.Y. Acad. Sci. 956: 426–429 (2002). © 2002 New York Academy of Sciences.

als. The study schedule for this investigation was divided into three phases: preflight, inflight and postflight. Three preflight data collection sessions were obtained 60, 30, and 15 days prior to flight (i.e., L-60, 30 and 15). Inflight pursuit data were recorded, when possible, on flight days (FD) 2, 5, 10, and 16. Postflight data collection began approximately 5 hours following the landing of the Space Shuttle (R+0), and was followed with additional data collections on R+1, 4, and 8. For the purposes of this paper, we only considered corrective saccades elicited during smooth pursuit tracking. In MATLAB, saccade endpoints during pursuit were manually selected for each trial. Saccade amplitude and peak velocity were then calculated from these points. The relationship between saccade amplitude and peak velocity were plotted as a *main sequence* for each phase of flight (i.e., combined preflight [L-60, 30 and 15]; each day inflight [FD2, 5, 10 and 16]; and each day postflight [R+0, 1, 4 and 8]. A linear regression analysis allowed us to determine the slope of each *main sequence* plot. The linear slopes were then combined for each flight phase (preflight, inflight, and postflight) for each individual subject. The linear regression slopes across flight phase allowed us to test for differences between preflight vs. inflight and preflight vs. postflight using an analysis of variance (ANOVA) for each subject.

RESULTS

FIGURE 1 shows the corrective saccade *main sequence* data comparing inflight (FD2, 5, 10 and 16 combined) and postflight (R+0, 1, 4 and 8 combined) behavior to that obtained prior to flight (L-60, 30 and 15 combined) for a single, representative subject. Panel A of FIGURE 1 shows this subject's *main sequence* data for the trials where smooth pursuit was executed with the EO to a 0.33-Hz target. Representative of panel A, panels B, C and D show the *main sequence* data for the EO at 1.0 Hz, the EH at 0.33 Hz, and the EH at 1.0 Hz, respectively. Note that the *main sequence* for both EO and EH trials at both the 0.33- and 1.0-Hz frequencies during flight show a reduction in saccade velocity and amplitude when compared to the preflight *main sequence*. This difference in the regression slopes between flight phase, head/eye condition (EO or EH), and pursuit target frequency was observed across all subjects and was statistically significant at the $p < 0.02$ level ($df = 2$ for phase of flight and $= 1$ for both the head/eye condition and target frequency) or better depending on the condition and subject. It is interesting to note that postflight there was an immediate recovery to the preflight *main sequence* across all conditions (smooth pursuit frequency and the EO and EH conditions). There were no significant differences observed between the preflight slopes for either head movement condition (EO vs. EH). When the immediate postflight (R+0) regression slopes were compared with the preflight slopes, there was a tendency (not significant) for both saccade amplitude and peak velocity to be higher during the postflight testing. This tendency had vanished by R+1. Of particular interest was the redistribution of saccades during the latter stages of the flight and immediately postflight in the EO condition. At the 1.0-Hz frequency the saccades tended to be clustered near the lowest target velocity. It was also interesting to note that gaze performance (eye in skull + head in space) was consistently better during the EH condition; a finding also observed by our Russian colleagues.[1]

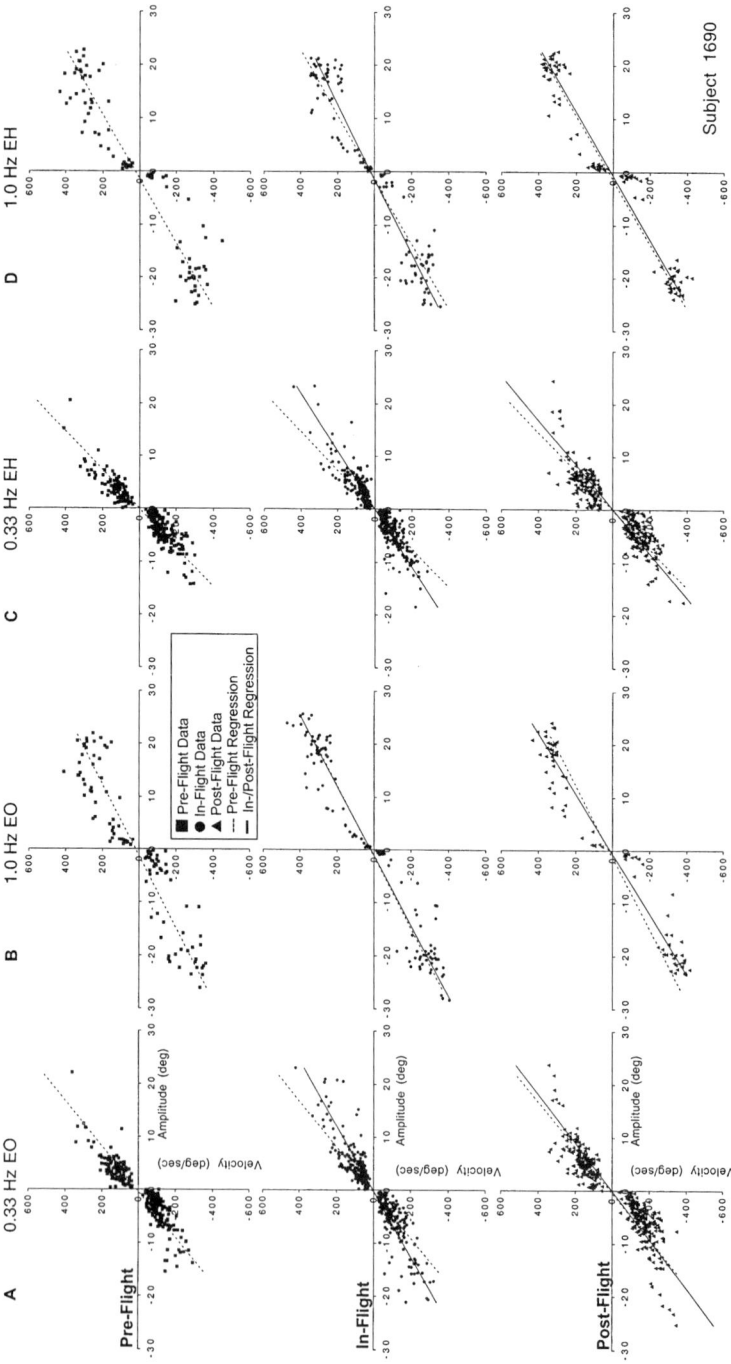

FIGURE 1. Shown are representative *main sequence* plots corresponding to the four test paradigms: (**A**) 0.33 Hz: eyes only; (**B**) 1.0 Hz: eyes only; (**C**) 0.33 Hz: eyes and head; and (**D**) 1.0 Hz: eyes and head. Each set of plots show inflight and postflight *main sequence* data compared to preflight data. In the case of the 0.33-Hz trials (**A** and **B**), saccade velocities during flight were lower than preflight values. For the 1.0-Hz trials (**C** and **D**), a saturation effect is evident. For both frequencies, head contributions do not seem to affect saccade velocities (panels **B** and **D** compared with **A** and **C**, respectively).

DISCUSSION

It is not clear what mechanism is responsible for the decreased peak saccadic velocity during flight unless the change is related to the control of retinal slip. For example, Bloomberg[4] has suggested, on the basis of other research, that saccades tend to initially undershoot their targets by about 10%; and are then followed, if vision is available, by a small augmenting corrective saccade.[5–9] It has been postulated that the functional significance of this undershooting tendency is to maintain the spatial representation of the target on the same side of the fovea (as opposed to racing across the fovea) and hence in the same cerebral hemisphere that initiated the primary saccade, thus minimizing delays caused by an intra-hemispheric transfer of information.[8,10] One could also speculate that with saccade velocities greater than normal, additional corrective saccades would be required to bring the target back on the fovea. A less plausible explanation of our findings could be fatigue.[3] Yet it seems unlikely that our subjects would show lower velocities on all inflight test days while showing increased saccade velocities immediately following space flight, where fatigue is usually the greatest. Finally, the redistribution effect noted late in flight is likely caused by adaptive changes. Overall, corrective saccades appeared to be used in maintaining gaze on target, reducing retinal slip, and assisting the astronauts in maintaining clear vision throughout the different phases of the space flight.

REFERENCES

1. KORNILOVA, L.N. et al. 1993. Vestibular function and sensory interaction in space flight. J. Vestib. Res. **3(3):** 219–230.
2. ANDRÉ-DESHAYS, C. et al. 1993. Gaze control in microgravity. 1. Saccades, pursuit, eye-head coordination. J. Vestib. Res. **3(3):** 331–343.
3. SCHMIDT, D. et al. 1979. Saccadic velocity characteristics: intrinsic variability and fatigue. Aviat. Space Environ. Med. **50:** 393–395.
4. BLOOMBERG, J.J. 1999. Personal communication based on the results of dissertation work.
5. BECKER, W. & A.F. FUCHS. 1969. Further properties of the human saccadic system: eye movements and correction saccades with and without visual fixation points. Vision Res. **9:** 1247–1258.
6. DEUBEL, H., W. WOLF & G. HAUSKE. 1982 Corrective saccades: effects of shifting the saccade goal. Vision Res. **22:** 353–364.
7. FROST, D. & E. POPPEL. 1976. Different programming modes of human saccadic eye movements as a function of stimulus eccentricity: indications of a functional subdivision of the visual field. Biol. Cybern. **23:** 39–48
8. HENSON, D.B. 1979. Investigation into corrective saccadic eye movements for refixation amplitudes of 10 degrees and below. Vision Res. **19:** 57–61
9. PRABLANC, C. & M. JEANNEROD 1975. Corrective saccades: dependence on retinal reafferent signals. Vision Res. **15:** 465–469.
10. HENSON, D.B. 1978. Corrective saccades: effects of altering visual feedback. Vision Res. **18:** 63–67.

Saccadic Palsy after Cardiac Surgery

Visual Disability and Rehabilitation

ROBERT L. TOMSAK,[a] BRUCE T. VOLPE,[b] JOHN S. STAHL,[a] AND R. JOHN LEIGH[a]

[a]*Department of Neurology, Case Western Reserve University, Cleveland, Ohio 44106, USA*

[b]*Burke Medical Research Institute, White Plains, New York, USA*

KEYWORDS: cerebellum; saccades; vergence; eye-head movements

Loss of voluntary gaze, especially saccades, is a recognized, but infrequently diagnosed, complication of cardiac surgery.[1,2] Patients' complaints are often dismissed as psychological, unless the clinician specifically examines horizontal and vertical saccades. Cases reported have had involvement of horizontal or vertical saccades, but reflexive quick phases of nystagmus are usually also involved, implying brainstem, rather than hemispheric, infarction. One patient who died of infection a month after the onset of his saccadic palsy showed midline pontine infarction.[1] The purpose of this report is to show that saccadic palsy after cardiac surgery can cause permanent visual disability, despite adaptive strategies.

Patient P1 was a 41-year-old woman, with no prior neurological problems, who underwent open heart surgery to repair patent ductus arteriosis in October 1999. Postoperatively, she complained of "blurred vision" and a scratchy voice (which was attributed to the endotracheal tube); later she was noted to have slurred speech. She reported difficulty walking, because her eyes would "lock" down or up during locomotion, depending on the predominant flow of visual information. She also noted mild clumsiness of her hands, and a tendency to develop uncontrollable laughter. On examination, visual acuity and fields were normal. Her speech was slow and dysprodic. Cognition, sensation, and tendon reflexes were normal. She was unable to make any voluntary vertical saccades or quick phases. However, vertical pursuit, optokinetic, and vestibular movements were preserved. Once her eyes were carried to an extreme up or down position, it was difficult for her to return them to the center position. She also noted that her eyes would "lock up" after a blink. In contrast, all horizontal eye movements were normal, except for square-wave jerks—small horizontal saccades that disrupted steady fixation.

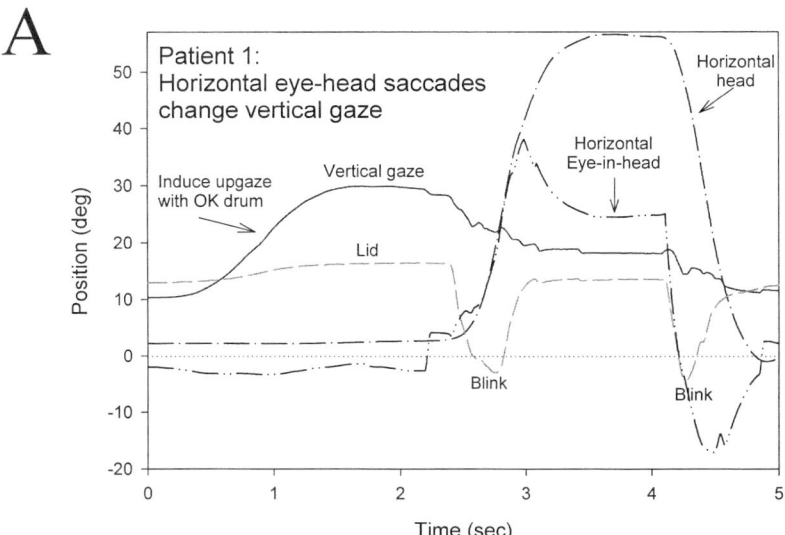

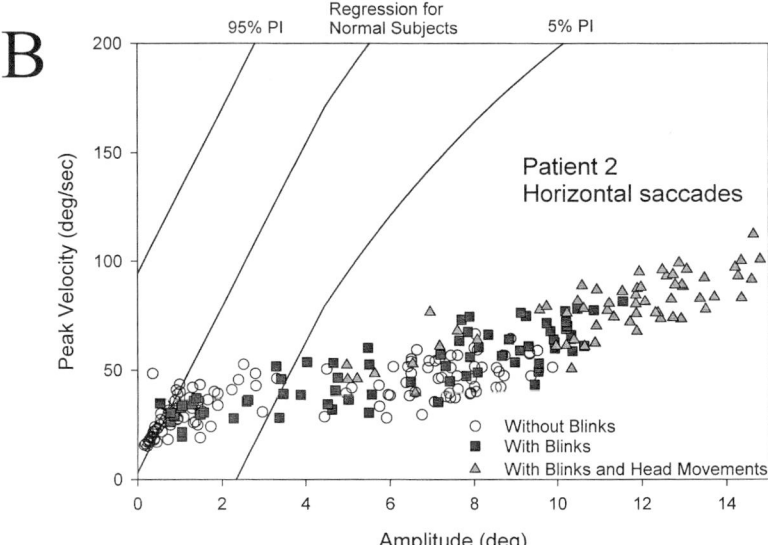

FIGURE 1. (**A**) Example of how a horizontal head rotation induced a vertical gaze shift in patient P1. Rightward deflections indicate rightward and upward rotations. (**B**) Demonstration of how blinks and head thrusts increased the size, but not speed, of saccadic gaze shifts in patient P2.

Patient P2 was 64-year-old man with an ascending aortic aneurysm, who underwent an aortic valve replacement with St. Jude valve in January 1998. Postoperatively, he complained of "blurry vision," unsteady gait, and slurred speech. A neurologist evaluated him six months later and diagnosed hysterical conversion reaction with blepharospasm. On examination, corrected visual acuity and fields were normal. Both horizontal and vertical saccades were small and slow. Vergence, pursuit, and the vestibulo-ocular reflex were preserved, horizontally and vertically.

We studied patient P1 seven months after her operation, and patient P2 two months, and again two years, after his operation. We measured horizontal and vertical movements of both eyes and head using the magnetic search coil technique. We tested fixation, horizontal and vertical saccades, smooth pursuit, and vestibulo-ocular reflex (VOR), vergence, and combined eye-head movements. We also monitored blink movements using a small search coil taped to one upper lid. Both patients gave informed, written consent of the protocol, which was approved by our Institutional Review Board. The goal of these measurements was to document their deficits and identify any adaptive behaviors that could be incorporated into rehabilitative strategies.

Patient P1 was completely unable to initiate vertical saccades. However, she had noted that whenever her eyes became "locked-up," making a large eye-head gaze shift in the horizontal plane would reset her eyes vertically. We confirmed this behavior, and established that a large horizontal eye movement, without a head movement, was ineffective. Accordingly, we developed a rehabilitation strategy in which she was required to make large, horizontal eye-head movements with the goal of improving her ability to voluntarily shift gaze vertically (FIG. 1A). To implement this constraint-induced therapy, she wore glasses that were opaque except for vertical slits. Despite three hours per week of rehabilitation for six weeks (supervised one hour per week), there was no improvement of her voluntary control of vertical saccades. She gave up her job as an evaluator of art, because she could no longer visually scan objects.

Patient P2 showed slow, hypometric horizontal and vertical saccades that appeared to be made more easily if combined with a head movement. Accordingly, he was instructed to practice gaze shifts between two stationary targets using combined eye-head movements. By the time of his second recording, he was able to increase the size of horizontal and vertical gaze-shifts by combining ocular saccades with head thrusts and blinks (FIG. 1B); however, the speed of these movements was not much changed. He also combined a transient convergence movement with each saccade. Despite these adaptations, his inability to voluntarily look around prevented him from driving. In conclusion: saccadic palsy is an underdiagnosed complication of cardiac surgery, which usually arises because of focal brain-stem infarction, and which causes visual disability that often persists, despite rehabilitation therapy.

ACKNOWLEDGMENTS

This work was supported by NIH Grant EY06717, Department of Veterans Affairs, and the Evenor Armington Fund. We thank Drs. J. Posner and P. Zweifach for referring patient P1 and Dr. G. Kosmorsky for referring patient P2.

REFERENCES

1. HANSON, M.R., M.A. HAMID, R.L. TOMSAK, et al. 1986. Selective saccadic palsy caused by pontine lesions: clinical, physiological, and pathological correlations. Ann. Neurol. **20:** 209–217.
2. DEVERE, T.R., A.G. LEE, M.B. HAMILL, et al. 1997. Acquired supranuclear ocular motor paresis following cardiovascular surgery. J. Neuro-ophthalmol. **17:**189–193.

Small Vertical Saccades Have Normal Speeds in Progressive Supranuclear Palsy (PSP)

L. AVERBUCH-HELLER,[a] C. GORDON,[a] A. ZIVOTOFSKY,[a,b] C. HELMCHEN,[c] H. RAMBOLD,[c] U. BÜTTNER,[d] J. BÜTTNER-ENNEVER,[e] AND R. J. LEIGH[f]

[a]*Department of Neurology, Rabin Medical Center, Tel Aviv University, Tel Aviv, Israel*

[b]*Brain Science Program, Bar Ilan University, Ramat Gan, Israel*

[c]*Department of Neurology, Medical University of Lübeck, Lübeck, Germany*

Departments of [d]Neurology and [e]Anatomy, Ludwig-Maximilians University, Munich, Germany

[f]*Department of Neurology, Case Western Reserve University, Cleveland, Ohio 44106, USA*

KEYWORDS: **midbrain; parkinsonism**

Slow vertical saccades are a cardinal feature of PSP that allows differentiation from other parkinsonian disorders.[1] The premotor signals for vertical saccades are generated by burst neurons in the rostral interstitial nucleus of the medial longitudinal fasciculus (riMLF).[2–4] Thus, characterization of the dynamic properties of vertical saccades in PSP presents an opportunity to understand better the pathogenesis of the saccadic palsy. The goal of this study was to follow up on the unpublished observation of Lea Averbuch-Heller, M.D. (1958–2000), that small saccades in PSP appeared to be of normal velocity.

We studied six patients with PSP; their ages ranged from 68–76 years, and mean duration of disease was 1.8 years. All patients gave informed consent. We compared saccadic properties with seven healthy controls (age range 62–75 years) previously studied.[1] We measured eye movements using the search coil technique (bandwidth 0–150 Hz) and defined saccade onset interactively using an eye velocity criterion of 15 deg/sec. We analyzed a total of 1054 vertical saccades from the PSP patients and fitted plots of saccadic amplitude and peak velocity with an equation of the form:

$$\text{Peak velocity} = V_{\max} * (1 - e^{-\text{Amplitude}/C})$$

where $V_{\max}$ is the asymptotic peak velocity, and C is a constant.

The results are plotted in FIGURE 1. Larger saccades made by PSP patients fell below the 95% confidence intervals for normal subjects, but most saccades <5 de-

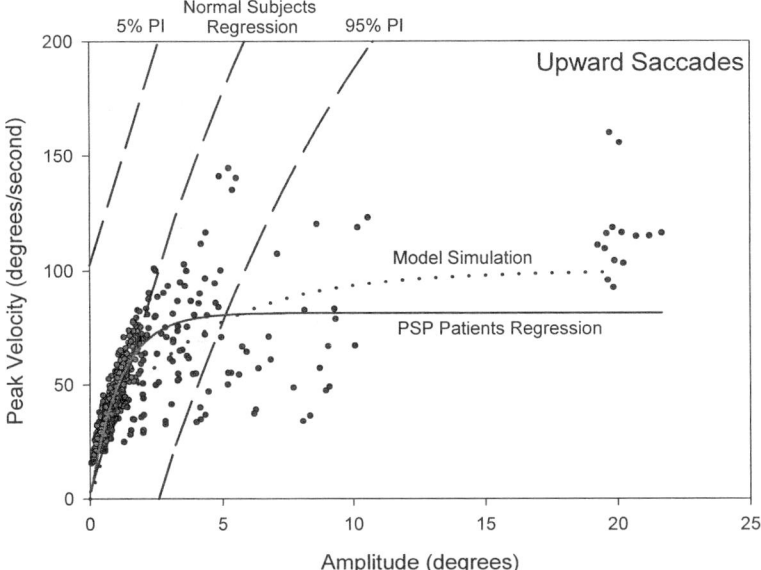
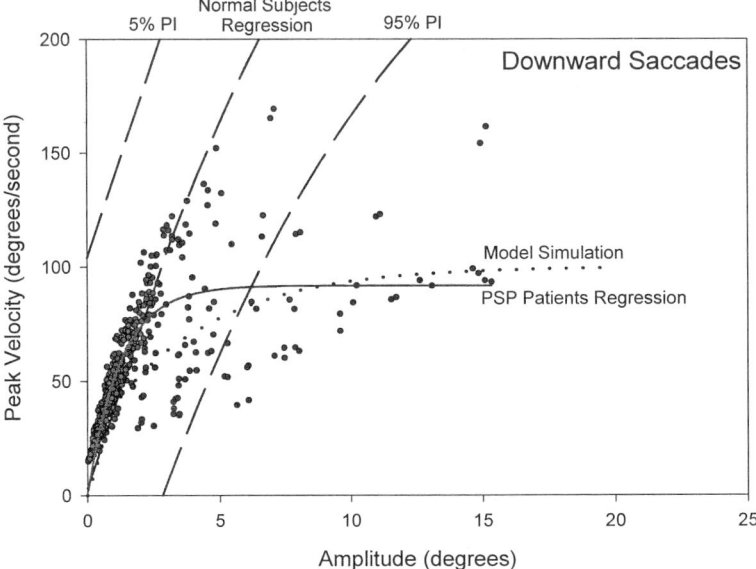

FIGURE 1. Plots of amplitude versus peak velocity of saccades from PSP patients (data points) and prediction intervals (PI) from age-matched normal subjects (*dashed lines*). *Solid lines* show regression fits for PSP patients; *dotted lines* represent model simulations (see text for details).

grees did not. A regression line for the saccades of PSP patients showed a similar slope to that of normal subjects up to 2 degrees. For upward saccades, V_{max} of the curve fit for upward PSP saccades was 81 deg/sec and for downward saccades was 92 deg/sec

We used these findings to test a current model for saccade generation.[5] In this model, the behavior of burst neurons is described by two exponential curves with an overlap region for saccades $< \pm 5$ deg. The physiological basis for this approach is that neurons in the paramedian pontine reticular formation (PPRF) discharge for small saccades in both horizontal directions.[6] The consequence of making burst neuron output the difference between these two curves is that the gain of the slope for small saccades is steep. In order to simulate slow saccades, we first set V_{max} to 100 deg/sec, but simulations with this modification made both large *and* small saccades slow. In order to simulate the normal-speed small saccades that we observed, we decreased the overlap region between the two slopes to ± 1 deg (to preserve the steep slope for small saccades) and decreased the constant C, which describes the rise of the exponential curve, from 10 to 4. Simulations of the model using these settings were similar to the observed data (dotted lines in FIGURE 1). Although our model accounts for our patients' saccadic behavior, other issues need to be addressed. Could it be that burst neurons in riMLF can fire normally for small saccades, but that their discharge saturates for large saccades (a frequency-dependent saturation)? Could the different speeds of small and large saccades be due to different subsets of burst neurons or reflect projections to riMLF from the PPRF region? Alternatively, could abnormal feedback control of burst neurons be responsible? Although burst neurons are likely to be involved, especially in the later stages of PSP, other structures, such as the superior colliculus, might be affected first. Inactivation of the superior colliculus is reported to cause slowing of saccades.[7] Finally, do small and large saccades show similar speed changes with peripheral ocular motor nerve palsies?

ACKNOWLEDGMENTS

This work was supported by the German-Israeli Foundation for Scientific Research, the Department of Veterans Affairs, NIH Grant EY06717, and the Evenor Armington Fund.

REFERENCES

1. ROTTACH, K.G. *et al.* 1996. Dynamic properties of horizontal and vertical eye movements in parkinsonian syndromes. Ann. Neurol. **36:** 129–141.
2. HORN, A.K.E. & J.A. BÜTTNER-ENNEVER. 1998. Premotor neurons for vertical eye-movements in the rostral mesencephalon of monkey and man: the histological identification by parvalbumin immunostaining. J. Comp. Neurol. **392:** 413–427.
3. VILIS, T. *et al.* 1989. On the generation of vertical and torsional rapid eye movements in the monkey. Exp. Brain Res. **77:** 1–11.
4. SUZUKI Y. *et al.* 1995. Deficits in torsional and vertical rapid eye movements and shift of Listing's plane after uni- and bilateral lesions of the rostral interstitial nucleus of the medial longitudinal fasciculus. Exp. Brain Res. **106:** 215–232.
5. ZEE, D.S., E.J. FITZGIBBON & L.M. OPTICAN. 1992. Saccade-vergence interactions in humans. J. Neurophysiol. **68:** 1624–1641.

6. VAN GISBERGEN, J.A.M., D.A. ROBINSON & S. GIELEN. 1981. A quantitative analysis of generation of saccadic eye movements by burst neurons. J. Neurophysiol. **45:** 417–442.
7. HIKOSAKA, O. & R. H. WURTZ. 1986. Saccadic eye movements following injection of lidocaine into the superior colliculus. Exp. Brain Res. **61:** 531–539.

Saccadic and Vestibular Abnormalities in Multiple Sclerosis

Sensitive Clinical Signs of Brainstem and Cerebellar Involvement

DEBORAH L. DOWNEY, JOHN S. STAHL, ROONGROJ BHIDAYASIRI,
JOY DERWENSKUS, NANCY L. ADAMS, ROBERT L. RUFF, AND R. JOHN LEIGH

Neurology Service, Veterans Affairs Medical Center, and Case Western University, Cleveland Ohio 44106, USA

KEYWORDS: internuclear ophthalmoplegia; saccadic dysmetria

Determining the effectiveness of treatments for multiple sclerosis (MS) is complicated because of two factors: (1) the clinical course is unpredictable; and (2) the disease may remain active, as evident on MRI, even though neither the patient nor the physician can detect progression.[1] Thus, there is a need to improve clinical methods for evaluating patients with MS that extend the system introduced by Kurtzke 40 years ago.[2,3] The goal of our study was to determine whether an examination that specifically tests saccades and vestibular eye movements is more sensitive than conventional clinical examinations in identifying brainstem and cerebellar dysfunction in MS.[3,4]

We examined 50 patients (8 female) with MS seen consecutively in our VA outpatient Neurology Clinic between August 1999 and April 2000. The study was approved by our Institutional Review Board. Kurtzke Functional Neurological Status (FSS) and Expanded Disability Status Scale (EDSS) scores in each patient were based on agreement between two examiners.[2,5] We assessed corrected, near visual acuity (Rosenbaum card, expressed as a decimal); visual fields by confrontation; color vision using Ishihara plates (correct responses expressed as decimal of total); pupillary size and reactions; and the presence of optic disc pallor by direct ophthalmoscopy. The eye movements examination consisted of: (1) observing range of movement and covering each eye in turn to test for static ocular alignment; (2) observation of fixation stability in central gaze, and eccentric horizontal, and eccentric vertical gaze; (3) observation of the speed, accuracy, and conjugacy of horizontal and vertical saccades (rapid eye movements) made between two stationary targets (a pencil tip and the examiner's nose); (4) observation of horizontal and vertical smooth

---

Address for correspondence: R. John Leigh, Department of Neurology, Case Western Reserve University and University Hospitals of Cleveland, 11100 Euclid Avenue, Cleveland, OH 44106. Voice: 216-421-3040; fax: 216-231-3461.
rjl4@po.cwru.edu

TABLE 1. Summary of ocular motor scoring system and frequency of abnormal findings

Abnormal finding	Score	Occurrence in AOM group ($n = 22$)
Misalignment of visual axes	1	7 (6 exotropia)
Gaze-evoked nystagmus[a]	1	8
Nystagmus in central position	1	4 (3 pendular, 1 downbeat)
Saccadic dysmetria[b]	1	20 in one plane, 13 in both
Internuclear ophthalmoplegia	1	15 (6 manifestly bilateral)
Smooth pursuit impaired[b]	0.5	7
VOR impaired[b]	1	8
Vergence impaired	1	5

ABBREVIATIONS: AOM, abnormal ocular motor; VOR, vestibulo-ocular reflex.
[a] For a score of 1, sustained bilateral gaze-evoked nystagmus must be present.
[b] Evaluated separately in horizontal and vertical planes; score doubled if abnormal in both planes.

pursuit of a moving target (inaccuracy being determined by the presence of corrective saccades); (5) observation of the vestibulo-ocular reflex (VOR) using small but rapid "head thrusts," and noting occurrence of corrective saccades;[4] and (6) observation of vergence in response to moving a target in the midsagittal plane towards the patient's nose. We developed an "ocular motor scoring system," which is summarized in TABLE 1. We formulated this system on the basis of knowledge that some normal subjects show nystagmus on horizontal eccentric gaze and many normal subjects have mild impairment of smooth pursuit.[4] Accordingly, patients were divided into two groups: those with a normal ocular motor (NOM) score of 2 or less, and those with an abnormal ocular motor (AOM) score of greater than 2. We then compared EDSS disability scores and FSS functional scores in the NOM and AOM groups. Statistical comparisons were made using the Mann-Whitney rank sum test.

We found that 22 of 50 patients (1 female) had AOM scores (median 4.5, mean 5.1), while 28 of 50 patients (7 female) had NOM scores (median 0, mean 0.6). The frequencies of occurrence of findings in the AOM group is summarized in the table. The commonest abnormality was saccadic dysmetria, taking the form of either overshoots (hypermetria) or an inappropriate vector (e.g., requiring a vertical correction following a horizontal saccade). With conventional testing, limitation of movement or ocular deviation was present in 7 patients, all in the AOM group; nystagmus was present in 14 patients, of whom 12 were in the AOM group. Five patients in the AOM group showed no limitation of movement and no nystagmus.

The age of patients in the AOM group (median 49 years) and NOM group (median 47 years) did not significantly differ; nor did the duration of disease (AOM median 22 years; NOM median 28 years). However, the Kurtzke EDSS scores in the AOM group (median 5.2) were significantly greater ($p = 0.02$) than the scores for the NOM group (median 3.5). Kurtzke FSS scores were greater in the AOM group for cerebellar functions ($p = 0.03$) and brainstem functions ($p < 0.01$). However, Kurtzke FSS scores for pyramidal, sensory, bowel/bladder, and mental functions were not different in the AOM and NOM groups. Visual acuity was significantly lower ($p < 0.01$) in the AOM group compared with NOM group (medians 0.65 versus 0.8), as was color vision (medians 0.79 versus 1.0). Of the NOM group, 13 of 28 had bilaterally normal optic discs compared with only 1 of 22 AOM patients. Relative afferent pupillary defects occurred in 3 of 28 NOM patients and 5 of 22 AOM patients.

In summary, we found that MS patients with abnormal eye movements were more disabled than patients with normal eye movements. Most of the signs that we elicited and scored are not part of the standard Kurtzke FSS scores of brainstem or cerebellar function, and five patients in the AOM group would not have been detected by conventional testing because they had full range of movements and no nystagmus. Clinical examination of eye movements, with attention to dynamic properties of saccades and the vestibulo-ocular reflex, takes only a few minutes to perform, but may provide better information concerning the presence of brainstem and cerebellar disease. Prospective studies are required to determine whether development of abnormalities with this testing is predictive of disease activity and progressive disability in MS.

ACKNOWLEDGMENTS

This work was supported by the Department of Veterans Affairs, NIH Grant EY06717, and the Evenor Armington Fund.

REFERENCES

1. SIMON, J.H. *et al.* 1999. A longitudinal study of brain atrophy in relapsing multiple sclerosis. Neurology **53:** 139–148.
2. KURTZKE, J.F. 1961. On the evaluation of disability in multiple sclerosis. Neurology **11:** 686–694.
3. LEIGH, R.J. & J.S. WOLINSKY. 2001. Keeping an eye on MS. Neurology **57:** 751–752.
4. LEIGH, R.J. & D.S. ZEE. 1999. The Neurology of Eye Movements, 3rd ed. Oxford University Press. New York.
5. KURTZKE, J.F. 1989. The disability status scale for multiple sclerosis. Neurology **39:** 291–302.

A Form of Inherited Cerebellar Ataxia with Saccadic Intrusions, Increased Saccadic Speed, Sensory Neuropathy, and Myoclonus

BARBARA E. SWARTZ,[a] MARGIT BURMEISTER,[b] JEFFREY T. SOMERS,[c] KLAUS G. ROTTACH,[d] IRINA N. BESPALOVA,[b] AND R. JOHN LEIGH[a]

[a]*Veterans' Affairs Medical Center and Case Western University, Cleveland 44106, Ohio, USA*

[b]*University of Michigan, Ann Arbor, Michigan, USA*

[c]*Johnson Space Center, Houston, Texas, USA*

[d]*87600 Kaufbeuren, Germany*

KEYWORDS: cerebellum; ataxia; saccades

Over the past five years, rapid progress has been made in genetically identifying different forms of spinocerebellar atrophy (SCA), for which several characteristic disorders of eye movements have been reported.[1,2] Nonetheless, the genetic disorder in some families has not yet been discovered, and this report concerns one such kinship. We studied a family of Slovenian descent in which 5 of 14 siblings presented with progressive ataxia. Two other siblings, who we did not study, may also show involvement, but neither parent and none of 27 children of the 14 siblings has been affected. We examined the five definitely affected patients several times over a 5-year period; electroencephalography, electromyography, and MRI brain imaging were performed. In Patient 5 (the proband), we carried out genetic screening (Athena Laboratories) for Friedreich's ataxia, spinocerebellar ataxia (SCA) 1-3 and 6-8, and Unverricht-Lundborg progressive myoclonic epilepsy (EMP1). We compared results of eye movement studies with a group of 10 normal subjects (2 female; median age 40 years, range 23–60). We measured horizontal and vertical eye movements using the magnetic search coil technique. We tested fixation, horizontal and vertical saccades, smooth pursuit, and vestibulo-ocular reflex (VOR), and vergence, as previously described.[3,4] All patients and subjects gave informed, written consent, as approved by our Institutional Review Board.

Clinical findings of the five affected sibs that we studied are summarized in TABLE 1. Disease onset was insidious; early symptoms could be remembered by all sibs during their early twenties, consisting of gait unsteadiness and difficulty read-

Address for correspondence: Barbara A. Swartz, M.D., Ph.D., Department of Neurology, University Hospitals of Cleveland, 11100 Euclid Avenue, Cleveland, OH 44106. Voice: 216-844-3714; fax: 216-844-5066.

Barbara.Swartz@uhhs.com

TABLE 1. Summary of clinical findings

Case/Sex	Age	Ataxia	Saccadic intrusions	UMN signs	LMN signs	Sensory loss	Disability	Myoclonus
1/M	54/39	+++	1.8/s	++	++	++	+++	+
2/M	52/36	++	3.1/s	+	+	++	++	+
3/F	50/28	+++	2.4/s	+	+	+	+++	+
4-P/M	49/23	++	3.1/s	+	+	++	++	+
5/M	47/34	+	2.0/s	++	−	+	−	+

NOTE: P = proband; age = present age/age at onset; UMN = upper motor neuron; LMN = lower motor neuron; − no disability; + mild disability; ++ ambulates with device; +++ wheel chair.

ing. Eventually, all patients showed gait, trunk, and limb ataxia, as well as pyramidal tract signs with increased reflexes and Babinski plantar responses. Myoclonic jerks were noted in all patients, but electroencephalography has shown no evidence of epilepsy. Fasciculations, impaired joint position sense, cerebellar dysarthria and mild pes cavus were uniform features. Nerve conduction studies showed findings consistent with axonal sensorimotor neuropathy, with evidence of active denervation in the distal muscles of the lower limbs. The optic fundi, visual fields, and corrected visual acuity were normal in all five patients.

The striking disturbance of eye movements was horizontal saccadic oscillations that were induced with each gaze shift (FIG. 1A). These oscillations about the position of the target showed inter-saccadic intervals (i.e., they were not flutter or opsoclonus, but macrosaccadic oscillations). Another abnormality was that the peak velocities of large, but not small, saccades were greater than 95% confidence limits for the control subjects (FIG. 1B). Patients 2 and 4 showed the greatest increase in saccadic velocities and these also showed the most frequent and largest saccadic oscillations with gaze shifts. There was no nystagmus in central or eccentric gaze, and smooth-pursuit and vestibular eye movements were normal.

We found a very consistent clinical picture in five sibs with SCA that has not yet been genetically identified. The pattern of inheritance suggests an autosomal recessive mode of inheritance. Although Friedreich's ataxia may cause a similar presentation, with saccadic intrusions,[5] this was excluded by genetic testing. Other forms of SCA that can be currently tested were also excluded, although the clinical picture —and especially the disorder of eye movements—is not typical for any of them. Linkage studies are proceeding, and should make it possible to determine whether any other families with the same defect have a similar phenotype.

Saccadic oscillations about the point of fixation following a gaze shift are a feature of midline cerebellar lesions (or experimental inactivation) involving the fastigial nucleus.[2,6] These macrosaccadic oscillations should be differentiated from opsoclonus and flutter, which consists of bursts of oscillations without an intersaccadic interval.[2] Flutter was absent in our patients, even when we tried to induce it using the combined saccadic-vergence movements.[4] A novel finding was increased speed of large saccades (FIG. 1B). This finding was most evident in the two patients with most marked macrosaccadic oscillations, suggesting that an increased gain of the saccadic system (due to fastigial nucleus dysfunction) was the cause of the oscil-

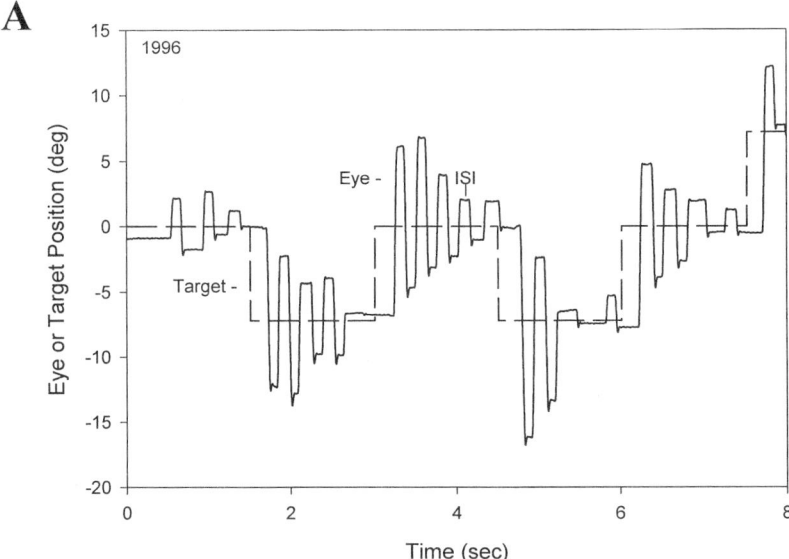

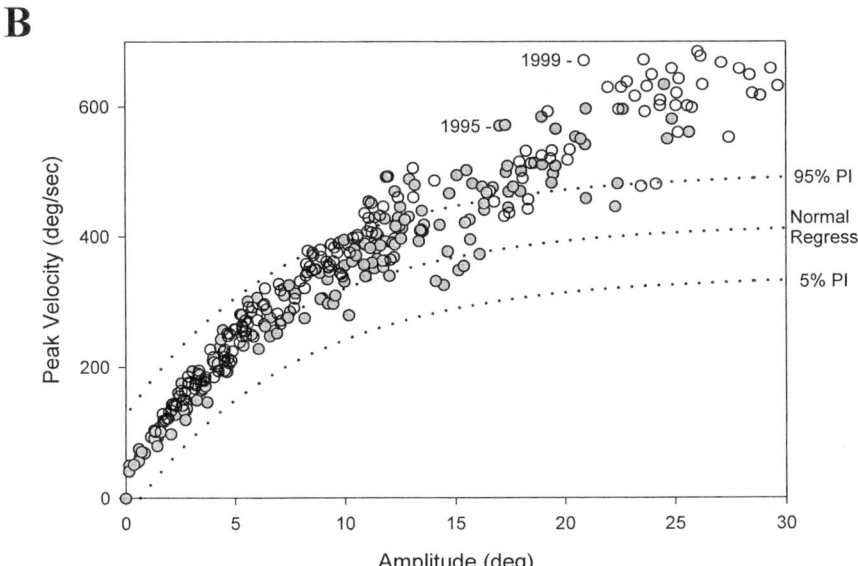

FIGURE 1. (**A**) Representative record of horizontal eye position from Patient 4, showing frequent, large saccadic intrusions during tracking of a stepping visual target; upward deflections indicate rightward movements. (**B**) Amplitude-peak velocity plot of horizontal saccades from Patient 4 during two sessions. Note that large saccades exceeded the 95% confidence interval for normal subjects.

lations. Remarkably, no other abnormalities of eye movements, such as nystagmus or impaired smooth pursuit that are common with vestibular cerebellar disease,[2] were present in our patients. The specificity of the disturbance of eye movements suggests a very restricted form of involvement of the cerebellum that affects the cells of the midline deep cerebellar nuclei.

ACKNOWLEDGMENTS

This work was supported by NIH Grant EY06717, the Department of Veterans Affairs, and the Armington Fund. We are grateful to Dr. Arthur Zinn for genetic advice, Dr. Eugene Dulaney for electromyography, and Drs. Lea Averbuch-Heller and Robert N. Sawyer, Jr. for assisting with evaluation of patients.

REFERENCES

1. EVIDENTE, V.G.H. *et al.* 2000. Hereditary ataxias. Mayo Clin. Proc. **75:** 475–490.
2. LEIGH, R.J. & D.S. ZEE. 1999. The Neurology of Eye Movements, 3rd ed. Oxford University Press. New York.
3. ROTTACH, K.G. *et al.* 1996. Dynamic properties of horizontal and vertical eye movements in parkinsonian syndromes. Ann. Neurol. **36:** 129–141
4. RAMAT, S. *et al.* 1999. Conjugate ocular oscillations during shifts of the direction and depth of visual fixation. Invest. Ophthalmol. Visual Sci. **40:**1681–1686.
5. SPIEKER, S. *et al.* 1995. Fixation instability and oculomotor abnormalities in Friedreich's ataxia. J. Neurol. **242:** 517–521.
6. ROBINSON, F.R., A. STRAUBE & A.F. FUCHS. 1993. Role of the caudal fastigial nucleus in saccade generation. II. Effects of muscimol inactivation. J. Neurophysiol. **70:** 1741–1758.

Anticompensatory Eye Position ("Contraversion") in Optokinetic Nystagmus

S. GARBUTT,[a] M.R. HARWOOD,[a] AND C.M. HARRIS[a,b]

[a]*Department of Ophthalmology, Great Ormond Street Hospital, and Visual Science Unit, Institute of Child Health, London, UK*

[b]*Plymouth Institute of Neuroscience, Plymouth University, Plymouth, UK*

KEYWORDS: optokinetic nystagmus; anticompensatory eye movement; gaze position; beating field; cerebellum

INTRODUCTION

The quick phases of full-field optokinetic nystagmus (OKN) not only reset the eyes, but also move them in an anticompensatory direction (that is, in the opposite direction to stimulus movement, "contraversion").[1–7] Although recognized as an oculomotor phenomenon, contraversion is poorly understood, and it has been suggested as a strategy for directing the line of sight into the visual field from which motion is originating.[5–7] We have previously observed extreme contraversion in patients with absent smooth pursuit and a leaky eye position integrator.[8] In this study we examined contraversion in healthy adults and propose a model for this behaviour based on observation from a clinical group.

METHODS

OKN was recorded from 10 healthy adults and an affected child from a family with a dominant vestibulocerebellar disorder.[8] Horizontal eye movements were measured using an infrared limbus eye tracker and bi-temporal dc-electro-oculography simultaneously. OKN was elicited by rotation of a full-field patterned curtain rightward and leftward at speeds of 2–30 deg/s, in steps of 2 deg/s. The direction and speed of the stimulus was randomized. Subjects were instructed to keep the curtain as clear as possible, but not to track any individual feature. Each step of optokinetic stimulation lasted 20 seconds followed by a 10-second fixation period.

The mean gaze position during stimulus motion was determined for each stimulus speed and direction. This was calculated from the start and end of each quick phase and then subtracted from the steady-state eye position prior to stimulus onset.

Address for correspondence: S. Garbutt, Department of Ophthalmology, Great Ormond Street Hospital, Great Ormond Street, London WC1N 3JH, UK. Voice: +44 207 405 9200, ext. 0283; fax: +44 207 829 8647.

s.garbutt@ich.ucl.ac.uk

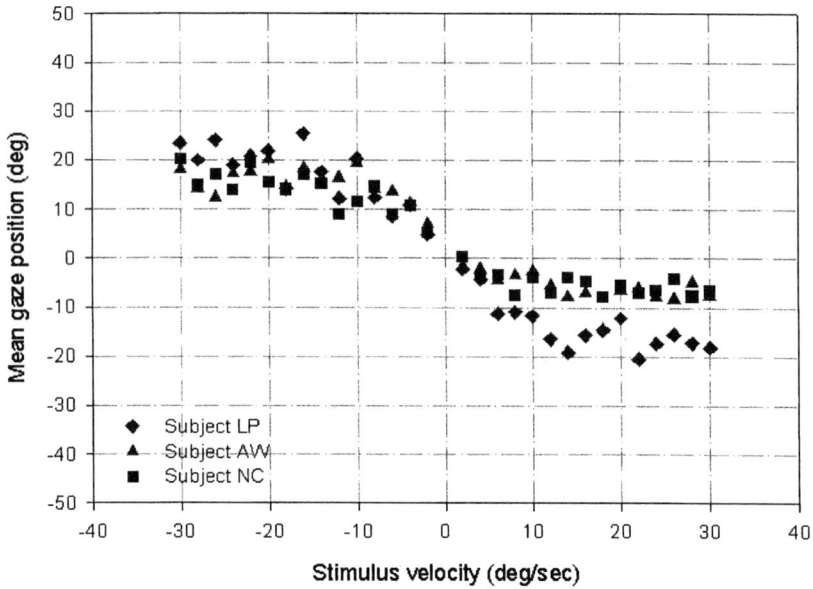
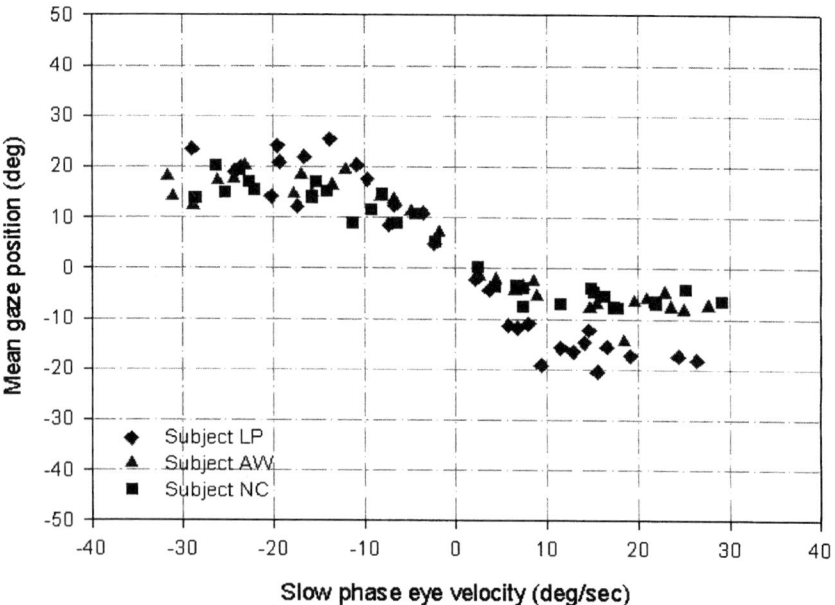

FIGURE 1, A and B. Representative responses in three normal subjects. (**A**) Mean gaze position plotted against stimulus velocity. (**B**) Mean gaze position plotted against slow phase eye velocity.

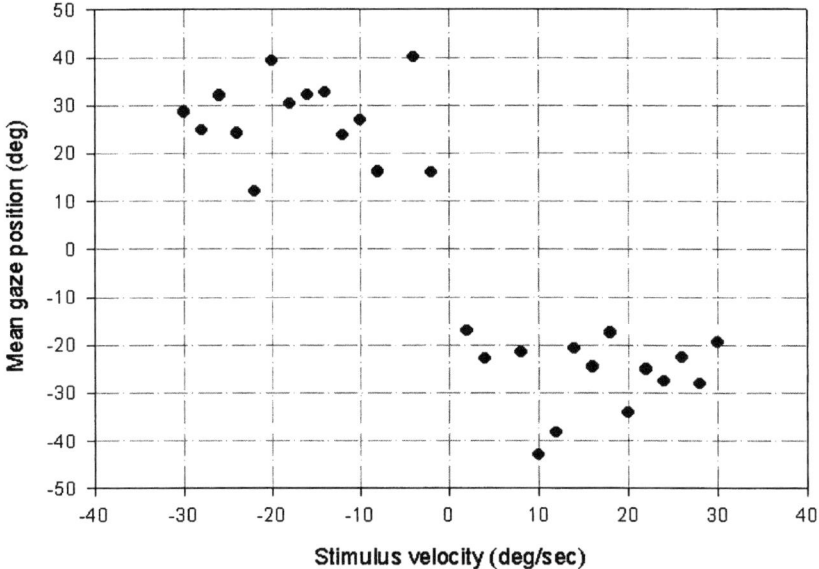

FIGURE 1C. Response from the child with a vestibular-cerebellar disorder. Mean gaze position plotted against stimulus velocity.

RESULTS

The mean gaze always shifted in the direction of the quick phase even at very low stimulus speeds. After onset of stimulus motion, usually two or more OKN quick phases were required to drive the subject's eyes away from the primary position into a new mean position of gaze. In normal subjects contraversion was *not* all-or-none, but showed a gradual increase with stimulus velocity up to about 10–15 deg/s, and saturated for higher velocities (FIG. 1A). In the normal subjects the mean saturated position shift was 12.5 degrees (FIGS. 1A and B). The vestibulocerebellar patient had poor smooth pursuit and on OKN stimulation immediately adopted a contraversive deviation of on average 24.5 degrees (FIG. 1C).

DISCUSSION

In normal subjects contraversion is not an all-or-none phenomenon, but increases with stimulus velocity up to about 10–15 deg/s before saturating. In patients with very poor/absent smooth pursuit (i.e., no fast OKN response) and a very leaky integrator, contraversion is a rapid all-or-none response. Thus, the stimulus for contraversion cannot be slow-phase eye velocity, but may be an internal reconstruction of stimulus velocity (e.g., efference copy) or retinal slip. We propose that the extreme contraversion seen in vestibulocerebellar patients reflects a "vestigial" system to match stimulus velocity with the subjects' own gaze-evoked nystagmus slow phases (pseudo-OKN) (rather than an optokinetic response seen in normals). A similar strategy to enhance velocity-matching could explain the less dramatic and graded contraversion in normal subjects, as velocity matching would be enhanced by even a slightly imperfect integrator. Thus, contraversion matches velocity when precise positioning is not needed or impossible. According to this hypothesis, contraversion, smooth pursuit (fast OKN) and the neural integrator are manifestations of the same function, namely, matching velocity and position as demanded by the visual apparatus or task.

REFERENCES

1. JUNG, R. & R. MITTERMAIER. 1939. Zur objectiven Registrierung und Analyse verscheidener Nystagmusformen: vestibularer, optokinetisher und spontaner Nystagmus in ihren Wechselbeziehungen. Archiv. Ohren. Nasen. Kehlkopfheilkd. **146:** 410–439.
2. MELVILL-JONES, G. 1964. Predominance of anticompensatory oculomotor response during rapid head rotation. Aerosp. Med. **35:** 965–968.
3. MIYOSHI, T., M. SHIRATO & S. HIWATASHI. 1978. Foveal and peripheral vision in optokinetic nystagmus. *In* Vestibular Mechanisms in Health and Disease. J.D. Hood, Ed.: 294–301. Academic Press. London
4. DUBOIS, M.F.W. & H. COLLEWIJN. 1979. Optokinetic reactions in man elicited by localised retinal motion stimuli. Vision Res. **19:** 1105–1115.
5. ABADI, R.V., I.P. HOWARD & M. OHMI. Gaze orientation during full-field and peripheral field passive optokinesis. Ophthalmic Physiol. Opt. **3:** 261–265.
6. THILO, K.V., M. GUERRAZ, A.M. BRONSTEIN, *et al.* 2000. Changes in horizontal oculomotor behaviour coincide with a shift in visual motion perception. Neuroreport **11:** 1987–1990.
7. WATANABE, K. 2001. Modulation of spatial attention with unidirectional field motion: an implication for the shift of the OKN beating field. Vision Res. **41:** 801–814.
8. HARRIS, C.M., J. WALKER, F. SHAWKAT, *et al.* 1993. Eye movements in a familial vestibulocerebellar disorder. Neuropediatrics **24:** 117-122.

Adaptive Control of Saccades in Children with Dancing Eye Syndrome

LAURA E. MEZEY[a,b] AND CHRISTOPHER M. HARRIS[a,c]

[a]*Department of Ophthalmology and Visual Sciences Unit, Great Ormond Street Hospital, and Institute of Child Health, London, UK*

[b]*Department of Psychology A19, University of Sydney, NSW 2006, Australia*

[c]*Plymouth Institute of Neuroscience, University of Plymouth, UK*

KEYWORDS: dancing eye syndrome; opsoclonus myoclonus; saccade; saccadic adaptation; cerebellum; vermis

INTRODUCTION

Dancing eye syndrome (DES), or opsoclonus-myoclonus, is a rare neurological syndrome characterized by opsoclonus (spontaneous chaotic involuntary bursts of back-to-back saccades in all directions) and myoclonus (jerky involuntary movements of the limbs). Onset is usually in childhood, under 3 years of age. Although there can be good recovery from the acute symptoms, long-term neurological sequelae are often found.[1,2] The etiology of DES is as yet unknown and may involve the brainstem and/or the cerebellum. A recent SPECT study of two children with DES has implicated the vermis of the cerebellum, in particular with the generation of opsoclonus.[3]

Shawkat *et al.* (1993) reported that visually guided reflexive saccades recorded from DES patients during the acute phase of their disease are hypermetric (gain >1).[4] Saccadic dysmetria in humans is often associated with lesions of the vermis of the cerebellum and/or the deep cerebellar nuclei.[5–7] Saccadic gain is under adaptive control (AC), which maintains the optimal accuracy of saccades with respect to their target, and it has been shown experimentally that the vermis of the cerebellum and fastigial nuclei in monkeys are necessary for AC.[8] Given possible neuropathology in the cerebellum, and specifically the vermis, it is plausible that patients with DES may have chronic saccade dysmetria consistent with the failure of the saccadic AC system. We set out to test this hypothesis in a group of children and young adults with DES. A gain-decreasing intra-saccadic target perturbation paradigm was used to test the active functioning of the AC system.

Address for correspondence: Laura Mezey, Department of Psychology A19, University of Sydney, NSW 2006, Australia. Voice: 61-2-9351-5953; fax: 61-2-9351-2603.
 lauram@psych.usyd.edu.au
 cmharris@plymouth.ac.uk

TABLE 1. Summary of patient details

Patient	Age (yr) at recording	Age (mo) at presentation	Treatment	Drug-free years	Opsoclonus now?	Learning difficulties?
1	10	~24	ACTH	~8	No	Yes
2	14	21	Pred	10	No	No
3	15	20	Pred	8	No	Yes
4	15	18	ACTH	7	No	Yes
5	17	20	ACTH	10	No	Yes
6	18	22	Unknown	13	No	Yes
7	19	27	ACTH	12	No	No

NOTE: Pred = prednisolone; ACTH = adrenocorticotropic hormone.

SUBJECTS

Seven DES patients (4 male and 3 female) and 7 control subjects were recruited. Informed consent was obtained. Long-term neurological deficits were reported in 5 of the 7 patients, in spite of the apparent recovery from the overt symptoms of DES. No opsoclonus was noted. See TABLE 1 for patient details.

METHODS

Binocular horizontal eye movements were recorded using an infra-red eye tracker (Skalar IRIS). Saccades were elicited to a red laser spot on a blank white screen. The experiment consisted of three phases presented in series, with no breaks: the preadaptive phase (25 trials); adaptive phase (200 trials); and postadaptive phase (25 trials). All testing took place to an initial target step of 10° to the right of the central fixation target. In the pre- and postadaptive phases there was no secondary intra-saccadic target step. These trials provided the baseline data for comparison of saccadic gain before and after adaptation. In the adaptive phase the target made an intra-saccadic target step of 2° backwards, triggered when the eye velocity crossed a 100°/sec threshold, so that the final position of the target was 8° to the right.

RESULTS

Control subjects had an average gain in the preadaptive phase of 0.94. The results from this group of DES patients produced a mixed picture. Baseline gain values from the preadaptive trials revealed that two patients (1 and 7) made saccades that were predominantly hypermetric; two (patients 4 and 6) made grossly hypometric saccades; while the remaining three patients had gain values within normal ranges, as determined from our control group. All control subjects showed a significant decrease in saccadic gain over the course of the adaptive phase of the experiment. The magnitude of gain decrease in control subjects ranged from −0.072 to −0.126 (aver-

TABLE 2. Average primary saccadic gain and percentage of overshooting saccades before and after adaptation, and the magnitude of the change in gain for each DES patient

Subject	Before adaptation (B)		After adaptation (A)		Difference (A−B)
	Gain	% o/s	Gain	% o/s	Gain
1	1.042	86.4	0.924	64.3	−0.118^a
2	0.941	0	0.854	0	−0.087^a
3	0.960	7.1	0.815	0	−0.145^a
4	0.632	0	0.674	0	+0.042
5	0.900	0	0.799	0	−0.101^a
6	0.800	0	0.693	0	−0.107^a
7	1.000	60.3	0.887	42.9	−0.113^a

NOTE: A negative value of gain difference indicates a decrease in gain.
^{a} Significance to level $p<0.01$.

age −0.098). All DES patients except for patient 4 were able to adaptively modify the gain of their saccades over the adaptive phase of the experiment. The magnitude of these gain changes was significant. The levels of gain before and after adaptation for all DES patients, along with the resultant magnitudes of adaptive gain change, are shown in TABLE 2.

CONCLUSIONS

The oculomotor outcome for patients with resolved DES is varied and unpredictable. Saccades may be hypometric, hypermetric, or normometric. This was not related to age or the presence of long-term cognitive impairment. However, adaptive capabilities were generally preserved in spite of chronic dysmetria. This was particularly surprising in the two patients who displayed chronic hypermetria in spite of demonstrating the ability to adaptively eliminate it in these experimental circumstances. This dissociation is consistent with the existence of a gain-setting mechanism that is functionally, and perhaps also anatomically, distinct from the gain-manipulating elements.

REFERENCES

1. POHL, K.R.E., J. PRITCHARD & J. WILSON. 1996. Neurological sequelae of the dancing eye syndrome. Eur. J. Pediatr. **155:** 237–244.
2. RUSSO, C. et al. 1997. Long term neurological outcome in children with opsoclonus-myoclonus. Med. Pediatr. Oncol. **28:** 284–288.
3. OGURO, K., et al. 1997. Opsoclonus-myoclonus syndrome with abnormal single photon emission computed tomography imaging. Pediatr. Neurol. **16:** 334–336.
4. SHAWKAT, F.S. et al. 1993. Eye movements in children with opsoclonus-polymyoclonus. Neuropediatrics **24:** 218–223.
5. SELHORST, J.B. et al. 1976. Disorders in cerebellar oculomotor control. I. Saccadic overshoot dysmetria: an oculographic control system, and clinico-anatomical analysis. Brain **99:** 497–508.

6. VAHEDI, K. *et al.* 1995. Horizontal eye movement disorders after posterior vermis infarctions. J. Neurol. Neurosurg. Psychiatry **58:** 91–94.
7. WAESPE, W. & E. MÜLLER-MEISSER. 1996. Directional reversal of saccadic dysmetria and gain adaptivity in a patient with a superior cerebellar artery infarction. Neuro-ophthalmology **16:** 65–74.
8. OPTICAN, L.M. & D.A. ROBINSON. 1980. Cerebellar-dependent adaptive control of the primate saccadic system. J. Neurophysiol. **44:** 1058–1076.

Integration of Motion Signals for Smooth Pursuit Eye Movements

RICHARD T. BORN AND CHRISTOPHER C. PACK

Department of Neurobiology, Harvard Medical School, Boston, Massachusetts 02115, USA

KEYWORDS: aperture problem; smooth pursuit; motion processing; area MT

How does the brain combine *local* motion measurements to form an accurate description of *object* motion? For example, if a vertically oriented bar moves upward and rightward at a constant velocity, a neuron with a small receptive field positioned along the length of the contour can measure only the rightward component of motion, since the upward component provides no time-varying information within its receptive field. In contrast, cells positioned at the endpoints of the contour can measure motion direction accurately. Because direction-selective neurons early in the visual pathways have small receptive fields, the visual system is constantly faced with this "aperture problem."[1] And since they provide the sole input to subsequent stages of cortical visual processing, which in turn inform the premotor circuitry used for making eye movements, the error in measuring local velocity could be perpetuated. How are these conflicting motion signals—the potentially erroneous signals measured along a contour and the correct signals originating from terminators—ultimately resolved in the visual cortex?

Microelectrode recordings from neurons in the middle temporal visual area (MT) of alert monkeys have shown that the earliest directional responses, beginning about 80 msec after the onset of stimulus motion, primarily encode the component of motion *perpendicular* to the orientation of a contour. That is, they are strongly affected by the ambiguous contour signals. However, the later responses (>140 msec after motion onset) encode the true direction of motion, irrespective of contour orientation. Thus the responses of MT neurons reflect the solution of the aperture problem for motion over a period of about 60 msec.[2]

Given the evidence that MT neuronal signals are important for the initiation of smooth pursuit eye movements,[3–5] we asked whether the time-evolving signals we observed in MT neurons had a behavioral correlate. We had subjects (monkeys and humans) track the center of a single moving bar whose orientation was varied with respect to its direction of motion. The bar was dim green (4.2 cd · m^{-2}; $u' = 0.28$, $v' = 0.59$), and its center was indicated by an isoluminant, red gaussian blob

---

Address for correspondence: Richard T. Born, M.D., Department of Neurobiology, Harvard Medical School, 220 Longwood Ave., Boston, MA 02115-5701. Voice: 617-432-1307; fax: 617-734-7557.

rborn@hms.harvard.edu

(4.2 cd · m^{-2}; $u' = 0.64$, $v' = 0.34$). The subject had to track the blob to within ±3° while eye movements were monitored using a scleral search coil. Eye position and velocity (analog differentiator: low pass, −3 dB at 50 Hz) were digitized at 1 kHz and stored to disk for off-line analysis.

In agreement with previous studies of both human perception[6] and the human ocular following response,[7] the initial direction of pursuit deviated in a direction *perpendicular* to the orientation of the bar, regardless of the true direction of bar motion. We then asked how the relative amounts of contour and terminator information affected the behavior by varying the *length* of the bar on randomly interleaved trials. Increasing bar length had the expected effect of increasing the magnitude of the deviation, as measured by the initial eye acceleration perpendicular to the direction of target motion. This effect of bar length was decreased if the targets were presented eccentrically, consistent with the predicted effects of larger receptive field sizes with increased eccentricity (magnification factor).

To study such scaling effects more thoroughly, we next used long bars of varying *widths*, which had the effect of symmetrically displacing the ambiguous contour signals various distances from the fovea. The "bars" for these experiments were actually parallelograms (i.e., end contour parallel to direction of bar motion), chosen so that we did not introduce potentially disambiguating contour signals at the ends of the bar. These results showed the same effect described above (i.e., smaller deviations with greater eccentricities) and also afforded us the opportunity to examine longer periods of pursuit uninterrupted by saccades. Interestingly, the earliest period (0 to 40 msec from the onset of pursuit) showed a relatively shallow, nearly *linear* decrement in perpendicular eye acceleration as a function of eccentricity, whereas the later period (40 to 80 msec) showed a much steeper, *exponential* decline. This differential sensitivity to eccentricity is reminiscent of a similar difference found for initial eye acceleration as a function of target position[8] and supports the idea that the early phase of pursuit initiation is driven by inputs that de-emphasize the central visual fields,[9] or, alternatively, have uniformly large receptive fields across the representation of visual space—such as those from the accessory optic system.[10] In this scheme, the later phase of pursuit initiation would be driven by the cortical pathway involving area MT, which inherits a considerable emphasis on foveal vision from the striate cortex and thus exhibits an exponential decline in magnification factor.[11]

Finally, given the robust nature of the "bottom-up" contour effects on pursuit, we thought it would be interesting to pit these signals against more "top-down" signals based on the predictability of target motion. We performed two types of experiments in both monkeys and humans. In the first, target motion on any given trial could be in one of four possible directions, but, on half of the trials, target onset was preceded by a cue that indicated the direction of motion of the upcoming target. In the second type of experiment, the direction of motion was exactly the same on every trial. For the monkeys, neither one of these manipulations affected the magnitude or timing of the contour-induced deviation; however, we as yet have no independent evidence that the monkey made use of the predictive signals. For human observers, predictability reduced the effect by about 50%.

In sum, our results indicate that the primate pursuit system initially derives a rough estimate of direction by *averaging* local motion signals obtained from neurons with relatively small receptive fields. The same effect of ambiguous contour motion has been observed in human perception,[6] the human ocular following response,[7] and the responses of MT neurons.[2] The initial integration takes place over a remarkably large spatial range—up to 35°—further blurring the distinction between smooth pursuit and other smooth eye movements, such as ocular following and the optokinetic response. The efficacy of spatial integration depends on both eccentricity and the period of pursuit measured, perhaps constituting a signature of early, subcortical influences followed by later, cortical signals. Lastly, in humans top-down signals, such as prior knowledge of true target direction, can clearly reduce the earliest contour-induced effect, but they cannot eliminate it completely.

REFERENCES

1. MARR, D. & S. ULLMAN. 1981. Directional selectivity and its use in early visual processing. Proc. R. Soc. Lond. B Biol. Sci. **211:** 151–180.
2. PACK, C.C. & R.T. BORN. 2001. Temporal dynamics of a neural solution to the aperture problem in visual area MT of macaque brain. Nature **409:** 1040–1042.
3. NEWSOME, W.T., R.H. WURTZ, M.R. DURSTELER & A. MIKAMI. 1985. Deficits in visual motion processing following ibotenic acid lesions of the middle temporal visual area of the macaque monkey. J. Neurosci. **5:** 825–840.
4. GROH, J.M., R.T. BORN & W.T. NEWSOME. 1997. How is a sensory map read out? Effects of microstimulation in visual area MT on saccades and smooth pursuit eye movements. J. Neurosci. **17:** 4312–4330.
5. BORN, R.T., J.M. GROH, R. ZHAO & S.J. LUKASEWYCZ. 2000. Segregation of object and background motion in visual area MT: effects of microstimulation on eye movements. Neuron **26:** 725–734.
6. LORENÇEAU, J., M. SHIFFRAR, N. WELLS & E. CASTET. 1993. Different motion sensitive units are involved in recovering the direction of moving lines. Vision Res. **33:** 1207–1217.
7. MASSON, G.S., Y. RYBARCZYK, E. CASTET & D.R. MESTRE. 2000. Temporal dynamics of motion integration for the initiation of tracking eye movements at ultra-short latencies. Vis. Neurosci. **17:** 753–767.
8. LISBERGER, S.G. & L.E. WESTBROOK. 1985. Properties of visual inputs that initiate horizontal smooth pursuit eye movements in monkeys. J. Neurosci. **5:** 1662–1073.
9. LISBERGER, S.G., E.J. MORRIS & L. TYCHSEN. 1987. Visual motion processing and sensory-motor integration for smooth pursuit eye movements. Annu. Rev. Neurosci. **10:** 97–129.
10. SIMPSON, J.I. 1984. The accessory optic system. Annu. Rev. Neurosci. **7:** 13–41.
11. DANIEL, P.M. & D. WHITTERIDGE. 1961. The representation of the visual field on the cerebral cortex in monkeys. J. Physiol. **159:** 203–221.

Visually Driven Eye Movements Elicited at Ultra-short Latency Are Severely Impaired by MST Lesions

AYA TAKEMURA,[a,b] YUKA INOUE,[c] AND KENJI KAWANO[a,b]

[a]*Neuroscience Research Institute, National Institute of Advanced Industrial Science and Technology (AIST), Tsukubashi, Ibaraki 305-8568, Japan*

[b]*CREST, JST, Ibaraki 305-8568, Japan*

[c]*National Institute for Physiological Science, Okazaki, Aichi 444-8585, Japan*

KEYWORDS: ocular following response; vergence eye movement; monkey; medial superior temporal (MST) area

Recent studies in primates have revealed three distinct oculomotor responses that can be elicited by visual stimuli at ultrashort latencies (~60 msec in monkeys). All three types of eye movement have a machinelike quality and are thought to help reduce the visual disturbances that we experience as we move around the environment.[1–4] One of these responses has been termed "ocular following" and is thought to deal with the visual stabilization problems confronting a moving observer who looks off to one side. Ocular following responses are conjugate (version) eye movements evoked by sudden drifting movements of a large-field visual stimulus.[1] The other two responses, termed "disparity vergence" and "radial-flow vergence," are thought to deal with the binocular fusion problems of the moving observer who looks in the direction of motion. These short-latency vergence eye movements are evoked by disparity and radial flow stimuli applied to large patterns.[2–4] In our recent experiments on monkeys, we have examined the discharges of single neurons in the medial superior temporal (MST) area associated with ocular following, disparity vergence, and radial-flow vergence. We have found ocular-following-related neurons that discharged in response to planar large-field motion.[5] Furthermore, some MST neurons responded to horizontal disparity steps[6] and some to radial flow stimuli. Most of these MST neurons increased their discharges before the associated eye movements, soon enough for them to have a causal role in producing even the earliest respective ocular responses. These electrophysiological experiments suggest that, despite their ultrashort latency, all three visually elicited eye movements are mediated by MST neurons. To further understand the role of the MST in eliciting ocular following, disparity vergence, and radial-flow vergence, we have examined the effects of injecting ibotenic acid into MST bilaterally in monkeys.

Address for correspondence: Aya Takemura, Neuroscience Research Institute, National Institute of Advanced Industrial Science and Technology (AIST), 1-1-1 Umezono, Tsukubashi, Ibaraki 305-8568, Japan. Voice: +81-298-61-3426; fax: +81-298-61-5849.

a.takemura@aist.go.jp

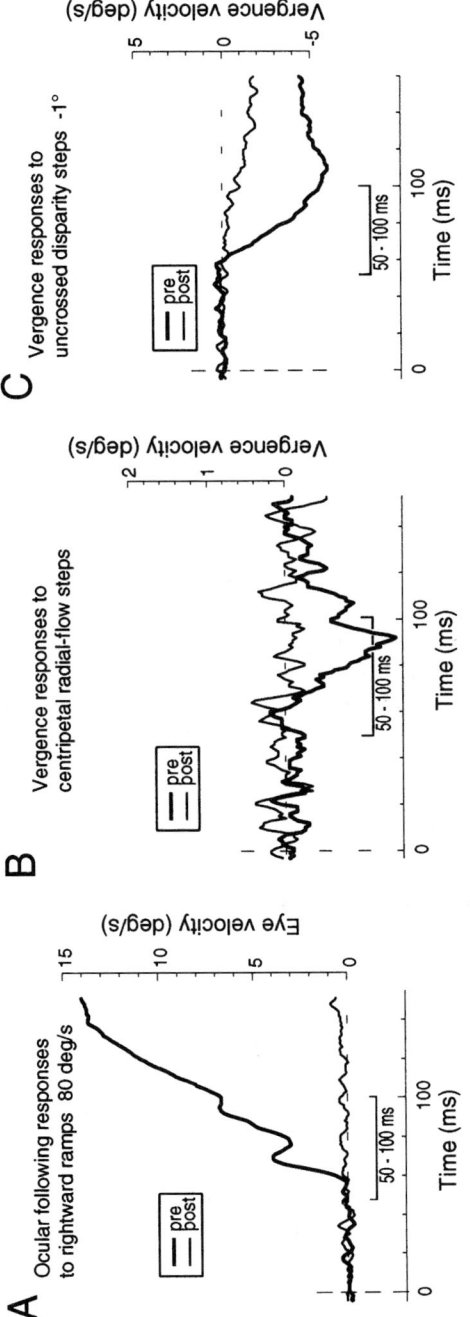

FIGURE 1. Sample mean eye-velocity profiles for ocular following response (**A**, $n = 88$; monkey Q), radial-flow vergence (**B**, $n = 200$; monkey R), and disparity vergence (**C**, $n = 100$; monkey Q). Preinjection (*thick lines*) and postinjection profiles (*thin lines*) are superimposed. Note: stimulus direction for ocular following response and convergence responses for vergence are shown by upward deflections.

METHODS

Ocular following responses and vergence eye movements were studied in three monkeys (*Macaca fuscata*). The monkey sat in a primate chair with its head secured in place, and faced a translucent screen (114 × 114 cm) located 50 cm away. The horizontal and vertical positions of both eyes were recorded using the electromagnetic search coil technique.[7] The monkey was trained to fixate stationary red target spots (LED), which were projected onto the pattern on the screen to elicit centering saccades. The monkey was given an occasional drop of fruit juice to remain alert and to facilitate brisk saccades.

We first recorded single-unit activity in association with ocular following and disparity vergence in order to locate the MST area in both hemispheres. Once this was accomplished, we recorded ocular following, disparity vergence, and radial-flow vergence responses to a standard set of stimuli. We then made lesions within the STS by injecting ibotenic acid bilaterally: 8 injections in the right hemisphere (~16 ml total) on the first day, and then 8 injections in the left hemisphere (~16 ml total) on the second day. Ibotenic acid (15 mg/ml in a basic saline solution) was pressure injected at a rate of 0.2 ml/min. The syringe was not raised for at least 15 min after completion of the injection. Then, on the third day, we recorded the eye movements elicited by the same standard set of visual stimuli.

RESULTS

Impairments were observed in ocular following, radial-flow vergence, and disparity vergence following bilateral ibotenic acid injections into STS, as shown by the sample records presented in FIGURE 1. On the day after the injections, there was almost total loss of ocular following (FIG. 1A) and a substantial decrease in both radial-flow vergence and disparity vergence (FIG. 1B and C). To quantify the effects on the ocular responses in each monkey, we measured the eye position change over the 50-msec time period starting 50 msec after the onset of ramps, corresponding to the so-called open-loop response period. On the basis of these measures, on average the injections reduced the ocular following responses (to 80 deg/sec motion in the four cardinal directions) by 81%, the disparity vergence responses (to crossed/uncrossed steps) by 63%, and the radial-flow vergence responses (to 2% expansion/contraction steps) by 73%.

DISCUSSION

Previous studies strongly suggested that middle temporal (MT) and MST areas were involved in visual motion processing by showing that deficits in smooth pursuit eye movements, saccades to moving targets, and optokinetic responses[8,9] followed punctate ibotenic acid lesions in these areas. In the present study, we examined the effect of ibotenic acid injections bilaterally in MST on the initial oculomotor responses elicited by visual stimuli at ultrashort latency. These chemical lesions severely attenuated initial ocular following responses and significantly attenuated the initial vergence responses elicited by both radial-flow and disparity steps. These results strongly support the hypothesis that MST is of primary importance in mediating all three short-latency visual tracking systems: ocular following responses, disparity vergence, and radial-flow vergence.

REFERENCES

1. MILES, F.A., K. KAWANO & L.M. OPTICAN. 1986. Short-latency ocular following responses of monkey. I. Dependence on temporospatial properties of visual input. J. Neurophysiol. **56:** 1321–1354.
2. BUSETTINI, C., F.A. MILES & R.J. KRAUZLIS. 1996. Short-latency disparity vergence responses and their dependence on a prior saccadic eye movement. J. Neurophysiol. **75:** 1392–1410.
3. BUSETTINI, C., G.S. MASSON & F.A MILES. 1997. Radial optic flow induces vergence eye movements with ultra-short latencies. Nature **390:** 512–515.
4. INOUE, Y. et al. 1998 Short-latency vergence eye movements elicited by looming step in monkeys. Neurosci. Res. **32:** 185–188.
5. KAWANO, K. et al. 1994. Neural activity in cortical area MST of alert monkey during ocular following responses. J. Neurophysiol. **71:** 2305–2324.
6. TAKEMURA, A. et al. 2001. Single unit activity in cortical area most associated with short-latency disparity-vergence eye movements: evidence for population coding. J. Neurophysiol. **85:** 2245–2266.
7. FUCHS, A.F. & D.A. ROBINSON. 1966. A method for measuring horizontal and vertical eye movement chronically in the monkey. J. Appl. Physiol. **21:** 1068–1070.
8. NEWSOME, W.T. et al. 1985 Deficits in visual motion processing following ibotenic acid lesions of the middle temporal visual area of the macaque monkey. J. Neurosci. **5:** 825–840.
9. DURSTELER, M.R. & R.H. WURTZ. 1988. Pursuit and optokinetic deficits following chemical lesions of cortical area MT and MST. J. Neurophysiol. **60:** 940–965.

Neuronal Activity in Monkey Cortical Area 6 during the Initial Phase of Smooth Pursuit against a Stationary Background

YASUSHI KODAKA,[a,b] SOHEI CHIMOTO,[a,b] AND KENJI KAWANO[a,b]

[a]*Neuroscience Research Institute, National Institute of Advanced Industrial Science and Technology (AIST), Tsukuba, Ibaraki 305-8568, Japan*

[b]*CREST, JST, Ibaraki 305-8568, Japan*

KEYWORDS: smooth pursuit; area 6; monkey frontal cortex

In many of the studies on pursuit eye movement, subjects tracked a moving spot in the dark. In the natural world, however, we usually track targets against a textured background. Laboratory experiments indicate that primates can pursue a target against a textured background, although some reduction in pursuit gain and/or increase in latency has been reported (for review, see Ilg[1]). Somewhere in the brain, there must be a system that can select the target from the background, perhaps involving a form of selective attention. In an attempt to understand some of the mechanisms involved, we have begun to study the effects of a stationary background on neuronal activity in area 6 of monkey frontal cortex during the initial phase of smooth pursuit.

METHODS

Data were collected from observations of a Japanese monkey (*Macaca fuscata*). Many of the procedures were the same as those used in previous studies.[2] The animal sat in a chair and faced a large tangent screen onto which a spot and a background could be projected. The background consisted of a stationary random dot pattern that filled the screen (90° × 90°) except in the paths traversed by the target spot (width of 1°). We measured eye movements and recorded neuronal activity under the following conditions. At the beginning of the trial, the target spot was positioned at the center, and the background could be either dark or patterned. After the animal had fixated on the stationary target, the target started to move at a constant speed in one of eight directions along one of four trajectories (horizontal, vertical, or diagonal). The animal was required to pursue the moving target. To quantify the initial open-loop neuronal responses, we computed the mean firing rate of the neuron over the 60-msec period starting 60 msec after the onset of target motion. The average initial

Address for correspondence: Yasushi Kodaka, Ph.D., AIST, Tsukuba Central 2, Neuroscience Research Institute, 1-1-1 Umezono, Tsukuba, Ibaraki, 305-8568, Japan. Voice: +81-298-61-5847; fax: +81-298-61-5849.

y.kodaka@aist.go.jp

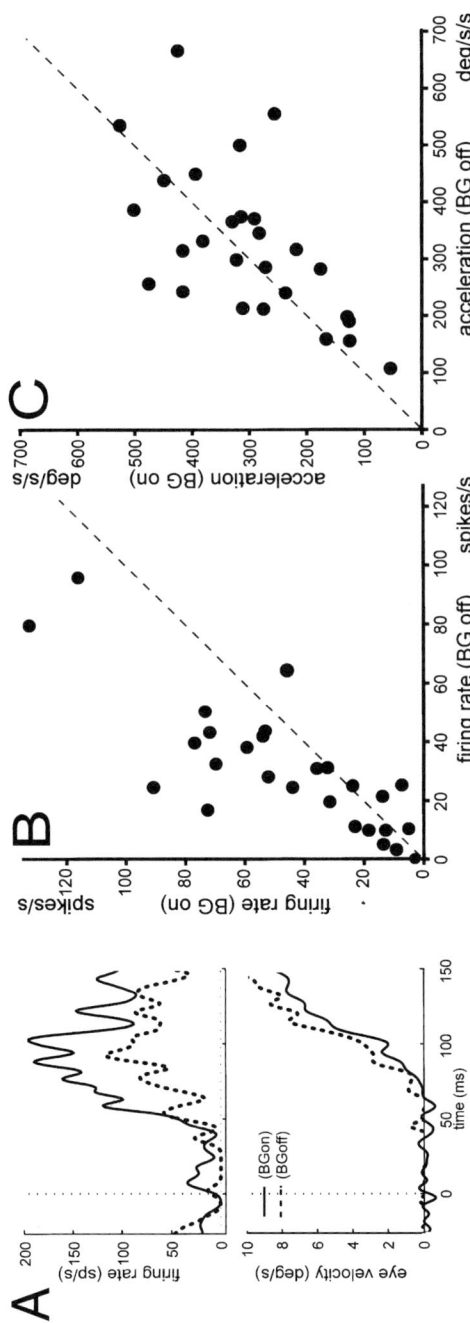

FIGURE 1. Initial neuronal and ocular tracking responses: dependence on background. (**A**) Neuronal discharges associated with smooth pursuit of a target moving left-upward at 20 deg/sec: dependence on the background. Superimposed mean spike density functions for a neuron in area 6 (*upper traces*) and mean eye velocity profiles (*lower traces*) when the background was patterned (*continuous line*) and dark (*broken line*). (**B**) Plot of mean discharge rate when background patterned against mean discharge rate when background dark for all 27 neurons characterized. (**C**) Plot of average initial eye acceleration when background patterned against when background dark (based on measurements of the pursuit eye movements while recording the neuronal responses plotted in B).

eye acceleration during the onset of pursuit was estimated by fitting a linear regression line to the mean eye velocity trace over the 50-msec period starting 80 msec after the onset of the target motion.

RESULTS

We recorded the activity of single neurons in the periarcuate cortex, close to the arcuate sulcus, in both hemispheres of a behaving monkey. We found neurons whose firing rates were modulated when the animal tracked a moving target, as reported previously,[3,4] and selected 27 neurons for further analysis. An example of a pursuit-related neuron is shown in FIGURE 1A. The neuron increased its firing rate before the animal began to pursue the moving target, whether the background was dark (broken line) or textured (continuous line). The preferred direction of movement for this neuron was similar irrespective of whether the background was patterned or dark. This was a consistent finding. Although the preferred directions were similar, the modulation of the firing rate was much more pronounced when the background was patterned than when the background was dark (FIG. 1A, upper traces). On the other hand, the initial eye acceleration was smaller when the background was patterned than when the background was dark (lower traces), though the difference was not statistically significant. FIGURE 1B compares the mean firing rates of the 27 neurons during the initiation of pursuit when the background was dark and patterned. It is clear that the majority of the neurons modulated their activity more when the background was patterned. However, no such trend was evident in the initial eye acceleration (FIG. 1C).

DISCUSSION

We have shown that more than half of the neurons in area 6 that have pursuit-related activity modulated their initial firing rate more when the track target was seen against a textured background than when it was seen against a dark background. Because we measured the neuronal responses during the initial open-loop period, we think that the difference is not secondary to the effects of the pursuit eye movements on retinal slip. There are studies reporting some effects of a structured background on the initiation of pursuit,[5,6] but we did not observe any consistent trends and no statistically significant effects. To this extent, the observed dependence of the neuronal responses on the background is not directly reflected in the associated tracking eye movements. We suggest that the background dependence of the neuronal activity in area 6 might reflect the operation of selective attention or some target selection process.

ACKNOWLEDGMENTS

We thank Dr. F.A. Miles for his valuable suggestions. This work was partly supported by the Cooperation Research Program of Primate Research Institute, Kyoto University.

REFERENCES

1. ILG, U.J. 1997. Slow eye movements. Progr. Neurobiol. **53:** 293–329.
2. KAWANO, K. *et al.* 1994. Neural activity in cortical area MST of alert monkey during ocular following responses. J. Neurophysiol. **71:** 2305–2324.
3. MACAVOY, M.G. *et al.* 1991. Smooth-pursuit eye movements representation in the primate frontal eye field. Cereb. Cortex **1:** 95–102.
4. TANAKA, M. & K. FUKUSHIMA. 1998. Neuronal responses related to smooth pursuit eye movements in the periarcuate cortical area of monkeys. J. Neurophysiol. **80:** 28–47.
5. KELLER, E.L. & N.S. KHAN. 1986. Smooth-pursuit initiation in the presence of a textured background in monkey. Vision Res. **26:** 943–955.
6. KIMMIG, H.G. *et al.* 1992. Effects of stationary textured backgrounds on the initiation of pursuit eye movements in monkeys. J. Neurophysiol. **68:** 2147–2164.

Perisaccadic Occipital EEG Changes Quantified with Wavelet Analysis

I. BODIS-WOLLNER, H. VON GIZYCKI, M. AVITABLE, Z. HUSSAIN, A. JAVEID, A. HABIB, A. RAZA, AND M. SABET

Departments of Neurology and Scientific Computing, State University of New York, Downstate Medical Center, Brooklyn, New York 11203, USA

KEYWORDS: visual cortex; oscillatory rhythms; activity-dependent plasticity

A role for the visual cortex in the dynamic interaction of saccades and visual processing is suggested by functional MRI studies, which reveal saccade-related striate (occipital) cortical activity.[1-6] Little is known, however, of the timing relationship between occipital changes and saccades. Occipital changes preceding the eye movement may be consistent with the notion of corollary discharge,[7,8] whereas occipital activity following the eye movement could signify the response of the visual cortex to a new, shifted visual field. In order to study the time course of occipital activity, we have studied electroencephalographic recordings in association with saccades. We concentrated on gamma range EEGs. It is known that gamma range activity represents local circuit binding of neurons engaged in the same task.[9] Our hypothesis was that there is a lawful relationship of gamma range occipital EEGs in relation to saccade timing and direction as the result of reorganization of neuronal binding in the visual cortex in association with saccades. Our hypothesis was based on the explicit suggestion of Morrone, Ross, and Burr[10] that visual space compression occurs during the short duration of a saccade. Previous studies of perisaccadic EEG changes[11,12] applied Fourier analysis, which is not adequate to quantify short-term EEG changes. Additionally, we concentrated on gamma range activity, as frequencies substantially lower could not be used by the brain to convey information of the chronology of neuronal reorganization during a saccade.

Ten normal subjects were studied in comparable age range, with the mean age being 38 years. The EOG and infrared eye scan recorded self-paced saccades and eye position. Permanent markers were located 78 cm from the observer, their distance subtending 60 degrees at the eye. Fourteen channels of EEGs were recorded over posterior occipital and parietal channels. Following a test session, the continuous recording EEG and eye movement recordings were visually inspected, and with our software, cursors were placed, bracketing acceptable saccades. Saccades were marked when their velocity exceeded 200 degrees/second, and they were free of ret-

---

Address for correspondence: Ivan Bodis-Wollner, M.D., D.Sc., Department of Neurology, SUNY, Downstate Medical Center, 450 Clarkson Ave., Brooklyn, NY 11203. Voice: 718-270-1482; fax: 718-270-3840.
bodisi01@bmec.hscbklyn.edu

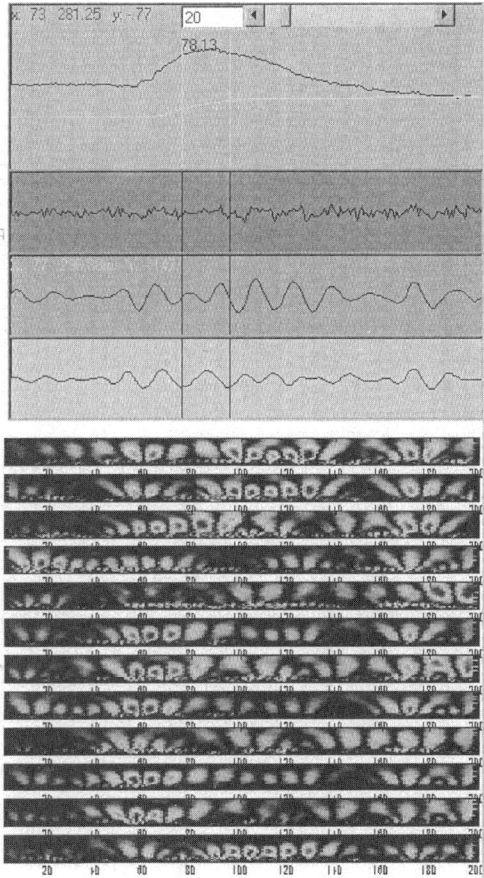

FIGURE 1. The *upper panel* shows the EOG and the ISCAN-recorded eye position for a rightward-directed, 30-degree voluntary saccade. Below are shown the occipitally recorded EEGs in the same time period. The first trace shows the midline EEG unfiltered and then left and right occipital EEGs filtered for 37 Hz. The color-coded panel shows from top to bottom the time and frequency distribution of the gamma wave obtained through continuous wavelet transform. The two cursors indicate the 75-msec time segment before new fixation. The order of channels is inion, and then left and right alternating: occipital 2–5, parietal 6–9, and parieto-temporal 10 and 11. All are referenced to a midfrontal electrode, except the last channel. The last color-coded row shows the results obtained from the occiput referenced to a temporal location. Hence, any response in channels 1–5 that corresponds to the activity in the last channel is due to occipital and not to frontal activity. Note the similarity of the second and last channels. Note also the phase reversal in the left and right occipital channels in this critical saccadic period.

rograde movements. The relevant EEG segments were analyzed with continuous wavelet transform[12–14] (wEEG) using MatLab tools. For each analyzed perisaccadic period, the wavelet coefficient of the 37-Hz band was entered into further analysis using standard statistics. The analyzed periods were 150 msec presaccade, 150 milliseconds intrasaccade, and 150 msec postsaccade. The intrasaccadic period was further subdivided into two 75-msec segments. The statistical analysis compared gamma range coefficient variation as a function of time (presaccade, intrasaccade, and new fixation) and hemisphere.

Side-to-side wEEG correlation (left vs. right) of the gamma band was low preceding the saccade, high during the saccade (preceding the new fixation), and then high once new fixation was achieved. All subjects showed time-dependent modulation of the perisaccadic gamma. Gamma occurred roughly 30–40 msec after the initiation of the saccade and before new fixation. Synchrony ceased roughly 60 msec after the eye came to rest (FIG. 1).

Our results suggest that perisaccadic cortical functional reorganization may be reflected in gamma band synchrony. High gamma (synchrony) is evident in all perisaccadic windows analyzed; however, hemispheric asymmetry reaches very high statistical significance before the time when new fixation is achieved. This result may represent new fixation-related preparatory neuronal binding, which is saccade direction (vector) dependent. At this point the occipital generator source of perisaccadic gamma is unknown. We do not know whether perisaccadic gamma *lateralizes* as the onset VEP,[15,16] or rather a recording over, say, the left occiput represents left occipital activity. Our data shows that the *timing* of high gamma synchrony is around 30–40 msec after the initiation of the saccade. This delay does allow for the possibility that it represents an input to the occipital cortex, arising not from initial eye command, but rather from eye position signals. Thus intrasaccadic gamma may possibly represent eye position information to the visual cortex. Only a few studies have concluded that eye position signals are cortically coded. Nevertheless, there is undeniable evidence of feedback signals reflecting eye position.[17] It remains to be seen whether functional imaging data and neurobehavioral and psychophysical measures can unequivocally define the functional significance of perisaccadic occipital gamma.

REFERENCES

1. BODIS-WOLLNER, I. *et al.* 1997. Functional MRI mapping of occipital and frontal cortical activity during voluntary and imagined saccades. Neurology **49:** 416–420.
2. BODIS-WOLLNER, I., S.F. BUCHER & K.C. SEELOS. 1999. Cortical activation patterns during voluntary blinks and voluntary saccades. Neurology **53:** 1800–1805.
3. LAW, I. *et al.* 1998. Parieto-occipital cortex activation during self-generated eye movements in the dark. Brain **121:** 2189–2200.
4. WENZEL, R. *et al.* 2000. Saccadic suppression induces focal hypooxygenation in the occipital cortex. J. Cereb. Blood Flow Metab. **20:** 1103–1110.
5. RAEMAEKERS, M., J.M. JANSMA, W. CAHN, *et al.* 2001. Neuronal substrate of the saccadic inhibition deficit in schizophrenia investigated with 3D event-related fMRI. NeuroImage **13:** S1091.
6. SUGIURA, M., R. KAWASHIMA, J. WANTANABE, *et al.* 2001. Human frontal and parietal cortices in memory-guided saccade: an event-related fMRI study. NeuroImage **13:** S365.
7. VON HOLST, E. & H. MILTELSTAEDT. 1950. Das Reafferenzprinzip. Naturwissenschaften **37:** 464–476.
8. SPERRY, R.W. 1950. Neural basis of the optokinetic response produced by visual inversion. J. Comp. Psychol. Physiol. **43:** 482–489.
9. SINGER, W. 1993. Synchronization of cortical activity and its putative role in information processing and learning. Annu. Rev. Physiol. **55:** 349–374.

10. MORRONE, M.C., J. ROSS & D.C. BURR. 1997. Apparent position of visual targets during real and simulated saccadic eye movements. J. Neurosci. **17:** 7941–7953.
11. SKRANDIES, W. & K. LASCHKE. 1997. Topography of visually evoked brain activity during eye movements: lambda waves, saccadic suppression, and discrimination performance. Int. J. Psychophysiol. **27:** 15–27.
12. MARI, Z. et al. 2000. Perisaccadic high frequency EEG changes in frontal and occipital regions are similar in light and dark. J. Physiol. (London) **25S:** 526.
13. DAUBECHIES, I. 1988. Orthonormal bases of compactly supported wavelets. Comm. Pure Appl. Math. **41:** 909–996.
14. DAUBECHIES, I. 1990. The wavelet transform, time-frequency localization and signal analysis. IEEE Trans. Inform. Theory **36:** 961–1004.
15. MALLAT, S. 1989. A theory for multi-resolution signal decomposition: the wavelet representation. IEEE Trans. Pattern Anal. Machine Intell. **11:** 674–693.
16. TZELEPI, A., T. BEZERIANOS & I. BODIS-WOLLNER. 2000. Functional properties of subbands of oscillatory brain waves to pattern visual stimulation in man. Clin. Neurophysiol. **111:** 259–269.
17. BODIS-WOLLNER, I. et al. 2001. Wavelet transform of the EEG reveals differences in low and high gamma responses to elementary visual stimuli. Clin. Electro. **32:** 139–144.
18. VELAY, J.L. et al. 1994. Eye proprioception and visual localization in humans: influence of ocular dominance and visual context. Vis. Res. **34:** 2169–2176.

Human Gaze Shifts to Acoustic and Visual Targets

L.C. POPULIN,[a] D.J. TOLLIN,[b] AND J.M. WEINSTEIN[c]

Departments of [a]Anatomy, [b]Physiology, and [c]Ophthalmology, University of Wisconsin, Madison, Wisconsin 53706, USA

KEYWORDS: orienting gaze shifts; acoustic targets; visual targets

INTRODUCTION

Eye and head contributions to orienting gaze shifts have been primarily studied using visual targets. Consequently, relatively little is known about the kinematics of eye and head movements in gaze shifts to acoustic targets. Although early work in nonhuman primates indicates that orienting responses to acoustic and visual targets are similar, suggesting that a common motor program is used for stimuli of both modalities,[1] supporting data have not followed.

Experiments using primates, which provide for concurrent behavioral and physiological measurements, suggest that it is difficult to train them to orient to acoustic targets.[2] Human subjects, on the other hand, do not require training and can be provided with specific instructions. In this study we characterize the kinematics of gaze shifts to acoustic targets using equivalent measures to visual targets for comparison.

METHODS

Five normal human subjects (4 males and 1 female) participated in the experiments. The experiments were carried out in a dimly illuminated, double-walled, soundproof chamber. Eye and head movements were measured with a phase angle search coil system (CNC Engineering). The eye coil (Skalar) was calibrated behaviorally by asking the subject to fixate on light-emitting diodes (LEDs) located at known positions. The head coil, having similar characteristics to the eye coil, was calibrated using a protractor and was firmly attached to the top of the subject's head with a headband.

Acoustic stimuli were 500-msec broadband (0.1–25 kHz) noise bursts presented well above detection threshold from one of 16 loudspeakers positioned in the frontal hemifield 84 cm away from the subject. Visual stimuli consisted of red/green LEDs attached to the front of the speakers. A dark cloth draped in front of the speakers occluded their view, but the LEDs were clearly seen as discrete spots.

Address for correspondence: Luis C. Populin, Ph.D., Department of Anatomy, University of Wisconsin, Madison, WI 53706. Voice: 608-265-6451; fax: 608-262-7306.
lpopulin@facstaff.wisc.edu

Ann. N.Y. Acad. Sci. 956: 468–473 (2002). © 2002 New York Academy of Sciences.

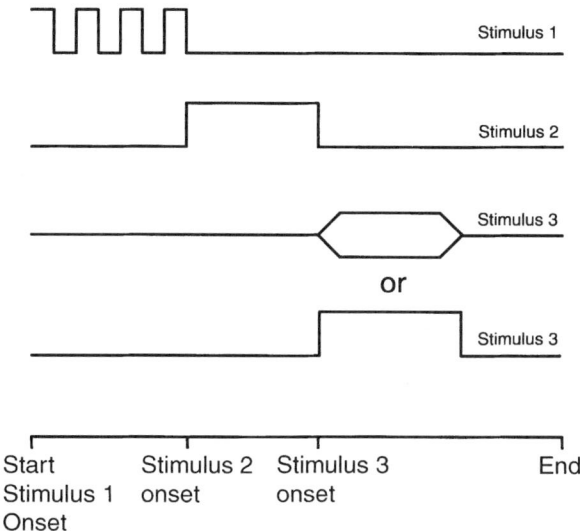

FIGURE 1. Dissociated saccade task.

Two experimental tasks were used. The *fixation task* (not shown) consisted of an acoustic or visual (red LED) target. The subject was instructed to look at the target. A blinking LED at the start of a trial indicated a *dissociated saccade task* trial (FIG. 1). In this task the subject was required to align his eyes and head to the blinking LED located at (±14°,0°) and (0°,0°). Once aligned, the blinking LED was extinguished, and a green LED was turned on at any of three positions (±14°,0°) and (0°,0°). The subjects were then required to look at the green LED without moving their heads. Coinciding with the offset of the green LED, an acoustic or visual target was presented from any of 16 positions. The subject was instructed to freely "look" to the target. This task allows us to independently manipulate the position of the eyes and the head relative to the position of the targets. A typical experimental session was composed of trials of both tasks presented in random order involving either visual or acoustic targets.

The onset of gaze, eye, and head movements was defined as the time at which the velocity exceeded two standard deviations of the mean baseline computed from a portion of the velocity trace starting 100 before to 30 msec after the onset of the stimulus; movement offset was defined as the time at which velocity returned to within two standard deviations of the mean baseline. The contribution of the head to gaze shifts was defined as the amplitude of the head displacement at the time of gaze offset.[3]

RESULTS

Overall, the relationships between velocity and movement amplitude of gaze, eye, and head movements to visual targets are consistent with previous reports.[4,5]

470   ANNALS NEW YORK ACADEMY OF SCIENCES

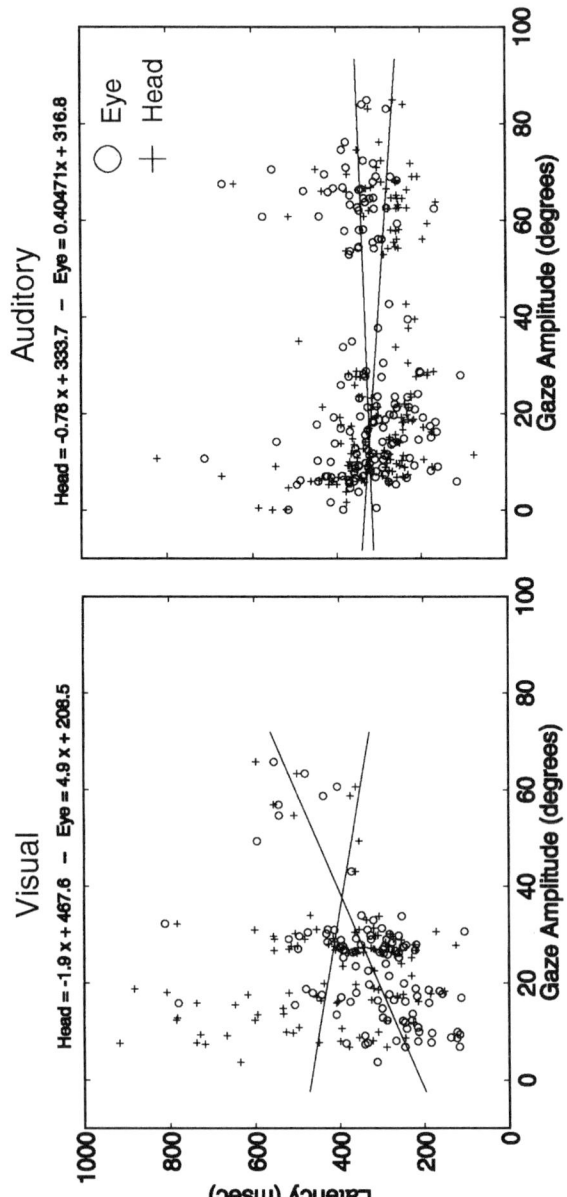

FIGURE 2. Eye movement initiates gaze shifts to visual and acoustic targets.

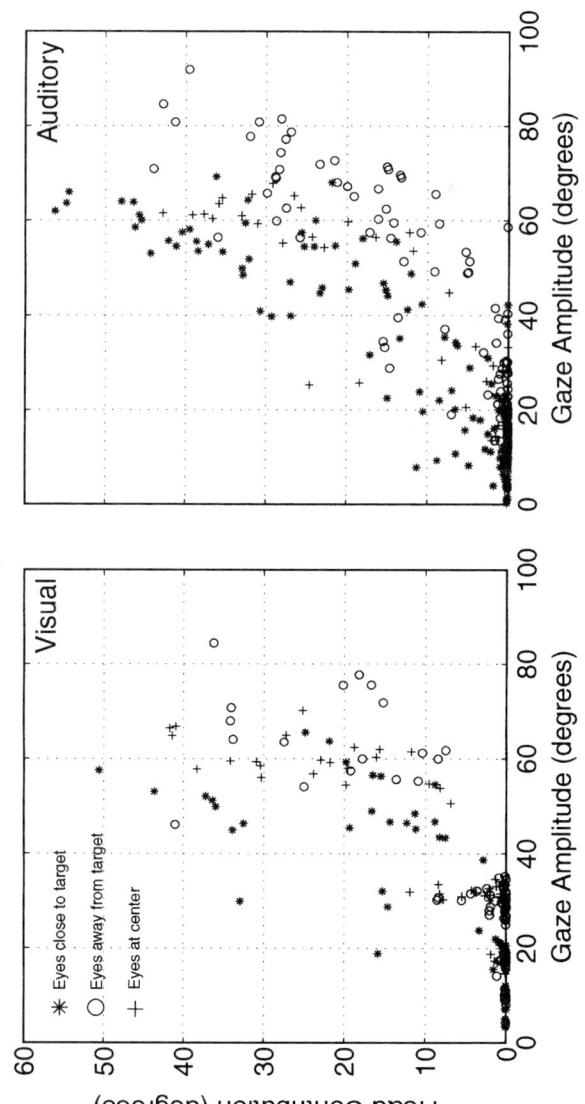

FIGURE 3. Contribution of the head to gaze shifts to acoustic and visual targets.

Eye and gaze velocity are saturating functions of amplitude, and head velocity is linearly related to head movement amplitude. Similar velocity–movement amplitude relationships were observed in auditory trials, but with slower eye and gaze velocities and higher head velocities.

FIGURE 2 illustrates that gaze shifts to visual and acoustic targets are generally initiated with an eye movement, but eye and head latencies in auditory trials varied differently with the amplitude of the gaze than in visual trials. In visual trials mean eye latency was shorter than head latency, and, consistent with head-fixed data,[6,7] eye latency increased for larger gaze shifts. Head latency, on the other hand, decreases slightly with larger gaze shifts. In auditory trials the eyes and head started to move at about the same time, and although head latency, like in visual trials, decreased slightly as a function of gaze amplitude, eye latency remained relatively constant, even for large gaze shifts.

The contribution of the head to gaze shifts to acoustic and visual targets is illustrated in FIGURE 3. Data from three different starting eye positions are shown: eye 14° in the direction of the target (o), eye 14° away from the targets (*), and eyes centered in the head (+). For both visual and acoustic targets, the contribution of the head to gaze shifts was small when the eyes were directed away from the target, increased when the eyes were at an intermediate position (centered in the orbit), and was large when the eyes were closer to the targets. However, the head contribution to gaze was slightly larger for small shifts to acoustic targets than for equivalent visual targets.

SUMMARY

1. Eye and gaze movements to acoustic targets tend to be slower than equivalent movements to visual targets. Head movements to acoustic targets are faster than to visual targets.

2. For small to moderately sized gaze shifts, the eyes moved before the head to both visual and acoustic targets, but the differences were smaller in auditory than visual trials.

3. When the initial eye position is close to the target, the head contribution to gaze is larger than when the initial eye position is away from the target. For small gaze shifts, the head contribution to gaze was greater for acoustic than equivalent visual targets.

4. With a few exceptions, head responses were composed of a single movement toward the target, but the eyes often exhibited numerous corrective saccades.

ACKNOWLEDGMENTS

This work was supported by National Institutes of Health (DC03693) and National Science Foundation (IBN-9904770) grants.

REFERENCES

1. WHITTINGTON, D.A. *et al.* 1981. Eye and head movements to auditory targets. Exp. Brain Res. **41:** 358–363.
2. LINDEN, L.F. *et al.* 1999. Responses to auditory stimuli in macaque lateral intraparietal Area II. Behavioral modulation. J. Neurophysiol. **82:** 343–358.
3. FREEDMAN, E.G. & D.L. SPARKS. 1997. Eye–head coordination during head-unrestrained shifts in rhesus monkeys. J. Neurophysiol. **77:** 2328–2348.
4. GUITTON, D. & M. VOLLE. 1987. Gaze control in humans: eye–head coordination during orienting movements to targets within and beyond the oculomotor range. J. Neurophysiol. **58:** 427–459.
5. VOLLE, M. & D. GUITTON. 1993. Human gaze shifts in which head and eyes are not initially aligned. Exp. Brain Res. **94:** 321–336.
6. ZAHN, J.R. *et al.* 1978. Audio-ocular response characteristics. Sens. Processes **2:** 32–37.
7. ZAHN, J.R. *et al.* 1979. The audioocular response: intersensory delay. Sens. Processes **3:** 60–65.

A Bayesian Approach to Change Blindness

MATTHIAS NIEMEIER,[a] J. DOUGLAS CRAWFORD,[b] AND
DOUGLAS B. TWEED[a,b]

[a]*Department of Physiology, University of Toronto, Toronto, Ontario M5S 1A8, Canada*
[b]*Centre for Vision Research, York University, Toronto, Ontario, Canada*

KEYWORDS: change blindness; Bayesian inference; neural network; transsaccadic integration

We use rapid eye movements called saccades to direct our foveae to different parts of our surroundings. But how do we combine information from successive fixations? Surprisingly, we are blind even to large changes in the visual display occurring during saccades.[1,2] For example, we are poor at recognizing shifts of a visual target when the shift coincides with the eye movement. This transsaccadic change blindness for displacement is commonly taken as evidence that our ability to integrate quantitative spatial information from successive fixations is poor. However, other studies suggest that under some conditions quantitative presaccadic information is still available after a saccade.[3,4] Here we argue for a new interpretation of change blindness. We suggest that the integration of information across saccades relies on principles of optimal—or Bayesian—inference, and that change blindness is a consequence of this strategy.

To interpret changes in the retinal image, the brain could employ visual and nonvisual information. From proprioceptive or motor information (e.g. efference copy), it could learn which changes in the retinal image might be due to eye motion rather than allocentric shifts, that is, movements of objects in the world. Visually it could identify common features in the retinal images before and after the saccade to try to deduce eye movements and allocentric shifts. Finally, the brain has sensitive visual motion detectors that respond to most allocentric motion, but that are switched off during saccades.[5]

We trained an artificial neural network to detect allocentric stimulus motion based on these three information sources: eye-movement signals, retinal position, and visual motion detectors. We took into account that (a) in normal life allocentric shifts seldom coincide precisely with saccades and (b) motion detection is quite accurate compared to the other two signals.

Using a backprop algorithm with momentum, the network learned to combine its three input signals into a statistically optimal estimate of the allocentric motion. As

Address for correspondence: Prof. Douglas B. Tweed, Department of Physiology, University of Toronto, 1 King's College Circle, Toronto M5S 1A8, Ontario, Canada. Voice: ++1 416 978 2603; fax: ++1 416 978 4940.
douglas.tweed@utoronto.ca

Ann. N.Y. Acad. Sci. 956: 474–475 (2002). © 2002 New York Academy of Sciences.

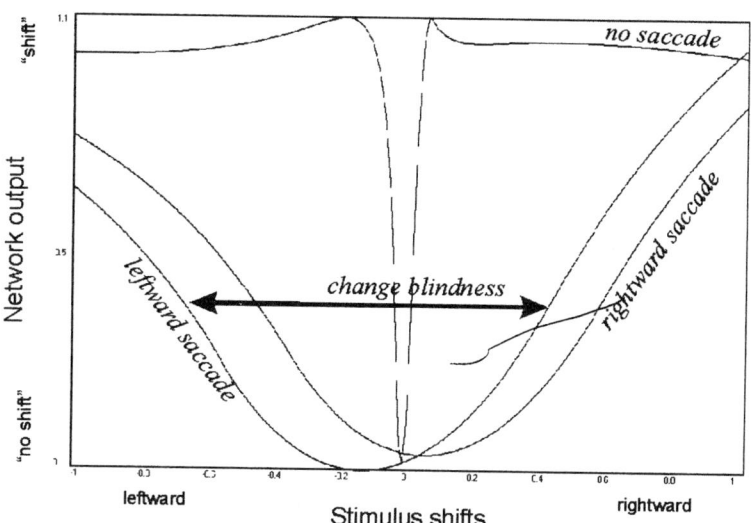

FIGURE 1. Network output after training. Shown is the probability that the network will detect an allocentric stimulus shift as a function of shift size in three situations: *no saccade* or a *leftward* or *rightward* saccade coinciding with the stimulus shift.

shown in FIGURE 1, the network matched important characteristics of human change blindness. It detected allocentric stimulus shifts when there was no saccade performed at the same time. During saccades the network was less likely to detect stimulus shifts. Detection was worse for smaller shifts than for larger shifts. And it was better for stimulus shifts opposite the saccade than for shifts in the same direction.

In further simulations we found that change blindness was reduced when saccades and allocentric stimulus shifts coincided more often during training. Moreover, when the network was provided with an unreliable visual motion signal, it showed no change blindness, but its detection of stimulus shifts was poor regardless of whether a saccade occurred at the same time. So change blindness depended on the statistical properties of the three simulated signals.

Our simulations suggest that transsaccadic change blindness for displacement may not imply poor transsaccadic memory but may be the result of optimal inference from visual and nonvisual signals available in the brain.

REFERENCES

1. BRIDGEMAN, B., D. HENDRY & L. STARK. 1975. Failure to detect displacement of the visual world during saccadic eye movements. Vision Res. **15:** 719–722.
2. SIMONS, D.J. 2000. Current approaches to change blindness. Visual Cognition **7:** 1–15.
3. HAYHOE, M., J. LACHTER & J. FELDMAN. 1991. Integration of form across saccadic eye movements. Perception **20:** 393–402.
4. DEUBEL, H., W.X. SCHNEIDER & B. BRIDGEMAN. 1996. Postsaccadic target blanking prevents saccadic suppression of image displacement. Vision Res. **36:** 985–996.
5. BURR, D.C., J. HOLT, J.R. JOHNSTONE & J. ROSS. 1982. Selective depression of motion selectivity during saccades. J. Physiol. **333:** 1–15.

Sensory Processing Delays Measured with the Eye-Movement Correlogram

J.B. MULLIGAN

NASA Ames Research Center, Moffett Field, California 94035, USA

KEYWORDS: smooth pursuit system; eye-movement correlogram; saccades

The smooth pursuit system causes the eyes to move to cancel "retinal slip" of a target and presumably uses the results of visual motion computations to control motor output. The overall latency of this system can be thought of as having two components: an input-processing component that depends on stimulus properties, and a motor output component that does not.

This poster presents a method for assessing pursuit latency that we call the *eye movement correlogram*. The subject is instructed to maintain fixation of a target that moves on a pseudo-random trajectory, designed to make prediction impossible.[1] The trajectories are computed by integrating a white noise velocity profile. Some low-pass filtering may be applied to the velocity signal before integration to smooth the trajectory. The subjects' eye movements are recorded, and the signals are differentiated to produce eye velocity. Saccades are detected using a velocity criterion, and values of the smooth velocity are interpolated in the neighborhoods of saccades. The resulting smooth velocity signal is cross correlated with the stimulus velocity. When averaged across a number of trials (each having a different random trajectory), an impulse response-like function is revealed, which is the correlogram. The time at which the peak of this function occurs can be interpreted as the latency of the pursuit system.

The method has the advantage of high sensitivity to weak signals, while being relatively insensitive to calibration artifacts. Because correlation is performed in the velocity domain, absolute positional calibration is not required, although rough calibration of velocity is required for proper function of the saccade-cutter.

There is ample evidence that the chromatic system is temporally sluggish.[2,3] Therefore, we expect that the pursuit latency for purely chromatic (isoluminant) targets will be long compared to that for luminance-defined targets. When the correlogram is measured for an isoluminant "red" target (modulated on the R-G axis, with constant S-cone stimulation), the responses are slightly weaker and delayed 50–100 milliseconds compared to the achromatic responses. More dramatic results are obtained with an S-cone-isolating "blue" stimulus. The subjective appearance of these stimuli is markedly different from both achromatic and red–green stimuli: the high

---

Address for correspondence: Jeffrey B. Mulligan, Ph.D., NASA Ames Research Center, Mail Stop 262-2, Moffett Field, CA 94035-1000. Voice: 650-604-3745; fax: 650-604-0255.
  jmulligan@mail.arc.nasa.gov

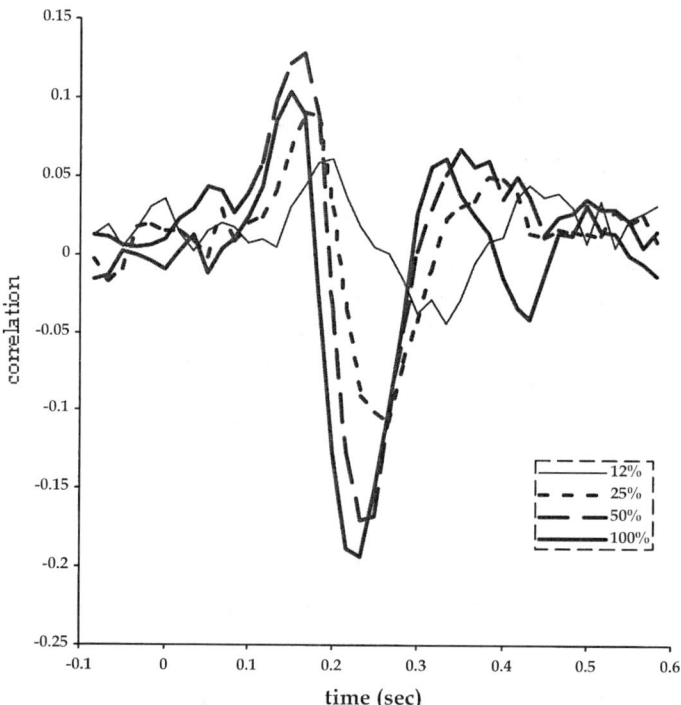

FIGURE 1. Average correlograms for tracking a small, bright Gaussian spot at 100% contrast (*heavy solid trace*), 50% contrast (*heavy dashed trace*), 25% contrast (*heavy dotted trace*), and 12% contrast (*light solid trace*). The latency of the response is reduced by approximately 20 milliseconds for each factor of 2 reduction in contrast, and as contrast is reduced below 50% a reduction in signal strength is seen.

temporal frequencies of the motion are virtually invisible. Nevertheless, the correlogram does reveal a small response, delayed 100–200 milliseconds relative to the achromatic response.

Proper equation of contrast and the exclusion of luminance artifacts are constant worries in the study of chromatic phenomena. To test the hypothesis that the chromatic responses might have been mediated by a sluggish response to a low-contrast luminance artifact, correlograms were measured for a series of reduced contrast achromatic targets. As contrast is reduced, the response becomes both weaker and slower, with the peak delayed by approximately 20 milliseconds for reach reduction by a factor of two (FIG. 1).

Current theories propose multiple mechanisms processing the motion of achromatic patterns: a "first-order" system that responds directly to luminance, a "second-order" system that responds to derived properties such as local contrast and flicker, and a "third-order" system that has very little in common with the other two.[4] We have compared correlograms for pursuit of a normal, luminance-defined spot and a second-order target defined by locally flickering a stationary random dot texture.

The response to the flicker-defined stimulus is significantly weaker and delayed by approximately 100 milliseconds.

The results described above for chromatic and flicker-defined targets were obtained from a single subject and have been reported in preliminary form previously.[5,6] These data were collected using a video ophthalmoscope, which provides excellent positional resolution and good rejection of head movement artifacts.[7] Unfortunately, this system requires use of a dental impression to stabilize the subject's head, and good images cannot be obtained from many subjects without dilating the pupil. Our recent work has therefore concentrated on replicating these findings using a simpler system in which the subject uses a chin/forehead rest, and cameras image the pupil and other anterior structures. Although the signal-to-noise ratio is somewhat poorer than in the ophthalmoscope system, correlograms for luminance-defined targets have been reliably obtained for three subjects.

In summary, the eye movement correlogram is a promising method for revealing the time course of the early stages of visual processing. The preliminary findings described here merely scratch the surface of the classes of stimuli that may be investigated with this technique.

REFERENCES

1. STARK, L. 1968. Neurological Control Systems: Studies in Bioengineering. Plenum Press. New York.
2. VAN DER HORST, G.J.C. & M.A. BOUMAN. 1969. Spatio-temporal chromaticity discrimination. J. Opt. Soc. Am. **59:** 1482–1488.
3. KELLY, D.H. 1983. Spatiotemporal variation of chromatic and achromatic contrast thresholds. J. Opt. Soc. Am. **73:** 742–750.
4. LU, Z. & G. SPERLING. 2001. Three-systems theory of human visual motion perception: review and update. J. Opt. Soc. Am. A **18:** 2331–2370.
5. MULLIGAN, J.B. 1998. Pursuit latency for chromatic targets. Invest. Ophthalmol. Visual Sci. (Suppl.) **39:** 444.
6. MULLIGAN, J.B. 1998. Ocular pursuit of flicker-define motion. Perception (Suppl.) **27:** 183.
7. MULLIGAN, J.B. 1997. Image processing for improved eye tracking accuracy. Behav. Res. Methods Instrum. Comput. **29:** 54–65.

Inhibitory Control of Saccade Generation Mediated by Various Oculomotor Regions in Response to Reading Difficulty

SHUN-NAN YANG AND GEORGE W. McCONKIE

Beckman Institute, University of Illinois at Urbana-Champaign, Urbana, Illinois 61801, USA

KEYWORDS: saccade inhibition; saccade generation; reading; eye movement control; superior colliculus; frontal eye field

Cognitive influences on saccade patterns during reading have been well documented, indicating that difficulties often result in prolonged fixation duration, shortened forward saccade length, and increased regressive saccades.[1] Nevertheless, the underlying neural mechanisms of the observed effects are not yet understood. A commonly accepted behavioral hypothesis suggests that the process of saccade generation in reading is triggered by the completion of certain psycholinguistic events, such as the completion of word identification or at least the execution of a familiarity check on the foveated words.[2]

Several studies conducted in our laboratory have shown that increasing the difficulty of identifying the fovated words (i.e., degrading word stimuli or replacing normal words with random letters) does not prolong the overall latency of all following saccades. Analyzing the frequency distributions of the durations of critical fixations (those on which the difficulty occurs) indicates that the cancellation or delay of saccades only begins at about 175–225 msec after the fixation onset.[3,4] Thus, the oculomotor mechanism for saccade generation actually proceeds normally, unaffected by cognitive difficulty, until some later time when the detected difficulty can exert an inhibitory influence on saccade generation.

In this report, we present results from two studies by Yang and McConkie,[5] using gaze-continent display change during saccades. In these studies, people read continuous text from a computer display while their eye movements were recorded. The normal text was replaced with an alternative stimulus pattern during randomly selected saccades, with normal text returning during the following saccade, thus occasionally producing an inappropriate stimulus pattern for a single fixation. We then observed the effect of the abnormal stimulus patterns on the time, direction, and length of the following saccade.

Address for correspondence: Shun-nan Yang, Beckman Institute, University of Illinois, 405 N. Mathews, Urbana, IL 61801. Voice: 217-333-0970; fax: 217-244-8371.
syang3@uiuc.edu

Ann. N.Y. Acad. Sci. 956: 479–481 (2002). © 2002 New York Academy of Sciences.

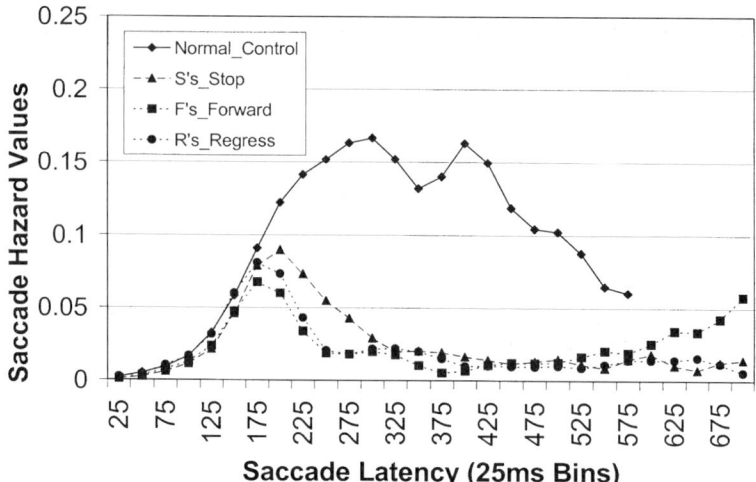

FIGURE 1. Hazard curves of forward saccades following the critical fixation in real reading, showing how the probability of making saccades is reduced by initial automated effects (at around 175 msec) and late cognitive control (at around 625 msec). In pseudo-reading (single-task conditions, not shown here), the initial inhibitory effects occur at about the same time, whereas the late effects occur much earlier (about 325 msec).

Hazard curves of the saccade onset times for forward and regressive saccades indicate changes over time in the momentary likelihood of making a saccade, for five conditions in the first study: normal text, random letters, segmented X strings, unsegmented string of Xs, and blank pages. (For the methods of calculating hazard values for saccade initiation, see Yang and McConkie.[5]) The unsegmented conditions produced an early, strong suppression of both forward and regressive saccades; whereas the segmented conditions produced a later suppression of only forward saccades, with saccade length being reduced and regression rate increased beginning 25–50 msec after the onset of this suppression.

The early, bilateral suppression of saccades observed in the unsegmented condition suggests an increase of fixation activity within the oculomotor system, probably mediated by the rostral region of the superior colliculus (SC), which also is critical for mediating inhibition-related signals from various oculomotor regions.[6] In addition, regions downstream of the SC have been shown to delay the timing of saccades without modifying the saccade program, different from the observed shortening of forward saccade length. The later inhibitory effects, which are directionally selective, may arise from another source, probably from higher-level oculomotor regions such as the frontal eye fields. This is consistent with the evidence that switching oculomotor responses based on visual events often involves the frontal eye fields.[7] Thus, our study is the first to show directionally selective inhibitory effects resulting from such a complex task as reading. We have also observed similar results when normal reading difficulties are encountered (lexical, syntactic, or semantic errors) without the occurrence of display changes.

In a second study, we examined the timing of intentional saccade control during reading. Subjects either pretended to read pseudo-text (strings of Xs broken into word-like units) or read normal text. Occasionally, for a single fixation, a change of letter identity for all letters would occur to indicate that the reader should make a specified response: S = stop the eyes, R = regressive saccade, F = forward saccade. Hazard curves of saccade onset times show that all signals produce a similar inhibitory response beginning 175–225 msec after the onset of the fixation; only much later do differences appear that are related to the response to be made. These latter differences occur later in real reading than in pseudo-reading, indicating the effect of the cognitive load produced by the reading activity. FIGURE 1 shows how the hazard values change over the period of the critical fixation, indicating that automated and effortful control of eye movements produce quite different effects on saccade initiation.

In these two studies, we observe three types of oculomotor response to stimulus abnormalities: an early, severe, saccadic suppression to unsegmented, homogeneous stimuli; a later, rather automated inhibitory response to processing difficulty or abnormality; and a much later cognitively mediated, intentional control. The second of these appears to involve higher-level brain regions but not the involvement of executive functions or working memory. In contrast, the intentional control of eye movements requires these cognitive resources, suggesting the involvement of prefrontal regions.[8]

These different categories of saccade control provide a foundation for developing a neurophysiological theory of eye movement control during reading, revealing how saccade generation is dynamically controlled by different neural circuits, depending on the nature of the task and momentarily perceived processing difficulty.

REFERENCES

1. RAYNER, K. 1998. Eye movements in reading and information processing: 20 years of research. Psychol. Bull. **124:** 372–422.
2. REICHLE, E.D., A. POLLATSEK, D.L. FISHER & K. RAYNER. 1998. Toward a model of eye movement control in reading. Psychol. Rev. **105:** 125–157.
3. MCCONKIE, G.W., M.R. REDDIX & D. ZOLA. 1992. Perception and cognition in reading: Where is the meeting point? *In* Eye Movement and Visual Cognition. K. Rayner, Ed.: 293–303. Springer-Verlag. New York.
4. MCCONKIE, G.W., N.R. UNDERWOOD, D. ZOLA & G.S. WOLVERTON. 1985. Some temporal characteristics of processing during reading. J. Exp. Psychol. Hum. Percept. Perform. **11:** 168–186.
5. YANG, S.-N. & G.W. MCCONKIE. 2001. Eye movements during reading: A theory of saccade initiation times. Vision Res. **41:** 3567–3585.
6. MUNOZ, D.P. & P.J. ISTVAN. 1998. Lateral inhibitory interactions in the intermediate layers of the monkey superior colliculus. J. Neurophysiol. **79:** 1193–1209.
7. HANES, D.P. 1998. Neural processes that regulate saccade production: role of the frontal eye fields. Ph.D. thesis, Vanderbilt University, Nashville, TN.
8. WALKER, R., M. HUSAIN, T.L. HODGSON, *et al.* 1998. Saccadic eye movement and working memory deficits following damage to human prefrontal cortex. Neuropsychologia **36:** 1141–1159.

Botulinum Toxin Therapy for Apraxia of Lid Opening

DAN BOGHEN,[a,c] VIORIKA TOZLOVANU,[b,c] ANDREEA IANCU,[b,c] AND ROBERT FORGET[b,c,d]

[a]*Neuro-ophthalmology Section, Centre Hospitalier de l'Université de Montréal (Hôtel-Dieu), Montréal, Québec, Canada*

[b]*Centre de Recherche Interdisciplinaire en Réadaptation du Montréal Métropolitain, Institut de Réadaptation de Montréal, Montréal, Québec, Canada*

[c]*Centre de Recherche en Sciences Neurologiques, Faculté de Médecine, Université de Montréal, Montréal, Québec, Canada*

[d]*École de Réadaptation, Université de Montréal, Montréal, Québec, Canada*

KEYWORDS: botulinum toxin therapy; apraxia of lid opening; orbicularis oculi; blepharospasm

BACKGROUND

Apraxia of lid opening (ALO) designates an intermittent inability to open the eyes voluntarily in the absence of apparent orbicularis oculi (OOc) contraction.[1] Botulinum toxin (BTX) is extensively used in the treatment of dystonic muscles and is presently the treatment of choice for blepharospasm.[2,3] Since previous studies have suggested that clinically undetectable, but recordable, persistent activity of the OOc plays an important role in ALO[4–6] and our previous work showed a strong relationship between this EMG persistence and the delays in lid opening in most patients with ALO,[6] we investigated whether BTX could be of benefit in patients with ALO.

OBJECTIVE AND METHODS

Our objective was to evaluate the effect of BTX injection in the pretarsal OOc on the eye opening delays, the ability to sustain lid opening, the OOc activity, and the relationship between lid movement and EMG activity in patients with ALO.

Twelve patients with ALO (mean age, 60 ± 11 years) and 12 healthy control subjects (mean age, 58 ± 10 years) participated in the study. One patient had isolated ALO, and 11 patients had ALO associated with essential blepharospasm. The patients were included in the study on the basis of a clinical diagnosis of ALO.[1] The duration of the disease ranged from 1 to 10 years.

Address for correspondence: Dan Boghen, Neurology Service, CHUM (Hôtel-Dieu), 3840 St-Urbain St., Montréal, Québec H3W 1T8, Canada. Voice: 514-890-8122; fax: 514-843-2665.
boghend@videotron.ca

The subjects were first asked to gently close their eyes and then, in response to a sound signal presented at random intervals, to open them as fast as possible and to keep them open until the end of the acquisition period (for 14 sec after the sound).

Lid movement (detected in an electromagnetic field)[7] and EMG of the septal and pretarsal portions of the OOc were simultaneously recorded. Lid movement and EMG recordings were performed twice before BTX treatment and once three to four weeks after the injection, at a time of expected maximum BTX effect. The time latencies to the onset of eye opening and to complete eye opening and the time during which eye opening was sustained were determined and correlated with OOc activity. The results of pre- and posttreatment evaluations were then compared.

RESULTS

Before treatment (1) the lid opening latencies and the lid movement duration were significantly increased in patients compared with control values, and (2) an abnormal persistence of OOc activity was present in 10 of the 11 patients with a delay in complete lid opening.

When compared with the pretreatment values, the patients' posttreatment values showed an improvement in lid movement latencies and lid movement duration, a decreased quantity of EMG during lid opening, and a shortened time from the signal to the inhibition of OOc activity. Even though the mean values for all lid movement measurements improved by 30 to 38% after treatment, they still differed significantly from those of the control subjects.

CONCLUSIONS

The results show that in patients with ALO, treatment with BTX leads to an improvement of lid movement metrics. This improvement is best explained by a decrease in the activity of the OOc muscle.

REFERENCES

1. BOGHEN, D. 1997. Apraxia of lid opening: a review. Neurology **48:** 1491–1503.
2. JANKOVIC, J. & M. HALLET. 1994. Therapy with Botulinum Toxin. Marcel Dekker. New York.
3. 1990. Assessment: the clinical usefulness of botulinum toxin-A in treating neurologic disorders. Report of the Therapeutics and Technology Assessment Subcommittee of the American Academy of Neurology. Neurology **40:** 1332–1336.
4. ELSTON, J.S. 1992. A new variant of blepharospasm. J. Neurol. Neurosurg. Psychiatry **55:** 369–371.
5. ARAMIDEH, M., B.W. ONGERBOER DE VISSER, J.H.T.M. KOELMAN, et al. 1995. Motor persistence of orbicularis oculi muscle in eyelid-opening disorders. Neurology **45:** 897–902.
6. TOZLOVANU, V., R. FORGET, A. IANCU, et al. 2001. Prolonged orbicularis oculi activity: a major factor in apraxia of lid opening. Neurology **57:** 1013–1018.
7. GUITTON, D., R. SIMARD & F. CODERE. 1991. Upper eyelid movements measured with a search coil during blinks and vertical saccades. Invest. Ophthalmol. Vision Sci. **32:** 3298–3305.

Distinguishing Progressive Supranuclear Palsy from Other Forms of Parkinson's Disease

Evaluation of New Signs

SANDRA KUNIYOSHI,[a] DAVID E. RILEY,[a] DAVID S. ZEE,[b] STEPHEN G. REICH,[b] CHRISTINA WHITNEY,[a] AND R. JOHN LEIGH[a]

[a]*Department of Neurology, Case Western University, Cleveland, Ohio 44106, USA*

[b]*Department of Neurology, Johns Hopkins University, Baltimore, Maryland, USA*

KEYWORDS: blink reflex; vestibulocollic reflex; saccades; progressive supranuclear palsy; Parkinson's disease

Distinguishing progressive supranuclear palsy (PSP) from other parkinsonian syndromes may be difficult, especially in the early stages of disease. Based on neuropathological criteria, some patients with PSP disease are misdiagnosed.[1] Characteristic eye movement abnormalities may be missed or delayed until late in the course of PSP.[2] Therefore, there is a need to identify additional clinical signs that may aid early, reliable diagnosis. The goal of this study was to determine the frequency of three clinical signs in patients with classic PSP or Parkinson's disease (PD).

We used currently accepted, clinical criteria to diagnose PSP[3] and PD.[4] Patients were diagnosed before entry into the study by a movement disorders specialist. Other neurologists, who were not informed of the diagnosis beforehand, conducted the study tests. We studied 12 patients (7 male) with PSP; their ages ranged from 68 to 80 years (median 75), duration of diagnosed illness ranged from 1 month to 11 years (median 4 years), and educational level ranged from 8 to 20 years (median 16). We also studied 28 patients (15 male) with PD; their ages ranged from 47 to 80 years (median 70), duration of illness ranged from 4 months to 20 years (median 6), and educational level ranged from 8 to 20 years (median 15). There were no significant differences in the duration of illness or educational level between PSP and PD patients.

We attempted to evaluate three clinical tests on all patients, although a few patients were unable to cooperate for some tests. The first consisted of attempting to induce a blink response by flashing a pen light repetitively into one open eye at 1–2 Hz. An abnormal response was defined as lack of habituation (continued blinking) after more than seven consecutive flash stimuli. The second test consisted of unidirectional rotation in an office chair for 45 seconds; an abnormal response consisted of a sustained head deviation.[5] For the third test, patients were requested to clap

---

Address for correspondence: R. John Leigh, Department of Neurology, University Hospitals of Cleveland, 11100 Euclid Ave., Cleveland, OH 44106. Voice: 216-421-3040; fax: 216-231-3461.

rjl4@po.cwru.edu

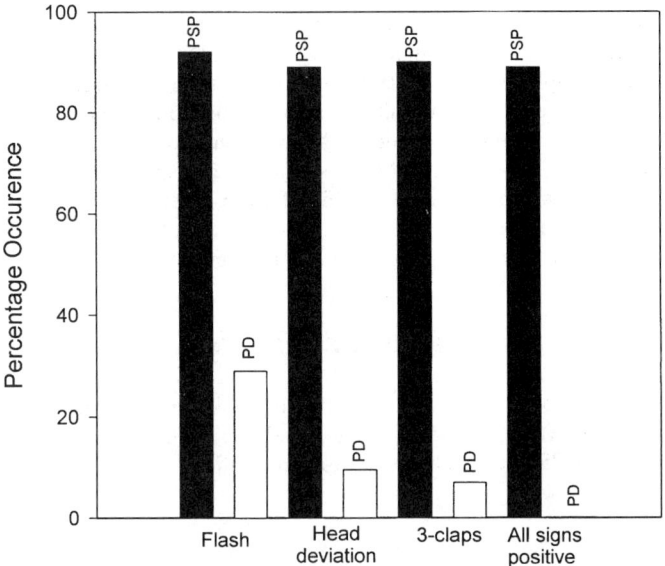

FIGURE 1. Summary of percentage occurrence of abnormal response for the three tests in the PSP and PD groups. "Flash" indicates nonhabituated, light-induced blinking. "Head deviation" refers to sustained deviation induced by body rotation. "3-claps" indicates an inability to perform this task correctly. "All signs positive" indicates the occurrence of all three abnormal signs combined.

quickly exactly three times. They were first asked verbally and then, after the patient had responded, with a demonstration. An abnormal response was defined as more than three claps.[6] Comparisons of the frequency of each finding in each group were made using Fisher's exact test.

We found that nonhabituating, light-induced blinking occurred in 11 out of 12 PSP patients but in only 9 out of 28 PD patients. Tonic head deviation occurred during rotation in 8 out of 9 PSP patients but in only 2 out of 21 PD patients tested. When shown or asked to clap exactly three times, 9 out of 10 PSP patients who could cooperate always exceeded three claps, some with sustained clapping—a positive "applause sign." In contrast, only 2 out of 28 PD patients exceeded three claps. The differences in responses between PSP and PD groups for each of the three tests were significant ($p < 0.01$). The combination of a positive light-blink test, vestibulocollic response, and applause sign was found in 8 out of 9 PSP patients but in 0 out of 24 PD patients. The percentage occurrence of abnormal signs for each of the three tests, and the three combined, in PSP and PD groups is summarized in FIGURE 1.

We selected patients who met accepted clinical criteria for either PSP or PD; all of our PSP patients had slow vertical saccades, and we excluded any patient with slow saccades from our PD group. Each of the tests that we evaluated, and especially the three combined, proved useful in distinguishing patients with PSP and PD. A more exacting evaluation would be to apply these tests to patients who are difficult to diagnose using current clinical criteria, then follow their clinical course, and, when possible, conduct clinical–neuropathological correlations.

The presence of each of these abnormal signs may also provide some insights into the underlying pathophysiology of PSP. Lack of habituation of light-induced blinking may reflect involvement of projections from the superior colliculus and pretectal nuclei to the facial nerve nuclei. Tonic neck deviation may reflect loss of corrective head saccades due to involvement of brainstem reticular structures, such as nucleus gigantocellularis in the rostral medulla. Finally, the inability to clap exactly three times may be one example of perseverative, learned motor behaviors in this disorder, reflecting subcortical frontal or striatopallidal dysfunction.

ACKNOWLEDGMENTS

This work was supported by National Institutes of Health Grants EY06717 and EY01849, Veterans Affairs, the Armington Fund, the Robert M. and Annetta J. Coffelt endowment for PSP research, the Evans Fund, the Kass Fund, and the estate of Nina Kramer.

REFERENCES

1. DANIEL, S.E., V.M. DE BRUIN & A.J. LEES. 1995. The clinical and pathological spectrum of Steele-Richardson-Olszewski syndrome (progressive supranuclear palsy). Brain **118:** 759–770.
2. LEIGH, R.J. & D.E. RILEY. 1999. Eye movements in parkinsonism: it's saccadic speed that counts. Neurology **54:** 1018–1019.
3. LITVAN, I., et al. 1996. Clinical research criteria for the diagnosis of progressive supranuclear palsy (Steele-Richardson-Olszewski syndrome) Neurology **47:** 1–9.
4. GELB, D.J., E. OLIVER & S. GILMAN. 1999. Diagnostic criteria for Parkinson disease. Arch. Neurol. **56:** 33–39.
5. JENKYN, L.R. et al. 1975. The nuchocephalic reflex. J. Neurol. Neurosurg. Psychiatry **38:** 561–566.
6. DUBOIS, B., B. DEFONTAINES, B. DEWEER, et al. 1997. Cognitive and behavioral changes in patients with focal lesions of the basal ganglia. In Behavioral Neurology of Movement Disorders. W.J. Weiner & A.E. Lang, Eds. Adv. Neurol. **65:** 29–41.

Binocular Eye Movement Responses to Dichoptically Presented Horizontal and/or Vertical Stimulus Steps

JOHANNES VAN DER STEEN AND RYOTA KANAI

Neuroscience Institute, Department of Physiology, FGG, Erasmus University Rotterdam, 3000 DR Rotterdam, The Netherlands

KEYWORDS: binocular eye movement; vertical stimulus; stereo vision; binocular oculomotor control; vergence responses

INTRODUCTION

Humans and primates with stereo vision depend on accurate binocular oculomotor control mechanisms to align their two eyes on corresponding points of a visual scene. Misalignments are sensed as the difference in the position of corresponding points of the two retinal images (binocular disparity) and are corrected by vergence eye movements. Vergence responses have been extensively investigated using dichoptically presented visual stimuli.[1-3] The dynamics of horizontal and vertical vergence eye movements are also known.[2,4,5] In general, vertical vergence dynamic properties are inferior to horizontal vergence responses. This suggests separate horizontal and vertical vergence centers with different dynamics. There is also scant evidence, however, for an interaction of the two systems.[6] Another important issue is whether the command signals that generate these vergence eye movements are entirely driven by binocular disparity or also depend on monocular input, that is, eye-of-origin information. Recent electrophysiological evidence suggests that eye movement commands are encoded monocularly for each eye in premotor pathways.[7,8] Here we provide behavioral evidence by using different combinations of horizontal and vertical vergence steps for an eye-dependent interaction between horizontal and vertical vergence signals in humans.

METHODS

Two identical random-dot patterns, one red and the other green, were projected onto a screen placed at 180 cm in front of the subjects. A wide range of conditions of orthogonal, pure horizontal/vertical, and diagonal steps were tested (FIG. 1). Each condition was presented in a pseudorandom order in 30-second trials. During each

---

Address for correspondence: Johannes van der Steen, Neuroscience Institute, Department of Physiology, FGG, Erasmus University Rotterdam, P.O. Box 1738, 3000 DR Rotterdam, The Netherlands. Voice: +31.10.4087572; fax: +31-10-408-9457.

vandersteen@fys.fgg.eur.nl

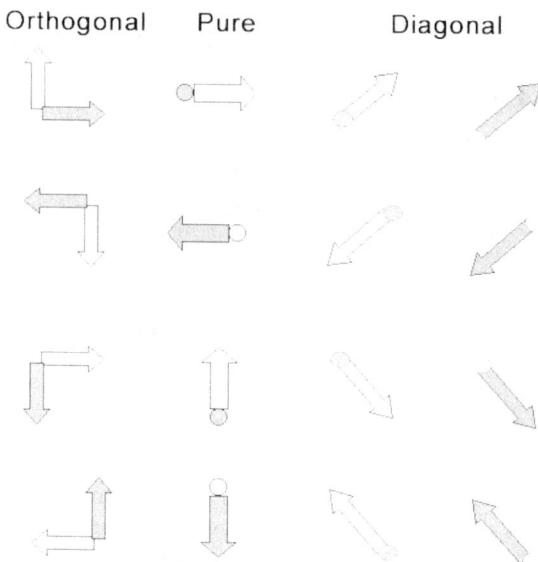

FIGURE 1. The range of combinations of stimulus steps used in our experiment. During pure and diagonal steps (A = 0.75°), one of the two random dot patterns (1000 dots, visual angle 30°) remained stationary. Light = right eye; dark grey = left eye.

trial one of the images stepped every three seconds. We tested six subjects with normal stereopsis (Titmus test) and normal binocular motility. Raw data were digitally low-pass filtered at a cut-off frequency of 125 Hz. For the latency analysis, periods of 1000 milliseconds of vergence position starting at stimulus onset were extracted from individual traces and normalized. Latency of saccadic vergence eye movements were estimated using velocity threshold analysis and a statistical minimum difference (p value) analysis. The latter method, presented here, was based on the average trace of stimulus and vergence response for each condition. The average traces were directly compared by systematically shifting them along the time axis so that a minimum difference of the two traces was obtained in a least-squares method.[9]

RESULTS

All six subjects made adequate vergence eye movements in response to the different combinations of stimulus steps.

Interactions of Horizontal and Vertical Vergence

FIGURE 2 (top panels) shows two collections of average vergence traces in four oblique directions. In both panels vergence had a horizontal and vertical component. In the left panel the vertical step was presented to one eye, and the horizontal step to the other eye (orthogonal stimulation). The right panel shows vergence in response

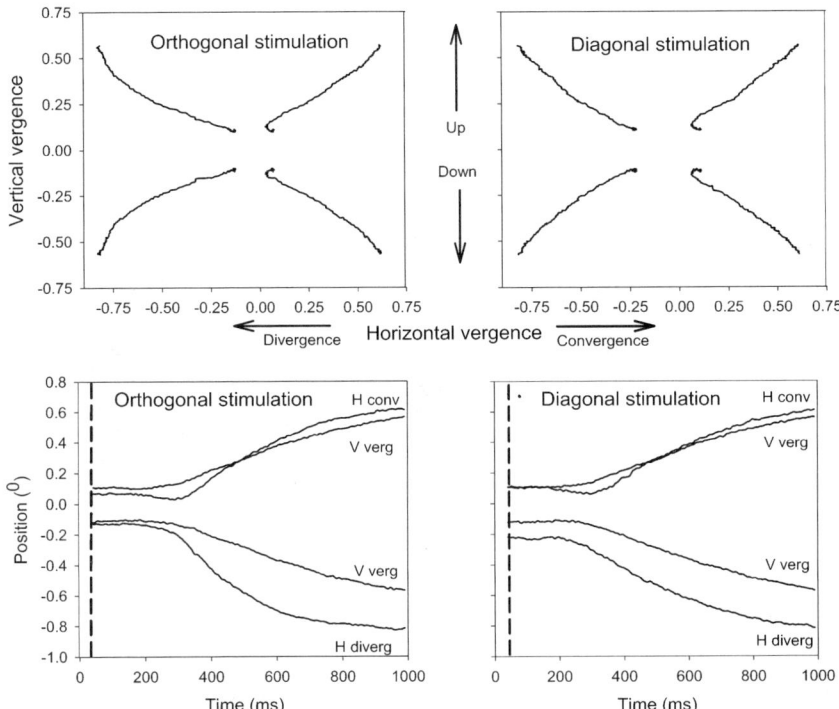

FIGURE 2. XY (*top panels*) and time series (*lower panels*) of averaged (*n* = 16) vergence traces of one subject. Note the transient divergence during convergence (*top panel*) and the decrease in horizontal vergence during diagonal stimulation.

to an oblique step in the same eye, whereas the stimulus remained stationary in the other eye (diagonal stimulation). Orthogonal stimulation resulted in curved oblique vergence paths. This was due to the relative independent dynamics of the horizontal and vertical vergence components in each eye. In contrast, the paths in the right panel were virtually straight. The left and right lower panels of FIGURE 2 demonstrate that, in the diagonal condition, the horizontal vergence is slowed down compared to the orthogonal condition. Thus, in this case horizontal and vertical vergence interact.

Latencies of Vergence Responses

Pure horizontal vergence latencies were 140 ± 21 msec during divergence and 162 ± 34 msec during convergence. This difference was due to a transient divergence. In combination with vertical vergence, horizontal latencies increased to 25 to 50 msec. Pure vertical vergence had longer latency values than horizontal vergence, but these were unaffected by the combination with horizontal vergence. The results are presented in TABLE 1 and FIGURE 3.

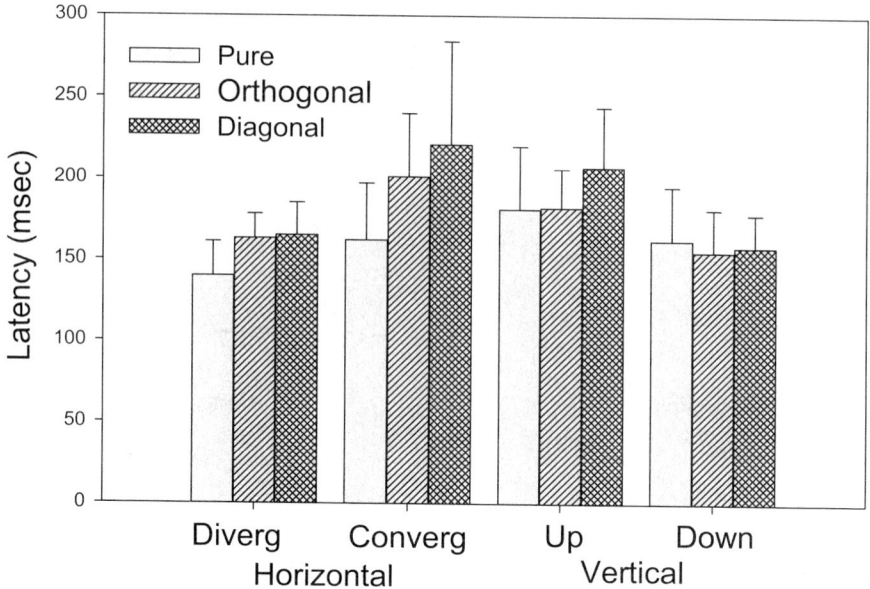

FIGURE 3. Latency histogram of horizontal convergence and divergence and vertical upward and downward vergence under three conditions: pure, orthogonal, and diagonal.

TABLE 1. Vergence latencies under pure, orthogonal, and diagonal conditions

Condition	Vergence			
	Horizontal		Vertical	
	Diverg	Converg	Up	Down
Pure	140 ± 21	162 ± 35	181 ± 39	162 ± 33
Orthogonal	163 ± 15	201 ± 39	182 ± 24	155 ± 26
Diagonal	165 ± 20	221 ± 63	207 ± 37	158 ± 20

CONCLUSIONS

Humans use both binocular disparity and monocular information to generate vergence eye movements in response to dichoptic stimulus steps. As with binocular saccades, a transient divergence occurs during convergent stimulus steps. Pure horizontal vergence responses have shorter latencies than vertical vergence. When combinations of horizontal and vertical vergence exist, both dynamics and timing of the horizontal vergence responses are attenuated.

REFERENCES

1. CUMMING, B.G. & S.J. JUDGE. 1986. Disparity-induced and blur-induced convergence eye movement and accomodation in monkey. J. Neurophysiol. **55:** 896–914.

2. ERKELENS, C.J. & H. COLLEWIJN. 1985. Eye movements and stereopsis during dichoptic viewing of moving random-dot stereograms. Vision Res. **25:** 1689–1700.
3. COLLEWIJN, H. & C.J. ERKELENS. 1990. Binocular eye movements and the perception of depth. Rev. Oculomot. Res. **4:** 213–261.
4. HOUTMAN, W.A., J.H. ROZE & W. SCHEPER. 1981. Vertical vergence eye movements. Doc. Ophthalmol. **51:** 199–207.
5. HOWARD, I.P, R.S. ALLISON & J.E. ZACHER. 1997. The dynamics of vertical vergence. Exp. Brain Res. **116:** 153–159.
6. PERLMUTTER, A. & A.E. KERTESZ. 1978. Human vertical fusional response under open and closed loop stimulation to predictable and unpredictable disparity presentations. IEEE Trans. Biomed. Eng. **29:** 56–61.
7. ZHOU, W. & W.M. KING. 1998. Premotor commands encode monocular eye movements. Nature **393:** 692–695.
8. KING, W.M. & W. ZHOU. 2000. New ideas about binocular coordination of eye movements: is there a chameleon in the primate family tree? Anat. Rec. **261:** 153–161.
9. RABNER, L.R. & B. GOULD. 1975. Theory and Application of Digital Signal Processing. Prentice-Hall. Englewood Cliffs, NJ.

Visual Processing in Disparity Vergence Control

SCOTT B. STEVENSON

University of Houston College of Optometry, Houston, Texas 77204-2020, USA

KEYWORDS: visual processing; disparity vergence control; luminance-modulated grating stimuli; contrast-modulated grating stimuli

Disparity vergence refers to the oculomotor control system that maintains horizontal, vertical, and torsional eye alignment based on retinal image disparity. It has both voluntary and involuntary aspects,[1–4] but voluntary control seems limited to the horizontal component.[2] Vertical disparity vergence appears to be completely reflexive: its gain is uninfluenced by attention or effort, and observers cannot normally detect the direction of vertical disparity. The control system for reflexive disparity vergence may represent a distinct aspect of visual processing with characteristics that differ from those measured with psychophysics or with voluntary eye movements. Vertical disparity vergence measurements provide a tool for studying this aspect of visual processing.

Our general research question can be stated informally as whether the reflexive and voluntary aspects of vergence "see" the same things. Specifically, we compare voluntary, horizontal vergence responses to reflexive, vertical vergence responses using a variety of visual stimuli. In this study we compared horizontal and vertical vergence responses to luminance-modulated ("first-order") and contrast-modulated ("second-order") grating stimuli.

METHODS

Subjects were persons from the research group with normal binocular vision. All procedures were approved by the University of Houston Committee on Human Subjects. Stimuli were sine-wave gratings defined either by luminance change (first-order stimuli) or by contrast change (second-order stimuli). The contrast-modulated stimuli were dynamic visual noise patterns with sinusoidal modulation of contrast, forming either horizontal or vertical bars (FIG. 1). Disparity modulation was sinusoidal in time at 0.25 Hz. Horizontal and vertical vergence responses were tested in separate blocks of trials. Subjects were instructed to fixate on the grating and follow

Address for correspondence: Scott B. Stevenson, Ph.D., University of Houston College of Optometry, Houston, TX 77204-2020. Voice: 713-743-1960; fax: 713-743-2053.
SBStevenson@uh.edu

Ann. N.Y. Acad. Sci. 956: 492–494 (2002). © 2002 New York Academy of Sciences.

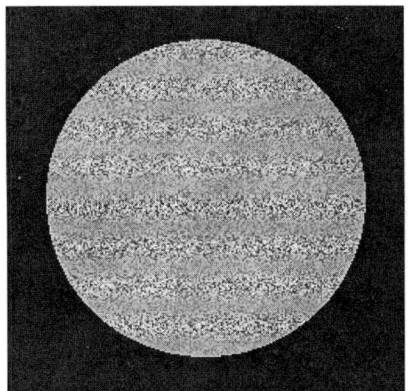 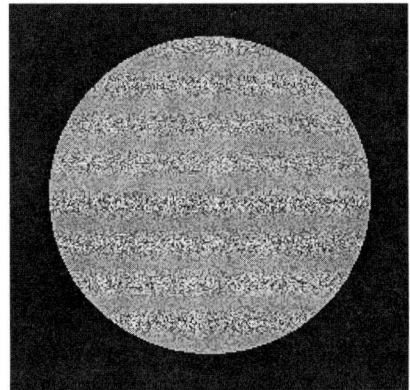

FIGURE 1. Contrast-defined second-order grating formed by multiplying a sine wave and visual noise. Left and right images were made with uncorrelated, dynamic noise to prevent spurious correlations from driving vergence. Field size in this case was 30 degrees, and no fixation or nonius marks were present. Bars were oriented horizontally for vertical disparity change, as shown in the figure. Bars were oriented vertically for horizontal disparity change (not shown). The contrast modulation pattern was a 0.5 cycle per degree sine-wave grating. Rectification or other nonlinearity in the visual system would make this stimulus effectively equivalent to a luminance grating of 0.5 cycle per degree.

its motion. Eye movements were recorded on a dual-Purkinje eye tracker and analyzed for gain and phase of tracking at the stimulus frequency.

RESULTS

Luminance modulation produced robust tracking for both horizontal and vertical disparity vergence. Tracking gain was systematically increased with increasing contrast of the grating. Contrast modulation produced robust tracking only for horizontal disparity. Subjects could clearly see the vertical disparity stimulus and its motion, but the reflex vergence system did not respond to the disparity change (FIG. 2).

DISCUSSION

Reflexive vergence responses are robust for luminance-defined stimuli, such as sine-wave gratings or binocularly correlated, dynamic visual noise. When uncorrelated noise is used to make a contrast-defined pattern, however, the reflex system appears to be blind to it. The voluntary vergence system is quite capable of tracking these second-order stimuli, underscoring the notion that the reflex and voluntary systems "see" different aspects of a visual stimulus. With respect to second-order stimuli, the nonlinearity that renders them visible in perception may occur relatively late in visual processing, past the stage of initial disparity extraction for eye alignment control. Alternatively, it may occur early in processing, but in a parallel pathway that does not contribute signals to reflex vergence control.

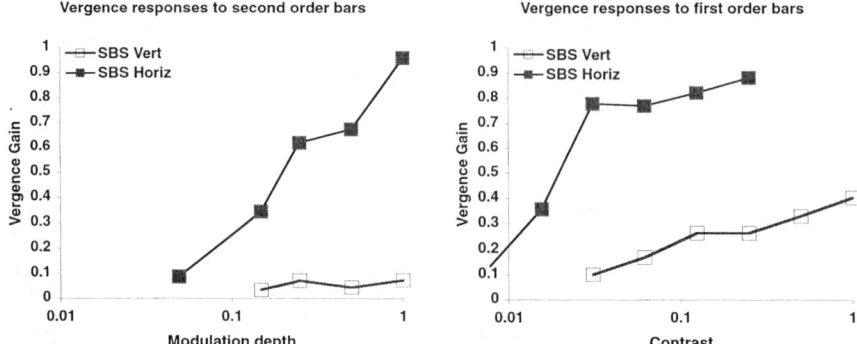

FIGURE 2. Vergence responses to luminance gratings (*right panel*) and to contrast-modulated, binocularly uncorrelated dynamic visual noise (*left panel*). When the second-order bars were vertical and the disparity change was horizontal, vergence responses were robust and depended systematically on the modulation depth. When the bars were horizontal and the disparity change was vertical, no reliable vergence responses were recorded. Luminance-modulated (first-order) stimuli produce robust tracking for both axes of disparity motion, although horizontal responses show higher gain than vertical. The vertical, reflexive vergence system seems to be blind to second-order patterns of this kind, even though the subjects are perceptually aware of them and can voluntarily track them with horizontal vergence.

ACKNOWLEDGMENTS

This work was supported by grants from the National Eye Institute.

REFERENCES

1. ERKELENS, C.J. & H. COLLEWIJN. 1991. Control of vergence: gating among disparity inputs by voluntary target selection. Exp. Brain Res. **87:** 671–678.
2. STEVENSON, S.B., L.A. LOTT & J. YANG. 1997. The influence of subject instruction on horizontal and vertical vergence tracking. Vision Res. **37:** 2891–2898.
3. ERKELENS, C.J. & H. COLLEWIJN. 1985. Motion perception during dichoptic viewing of moving random-dot stereograms. Vision Res. **25:** 583–588.
4. ERKELENS, C.J. & H. COLLEWIJN. 1985. Eye movements and stereopsis during dichoptic viewing of moving random-dot stereograms. Vision Res. **25:** 1689–1700.
5. STEVENSON, S.B., P.E. REED & J. YANG. 1999. The effect of target size and eccentricity on reflex disparity vergence. Vision Res. **39:** 823–832.
6. HOWARD, I.P., X. FANG, R.S. ALLISON & J.E. ZACHER. 2000. Effects of stimulus size and eccentricity on horizontal and vertical vergence. Exp. Brain Res. **130:** 124–132.
7. ALLISON, R.S., I.P. HOWARD & X. FANG. 2000. Depth selectivity of vertical fusional mechanisms. Vision Res. **40:** 2985–2998.
8. HOWARD, I.P., R.S. ALLISON & J.E. ZACHER. 1997. The dynamics of vertical vergence. Exp. Brain Res. **116:** 153–159.

Anticipatory Saccadic–Vergence Responses in Humans

ARUN N. KUMAR,[a] YANNING HAN,[a] STEFANO RAMAT,[b] AND
R. JOHN LEIGH[a]

[a]*Departments of Biomedical Engineering and Neurology, Case Western Reserve University, Cleveland, Ohio 44106, USA*

[b]*Dipartimento di Informatica e Sistemistica, University of Pavia, Pavia, Italy*

KEYWORDS: **saccades; vergence movements; burst neurons; omnipause neurons; brainstem**

Under natural conditions, most shifts of our point of visual fixation are between objects that lie in different directions and different distances in the environment. These shifts are achieved by combined saccadic and vergence movements.[1] Saccades and vergence movements are each generated by separate populations of "burst neurons," which lie in the brainstem.[2,3] Saccadic burst neurons are inhibited by omnipause neurons, which lie in the pons of the brainstem.[4] Recent studies have suggested that omnipause neurons gate both saccadic and vergence burst neurons, and thus act as a premotor switch that allows gaze shifts.[1,5] Omnipause neurons receive motor commands from the frontal eye fields and the superior colliculus[6] and also respond to visual stimuli.[7] We investigated whether it was the visual stimulus, the saccadic command, or the vergence command that triggered the omnipause "switch" to permit a shift in the point of fixation.

We studied 10 normal subjects, all of whom gave informed consent. We measured movements of each eye using the magnetic search-coil technique. Subjects alternately switched their point of fixation between near and far targets, both aligned on the midline. The far visual stimulus was a red laser spot at 1.2 m, and the near target was a green LED at 20 cm. Each target was alternately illuminated in a predictable sequence, every 1.25 sec. Digitized signals were filtered and differentiated as previously described.[8] Our experimental strategy was to look for small, high-frequency, conjugate oscillations, which are a behavioral marker that omnipause neurons are turned off.[8] We used our data to test a current model for combined saccade–vergence movements.[1]

All subjects showed some anticipatory responses, each consisting of a vergence drift, accompanied by a small saccade, that preceded target jumps (FIG. 1A). Overall, predictive responses occurred during 95% of divergence and 85% of convergence movements. In all trials, an initial vergence movement preceded the saccade, and this

---

Address for correspondence: R. John Leigh, M.D., Department of Neurology, University Hospitals of Cleveland, 11100 Euclid Avenue, Cleveland, OH 44106. Voice: 216-421-3040; fax: 216-231-3461.

rjl4@po.cwru.edu

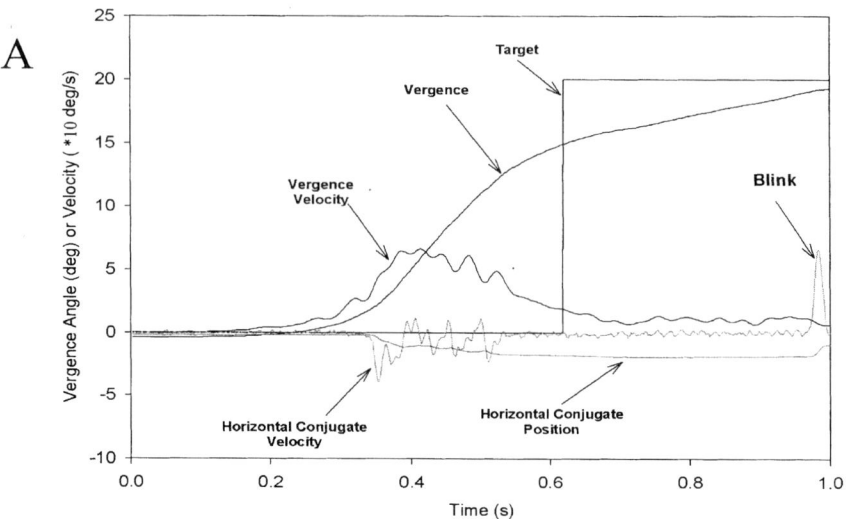

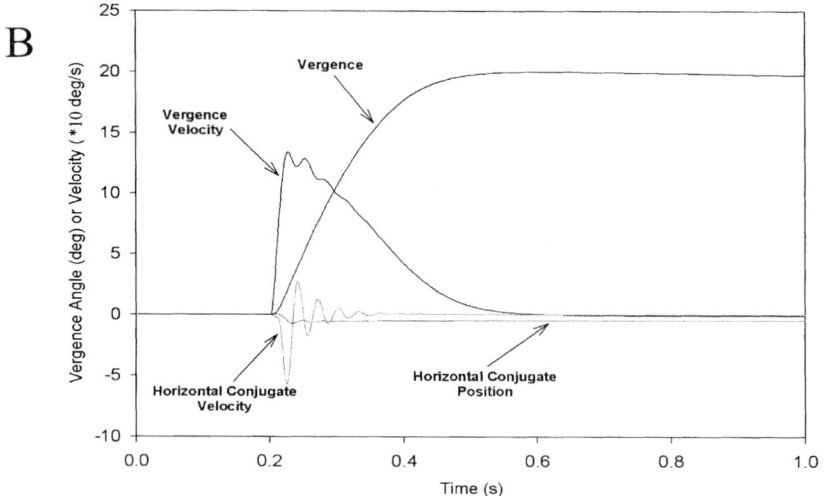

FIGURE 1. (A) Representative records of a convergence response. **(B)** Results from the simulation of the saccade–vergence model by Zee *et al.*[1] To produce the conjugate oscillations, it was necessary to increase the gain of the saccadic burst neurons. Note different scales for eye position and velocity.

was always free of high-frequency oscillations. At the onset of the saccadic component, conjugate oscillations at around 30 Hz commenced, and these were sustained during the part of the vergence movement that continued after the saccade was completed (FIG. 1A). High-frequency oscillations did not occur during the vergence movement that preceded the saccade; they commenced after the initial saccade, and occurred before the visual target jumped to its new position. Thus, these results indicate that the saccadic command causes omnipause neurons to be turned off, not the vergence command and not the visual stimulus for the gaze shift.

We used these data to test a current model for saccade–vergence interactions.[1] We did not attempt to simulate the presaccadic vergence response, nor did we allow an interaction between vergence and saccadic burst neurons. Our focus of interest was whether conjugate oscillations, such as we observed (FIG. 1A), were generated by the model, since these are the behavioral marker for the brainstem saccadic switch. We found that to produce these oscillations (FIG. 1B), we needed to increase the gain of saccadic burst neurons for small movements. This result suggests that the gain of the saccadic system for small motor errors (desired change in direction of fixation) is larger than Zee and colleagues suggested and that if omnipause neurons are turned off, but no saccadic command is present (after the saccade), then small conjugate oscillations at about 30 Hz occur. Some normal subjects can voluntarily induce these oscillations using a vergence effort gain ("voluntary nystagmus").[9] In summary, our data indicate that a small saccade is needed to turn off the omnipause switch and that a high gain of burst neurons is required to sustain the oscillations. We have shown that a behavioral marker of a brainstem motor switch can be used to detect the nature of the sensory and motor commands that drive ballistic eye movements. We suggest a modification to a current model for combined saccade–vergence shifts of the point of fixation.

ACKNOWLEDGMENTS

This work was supported by National Institutes of Health Grant EY06717 and grants from Veterans Affairs and the Evenor Armington Fund.

REFERENCES

1. ZEE, D.S., et al. 1992. Saccade–vergence interactions in humans. J. Neurophysiol. **68:** 1624–1641.
2. VAN GISBERGEN, J.A.N., D.A. ROBINSON & S. GIELEN. 1981. A quantitative analysis of generation of saccadic eye movements by burst neurons. J. Neurophysiol. **45:** 417–442.
3. MAYS, L.E., J.D. PORTER, P.D.R. GAMLIN & C.A. TELLO. 1986. Neural control of vergence eye movements: neurons encoding vergence velocity. J. Neurophysiol. **56:** 1007–1021.
4. BÜTTNER-ENNEVER, J.A., et al. 1988. Raphe nucleus of pons containing omnipause neurons of the oculomotor system in the monkey, and its homologue in man. J. Comp. Neurol. **267:** 307–321.
5. MAYS, L.E. & P.D.R. GAMLIN. 1995. A neural mechanism subserving saccade–vergence interactions. In Eye Movement Research: Mechanisms, Processes and Applications. J.M. Findlay, R. Walker & R.W. Kentridge, Eds.: 215–223. Elsevier. Amsterdam.

6. BÜTTNER-ENNEVER, J.A. & A.K.E. HORN. 1997. Anatomical substrates of oculomotor control. Curr. Opin. Neurobiol. **7:** 872–879.
7. EVINGER, C., C.R.S. KANEKO & A.F. FUCHS. 1982. Activity of omnipause neurons in alert cats during eye movements and visual stimuli. J. Neurophysiol. **47:** 825–844.
8. RAMAT, S., J.T. SOMERS, V.E. DAS & R.J. LEIGH. 1999. Conjugate ocular oscillations during shifts of the direction and depth of visual fixation. Invest. Ophthalmol. Vis. Sci. **40:** 1681–1686.
9. HOTSON, J.R. 1994. Convergence-initiated voluntary flutter: a normal intrinsic capability in man. Brain Res. **294:** 299–304.

Vergence Eye Movements in Strabismus Patients

HORACE R. LEE[a,b] AND MOSHE EIZENMAN[a,b,c]

[a]*Edward S. Rogers Sr. Department of Electrical and Computer Engineering,* [b]*Institute of Biomaterials and Biomedical Engineering,* [c]*Department of Ophthalmology, University of Toronto, Toronto, Ontario M5S 3G9, Canada*

> KEYWORDS: vergence eye movements; disparity vergence stimuli; strabismus; infantile esotropia

Fusional vergence eye movements in response to step disparity vergence stimuli were recorded and analyzed in 10 children with infantile esotropia. Unlike children with normal binocular vision who generated smooth vergence eye movements (see FIG. 1A), to fuse the disparity stimuli, children with infantile esotropia exhibited responses with a combination of unequal saccades, monocular drifts, and/or disconjunctive eye movements (see FIG. 1B,C). The end result of the combined sequence of eye movements achieved partial to full motor compensation for the disparity vergence stimuli.

To better understand the set of responses, the effects of monocular suppression scotomas on responses to disparity vergence stimuli were explored. For this purpose, eye movements of subjects with normal binocular vision were studied, using disparity vergence stimuli with artificial monocular scotomas.

To pursue this study, we developed an integrated system that includes a binocular eye-tracking system, VISION 2000 (EL-MAR Inc. Toronto, Ontario), and a stereoscopic display that can present time-multiplexed images to the two eyes. With the display, we generated disparity stimuli with artificial monocular scotomas of varying sizes (1–15 degrees).

Responses to disparity vergence stimuli with small monocular scotomas (1 degree) were slower than vergence eye movements that were observed in the absence of a scotoma (FIG. 2A). Symmetric fusional vergence eye movements fully compensated for the stimulus disparity. With large (15 degrees) scotomas (FIG. 2C), the responses consisted of symmetric saccades followed by small monocular drifts in the eye with the scotoma. The motor response only partially compensated for the induced disparity (FIG. 2C, right). Scotomas that fall between 1 and 15 degrees consisted of a combination of vergence and saccadic eye movements (FIG. 2B). The

---

Address for correspondence: Horace R. Lee, Institute of Biomaterials and Biomedical Engineering, University of Toronto, Rosebrugh Building, 4 Taddle Creek Road, Toronto, Ontario M5S 3G9, Canada. Voice: 416-978-2255; fax: 416-978-4317.

horace.lee@utoronto.ca

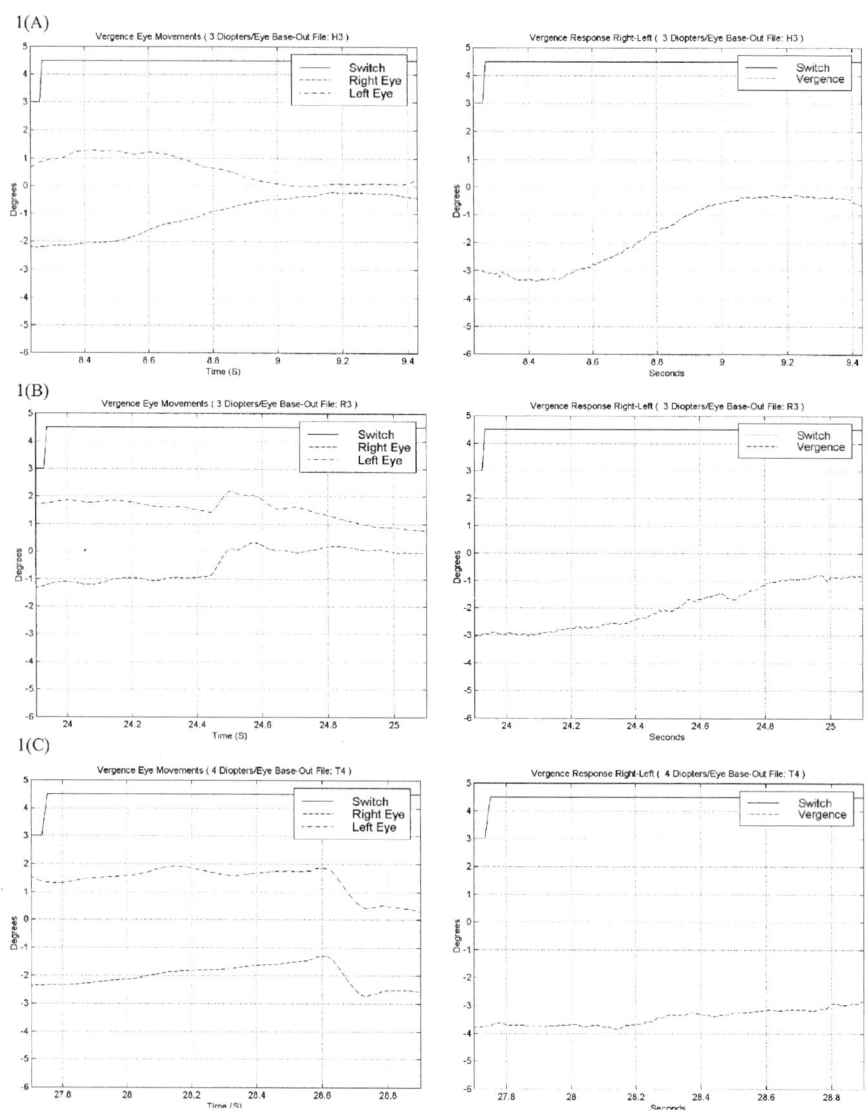

FIGURE 1. (**A**) Typical eye movement response of a subject with normal binocular vision to 3-diopter/eye prisms. *Right*: Vergence response (right eye–left eye). (**B,C**) Examples of disparity vergence eye movements in infantile esotropes. (**B**) Asymmetric saccades followed by a monocular drift. (**C**) Saccade movements with little or no vergence.

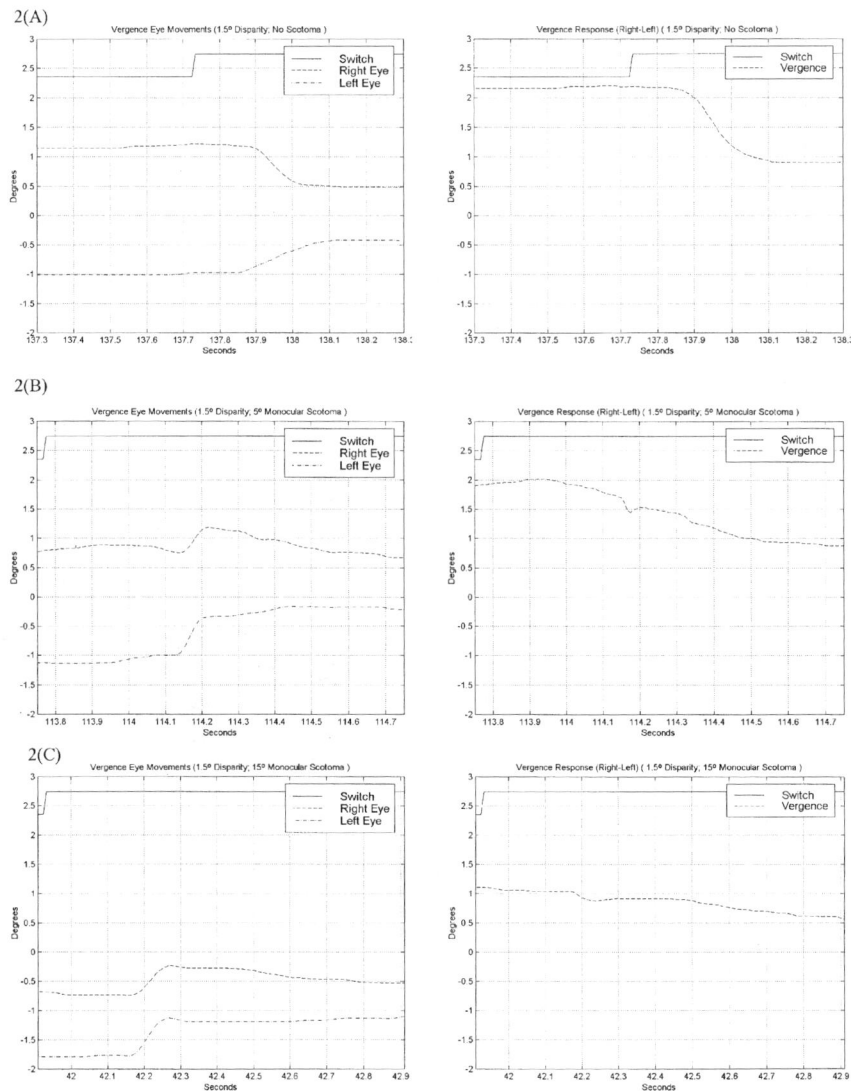

FIGURE 2. Vergence eye movements of a subject with normal binocular vision in response to 1.5-degree disparity stimuli with an artificial monocular scotoma presented to the right eye. **(A)** Vergence eye movements in response to stimuli without a scotoma. **(B)** Response to stimuli with a 5-degree scotoma. **(C)** Response to stimuli with a 15-degree scotoma.

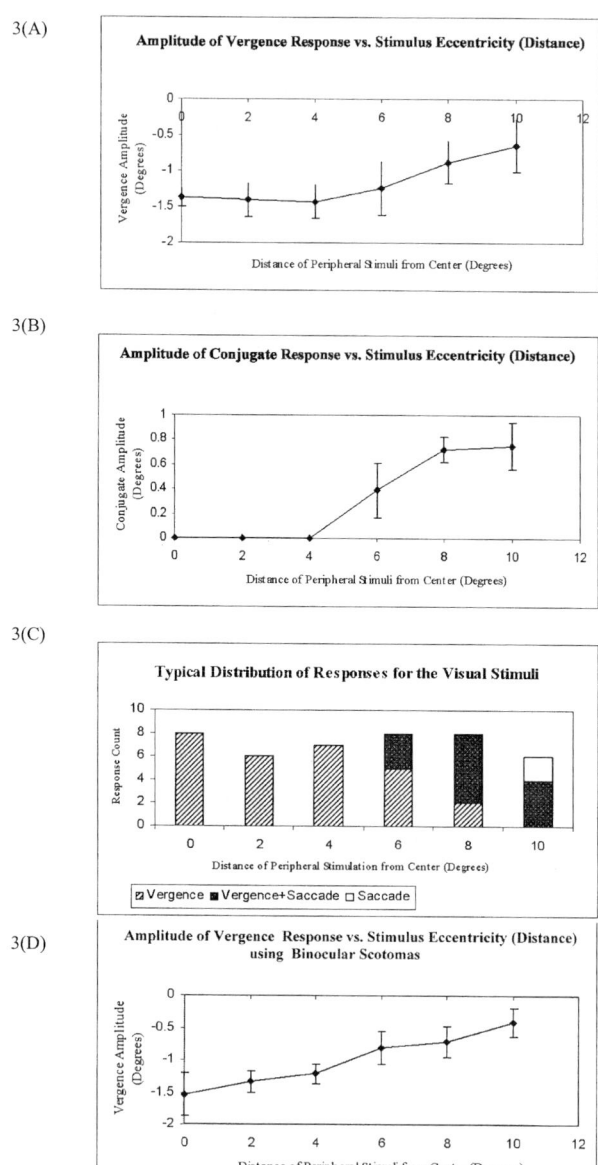

FIGURE 3. (**A**) Summary of relationships between vergence amplitude and stimulus eccentricity; (**B**) conjugate amplitude and stimulus eccentricity; (**C**) nature of the response and stimulus eccentricity; and (**D**) vergence amplitude and stimulus eccentricity with binocular scotomas.

range of responses obtained in this study was similar to those in the study with infantile esotropes.

The above results suggest that the observed response to disparity vergence stimuli in children with infantile esotropia is a combination of two components. The visual stimuli from the portion of the field that is seen monocularly stimulates saccadic eye movements, while the visual stimuli from the portion of the field that is seen binocularly stimulates fusional vergence eye movements. The observed response is dependent on the relative contribution of these two systems.

Studies of stimulus parameters that affect eye movement responses to vergence disparity stimuli in the presence of monocular scotomas demonstrate that the magnitude of vergence responses depends on the eccentricity of the binocular disparity vergence stimulus (FIG. 3) and is independent of the ratio of information that is contained in monocular and binocular portions of the stimulus. In these studies, the visual stimuli consisted of a set of step-symmetric disparity vergence stimuli with different sizes of monocular scotomas. Unlike the previous study, the content of the monocular portion of the stimuli did not change with the size of the scotoma (creating a constant saccadic demand), and the binocular portion of the stimuli changes in eccentricity but not in size. FIGURE 3A demonstrates that the vergence amplitude decreases as the eccentricity of the binocular vergence stimuli increases, whereas the conjugate saccadic amplitude increases (FIG. 3B). Furthermore, as the eccentricity of the binocular portion of the stimulus increases, the probability of observing pure saccadic or combined saccadic and vergence responses increases (FIG. 3C). To learn more about the interaction between the saccadic and vergence systems, we used the same set of stimuli, but with binocular central scotomas in the two eyes. The results in FIGURE 3D show that even in the absence of monocular saccadic stimuli, motor vergence responses do not fully compensate for the disparity introduced by the visual stimuli when the scotoma sizes are larger than 4 degrees.

The above results suggest that when the amplitude (or more precisely the initial velocity) of vergence eye movement in response to disparity vergence stimuli is large enough to compensate for the disparity, the probability of observing saccades in the responses is low. When the initial vergence response to the disparity stimuli is inadequate (i.e. too slow) saccadic eye movements are frequently superimposed on the vergence response so as to foveate the image that is presented to the eye without the artificial scotoma.

Vergence Disorders in Progressive Supranuclear Palsy

KITTISAK KITTHAWEESIN, DAVID E. RILEY, AND R. JOHN LEIGH

Neurology Service, Veterans Affairs Medical Center and Case Western Reserve University, Cleveland, Ohio 44106, USA

KEYWORDS: convergence; saccades; progressive supranuclear palsy; vergence disorders

Slow vertical saccades are a cardinal feature of progressive supranuclear palsy (PSP), but vergence movements have not been formally evaluated.[1] Most natural shifts of our point of visual fixation are made between objects lying in different directions and at different distances, requiring combined saccadic–vergence movements. The midbrain, which houses the premotor machinery for both vertical saccades and vergence, is involved in PSP.[2] Therefore, we measured the dynamic properties of vergence components of saccadic–vergence movements in PSP.

We studied five women with probable PSP[3]; ages ranged from 68 to 76 years, and disease duration from 2 to 7 years (median 2.5). We also studied 12 healthy control subjects, ages ranging from 25 to 62 years. All patients and subjects gave informed consent. Horizontal and vertical gaze were measured using the magnetic search coil technique. A red laser spot served as a "far target." It was projected onto a tangent screen at a viewing distance of 1.2 meters, 10 degrees above the zero position. A green LED served as a the "near target." It was located at vertical zero position and at a distance, calculated for each patient, to require 10 degrees of vergence (typically at about 35 cm). Thus, each shift of fixation between the far and near targets required both a 10-degree vertical saccade and a 10-degree vergence movement. We ran three types of trials, in which the LED was aligned with the far target for either the right or the left eye (Müller paradigm) or in the subject's midline. Each gaze shift was prompted by illumination of either the near or far targets, with a 100-msec gap; each target light remained illuminated for 2.4 sec to allow time for subjects to fixate it. Three subjects also made vergence movements ranging from 1 to 8 degrees, with vertical saccades. Coil signals were filtered, digitized, and differentiated, as previously described.[1] Vergence was defined as left horizontal gaze minus right horizontal gaze. All responses were analyzed interactively. To compare the amplitude-peak velocity relationships of vergence paradigms, we fit data for convergence and divergence separately with equations of the form:

Address for correspondence: R. John Leigh, Department of Neurology, University Hospitals of Cleveland, 11100 Euclid Ave., Cleveland, OH 44106. Voice: 216-421-3040; fax: 216-231-3461.

rjl4@po.cwru.edu

Ann. N.Y. Acad. Sci. 956: 504–507 (2002). © 2002 New York Academy of Sciences.

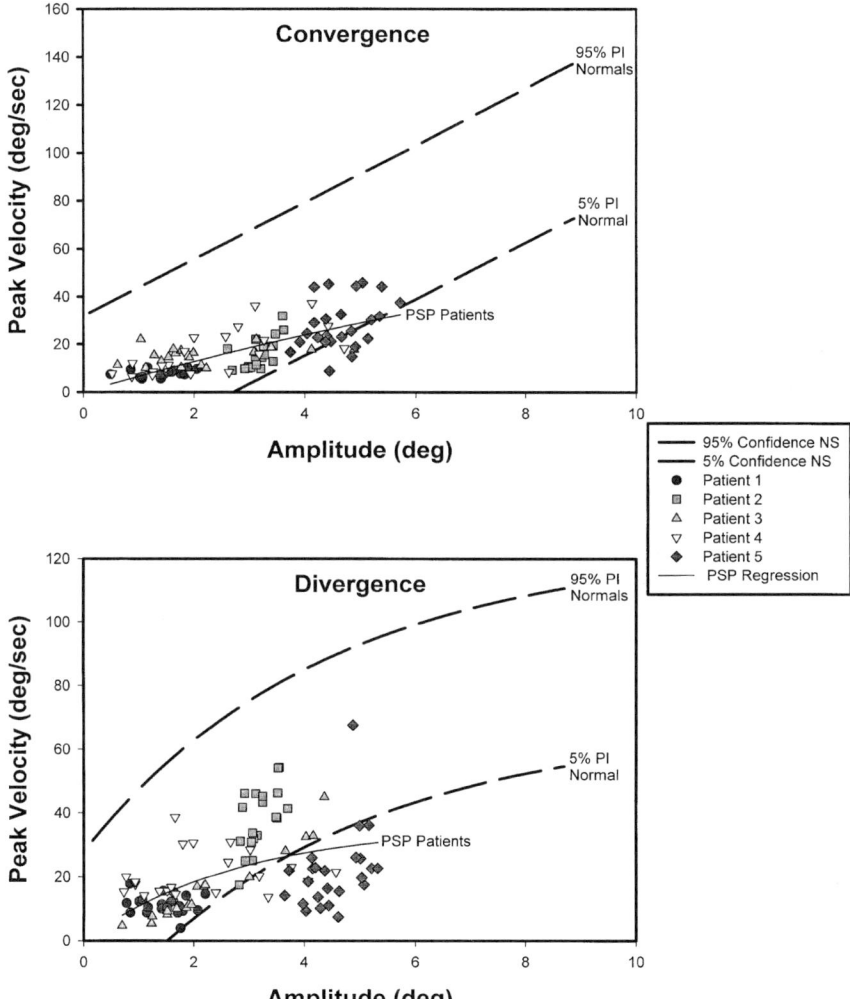

FIGURE 1. Summary of relationships between amplitude and peak velocity of vergence movements showing prediction intervals (PI) for the group of normal subjects (*dashed lines*) and individual responses from PSP patients (*data points*) with regression curve.

$$\text{Peak velocity} = V_{\max} * (1 - e^{\text{amplitude}/C})$$

where $V_{\max}$ is the asymptotic peak velocity, and C is a constant.

For normal subjects, mean (±SD) amplitudes of convergence responses were 6.66 ± 1.56° for asymmetric (Müller) stimuli, and 6.24 ± 1.82° for symmetrical stimuli. Mean amplitudes of divergence responses to asymmetrical stimuli were 6.61 ± 1.60°, and 6.17 ± 1.85° for symmetric stimuli. Using the equation above, we defined relationships between amplitude and peak velocity of convergence and divergence (line plots in FIG. 1), and calculated 95% prediction intervals. All five patients with PSP made smaller vergence movements than controls ($p < 0.01$). Mean amplitudes of convergence responses were $2.89 \pm 1.43°$ for asymmetric stimuli and $2.58 \pm 1.28°$ for symmetrical stimuli. Mean amplitudes of divergence responses to asymmetrical stimuli were $2.89 \pm 1.39°$ and $2.56 \pm 1.22°$ for symmetric stimuli. The regression slope of the relationship between amplitude and peak velocity vergence movements for PSP patients (FIG. 1) was lower than for normal subjects, but the peak velocities of smaller vergence movements (4 degrees or smaller) were within 95% prediction intervals for normal subjects.

Current evidence suggests that the vertical saccadic component is generated by burst neurons in the riMLF,[4] and the vergence component is generated by vergence burst neurons, lying in the mesencephalic reticular formation, dorsolateral to the oculomotor nucleus.[5] Electrophysiological evidence suggests that both saccadic and vergence burst neurons may be gated by omnipause neurons that lie in the nucleus raphe interpositus.[6] We found that vergence components of combined responses in PSP patients were smaller than control subjects, and that the regression slope of their amplitude–peak velocity relationship was lower than for control subjects. However, vergence movements less than 4 degreees were within 95% prediction intervals for normal subjects. It is also reported that small vertical saccades are similar to controls in PSP[7]; only larger movements are slowed. Thus, it seems possible that a similar disturbance affects vertical saccades and vergence movements in PSP. The observed disturbance of vertical saccades and vergence movements in PSP may indicate either a similar abnormality of saccadic and vergence burst neurons, or a common disturbance of feedback control of their discharge.

ACKNOWLEDGMENTS

This work was supported by Department of Veterans Affairs, National Institutes of Health Grant EY06717, the Evenor Armington Fund Fund, the Kass Fund, and the estate of Nina Kramer.

REFERENCES

1. BHIDAYASIRI, R., *et al.* 2001. Pathophysiology of slow vertical saccades in progressive supranuclear palsy. Neurology **57:** 2070–2071.
2. LITVAN, I. *et al.* 2000. Research goals in progressive supranuclear palsy. Mov. Disord. **15:** 446–458.
3. LITVAN, I. *et al.* 1996. Clinical research criteria for the diagnosis of progressive palsy (Steel-Richardson-Olszewski syndrome). Neurology **47:** 1–9.
4. HORN, A.K.E. & J.A. BÜTTNER-ENNEVER. 1998. Premotor neurons for vertical eye-movements in the rostral mesencephalon of monkey and man: the histological identification by parvalbumin immunostaining. J. Comp. Neurol. **392:** 413–427.
5. MAYS, L.E., J.D. PORTER, P.D.R. GAMLIN & C. TELLO. 1986. Neural control of vergence eye movements: neurons encoding vergence velocity. J. Neurophysiol. **56:** 1007–1021.
6. MAYS, L.E. & P.D.R. GAMLIN. 1995. A neural mechanism subserving saccade–vergence interactions. *In* Eye Movement Research: Mechanisms, Processes and

Applications. J.M. Findley, R. Walker & R.W. Kentridge, Eds.: 215–223. Elsevier. Amsterdam.
7. AVERBUCH-HELLER, L., et al. 2002. Small vertical saccades have normal speeds in progressive supranuclear palsy (PSP). Ann. N.Y. Acad. Sci. This volume.

A New Way to Find the Primary Position in Three-Dimensional Measurements

J. YGGE,[a] M. BENASSI,[b] AND R. BOLZANI[b]

[a]*Department of Ophthalmology, Karolinska Institutet, SE-141 86 Stockholm, Sweden*
[b]*Department of Psychology, University of Bologna, Bologna, Italy*

KEYWORDS: primary position; 3-D eye movement recording; ocular torsion

BACKGROUND

The eye can rotate around an infinite number of axes, all but one located within Listing's plane, which in most instances corresponds to the equatorial plane. Listing's plane is thus perpendicular to the visual axes when the eyes are in primary position. Ocular torsion occurs around an axis parallel to the line of sight and corresponds to the angle between eye vertical axis and the objective vertical direction in Fick's coordinates. The ocular torsion in tertiary positions can be explained by using Listing's law. Listing's law is valid for saccadic and smooth pursuit movements, but it can be violated by vestibular eye movements like head rotations in roll. In these cases the real primary position (PP) coordinates are required to evaluate the theoretical torsion from the Listing's law and to compare the theoretical to the measured torsion values. For torsion we need to refer the measured angle to the PP, as defined by Listing's law, from which pure horizontal and vertical movements do not produce any torsional change. Many methods have been proposed to evaluate the PP. The aim of this work has been to set up and test a new method for the evaluation of the PP based on two-dimensional fitting curves.

METHODS

Ten normal subjects took part in this experiment. A search coil system (Skalar) was used to measure the 3-D eye position. The stimulus target, a red spot on a dark screen, moved from the center position to 20° left-down to 20° right-up and again 20° left-down (FIG. 1). The cycle was repeated ten times. Then the target moved from the center to 20° right-down to 20° left-up and 20° right-down. This cycle was also repeated ten times. Thus an X-shaped eye movement pattern was performed that was analyzed off-line by computer software. Using a polynomial fitting function, two fitting curves were achieved (one for the movement from left-down to right-up and the other from right-down to left-up). These curves represent the torsional values

Address for correspondence: J. Ygge, Department of Ophthalmology, Karolinska Institutet, SE-141 86 Stockholm, Sweden. Voice: 46-8-58581496; fax: 46-8-7793506.
jan.ygge@klinvet.ki.se

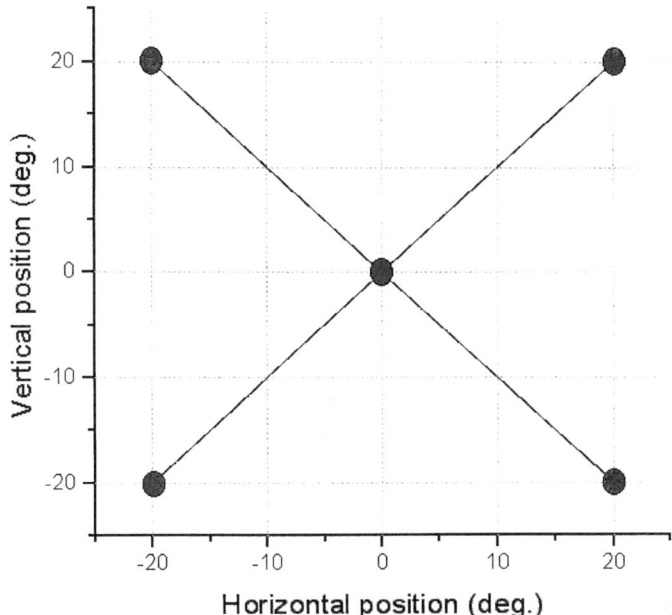

FIGURE 1. Target positions as seen by the subject. The stimuli target moved from the center position to 20° left-down to 20° right-up and again to 20° left-down. The cycle was repeated ten times. Thereafter the stimuli moved from center to 20° right-down to 20° left-up and again to 20° right-down. This cycle was also repeated ten times. Thus an X-shaped movement pattern was created.

related to the vertical and horizontal position (FIGS. 2A and B). These graphs, when adjusted to zero position, had two crossing points. The zero degree crossing point corresponds to the central target position. The other crossing point corresponds to the real PP in the Listing's law sense, and was related to the intersection between the curves and the vertical axis started from the PP. Following Listing's law, the pure vertical movements from the PP do not change the torsional value, conventionally assumed to be zero in the PP. In this way the horizontal displacement of the PP was determined.

RESULTS

The method proposed here allowed us to estimate the PP for all subjects. The R^2 goodness-of-fit ranged from 0.49 to 0.99. The PP horizontal displacement ranged from −10° to +7°, while the vertical ranged from −5° to +2°. The PP displacement was used to correct the horizontal and vertical eye position. The theoretical torsion angle was then evaluated by means of the Listing formula. The comparison between the theoretical and the measured torsion showed a good agreement in XY graph and correlation coefficients were close to 0.9.

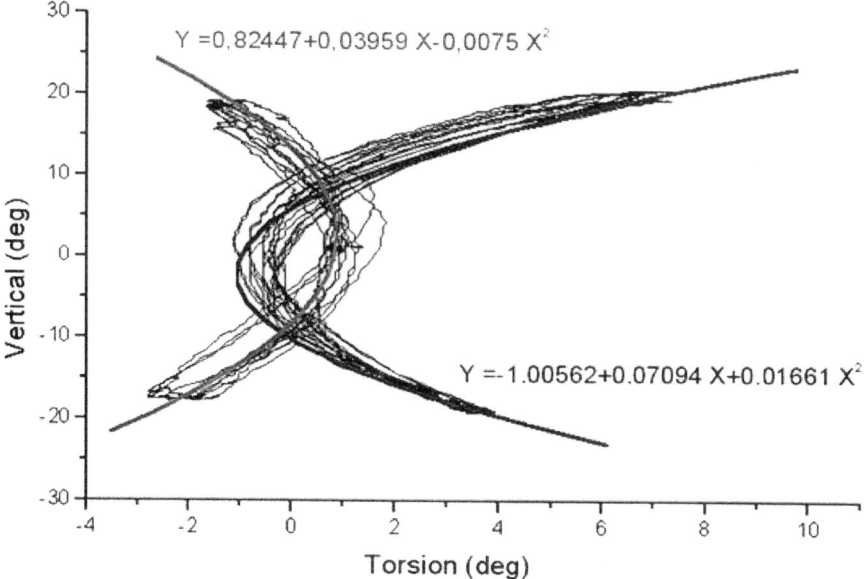

FIGURE 2A. Eye movement recorded raw data and superimposed fitted curve of vertical eye position versus torsion of the right eye when following the target from left-down to right-up (right convex curve) and from right-down to left-up (left convex curve).

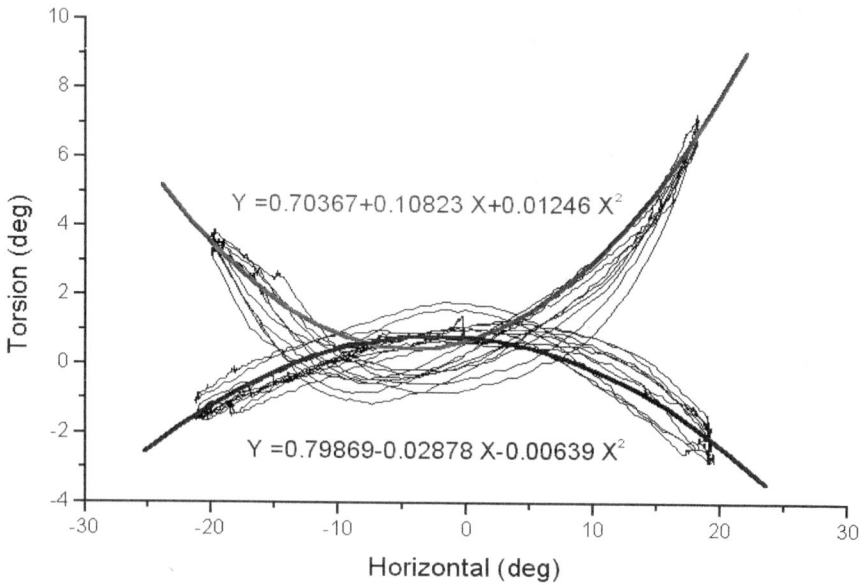

FIGURE 2B. Eye movement recorded raw data and superimposed fitted curve of horizontal eye position versus torsion of the right eye when following the target from left-down to right-up (down convex curve) and from right-down to left-up (up convex curve).

CONCLUSION

The PP evaluation can be obtained using different mathematical methods which are based on the implications of Listing's and Donder's laws.

Using the method proposed in this study, which applies fitting function to the recorded data, we found that the PP coordinates are simply available and very reliable. Moreover, the reliability obtained by this method depends on the number of recorded points. Even analyzing only a single movement in the two main diagonals is sufficient to evaluate the PP, but having many repetitions of the diagonal movements would perform a better fitting coefficient.

REFERENCE

1. ROBINSON, D.A. 1975. A quantitative analysis of extraocular muscle cooperation and squint. Invest. Ophthalmol. **14:** 801–825.

Neural Mechanisms of Three-Dimensional Eye and Head Movements

ELIANA M. KLIER, HONGYING WANG, AND J. DOUGLAS CRAWFORD

CIHR Group for Action and Perception, York Centre for Vision Research, and Departments of Psychology, Biology and Kinesiology & Health Sciences, York University, Toronto, Ontario, Canada

KEYWORDS: superior colliculus; gaze; stimulation; Donders' law; primate

Previous researchers have identified the superior colliculus (SC) as a structure encoding two-dimensional (2-D) gaze commands.[1,2] In a recent study, we showed that these SC output commands are organized in oculocentric coordinates.[3] Given these findings, it is expected that any three-dimensional (3-D) aspects of gaze control and eye/head kinematics should occur downstream from the SC. In this preliminary study, we addressed this hypothesis by examining eye/head movements evoked by head-free stimulations of the SC and comparing them with normal, head-free behavior. If the 3-D aspects of both these movement types are similar, then one can conclude that the 3-D properties of movement are implemented after the SC. This is because SC stimulation bypasses all neural structures upstream from the SC and thus highlights the functions of downstream structures associated with gaze generation.

Two monkeys (*Macaca fascicularis*) were fitted with (1) recording chambers (laterally across the midline and 5 mm anteriorly), (2) scleral search coils (two in one eye to record 3-D eye movements), and (3) one 3-D head search coil. The SC was first identified by recording oculomotor burst activity. Stimulations were then delivered, using tungsten microelectrodes, with pulse trains (50 µA, 500 Hz) of 200 msec, while the monkeys made natural, unencumbered, head-free gaze shifts in both the dark and in dim light. The animals were encouraged to use their entire eye/head motor ranges by the presentation of novel objects (in dim light) and sounds (in the dark) at eccentric positions. This allowed us to obtain a large variety of different initial eye and head positions. The resultant 3-D eye and head positions were recorded off-line and analyzed using the quaternion method.[4]

Movements of the eye-in-space (Es), head-in-space (Hs), and eye-in-head (Eh) occur about three axes—horizontal, vertical, and torsional. During normal behavior, the Es, Hs, and Eh obey Donders' law. It states that that the eyes and head use only one, unique torsional orientation for every possible direction of visual gaze. Furthermore, the Eh obeys a specific form of Donders' law, named Listing's law, which holds that the torsional component of the Eh is always zero. Typically, one can assess

---

Address for correspondence: J.D. Crawford, Department of Psychology, York University, 4700 Keele Street, Toronto, Ontario, Canada M3J 1P3. Voice: 416-736-2100, ext. 88621; fax: 416-736-5814.

jdc@yorku.ca

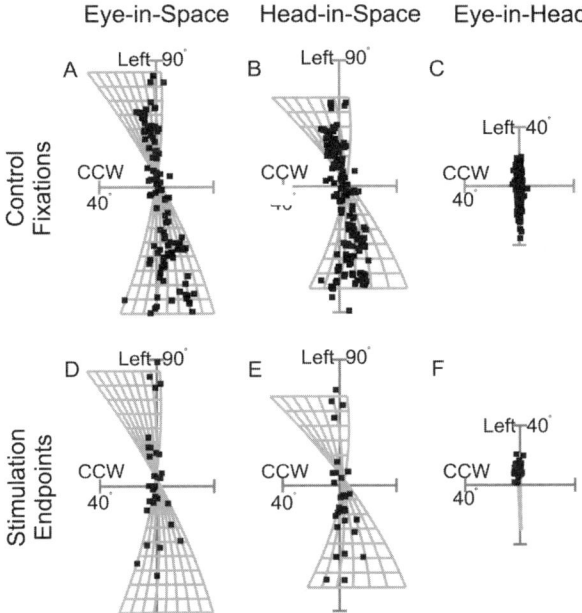

FIGURE 1. Movement endpoints relative to control planes. Best-fit surfaces drawn to random fixation points are sketched in the background, while either control fixation points (**A-C**) or SC stimulation-induced endpoints (**D-F**) are superimposed. Measures of Tsd reveal whether Donders' law is obeyed.

how well Donders' and Listing's laws hold by fitting a surface to the collected data and then computing the torsional standard deviation (Tsd) of these data to this best-fit surface (i.e., to determine how well the data adhere to the best-fit surface).[5]

We first conducted this analysis for normal gaze fixations (when both eye and head velocity were <10°/sec) taken from a random control paradigm. We then took the stimulation-induced gaze end-points and computed their Tsd from the best-fit planes of the random controls. In our preliminary data, there appeared to be no difference in Tsd between normal and stimulation-induced fixation points (FIG. 1).

As a second quantification, we computed second-order surfaces for the stimulation-induced gaze end-points and compared the parameters describing the resultant surfaces to the surface parameters taken from normal gaze fixations. Once again, the two did not appear to be different.

Thus, our preliminary 3-D analysis demonstrated that SC stimulation-induced movements seem to be nearly identical to movements found in normal head-free behavior. The Es and Hs obey Donders' law, while the Eh obeys Listing's law. These kinematics occur with great precision despite widely varying 2-D gaze trajectories and initial eye and head positions. This suggests that the SC encodes 2-D gaze in eye coordinates, with eye-head reference frame transformations occurring downstream in the brainstem.

ACKNOWLEDGMENTS

E.M. Klier is supported by NSERC and OGS scholarships. J.D. Crawford holds a CIHR operating grant and is a Canada Research Chair.

REFERENCES

1. VAN OPSTAL, A.J. *et al.* 1991. Two- rather than three-dimensional representation of saccades in monkey superior colliculus. Science **252:** 1313–1315.
2. FREEDMAN, E.G. & D.L. SPARKS. 1997. Activity of cells in the deeper layers of the superior colliculus of the rhesus monkey: evidence for a gaze displacement command. J. Neurophysiol. **78:** 1669–1690.
3. KLIER, E.M., W. WANG & J.D. CRAWFORD. 2001. The superior colliculus encodes gaze commands in retinal coordinates. Nat. Neuro. **4:** 627–632.
4. TWEED, D.B., W. CADERA & T. VILIS. 1990. Computing three-dimensional eye position quaternions and eye velocity from search coils. Vision Res. **30:** 97–110.
5. RADAU, P., D.B. TWEED & T. VILIS. 1994. Three-dimensional eye, head and chest orientations after large gaze shifts and the underlying neural strategies. J. Neurophysiol. **72:** 2840–2852.

The Visuomotor Transformation for Arm Movement Accounts for 3-D Eye Orientation and Retinal Geometry

D.Y.P. HENRIQUES,[a] J.D. CRAWFORD,[a] AND T. VILIS[b]

[a,b]*CIHR Group for Action and Perception, and* [a]*Departments of Biology, Psychology, and Kinesiology and Health Sciences, Center for Vision Research, York University, Toronto, Canada*

[a]*Department of Physiology, University of Western Ontario, London, Canada*

> KEYWORDS: **visuomotor; spatial vision; eye position; retina; three-dimensional geometry; pointing; arm movements**

To point or reach to a visual target, you need to know its direction relative to your shoulder. That direction can be computed from the retinal image, if the brain also knows the orientation of the eyeball, head, and clavicle. To be geometrically exact, the neural computation would have to involve rotary operations.[1] When the eye was turned 30° up, for instance, the brain would take the locations of all objects relative to the eye and rotate them 30° down to find the locations relative to the head. Some theories[2,3,5–7] suggest that the brain might approximate that rotation by a simpler vector addition, merely shifting all retinal locations by the same vector. But that strategy would lead to marked errors in some situations. FIGURE 1A shows five earth-fixed, horizontal arcs wrapped around a cylinder centered on an eyeball. In front of the eye are two spheres and 90° to the right are two cubes, one sphere-cube pair at eye level and the other 30° up. When the eye fixates the eye-level sphere, the retinal image of the eye-level cube falls on the eye's horizontal meridian (FIG. 1B). But when the eye fixates the sphere at 30° up, the retinal image of the corresponding cube falls well below the horizontal meridian (FIG. 1C). That is, both pairs of spheres and cubes are horizontally separated, but in one case their retinal images are not, owing to the rotation of the eye. If the brain tried to compute these objects' locations relative to the head or the shoulder simply by shifting all their retinal locations by a fixed vector, it would misestimate the elevations of at least some objects, and would misaim its eye and arm movements. Here we study whether the visuomotor transformation for arm movements uses a vector-additive strategy or correctly accounts for the rotary geometry of retinal projection.

Address for correspondence: J.D. Crawford, Department of Psychology, York University, 4700 Keele Street, Toronto, Ontario, Canada, M3J 1P3. Voice: 416-736-5121, ext. 88621; fax: 416-736-5814.

jdc@yorku.ca

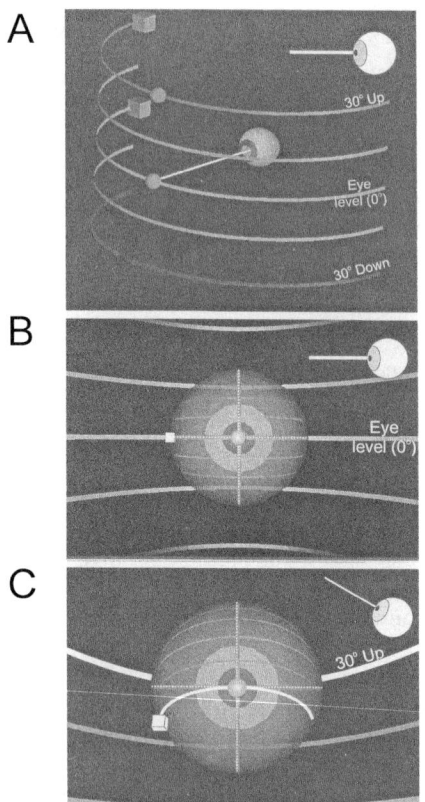

FIGURE 1. Simulated retinal projections for two horizontally paired targets as a function vertical eye position. **A:** Five horizontal lines (wrapped around the eye) with paired targets at different elevations. **B** and **C:** Behind view of the semi-transparent eye and retinal projections of the paired targets while the eye is looking straight ahead (**B**) and while the eye is rotated 30° up to view the top paired stimuli (**C**).

METHODS

Six subjects took part. With the head fixed, the subject looked at one of five target lights at different elevations in an otherwise dark room. Another LED flashed for 500 msec, 40°–80° right of the fixated light. The fixation light went out at the same time, and in complete darkness the subjects kept their eyes on the remembered location, but pointed their arm to the location of the flash. All targets lay on a vertical cylinder of diameter 1.1 m, centered on the right eye. We measured the orientation of the right eye and arm with search coils, and calculated the flashed target's location in eye coordinates (also known as retinal error).

RESULTS

FIGURE 2 shows the type of pointing error that would result if the brain used a vector-additive computation to process retinal target locations. Final pointing directions

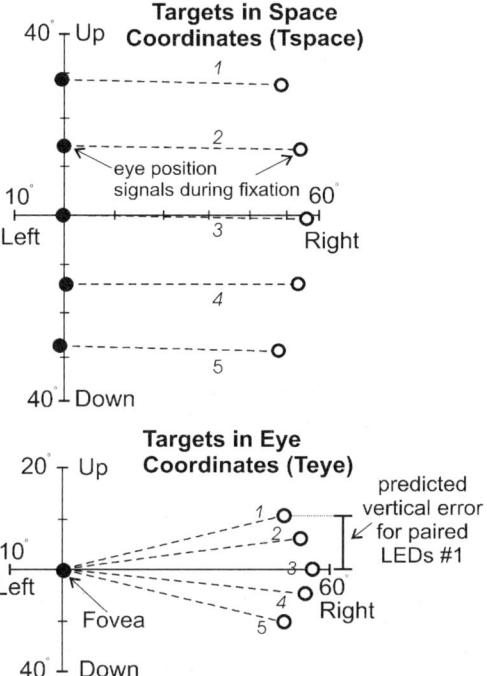

FIGURE 2. Target locations in space (*top*) and eye (*bottom*) coordinates. *Solid circles:* LED location for ocular fixation. *Open circles:* location of final pointing direction as derived from eye position signals. Five horizontal pairs are shown. To get the Teye from Tspace, we rotate the target vectors from the latter by the inverse of initial 3-D eye position (*toward solid circles*).[1,4]

would "fan out," yielding vertical errors that varied systematically as a function of vertical eye position. Only a rotary computation would allow accurate pointing to all targets.

Do subjects make the systematic fanning-out errors predicted by the vector-additive theory? FIGURE 3A shows mean vertical pointing errors for one typical subject for 24 trials. Actual vertical errors are plotted versus the vertical errors predicted by the vector-additive theory. Solid lines are regression fits to the data for the one subject and for all subjects (FIG. 3B). A slope of unity (dashed line) would suggest that the brain performed a purely vector-additive computation. A slope of zero would suggest a rotary computation. Actual slopes were near zero: subjects showed little or no tendency to make the errors predicted by the vector-additive theory, but instead closely approximated the exact, rotary transformation.

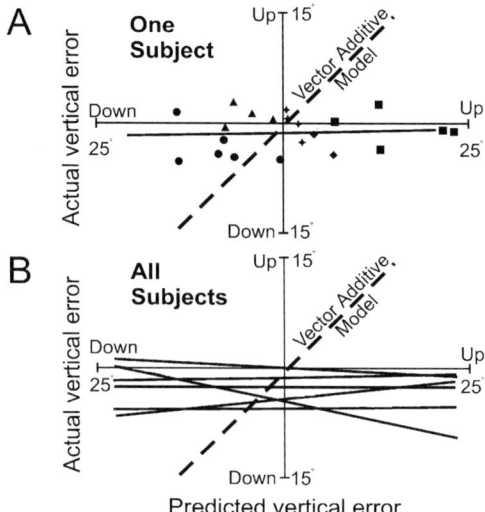

FIGURE 3. Actual vertical pointing errors plotted as a function of errors predicted by the vector-additive theory for one subject (**A**). *Solid lines* indicate the regression fit to the data for the one subject (**A**) and for all six subjects (**B**). The vector-additive model predicts a slope of one (*dashed line*) whereas correction for the rotary geometry of retinal projection predicts a slope of zero.

CONCLUSION

We have shown that the brain closely approximates the rotary computations that are needed to convert retinal images into accurate arm movements. In 1998 Klier and Crawford, studying the rapid eye movements called saccades, showed that they are accurately directed even when the retinal image of the target is rotated owing to torsion of the eye about its own line of sight. We have generalized that result to show that the visuomotor processing underlying arm movements also employs rotary computation, and that the brain also corrects for the retinal effects of vertical eye motion.

REFERENCES

1. CRAWFORD, J.D. & D. GUITTON. 1997. Visuomotor transformation required for accurate and kinematically correct saccades. J. Neurophysiol. **78:** 1447–1467.
2. HALLET, P.E. & A.D. LIGHTSTONE. 1976. Saccadic eye movements to flashed targets. Vision Res. **16:** 107–114.
3. FLANDERS, M., S.I. HELMS TILLERY & J.F. SOECHTING. 1992. Early stages in a sensorimotor transformation. Behav. Brain Sci. **15:** 309–362.
4. KLIER, E.M. & J.D. CRAWFORD. 1998 The human oculomotor system accounts for 3-D eye orientation in the visual-motor transformation for saccades. J. Neurophysiol. **80:** 2274–2294.

5. MAY, L.E. & D.L. SPARKS. 1980. Saccades are spatially, not retinotopically coded. Science **208:** 1163–1164.
6. MILLER, J.M. 1996. Egocentric localization of a perisaccadic flash by manual pointing. Vision Res. **36:** 837–851.
7. ZIPSER, D. & R.A. ANDERSEN. 1988. A back-propagation programmed network that simulates response properties of a subset of posterior parietal neurons. Nature **331:** 679–684.

Implementation of Listing's Law in Patients with Unilateral Sixth Nerve Palsy

AGNES M.F. WONG,[a,b] DOUGLAS TWEED,[a,c] AND JAMES A. SHARPE[a,b]

From the [a]Division of Neurology, and Departments of [b]Ophthalmology and [c]Physiology, University of Toronto, and University Health Network–Toronto Western Hospital, Toronto, Ontario, Canada

KEYWORDS: Listing's law; sixth nerve palsy; adaptation

INTRODUCTION

During fixation and saccades, human eye movements obey Listing's law, which specifies the torsional eye position for each combination of horizontal and vertical eye position.[1,2] To study the mechanisms that implement Listing's law, we investigated whether it was violated in peripheral and central unilateral sixth nerve palsy.

METHODS

Twenty-seven patients with unilateral sixth nerve palsy and 10 normal subjects were studied. Informed consent was obtained from each subject. Patients with diplopia of less than 4 weeks' duration were classified as having *acute* palsy; all others were designated here as having *chronic* palsy.

MRI with enhancement was performed for patients under 50 years of age and for those with neurologic signs. CT with contrast was obtained in patients with ischemic risk factors and those over 50 years of age. If the CT was normal, patients were followed at about 3 months. Those without improvement at 3 months and those with an abnormal CT were further investigated with MRI.

Eye positions were measured with magnetic search coils while patients followed a laser target at 1 m with one eye covered. The laser was programmed to appear in nine different target positions, arranged in a 3 × 3 square. With head immobile, subjects made saccades to a target that moved between straight ahead and 8 eccentric positions. At each target position, fixation was maintained for 3 seconds before the next saccade.

Using fitted functions, we quantified violations of Listing's law by comparing the ocular torsion in each recorded eye position to the torsion predicted by the law. The standard deviation of the differences between the predicted and measured torsion was called *Listing deviation*.

Address for correspondence: Dr. Agnes Wong, 5-440 WW, Toronto Western Hospital, 399 Bathurst Street, Toronto, Ontario, Canada M5T 2S8. Voice: 416-501-5100; fax: 416-603-5596.
agnes.wong@utoronto.ca

Ann. N.Y. Acad. Sci. 956: 520–522 (2002). © 2002 New York Academy of Sciences.

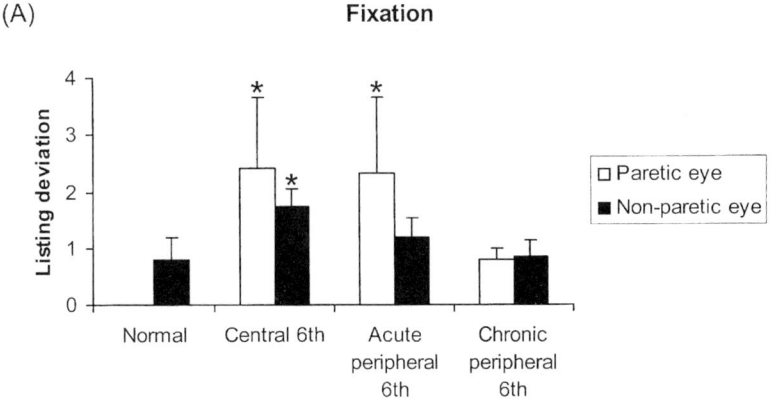

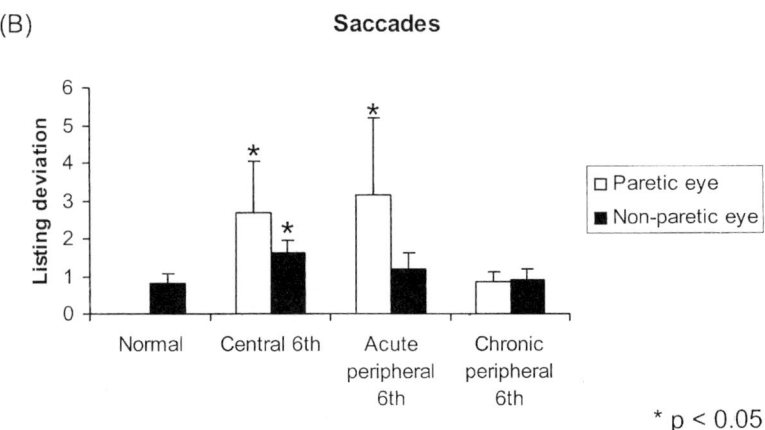

FIGURE 1. Listing deviation in patients with peripheral (chronic and acute) and central sixth nerve palsy caused by brainstem lesions, as compared with normal controls, during fixation (**A**) and saccades (**B**).

In all 27 patients, Listing deviation in both the paretic and non-paretic eyes did not differ during paretic or non-paretic eye viewing. In what follows, we report only Listing deviation during non-paretic eye viewing; deviations during paretic eye viewing were similar. Statistical analysis was performed using analysis of variance. Values were defined as significant when $p < 0.05$.

RESULTS

Patients with *central* sixth nerve palsy had abnormal ocular torsion in both the paretic and non-paretic eyes, violating Listing's law (FIG. 1). During fixation, Listing deviation averaged 2.4° in the paretic eye, and 1.7° in the non-paretic eye, compared to 0.8° in normal controls ($p < 0.05$). During saccades, Listing deviation averaged 2.7° in the paretic eye, and 1.6° in the non-paretic eye, compared to 0.8° abnormal ocular torsion only in the *paretic* eye, but not the non-paretic eye (FIG. 1). Listing deviation of the paretic eye averaged 2.3° during fixation and 3.2° during saccades ($p < 0.05$). Patients with *chronic peripheral* sixth nerve palsy obeyed Listing's law during both fixation and saccades (FIG. 1).

DISCUSSION

In *acute peripheral* sixth nerve palsy, Listing's law was violated in the paretic eye, presumably because the lateral rectus muscle was paretic and perhaps also because its pulley was abnormally positioned. In *chronic peripheral* palsy, both eyes obeyed Listing's law, even though the lateral rectus was still markedly weak. This recovery shows that the neural circuitry underlying Listing's law is adaptive, restoring the law despite a palsied muscle and possibly a disrupted pulley system. Neural adaptation must work by readjusting the innervations to the remaining extraocular muscles; it may also adjust their pulleys, though theoretically Listing's law could be restored with or without a new pattern of pulley placement and motion.

All patients with *central* fascicular sixth nerve palsy caused by brainstem lesions had abnormal ocular torsion in both the paretic and non-paretic eyes. Evidently the neural adaptive mechanisms underlying Listing's law are unable to restore it after brainstem lesions. Our results also indicate that the pontomedullary region is an element of the neural pathway that enforces Listing's law.

ACKNOWLEDGMENTS

This work was supported by the E.A. Baker Foundation (Canadian National Institute for the Blind), the Vision Science Research Program (University of Toronto), and Canadian Institutes of Health Research Grants MT 15362 and ME 5504.

REFERENCES

1. TWEED, D. *et al.* 1992. Three-dimensional properties of human pursuit eye movements. Vision Res. **32:** 1225–1238.
2. TWEED, D. & T. VILIS. 1990. Geometric relations of eye position and velocity vectors during saccades. Vision Res. **30:** 111–127.

Hyperdeviation and Static Ocular Counterroll in Unilateral Abducens Nerve Palsy

AGNES M.F. WONG,[a,b] DOUGLAS TWEED,[a,c] AND JAMES A. SHARPE[a,b]

From the Division of [a]Neurology, and Departments of [b]Ophthalmology and [c]Physiology, the University of Toronto, and University Health Network – Toronto Western Hospital, Toronto, Ontario, Canada

KEYWORDS: abducens nerve palsy; vertical strabismus; ocular counterroll

INTRODUCTION

When a vertical strabismus accompanies sixth nerve palsy, multiple cranial nerve palsy or skew deviation should be considered. The purpose of this study is to detect and determine the magnitude of vertical deviation in patients with unilateral sixth nerve palsy.

METHODS

Twenty-seven patients with unilateral sixth nerve palsy and ten normal subjects were studied. The range of ductions was examined and the degree of abduction defect was recorded. The amount of horizontal and vertical deviations was measured in nine diagnostic positions by the prism and cover test, Maddox rod and prism test, and magnetic scleral search coils.

MRI with enhancement was performed for patients under 50 years of age and those with neurologic signs. CT with contrast was obtained in patients with ischemic risk factors and those over 50 years of age. If CT was normal, patients were followed at about 3 months. Those without improvement at 3 months and those with an abnormal CT were further investigated with MRI. We report here the primary deviations in our subjects; the results for secondary deviations were similar. ANOVA was used for statistical analysis.

Address for correspondence: Dr. Agnes Wong, 5-440 WW, Toronto Western Hospital, 399 Bathurst Street, Toronto, Ontario, Canada M5T 2S8. Voice: 416-501-5100; fax: 416-603-5596.
agnes.wong@utoronto.ca

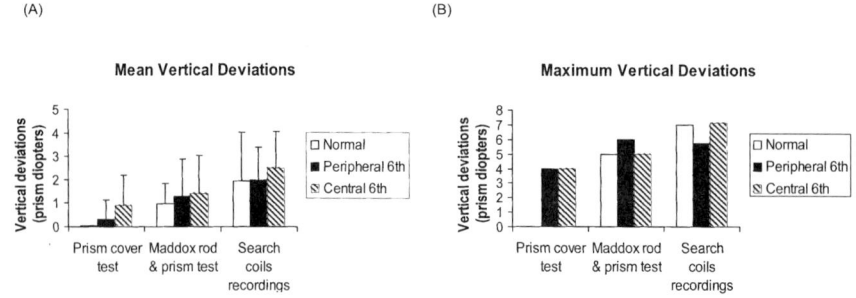

FIGURE 1. (**A**) Mean and (**B**) maximum vertical deviations in patients and normal control subjects.

RESULTS

Vertical Deviations in Nine Diagnostic Positions

Mean vertical deviations, for all positions of gaze, in peripheral palsy ($n = 20$) were: 0.3 ± 0.8 prism diopters (PD) by prism-cover test, 1.3 ± 1.6 PD by Maddox test, and 2.0 ± 1.4 PD by coil recordings. Mean vertical deviations in normal subjects were: 0.0 ± 0.0 PD by prism-cover test, 1.0 ± 0.9 PD by Maddox test, and 1.9 ± 2.1 PD by coil recordings. Therefore, peripheral palsy did not cause abnormal vertical deviation. In central palsy ($n = 7$), mean vertical deviations were: 0.9 ± 1.3 PD by prism-cover test, 1.4 ± 1.6 PD by Maddox test, and 2.5 ± 1.6 PD by coil recordings; they were not different from normal values. There was no difference in mean or maximum vertical deviations between patients and normal controls (FIG. 1). The vertical deviations were comitant in individual patients and in normal subjects, as were the group means. The maximum difference between hyperdeviation on up- and down-gaze in any individual patients was 3 PD.

Vertical Deviations during Static Head Roll

Eighteen of 20 patients (90%) with *peripheral palsy* exhibited a right hyperdeviation (mean = 4.3 PD) on static lateral head tilt to the right shoulder (right head tilt), and a left hyperdeviation (mean = 2.9 PD) on lateral tilt toward the left shoulder (left head tilt), regardless of the side of palsy (FIG. 2) ($p < 0.001$). In contrast, four of seven patients (57%) with *central* palsy had hyperdeviation that remained on the same side during static lateral head tilt to either side (FIG. 2). Mean deviations were 2.1 PD right hyperdeviation on *right* head tilt, and 1.3 PD right hyperdeviation on *left* head tilt.

DISCUSSION

A small hypertropia can be detected in patients with peripheral and central sixth nerve palsy. This hypertropia falls within the normal range of hyperphoria seen in

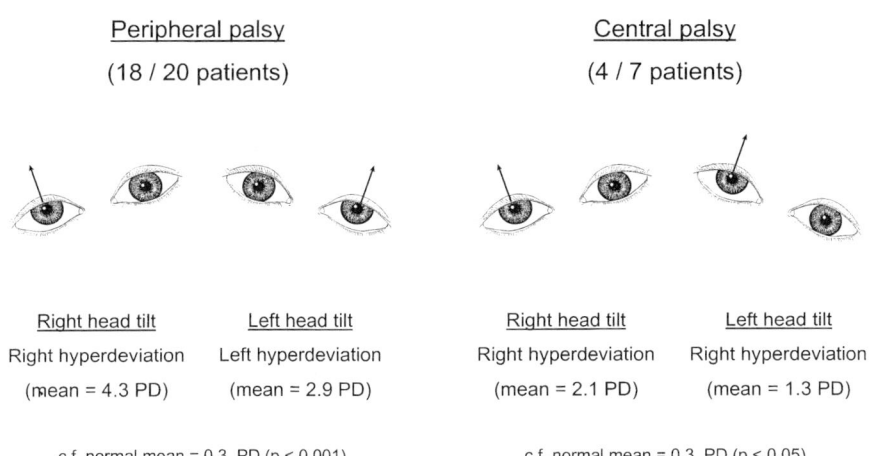

FIGURE 2. Mean vertical deviations on static lateral head roll in peripheral and central palsy.

healthy subjects, indicating that it is a normal hyperphoria that becomes manifest in the presence of esotropia. In normal subjects, the mean vertical deviation in the straight-ahead position is 1.52 ± 1.49 PD. Thus, in patients with sixth nerve palsy, if a hypertropia is detected in the straight-ahead position, which is less than or equal to 5 PD (normal mean + 2 SD), investigation for multiple cranial nerve palsy or skew deviation is not indicated.

Static head tilt stimulates otolith receptors, leading to ocular counterroll and a small change in vertical alignment in normal subjects.[1] However, when the otolith–ocular reflex pathway is disrupted, ocular torsion and skew deviation are observed.[2] This indicates that under normal circumstances, the otolith–ocular reflex is symmetrical and balanced; it is also suppressed during static head roll. Disruption of binocular vision may remove the suppression on the otolith–ocular reflex, and lead to the pattern of right hyperdeviation on right head tilt and left hyperdeviation on left head tilt observed in *peripheral* palsy. In contrast, in *central* fascicular sixth nerve palsy, unilateral brainstem lesions can also disrupt the balance of the otolith–ocular reflex and lead to the pattern of vertical deviation that remains on the same side regardless of the direction of head roll. This pattern of hyperdeviation induced by lateral head tilt may warrant investigation for a brainstem lesion as the cause of paretic abduction.

ACKNOWLEDGMENTS

This work was supported by the E.A. Baker Foundation (Canadian National Institute for the Blind), the Visual Science Research Program (University of Toronto, Ontario, Canada), and the Canadian Institutes of Health Research (Grants MT 15362 and ME 5504).

REFERENCES

1. AVERBUCH-HELLER, L. *et al.* 1997. Torsional eye movements in patients with skew deviation and spasmodic torticollis: responses to static and dynamic head roll. Neurology **48:** 506–514.
2. BRANDT, T. & M. DIETERICH. 1993. Skew deviation with ocular torsion: a vestibular brainstem sign of topographic diagnostic value. Ann. Neurol. **33:** 528–534.

Vestibulo-Ocular Responses during Mirror-Viewing

Y. HAN,[a] A. KUMAR,[b] J. SOMERS,[c] J.I. KIM,[d] AND R.J. LEIGH[a,b]

[a]*Department of Neurology and* [b]*Biomedical Engineering, Veterans Affairs Medical Center, and Case Western Reserve University, Cleveland, Ohio, USA*

[c]*Johnson Space Center, Houston, Texas, USA*

[d]*Department of Neurology, Dankook University Hospital, Cheonan, Korea*

> KEYWORDS: vestibulo-ocular responses; mirror-viewing; compensatory eye movements

When we view a target located at optical infinity, eye rotations will compensate for head rotations if they are equal and opposite in direction. During viewing of a near target, the situation is more complicated because the eyes do not lie at the center of head rotation and are separated by several centimeters. In general, the gain of compensatory eye movements is increased during near viewing and the gain is greater for the eye that is closer to the object of interest.[1]

We have recently shown that, during close viewing of one's own image in a mirror, the visual demands during head rotation differ from when viewing a real, near target.[2] This is mainly because translations of the subject's head in a plane parallel to that of a mirror are matched by translations of the image. Our calculations indicated that the main factor determining gain differences between mirror-viewing and near-viewing condition is the eccentricity of the target, which is zero during near-viewing, but varies as a function of head rotation in the mirror-viewing condition. Vergence angle was not the main determinant of VOR gain because it was similar during near- and mirror-viewing conditions, but gain values differed substantially under the two test conditions.[2] Overall, median vergence angles differed by only 0.8 degrees (4%) under the two viewing conditions, but median gain was 65% greater during near-viewing than mirror-viewing.[2]

The mirror paradigm might be regarded as a combined vestibulo-ocular reflex (VOR)–smooth pursuit task in which the pursuit stimulus (the virtual image of bridge of the subject's nose) moves as a linear function of head rotation. To investigate this possibility, we used the experimental strategy of switching to darkness and measuring VOR gain while the eyes were still converged but visual feedback was not possible.

Address for correspondence: R. John Leigh, Department of Neurology, University Hospitals of Cleveland, 11100 Euclid Avenue, Cleveland, OH 44106. Voice: 216-421-3040; fax: 216-231-3461.
rjl4@po.cwru.edu

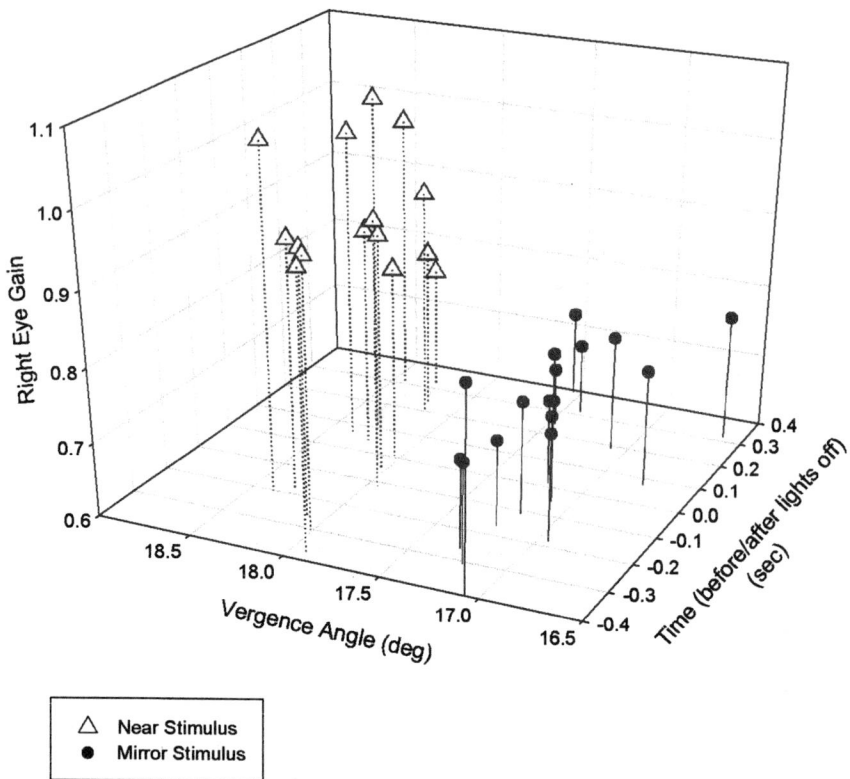

FIGURE 1. Example of the effects of switching to darkness during near-viewing or mirror viewing in subject 2, shown as a plot of gain, vergence angle, and time. The lights were turned off at time zero. The subject showed significantly greater gain values during viewing of the near target than viewing her image in the mirror, even though the vergence angle differed by only about one degree during the two conditions.

We studied six healthy human subjects; all subjects gave informed consent. Subjects sat in a vestibular chair and firmly braced their heads against the headrest of the chair. Eye and the head rotations were measured using the magnetic search coil technique, as previously described.[2] Two visual stimuli were used: (1) a black cross drawn on a circular piece of white paper, diameter 16 cm, aligned at the subject's eye level at a viewing distance close to the near point of accommodation for each subject ("near target"); (2) a circular mirror diameter 16 cm, mounted at the subject's eye level at a viewing distance of about half the distance of the near target ("mirror"); subjects viewed a small dark ink-spot made on the bridge of their nose. The near stimuli and mirror stimuli were positioned at the near point of accommodation for each subject, and vergence angle was monitored so that it was similar during these two experimental conditions. Both the near stimulus and the mirror were illuminated by a surrounding array of bright LEDs. In the otherwise darkened experimental room, the LED array made the near target and the subject's own face in the mirror easily visible, but could be turned on and off in a few msec. During each test paradigm, which lasted 40 sec, the subject was sinusoidally oscillated in the chair at 2.0 Hz with a peak velocity of 30 deg/sec. The test paradigms were: (1) attempted fixation of the near target as lights were alternately switched on for 6 sec and off for periods of 6 sec; and (2) attempted fixation of the bridge of the subject's own nose in the mirror as lights were alternately switched on and off for periods of 6 sec.

As previously reported,[2] all subjects showed substantially smaller gain values, for each eye, during mirror-viewing compared with the near-target condition, even though vergence angles during the two conditions were similar for each subject. After switching to darkness, vergence angle declined over several seconds. In three of the six subjects studied, VOR gain measured 0.1–1.0 sec after switching to darkness, remained significantly greater after near-viewing than after mirror-viewing ($p < 0.001$ using Mann-Whitney rank sum test), during which vergence angles remained similar for the two conditions. An example from one of these three subjects of gain measurements before and after switching to darkness is shown in FIGURE 1. In the other three subjects, a similar trend was noted but differences were not significantly different.

Evidence from our prior study suggested that difference in the gain of compensatory eye movements during the two test conditions mainly reflected parametric changes in VOR gain rather than the effects of smooth pursuit[2] The present results provide further evidence to support this view, since differences in gain during the two conditions were still evident 0.1–1.0 sec after switching to darkness in three subjects, when the effects of visual feedback on the response would be minimal.

REFERENCES

1. VIIRRE, E., D. TWEED, K. MILNER & T. VILIS. 1986. A reexamination of the gain of the vestibuloocular reflex. J. Neurophysiol. **56:** 439–450.
2. HAN, Y., J.T. SOMERS, J.-I. KIM, A.N. KUMAR & R.J. LEIGH. 2002. Ocular responses to head rotation during mirror-viewing. J. Neurophysiol. In press.

The Effects of Horizontal Head Position (Yaw Axis) and Step Velocity on the Vestibulo-Ocular Reflex

PHILLIP KRAMER,[a] ELLIOT M. FROHMAN,[b] DANIELE NUTI, AND DAVID S. ZEE[d]

[a]*The New Jersey Neuroscience Institute, Edison, New Jersey, USA*

[b]*Department of Neurology and Ophthalmology, University of Texas Southwestern Medical Center, Dallas, Texas, USA*

[c]*Department of Otolaryngology, University of Siena, Siena, Italy*

[d]*Department of Neurology, Neurosciences, Otolaryngology and Ophthalmology, The Johns Hopkins Hospital, Baltimore, Maryland, USA*

KEYWORDS: vestibulo-ocular reflex; gain; time constant; head position; velocity; beating frequency; beating span

We looked at two data sets to determine the effects of horizontal head position and step velocity on the vestibulo-ocular reflex (VOR).

EXPERIMENT 1

We studied six normal subjects to investigate whether horizontal head-on-body position and the velocity of step rotations influence the vestibulo-ocular reflex (VOR). Seated subjects were rotated using left and right velocity steps of 60, 120 and 180 degrees/sec. For each velocity step the head was stabilized with a bite bar at 60 deg left of center, centered, and 60 deg right of center for a total of 18 rotations done in a Latin square order. Eye movements were recorded at 500 Hz using scleral search coils. The eye movements during rotation were analyzed for slow phase velocity, gain, time constant (Tc), mean beating span (MBS, the mean of the number of degrees the eyes travel during each slow phase), mean beating frequency (MBFq, the mean of the inverse of time between fast phases), and mean beating field (MBF, the mean of the eye position at the center of each slow phase). Leftward and rightward rotations were averaged and head position was considered either ipsi- or contralateral to the direction of rotation. Analysis of variance indicated that head position did not influence any of these parameters, but that step velocity affected all except gain.

As step velocity increased from 60 to 180 deg/sec; Tc decreased (11.5 to 9.0 sec, $p = 0.02$), MBS increased (5.8 to 10.7 deg, $p = 0.001$), and MBFq increased (2.8 to 3.2 Hz, $p = 0.007$). While the change in MBF was statistically significant ($p = 0.04$), it was not proportional to step velocity (nearly midline at 60 and 180 deg/sec and 5 degrees contralateral to the direction of chair rotation at 120 deg/sec).

EXPERIMENT 2

We retrospectively examined the effect of step velocity on the vestibulo-ocular reflex (VOR) during rotary chair testing of 896 "dizzy" patients. Patients were exposed to four velocity step profiles (60 deg/sec left, 60 deg/sec right, 240 deg/sec left, and 240 deg/sec right) with their head maintained in the midline position. The 60 deg/sec rotation always preceded the 240 deg/sec rotations. Eye movements were recorded during both the per and post portions of the rotation for a total of eight accelerations. At the Johns Hopkins Hospital (JHH) eye movements were recorded using electro-oculography at 100 Hz, and slow phases were identified automatically but verified manually. At the New Jersey Neuroscience Institute (NJNI) eye movements were recorded using infrared video-oculography at 60 Hz and slow-phase velocities were calculated with the use of a heuristic filter and by determining the modal velocity every 0.25 sec. The resulting slow-phase velocity (SPV) data was fitted to the form

$$SPV = A * \exp(-t/Tc),$$

where t is time and Tc is the time constant of the exponential decay. The gain of the VOR was calculated by dividing the calculated peak SPV (A) by the peak velocity of chair rotation.

The records of 985 patients were examined (658 from JHH and 327 from NJNI) and 89 were eliminated due to incomplete testing, noise or excessive spontaneous nystagmus leaving 896 patients.

We made seven comparisons and found them all to be statistically significant $p \ll 0.0001$. Gains were higher and time constants were longer for rotations at 60 deg/sec than for rotations at 240 deg/sec. For both 60 and 240 deg/sec the gain during the per portion of the rotation exceeded the gain during the post portion of the rotation. For both 60 and 240 deg/sec the Tc during the post portion of the rotation exceeded the Tc during the per portion of the rotation. Finally directional preponderance was higher for rotations at 60 deg/sec than at 240 deg/sec. These results are shown in TABLE 1. For instance the first line of TABLE 1 is read as: the gain of the VOR for step rotations to 60 deg/sec exceeded the gain of the VOR for step rotations to 240 deg/sec 71% of the time (p much less than 0.0001). The mean gain for 60 deg/sec steps was 0.58 and for 240 deg/sec steps was 0.47.

Since the 60 deg/sec rotations were performed before the 240 deg/sec rotation, it is possible that the 60 vs. 240 deg/sec results were influenced by the order of testing. To examine this, an additional 15 patients were tested at NJNI using the same protocol, except the 240 deg/sec rotations were performed before 60 deg/sec rotations. For these subjects the results were similar (TABLE 2). Gains were higher and time constants were longer for rotations at 60 deg/sec than for rotations at 240 deg/sec. Therefore the order of testing does not seem to be a factor.

TABLE 1. Step rotations to 60 deg/sec preceded step rotations to 240 deg/sec: 896 patients

Parameter	How often	p	Mean	Mean
Gain: 60 deg/sec > 240 deg/sec	71%	≪0.0001	60 deg/sec: 0.58	240 deg/sec: 0.47
Tc: 60 deg/sec > 240 deg/sec	71%	≪0.0001	60 deg/sec: 16.5 sec	240 deg/sec: 13.2 sec
Gain: 60 deg/sec per>post	63%	≪0.0001	per: 0.61	post: 0.55
Gain: −240 deg/sec per>post	60%	≪0.0001	per: 0.50	post: 0.47
Tc: 60 deg/sec per<post	60%	≪0.0001	per: 15.9 sec	post: 17.0 sec
Tc: 240 deg/sec per<post	66%	≪0.0001	per: 12.5 sec	post: 13.9 sec
DP: 60 deg/sec > 240 deg/sec	59%	≪0.0001	60 deg/sec: 10.5%	240 deg/sec: 7.8%

TABLE 2. Step rotations to 240 deg/sec preceded step rotations to 60 deg/sec: 15 patients

Parameter	How often	p	Mean	Mean
Gain: 60 deg/sec > 240 deg/sec	72%	0.002	60 deg/sec: 0.49	240 deg/sec: 0.43
Tc: 60deg/sec > 240 deg/sec	77%	<0.0001	60 deg/sec: 17.6 sec	240 deg/sec: 14.2 sec

DISCUSSION

From these data it appears that horizontal head position during step rotations does not affect the VOR. Increasing velocity of rotation affects the VOR by decreasing the Tc, increasing the MBS and the MBF. Varying step velocities between 60 and 180 deg/sec does not seem to affect the gain of the VOR, but a step velocity of 240 deg/sec results in a lower gain. A possible explanation for the decrease in the gain of the VOR at 240 deg/sec is that this velocity may exceed the saturation velocity of the VOR in humans.

The Vertical Vestibulo-Ocular Reflex, and Visual-Vestibular Interaction during Active Head Motion

JI SOO KIM[a,b] AND JAMES A. SHARPE[a]

[a]*Division of Neurology, University Health Network, University of Toronto, Canada*
[b]*Department of Neurology, College of Medicine, Cheju National University, Cheju, Korea*

KEYWORDS: vertical vestibulo-ocular reflex; visual-vestibular interaction; head motion

INTRODUCTION

The dynamics of vertical eye motion are reported to differ from those of horizontal motion.[1,2] The characteristics of vestibular smooth eye motion during passive whole-body rotation may differ from those during active head rotation.[3] The vertical vestibulo-ocular reflex (VOR), its interactions with vision during active head motion, and effects of aging on them had not been investigated.

METHODS

We measured smooth pursuit, combined eye-head tracking, the VOR, and its visual enhancement and cancellation during active head motion in pitch using a magnetic search coil technique in 21 younger (age < 65; mean, 36.4 ± 13.7) and 10 elderly (age 65; mean, 77.6 ± 7.2) subjects. With the head immobile, subjects pursued a target moving sinusoidally with a frequency range of 0.125 to 2.0 Hz, and with peak target accelerations (PTAs) ranging from 12 to 789°/sec^2. Combined eye-head tracking, the VOR in darkness, and its visual enhancement during fixation of an earth-fixed target (VVOR) were measured during active sinusoidal head rotation with a peak-to-peak amplitude of 20° at frequencies of 0.25, 0.5, 1.0 and 2.0 Hz. The efficacy of VOR cancellation was determined from VOR gains during combined eye-head tracking.

Address for correspondence: James A. Sharpe, M.D., Division of Neurology, University Health Network, 399 Bathurst St., ECW 5-042, TWH, Toronto, ON, M5T 2S8, Canada. Voice: 1-416-603-5950; fax: 1-416-603-5596.
sharpej@uhnres.utoronto.ca

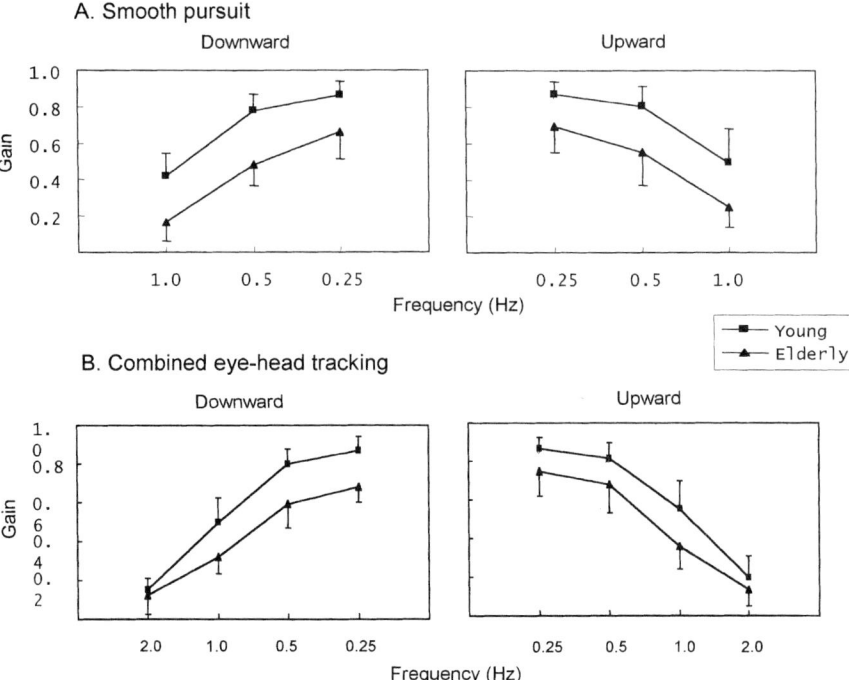

FIGURE 1. (**A**) Smooth pursuit gains are plotted for a peak-to-peak target amplitude of 20°. (**B**) Combined eye-head tracking. Gaze (eye plus head) gain declined with increasing frequency in both age groups ($p < 0.0001$, repeated measures of ANOVA). Gains are greater in the young than in the elderly. Error bars indicate 1 standard deviation.

RESULTS

Vertical smooth pursuit. Gains of upward and downward smooth pursuit declined with increasing frequency ($p < 0.0001$, ANOVA) (FIG. 1A). Both upward and downward smooth pursuit gains decreased significantly with advancing age (Pearson correlation coefficient, $-0.80 < r < -0.63, p < 0.05$).

Vertical eye-head tracking. Gaze gain decreased significantly as target frequency increased from 0.25 to 2.0 Hz ($p < 0.0001$, ANOVA) (FIG. 1B). Both upward and downward gaze (head plus eye motion) gains decreased with advancing age at frequencies from 0.25 to 1.0Hz (Pearson correlation coefficient, $-0.48 < r < -0.75, p < 0.005$).

Vertical vestibulo-ocular reflex in darkness (VOR). The VOR gain or phase did not vary with increasing age (Pearson correlation coefficient, $-0.12 < r < 0.19$, $p > 0.5$), except for downward movement at 0.25 Hz (FIG. 2A) where it decreased in the elderly ($p < 0.05$). Group mean gains did not vary with frequency or with head velocity. Vestibular smooth eye movements were approximately 180° out of phase with head movements.

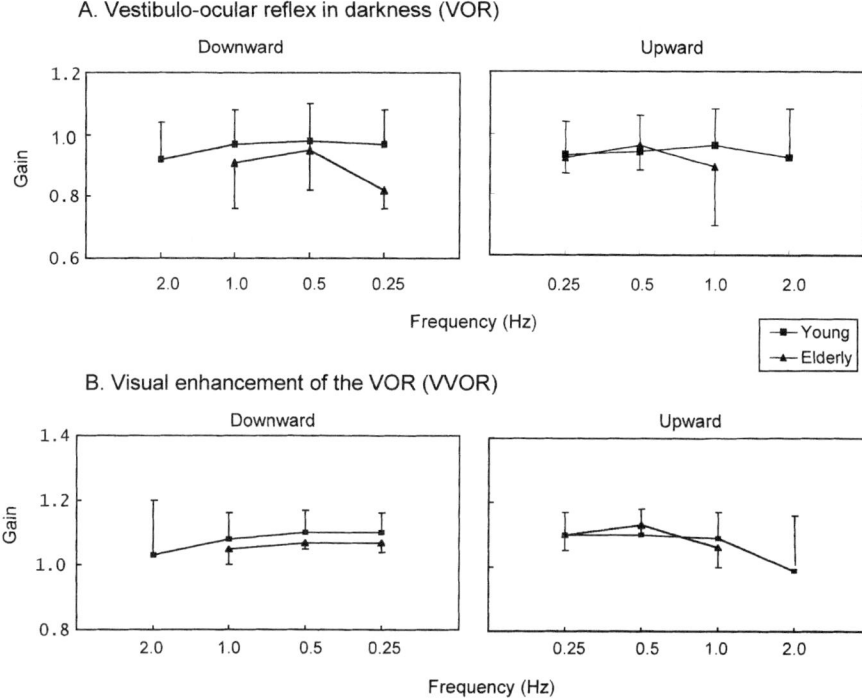

FIGURE 2. (A) Vestibulo-ocular reflex (VOR) gain. (B) Visual enhancement of VOR (VVOR). VOR and VVOR gains did not differ between younger and elderly groups and did not vary with frequency. Error bars indicate 1 standard deviation.

Visual enhancement of the VOR (VVOR). No change in VVOR gain occurred with advancing age (Pearson correlation coefficient, $-0.40 < r < 0.32, p > 0.5$). Few elderly subjects could sustain tracking at 2.0 Hz, and in the younger group VVOR gain decreased significantly at 2. 0Hz.

Visual cancellation of the VOR. VOR cancellation declined as target frequency increased ($p < 0.001$, ANOVA). During combined eye-head tracking, cancellation of the vertical VOR deteriorated with advancing age at lower frequencies (0.25 and 0.5 Hz) (Pearson correlation coefficient, $-0.80 < r < 0.65, p\ 0.001$).

DISCUSSION AND CONCLUSIONS

Quantitative investigation of vertical smooth pursuit, head-free tracking, the VOR in darkness, and visual enhancement of the VOR (VVOR) during active head motion showed that vertical smooth pursuit, combined eye-head pursuit, and VOR cancellation have common features of lowered gains at high target accelerations and senescent degradation. The angular vertical VOR and VVOR vary little with advancing age. VOR and VVOR performance in the elderly implicates relative preservation of neural structures subserving vertical vestibular smooth eye motion in senescence.

ACKNOWLEDGMENTS

This work was supported by an Elizabeth Barford Award, University of Toronto (JSK), and by Canadian Institutes of Health Research (CIHR) Grants ME 5509 and MT 15362 (JAS).

REFERENCES

1. ROTTACH, K.G. *et al.* 1996. Comparison of horizontal, vertical and diagonal smooth pursuit eye movements in normal human subjects. Vision Res. **36:** 2189–2195.
2. BALOH, R.W. *et al.* 1983. The dynamics of vertical eye movements in normal human subjects. Aviat. Space Environ. Med. **54:** 32–38.
3. DEMER, J.L., J.G. OAS & R.W. BALOH. 1993. Visual-vestibular interaction in humans during active and passive, vertical head movement. J. Vestibular Res. **3:**101–114.

A Three-Channel Model for Generating the Vestibulo-Ocular Reflex in Each Eye

LAURENCE R. HARRIS,[a] KARL A. BEYKIRCH,[b] AND MICHAEL FETTER[c]

[a]*Department of Psychology, York University, Toronto, Canada*

[b]*Department of Neurology, University of Tübingen, Tübingen, Germany*

[c]*Department of Neurologie, Klinikum Karlsbad-Langensteinbach, Karlsbad-Langensteinbach, Germany*

KEYWORDS: vestibulo-ocular reflex; three-dimensional eye movements; channels; passive rotation

Coding head movement involves representing the head's velocity and axis of rotation. The neural representation can then be used to inform perceptual and motor processes. An important motor response to head movement is the compensatory eye movements evoked, one component of which is the vestibulo-ocular reflex (VOR). Historically a three-neuron arc has been described as the core of the neural mechanism underlying the generation of the VOR.[1,2] Such a direct line between sensor (the canals) and effector (the eye muscles) implies independent processing of the geometric components of the three-dimensional VOR.[3] A more flexible and robust representation of the movement involves an interactive process in which the activity coding movement in each direction is interpreted in the context of the activity of the others. Many sensory attributes are coded by the activity of a small set of channels,[4] and the closely constrained three-dimensional movement of the head could be efficiently represented by such a system. Psychophysical methods have been developed to investigate channel systems among which is adaptation. After adapting the response to a particular stimulus, the effect on the responses to closely related stimuli can often reveal a channel-coding system.[5,6] Here we use an adaptation technique to provide evidence for a three-channel model underlying the representation of head rotation and generating the vestibulo-ocular reflex of each eye. These channels are conceptually different from those proposed for coding head velocity ranges,[7] as discussed elsewhere.[8]

Address for correspondence: Laurence R. Harris, Department of Psychology, York University, 4700 Keele Street, Toronto, Ontario M3J 1P3, Canada. Voice: 416-736-2100, ext. 66108; fax: 416-736-5814.

harris@yorku.ca

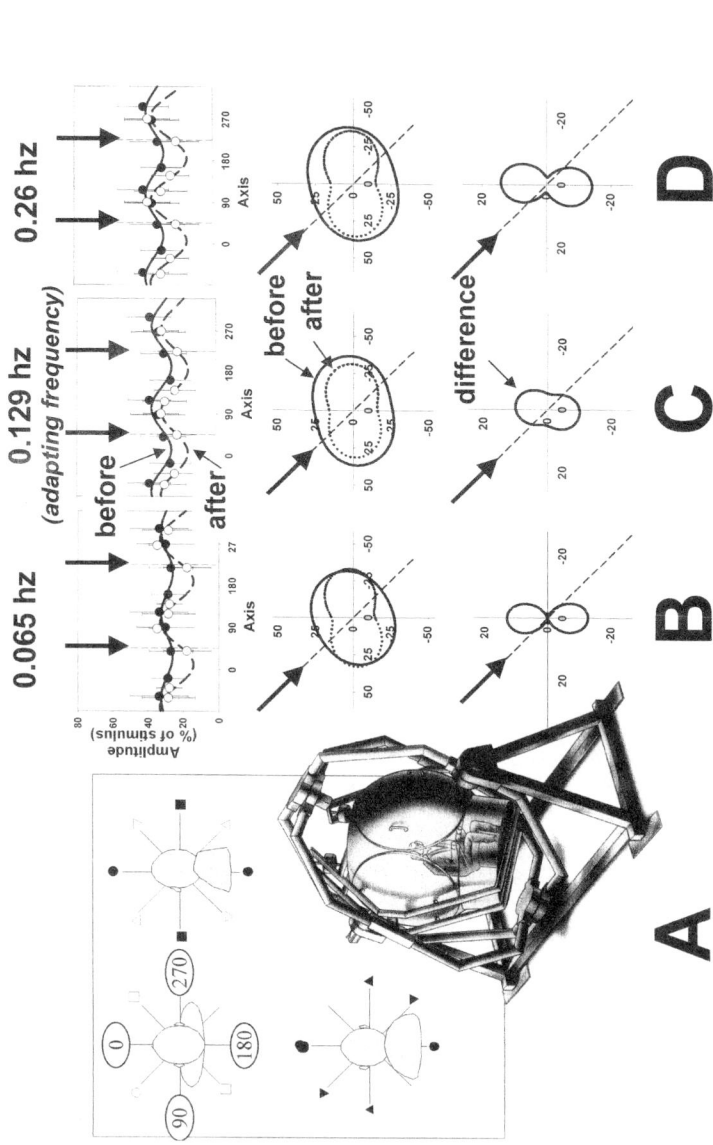

FIGURE 1. The VOR measured before and after adaptation. **A** shows the axes tested and the Tübingen Vestibular Testing Facility. **B-D** show the amplitude and orientation in the horizontal plane of the slow-phase component of the VOR evoked in response to each of three frequencies of rotation around four axes in the horizontal plane before and after adaptation around the axis shown by the fat *black arrow*. The *top row* plots the amplitude and direction of the responses before (*solid circles*) and after (*open circles*) adaptation plotted on a linear scale. The lines are best-fit sines. The next row plots the best-fit sines in polar coordinates. The *bottom row* plots the differences between the conditions before and after. The largest effects are around the roll axis, some 45 degrees from the adapted axis.

METHODS

The experiment was run in two sessions separated by at least two days. The first session evaluated the eye movements evoked by rotation around eight test axes. Using the Tübingen Vestibular Research Stimulator and starting from upright, we rotated seven subjects in yaw, pitch, roll and axes 45 degrees in between (FIG. 1A) in the dark and measured the evoked eye movements with 3-D scleral search coils on the left eye. We used 100 sec of sum-of-sines (0.032, 0.065, 0.13, 0.26 Hz, ± 20 degrees). In the second session, subjects were adapted to subject-stationary vision viewed at 1 m during physical rotation (0.13 Hz; ± 100 degrees) in the plane of the right-anterior, left-posterior (RALP) canal-pair. Under these conditions the vestibulo-ocular reflex is suppressed[9] and the reduced gain persists when performance is measured subsequently in darkness.[10] After the 30-minute adaptation procedure, eye coils were placed on the left eye, and the VOR evoked by rotation about the test axes measured in the dark. The data analysis method has been described elsewhere.[11]

RESULTS

The slow-phase component of the VOR induced by rotation before adaptation was not always aligned with the stimulating axis and showed a variation in amplitude and deviation that varied from axis to axis[11] (FIG. 1B-D [solid circle and solid lines] and FIG. 2 ["pre"]). After visually driven VOR gain reduction by stabilized vision around the RALP axis, the response in the dark to rotation around the adapting axis (indicated by thick arrows in FIG. 1) was reduced. However, changes in both amplitude and alignment were also seen in response to rotation around other axes, most notably roll. The differences between the before and after conditions are illustrated in the bottom row of FIG. 1 B-D: clearly the largest effect is in response to roll rotation. Notice that the response to pitch rotation, which, like roll, was only 45 degrees away from the adaptation axis, was not significantly altered.

MODELING

The data were modeled with a three-channel model to code the orientation and velocity of head rotation. Each of the three channels of this model represents a component of the movement. They are not based on any anatomical features but are proposed simply as an information-processing system. The head rotation is represented as a three-dimensional vector of a length representing velocity. This vector is then projected orthogonally onto each of the "channels." Each channel has a gain by which the projected component is multiplied and an orientation. To recover the head movement, the activity in the three channels are taken as the coordinates. The output is the vector sum of the three channels' activity.

In order to model the VOR before any adaptation, the orientations and gains of the three channels were configured as free variables and the output of the model, with a random initial configuration, was compared to the actual response for each subject around each axis. Gain and orientations were varied systematically and the process repeated until the output of the model best matched the data. The orientation

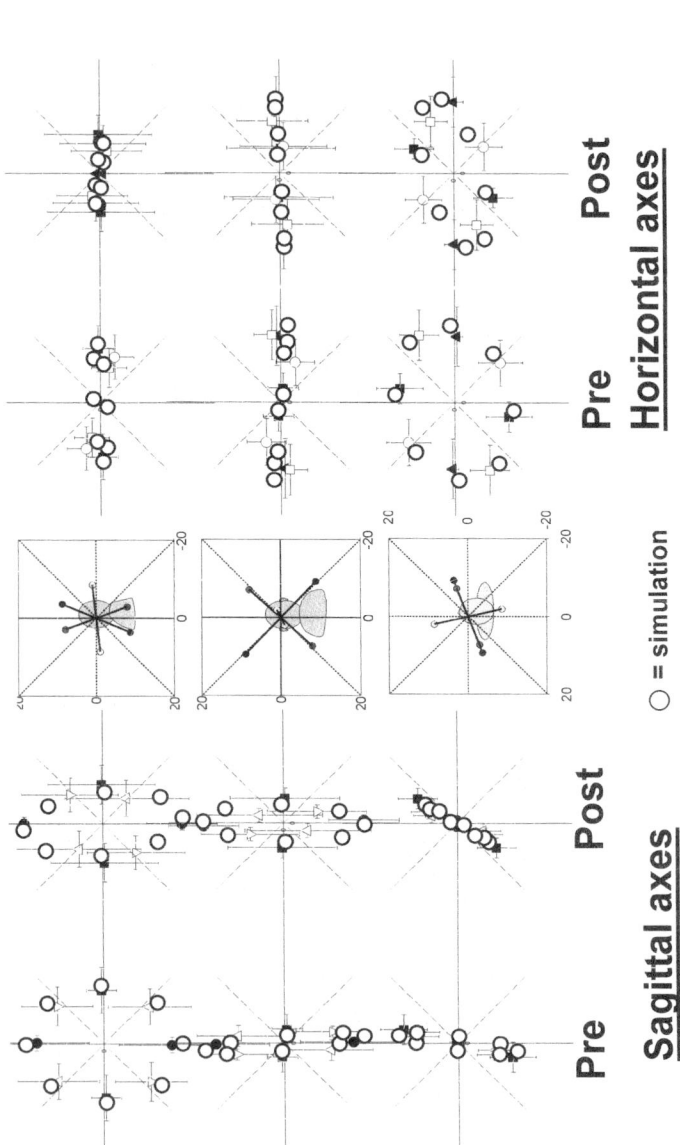

FIGURE 2. Simulating the VOR for the left eye. The orientation and relative length of the hypothetical channels are shown in the center column seen from the side (*top row*), back (*middle row*) and above (*bottom row*). For clarity the results are divided into those obtained from rotation around axes in the sagittal plane and the horizontal plane. The average data are shown with standard deviations as symbols (use FIG. 1A as key). Superimposed are the output of the three-channel model for the pre- and post-adaptation conditions (*large open circles*). All the major features of the data are reproduced by this model.

and gains of the set of three channels that optimally reproduced all the major features of the pre-adapt VOR, are shown in FIGURE 2. The prediction of this model of the response to the eight axes used are shown as large open symbols on either side in the columns labelled "pre." There is a good correspondence.

Next the model was "adapted." First the relative hypothetical activity of each of the channels expected in response to the physical motion component of the adapting stimulus was calculated given the gains and orientations of the channels obtained from the best fit to the pre-adapted data.The relative gains of each channel were then adjusted by an amount proportional to their relative activity during the adaptation experience to produce an "adapted" model. The response of the "adapted" model was simulated by projecting each head rotation onto the "adapted" channels (FIG. 2). The output of the adapted model showed an excellent fit to the adapted data, reproducing all the major features.

The properties of the proposed channels (FIG. 2, center) do not correspond to canal or roll/pitch/yaw coordinates. They are far away from the planes of individual canals, with one close to roll and the others forming an X approximately in Listing's plane,which can be expected to be tilted outwards since the eyes were likely verged in dark at a distance roughly corresponding to the screen.[12] A separate set of channels is needed for each eye since they are not in a plane orthogonal to the straight ahead but are instead tilted outwards by about 20 degrees.

This study shows that the vestibulo-ocular reflex can be elegantly modeled by a three-channel system. The proposed location of the channels correspond to emerging studies indicating a neural coordinate system involving an axis close to roll[13] and Listing's plane.[14–16]

ACKNOWLEDGMENTS

This work was supported by the Deutsche Forschungsgemeinschaft; Natural Science & Engineering Council (NSERC), Canada; and the Centre for Research in Earth and Space Technology (CRESTech), Ontario.

REFERENCES

1. LORENTE DE NÓ, R. 1933. Vestibulo-ocular reflex arc. Ann. Neurol. Psychiat. **30**: 245–291.
2. SZENTÁGOTHAI, J. 1950. The elementary vestibulo-ocular reflex arc. J. Neurophysiol. **13**: 395–407.
3. VILIS, T. & D. TWEED. 1988. A matrix analysis for a conjugate vestibulo-ocular reflex. Biol. Cybernet. **59**: 237–245.
4. BLUM, B. 1991. Channels in the Visual Nervous System: Neurophysiology, Psychophysics and Models. Freund. London.
5. CAMPBELL, F.W. & R.W. TEGEDER. 1991. A survey of channels and challenges, of information and meaning. *In* Channels in the Visual Nervous System: Neurophysiology, Psychophysics and Models. B. Blum, Ed.: 1–10. Freund. London.
6. GRAHAM, N. 1989. Visual Pattern Analyzers. Oxford University Press. Oxford.
7. LISBERGER, S.G., F.A. MILES, *et al.* 1983. Frequency-selective adaptation: evidence for channels in the vestibulo-ocular reflex? J. Neurosci. **3**: 1234–1244.
8. HARRIS, L.R. 1997. The coding of self motion. *In* Computational and Psychophysical Mechanisms of Visual Coding. L.R. Harris & M. Jenkin, Eds.: 157–183. Cambridge University Press. Cambridge.

9. BARNES, G.R. 1982. Visual factors affecting suppression of the vestibulo-ocular reflex. *In* Functional Basis of Ocular Motility Disorders. G. Lennerstrand *et al.* Eds.: 387–389. Pergamon. Oxford and New York.
10. BERTHOZ, A. & G. MELVILL JONES. 1985. Adaptive Mechanisms in Gaze Control. Elsevier. New York.
11. HARRIS, L.R. & K. BEYKIRCH, *et al.* 2001. The visual consequences of deviations in the orientation of the axis of rotation of the human vestibulo-ocular reflex. Vision Res. **41:** 3271–3281.
12. MOK, D. & A. RO, *et al.* 1992. Rotation of Listing's plane during vergence. Vision Res. **32:** 2055–2064.
13. CRAWFORD, J.D. & W. CADERA, *et al.* 1991. Generation of torsional and vertical eye position signals by the interstitial nucleus of Cajal. Science **252:** 1551–1553.
14. CRAWFORD, J.D. & T. VILIS. 1991. Axes of eye rotation and Listing's law during rotations of the head. J. Neurophysiol. **65:** 407–423.
15. SMITH, M.A. & J.D. CRAWFORD. 1998. Neural control of rotational kinematics within realistic vestibuloocular coordinate systems. J. Neurophysiol. **80:** 2295–2315.
16. SMITH, M.A. & J.D. CRAWFORD. 2001. Self-organizing task modules and explicit coordinate systems in a neural network model for 3-D saccades. J. Comput. Neurosci. **10:** 127–150.

Rectified Cross-Axis Adaptation of the Vestibulo-Ocular Reflex in Rhesus Monkey

M.F. WALKER AND D.S. ZEE

Departments of Neurology and Ophthalmology, Johns Hopkins University Hospital, Baltimore, Maryland 21287, USA

KEYWORDS: vestibulo-ocular reflex; cross-axis adaptation

In order to maintain perfect gaze stability, the vestibulo-ocular reflex (VOR) must be appropriately calibrated in three dimensions. Prior experiments have shown that not only the gain but also the direction of the VOR can be adaptively modified. Directional (cross-axis) VOR plasticity is dependent upon integrity of the vestibulo-cerebellum.[1] The usual paradigm for cross-axis adaptation pairs a sinusoidal head rotation about one axis (e.g., yaw) with an orthogonally moving visual stimulus (e.g., vertical) that has the same frequency and phase. After prolonged exposure to this stimulus, the response to yaw-axis rotation alone includes a vertical component. This process was modeled by Robinson using a 3×3 brainstem matrix to map inputs from semicircular canal pairs to pairs of eye muscles.[2]

In patients with cerebellar disease, we have observed a pattern of cross-coupling in which there is an inappropriate upward eye velocity during yaw-axis rotation in either direction.[3] This differs from the typical cross-axis adaptation experiment, in which the vertical component changes direction with the horizontal component. For example, when the head moves to the right, the stimulus moves up, and when the head moves to the left, the stimulus moves down. We asked whether the normal VOR could be trained to produce the response pattern seen in patients.

In three rhesus monkeys, we used the magnetic field search coil technique to record responses to yaw-axis rotation before, during, and after adaptation. Animals were rotated (0.5 Hz, 30–62 degrees/sec peak velocity) about the yaw axis for 35–120 minutes. During rotation, they viewed (and followed) a random dot optokinetic stimulus (also 0.5 Hz) that was in phase with the chair, except that it moved upward during each half-cycle of rotation. We term this "rectified cross-axis adaptation" because the velocity profile of the visual stimulus is a rectified sine. Before and after this adaptation, the response to yaw axis rotation was recorded, both in the dark and with viewing of a space-fixed center target (to maintain orbital position). Analysis was performed on the latter data because visual fixation did not suppress the effect of adaptation.

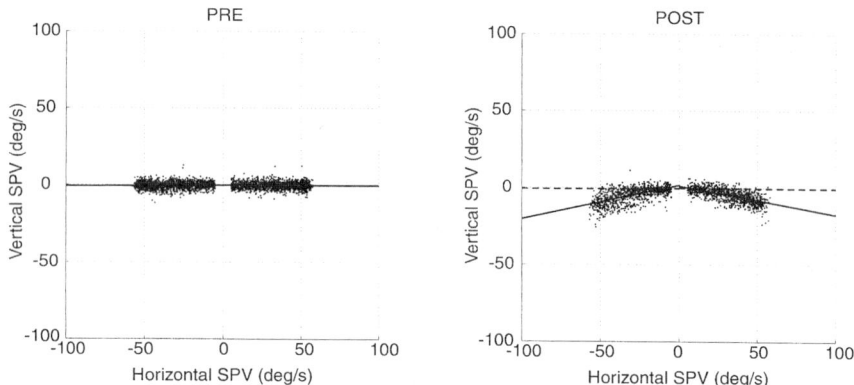

FIGURE 1. Vertical and horizontal components of slow-phase eye velocity before (PRE) and after (POST) adaptation in monkey M2. The *points* represent instantaneous slow-phase eye velocity (SPV), and the *solid lines* are the results of the linear regressions. For comparison, the regression from the pre-adaptation data is replotted as the *dashed line* on the right-hand panel. Positive values indicate left and down rotations.

After adaptation, yaw rotation consistently produced a mixed horizontal-vertical response in which the eyes moved upward for each direction of rotation (FIG. 1). We performed a linear regression of the vertical to the horizontal component of instantaneous eye velocity. The arctangent of the slope of this regression gives the axis of eye velocity in the horizontal–vertical plane. In 10 experiments, there was upward shift of eye velocity axis of 6.9 ± 4.7 (SD) degrees for rightward rotation and 6.5 ± 3.6 degrees for leftward rotation (left eye). Vertical eye velocity modulated clearly with chair (head) velocity and thus could not be attributed merely to an increase in the slow-phase velocity of a spontaneous downbeat nystagmus.

This rectified cross-axis adaptation paradigm is able to reproduce in normal animals a typical feature of VOR cross-coupling seen in patients with cerebellar disease: an upward vertical eye velocity during yaw rotation in both directions. An important characteristic of this cross-axis response is its directional nonlinearity: opposite inputs (right and left rotations) produce the same vertical response. This behavior cannot be explained by the classic linear push-pull model of the VOR, including Robinson's 3 × 3 matrix formulation. Instead, these models predict that the vertical response should change direction with a change in the direction of rotation. Yet the rectified response was easily achieved, even with a stimulus whose frequency and amplitude were well within what is generally thought to be the linear range of operation of the VOR.

The exact site responsible for this adaptation is unknown, but the vestibulocerebellum is likely to be involved, as vestibulocerebellar ablation has been shown to abolish cross-axis plasticity.[1] In this case, the nonlinearity of the adapted response suggests that it may not result from a simple modification of primary VOR pathways. Another possibility would be an axis shift within the velocity storage system. The velocity storage system has been shown to reorient the axis of eye velocity to align it more closely with the gravito-inertial axis; lesions to the cerebellar nodulus and ventral uvula abolish this reorientation.[4,5] Further studies will be required to elucidate the dynamics of this rectified cross-axis adaptation and the effects of specific cerebellar lesions.

ACKNOWLEDGMENTS

This work was supported by NIH Grants K23-EY00400 and R01-EY01849, and by the Arnold-Chiari Foundation.

REFERENCES

1. SCHULTHEIS, L.W. & D.A. ROBINSON. 1981. Directional plasticity of the vestibulo-ocular reflex in the cat. Ann. N.Y. Acad. Sci. **374:** 504–512.
2. ROBINSON, D.A. 1982. The use of matrices in analyzing the three-dimensional behavior of the vestibulo-ocular reflex. Biol. Cybern. **46:** 3–66.
3. WALKER, M.F. & D.S. ZEE. 1999. Directional abnormalities of vestibular and optokinetic responses in cerebellar disease. Ann. N.Y. Acad. Sci. **871:** 205–220.
4. ANGELAKI, D.E. & B.J. HESS. 1995. Inertial representation of angular motion in the vestibular system of rhesus monkeys. II. Otolith-controlled transformation that depends on an intact cerebellar nodulus. J. Neurophysiol. **73:** 1729–1751.
5. WEARNE, S.T., T. RAPHAN & B. COHEN. 1998. Control of spatial orientation of the angular vestibulo-ocular reflex by the nodulus and uvula. J. Neurophysiol. **79:** 2690–2715.

Three-Dimensional Eye-Movement Responses to Surface Galvanic Vestibular Stimulation in Normal Subjects and in Patients

A Comparison

H.G. MACDOUGALL,[a] A.E. BRIZUELA,[a] I.S. CURTHOYS,[a] AND G.M. HALMAGYI[b]

[a]*Department of Psychology, University of Sydney, NSW, Australia*
[b]*Department of Neurology, Royal Prince Alfred Hospital, NSW, Australia*

KEYWORDS: galvanic vestibular stimulation (GVS); ocular torsion, labyrinth; eye movement

Vestibular stimulation produces characteristic eye movements or vestibular ocular reflexes (VORs). The spatial components of VORs have been attributed to activation of specific vestibular sensory regions:[1] stimulation of a semicircular canal (SCC) mainly produces nystagmus around an axis which is roughly perpendicular to the plane of that canal, whereas stimulation of the otoliths mainly produces changes in ocular torsional position (OTP). Galvanic vestibular stimulation (GVS) of human subjects by currents of 5mA delivered through large-surface-area electrodes on the mastoids is painless and produces a range of vestibular responses,[2,3] including characteristic eye movements. By recording human eye movements using 3-D video during prolonged GVS, we have observed that normal subjects show eye-movement response patterns that could be expected from stimulation of a combination of vestibular sensory regions. Reliable characteristic patterns of these eye-movement components suggest that in humans surface GVS acts on otoliths and SCCs in an idiosyncratic fashion.

Similarities in the patterns of results between normal subjects provide the basis of a heuristic model that describes eye-movement responses to GVS as the weighted sum of inputs from all vestibular end-organs. In our work, the eye-movement response to GVS is modelled so that stimulation of the otoliths produces mainly OTP changes (upper poles of the eyes rotate toward the anode/away from the cathode). Stimulation of the horizontal canals mainly produces horizontal nystagmus with the slow phases directed towards the anode/away from the cathode. Stimulation of the anterior SCC produces downward eye movement, and stimulation of the posterior SCC produces upward eye movement, with a torsional velocity component. In nor-

Address for correspondence: H.G. MacDougall, Department of Psychology (A19), University of Sydney, Sydney, NSW, 2006, Australia. Voice: +612 9351 5953; fax: +612 9351 2603.
hamish@psych.usyd.edu.au

Ann. N.Y. Acad. Sci. 956: 546–550 (2002). © 2002 New York Academy of Sciences.

mal healthy subjects, during GVS, the downward eye movement produced by the anterior SCC is opposed by upward eye movement produced by the posterior SCC, so that the simultaneous stimulation of both vertical SCCs produces mainly torsional nystagmus with almost no vertical components (FIG. 1A).

Using the foregoing principles, we have shown that individual characteristic eye-movements to GVS can be modelled by varying the effective activation and inhibition of each vestibular sensory region. Such variability in the pattern of effective stimulation may reflect the variability of inner-ear morphology and impedance paths. This model also allows predictions about the eye-movement response characteristics in patients with specific vestibular dysfunction. The present study investigates whether the responses of patients with known vestibular dysfunction are consistent with these predictions.

BILATERAL VESTIBULAR LOSS

The model predicts that a patient with non-surgical bilateral vestibular loss should not show a substantial change in any eye-movement component in response to GVS delivered to either side. We tested a patient with non-surgical bilateral vestibular dysfunction (total loss on left, 80% paresis on right) as a result of sequential neuritis. The subject's data show small horizontal and vertical nystagmus, but otherwise minimal oculomotor responses to GVS delivered to either side (FIG. 1B).

UNILATERAL VESTIBULAR LOSS

The model predicts that a patient with surgical unilateral vestibular loss should show no response to GVS delivered to the affected side, whereas the eye-movement response to GVS on the healthy side should be normal. We tested a patient who underwent left unlilateral vestibular neurectomy 15 years earlier. The subject has some hearing preserved on the left, suggesting that some response to GVS might be possible on that side. The subject's data show marked reduction in all oculomotor responses to GVS delivered on the affected side, compared to a relatively normal response to GVS on the healthy side (FIG. 1C).

INFERIOR VESTIBULAR NEURITIS

The model predicts that a subject with inferior division neuritis (posterior SCC and saccule affected) should show vertical nystagmus in response to GVS: with anodal current delivered to the affected side producing slow phases directed down as a result of the imbalance in activity of the vertical canals. We tested a patient with profound deafness, absent VEMPs, and absent caloric responses on the left but preserved function on the right. The subject's data show vertical nystagmus (slow phases: down) in response to GVS delivered on the right side, compared to a relatively normal response to GVS on the left side. There is no reduction in OTP response to GVS on the right side compared to the left side, which is consistent with dysfunction of the saccule but preserved function of the utricle (FIG. 1D).

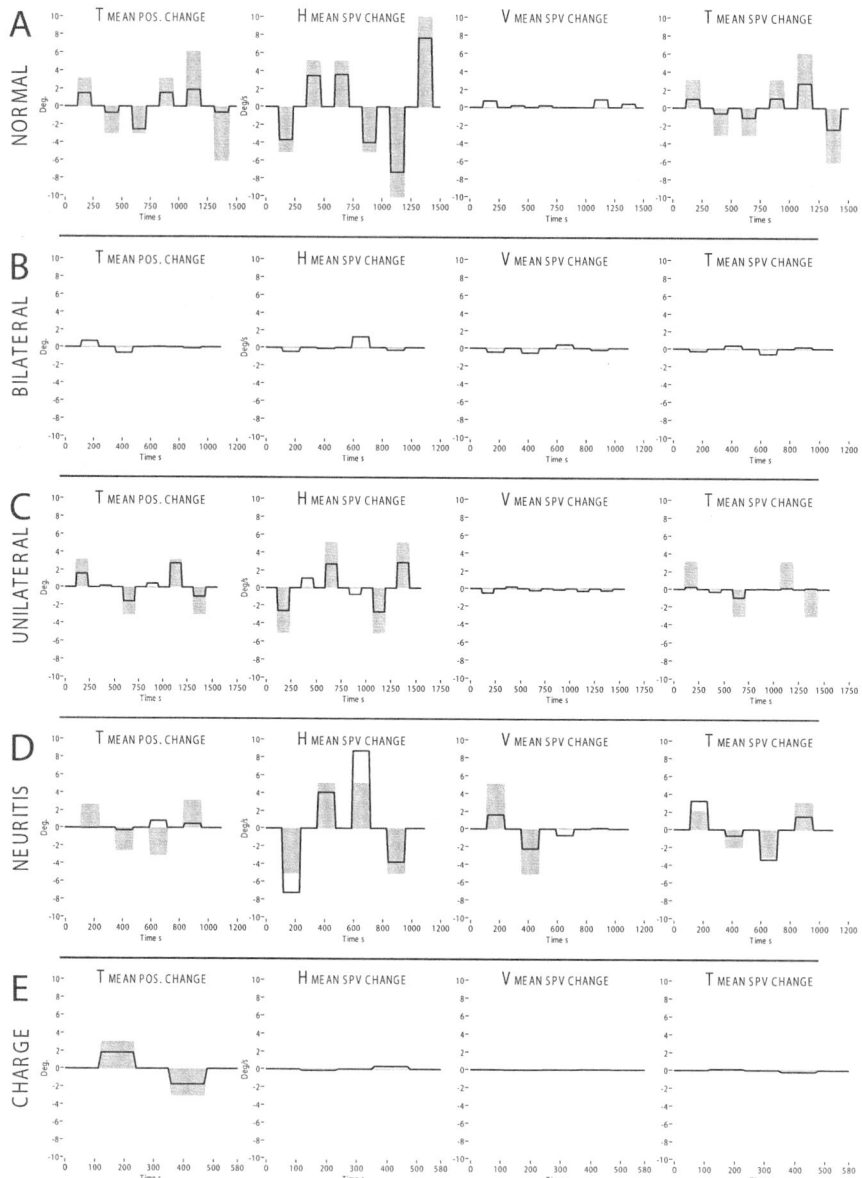

FIGURE 1. Predicted range (*grey bars*) and observed magnitude (*solid black trace*) of OTP and horizontal, vertical and torsional slow-phase velocity (SPV) responses to GVS for five subjects. Row **A** shows the results of a subject without diagnosed vestibular dysfunction (normal); row **B** shows a subject with non-surgical bilateral vestibular loss; row **C** shows a subject with surgical (left) unilateral vestibular loss; row **D** shows a subject whose response is consistent with (right) inferior vestibular neuritis; and row **E** shows a subject with CHARGE syndrome.

CHARGE SYNDROME

The model predicts that a patient with absent SCCs associated with the CHARGE syndrome (a combination of various congenital abnormalities, including ear anomalies)[4] should not show substantial eye-velocity response components to GVS delivered to either side, but should show normal otolith-related eye-movement response components. We tested a patient with CHARGE syndrome with absent SCCs (visualized on scans) and absent responses to rotations but preserved VEMPs. The subject's data show absent velocity responses but preserved OTP changes in response to GVS; however, it should be noted that testing was carried out with fixation present in this instance, and this would tend to suppress velocity responses (FIG. 1E).

CONCLUSION AND DISCUSSION

The predictions generated by the present model of eye-movement response to GVS are consistent with the responses obtained by testing patients with known vestibular dysfunction. The results of the present study lend weight to the validity of the model and thereby support the argument that surface GVS tends to stimulate all end organs in an idiosyncratic, yet predictable, fashion. In each case, the response predicted by the model was derived by modifying the sensitivity of the sensory regions affected by the diagnosed dysfunction. The idealized response could then be modelled on the subject's observed response by making minor modifications to the sensitivity of sensory regions that was consistent with the variability in sensitivity seen in a normal population (MacDougall *et al.*, submitted for publication). It is important to bear in mind that patients are unlikely to show any GVS response from end-organs which have been surgically ablated unless some nerve endings remain (in that GVS is thought to act on the spike trigger zone of primary afferents[5]). Patients with vestibular loss which has been inferred from other tests might show responses to GVS; thus any diagnosis is likely to remain a question of evidence from a number of sources. Developments in 3-D eye-movement recording systems and GVS delivery methods are making it possible to understand the variability and apparent complexity of the responses to this stimulus. GVS may be more reliable than might be inferred from recent findings[6–8] and may potentially have some useful application in a clinical, diagnostic, or therapeutic setting.

ACKNOWLEDGMENTS

Mr. Hamish MacDougall was supported by a Research Scholarship provided by the Garnett Passe and Rodney Williams Memorial Foundation during the period of this study. This work was also supported by a NHMRC (of Australia) Clinical Excellence Grant. We gratefully acknowledge the assistance of Imelda Hannigan, Robbie Yavor, Ann Burgess, Laura Mezey, Americo Migliaccio, and the subjects.

REFERENCES

1. SUZUKI, J.I. & B. COHEN. 1966. Integration of semicircular canal activity. J. Neurophysiol. **29:** 981–995.
2. WATSON, S.R. *et al.* 1998. Maintained ocular torsion produced by bilateral and unilateral galvanic (DC) vestibular stimulation in humans. Exp. Brain Res. **122:** 453–458.
3. WATSON, S.R., P. FAGAN & J.G. COLEBATCH. 1998. Galvanic stimulation evokes short-latency EMG responses in sternocleidomastoid which are abolished by selective vestibular nerve section. Electroencephalogr. Clin. Neurophysiol. **109:** 471–474.
4. WIENER-VACHER, S.R. *et al.* 1999. Vestibular function in children with the CHARGE association. Arch. Otolaryngol. Head Neck Surg. **125:** 342–347.
5. GOLDBERG, J.M., C.E. SMITH & C. FERNANDEZ. 1984. Relation between discharge regularity and responses to externally applied galvanic currents in vestibular nerve afferents of the squirrel monkey. J. Neurophysiol. **51:** 1236–1256.
6. KLEINE, J.F., W.O GULDIN & A.H. CLARKE. 1999. Variable otolith contribution to the galvanically induced vestibulo-ocular reflex. Neuroreport **10:** 1143–1148.
7. LOBEL, E., J.F. KLEINE, D.L. BIHON, A. LEROY-WILLIG & A. BERTHOZ. 1998. Functional MRI of galvanic vestibular stimulation. J. Neurophysiol. **80:** 2699–2709.
8. SCHNEIDER, E., S. GLASAUER & M. DIETERICH. 2000. Central processing of human ocular torsion analyzed by galvanic vestibular stimulation. Neuroreport **11:** 1559–1563.

Translational VOR Responses to Abrupt Interaural Accelerations in Normal Humans

S. RAMAT[a,b] AND D.S. ZEE[a]

[a]*Department of Neurology, The Johns Hopkins University School of Medicine, Baltimore, Maryland, USA*

[b]*Dipartimento di Informatica e Sistemistica, Università di Pavia, Pavia, Italy*

KEYWORDS: translational vestibulo-ocular reflex; saccade amplitude

INTRODUCTION AND METHODS

We investigated the translational vestibulo-ocular reflex (tVOR) in response to abrupt high-acceleration stimuli in normal humans. Using the search coil technique we recorded eye responses to head translation in six normal subjects while they viewed an LED at 15 cm or 30 cm distance. Head position was monitored with a six-degrees-of-freedom miniBIRD device, and a linear accelerometer detected the onset of head motion. Subjects looked at the LED in dim illumination while abrupt interaural head translations (2–3 cm amplitude, 0.4–1.4 g max acceleration) were delivered manually using a "head sled'"device,[3] which has the advantage of a low inertia compared to whole-body sleds. Head movements had an average duration of 190 msec, reaching an average peak head velocity of 25 cm/sec at 86 msec (mean) from the onset of head motion. The effect of monocular versus binocular viewing was also assessed for the 15-cm target. On the basis of geometrical relationships and individual subject measurements, we computed the ideal eye movement needed to maintain target fixation during head translation. Position gains were calculated as recorded/ideal eye rotation from onset of head movement to the time of peak head velocity.

RESULTS

The responses had qualitatively the same characteristics at 15- and 30-cm distance and a typical response in shown in FIGURE 1. During both eyes viewing (BEV) of the 15-cm target the average (± SD) latency of the slow-phase response was 36 (± 15) msec and was supplemented by a saccade, with an amplitude of 3.2 (±1) deg and occurring at 150 (± 44) msec after the onset of head motion (FIG. 1). For the 30-cm target, there were statistically significant ($p < 0.01$) increases in both the latency of the slow-phase response (average 41 ± 27 msec) and the latency of the saccadic correc-

Address for correspondence: Dr. Stefano Ramat, Department of Neurology, Johns Hopkins Hospital, Pathology 2-210, 600 N. Wolfe Street, Baltimore, MD 21287. Voice 410-955-3319; fax: 410-614-1746.

stefano@dizzy.med.jhu.edu

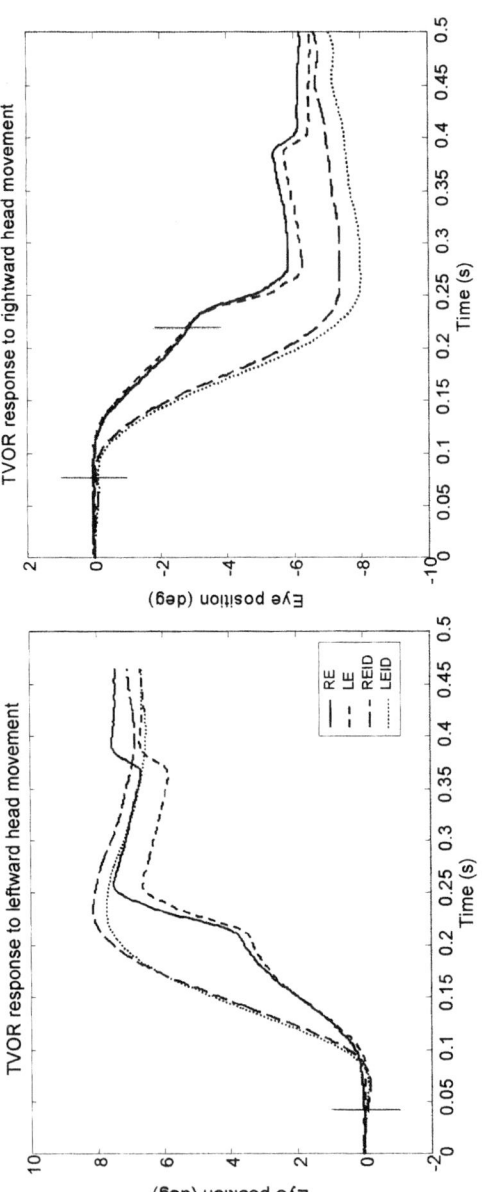

FIGURE 1. Overall responses of one subject to rightward and leftward head accelerations. The *continuous* and the *short dashed lines* represent the recorded right eye and left eye, respectively. The *long dashed* and the *dotted lines* represent the ideal eye movement computed for the given head movement, the target distance, and the anatomical measurements of the subject.

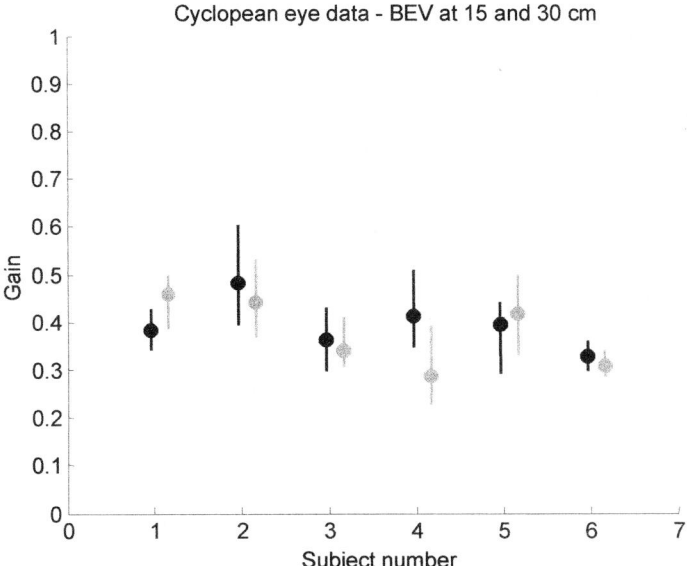

FIGURE 2. Median (*circle*) and vertical bar extending from the 25th to the 75th percentile of each subject's gains computed at the time of peak head velocity in BEV at 15 cm (*black symbols*) and at 30 cm (*gray symbols*).

tion (165 ± 60 msec), and a decrease in saccade amplitude (1.8 ± 0.7 deg). However, no statistically significant difference was found between the gains at the two viewing distances (FIG. 2). Medians (and 95% confidence intervals) for subject gains were 0.38 (0.25 to 0.62) at the 15-cm and 0.37 (0.15 to 0.62) at the 30-cm distance.

For monocular viewing of the 15-cm target, three of six subjects showed a significant ($p < 0.05$) decrease in gain in at least one monocular viewing condition (comparing the right eye during right-eye viewing or the left eye during left-eye viewing, with the same eye during binocular viewing). All subjects showed statistically significant changes in the vergence angle (measured at the onset of the head movement) between binocular and monocular viewing conditions ($p < 0.05$). During monocular viewing, vergence angle and tVOR gain were loosely related. The decrease in tVOR gain was greatest (44%) in the subject with the largest change in vergence angle (exophoria during monocular viewing), and was smallest (4%) in the subject that showed the least change in vergence angle (6%). One subject showed a decrease of gain with a relative esophoria relative to the alignment with BEV. Three subjects showed significantly greater gains for adduction (temporal to nasal) of the viewing eye in one or both monocular viewing conditions.

DISCUSSION

A major finding of this study is that in response to abrupt, high-acceleration stimuli the gain of the slow-phase component of the tVOR is relatively low (about 40%)

and relatively independent of target distance. This finding suggests that the tVOR is tuned to compensate for a relatively small proportion of head translation, no matter what the required velocity or amplitude of the compensatory slow phase.[1,2] We also found that the average size of the first saccadic correction for 30-cm target viewing was significantly smaller than for 15-cm target viewing. Saccade amplitude was reduced by about 44% compared to the amplitude for the 30-cm target, which roughly corresponds to the reduction in eye-movement requirement for the same amount of head movement (2.5 cm). Clearly, rapid, position-correcting movements also play a role in the tVOR response in normal subjects. Finally, our findings also indicate that the performance of the tVOR is subject to factors related to binocular versus monocular viewing, and to direction of eye motion.

ACKNOWLEDGMENTS

This work was supported by NIH Grants EY01849, DC02849, and NASA/NSBRI NCC 9-58 and by the Robert M. and Annetta J. Coffelt endowment for PSP research.

REFERENCES

1. CRANE, B.T. *et al.* 2000. Initiation of the unilaterally deafferented human otolith-ocular reflex, Soc. Neurosci. Abstr. **26:** 6.
2. PAIGE, G.D. *et al.* 1998 Human vestibuloocular reflex and its interactions with vision and fixation distance during linear and angular head movement. J. Neurophysiol. **80:** 2391–2404.
3. RAMAT, S., D.S. ZEE & L.B. MINOR. 2001 Translational vestibulo-ocular reflex evoked by a "head heave" stimulus. Ann. N.Y. Acad. Sci. **942:** 95–113.

Rapid Adaptation of Translational Vestibulo-Ocular Reflex: Time Course, Consolidation, and Specificity

W. ZHOU,[a,b,c] P. WELDON,[b] B. TANG,[b] AND W.M. KING[b,c]

Departments of Surgery/ENT,[a] Neurology,[b] and Anatomy,[c] University of Mississippi Medical Center, Jackson, Mississippi 39216, USA

KEYWORDS: motor learning; VOR; monkey; sensory-motor transformation; synaptic plasticity; Hebbian learning

When monkeys were subjected to repeated trials of brief linear head acceleration and were required to maintain fixation on a visual target, the initial eye speed of their translational vestibulo-ocular reflex (TVOR) was systematically increased if the target was world-fixed (TVOR condition), or systematically decreased if the target was head-fixed (TVORC condition). To study the time course of adaptation and its consolidation, monkeys were given repeated TVOR trials for several days (1000 trials/day) in order to maximize their initial eye speed. After eye speed was stable (high gain state), the monkeys were given TVORC trials for several days to minimize eye speed (low gain state). TVOR responses were computed by averaging eye speed over multiple trials in the high and low gain states. Using this paradigm, we identified three temporally distinct components in the initial 80 msec of the TVOR. The earliest component began 7.7 msec after the onset of head motion and showed minimal modification even after several days of high or low gain training. The intermediate component began 20 msec after the onset of head motion and could be systematically modified by training trials. The late component was the most labile of the three components. It began 35 msec after the onset of head motion and was parametrically modified after a single trial of new context. Neither the early nor the intermediate latency component was parametrically influenced by a change in task context. Furthermore, these learned changes were consolidated and could be observed several days after training without any reinforcement, indicating that the adaptive changes are due to long-term synaptic plasticity and not to a simply learned strategy. We hypothesize that the initial component of the TVOR is mediated solely by brainstem circuits that include central vestibular and oculomotor neurons. The second component of the TVOR response is hypothesized to involve the vestibulo-cerebellum (flocculus and ventral paraflocculus). The third component is hypothesized to involve cortical circuits that are sensitive to behavioral context.

Address for correspondence: Wu Zhou, Ph.D., Department of Surgery/ENT, University of Mississippi Medical Center, 2500 North State Street, Jackson, MS 39216. Voice: 601-984-5491; fax: 601-984-5503.

wuz@vor.umsmed.edu

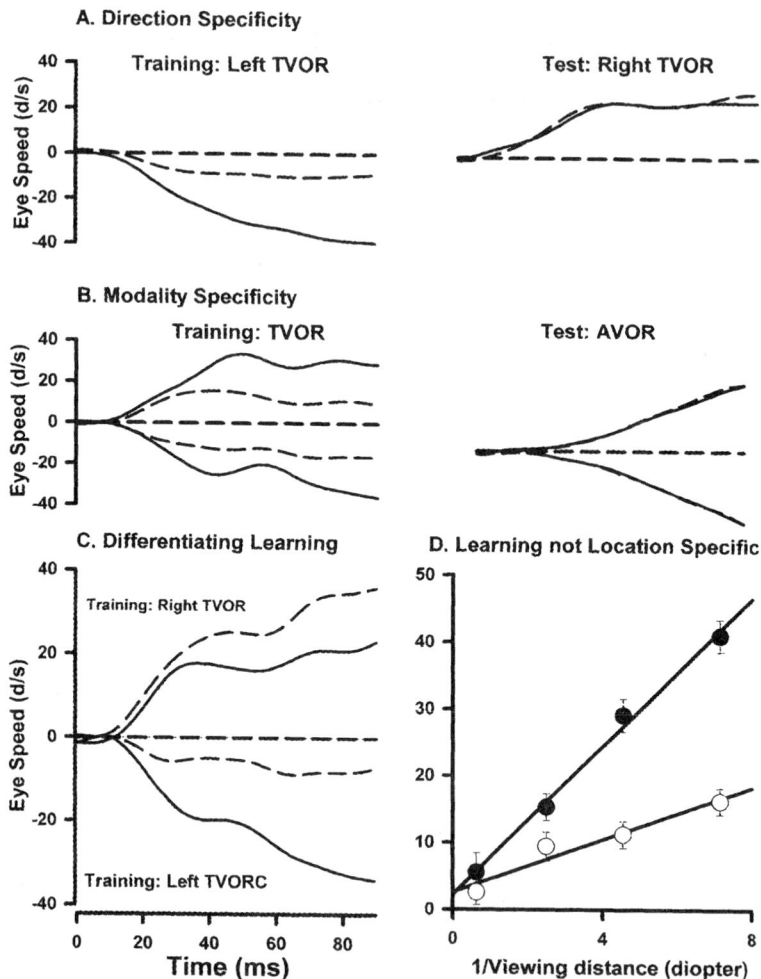

FIGURE 1. Specificity of TVOR motor learning. (**A**) TVOR motor learning is direction-specific. After training, the ocular response to rightward translation was decreased (*dotted line*), while the ocular response to leftward translation was unaffected. *Solid line* = before learning. (**B**) TVOR motor learning is modality-specific. After training with translation, TVOR ocular responses were decreased, but AVOR ocular responses were unaffected. (**C**) Directional learning. Monkeys were trained with TVOR condition associated with leftward translations and the TVORC condition associated with rightward translations. After training, the ocular responses to rightward translations were decreased, but ocular responses to leftward translations were increased. (**D**) Motor learning is not viewing distance–specific. The ocular response to translation was measured before and after learning at a particular target location (22cm).

To gain insights into the neural mechanisms underlying the TVOR adaptation, we carried out a series of experiments to characterize the specificity of the TVOR motor learning. FIGURE 1A shows that the learned changes were specific to the direction of head translation. More interestingly, monkeys could increase their responses in one direction and decrease them in the opposite direction in the same training session (FIG. 1C). FIGURE 1B shows that motor learning induced with translational head movements did not transfer to compensatory responses elicited by rotary head movements (angular VOR) or *vice versa*. These data suggest that the neural substrates for translational and angular VOR motor learning are functionally distinct. Finally, FIGURE 2D shows the learned changes transferred to other viewing distances. Solid circles showed that eye speed (the first 80 msec) and inverse viewing distance are linearly related.[1,2] The regression line has a slope of 5.54 degrees/sec per diopter. After the TVOR motor learning, initial eye speed remained inversely proportional to viewing distance (open circles); however, the slope of the linear relationship was reduced to 1.95 degrees/sec per diopter ($p < 0.001$). Furthermore, the learned changes were not gaze-direction-specific. In summary, the TVOR adaptation is highly specific for the combination of the head motion and the evoked eye-movement response. These results suggest that the TVOR learning does not occur in the sensory limb of the reflex or the motor limb of the reflex, but in the sensory-motor transformation stage of the reflex. The dependence of the adaptation on both the head motion and the evoked eye movement suggests that Hebbian learning may be one of the underlying cellular mechanisms.

ACKNOWLEDGMENTS

This work was supported by PHS grant (EY04045) and a NSBRI grant to W.M. King and a NSBRI/NASA grant and a UMC intramural grant to W. Zhou.

REFERENCES

1. PAIGE, G.D. & D.L. TOMKO. 1991. Eye movement responses to linear head motion in the squirrel monkey. I. Basic characteristics. J. Neurophysiol. **65:** 1170–1182.
2. SCHWARZ, U., C. BUSETTINI & F.A. MILES. 1989. Ocular responses to linear motion are inversely proportional to viewing distance. Science **245:** 1394–1396.

Rapid Adaptation of Translational Vestibulo-Ocular Reflex: Independence of Retinal Slip

W. ZHOU,[a,b,c] P. WELDON,[b] B. TANG,[b] AND W.M. KING[b,c]

Departments of Surgery/ENT,[a] Neurology,[b] and Anatomy,[c] University of Mississippi Medical Center, Jackson, Mississippi 39215, USA

KEYWORDS: motor learning; VOR; monkey; sensory-motor transformation; synaptic plasticity

Recently, we developed a novel paradigm to induce rapid adaptation of the translational vestibulo-ocular reflex (TVOR).[1,2] When monkeys were given repeated trials of brief head translation (0.4 second duration, 0.2 g peak acceleration) and were required to maintain fixation on a visual target, the initial eye speed of their TVOR (as early as 20 ms after the onset of head translation) systematically increased if the target's position was world-fixed (TVOR condition) or systematically decreased if the target's position was head-fixed (TVORC condition). The time course of adaptation is exponential (time constant less than 100 trials) and the adaptation is specific for stimulus direction and modality (linear or angular). Retinal slip has been implicated as an essential signal for angular VOR gain adaptation.[3] Here, we examine a possible role of retinal slip in TVOR adaptation. In darkness, monkeys were given repeated TVORC training trials in two conditions: with or without retinal slip during head translation. In the without-retinal-slip condition, the target was turned off before head translation; however, when the head movement was completed, the target was re-illuminated. Each trial lasted about 3 s with an inter-trial interval of 1 s. During the inter-trial interval, photographers' safe lights were turned on to minimize dark adaptation. Search coils were surgically implanted in both eyes to record binocular eye movements.[4] FIGURE 1 shows that initial eye speed (i.e., open-loop interval of the TVOR: ~first 90 ms) decayed exponentially (adapted) with successive trials ($p < 0.001$) in both conditions (FIG. 1A, with retinal slip; FIG. 1B, without retinal slip). In the with-retinal-slip condition, the initial eye speed was reduced by $52 \pm 3\%$. In the without-retinal-slip condition, the initial eye speed was reduced by $45 \pm 3\%$, similar to the with-retinal-slip condition. The time constants of adaptation were 64 ± 4 trials (without retinal slip) and 76 ± 3 trials (with retinal slip). These

---

Address for correspondence: Wu Zhou, Ph.D., Department of Surgery/ENT, University of Mississippi Medical Center, 2500 North State Street, Jackson, MS 39216. Voice: 601-984-5491; fax: 601-984-5503.

wuz@vor.umsmed.edu

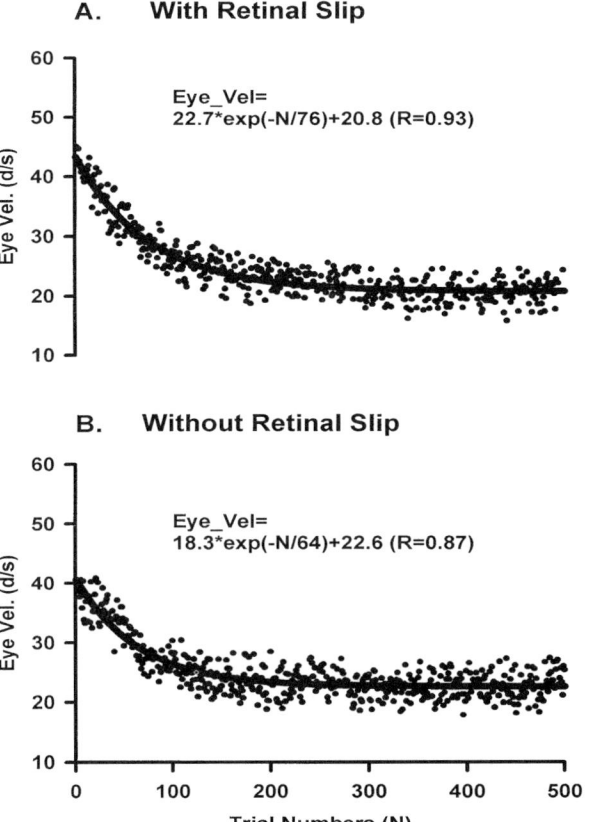

FIGURE 1. Rapid TVOR adaptation with retinal slip (**A**) or without retinal slip (**B**). In both conditions, the data are best-fit by a single exponential [eye speed = $a*\exp(-\text{trialN}/b)+c$].

data provide evidence that retinal slip is not necessary for rapid TVOR adaptation in this paradigm.

ACKNOWLEDGMENTS

This work was supported by PHS grant EY04045 to W.M. King and a NSBRI/NASA grant and a UMC intramural grant to W. Zhou.

REFERENCES

1. ZHOU, W., B.F. TANG & W.M. KING. 1999. Rapid motor learning in vestibulo-ocular reflex guided by context cues. Soc. Neurosci. Abst. **25:** 661.
2. ZHOU, W., P. WELDON, B.F. TANG & W.M. KING. 2001. Rapid adaptation of translational vestibulo-ocular reflex: time course, consolidation, and specificity. Ann. N.Y. Acad. Sci. This volume.

3. DU LAC, S., J.L. RAYMOND, T.J. SEJNOWSKI & S.G. LISBERGER. 1995. Learning and memory in the vestibulo-ocular reflex. Annu. Rev. Neurosci. **18:** 409–441.
4. ROBINSON, D.A. 1963. A method of measuring eye movement using a scleral search coil in a magnetic field. IEEE Trans. Biomed. Eng. **10:** 137–145.

Effects of a Large-Field Visual Scene on the Vergence Response to Naso-Occipital Linear Motion in Monkeys

YOSHIRO WADA,[a,b] YASUSHI KODAKA,[a,c] AND KENJI KAWANO[a,c]

[a]Core Research for Evolutional Science and Technology (CREST), Japan Science and Technology Corporation (JST), Ibaraki, Japan

[b]Department of Physiology, Nara Medical University, Kashihara, Nara, 634-8521 Japan

[c]Neuroscience Research Institute, National Institute of Advanced Industrial Science and Technology, Tsukuba, Ibaraki, 305-856 Japan

KEYWORDS: vergence; vestibular; visual; monkey; linear motion

INTRODUCTION

Vergence eye movements are elicited by visual stimuli[1,2] as well as naso-occipital (NO) linear head motion.[3] Therefore, vergence eye movements during NO linear motion can be influenced by visual stimuli. As the first step in understanding the role of visual stimuli in the vestibular vergence response, we studied the effects of a large-field visual scene on the oculomotor response to forward linear motion in monkeys.

METHODS

We used three alert Japanese monkeys (*Macaca fuscata*, weighing 5.5–10.5 kg) previously trained to fixate on small targets. During recording sessions, the monkey was seated in a chair with its head secured by a head holder. We measured the horizontal and vertical eye positions binocularly using an electromagnetic search-coil system (Enzanshi-Kogyo). The monkey chair was mounted on an 80-cm-long linear sled riding on linear bearings and driven by a lead screw connected to an AC servomotor (Nippon Thompson Co., Ltd). The linear sled generated sinusoidal oscillations along the NO-axis at 4, 2, 1 and 0.67 Hz. A random-dot pattern (=RD, 69° × 69°) large-field visual scene and/or a red center fixation target (=FT, 0.5° × 0.5°) were/was back-projected onto a screen 35 cm away from the monkey. The presentation of the stimuli and the collection, storage, and display of data

---

Address for correspondence: Yoshiro Wada, Department of Physiology, Nara Medical University, 840 Shijo-cho, Kashihara, Nara, 634-8521 Japan. Voice: +81-744-29-8827; fax: +81-744-29-0306.

wada@naramed-u.ac.jp

Ann. N.Y. Acad. Sci. 956: 561–563 (2002). © 2002 New York Academy of Sciences.

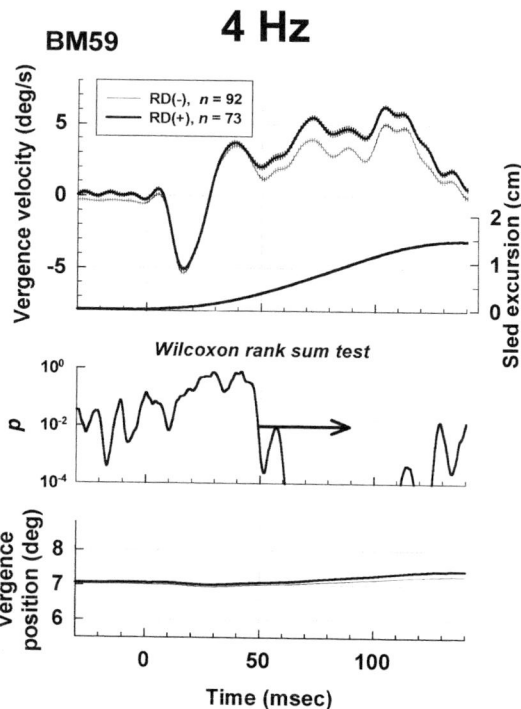

FIGURE 1. Effects of a large-field visual scene on the vergence response to initial forward linear motion at 4 Hz without FT. **Top traces:** the means ± SEs of vergence velocities and sled excursion. **Middle traces:** the results (probabilities) of Wilcoxon rank-sum test for the vergence velocities with and without RD. **Bottom traces:** vergence positions with and without RD.

were controlled by the REX system.[4] We studied the early effects (within 140 msec) of a large-field visual scene on the vergence response to forward linear motion.

RESULTS

FIGURE 1 summarizes the results of monkey BM59 during 4 Hz NO linear motion without FT. The top traces illustrate the vergence velocities (mean ± SE for every 1 msec) with/without RD and the sled excursion. The middle trace shows the probability (p) of Wilcoxon rank-sum test for the vergence velocities with and without RD. These results demonstrate that the vergence velocities with RD were significantly ($p < 0.01$) larger than those without RD from 49.0 msec after the onset of sled motion. The means ± SDs of these RD effect latencies without FT in three monkeys were 55.6 ± 10.7 msec at 4 Hz and 59.9 ± 7.8 msec at 2 Hz. Those with FT were 55.5 ± 10.0 msec at 4 Hz and 60.8 ± 4.3 msec at 2 Hz. The bottom traces illustrate the vergence positions, which did not exhibit large changes. These data reveal that the monkey could maintain the fixation distance for 140 msec even if FT was not presented.

CONCLUSION

This study showed that a large-field visual scene significantly increases the vergence response to initial forward motion with short latencies (<56 msec) in monkeys. In addition, these effects were not influenced by the fixation target.

ACKNOWLEDGMENT

This work was supported by Japan Space Forum.

REFERENCES

1. MILES, F.A. 1999. Short-latency visual stabilization mechanisms that help to compensate for translational disturbances of gaze. *In* Otolith Function in Spatial Orientation and Movement. B. Cohen & B.J.M. Hess, Eds. Ann. N.Y. Acad. Sci. **871:** 260–271.
2. INOUE, Y.A. *et al.* 1998. Short-latency vergence eye movements elicited by looming step in monkeys. Neurosci. Res. **32:** 185–188.
3. PAIGE, G.D. & D.L. TOMKO. 1991. Eye movement responses to linear head motion in the squirrel monkey. II. Visual-vestibular interactions and kinematic considerations. J. Neurophysiol. **65:** 1183–1196.
4. HAYS, A.V., B.J. RICHMOND & L.M OPTICAN. 1982. A UNIX-based multiple process system for real-time data acquisition and control. WESCON Conference Proceedings **2/1:** 1–10.

Dynamic Changes of Eye Cyclo Position during Head Tilt

TONY PANSELL,[a] JAN YGGE,[a] AND HERMANN D. SCHWORM[a,b]

[a]*Karolinska Institutet, Department of Clinical Science, Division of Ophthalmology, Huddinge University Hospital, 141 86 Stockholm, Sweden*

[b]*Universitätsklinikum Hamburg-Eppendorf, Hamburg, Germany*

KEYWORDS: torsion dynamics; head tilt; vestibulo-ocular reflex; otoliths; semicircular canals

INTRODUCTION

In maintaining steady vision during locomotion the eyes have to make compensatory rotations to the movements of the head, that is, the vestibulo ocular reflex (VOR). The semicircular canals detect rotary (r-VOR) and the otoliths detect linear (t-VOR) movements of the head.[1] When a subject performs head tilt (roll), the ipsilateral eye elevates and the contralateral eye depresses. An ocular counter-rolling (OCR) in the opposite direction to head tilt is also induced. The dynamic shift in torsional position is usually attributed to the semicircular canals and the stable post-tilt OCR is attributed to the otoliths. Recent findings[2] suggest that the otoliths contribute to the dynamic changes of eye position as well. Our aim with this study was to evaluate the dynamic changes of eye position in response to head tilt, and to try to differentiate the input from the canals and the otoliths to provide a better understanding of the r-VOR dynamics.

METHODS

Ocular torsion in 20 normal subjects was evaluated binocularly using the three-dimensional video-oculography technique (3D-VOG, SMI, Germany) during stepwise tilting of the head (FIG. 1) in the roll plane to 15, 30, and 45 degrees both to the right and left shoulder. A visual stimulus (a photographic picture of a castle with added spatial cues) was displayed on a screen 1.5 m in front of the subject. The tests were performed while the subjects were viewing binocularly. During the experiment, the subject's head was maneuvered with a specially designed tiltable chin rest with a bite-board (FIG. 1).

Address for correspondence: Tony Pansell, Ogonkliniken, Karolinska Institutet, Huddinge sjukhus, B52, 141 86 Huddinge, Sweden. Voice: +46 8 585 879 89; fax +46 8 779 35 06.

tony.pansell@klinvet.ki.se

FIGURE 1. The head was tilted from head-straight position to 15, 30, and 45 degrees to both right and left shoulder. A specially designed tiltable chin-rest with a bite-board was used.

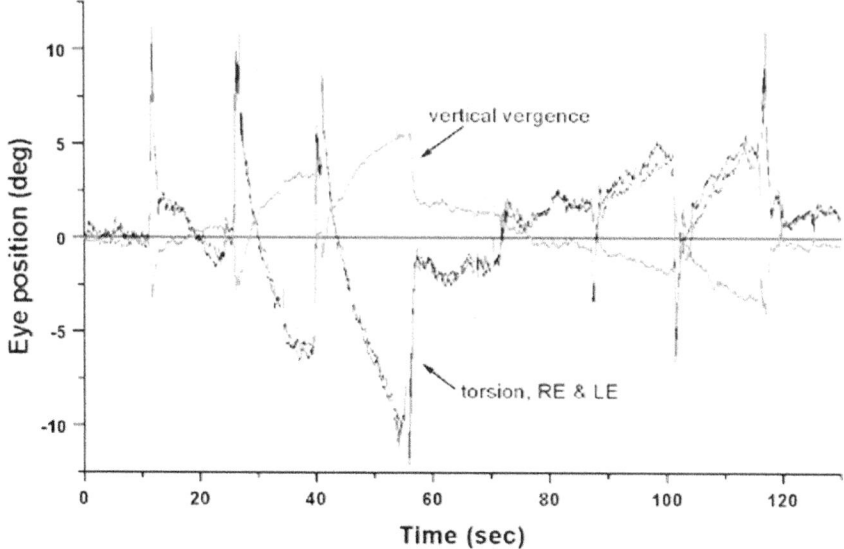

FIGURE 2. Graph displaying binocular torsion and the vertical vergence from one representative subject. Notice the similar characteristics of the torsion and the vertical eye dynamics. Positive torsion is counter-clockwise from the subject's point of view. Positive vertical vergence is right eye over left eye.

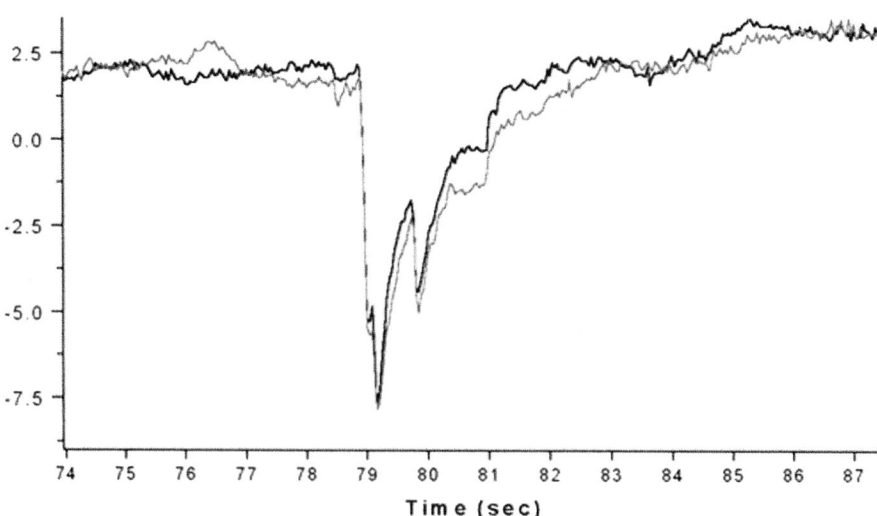

FIGURE 3. A magnification of the rapid torsion movement. Nystagmus beats were seen superimposed on the return-movement.

RESULTS

A rapid torsional movement in the same direction as the head tilt was a consistent finding in all subjects (FIG. 2) The amplitude of this rapid torsional change varied between 3 and 10 degrees and the peak velocity to approximately 80 degrees/sec. This was followed instantly, with no inter-saccadic interval, by a slower torsional return of the eye towards the initial position. The peak velocity of the return movement was about half the above value. The torsional change was superimposed by nystagmus beats (~3 Hz) with the fast phase in the same direction as the initial rapid part of the movement (FIG. 3). The eye position data were used for calculating the vertical vergence (LE-RE). A saccade-like change in vertical vergence position was found to be synchronous with the torsional rapid change (FIG. 2). Initially a left eye over right eye was induced in right head tilt. This vertical vergence change had a peak velocity of approximately 20 degrees/sec and like the torsional rapid change it was directed in the opposite direction to the final position after the head tilt. The return movement of the vertical vergence showed a similar pattern as the torsional return movement.

CONCLUSIONS

Both the torsion and vertical vergence rapid changes are in the opposite direction to their final position when the head is held in a tilted position. A physiological skew deviation is thus induced during the first part of the head tilt. These dynamic changes of eye position are initiated with a short latency compared to the change in head position. Inertia of the utricular mass results in a bending of the receptor cilias in the otolith macula in an opposite direction with respect to the final position after the head tilt. This could presumably explain the skew deviation found in the present study. If the rapid changes in torsional and vertical vergence position place the eye in a favorable position in the orbit due to a slow phase VOR cannot be explained at present.

REFERENCES

1. LEIGH, R.J. & D.S. ZEE. 1999. The neurology of eye movements, 3rd ed. Contemporary Neurology Series.: 21–22. Oxford University Press. New York.
2. MORROW, M.J. & J.A. SHARPE. 1993. The effects of head and trunk position on torsional vestibular and optokinetic eye movements in humans. Exp. Brain Res. **95:** 144–150.

The Myth of Static Ocular Counter-Rolling
The Response of the Eyes to Head Tilt

ROBERT S. JAMPEL

Kresge Eye Institute, Detroit, Michigan 48201, USA

KEYWORDS: head tilt; ocular counter-rolling

INTRODUCTION

Static ocular counter-rolling (OCR) has been accepted as a foundation or as a variable in many scientific studies in vestibular and ocular motor physiology, including space-motion sickness, and as a precept in the interpretation of clinical ocular motor defects.[1–9] The idea of OCR appears so ingrained that there is a published history of this phenomenon.[10] A law was proposed in 1871 stating that any degree of tilt of the head towards the shoulder produced a torsional rotation of both eyes in the opposite direction, equal to approximately one-sixth of the head tilt.[11] Subsequent investigators have accepted the existence of static OCR, but believe that it is highly variable and asymmetrical in every position of head tilt. All agree that its magnitude is related to the magnitude of the head tilt.[12–14]

METHODS

Video-oculography (VOG) was carried out by suspending two miniature video cameras, a fiberoptic light source, and a fixation target from a headband so that they moved in synchrony with the head. One camera video-recorded eye landmarks, the other recorded head movement and position. The video frames were digitized and clips were grabbed and then analyzed with computer algorithms. The eye and head movements of 15 subjects (7 male, 8 female) were recorded. All were healthy adults who ranged in age from 21 to 74 years. Visual acuity in all subjects was 20/25 or better without correction, with contact lenses, or after refractive surgery. All had clinically normal-appearing conjunctivas, irises, and retinas and clinically normal eye and head movements.

RESULTS

No OCR was measured in 15 normal subjects when they gazed in any direction with the head held in any position in space (FIGS. 1 and 2). During head tilt of about

---

Address for correspondence: Robert S. Jampel, M.D., Ph.D., Kresge Eye Institute, 4717 St. Antoine, Detroit, Michigan 48201-1423. Voice: 313-577-6151; fax: 248-646-8677.
rsjampel@mediaone.net

Ann. N.Y. Acad. Sci. 956: 568–571 (2002). © 2002 New York Academy of Sciences.

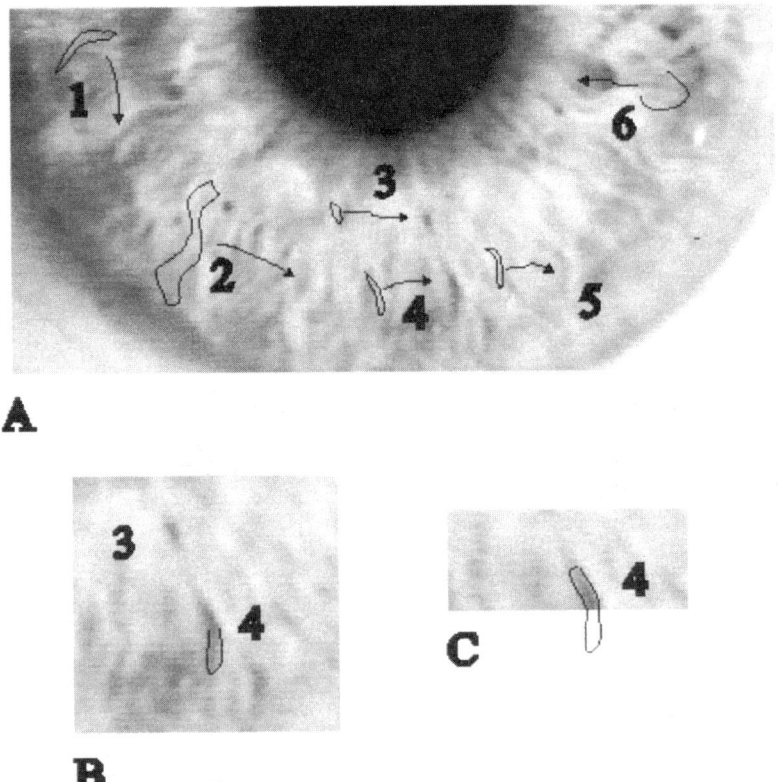

FIGURE 1. Tracings of six iris landmarks in a greatly magnified iris segment (**A**). The tracings are drawn in separate moveable layers of Adobe PhotoShop 5.5™ and can be superimposed with micrometer-like precision over the landmarks (± 10 arc min) (**B** and **C**).

30 deg/s there were periodic tonic torsional eye movements counter to the head tilt that were followed by torsional saccades in the direction of the head tilt, which caused the eyes to catch up to the head and reset to the primary position. The amplitudes of the torsional saccades increased and their number decreased as the head tilted faster and faster. When the head tilted slowly, the torsional eye movements required magnification to be seen.

CONCLUSIONS

There is no static OCR. The eyes (retinas) are oriented to the brain and not the horizon. When the head stabilizes in any tilted position, the retinas attain the same dynamic state of equilibrium with the brain that they attain in the head-erect position. The oblique muscles function to stabilize the retinas in regard to a brain plane in all stable head and gaze positions. During head tilt there are involuntary torsional saccades

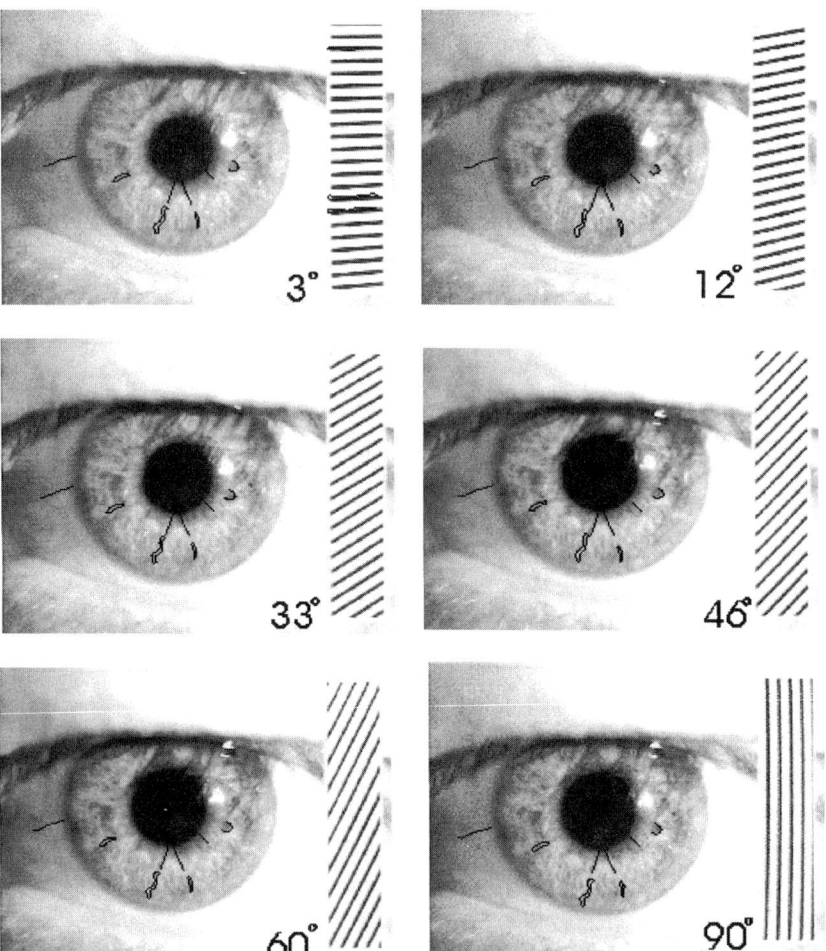

FIGURE 2. Six video frames of the iris are shown with the head held tilted to the shoulder and horizon from 3 to 90 degrees. Superimposition was precise and the same in all frames. There was no OCR.

around the visual line that appear to anticipate the final brain position and to dampen the effect of head movement on the retinas. The difference between how the eyes respond when the head is maintained in a stable tilted position (actually any head position in space) and how the eyes move during head tilt (actually any head movement) explains some of the confusion in the literature concerning ocular torsion.

REFERENCES

1. GRAYBIEL, A. & R.C. WOELLNER. 1959. A new and objective method for measuring ocular torsion. Am. J. Ophthalmol. **47**: 349–352.

2. MILLER, E.F. 1962. Counterrolling of the human eyes produced by head tilt with respect to gravity. Acta Otolaryngol. **54**: 479–501.
3. KREJCOVA, H., S. HIGHSTEIN & B. COHEN. 1971. Labyrinthine and extra-labyrinthine effects on ocular counter-rolling. Acta Otolaryngol. **72**: 165–171.
4. DIAMOND, S.G., C.H. MARKHAM & N. FURUYA. 1982 Binocular counterrolling during sustained body tilt in normal humans and in a patient with unilateral vestibular nerve section. Ann. Otol. Rhinol. Laryngol. **91**: 225–229.
5. YOUNG, L.R. *et al.* 1981 Ocular torsion on earth and in weightlessness. Ann. N.Y. Acad. Sci. **374**: 80–92.
6. DIAMOND, S.G. & C.H. MARKHAM. 1983. Ocular counterrolling as an indicator of vestibular otolith function. Neurology **33**: 1460–1469.
7. NELSON, J.R. & W.F. HOUSE. 1971. Ocular countertorsion as an indicator of otolith function: effects of unilateral vestibular lesions. Trans. Am. Acad. Ophthalmol. Otolaryngol. **75**: 1313–1321.
8. MILLER, E.F. & A. GRAYBIEL. 1971. Effect of gravitoinertial force on ocular counterrolling. J. Appl. Physiol. **5**: 697–700.
9. DIAMOND, S.G. & C.H. MARKHAM. 1991. Prediction of space motion sickness susceptibility by disconjugate eye torsion in parabolic flight. Aviat. Space Environ. Med. **62(3)**: 201–205.
10. SIMONSZ, H.J. 1985. The history of the scientific elucidation of ocular counterrolling. Doc. Ophthalmol. **61**: 183–189.
11. NAGEL, A. 1871. On the occurrence of true rolling of the eye about the line of sight. Arch. Ophthalmol. **17**: 237–264. [Translated from German by H.J. Simonsz. 2000. Strabismus **8**: 33–38.]
12. KUSHNER, B.J. & S. KRAFT. 1983. Ocular torsional movements in normal humans. Am. J. Ophthalmol. **95**: 752–762.
13. GROEN, E., J.E. BOS, P.F. NACKEN & B. DE GRAAF. 1996 Determination of ocular torsion by means of automatic pattern recognition. IEEE Trans. Biomed. Eng. **43**: 471–479.
14. KUSHNER, B.J., S.E. KRAFT & M. VRABEC. 1984. Ocular torsional movements in humans with normal and abnormal ocular motility. Part I: Objective measurements. J. Pediatr. Ophthalmol. Strabismus **21**: 172–177.

Otolith Effect on Torsional Quick Phases of Vestibular Nystagmus in Humans

A.A. KORI, A. SCHMID-PRISCOVEANU, AND D. STRAUMANN

Neurology Department, Zurich University Hospital, Zurich, Switzerland

KEYWORDS: vestibulo-ocular reflex; otoliths

The vestibulo-ocular reflex (VOR) ensures the stabilization of the foveal image during head movements. The VOR can be characterized by the gains in both the velocity (eye velocity divided by head velocity) and position (eye position divided by head position) domains. In the absence of quick phases, velocity and position gains are equivalent. Vestibular slow phases are usually interrupted by anticompensatory quick phases, which move the fovea away from the target in a direction opposite to the slow phase, thus reducing the position gain of the VOR. In the upright position, both semicircular canal and otolith organs contribute to the velocity gain of the torsional VOR. During this off-vertical axis rotation, the otoliths provide an additional positional signal as they sense the angle between the head and gravity vectors. This signal can be used to increase the position gain of the VOR by modifying the anticompensatory quick phases. A previous study has shown that, in humans, the otoliths decrease the frequency of quick phases during torsional oscillatory VOR stimulation. To investigate which parameters of the quick phases are influenced by the otoliths, we studied the torsional quick phases elicited during torsional whole-body position steps (amplitude 10°, peak acceleration of $900/s^2$) in the upright (with otolith input) and supine (without otolith input) body positions in healthy human subjects. Eye movements were recorded binocularly in three dimensions with dual search coils. The number of position steps that evoked quick phases was significantly greater in the supine (average: 14.2 ± 4.4 SD of 20 trials) than in the upright position (7.0 ± 5.3 of 20 trials). The latencies from the beginning of the turntable movement to the appearance of a quick phase were significantly shorter in the supine (181 ms ± 50.99) than in the upright position (250 ms ± 67.69). No significant differences were seen in the amplitude, duration, and peak velocity between the two body positions. However, a comparison of the main sequence of the torsional quick phases showed that the peak velocities relative to amplitudes were higher in the upright than in the supine position. These results indicate that the otoliths increase the positional stability of the eye-in-space during torsional head movements through two

Address for correspondence: Adriana A. Kori, M.D., Vestibulo-Oculomotor Laboratory, Zurich University Hospital, Frauenklinikstrasse 26, CH-8091 Zurich, Switzerland. Voice: 41-1-255-5591; fax: 41-1-240-0075.
adrianakori@yahoo.co.uk

mechanisms: by inhibiting the generation of torsional quick phases and by modifying their dynamics.

ACKNOWLEDGMENTS

This work was supported by the Swiss National Science Foundation (3231-051938.97/3200-052187.97) and the Betty and David Koetser Foundation for Brain Research.

Impaired Linear Vestibulo-Ocular Reflex Initiation and Vestibular Catch-Up Saccades in Older Persons

JUN-RU TIAN,[a,d] BENJAMIN T. CRANE,[b] GERALD WIEST,[c] AND JOSEPH L. DEMER[a,c,d]

Departments of [a]Ophthalmology, [b]Otolaryngology, and [c]Neurology, and the [d]Jules Stein Eye Institute, University of California, Los Angeles. California 90095-7002, USA

KEYWORDS: linear vestibulo-ocular reflex; aging; otolith; saccades

The vestibulo-ocular reflex, which stabilizes gaze to reduce retinal slip during head perturbations, has two components: the angular (AVOR), mediated by the semicircular canals, and the linear (LVOR), mediated by the otoliths. Progressive losses in vestibular sensory cells and primary neurons begins at about age 40 yrs,[1] and are associated with age-related impairments in the steady-state[1-3] and transient AVOR.[4] However, little is known about possible age-related impairments of the LVOR. Vestibular catch-up saccades (VCUS) cued by the semicircular canals during transient rotations can supplement the hypometric AVOR to assist people with vestibular deafferentation to stabilize gaze.[5] However, VCUS cued by otolith organs have not been previously studied, especially in terms of a possible association with aging. This study sought to characterize LVOR initiation and VCUS in older humans.

We investigated nine younger subjects 24 ± 5 years of age (mean ± SD, range 18–31) and eight older subjects 65 ± 7 years of age (range 56–75) who gave written informed consent to a protocol approved by the UCLA Human Subject Protection Committee. Random lateral ("heave") translations were delivered by a pneumatic position servo at peak acceleration of 0.5 G over a distance of ± 25 cm. Subjects were firmly secured to a chair-mounted head-holder. Eye and head movements were sampled at 1200 Hz using binocular magnetic search coils and a cranial accelerometer. Ten responses were averaged for each condition. Subjects fixed targets at 200, 50, or 15 cm distant immediately before unpredictable onset of randomly directed translation in dark (LVOR) or light (visually enhanced LVOR, LVVOR).

All older subjects maintained ideal vergence of 1.5–2 deg for the 200-cm target, 6–8 deg for the 50-cm target, and 21–26 deg for the 15-cm target, with actual vergences depending on individual interpupillary distances. Search coil recording of head rotation showed it to be negligible (< 0.5°) for the first 250 ms after onset of head translation, excluding a role for the AVOR in the responses studied. The typical

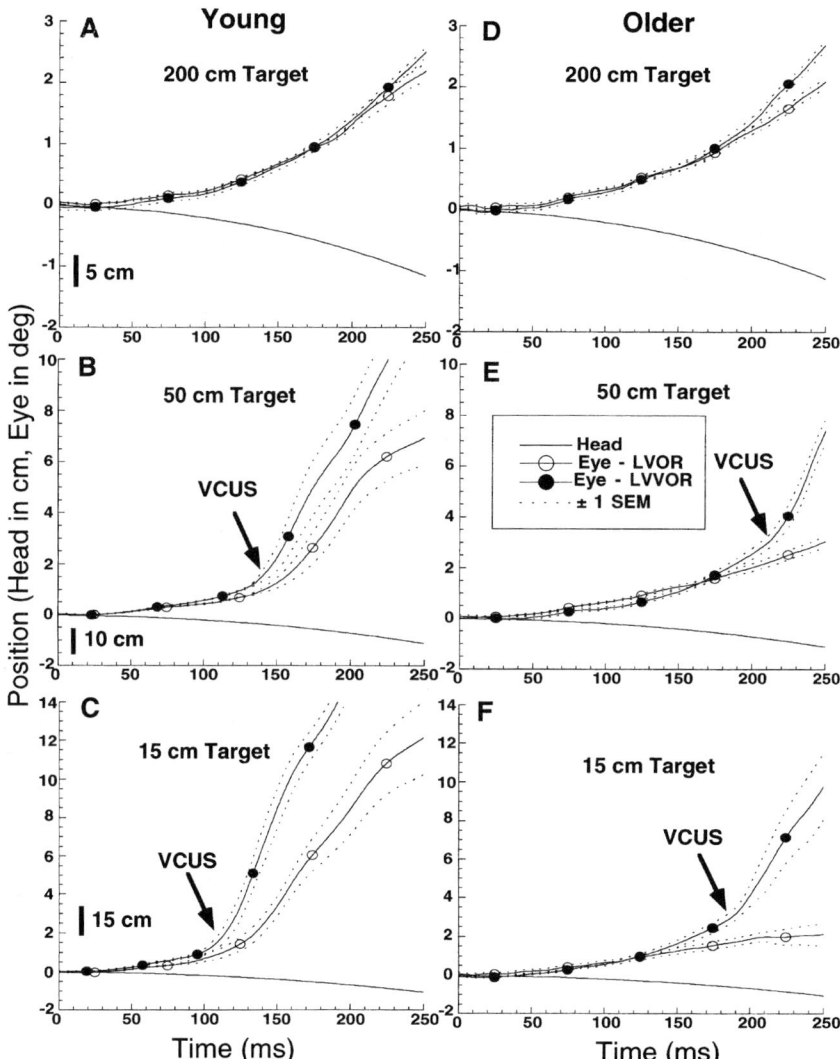

FIGURE 1. LVOR and LVVOR in a younger and an older subject undergoing transient translation to the left after viewing targets at multiple distances. Data were sampled at 1200 Hz from onset of head translation at time zero and averaged over 10 trials. In darkness, vestibular catch-up saccades (VCUS) were employed by seven of nine younger subjects, but only three of eight older subjects.

LVOR response to lateral translation was an oppositely directed eye rotation occurring after a latency (FIG. 1). Mean LVOR latency was 42 ± 3 ms (mean ± SD) for younger and 62 ± 3 ms for older subjects ($p < 0.0001$). The peak of the latency distribution was from 20–60 ms for younger subjects, and 60–100 ms for older subjects.

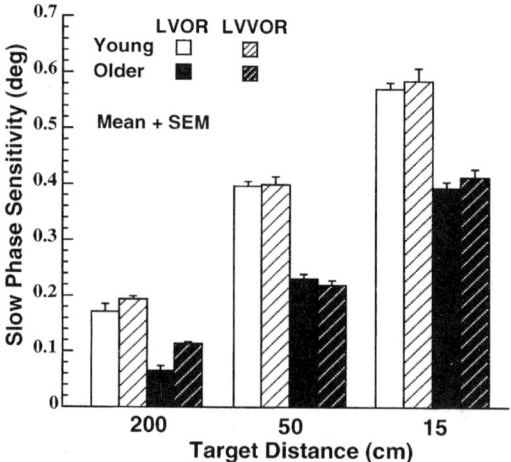

FIGURE 2. Slow phase sensitivity of LVOR and LVVOR in nine young and eight older subjects averaged during the interval of 70–80 ms from onset of head translation. Responses increased with decreasing target distance, but were significantly reduced for older subjects ($p < 0.0001$).

During the early interval 70–80 ms from head motion onset prior to any saccades, all subjects had significantly enhanced LVOR with decreasing target distance (FIG. 2). In this interval, the LVOR position amplitude of younger subjects was $0.17 \pm 0.01°$, $0.40 \pm 0.01°$, $0.57 \pm 0.01°$ (mean ± SE), respectively, in descending order of target distance. Early sensitivities were significantly reduced for older subjects to $0.07 \pm 0.01°$, $0.23 \pm 0.01°$, $0.40 \pm 0.01°$ ($p < 0.0001$). There was no significant difference between the LVOR and LVVOR in either group during the first 110 ms of the response ($p > 0.05$), excluding a contribution from visual tracking. Visual-otolith interaction was mainly reflected by saccades, not on the vestibular slow phase. These compensatory saccades were in the same direction as the LVOR slow phase, and are termed VCUS (FIG. 1).

There was a significant correlation between VCUS amplitude and gaze position error, the difference between ideal eye position at the end of the VCUS and actual eye position before the VCUS (FIG. 3, $r = 0.63$, $p < 0.001$). The LVOR slow phase only accounted for 19–70% and 16–57% of ideal eye position required to maintain the target on the fovea, for the younger and older subjects, respectively, and was less for nearer targets. Gaze position errors were systematically corrected by VCUS whose prevalence and speed increased, and latency decreased, with decreasing target distance. The ability to generate VCUS was significantly impaired in older subjects. For the 15-cm target distance, older subjects only made VCUS in 28% of trials, compared with 64% for younger subjects (FIG. 4, $p < 0.0001$). However, when the target remained visible (LVVOR), VCUS were more common than in darkness, with both groups of subjects making VCUS in most trials. For the 15-cm target distance, the frequency of VCUS was doubled in older subjects to 56% ($p < 0.0001$), albeit less than the 81% frequency in younger subjects. VCUS latency was significantly corre-

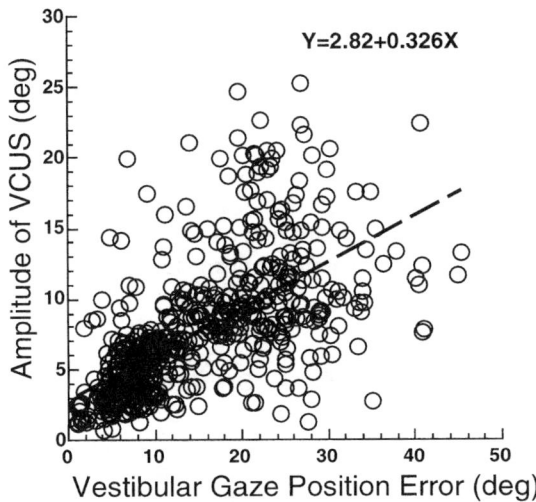

FIGURE 3. VCUS amplitude was correlated with gaze position error ($r = 0.63$, $p < 0.001$), the difference between actual eye position before, and ideal eye position at the end of the VCUS.

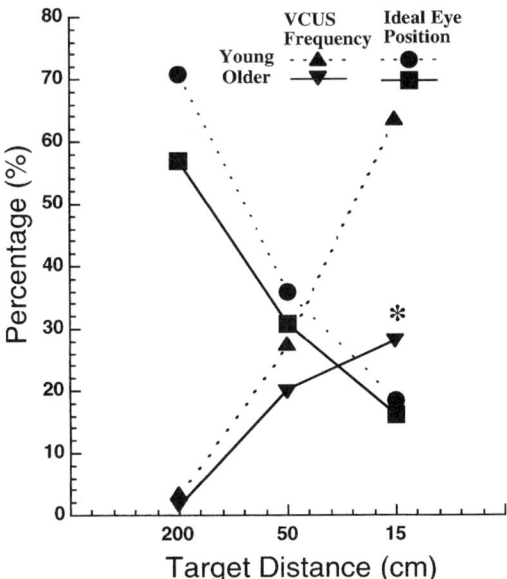

FIGURE 4. VCUS were more frequent when the LVOR slow phase fell increasingly short of ideal eye position. The LVOR slow phase accounted for only 19–70% of ideal eye position in young subjects, and 16–57% of ideal eye position in older subjects, being less for nearer targets. For the 15-cm target, older subjects made VCUS significantly less often than that in young subjects (* $p < 0.0001$).

lated with gaze position error in both groups ($r = 0.8$–0.9, $p < 0.001$). The peak of the VCUS latency distribution during LVVOR was about 200 ms for older subjects, significantly later than the 160-ms peak for younger subjects. Mean VCUS latency was also significantly prolonged to 216 ± 34 (mean $\pm$ SD) and 183 ± 35 ms in older subjects, from 188 ± 36 and 145 ± 32 ms in younger subjects, for target distances of 50 and 15 cm, respectively.

A significant effect of aging on the initial human lateral LVOR is to prolong latency, reduce early sensitivity and reduce occurrence of VCUS in darkness. Otolith-mediated VCUS calibrated to target distance can assist LVOR slow-phase eye movements to stabilize gaze. This normal mechanism is impaired in older subjects, especially under conditions of increased demand posed by near targets.

ACKNOWLEDGMENTS

This work was supported by United States Public Health Service Grant DC-02952 and an unrestricted grant from Research to Prevent Blindness. J.L.D. is Laraine and David Gerber Professor of Ophthalmology.

REFERENCES

1. BALOH, R.W., K.M. JACOBSON & T.M. SOCOTCH. 1993. The effect of aging on visual-vestibuloocular responses. Exp. Brain Res. **95:** 509–516.
2. PAIGE, G.D. 1991. The aging vestibulo-ocular reflex (VOR) and adaptive plasticity. Acta Otolaryngol. Suppl. **481:** 297–300.
3. PAIGE, G.D. 1992. Senescence of human visual-vestibular interactions I. Vestibulo-ocular reflex and adaptive plasticity with aging. J. Vestib. Res. **2:** 133–151.
4. TIAN, J.-R., I. SHUBAYEV & J.L. DEMER 2001. Impairments in the initial horizontal vestibulo-ocular reflex of older humans. Exp. Brain Res. **137:** 309–322.
5. TIAN J.-R., B.T. CRANE & J.L. DEMER 2000. Vestibular catch-up saccades in labyrinthine deficiency. Exp. Brain Res. **131:** 448–457.

A New Differential Diagnosis for Spontaneous Nystagmus

Lateral Canal Cupulolithiasis

ALEXANDRE R. BISDORFF AND DAMIEN DEBATISSE

Department of Neurology, Hôpital de la Ville, L-4240 Esch-sur-Alzette, Luxembourg

KEYWORDS: spontaneous nystagmus; neuro-otology; positional nystagmus; lateral canal cupulolithiasis

In two patients with a history of positional vertigo, a spontaneous nystagmus was observed in the dark while sitting and lying down. When supine with either ear turned downwards, they had a permanent apogeotropic nystagmus (beating to the upper ear) compatible with a recently recognized variant of benign positional vertigo: horizontal canal cupulolithiasis.[1] Neither patient had evidence of underlying neurological disease, and they entered remission within days to weeks, either spontaneously or after some sessions of head shaking and general mobility exercises.

The modulation of their nystagmus in pitch and yaw while supine revealed a particular and reproducible behavior.[2] In pitch, the nystagmus stopped when the head was slightly tilted forwards and reversed direction when the face was turned downwards. While supine with the head in the sagittal plane, the spontaneous nystagmus was similar to that when in the sitting position.

In yaw there was a permanent apogeotropic nystagmus with either ear turned down. A null position for the nystagmus in the supine position was found when the head was slightly turned to one side.

This behavior of the nystagmus can be explained by the orientation of the plane of the lateral canals in the skull and the orientation of the cupula in the lateral canal.[3] The cupula of the lateral canal forms an angle with the sagittal plane of the head with its base pointing medially. A null point where the nystagmus stopped while supine was found when the head was slightly turned ipsilaterally. This aligned the cupula with gravity such that hyperdense debris attached to it could not cause a deflection. In pitch the null point for the nystagmus was reached with the head slightly tilted forwards; in this position the lateral canals were perpendicular to gravity, and again hyperdense debris attached to the cupula could not cause a deflection.

Address for correspondence: Alexandre Bisdorff, M.D., Department of Neurology, Hôpital de la Ville, Rue Emile Mayrisch, L-4240 Esch-sur-Alzette, Luxembourg. Voice: 352 5444491; fax: 352 547886.

alexbis@pt.lu

In patients with a history of positional vertigo and spontaneous nystagmus, the differential diagnosis of lateral canal cupulolithiasis should be considered, and the modulation of the nystagmus in pitch and yaw while supine examined.

REFERENCES

1. BALOH, R.W., Q. YUE, K.M. JACOBSON & V. HONRUBIA. 1995. Persistent direction-changing positional nystagmus: another variant of benign positional nystagmus? Neurology **45:** 1297–1301.
2. BISDORFF, A.R. & D. DEBATISSE. 2001. Localizing signs in positional vertigo due to lateral canal cupulolithiasis. Neurology **57:** 1085–1088.
3. CURTHOYS, I.S. & C.M. OMAN. 1987. Dimensions of the horizontal semicircular duct, ampulla and utricle in the human. Acta Otolaryngol. (Stockh.) **103:** 254–261.

Changes in the Angular Vestibulo-Ocular Reflex after a Single Dose of Intratympanic Gentamicin for Ménière's Disease

J.P. CAREY,[a] T. HIRVONEN,[a] G.C.Y. PENG,[b] C.C. DELLA SANTINA,[a] P. D. CREMER,[c] T. HASLWANTER,[d] AND L.B. MINOR[a]

Departments of Otolaryngology[a] and Neurology,[b] The Johns Hopkins University School of Medicine, Baltimore, Maryland 21287, USA

[c]Eye and Ear Research Unit, Institute of Clinical Neurosciences, Royal Prince Alfred Hospital, Sydney, Australia

[d]Department of Neurology, Zürich University Hospital, Zürich, Switzerland

KEYWORDS: Ménière's disease; angular vestibulo-ocular reflex; gentamicin; intractable vertigo

Ménière's disease is an inner ear disorder that causes episodic vertigo, hearing loss, tinnitus, and fullness in the ear. The clinical disorder has been associated with hydrops, a distension of the endolymphatic fluid compartment of the inner ear. The majority of patients can successfully control their attacks of vertigo with diuretics and restriction of sodium intake. However, about 30% of patients with Ménière's disease have intractable vertigo that does not respond to these measures. The intratympanic (middle ear) injection of gentamicin is now an alternative to ablative surgery for these patients. Gentamicin is an aminoglycoside antibiotic that is toxic to the hair cells of the inner ear, with somewhat greater vestibular toxicity than cochlear toxicity. We have treated 102 patients with unilateral Ménière's disease with intratympanic gentamicin. Attacks of vertigo have been completely controlled in 83%, and substantially controlled (>60% reduction in frequency) in 12%. Profound hearing loss occurred as a result of treatment in only 2%, and the overall rate at which hearing declined after treatment (19%) did not differ from the expected decline in hearing with active Ménière's disease.

Recent evidence suggests that a single dose of intratympanic gentamicin can be effective in controlling vertigo in Ménière's disease.[1] This study was undertaken to determine the effect of a single intratympanic injection of gentamicin on the function of the human angular vestibulo-ocular reflex (AVOR) in subjects with Ménière's disease.

Address for correspondence: John P. Carey, M.D., Otolaryngology—Head and Neck Surgery, The Johns Hopkins University School of Medicine, 601 N. Caroline St., 6th floor, Baltimore, MD 21287-0910. Voice: 410-955-3403; fax: 410-955-0035.
jcarey@jhmi.edu

TABLE 1. AVOR gain values for head thrusts that excited the indicated semicircular canals in subjects with Ménière's disease pre- and post-gentamicin treatment and for subjects with SUVD

	Ipsilateral canals			Contralateral canals		
	HC	AC	PC	HC	AC	PC
Gent-Pre	0.94 ± 0.18	0.80 ± 0.21	0.84 ± 0.10	0.93 ± 0.09	0.89 ± 0.05	0.78 ± 0.18
Gent-Post	0.39 ± 0.11**	0.41 ± 0.11*	0.35 ± 0.15**	0.83 ± 0.08**	0.86 ± 0.11	0.72 ± 0.21
SUVD	0.26 ± 0.10*	0.24 ± 0.10*	0.26 ± 0.07	0.70 ± 0.14*	0.78 ± 0.09	0.66 ± 0.12

NOTE: Figures given are mean ± S.D. Asterisks indicate the significance of the difference of the given mean from the one in the row above it: ** $p \leq 0.001$; * $p \leq 0.01$.

The three-dimensional AVOR responses elicited by rapid rotary head thrusts were studied in 12 subjects with unilateral Ménière's disease before and 2 to 10 weeks after a single treatment with intratympanic gentamicin and in 13 subjects after surgical unilateral vestibular destruction (SUVD). Each head thrust was in the horizontal plane or in either diagonal plane of the vertical semicircular canals (left anterior/right posterior [LARP] or right anterior/left posterior [RALP] planes). Thus, each head thrust effectively stimulated only one pair of canals. The AVOR gains (eye velocity/head velocity during the 30 msec before peak head velocity) for the head thrusts exciting each individual canal were averaged and taken as a measure of the function of that canal.

Before treatment with intratympanic gentamicin, subjects with unilateral Ménière's disease usually had AVOR gains similar to values reported in normal subjects (FIG. 1, TABLE 1).[2-5] There were minimal asymmetries between responses for head thrusts expected to excite canals on the ipsi- versus contralateral sides (TABLE 1). Responses after a single intratympanic gentamicin injection demonstrated marked gain reductions for head thrusts expected to excite canals on the treated side (FIG. 1). On average, gains for excitation of the treated horizontal canal (HC) decreased by 59%, for the treated anterior canal (AC) by 49%, and for the treated posterior canal (PC) by 58% (TABLE 1). Gains for head thrusts that excited the contralateral canals decreased by much smaller percentages: HC, 11%; AC, 4%; and PC, 8%. In comparisons of the gains of the subjects with Ménière's disease after treatment with intratympanic gentamicin and the gains of the SUVD subjects, the gains were lower for the SUVD subjects for those head thrusts that excited the HC and AC on the lesioned (ipsilateral) side and the HC on the contralateral side. These significant differences are indicated by asterisks in TABLE 1.

Ménière's disease itself did not reduce the AVOR gain values for the affected semicircular canals in most of our subjects with the disorder. Tsuji et al. examined temporal bones from 24 patients with longstanding Ménière's disease.[6] While the loss of type II hair cells and cells of Scarpa's ganglion were greater than expected for age, they rarely observed >50% reduction in either type of hair cell or their afferent neurons. Our data suggest that the preservation of these structures is adequate to preserve AVOR function in response to rapid rotary head thrusts in subjects with symptomatic Ménière's disease.

In the cases presented here, a single treatment with intratympanic gentamicin controlled the attacks of vertigo due to Ménière's disease during the initial

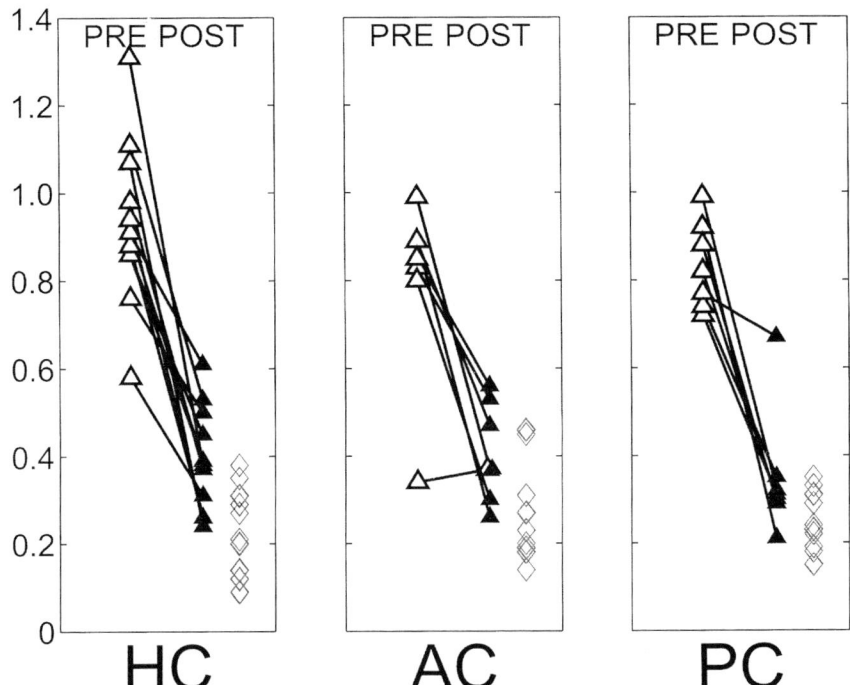

FIGURE 1. Comparison of gain values measured in 10 subjects with unilateral Ménière's disease for head thrusts that excited the semicircular canals on the affected side before intratympanic gentamicin treatment (PRE) and following treatment (POST). Triangles represent mean gain values from individual subjects with Ménière's disease, and each line connects pre- and post-gentamicin values for one subject. Filled triangles in the POST data indicate decreased gains compared to PRE values ($p < 0.01$). The open diamonds show gain values for head thrusts toward the lesioned side in SUVD subjects.

(6 months) follow-up period. In every case the single dose of gentamicin markedly reduced AVOR gains for the semicircular canals on the treated side. The effect of intratympanic gentamicin was specific to the treated ear, as indicated by the more modest reduction in AVOR gains for excitation of canals on the contralateral side. The gain reduction for the ipsilateral horizontal and anterior canals caused by intratympanic gentamicin treatment was not as great as that caused by SUVD.

These results suggest that a single treatment with intratympanic gentamicin might not cause complete hair cell destruction. Indeed, partial hair cell loss in the labyrinth has been noted in the chinchilla after a single intraotic application of gentamicin.[7,8] The hair cells remaining after intratympanic gentamicin treatment may provide some ipsilateral excitatory component to the response that is missing in the SUVD subjects. An alternative or adjunctive mechanism for the improved response after intratympanic gentamicin may be the preservation of baseline afferent discharge on the treated side. Such preservation of the afferent firing would maintain the symmetry in afferent activity reaching the central vestibular nuclei. During a rapid head thrust exciting the canals on the treated side, the balanced central vestibular activity might improve the response generated by inhibition of afferents from the contralateral side.

ACKNOWLEDGMENTS

This work was supported by NIDCD Grants K23 DC00196 and R01 DC02390.

REFERENCES

1. HARNER, S.G. *et al.* 2001. Long-term follow-up of transtympanic gentamicin for Ménière's syndrome. Otol. Neurotol. **22:** 210–214.
2. AW, S.T. *et al.* 1996. Three-dimensional vector analysis of the human vestibulo-ocular reflex in response to high-acceleration head rotations. I. Responses in normal subjects. J. Neurophysiol. **76:** 4009–4020.
3. CREMER, P.D. *et al.* 1998. Semicircular canal plane head impulses detect absent function of individual semicircular canals. Brain **121:** 699–716.
4. TABAK, S. *et al.* 1997. Gain and delay of human vestibulo-ocular reflexes to oscillation and steps of the head by a reactive torque helmet. I. Normal subjects. Acta Otolaryngol. (Stockh.) **117:** 785–795.
5. TIAN, J.-R. *et al.* 2001. Impairments of the initial horizontal vestibulo-ocular reflex in older humans. Exp. Brain Res. **137:** 309–322.
6. TSUJI, K. *et al.* 2000. Temporal bone studies of the human peripheral vestibular system. 4. Ménière's disease. Ann. Otol. Rhinol. Laryngol. **109:** 26–31.
7. LOPEZ, I. *et al.* 1997. Quantification of the process of hair cell loss and recovery after gentamicin treatment. Int. J. Dev. Neurosci. **15:** 447–461.
8. LOPEZ, I. *et al.* 1998. Hair cell recovery in the chinchilla crista ampullaris after gentamicin treatment: a quantitative approach. Otolaryngol. Head Neck Surg. **119:** 255–262.

Unilateral Rebound Nystagmus

One Manifestation of Two Different Pathologic Processes

M.L. ROSENBERG[a] AND D.S. ZEE[b]

[a]*New Jersey Neuroscience Institute, Edison, New Jersey 08818, USA*

[b]*Johns Hopkins University, Baltimore, Maryland 21287, USA*

KEYWORDS: rebound nystagmus; vestibulo-ocular reflex

Hood described a series of patients with cerebellar disease who had no nystagmus in primary position except after returning from a period of prolonged eccentric gaze.[1] In all of his patients, this would occur after a gaze to either side. Bilateral rebound nystagmus is now generally recognized as a sign of cerebellar disease. The underlying pathology is unclear.

We report two patients with rebound nystagmus evident only after sustained eccentric gaze in one direction. Results of vestibular testing suggest that there are least two different pathophysiologic mechanisms underlying rebound nystagmus.

METHODS

Two patients were identified clinically as having a primary position nystagmus present only after sustained eccentric gaze in one direction. Vestibular testing included testing for vestibulo-ocular reflex (VOR) gain and rebound after sustained eccentric gaze with and without fixation. VOR gain was tested in a rotary chair at a rotation velocity of 60 degrees per second. Rebound was elicited by having the patient maintain extreme eccentric gaze for 15 seconds in the light and then returning to primary position. At this point the room lights were turned off, but a small fixation light in primary position was turned on for 0.2 seconds every second. Eye position was recorded until the eyes were visibly stable while viewed via infrared camera.

PATIENTS

Patient 1: During an evaluation for headache, a 31-year-old woman was found to have several neuro-ophthalmologic abnormalities including a left relative afferent

---

Address for correspondence: Michael L. Rosenberg, M.D., New Jersey Neuroscience Institute, 65 James St., Edison, NJ 08818. Voice: 732-321-7950; fax: 732-632-1671.
 mrosenberg@solarishs.org

TABLE 1. Results of vestibular testing

Patient No.	Rebound from	VOR gain		Maximum slow-phase velocity (deg/sec)			
				Rebound in dark		Rebound in light	
		LT	RT	LT	RT	LT	RT
1	RT	0.56	0.68	13.7	15.4	5.1	26.3
2	RT	0.39	0.63	9.2	41.5	0.2	23.1

pupillary defect, gaze-evoked nystagmus that was worse on left gaze, and rebound nystagmus only after prolonged left gaze. The VOR as judged by head thrusts was normal, and head shaking resulted in no post-headshaking nystagmus. Work-up confirmed a diagnosis of multiple sclerosis.

Patient 2: A 30-year-old man with a known right optic nerve glioma developed secondary limitation of oculomotor ductions during follow-up. Vestibular abnormalities appearing on exam included small-amplitude primary position nystagmus only with fixation removed, decreased VOR to the right by head thrusts, left beating nystagmus after head shaking, and unilateral rebound nystagmus after sustained eccentric gaze to the right.

RESULTS

The results of the testing are shown in TABLE 1. Patient 1 had an underlying vestibular imbalance with a decreased VOR gain and asymmetric rebound in both the light and the dark. Patient 2 had symmetric VOR gain and symmetric rebound in the dark but asymmetric rebound in the light.

DISCUSSION

Rebound nystagmus has been found routinely in normal subjects when tested without fixation.[2] Prolonged eccentric gaze to one side evokes a tendency to tonic deviation toward that side after return to primary position. In the light, normal subjects can use fixation to suppress this tonic deviation, but patients with rebound nystagmus cannot. This could be due to a tonic deviation that is greater than that normally present and outside of the suppressible range. Another possible cause is an inability to suppress even the normal tonic deviation produced by eccentric gaze.

Patients with unilateral rebound nystagmus presented an ideal opportunity to distinguish these processes. If the rebound nystagmus generated in the dark was symmetric and similar to normal subjects, one could assume the abnormality was in the patient's ability to suppress the nystagmus from that side. Patient 1 fits this profile. In contrast, patient 2 had asymmetric rebound in both the dark and the light. The slow-phase velocity in the dark was greater than normal after gaze in one direction. One could assume that there was a normal ability to suppress nystagmus, and propose that the tonic deviation and secondary nystagmus was too great to suppress in this patient.

These same mechanisms may be present in patients with bilateral rebound. As these two pathophysiologic processes may have different underlying neurologic substrates, future attempts at localization of either unilateral or bilateral rebound nystagmus should include a determination of the underlying physiologic process.

ACKNOWLEDGMENT

This work was supported in part by National Institutes of Health Grant EY 01849.

REFERENCES

1. HOOD, J.D., A. KAYAN & J. LEECH. 1973. Rebound nystagmus. Brain **96:** 507–526.
2. ZEE, D.S., R.D. YEE, D.G. COGAN, *et al.* 1976. Oculomotor abnormalities in hereditary cerebellar ataxia. Brain **99:** 207–234.

Delayed-Onset Seesaw Nystagmus Posttraumatic Brain Injury with Bitemporal Hemianopia

ERIC R. EGGENBERGER

Michigan State University Department of Neurology and Ophthalmology, A217 Clinical Center, 138 Service, East Lansing, Michigan 48824-1313, USA

KEYWORDS: seesaw nystagmus; posttraumatic brain injury; chiasma

INTRODUCTION

Delayed neurologic syndromes are relatively uncommon. We report two cases of delayed-onset seesaw nystagmus (SSN) 21 and 37 years post trauma. An immediate chiasmal lesion with bitemporal hemianopia characterized both original injuries. The delayed onset of SSN following chiasmal trauma suggests a chiasmal region link to the brainstem and cerebellar structures involved in long-term control of ocular oscillations.

CASE REPORTS

Case 1: A 45-year-old man originally developed oscillopsia in 1996. Past medical history was notable for a snowmobile accident with closed head injury in 1975, 21 years before the onset of his oscillopsia, resulting in diplopia and bitemporal hemianopia. Following the onset of oscillopsia, a neuro-ophthalmologic examination in 1997 demonstrated visual acuity of 20/20 in the right eye and 20/40 in the left. Binocular color vision was preserved with correct identification of 9 out of 10 Hardy-Rand-Rittler plates, and normal pupils were without a relative afferent pupillary defect. Complete bitemporal hemianopia was present. A 4-prism diopter right hypertropia was present in left gaze, with a 4-prism diopter left hypertropia in right gaze and approximately 20-prism diopters exotropia. Seesaw nystagmus (elevation with incyclotorsion, and depression with excyclotorsion) was noted; this increased notably following a blink. Horizontal vestibular ocular reflex suppression was normal bilaterally. The remainder of the neurologic examination was unremarkable. Gabapentin and baclofen monotherapy failed to improve oscillopsia; however, clonazepam and baclofen in combination produced moderate improvement that was

---

Address for correspondence: Eric R. Eggenberger, D.O., Michigan State University Department of Neurology and Ophthalmology, A217 Clinical Center, 138 Service, East Lansing, Michigan 48824-1313. Voice: 517-353-8122; fax: 517-432-3713.
eric.Eggenberger@ht.msu.edu

maintained with clonazepam monotherapy. MRI demonstrated bilateral subfrontal encephalomalacia, and no images demonstrated clear chiasmal discontinuity.

Case 2: A 58-year-old man originally developed oscillopsia in March 2001. His past medical history was notable for a motor vehicle accident in 1964, with traumatic brain injury including loss of consciousness for approximately 3 days, left infraorbital fracture, and bitemporal hemianopia. The patient recalled decreased acuity in both eyes that spontaneously improved over approximately 2 months. He noted no new ocular symptoms until the development of oscillopsia. A neuro-ophthalmologic examination in May 2001 demonstrated visual acuity of 20/30 in the right eye and $20/60^{+1}$ in the left. Binocular color vision was impaired, with 0 out of 10 Hardy-Rand-Rittler plates correctly identified. A complete temporal hemianopia was present in the right eye, with relative temporal hemianopia on the left. Both pupils were 4 mm and showed a slight reaction to light. Small-amplitude see-saw nystagmus was present. Horizontal vestibular ocular reflex suppression was normal bilaterally. MRI revealed an empty sella and encephalomalacia of the left subfrontal and left anterior temporal lobes; the chiasm was not well visualized (integrity was not demonstrated). The patient deferred any medical attempts at therapy.

DISCUSSION

Seesaw nystagmus (SSN) has rarely been reported since its first report in 1913.[1] SSN has been reported in association with several conditions (TABLE 1), but rarely in a delayed fashion following trauma. Frisén and Wikkelsø described a patient who developed seesaw nystagmus 4 months after traumatic injury inclusive of bitemporal hemianopia.[2] Their patient was unique because of relatively minor head trauma (although basilar skull fracture was noted), development of hydrocephalus with nystagmus, and positive response of SSN to alcohol. Association with hydrocephalus also characterized the case reported by Sano *et al.*[3] Cochin *et al.* reported a patient who developed intermittent seesaw nystagmus, causing bitemporal visual defects, 5 years after head injury.[4] All reported cases following trauma appear to develop SSN in a delayed fashion, typically weeks to months following injury.[5,6]

Several delayed-onset neurologic syndromes have been described, including myoclonus, dystonia, nystagmus, and ataxia following initial injuries such as hypoxic-ischemic injury, infarction, or trauma. The most common delayed neuro-ophthalmologic syndrome is oculopalatal myoclonus, which involves cerebellar connections and occasionally has a delayed ataxia component.[7] Delayed neurologic sequelae of head trauma often include movement disorders following relatively severe brain injury.[8] Most of these delayed neurologic syndromes share a common link involving the cerebellum and its connections.

Posttraumatic optic neuropathy is well known; however, traumatic chiasmal injury is less well recognized. Histopathology of traumatic chiasmal lesions appears to involve contusion hemorrhage, contusion necrosis, and contusion tears.[9] Savino and colleagues reported 11 patients with permanent chiasmal injuries following trauma, while Domingo and de Villiers presented an additional 10 such patients including MRI findings; however, seesaw nystagmus was not observed.[10,11] Accordingly, traumatic chiasmal injury alone does not appear to be sufficient for the development of seesaw nystagmus. Barton reported intermittent low-amplitude seesaw post blink-

TABLE 1. Conditions associated with see-saw nystagmus

Parasellar tumors with extension[18]
Upper brainstem infarction[19]
Trauma[5,20,21]
Achiasma and hemichiasma[22]
Syringobulbia[23]
Septo-optic dysplasia[24]
Arnold Chiari malformation[25]
MS[26]
Vision loss with RP or rod cone dystrophy[27,28]
Midbrain tumor[29]
Congenital

ing.[12] Our patient 1 exhibited increased seesaw nystagmus following a blink, and blinking may represent a modifying factor in patients with SSN. The delayed onset of SSN following chiasmal trauma suggests a chiasmal region link to the brainstem and cerebellar structures involved in long-term control of ocular oscillations. Seesaw nystagmus appears to require an intact interstitial nucleus of Cajal and vestibular connections.[3,13,14] Head roll normally produces small-amplitude seesaw rotation of the eyes, and impaired calibration of this vestibular response involving cerebellar input may operate in acquired seesaw nystagmus.[15–17]

REFERENCES

1. MADDOX, E.E. 1913. See-saw nystagmus with bitemporal hemianopia. Proc. R. Soc. Med. **7:** 12–13.
2. FRISEN, L. & C. WIKKELSØ. 1986. Posttraumatic seesaw nystagmus abolished by ethanol ingestion. Neurology **36:** 841–844.
3. SANO, K., H. SEKINO, N. TSUKAMOTO, *et al.* 1972. Stimulation and destruction of the region of the interstitial nucleus in cases of torticollis and see-saw nystagmus. Confin. Neurol. **34:** 331–338.
4. COCHIN, J.P., D. HANNEQUIN, C. DO-MARCOLINO, *et al.* 1995. Intermittent see-saw nystagmus successfully treated with clonazepam. Rev. Neurol. **151:** 60–62.
5. SCHMIDT, D. & G. KOMMERELL. 1969. Seesaw nystagmus with bitemporal hemianopia following head traumas. Albrecht Von Graefes Arch. Klin. Exp. Ophthalmol. **178:** 349–366.
6. ARNOTT, E.J. & S.J.H. MILLER. 1970. See saw nystagmus. Trans. Ophthalmol. Soc. UK **90:** 491–496.
7. EGGENBERGER, E.R., W. CORNBLATH & D.H. STEWART. 2001. Oculopalatal tremor with tardive ataxia. J. Neuroophthalmol. **21:** 83–86.
8. KRAUSS, J.K., R. TRÄNKLE, K.-H. KOPP, *et al.* 1996. Post-traumatic movement disorders in survivors of severe head injury. Neurology **47:** 1488–1492.
9. LINDBERG, R., F.B. WALSH & J.G. SACKS. 1973. Neuropathology of Vision: An Atlas. Lea & Febiger. Philadelphia.
10. SAVINO, P.J., J.S. GLASER & N.J. SCHATZ. 1980. Traumatic chiasmal syndrome. Neurology **30:** 963–970.
11. DOMINGO, Z. & J.C. DE VILLIERS. 1993. Post-traumatic chiasmatic disruption. Br. J. Neurosurg. **7(2):** 141–147.

12. BARTON, J.J.S. 1995. Blink- and saccade-induced seesaw nystagmus. Neurology **45:** 831–833.
13. SUZUKI, Y., J.A. BÜTTNER-ENNEVER, D. STRAUMANN, et al. 1995. Deficits in torsional and vertical rapid eye movements and shift of Listing's plane after uni- and bilateral lesions of the rostral interstitial nucleus of the medial longitudinal fasciculus. Exp. Brain Res. **106:** 215–232.
14. RAMBOLD, H., C. HELMCHEN, A. STRAUBE & U. BÜTTNER. 1998. Seesaw nystagmus associated with involuntary torsional head oscillations. Neurology **51:** 831–837.
15. LEIGH, R.J. & D.S. ZEE. 1999. The Neurology of Eye Movements. 3rd edit. Oxford University Press. New York. p. 428.
16. DELL'OSSO, L.F. & R.B. DAROFF. 1998. Two additional scenarios for see-saw nystagmus: achiasma and hemichiasma. J. Neuroophthalmol. **18(2):** 112–113.
17. NAKADA, T. & I.L. KWEE. 1981. Seesaw nystagmus. Role of visuovestibular interaction in its pathogenesis. J. Clin. Neuroophthalmol. **8:** 171–177.
18. DAROFF, R.B. 1965. See-saw nystagmus. Neurology **15:** 874–877.
19. KANTER, D.S., R.L. RUFF, R.J. LEIGH & M. MODIC. 1987. See-saw nystagmus and brainstem infarction. MRI findings. Neuroophthalmology **7:** 279–283.
20. KEANE, J. 1978. Intermittent see-saw eye movements. Report of a patient in coma after hyperextension neck injury. Arch. Neurol. **35:** 173–174.
21. FISHER, N.F., A. JAMPOLSKY, A.B. SCOTT, et al. 1968. Traumatic bitemporal hemianopsia. Am. J. Ophthalmol. **65:** 578–581.
22. DELL'OSSO, L.F. & R.B. DAROFF. 1998. Two additional scenarios for see-saw nystagmus: achiasma and hemichiasma. J. Clin. Neuroophthalmol. **18(2):** 112–113.
23. FEIN, J.M. & R.D.B. WILLIAMS. 1969. See-saw nystagmus. J. Neurol. Neurosurg. Psychol. **32:** 202–207.
24. DAVIS, G.V. & J.P. SCHOCK. 1975. Septo-optic dysplasia associated with see-saw nystagmus. Arch. Neurol. **93:** 137–139.
25. ZIMMERMAN, C.F., E.S. ROACH & B.T. TROOST. 1986. See-saw nystagmus associated with Chiari malformation. Arch. Neurol. **43:** 299–300.
26. SAMKOFF, L.M. & C.R. SMITH. 1994. See-saw nystagmus in a patient with clinically definite MS. Eur. Neurol. **34:** 228.
27. MAY, E.F. & A.R. TRUXAL. 1997. Loss of vision may result in seesaw nystagmus. J. Neuroophthalmol. **17:** 84.
28. BERGIN, D.J. & J. HALPERN. 1986. Congenital see-saw nystagmus associated with retinitis pigmentosa. Ann. Ophthalmol. **18:** 346–349.
29. HALMAGYI, G.M. & W.F. HOYT. 1991. See-saw nystagmus due to unilateral mesodiencephalic lesion. J. Clin. Neuroophthalmol. **11(2):** 79–84.

Main Sequence of Convergence Retraction Nystagmus Indicates a Disorder of Vergence

HOLGER RAMBOLD, DETLEF KÖMPF, AND CHRISTOPH HELMCHEN

Department of Neurology, Medical University of Lübeck, D-23538 Lübeck, Germany

KEYWORDS: convergence retraction nystagmus; convergence; midbrain; mesencephalic lesion

Convergence retraction nystamus (CRN) is characterized by fast convergence and retraction of the eyes with a subsequent slow divergence and forward movement of the eye. CRN is found in midbrain disease, but its pathology is unknown. To test the hypothesis whether CRN is a disorder of (1) asynchronous opposing saccades[1] or (2) the vergence system, we investigated two patients with CRN who suffered from unilateral midbrain lesions.

Patient 1: A 62-year-old patient—as previously reported[2]—presented with a vertical gaze palsy, a positive torsional nystagmus, and an ocular tilt reaction to the right. CRN was elicited during attempted upgaze. On MRI scan a left paramedian thalamic infarction extending in the midbrain was found.

Patient 2: A 60-year-old patient presented with CRN on attempted upgaze, a vertical gaze palsy, and vertical gaze–evoked nystagmus. An ocular tilt reaction to the right was found. MRI scan of this patient revealed a right-sided mesencephalic hemorrhage.

After both patients and five healthy control subjects (age 30–40 years) gave their informed consent, binocular three-dimensional eye movements were recorded under head-fixed condition using the search coil system (three magnetic fields, Skalar; CNC, Seattle). Visually guided eye movements including vergence and saccade–vergence interaction, elicited by a LED and a laser target, were recorded.[2,3] Vergence was calculated of left minus right horizontal eye position.

CRN was found in both patients independent of saccades. Quick and slow phases of both eyes were synchronous in both patients (FIG. 1A,B). The main sequence (eye movement amplitude/eye movement peak velocity) of the control subjects (dashed line) was not different from the horizontal saccades in both patients (FIG. 1C) but differed from the adducting CRN fast phases. CRN fell on the main sequence of vergence (patient 1; FIG. 1D) or saccade-enhanced vergence eye movements (patient 2; FIG. 1D). Peak velocity of abducting was slower than of the adducting phases. Ab-

---

Address for correspondence: Holger Rambold, M.D., Department of Neurology, Medical University Lübeck, Ratzeburger Allee 160, D-23538 Lübeck, Germany. Voice: +49-451-500-3709; fax: +49-451-500-2489.
rambold_h@neuro.mu-luebeck.de

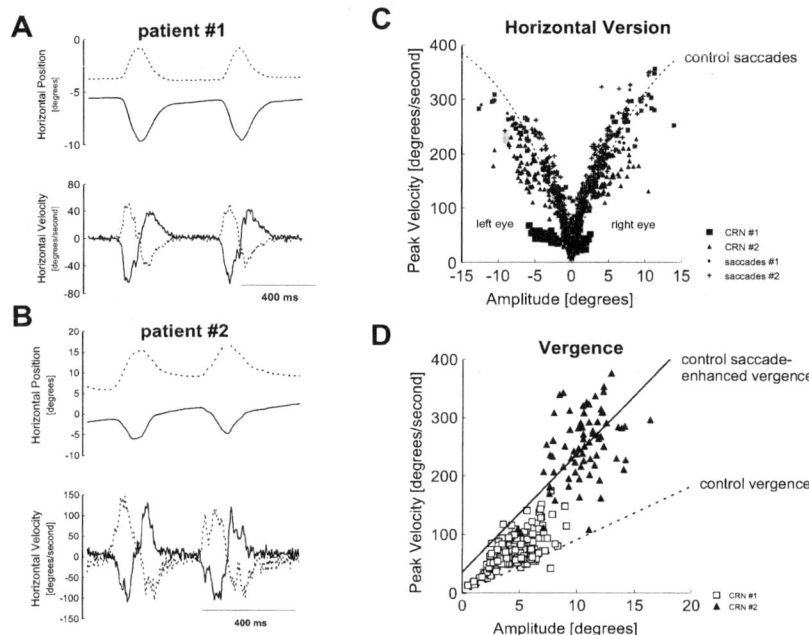

FIGURE 1. Position and velocity traces of two CRN phases for the right (*solid line*) and left (*dashed line*) eyes are presented for patients #1 (**A**) and #2 (**B**). The main sequence (eye amplitude to peak eye velocity) of saccades of control subjects (**C**, *dashed line*) is compared with the saccades of the patients (saccades #1, saccades #2) and with the adducting phases of CRN (CRN #1 and #2) of the right and left eye separately. In **D** the main sequence of control vergence (*dotted line*) and saccade-enhanced vergence eye movements of control subjects (*solid line*) are compared with CRN-adducting phases of patients #1 (CRN #1) and #2 (CRN #2).

ducting phases had an exponential time constant of 70 msec (patient 1) and of 75 msec (patient 2) in both eyes.

In contrast to our first hypothesis, CRN of the right and left eyes was not asynchronous (in time) and not linked to the occurrence of saccades as proposed for opposing adducting saccades.[1] Neither adducting nor abducting phases of CRN fell on the main sequence of saccades, but rather on the main sequence of vergence or saccade-enhanced vergence eye movements. Thus our first hypothesis was not supported by our recordings, which rather showed evidence for the second hypothesis. The increase in the vergence main sequence in patient #2 could be explained by a simultaneous vertical gaze–evoked nystagmus, which could enhance the peak vergence velocity.[3] Thus the convergence phases of CRN fell on the main sequence of control vergence or saccade-enhanced vergence eye movements, suggesting that CRN is a disorder of vergence. At present, however, we cannot exclude the possibility that co-contraction of the eye muscles during eye retraction contributes to the slowing of the ab- and adducting phases of CRN.

ACKNOWLEDGMENTS

This work was supported by Deutsche Forschungsgemeinschaft (DFG) and German–Israel Foundation (GIF).

REFERENCES

1. OCHS, L., L. STARK, W.F. HOYT & D. D'AMICO. 1979. Opposed adducting saccades in convergence retraction nystagmus. A patient with sylvian aquaeduct syndrome. Brain **102:** 497–508.
2. RAMBOLD, H., D. KÖMPF & C. HELMCHEN. 2001. Convergence retraction nystagmus: a disorder of vergence? Ann. Neurol. **50:** 677–681.
3. COLLEWIJN, H., C.J. ERKELENS & R.M. STEINMAN. 1995. Voluntary binocular gaze-shifts in the plane of regard: dynamics of version and vergence. Vision Res. **35:** 3335–3358.

Effects of Intravenous Opioids on Eye Movements in Humans

Possible Mechanisms

K.G. ROTTACH,[a] A.E. DZAJA,[b] W.A. WOHLGEMUTH,[c]
T. EGGERT,[d] AND A. STRAUBE[d]

[a]*Department of Psychiatry, Klinikum Augsburg, Germany*

[b]*Max Planck Institute of Psychiatry, Munich, Germany*

[c]*Department of Radiology, Klinikum Augsburg, Germany*

[d]*Department of Neurology, University of Munich, Munich, Germany*

KEYWORDS: opioids; eye movements; downbeat nystagmus; saccades

INTRODUCTION

Downbeat nystagmus (DBN) is a sign of a central vestibular disorder. DBN can be caused by bilateral lesions of the vestibulocerebellum or lesions of the brainstem close to the midline. It has also occurred as a side effect of several pharmacological substances. After observing DBN in several patients who had received i.v. pethidine, we performed a study on the clinical effects of intravenous opioids (pethidine and fentanyl) on particular eye movement symptoms.

METHODS

Three normal subjects underwent a clinical neurological examination before and after an injection of 100 mg pethidine. The eye movements of four normal subjects were recorded using the scleral search coil technique before and after an i.v. injection of 100 mg pethidine. The eye movements of one normal subject were recorded before and after an i.v. injection of 0.1 mg fentanyl.

RESULTS

Clinical examination: About 1 minute after the pethidine injection, all three subjects developed a mild cerebellar syndrome with ataxia, dysarthria, intention tremor, and a severe disturbance of the eye movements. They showed DBN, brief episodes

---

Address for correspondence: Klaus G. Rottach, Department of Psychiatry, Klinikum Augsburg, Dr-Mack-Str. 1, 86156 Augsburg, Germany. Voice: 49-177-2007054; fax: 49-8341-100037.
Klaus.Rottach@t-online.de

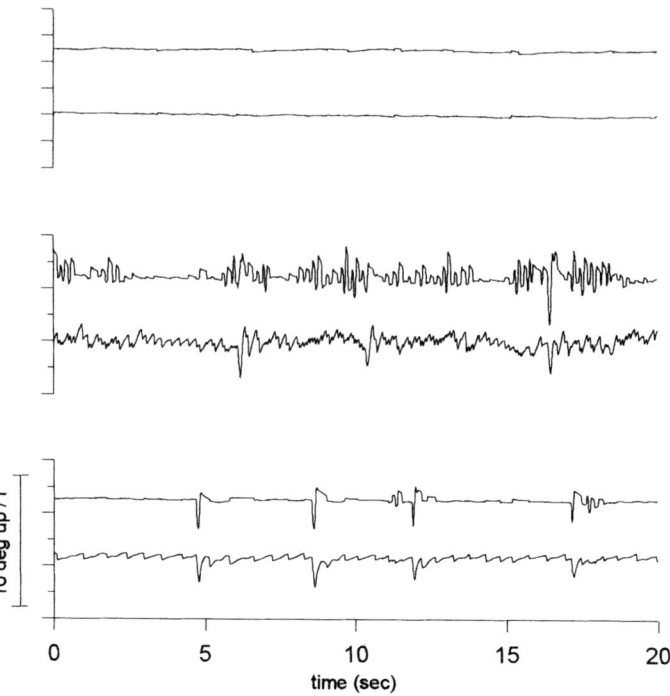

FIGURE 1. *Upper panel*: fixation before the application of any drug. *Middle panel*: the maximum of the fixation disturbance after the administration of pethidine. *Lower panel*: the maximum of the fixation disturbance after the administration of fentanyl (five weeks after the trail with pethidine). The upper trace in each panel shows the horizontal gaze position, the lower trace the vertical gaze position.

of ocular flutter, and a reduced smooth pursuit and vestibulo-ocular reflex (VOR) gain. The symptoms disappeared after about 20 minutes.

Eye movement recording (FIG. 1): All four subjects developed DBN within less than 1 minute after the pethidine injection. Furthermore, there was a dysfunction of horizontal fixation: square wave jerks, groups of ocular flutter, and brief episodes of opsoclonus were observed. After 12–20 minutes, normal fixation was recovered. Blinks elicited groups of square wave jerks and ocular flutter. However, lid closure suppressed all the fixation symptoms. In addition to these malfunctions of fixation, the other ocular motor subsystems were affected as well. Refixation saccades were slowed; the median peak velocity was decreased by 20% for horizontal and by 25% for vertical saccades. The gain of the horizontal and vertical vestibulo-ocular reflex was decreased by about 10%. The gain of the horizontal and vertical smooth pursuit was also markedly decreased.

Fentanyl caused basically the same effects as pethidine. But the effects were less pronounced in the intraindividual comparison.

DISCUSSION

Both pethidine and fentanyl caused the same symptoms. The i.v. administration of morphine has also been reported to cause transient DBN.[1] Thus, it can be assumed that the i.v. administration of opioids generally leads to transient ocular motor symptoms. The role of opioids and opioid receptors within the ocular motor system is still unknown. Fixation symptoms like those caused by the opiates can be observed in patients with cerebellar malfunctions.[2] A dysfunction of the brainstem omnipause neurons should also be considered in horizontal fixation symptoms.[3] Arguments for this possibility are the marked slowing of saccades[4] and the provocation of horizontal fixation symptoms by blinks.[5] However, a malfunction of the omnipause neurons does not cause DBN. Since the cerebellar nuclei also project to the omnipause neurons, all the observed symptoms could be explained by an alteration of cerebellar functions. Recently, opioid receptors have also been detected within the cerebellum.[6]

CONCLUSION

The main effect of i.v. administration of opioids on the ocular motor system is a pronounced fixation disorder consisting of DBN, square wave jerks, and ocular flutter. The underlying mechanism is probably a diminished activation of the Purkinje cells of the cerebellum.

REFERENCES

1. HENDERSON, R.D., *et al.* 2000. Downbeat nystagmus associated with intravenous patient-controlled administration of morphine. Anesth. Analg. **91:** 691–692.
2. BÜTTNER, U., *et al.* 1995. The localizing value of nystagmus in brainstem disorders. Neuroophthalmology **15:** 283–290.
3. AVERBUCH-HELLER, L., *et al.* 1996. Dysfunction of pontine omnipause neurons causes impaired fixation: macrosaccadic oscillations with a unilateral pontine lesion. Neuroophthalmology **16:** 99–106.
4. KANEKO, C.R.S., *et al.* 1996. Effect of ibotenic acid lesions of the omnipause neurons on saccadic eye movements in rhesus macaques. J. Neurophysiol. **75:** 2229–2242.
5. ROTTACH, K.G., *et al.* 1998. Properties of saccades accompanied by blinks. J. Neurophysiol. **79:** 2895–2902.
6. SCHADRACK, J., *et al.* 1999. Opioid receptors in the human cerebellum: evidence from [11C]diprenorphine PET, mRNA expression and autoradiography. Neuroreport **10:** 619–624.

Evaluation of Current Optical Methods for Treating the Visual Consequences of Nystagmus

STACY S. YANIGLOS,[a] JOHN S. STAHL,[b] AND R. JOHN LEIGH[b]

[a]*Optometry Service and* [b]*Neurology Service, Veterans Affairs Medical Center and Case Western University, Cleveland, Ohio 44106, USA*

KEYWORDS: **retinal image stabilization; prisms**

Clear vision requires that images must be held quite still upon the retina, especially the foveal region.[1] Pathological nystagmus causes excessive motion of images on the retina, which degrades vision and may lead to oscillopsia. Any therapeutic approach to treat nystagmus and its visual consequences must take into consideration the normal functions of eye movements, which evolved to ensure clear, stable, single binocular vision. Thus, for example, the approach of weakening the extraocular muscles with botulinum toxin suffers the drawback that not only nystagmus, but also normal eye movements such as vergence may be abolished—so that overall vision is not improved.[2] An alternative approach is to negate the visual effects of eye movements using optical devices. Over the past 15 years, we have tried three different optical approaches to treat visual consequences of nystagmus, which we will summarize and compare here.

In patients whose nystagmus suppresses during viewing of either a near or distant target, prisms may improve vision. This is especially true for certain patients with congenital nystagmus that suppresses with convergence; base-out prisms may improve vision so that they can qualify for a driving license.[3] Occasional patients with acquired nystagmus also show substantial modulation of their nystagmus with vergence angle. We have encountered a few patients with downbeat or acquired pendular forms of nystagmus who have gained sustained benefit from wearing prisms.

Another distinct approach is to use optical devices that negate the effects of nystagmus without seeking to modulate the oscillation itself. One such device consists of a high-power plus spectacle lens worn in combination with a high-power minus spectacle lens.[4] Although the system can produce up to 90% stabilization of retinal images, this requires hard contact lenses that are uncomfortable to wear.[5] With gas-permeable or soft contact lenses, it is possible to achieve up to about 50% stabilization, which is often adequate to negate oscillopsia and achieve vision that is clear

---

Address for correspondence: R. John Leigh, M.D., Department of Neurology, University Hospitals of Cleveland, 11100 Euclid Avenue, Cleveland, OH 44106. Voice: 216-421-3040; fax: 216-231-3461.

rjl4@po.cwru.edu

enough to view a movie.[6] However, two main drawbacks limit the usefulness of this device. First, the device not only negates the visual consequences of nystagmus but also of all normal eye movements. Thus, vergence is ineffective and monocular viewing is necessitated; further, the vestibulo-ocular reflex is disabled, and vision blurs whenever the head moves (such as during locomotion). Second, many patients with neurological disorders causing nystagmus, such as multiple sclerosis, have difficulty inserting and caring for contact lenses. Thus, few of our patients have persevered with this type of optical device.

A new approach has been to use an electro-optical device to selectively negate the visual consequences of nystagmus, but not of normal eye movements.[7] The system is most applicable to pendular forms of nystagmus. Eye movements are measured using an IR sensor and fed to a phase-locked loop that generates a sinusoidal signal similar in frequency and amplitude to the nystagmus; the system is insensitive to other eye movements, such as saccades. The sinusoidal electronic signal is then used to rotate Risley prisms, through which the patient views the world. When the Risley prisms rotate in synchrony with the patient's nystagmus, they negate the visual effects of the ocular oscillations. A prototype system has been tested in five patients with acquired pendular nystagmus.[7] All five reported decreased oscillopsia, and visual acuity improved in four. At present, the main drawbacks of the prototype system are that it is cumbersome and only responds to pendular forms of nystagmus. Miniaturization of the device so that the prisms can be fitted to spectacle frames should make the device more portable. Development of software filters, perhaps using neural networks, may allow recognition of nonsinusoidal nystagmus waveforms that can be used to rotate the prisms, appropriately negating the visual effects of nystagmus but not of normal eye movements.

In summary, in studying approximately 30 patients during a 15-year period, we have experienced limited success using simple optical treatments for nystagmus, but greater promise is held using an electro-optical device. The latter controls the performance of optical components using computer programs that can distinguish abnormal from normal eye movements. Despite this limited success, one outcome of our research effort has been to clarify the visual requirements of eye movements and better understand how nystagmus degrades vision.

ACKNOWLEDGMENTS

This work was supported by the Office of Research and Development, Medical Research Service, Department of Veterans Affairs; National Institutes of Health Grants EY00356 and EY06717; and the Evenor Armington Fund.

REFERENCES

1. LEIGH, R.J. & D.S. ZEE. 1999. The Neurology of Eye Movements. Third edit. Oxford University Press. New York.
2. TOMSAK, R.L. *et al.* 1995. Unsatisfactory treatment of acquired nystagmus with retrobulbar botulinum toxin. Am. J. Ophthalmol. **119:** 489–496.
3. DELL'OSSO, L.F. 1973. Improving visual acuity in congenital nystagmus. *In* Neuro-ophthalmology. J.L. Smith & J.S. Glaser, Eds. Vol. 7: 98–106. C.V. Mosby. St. Louis, MO.

4. RUSHTON, D. & N. COX. 1987. A new optical treatment for oscillopsia. J. Neurol. Neurosurg. Psychiatry **50:** 411–415.
5. LEIGH, R.J. *et al.* 1988. Effects of retinal image stabilization on acquired nystagmus due to neurological disease. Neurology **38:** 122–127.
6. YANIGLOS, S.S. & R.J. LEIGH. 1992. Refinement of an optical device that stabilizes vision in patients with nystagmus. Optom. Vis. Sci. **69:** 447–450.
7. STAHL, J.S. *et al.* 2000. Prospects for treating acquired pendular nystagmus with servo-controlled optics. Invest. Ophthalmol. Vis. Sci. **41:** 1084–1090.

Latency of Dynamic and Gaze-Dependent Optotype Recognition in Patients with Infantile Nystagmus Syndrome versus Control Subjects

RICHARD W. HERTLE,[a] MITRA MAYBODI,[a] GEORGE F. REED,[b] AMIR H. GUERAMI,[a] DONGSHENG YANG,[a] AND EDMOND J. FITZGIBBON[a]

[a]*Pediatric Ophthalmology, Strabismus and Eye Movement Disorders, The Laboratory of Sensorimotor Research,* and [b]*Division of Biometry and Epidemiology, The National Eye Institute, The National Institutes of Health, Bethesda, Maryland 20892, USA*

KEYWORDS: dynamic optotype recognition; gaze-dependent optotype recognition; infantile nystagmus syndrome; visual acuity

INTRODUCTION AND PURPOSE

The quality of visual experience in patients with disorders of the ocular motor system is often not represented by resolution or recognition acuity measures.[1–4] Patients with infantile nystagmus syndrome (INS) who have had surgery for strabismus, gaze repositioning, or artificial divergence report a subjective improvement in visual function separate from that which is quantified by optotype testing. We have developed and tested a dynamic measure of visual function in controls and patients with INS.

METHODS

The investigations were performed according to the guidelines of the Declaration of Helsinki and approved by the Institutional Review Board at the National Eye Institute. The presentation of stimuli and the acquisition, display, and storage of data were controlled by a PC. Patients and control subjects were seated with their heads stabilized by means of a chin cup, headrest, and headstrap. A joy-stick with a four-directional switch was used by the subject's dominant hand while being asked to move the joy-stick in the direction of a "tumbling," single, surrounded "E" optotype as soon as it was recognized. The stimulus was presented on a 42-inch plasma screen at 1 meter where the following two tasks were performed by all patients and controls.

Static task: Subjects were asked to binocularly view four different logMAR levels (1.58, 1.04, 0.944, and 0.602) of a tumbling, single, surrounded "E" optotype at varying horizontal gaze angles between 22 degrees to the right and 22 degrees to the

---

Address for correspondence: Richard W. Hertle, M.D., Pediatric Ophthalmology, Strabismus and Eye Movement Disorders, The Laboratory of Sensorimotor Research, The National Eye Institute, NIH, Building 49, Room 2A50, Bethesda MD 20892. Voice: 301-402-9375; fax: 301-402-2279.

rwh@lsr.nei.nih.gov

left. The stimulus angle varied from one trial to the next, from 0 to +5 (to the right), −5 (to the left), +10, −10, +15, −15, +20, −20, +22, −22 degrees. Each gaze angle was tested four times, resulting in a total of 44 trials at each optotype level.

Dynamic task: Subjects were asked to binocularly view three different logMAR levels (1.04, 0.944, 0.602) of a tumbling, single, surrounded "E" optotype moving horizontally to the right (+) or left (−) at 10, 20, and 30 deg/sec. The stimulus velocity varied from one trial to the next in a pseudo-random order. Each velocity was tested 16 times, resulting in a total of 96 trials at each optotype level.

Data collected included (1) time to recognition ("latency"), in milliseconds, and (2) percent of correct responses ("accuracy") of the optotype at each position of gaze and at each velocity trial, collected at each logMAR optotype level. An "accurate" response was defined as a minimum of 75% correct responses (three out of four at each gaze angle in the static task and 12 out of 16 at each gaze angle in the dynamic task). A 75% success rate was chosen as the level of accuracy because at this level there is only a 3–4% probability that the subject's response would be correct as the result of chance. Statistical analysis was performed using the Student's t-test. The significance level chosen for each optotype level was based on the Bon Ferroni principle. Thus, the overall significance level of 0.05 for each optotype level was determined by dividing 0.05 by the number of gaze angles ($0.05/11 = 0.00455$) in the static task and by the number of stimulus velocities ($0.05/6 = 0.00833$) in the dynamic task.

RESULTS

We studied 23 control patients (ages 7–50 years, mean 23 years) and 10 patients with INS (ages 9–61 years, mean 32 years). A comparison of control and patient characteristics revealed no significant difference in the age or male-to-female ratio, but did reveal a significant difference in visual acuity. In the test–retest analysis of the controls, there was no significant difference in the mean latencies at each optotype level when comparing the first to the second test. Comparison of average latencies between all controls for the static and dynamic tasks showed no significant differences at various gaze angles, velocities, or logMAR steps. The INS patients differed from controls at the smallest logMAR optotype size both in the static and in the dynamic tasks ($p < 0.0083$). These differences were significant for both gaze angle and velocity measurements.

CONCLUSIONS

Our study, testing static and dynamic latencies in the INS population, has shown that these latencies can be a useful measure of visual function. With further development of this new technique, this measure may be useful in clinical studies.

REFERENCES

1. HILLMAN, E.J., *et al*. 1999. Dynamic visual acuity while walking in normals and labyrinthine-deficient patients. J. Vestib. Res. **9:** 49–57.

2. HAARMEIER, T. & P. THEIR. 1999. Impaired analysis of moving objects due to deficient smooth pursuit eye movements. Brain **122:** 1495–1505.
3. SCHMAL, F., R. KUNZ & W. STOLL. 2000. Dynamic visual acuity during linear acceleration along the inter-aural axis. Eur. Arch. Otorhinolaryngol. **257:** 193–198.
4. OKADA, T., E. GRUNFELD, J. SHALLO-HOFFMANN & A.M. BRONSTEIN. 1999. Vestibular perception of angular velocity in normal subjects and in patients with congenital nystagmus. Brain **122:** 1293–1303.

A Robust, Normal Ocular Motor System Model with Latent/Manifest Latent Nystagmus and Dual-Mode Fast Phases

JONATHAN B. JACOBS[a,c] AND LOUIS F. DELL'OSSO[a,b,c]

Ocular Motor Neurophysiology Lab, VAMC[a] and Departments of Neurology[b] and Biomedical Engineering,[c] Case Western Reserve University, Cleveland, Ohio 44106, USA

KEYWORDS: ocular motor system; nystagmus; saccades

INTRODUCTION

Latent/manifest latent nystagmus (LMLN) is a specific type of infantile nystagmus that occurs subsequent to strabismus in some patients.[1,2] The amplitude of LMLN usually follows Alexander's law, increasing as the fixating eye moves into abduction and decreasing in adduction.[1] The slow phases of LMLN may be either linear or of decreasing velocity in the same patient.[3] Depending on the slow-phase velocity, LMLN fast phases may cause the target image to fall either within (foveating) or outside (defoveating) the foveal area.[3] Higher slow-phase velocities precipitate defoveating fast phases.[4] Also, as presaccadic slow-phase velocities grow, fast-phase amplitudes follow.

We present a dual-mode, ocular motor system model capable of producing normal saccades and both foveating and defoveating fast phases in LMLN. Additionally, the model contains a mechanism by which linear slow phases undergo the transition to decreasing velocity slow phases. The model includes programmable Alexander's law behavior and fixation conditions (binocular, right- and left-eye viewing), allowing simulation of the idiosyncratic characteristics of a broad spectrum of individuals with LMLN.[5]

MATERIALS AND METHODS

The model was built using the Simulink component of MATLAB (The MathWorks, Natick, MA) and is of modular design, consisting of subsystems thought to be required for accurate ocular motor control. The model also contains distributed delays that duplicate those known to exist from neurophysiological studies.

Address for correspondence: Jonathan B. Jacobs, Ph.D., Ocular Motor Neurophysiology Lab (W151), Louis Stokes Cleveland VA Medical Center, 10701 East Blvd., Cleveland, OH 44106. Voice: 216-421-3224; fax: 216-231-3461.
  jxj24@po.cwru.edu

The saccadic system responds to abrupt changes in target position and is capable of making short-latency (130-msec) corrective saccades, based on efference copy of position motor commands.

The internal monitor (IM) is essential for this model; the functions it performs have been required by all our past models of ocular motor function and *dysfunction*.[6-11] It uses afferent signals from the retina and efferent signals from the brainstem to accurately reconstruct target position and velocity and to distinguish them from eye position and velocity in the presence of motor instabilities. It calculates saccadic motor commands for voluntary and corrective saccades and for fast phases, and controls what portion of the saccadic pulse should be integrated. Provision is also made for Alexander's law variation of nystagmus slow phases.[9,11] Without such abilities, we contend that the ocular motor system could not function in the presence of either nystagmus or saccadic instabilities, let alone under normal circumstances.

To generate a *foveating* fast phase, the output of the neural integrator is compared with a desired eye-position signal, and the difference is compared to a position–signal error threshold. If this error exceeds the threshold, a saccade proportional to the error is generated. When the slow-phase velocity exceeds the velocity threshold (4°/sec), a *defoveating* fast phase is generated. The transition from foveating to defoveating saccades in the model is based on phase-plane data from LMLN subjects that show a significant difference in the presaccadic velocities for the foveating and defoveating cases. Some subjects have a region of overlapping slow-phase velocities where either foveating or defoveating fast phases can occur. This can be simulated by a change in the position-error threshold.

Previous studies correlated fast-phase magnitude with pre- and post-saccadic velocity.[4,12] The linear relationship between the magnitude and post-saccadic velocity suggested that an unintegrated, or "stepless," pulse was being employed. The post-saccadic velocities indicated that the pulse was not totally unintegrated, and the data suggested that the fast-phase generator produces a pulse width and height for a saccade of a relatively small size. To generate the decreasing velocity profiles of LMLN slow phases additional mechanisms were required: increasing the pulse to produce saccades greater than necessary to foveate the target leads to a larger unintegrated pulse, which is summed with the output from the neural integrator, producing a decreasing velocity slow phase.

RESULTS

The nystagmus of persons with MLN (both eyes open) contains linear slow phases and foveating fast phases throughout most gaze angles. When the model simulates a small gaze-angle Alexander's law effect, even though slow–phase velocity increases as the fixating eye abducts, the fast phases remain foveating until the eye is far into abduction. With a larger gaze-angle effect, slow-phase velocity increases more rapidly as the eye moves into abduction. When the velocity exceeds 4°/sec, the fast phases become larger and defoveating, and the slow phases exhibit a decreasing velocity profile. Importantly, neither type of MLN interferes with the accuracy of the saccadic subsystem, which can still make corrective saccades as necessary.

The nystagmus of individuals with LN (one eye occluded) contains decreasing velocity slow phases and defoveating fast phases throughout most gaze angles.

When a small gaze-angle Alexander's law effect is simulated, even though slow–phase velocity decreases as the fixating eye adducts, the fast phases remain defoveating except in extreme adduction. With a larger gaze-angle effect, the slow-phase velocity decreases more rapidly as the fixating eye adducts. As a result, the slow phases drop below 4°/sec at a more central gaze angle, causing smaller foveating fast phases and linear slow phases. As before, normal saccadic behavior is preserved.

CONCLUSIONS

We constructed a computer model of the normal ocular motor system that can also simulate LMLN. We demonstrated that an internal monitor could make use of afferent retinal and efferent motor information to detect changes in target position and to accurately differentiate target position and velocity from internally generated eye position and velocity (e.g., resulting from LMLN). In addition, we demonstrated that when slow-phase velocity exceeded 4°/sec, the foveating fast phases became defoveating, and the resulting slow phases decreased in velocity due to unintegrated portions of the fast-phase pulses. Finally, we demonstrated that if slow-phase velocity increased as gaze was directed in the abducting direction of the fixating eye (due to Alexander's law), that would ultimately cause the switch from foveating to defoveating fast phases.

ACKNOWLEDGMENTS

This work was supported by the Veterans Affairs Merit Review.

REFERENCES

1. DELL'OSSO, L.F., D. SCHMIDT & R.B. DAROFF. 1979. Latent, manifest latent and congenital nystagmus. Arch. Ophthalmol. **97:** 1877–1885.
2. DELL'OSSO, L.F., S. TRACCIS & L.A. ABEL. 1983. Strabismus—a necessary condition for latent and manifest latent nystagmus. Neuroophthalmology **3:** 247–257.
3. DELL'OSSO, L.F. et al. 1995. Two types of foveation strategy in "latent" nystagmus. Fixation, visual acuity and stability. Neuroophthalmology **15:** 167–186.
4. ERCHUL, D.M., L.F. DELL'OSSO & J.B. JACOBS. 1998. Characteristics of foveating and defoveating fast phases in latent nystagmus. Invest. Ophthalmol. Visual Sci. **39:** 1751–1759.
5. JACOBS, J.B. & L.F. DELL'OSSO. 1999. A dual-mode model of latent nystagmus. Invest. Ophthalmol. Visual Sci. **40:** S962.
6. DELL'OSSO, L.F. 1968. A dual-mode model for the normal eye tracking system and the system with nystagmus. Ph.D. thesis, University of Wyoming, Laramie, WY.
7. WEBER, R.B. & R.B. DAROFF. 1972. Corrective movements following refixation saccades: type and control system analysis. Vision Res. **12:** 467–475.
8. ABEL, L.A., L.F. DELL'OSSO & R.B. DAROFF. 1978. Analog model for gaze-evoked nystagmus. IEEE Trans. Biomed. Eng. **25:** 71–75.
9. DOSLAK, M.J., L.F. DELL'OSSO & R.B. DAROFF. 1979. A model of Alexander's law of vestibular nystagmus. Biol. Cybern. **34:** 181–186.
10. ABEL, L.A. et al. 1980. Myasthenia gravis: analogue computer model. Exp. Neurol. **68:** 378–389.

11. DOSLAK, M.J., L.F. DELL'OSSO & R.B. DAROFF. 1982. Alexander's law: a model and resulting study. Ann. Otol. Rhinol. Laryngol. **91:** 316–322.
12. ERCHUL, D.M., J.B. JACOBS & L.F. DELL'OSSO. 1996. Latent nystagmus fast-phase generation. Invest. Ophthalmol. Visual Sci. **37:** S277.

A Hypothetical Fixation System Capable of Extending Foveation in Congenital Nystagmus

JONATHAN B. JACOBS[a,c] AND LOUIS F. DELL'OSSO[a,b,c]

Ocular Motor Neurophysiology Lab, VAMC[a] and Departments of Neurology[b] and Biomedical Engineering,[c] Case Western Reserve University, Cleveland, Ohio 44106, USA

KEYWORDS: congenital nystagmus; saccades; ocular motor system

INTRODUCTION

Good visual acuity in the presence of the high-velocity oscillations of pendular nystagmus requires that the ocular motor system be capable of reliable and repeatable foveation of the target. The ability of the ocular motor system to acquire a target, despite these ongoing oscillations, requires seamless interaction between the smooth pursuit (SP) and saccadic subsystems. In a previous model[1] we proposed that this cooperation is mediated by the "internal monitor" (IM), a hypothetical grouping of functions that calculates the necessary control signals based on efference copy of position and velocity commands issued by the ocular motor system (OMS). Here we extend this model, adding the ability to extend foveation duration beyond the period of low-velocity motion that occurs naturally at the reversal of eye direction during pendular waveforms.

MATERIALS AND METHODS

The model was designed and implemented using the Simulink component of MATLAB (The MathWorks, Natick, MA). The model is of modular, hierarchical design. The SP subsystem is based on Robinson,[2] chosen for its relative simplicity and ability to be induced into instability, yielding a sinusoidal oscillation characteristic of pendular nystagmus. The saccadic system was based around a resettable neural integrator[3,4] (RNI) with pulse-height and -width nonlinearities. The RNI is part of the circuit that determines saccade duration; when the RNI resets, the saccade ends.

The IM, which is the "brains" of the model, has a long history in ocular motor models.[3–8] The IM makes use of visual signals from the retina, as well as position and velocity efference signals available in the brainstem, to reconstruct and properly respond to changes in target position and velocity. It can then use this information to coordinate interaction between the SP and saccadic and fixation subsystems, despite

---

Address for correspondence: Jonathan B. Jacobs, Ph.D., Ocular Motor Neurophysiology Lab (W151), Louis Stokes Cleveland VA Medical Center, 10701 East Blvd., Cleveland, OH 44106. Voice: 216-421-3224; fax: 216-231-3461.

jxj24@po.cwru.edu

the confounding "noise" of the nystagmus. The operational details of these systems can be found elsewhere.[9]

We then tested two separate designs for fixation subsystems. The first approach calculated a "counter-signal," based on the reconstructed nystagmus oscillation, to be summed destructively with (i.e., subtracted from) the commands sent by the SP subsystem to the OMN. The second approach also relied on the reconstructed nystagmus signal, using it as a variable gain to modulate the SP commands to the OMN. After these separate fixation systems were tested in isolation, they were integrated into the full model to determine whether they could utilize position and velocity efference signals to extend foveation.

RESULTS

FIGURE 1 shows the difference between model output for the pseudopendular with foveating saccades (PP_{fs}) waveform when the counter-signal fixation system is disabled (A) and enabled (B). Compare the portion of the slow phase immediately following the foveating saccade (arrow). Without the effect of the fixation system, eye velocity is below 4°/sec for only 18 msec. When fixation is enabled, that duration rises to 40 msec, consistent with better visual acuity. Also, eye position is noticeably

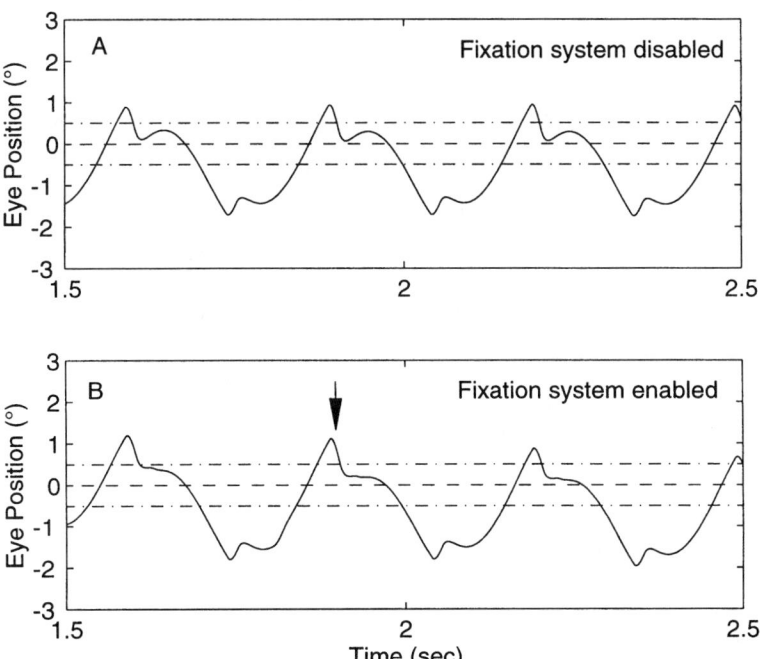

FIGURE 1. Difference between model output for the pseudopendular with foveating saccades (PP_{fs}) waveform when the counter-signal fixation system is disabled (**A**) and enabled (**B**). Compare the portion of the slow phase immediately following the foveating saccade (*arrow*).

more constant during the period of low velocity when the fixation system is active. Note that the slow phase following the leftward, or braking, saccades is not affected by the fixation system, consistent with actual patient data, indicating that the effort of foveation is a necessary condition for fixation extension.

CONCLUSIONS

Comparative testing of the two possible fixation system designs that we examined favors the counter-signal approach over that of variable gain, as the latter can adversely affect legitimate pursuit signals passing through the OMN, whereas the former acts to remove only the nystagmus oscillation, leaving the true pursuit commands intact. Therefore, the model provides a possible mechanism for a fixation system that acts effectively in the presence of the high-velocity oscillations of the smooth pursuit system typical in CN, slowing the eye sufficiently so that a more useful period of low-velocity foveation is available to the visual system, allowing for greater visual acuity.

ACKNOWLEDGMENTS

This work was supported by the Veterans Affairs Merit Review.

REFERENCES

1. JACOBS, J.B. & L.F. DELL'OSSO. 2000. A model of congenital nystagmus (CN) incorporating braking and foveating saccades. Invest. Ophthalmol. Visual Sci. **41:** S701.
2. ROBINSON, D.A., J.L. GORDON & S.E. GORDON. 1986. A model of smooth pursuit eye movements. Biol. Cybern. **55:** 43–57.
3. ABEL, L.A., L.F. DELL'OSSO & R.B. DAROFF. 1978. Analog model for gaze-evoked nystagmus. IEEE Trans. Biomed. Eng. **25:** 71–75.
4. ABEL, L.A., L.F. DELL'OSSO, D. SCHMIDT, *et al.* 1980. Myasthenia gravis: analogue computer model. Exp. Neurol. **68:** 378–389.
5. DELL'OSSO, L.F. 1968. A dual-mode model for the normal eye tracking system and the system with nystagmus. Ph.D. dissertation, University of Wyoming, Laramie, WY.
6. WEBER, R.B. & R.B. DAROFF. 1972. Corrective movements following refixation saccades: type and control system analysis. Vision Res. **12:** 467–475.
7. DOSLAK, M.J., L.F. DELL'OSSO & R.B. DAROFF. 1979. A model of Alexander's law of vestibular nystagmus. Biol. Cybern. **34:** 181–186.
8. DOSLAK, M.J., L.F. DELL'OSSO & R.B. DAROFF. 1982. Alexander's law: a model and resulting study. Ann. Otol. Rhinol. Laryngol. **91:** 316–322.
9. JACOBS, J.B. 2001. An ocular motor system model that simulates congenital nystagmus, including braking and foveating saccades. Ph.D. dissertation, Case Western Reserve University, Cleveland, OH.

Congenital Periodic Alternating Nystagmus
Response to Baclofen

DAVID SOLOMON,[a] NEIL SHEPARD,[b] AND ANUPAM MISHRA[c]

[a]*Departments of Neurology and* [b]*Otorhinolaryngology, University of Pennsylvania, Philadelphia, Pennsylvania 19104, USA*

[c]*K.G. Medical College, Lucknow, India*

KEYWORDS: periodic alternating nystagmus; baclofen; posterior cerebellar vermis

INTRODUCTION

Acquired periodic alternating nystagmus (PAN) has been associated with lesions of the posterior cerebellar vermis, particularly the nodulus and uvula.[1] It is usually a horizontal spontaneous jerk nystagmus present in straight-ahead gaze, which reverses its quick phase direction approximately every 2 minutes.[2] Baclofen, a $GABA_B$ agonist, is often effective in reducing PAN in patients with acquired cerebellar lesions.[3] PAN in these cases is thought to be due to disruption of inhibitory control of the velocity storage integrator by GABAergic cerebellar Purkinje cells projecting from the posterior vermis to the vestibular nuclei.[4] This is believed to be the substrate for adaptive control by the cerebellum of central vestibular processing and is responsible for secondary afternystagmus.[5] Dysfunction in this pathway could cause oscillatory behavior from an unstable integrator. This is supported by findings that vestibular stimulation can alter the cycle of PAN.[6]

PAN is probably underrecognized in congenital nystagmus patients[7] and may be a component of the nystagmus in up to 9% of these patients.[8] In congenital or acquired cases, the reversals in direction may occur with irregular intervals, so-called aperiodic alternating nystagmus.[9] Often patients have anomalous head postures. Baclofen has not been found as effective in patients with congenital PAN[3,8] or in acquired pendular nystagmus (APN), but may be effective in some cases of downbeat nystagmus.[10] One case of PAN "probably of the congenital type" that responded to baclofen has been reported[11] in a patient with cerebral palsy.

PATIENT HISTORY

RK is a 73-year-old Caucasian ambidextrous male (retired chemist) with a diagnosis of "congenital nystagmus." At age 30 he underwent strabismus surgery on the

---

Address for correspondence: David Solomon, M.D., Ph.D., Department of Neurology, University of Pennsylvania, 3 West Gates Building, 3400 Spruce Street, Philadelphia, PA 19104-4283. Voice: 215-614-0233; fax: 215-573-2107.

Dsolomon@mail.med.upenn.edu

right eye (details not available), with some improvement in his "ability to focus." Lifelong complaints of imbalance worsened recently with head-movement and positional provoked episodic dizziness. Episodes of prolonged vertigo occurred twice in the past 10 years, with more frequent bouts of lightheadedness/imbalance lasting hours occurring every 1 to 3 months. Fluctuating hearing loss, aural fullness, and tinnitus occurs in the left ear.

Family history is negative for any ataxia, hearing loss, or ocular motor problems; the mother had a "mild stroke." At the time of presentation (10/98), only an aspirin and vitamins were being taken. Past medical history was remarkable only for eye muscle surgery, appendectomy, and hernia repair. No tobacco was used, and one glass of red wine was consumed daily. Previous workup included a normal MRI of brain and cervical spine. No asymmetry was found with caloric testing. An audiogram demonstrated a left-sided low-frequency sensorineural hearing loss and a symmetric-sloping moderate to severe high-frequency SNHL.

EXAMINATION

Spontaneous nystagmus was readily observed with fixation present. Behind Frenzel lenses, with several minutes of observation, the horizontal nystagmus changed from its predominantly right beating direction to left beating, and back to right beating. Superimposed on the horizontal nystagmus, typical left-sided posterior semicircular canal benign paroxysmal positional nystagmus and symptoms were elicited in the left Dix-Hallpike position. There was no evidence of a latent nystagmus with monocular viewing. Pigmentation of the iris was normal. There was no evidence of rebound nystagmus. Pursuit responses were poor. No null position could be identified. Cerebellar testing of the limbs was normal. No anomalous head posture was ob-

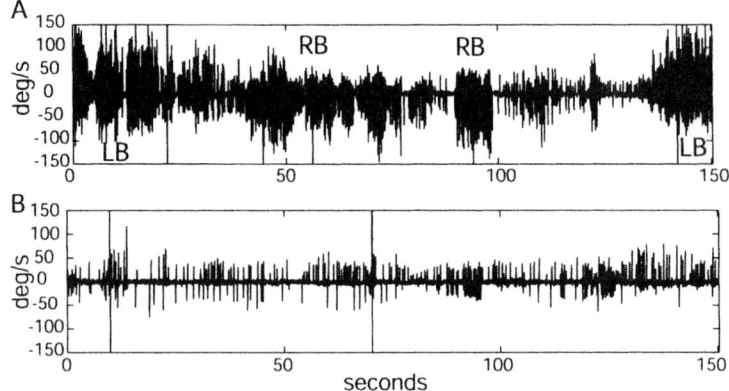

FIGURE 1. Eye velocity during binocular viewing in dim light before (**A**) and after (**B**) a dose of baclofen. In **A**, nystagmus direction alternates (RB: right beating; LB: left beating), and reaches peak slow-phase eye velocity up to 80 deg/sec. Following treatment, nystagmus slow-phase eye velocity is substantially reduced, and reversals are either very frequent or absent.

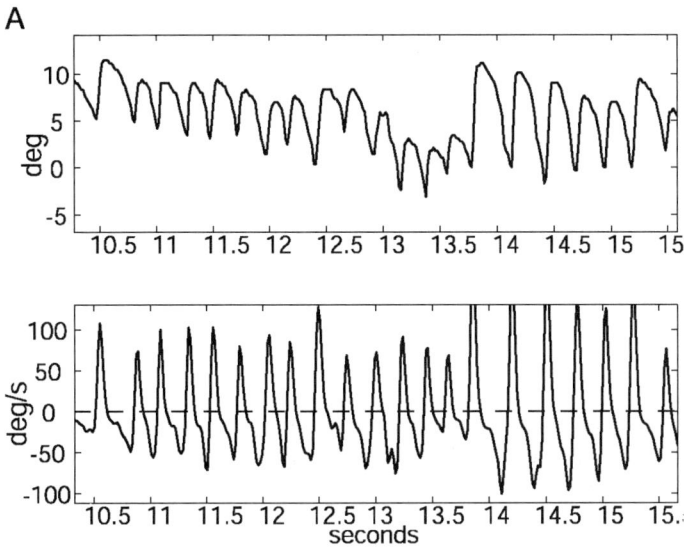

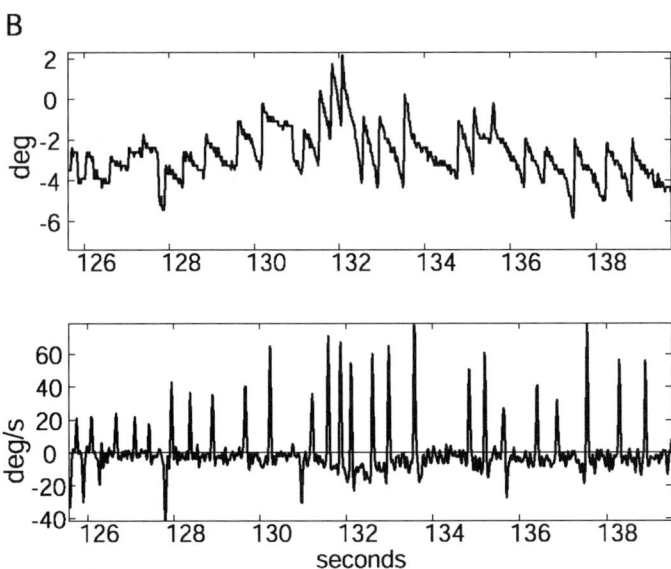

FIGURE 2. Expanded traces from records shown in FIGURE 1. (**A**) Before the baclofen dose, a typical congenital waveform nystagmus is recorded, with fixation intervals and accelerating slow phases (*top trace*, eye position; *bottom trace*, eye velocity). (**B**) 150 minutes following a 20-mg baclofen dose. The waveform has changed to one with more linear or decelerating slow phases. Nystagmus amplitude is greatly decreased. (Note difference in position and velocity scales.)

served, and recordings of yaw head position throughout reversals of the horizontal nystagmus showed no angular deviations.

METHODS

The patient signed a consent form approved by the local Institutional Review Board. Eye movements were recorded usng video-oculography (Micromedical), with a camera sampling rate of 60 Hz. Eye position signals were smoothed with a Gaussian filter before digital differentiation. Conjugacy was confirmed with binocular recording during binocular viewing. No significant vertical nystagmus components were observed.

CLINICAL COURSE

The patient's positional vertigo resolved with a canalith repositioning procedure, and he has remained free of BPPV while continuing to perform daily Brandt-Daroff exercises.[12] He has had no further episodes of prolonged vertigo, hearing fluctuation, or aural discomfort since beginning a salt-restricted diet with increased water intake. He noted a significant decrease in oscillopsia and nystagmus (FIG. 1) and an improved ability to read after starting baclofen. He has been taking 40 mg/day for three years with continued benefit. He still has complaints of generalized imbalance, though he leads walking tours through historic Philadelphia throughout each summer.

DISCUSSION

The diagnosis of congenital periodic alternating nystagmus in this case is supported by accelerating slow-phase velocity waveform, fixation intervals (FIG. 2A), horizontal direction of nystagmus, and reversal of the optokinetic response.[13] Horizontal shifts in orbital eye position (periodic alternating gaze deviation) have been associated with cerebellar damage and PAN[14] but are also reported in reversible conditions associated with metabolic derangements affecting GABAergic transmission[15] in the absence of any structural lesion. Improvement in the patient's congenital nystagmus with pharmacotherapy suggests a possible underlying defect in neurotransmission as an etiologic factor.

REFERENCES

1. WAESPE, W., B. COHEN & T. RAPHAN. 1985. Dynamic modification of the vestibuloocular reflex by the nodulus and the uvula. Science **228**: 199–202.
2. BALOH, R.W., V. HONRUBIA & H.R. KONRAD. 1976. Periodic alternating nystagmus. Brain **99**: 11–26.
3. HALMAGYI, G.M., P. RUDGE, M.A. GRESTY, et al. 1980. Treatment of periodic alternating nystagmus. Ann. Neurol. **8**: 609–611.
4. FURMAN, J.M., C. WALL III & D.L. PANG. 1990. Vestibular function in periodic alternating nystagmus. Brain **113**: 1425–1439.

5. FURMAN, J.M., T.C. HAIN & G.D. PAIGE. 1989. Central adaptation models of the vestibulo-ocular and optokinetic systems. Biol. Cybern. **61:** 255–264.
6. LEIGH, R.J., D.A. ROBINSON & D.S. ZEE. 1981. A hypothetical explanation of periodic alternating nystagmus: instability in the optokinetic-vestibular system. Ann. N.Y. Acad. Sci. **374:** 619–635.
7. SHALLO-HOFFMANN, J., M. FALDON & R.J. TUSA. 1999. The incidence and waveform characteristics of periodic alternating nystagmus in congenital nystagmus. Invest. Ophthalmol. Vis. Sci. **40:** 2546–2553.
8. GRADSTEIN, L., R.D. REINECKE, S.S. WIZOV & H.P. GOLDSTEIN. 1997. Congenital periodic alternating nystagmus. Diagnosis and management. Ophthalmology **104:** 918–928; discussion 928–929.
9. NUTI, D., G. CIACCI, F. GIANNINI, et al. 1986. Aperiodic alternating nystagmus: report of two cases and treatment by baclofen. Ital. J. Neurol. Sci. **7:** 453–459.
10. AVERBUCH-HELLER, L., R.J. TUSA, L. FUHRY, et al. 1997. A double-blind controlled study of gabapentin and baclofen as treatment for acquired nystagmus. Ann. Neurol. **41:** 818–825.
11. ISAGO, H., R. TSUBOYA & A. KATAURA. 1985. A case of periodic alternating nystagmus: with a special reference to the efficacy of baclofen treatment. Auris Nasus Larynx **12:** 15–21.
12. BRANDT, T. & R.B. DAROFF. 1980. Physical therapy for benign paroxysmal positional vertigo. Arch. Otolaryngol. **106:** 484–485.
13. HALMAGYI, G.M., M.A. GRESTY & J. LEECH. 1980. Reversed optokinetic nystagmus (OKN): mechanism and clinical significance. Ann. Neurol. **7:** 429–435.
14. KENNARD, C., G. BARGER & W.F. HOYT. 1981. The association of periodic alternating nystagmus with periodic alternating gaze. A case report. J. Clin. Neuroophthalmol. **1:** 191–193.
15. AVERBUCH-HELLER, L. & Z. MEINER. 1995. Reversible periodic alternating gaze deviation in hepatic encephalopathy. Neurology **45:** 191–192.

Index of Contributors

Adams, N.L., 438–440
Andrade, F.H., 391–393
Arai, Y., 190–204
Averbuch-Heller, L., 434–437
Avitable, M., 464–467
Aw, S.T., 306–313
Ayers, A., 111–129

Baloh, R.W., 338–345
Barton, E.J., 85–98
Barton, J.J.S., 250–263
Benassi, M., 508–511
Bense, S., 230–241
Berthoz, A., 426–429
Bespalova, I.N., 441–444
Beykirch, K.A., 537–542
Bhidayasiri, R., 438–440
Bisdorff, A.R., 579–580
Bisley, J., 205–215
Bodis-Wollner, I., 464–467
Boghen, D., 482–483
Bolzani, R., 508–511
Boothe, R.G., 346–360
Born, R.T., 453–455
Brandt, T., 230–241
Brizuela, A.E., 546–550
Burmeister, M., 441–444
Büttner, U., 99–110, 434–437
Büttner-Ennever, J.A., 75–84, 99–110, 434–437

Carey, J.P., 324–337, 581–584
Chen, B., 405–408
Cherkasova, M.V., 250–263
Chimoto, S., 460–463
Cohen, B., 190–204
Colnaghi, S., 401–404
Cosi, V., 401–404
Crane, B.T., 574–578
Crawford, J.D., 474–475, 512–514, 515–519
Cremer, P.D., 581–584
Curthoys, I.S., 306–313, 546–550

Dai, M., 190–204
Daroff, R.B., 1–6
Das, V.E., 346–360, 380–390
Debatisse, D., 579–580
Dell'Osso, L.F., 361–379, 604–607, 608–610
Della Santina, C.C., 581–584
Demer, J.L., 17–32, 574–578
DePalma, S., 111–129
Derwenskus, J., 438–440
Deutschländer, A., 230–241
Dieterich, M., 230–241
Dill, N., 409–413
Downey, D.L., 438–440
Dzaja, A.E., 595–597

Eggenberger, E.R., 588–591
Eggert, T., 595–597
Eizenman, M., 499–503
Engle, E.C., 55–63

Fetter, M., 537–542
Fitzgibbon, E.J., 601–603
Forget, R., 482–483
Frohman, E.M., 530–532
Fuchs, A.F., 155–163

Gamlin, P.D.R., 264–272
Gandhi, N.J., 85–98
Garbutt, S., 445–448
Gaymard, B., 216–229
Gibson, J.M., 421–425
Glasauer, S., 230–241
Goff, D.C., 250–263
Goldberg, M.E., 205–215
Gordon, C., 434–437
Gottlieb, J., 205–215
Graf, E., 297–305
Graf, W., 75–84
Guerami, A.H., 601–603

Habib, A., 464–467
Halmagyi, G.M., 306–313, 546–550
Han, Y., 495–498, 527–529

Harris, C.M., 414–417, 445–448, 449–452
Harris, L.R., 537–542
Harwood, M.R., 414–417, 445–448
Haslwanter, T., 33–41, 581–584
Helmchen, C., 99–110, 434–437, 592–594
Henriques, D.Y.P., 515–519
Hertle, R.W., 601–603
Hirvonen, T., 581–584
Horn, A.K.E., 75–84
Hussain, Z., 464–467

Iancu, A., 482–483
Inoue, Y., 456–459
Intriligator, J.M., 250–263

Jacobs, J.B., 604–607, 608–610
Jampel, R.S., 568–571
Javeid, A., 464–467
Jen, J.C., 338–345
Jennings, D.W., 421–425

Kaminski, H.J., xii–xiii, 42–54, 397–398, 399–400
Kanai, R., 487–491
Karlberg, M., 306–313
Kawano, K., 284–296, 456–459, 460–463, 561–563
Keller, E.L., 130–142
Kennard, C., 242–249
Khanna, S., 394–396
Kim, J.I., 527–529
Kim, J.S., 143–154, 533–536
King, W.M., 273–283, 555–557, 558–560
Kitthaweesin, K., 504–507
Klier, E.M., 512–514
Kodaka, Y., 460–463, 561–563
Kömpf, D., 592–594
Kori, A.A., 572–573
Kornylo, K., 409–413
Kramer, P., 530–532
Krauzlis, R.J., 409–413
Kumar, A.N., 495–498, 527–529
Kunin, M., 190–204
Kuniyoshi, S., 484–486
Kusner, L.L., 42–54
Kusunoki, M., 205–215

Lasker, D.M., 324–337
Lavery, K., 421–425
Lee, H.R., 499–503
Leigh, J., 380–390
Leigh, R.J., xii–xiii, 430–433, 434–437, 438–440, 441–444, 484–486, 495–498, 504–507, 527–529, 598–600
Lindgren, K., 250–263

MacDougall, H.G., 546–550
Manoach, D.S., 250–263
Maxwell, J.S., 297–305
May, P.J., 405–408
Maybodi, M., 601–603
McCandless, J., 297–305
McConkie, G.W., 479–481
McGivern, R.C., 421–425
McPeek, R.M., 130–142
Merriam, A.P., 391–393
Mezey, L.E., 449–452
Miles, F.A., 284–296
Minor, L.B., 324–337, 581–584
Mishra, A., 611–615
Mitsumoto, H., 42–54
Montgomery, C., 421–425
Mulligan, J.B., 476–478
Müri, R.M., 216–229
Mustari, M.J., 346–360

Nelson, J., 85–98
Niemeier, M., 474–475
Noto, C.T., 155–163
Nuti, D., 530–532

Olivier, E., 405–408
Optican, L.M., 164–177

Pack, C.C., 453–455
Paige, G.D., 314–323
Pansell, T., 564–567
Pathmanathan, J., 111–129
Peng, G.C.Y., 581–584
Pierrot-Deseilligny, Ch., 216–229
Ploner, C.J., 216–229
Populin, L.C., 468–473
Porter, J.D., 7–16, 391–393, 394–396

INDEX OF CONTRIBUTORS

Powell, K.D., 205–215
Presnell, R., 111–129

Quaia, C., 164–177, 284–296

Ramat, S., 178–189, 324–337,
  495–498, 551–554
Rambold, H., 99–110, 434–437,
  592–594
Raphan, T., 190–204
Raza, A., 464–467
Reed, G.F., 601–603
Reich, S.G., 484–486
Reschke, M.F., 426–429
Richmond, F.J.R., 405–408
Richmonds, C.R., 42–54, 397–398,
  399–400
Riley, D.E., 484–486, 504–507
Rivaud-Péchoux, S., 216–229
Robinson, F.R., 155–163
Rosenberg, M.L., 585–587
Rottach, K.G., 441–444, 595–597
Ruff, R.L., 438–440

Sabet, M., 464–467
Sandrini, G., 401–404
Schmid-Priscoveanu, A., 572–573
Schor, C.M., 297–305
Schworm, H.D., 564–567
Seidman, S.H., 380–390
Sharpe, J.A., 143–154, 520–522,
  523–526, 533–536
Shepard, N., 611–615
Solomon, D., 611–615
Somers, J.T., 426–429, 441–444,
  527–529
Sparks, D.L., 85–98
Stahl, J.S., 64–74, 418–420, 430–433,
  438–440, 598–600
Stephan, T., 230–241
Stevenson, S.B., 492–494
Straube, A., 595–597
Straumann, D., 572–573
Suzuki, J.-I., 190–204
Swartz, B.E., 441–444

Takemura, A., 284–296, 456–459
Tang, B., 555–557, 558–560
Taylor, L.C., 426–429
Tian, J.-R., 574–578
Todd, M.J., 306–313
Tollin, D.J., 468–473
Tomsak, R.L., 430–433
Tozlovanu, V., 482–483
Tusa, R.J., 346–360
Tweed, D.B., 474–475, 520–522,
  523–526

Ugolini, G., 75–84

Van der Steen, J., 487–491
Versino, M., 401–404
Vilis, T., 515–519
Volpe, B.T., 430–433
von Gizycki, H., 464–467

Wada, Y., 561–563
Waitzman, D.M., 111–129
Walker, M.F., 178–189, 543–545
Wang, H., 512–514
Warren, S., 405–408
Weinstein, J.M., 468–473
Weldon, P., 555–557, 558–560
Whitney, C., 484–486
Wiest, G., 574–578
Wohlgemuth, W.A., 595–597
Wong, A.M.F., 520–522, 523–526

Yakushin, S.B., 190–204
Yang, D., 601–603
Yang, S.-N., 479–481
Yaniglos, S.S., 598–600
Ygge, J., 508–511, 564–567
Yousry, T.A., 230–241

Zee, D.S., 178–189, 484–486,
  530–532, 543–545, 551–554,
  585–587
Zhou, W., 273–283, 555–557, 558–560
Zivotofsky, A., 434–437